DRILLING WASTES

PROCEEDINGS OF THE 1988 INTERNATIONAL
CONFERENCE ON DRILLING WASTES
CALGARY, ALBERTA, CANADA, 5–8 APRIL

DRILLING WASTES

Edited by

F. R. ENGELHARDT
Canada Oil and Gas Lands Administration,
Ottawa, Ontario, Canada

J. P. RAY
Shell Oil Company, Houston, Texas, USA

and

A. H. GILLAM
CBR International Corp., Sidney,
British Columbia, Canada

CRC Press
Taylor & Francis Group
Boca Raton London New York

CRC Press is an imprint of the
Taylor & Francis Group, an **informa** business
A TAYLOR & FRANCIS BOOK

CRC Press
Taylor & Francis Group
6000 Broken Sound Parkway NW, Suite 300
Boca Raton, FL 33487-2742

First issued in paperback 2019

© 1990 by Taylor & Francis Group, LLC
CRC Press is an imprint of Taylor & Francis Group, an Informa business

ISBN-13: 978-1-85166-278-4 (hbk)
ISBN-13: 978-0-367-86536-8 (pbk)

British Library Cataloguing in Publication Data

International Conference on Drilling Wastes
(1988: Calgary, Canada).
Drilling wastes.
1. Petroleum deposits. Extraction.
Environmental aspects
I. Title   II. Engelhardt, F.R. (F. Rainer)
III. Ray, J.P.   IV. Gillam, A.H.
622′.3382

Library of Congress Cataloging in Publication Data

Internal Conference on Drilling Wastes (1988:
Calgary, Alta)
Drilling wastes.

Bibliography: p.
Includes index.
1. Oil well drilling—Environmental aspects—
Congresses.   2. Drilling muds—Environmental aspects—
Congresses.   3. Oil well drilling—Waste disposal—Congresses.
I. Engelhardt, F.R. (F. Rainer)
II. Ray, J. P.   III. Gillam, A. H.   IV. Title.
TD195.P4156 1988   622′.3382   88-24436

Typeset by Lasertex Ltd, Stretford, Manchester

Visit the Taylor & Francis Web site at
http://www.taylorandfrancis.com

and the CRC Press Web site at
http://www.crcpress.com

Preface

The scientific and technical papers which follow are a product of the 1988 International Conference on Drilling Wastes, held in Calgary, Alberta, Canada on April 5–8 of that year. The papers represent the results of recent and current studies on drilling wastes. In general, these papers fall into categories of case histories, treatment and control methods, discharge and effects monitoring, and testing procedures, for both land-based and offshore marine drilling activities. More specifically, there are reports on land spreading of drilling wastes, desalination as a treatment method, biodegradation of discharged wastes, the tainting of fish, the utility of compliance testing protocols, the diesel pill issue, monitoring of fate and effects around production platforms, and other current topics. The last section of these proceedings presents two summary papers, written from the perspective of their authors who were invited to close the Drilling Wastes Conference with such overviews.

Some of the results of the studies reported here address areas of controversy, for example, the question of tainting of fish from the discharge of oily drilling wastes or the effectiveness of treatment of salt water mud discharges. Others may give rise to controversy, such as the interpretations by the Norwegian regulator of hydrocarbon levels and of benthic population changes recorded in a study in the North Sea. We have not stayed away from controversial papers in the view that it is important to bring such issues into the open forum. In instances where we felt the reader might benefit by some guidance to additional information, we have added an Editors' Note to the paper. It is expected that these papers will stimulate further research into the effects of the discharge of drilling wastes and the further development of control methodologies. As well, the results from these studies may help to focus regulatory approaches to the control and discharge of drilling wastes in both the land and marine environment.

The Calgary Conference was the first time since 1980, at the Lake Buena Vista, Florida Symposium 'Research on Environmental Fate and Effects of Drilling Fluids and Cuttings', that a broad evaluation of the subject was carried out in a meeting forum. It was clear that there had been many advances since 1980 in the treatment and control of drilling wastes and in our understanding of their environmental fate and effects. Further, changes have occurred in the methods of drilling for petroleum reserves, such as an increasing use of low-toxicity or alternate oil-based drilling muds. Advances have occurred in both land and offshore operations of the petroleum drilling sector.

The proposal for an international conference on drilling wastes was received with great interest in the petroleum sector, and led to joint sponsorship agreements from petroleum industry associations and provincial and national government organizations. The following lists the American, Norwegian and Canadian sponsors, whose financial contributions and additional help in organizing the conference is gratefully acknowledged:

— American Petroleum Institute
— Canada Oil and Gas Lands Administration/Department of Energy, Mines and Resources
— Canadian Petroleum Association
— Department of Energy and Mines, Saskatchewan
— Department of Environment, Canada
— Department of Environment, Alberta
— Department of Fisheries and Oceans, Canada
— Department of Indian Affairs and Northern Development, Canada
— Energy Resources Conservation Board, Alberta
— Ministry of Energy, Mines and Petroleum Resources, British Columbia
— US Department of Energy
— US Department of the Interior, Minerals Management Service
— US Environmental Protection Agency
— State Pollution Control Authority, Norway

This is also an opportunity to acknowledge the hard work and expertise of the following individuals who participated in the steering, program and review committees necessary to help us realize the conference and produce these proceedings: Bob Ayers (Exxon Production Research), Evan Birchard (Esso Resources Canada), Robert Bisson (Canada Oil and Gas Lands Administration), Paul Chénard (Canada Oil and Gas Lands Administration), Jim Cimato (US Minerals Management Service), Ruth Fry (US Department of Energy), Andrew Gillam (CBR International Corp.), Jerry Gossard (Saskatchewan Energy and Mines), Jack Gould (American Petroleum Institute), Ken Hindmarch (British Columbia Ministry of Energy, Mines and Petroleum Resources), Aston Hinds (International Drillers Association), Nancy Johnson (US Department of Energy), Fred Lepine (Canada Oil and Gas Lands Administration), David Lloyd (Alberta Forestry, Lands & Wildlife), Doug Mead (Shell Canada Limited), Jim Osborne (Environment Canada), Rod Paterson (Fisheries and Oceans Canada), Bob Peters (Argonne National Laboratory), Jim Ray (Shell Oil Company), Glen Singleton (Alberta Environment), David Stone (Indian and Northern Affairs Canada), Bill Telliard (US Environmental Protection Agency, and Ivan Weleschuk (Energy Resources Conservation Board, Alberta). In addition to the commitments of these individuals and their agencies, we the editors would like to thank Peter Faloon and his staff of Info-Tech, Calgary, who provided truly expert and efficient management services for the conference in its planning and execution, as well as for the preparation of these proceedings.

More than twice as many papers as could be accommodated were received from practitioners in the field of drilling wastes. The papers accepted by the conference review committee for presentation and publication have additionally undergone extensive peer review. I would like to take this opportunity to thank the many scientists and engineers in the field of drilling muds and wastes who have contributed much time and effort in commenting on these papers, recommending changes or additions where needed based on their experience.

The conference was attended by several hundred registrants, who arrived from a dozen different countries. The real success of a meeting, however, is measured by the quality of the information exchanged, and our high expectations were met as shown by the quality of the papers in these proceedings.

F. RAINER ENGELHARDT
Ottawa, Canada

Contents

PART II: Treatment Practices

PART III: Land Disposal and Effects

PART I
Case Histories

1

Trends in Sediment Trace Element Concentrations around Six Petroleum Drilling Platforms in the Northwestern Gulf of Mexico

P. N. BOOTHE and B. J. PRESLEY

Department of Oceanography, Texas A&M University, College Station, Texas 77843-3146, USA

ABSTRACT

Sediments within 500 m of 6 offshore drilling sites in the northwestern Gulf of Mexico were studied. The objective was a detailed characterization of sediment gradients of elements known to be major constituents of drilling fluids e.g. Ba, and of trace elements of environmental concern, e.g. Cd, Cr, Cu, Hg, Pb, Zn, which may be released during drilling. Exploration, development and production sites in both shallow and deep water were studied to determine how the amount of drilling, environment and time since last drilling influence ambient sediment characteristics.

Using a high sampling density (36 samples within 500 m), complete discharge history and analysis of subsurface sediments, a detailed three-dimensional barium mass balance was calculated which showed that only a small fraction of the total barium used, i.e. <1·5% nearshore, <12% offshore, and presumably similarly behaving drilling mud components, are retained in near-site sediments. The more pervasive sediment perturbations attributable to drilling activities are largely restricted to deep water development and production sites which had by far the largest sediment Ba retention among the 6 study sites.

Statistically significant elevations in surficial sediment Hg and Zn concentrations, i.e. within 125 m of the site, Hg 4–7 fold above background, Zn 5–10 fold, were observed at these two deep water sites. Significant elevations, e.g. 3–5 times background, in other trace metals (Cd, Cu, Pb,

3

Zn) were not common and generally restricted to sediments within 125 m of the drilling site. Chromium showed few elevations above background levels.

1. INTRODUCTION

Offshore petroleum recovery is one of the major human activities in the oceans of the continental margin. Assessing the environmental ramifications of this activity has been a highly funded research topic in continental shelf oceanography over the past decade (Gettleson & Laird, 1980; Menzie, 1982; National Research Council, 1983). Much of this research has dealt with short-term effects of one or a few exploratory wells drilled in a single area. The long-term, cumulative effects of development and production drilling have received much less attention. The study discussed here was conducted to address this need.

The goal of this study was to determine typical concentrations of drilling mud residuals in surface and subsurface sediments within 500 m of drilling sites of various types and ages. The residuals chosen were elements known to be major constituents of drilling fluids, e.g. barium, and other trace elements of environmental concern, i.e. cadmium, chromium, copper, mercury, lead, zinc. The detailed characterization of these sediments was emphasized because the highest elevations of drill mud components such as barium typically occur close to the drilling site. Also, in previous studies, insufficient surface and subsurface samples were taken within 500 m to adequately characterize the concentration gradients of drill mud components in these sediments. A detailed knowledge of these concentrations is important when extrapolating the results of drilling fluid bioassay studies conducted in the laboratory to actual field conditions. The data from this study can also be used to estimate the area of the sea floor containing elevated levels of discharged materials and to evaluate the persistence of these materials in the sediments.

Three types of drilling sites were studied: exploratory sites as isolated as possible from other wells, developmental sites with multiple, recently completed wells, and production sites where considerable time had elapsed since drilling was completed. For each of the three types, a location was chosen in shallow water (about 30 m) and in deep water (about 100 m). Figure 1 shows the locations of the 6 study sites in the northwestern Gulf of Mexico. A detailed, descriptive comparison of the sites is given in Table 1. The data on drilling mud components was compiled from actual mud records provided by the companies which drilled each well. This sampling

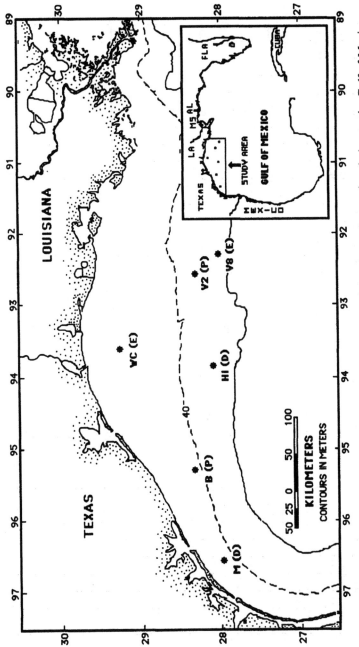

FIG. 1. Locations of the 6 drilling sites used in this study. Inset shows the study area in relation to the Gulf of Mexico.

TABLE 1
Descriptive summary of the six drilling sites used in this study

Characteristic	West Cameron (ES)[a]	Vermilion 381 (ED)	Matagorda 686 (DS)	High Island A-341 (DD)	Brazos A-1 (PS)	Vermilion 321 (PD)
Type of drilling activity	Exploration	Exploration	Development	Development	Production	Production
Number of wells	2	2	8	8	4	25
Water depth at site (m)	13	102	29	76	34	79
Sediment texture (% sand/silt/clay)	92/3/5	1/39/60	7/32/61	64/18/18	67/14/19	15/38/47
Elapsed time between last drilling activity and time of sampling (years)	In progress	In progress	In progress	0·48	7·6	5·5
Total well depth (sum all wells, m)	8845	4693	24938	23943	10919	77890
Total volume of cuttings discharged[b] (sum all wells, cubic meters)	1110	590	3130	3000	1370	9780
Drill mud components used (sum all wells, 1000 kg[c])	5667	651	5334	3474	2244	11935
Total barite used (sum all wells in 1000 kg, % total components used)	4702 (83)	446 (69)	4547 (85)	2957 (85)	2028 (90)	9671 (81)
Total Ba used (sum all wells, 1000 kg)[d]	2414	229	2334	1518	1041	4964
Number of surface (0-2 cm) sediment samples analyzed	30	30	39	37	29	40
Number of subsurface sediment samples analyzed (number of cores analyzed)	NA	18(3)	24(5)	28(7)	17(4)	47(8)
Depth interval range covered by sub-surface sediment samples analyzed (cm)	NA	2-31	2-31	2-31	3-22	2-31

[a] Well type designation: Exploratory (E), Development (D), Production (P)/Shallow (S), Deep (D).
[b] Estimated as 1:1 times total volume all wells.
[c] Mud systems used:

West Cameron: Seawater/gel (to 1 280 m), Lignosulfonate (below 1 280 m)
Vermilion 381: Seawater/gel (to 395 m), Freshwater/Gel/Caustic (to 1 415 m), lignosulfonate (to 3 715 m)
Matagorda: Seawater/gel (to 900 m), Lignite/lignosulfonate (below 900 m)
High Island: Seawater/gel (to 1 667 m), Lignosulfonate (below 1 667 m)
Brazos: Seawater/lignosulfonate
Vermilion 321: Salt/lime/lignosulfonate. Oil-base mud used in 6 wells below an average of 3470 m.

[d] Assuming the barite used was 87% $BaSO_4$ and that the percent composition of Ba in $BaSO_4$ is 58·8%.

design made it possible to determine how the amount of drilling (up to 25 wells at Vermilion 321), water depth and elapsed time between cessation of drilling and sampling i.e. 0–7·6 years, influenced the characteristics of surrounding sediments (≤ 500 m).

This paper focuses on trace element concentration gradients both with distance from the drilling site and depth in the sediment column. Identification of those gradients attributable to drilling discharges will be made on the basis of significant correlations (Spearman $p < 0.05$) between distance and barium and significant elevations above both surface and subsurface control sediments.

2. METHODS

2.1. Field Sampling

Sediment samples were collected once at each drilling site during the period from 19 November to 7 December 1980. The samples were collected from shipboard using a 0.1 m^2 stainless steel box core from which subcores were taken for analysis. Because of the sandy sediment at the West Cameron site, no subsurface samples could be taken. Sediment was collected at 40 stations around each drilling site using a circularly and radially symmetrical pattern around the drilling site (Fig. 2). This pattern also represents a 2–3 fold improvement in sampling density compared to earlier studies (Gettleson, 1980; Gettleson and Laird, 1980).

Selected depth intervals were taken from subcores at each station and homogenized by manual stirring. For all elemental analyses except Hg, samples were washed with distilled water to remove sea salts, freeze dried, and ground to a uniform powder in a porcelain ball mill. Mercury analyses were performed on wet or carefully oven dried (60°C) aliquots to avoid any losses. The total number of surface and subsurface samples analyzed from each site are summarized in Table 1.

2.2. Elemental Analysis

Barium, Cr and Fe were determined by instrumental neutron activation analysis (INAA). Samples (0.3–0.6 g) were irradiated in polyethylene vials for 14 h at a thermal neutron flux of 1×10^{13} cm^{-2} s^{-1} and allowed to decay for 10–12 days prior to counting on an Ortec Ge(Li) detector for 20–30 min. Concentrations were calculated with a proprietary Canberra NAA program using primary element standards and standard reference materials as comparator standards. Accuracy and precision of INAA,

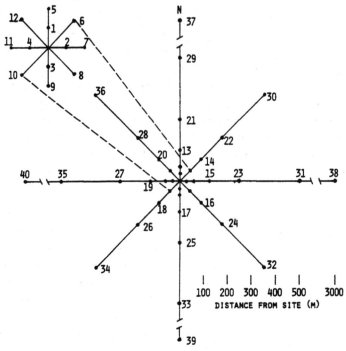

FIG. 2. Sampling pattern used at the 6 drilling sites in this study. Sampling radii are 30, 60, 125, 250, 500 and 3000 m. See Table 1 for number of surface samples, cores and subsurface samples analyzed from each site.

determined by repetitive analysis of reference materials with known concentrations of the elements of interest, was $\pm 10\%$ for Ba and Cr and $\pm 5\%$ for Fe.

Total Cd, Cu, Pb and Zn were determined by atomic absorption spectrophotometry (AAS). Samples (1 g) were digested in Teflon beakers using a mixture of hydrofluoric and perchloric acids. The resulting solutions were analyzed by flame AAS using operating parameters recommended by the instrument manufacturers. Because of its low concentration in the sediments, Cd was analyzed using a graphite furnace atomizer (flameless AAS). In all cases, background correction was performed using either a deuterium lamp or nonabsorbing lines near the analytical wavelength. Total Hg was determined in 0·5 g samples leached for 20 min in boiling aqua regia using the cold vapor method (Hatch & Ott, 1968). Accuracy and precision of AAS, determined by analyzing ≥ 4

replicate aliquots of reference materials with known concentrations, was $\pm 3\%$ for Cu, $\pm 7\%$ for Cd and Zn, $\pm 10\%$ for Hg and $\pm 13\%$ for Pb.

3. RESULTS AND DISCUSSION

All surface (0–2 cm) sediment samples (total = 205) were analyzed for Ba, Cr, Fe, Hg and Pb. In addition, a total of 60 selected samples among all 6 sites were analyzed for Cd, Cu and Zn. Only Ba, Cr and Fe were determined in subsurface samples. Because of the high sampling density used, systematic analyses of replicate samples at each station were not feasible. Instead variability estimates were made by grouping stations together. Since we were studying the effects of point source discharges, the most common grouping was by sampling distances, i.e. 1–8 samples per sampling radii, see Fig. 2. However, many of the sediment parameters showed considerable directional variability depending on the orientation of the prevailing water currents at the study site. Thus the grouping of samples by sampling distance often resulted in considerable variability especially near the drilling sites where the levels of residual discharged materials was the highest. This effect is well demonstrated in Figs 3A–3C where the variability in surficial sediment Ba decrease drastically, i.e. data point box size decreases, as a function of distance from the site.

To analyze these grouped data, a conservative statistical approach was taken which emphasized the use of non-parametric techniques. These techniques included Spearman rank-order correlations to measure the strength of the relationship between variable pairs, and Wilcoxon rank-sum (two-sample comparisons) or Kruskal–Wallis (k-sample comparisons) tests to evaluate the statistical significance of differences among sample groups. These techniques, although not as powerful as parametric methods, make minimal assumptions concerning the underlying distribution of the data and are especially well suited to nonlinear, discrete groupings of data such as those in this study. This conservative approach allowed us to clearly distinguish major trends and effects from variable grouped data. The probability of mis-identifying a trend as significant using this approach is minimal. However, these techniques may well have failed to identify more subtle trends in the data. Considering this trade-off and the overall variability of the data set, the conservative approach used is certainly the prudent choice.

Since trace elements in sediments are predominantly associated with the fine fraction, some portion of any gradient observed can be caused

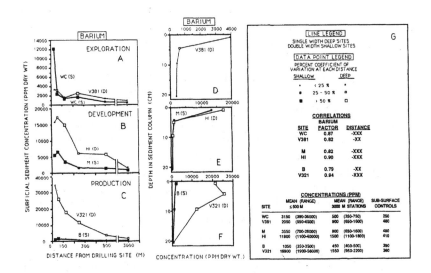

FIG. 3. Surface sediment Ba (0–2 cm, A–C) as a function of distance and subsurface (D–F) Ba as a function of depth at the 6 drilling sites. Each surface Ba data point represents the average of 1–8 samples collected at each sampling radii indicated. Each subsurface data point represents one sample analyzed from the depth indicated. Percentage coefficient of variation, a measure of data variability, is the ratio of the standard deviation to the mean expressed as a percentage (G). Inset (G) summarizes factor analysis loadings (see Table 3) and Spearman correlations for surface Ba data. The significance of the negative or positive distance correlations is indicated by one ($p < 0.01$), two ($p < 0.001$) or three ($p < 0.0001$) Xs. Inset (G) also lists the means (and ranges) of surface Ba concentrations within 500 m, at the 3000 m control stations and the mean of the subsurface control samples. Total excess barium within 500 m at each site is given in Table 2.

by natural variations in sediment texture and not drilling discharges. One way to remove the effects of natural variability is to use elemental concentrations normalized (ratioed) to iron in the statistical analyses. Iron is highly correlated with sand and clay because Fe is a major component of aluminosilicates which are the dominant constituents in clay-rich sediments but which are in very low concentrations in sand. In this study, no gradient was considered significant unless it was significant using concentration data which were both un-normalized and normalized to iron.

To link trends observed in sediment trace element levels to drilling discharges three criteria were used. First was a significant correlation with Ba (Spearman, $p < 0.05$). Barium sulfate (barite) is the major component of drilling muds (up to 90% by dry weight, see Table 1) and because of its low water solubility, chemical inertness and low background concentrations in undisturbed sediment, it is an excellent tracer of the settleable, insoluble components of discharges from drilling operations. Any trace element associated with the settleable drilling mud fraction would show a high level of covariance with Ba. Second was a significant Spearman correlation with distance from the drilling site, i.e. discharge point. Finally, discharged trace elements would be present in near-rig sediments at levels above those in control (background) sediments. Background concentrations were determined by analyzing sediments from 4 control stations located 3000 m from each drilling site (see Fig. 2) as well as subsurface sediments located well below the possible influence of surface discharges, i.e. 60 sub-surface samples, 4–31 cm depth. A control sample was used only if it exhibited an appropriate Gulf of Mexico (GOM) background Ba concentration, which ranges from < 200 to 700 ppm dry weight, depending on sediment texture (Boothe & Presley, 1985). Barium and Cr background concentrations are summarized in Figs 3(G) and 4(G), respectively.

3.1. Barium

Figure 3 shows the distribution of Ba in surface and subsurface sediments at the 6 study sites. Because of high sampling density, complete discharge history and subsurface analyses, a detailed, three-dimensional mass balance of discharged (excess) barium could be determined (Boothe & Presley, 1985). This three-dimensional approach estimates all excess barium present in the sediment column within 500 m of each study site.

All Ba concentration data used in the mass balance computations were corrected for background Ba levels at each site (Fig. 3(G)). For each core analyzed, a profile of excess Ba above background (BAEX) versus depth was plotted and integrated to determine the total excess Ba present (BAEXAC in units of g Ba/m^2 of seafloor). For stations where no subsurface sediment was analyzed, BAEXAC was estimated using the surface Ba concentration observed at the station and assuming a Ba depth profile similar to the profile at a similar, neighboring station with subsurface data. Since BAEX was generally observed to a depth of only 5 cm at more stations. the error associated with this estimate should be

P. N. Boothe and B. J. Presley

TABLE 2

Mass balance of total excess barium in sediments within 500 m of the six drilling sites used in this study

Drilling site	Total barium used (TBU) in drilling activities (1000 kg)[a]	Mean BAEXAC 0–500 m radius (g/m²)[b]	Total excess barium (TEB) in sediments (1000 kg) within 500 m[c]	Percentage of TBU within 500 m
West Cameron	2414	25.8	20.3	0.84
Vermilion 381	229	28.0	22.0	9.6
Matagorda	2334	27.5	21.6	0.93
High Island	1518	173.0	136.0	9.0
Brazos	1041	19.2	15.1	1.5
Vermilion 321	4964	759.0	596.0	12.0

[a] From Table 1.

[b] Mean total excess barium in the sediment column areal concentration (BAEXAC) within a 500 m radius of the drilling site = $TEB_{500}/2 \pi(500)^2$.

[c] Number of surface and subsurface samples used to make this estimate are given in Table 1.

small. For each site BAEXAC values for all stations were plotted and contoured. Each contour interval was integrated to yield total excess Ba (TEB, units kg Ba) in the sediment column within the area of seafloor enclosed by that interval. Finally, TEB values for all intervals at each site were summed to yield a single TEB for the entire sediment column within 500 m. These data are given in Table 2.

The total error associated with the Ba mass balance calculations cannot be explicitly determined but should not be enough to change the trend observed. The sampling density used in this study represents at least a two-fold improvement over other studies. Whether this density was sufficient to adequately define the patchy distribution of discharged Ba, or that enough significant patches were missed to change the results observed, is not known. However, the similarity of the percentage Ba retention within the shallow (1%) and deep water (10%) groups suggests the density was adequate. If such patches were being missed, the variability among similar sites should be greater than observed. The seafloor directly under the platform could not be sampled. This situation should not be a source of much error for deep water sites since most of the discharged materials would be carried by currents into the sampling area before reaching bottom. Accumulations under the platform could be a source of error for the shallow water sites. However, such accumulations persist for only a few months (summarized in Gettleson, 1980) and most of the drilling at the shallow water sites had occurred at least a year before sampling (Boothe & Presley, 1985). As discussed above, the error associated with determining BAEXAC values at each station should be low. Finally, the error associated with determining the total amount of Ba (TBU) used at each site is low. The TBU data were determined from actual mud logs and accounting records provided by the drilling companies. These records represent the best discharge history available to any rig monitoring study.

The mass balance shows that only a small fraction of the total barium used, i.e. $<1.5\%$ nearshore, $<12.0\%$ offshore, and presumably similarly behaving drilling mud components, are present in near-site sediments. This nearshore–offshore dichotomy is also apparent from the surface Ba data (Fig. 3(A–C)). Significant resuspension and removal of bottom sediments in the high current (high energy) nearshore environment appears to be responsible for this low percentage of residual barium at the shallow water sites. The higher barium levels at deep water sites probably come from barite associated with heavier cuttings and other aggregates which settle at a much faster rate than the normal bulk mud. The differences in length of time between cessation of drilling and sampling among the 6 sites (Table 1) has little effect on the percentage of the total barium used in drilling activities which was present in near-site sediments.

The TEB within 500 m (Table 2) is highly correlated with the total barium used at a site ($p < 0.005$). This means that in terms of total excess barium, the effect of multiple wells on near-site sediments is directly additive. As the number of wells, i.e. mass of materials discharged, at a site increases, the mass of residual barium, and presumably similarly behaving drilling mud components, present in the near-site sediments also increases. This increase represents about 1% of the new barium discharged at nearshore sites and about 10% at offshore sites. This additive process is very constant as shown by the fact that the percentage of the total barium used, which is present within 500 m of a site (Fig. 3(G)), is surprisingly constant within nearshore and offshore groups despite the large differences in number and age of wells among the sites in each group.

The distribution of Ba in the sediment column at each site (except West Cameron) is shown in Fig. 3(D–F). The excess Ba is generally restricted to the upper 5 cm except at the Vermilion 321 site where the excess extends down to 15–20 cm. This is a production site which had the most drilling (25 wells) and is in deeper water where Ba retention in the sediments was higher.

3.2. Trends in Trace Elements Distributions

An important goal of this study was to describe gradients of potentially environmentally important trace elements (Ba, Cr, Hg, Pb, Cd, Cu, Zn) in near-rig sediments and identify those perturbations attributable to drilling activities. The Ba mass balance data suggest that only a small fraction of the total barium used, and presumably other drilling fluid components, was present in near-rig sediments and thus could have contributed to the gradients observed. Presumably other drilling fluid components would behave as Ba did, and thus not accumulate appreciably near the drilling sites. Nevertheless, the sediments surrounding drilling sites exhibit the highest concentrations of drilling related materials found and it is important to document and evaluate these trends.

Factor analysis was used to identify those variables whose distributions were highly correlated with Ba and therefore consistent with a drilling discharge origin. Factor analysis is a multivariate statistical technique by which underlying relationships in a data set are reduced to a small set of factors which account for most of the variance in the data. To reduce the effects of any natural (nondrilling) sediment variability, factor analysis was run using both unnormalized and normalized (ratioed to Fe) trace element concentration data. Both approaches gave similar results. For each drilling site, factor analysis of surficial sediment data yielded 3 factors which accounted for an average of 77% of the variance among the data sets. One factor had high loadings (mean = 0·71) for sediment texture parameters (sand, silt, clay) and Al, Fe and Cr. This factor depends on the sediments at the sites including the clay-rich aluminosilicates. A second factor exhibited high loadings (mean = 0·94) for $CaCO_3$ and represents biological input to the sediments. A third factor was characterized by high correlations with Ba accounting for an average of 74% of the total variance in that variable at all sites studied.

Table 3 gives the results of the factor analysis for the barium factor. It is important to note that the 2 sites with the highest variable loading on this factor, i.e. High Island and Vermilion 321, deep water sites, are also the 2 sites with the largest amounts of excess Ba in their sediments (Table 2, 6 and 26 times more than any other site). The nearshore sites, where very little of the drilling muds discharged remained in the vicinity of the site, showed much lower variable loadings. This observation suggests that the number of sediment parameters showing significant correlations with Ba, i.e. Spearman $p < 0.05$, similar trends of contamination, is directly related to the amount of excess Ba in the sediments. This observation also shows that the more pervasive sediment perturbations attributable

TABLE 3

Iron and trace element loadings (correlations) on the barium factor for the six drilling sites studied[a]

Variable measured	West Cameron (ES)	Variable: factor loading by drilling sites[b]				
		Vermilion 381 (ED)	Matagorda (DS)	High Island (DD)	Brazos (PS)	Vermilion 321 (PD)
Distance from drilling site	−0·81	—	−0·65	−0·89	−0·76	−0·86
Barium	0·87	0·82	0·82	0·90	0·79	0·94
Chromium	—	—	—	—	—	—
Iron	—	—	—	—	—	−0·51
Lead	—	0·72	—	0·97	—	—
Mercury	0·50	—	0·86	0·92	0·45	0·82
Proportion of total variance[c]	16·9	21·5	18·0	35·4	16·5	32·2

[a] Insufficient samples were analyzed for Cd, Cu and Zn to be included in this analysis.

[b] Table represents that portion of the factor analysis structure matrix describing the barium factor. Data given (except proportion of total variance) are pairwise correlation coefficients of the variables measured with the barium factor. Only coefficients >0·45 i.e. >20% of the variance accounted for, are listed.

[c] The proportion of the total variance in the data accounted for by the barium factor. In this case the value is a relative measure of the number of variables that load strongly on this factor among the drilling sites studied.

to drilling activities are largely restricted to deep water development and production sites.

3.2.1. Chromium

Based on drill mud logs from each site, Cr was an important component of many drilling mud systems used at the sites studied (Boothe & Presley, 1985). Because of its toxicity to marine organisms (hexavalent form), there has historically been considerable concern over the discharge of Cr during drilling operations. Figure 4 shows the trends in surface and subsurface sediment Cr concentrations at the 6 study sites. At the Matagorda and Brazos sites (Figs 4(B) and 4(C)), mean sediment Cr levels were greater at 125 m from the site than at 30 m. This difference is significant (Wilcoxon, $p < 0.01$) and reflects a similar increase in clay content over that interval. Chromium is highly correlated with both clay and Fe in undisturbed, clayey marine sediments. Vermilion 321 is the only site exhibiting a significant change in Cr with depth in the sediment (Kruskal–Wallis

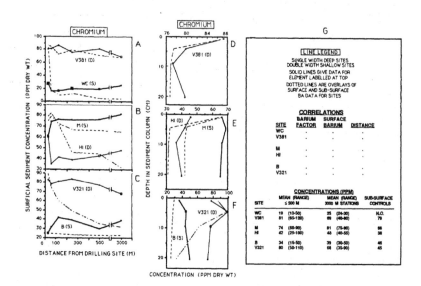

FIG. 4. Surface sediment Cr (0–2 cm, A–C) as a function of distance and subsurface Cr (D–F) as a function of depth at the 6 drilling sites. Each surface Cr data point represents the average of 1–8 samples collected at each sampling radii indicated. Each subsurface data point represents one sample analyzed from the depth indicated. No subsurface samples could be collected at the West Cameron site. Data point size is as explained in Fig. 3. Dotted surface and subsurface Ba overlays are from Fig. 3(A–F). Inset G summarizes the surface Cr data as described in Fig. 3 for the Ba data. A dot indicates no significant ($p < 0.05$) loading or correlation of Cr with the variable indicated was observed at a given site.

$p < 0.05$, Fig. 4(F)). At this site, Cr is highly correlated ($p < 0.0001$) with Ba which is consistent with a drilling mud origin. The maximum Cr concentration (190 ppm, 4–5 cm below the sediment surface) represents about a four-fold elevation over control levels (Fig. 4(G)). Elevations in near-rig sediment Cr compared to control stations are infrequent (patchy) and generally less than twice expected background levels. Also Cr levels show no correlation with the amount of chrome-containing materials used at a given site.

3.2.2. Mercury
High Island and Vermilion 321 show elevated concentrations of sediment Hg which are attributable to drilling activities. Mercury shows a significant

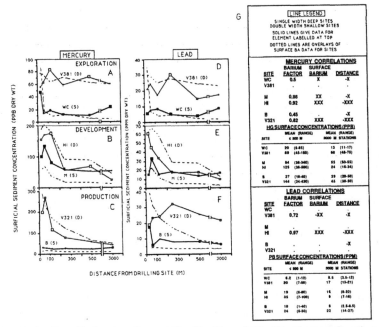

FIG. 5. Surface (0–2 cm) sediment Hg (A–C) and Pb (D–F) as a function of distance from the 6 drilling sites. Each data point represents the average of 1–4 samples collected at each sampling radii indicated. Data point size is as explained in Fig. 3 except here the smallest box represents only one sample analyzed at the sampling radii indicated. Dotted surface Ba overlays are from Fig. 3(A–C). Inset G summarizes the surface Hg and Pb data as described in Fig. 3 for the Ba data. A dot indicates no significant ($p < 0.05$) loading or correlation. No subsurface samples were analyzed for Hg or Pb.

decrease with distance from the platform at both sites (Figs 5(B) and 5(C), Kruskal–Wallis test $p < 0.01$). These trends are the result of 7–10 samples having ≥ 100 ppb Hg within 125 m of both sites. The mean Hg elevation within 125 m of the site compared to control stations is 6·9-fold at High Island and 4·5 fold at Vermilion 321. No Hg concentrations were determined in subsurface samples.

Where Hg levels were elevated, it was also strongly correlated with Ba at these sites ($p < 0.0001$). This observation suggests that Hg may be present as a trace contaminant of the barite used. The ratio of Ba to Hg in sediments at these sites (Ba/Hg = 100 000) is in the range of that found in veined barite deposits by Kramer *et al.* (1980) although the exact type or source of barite used at the study sites is unknown. Also High Island

and Vermilion 321 are the two sites with the largest amounts of excess Ba in near-site sediments (Table 2). The concentration of Hg in the barite required to cause the elevated sediment Hg levels observed is only 1–3 μg Hg/g of barite. Data from the Matagorda site supports the hypothesis that barite is the source of the elevated Hg. The elevation in sediment Hg levels within 125 m of the Matagorda site is significant (Wilcoxon test $p < 0.05$) compared to control stations (Fig. 5(B)) and mercury is highly correlated with Ba ($p < 0.001$). However, the amount of Hg contamination is much less at this site which corresponds with the fact that the total amount of excess Ba in sediments surrounding the Matagorda site is much less (6 and 26 fold) than that at the High Island and Vermilion 321 sites (Table 2). The trend with distance at the Matagorda site is due to 3 high samples and the mean Hg elevation within 125 m compared to controls is 1·9 fold.

3.2.3. Lead

Figures 5(D–F) show the trends in sediment Pb concentrations as a function of distance from the drilling sites studied. High Island is the only site which shows a consistent and significant Pb gradient related to the drilling platform. This trend is significant (Kruskal–Wallis test $p < 0.03$). The mean Pb concentration within 500 m of the site was significantly greater (Wilcoxon test $p < 0.02$) than that at the 3000 m control stations, i.e. mean = 8 ppm dry weight. At High Island Pb was strongly correlated with Ba ($p < 0.0001$) which is consistent with a drilling discharge origin. However, unlike Hg, the source of the Pb observed did not appear to be the barite. The elevations in sediment Pb levels were much greater than could be accounted for by the Pb concentrations found in natural barite (Kramer *et al.*, 1980). Also the Vermilion 321 site, which has 4·2 times more excess Ba in near-site sediments than the High Island site, did not show a similar Pb gradient. The other study sites exhibit a patchy distribution of elevated sediment Pb concentrations but no consistent trend with distance from the site. The highest Pb concentrations observed are generally <2 times control levels.

3.2.4. Zinc and Other Trace Metals

The concentrations of Cd, Cu and Zn were determined in 9–10 surficial sediment samples at each study site. Statistical evaluation of any trends observed is difficult with such small sample sizes. More specifically, not every distance is sampled and the level of each metal observed at any given distance from the site is based on only 1–2 samples. Figure 6 shows

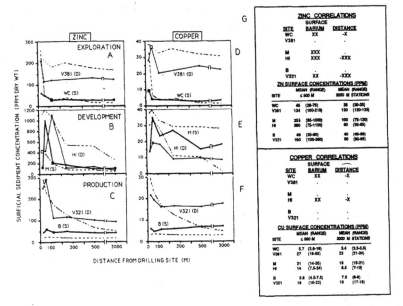

FIG. 6. Surface (0–2 cm) sediment Zn (A–C) and Cu (D–F) as a function of distance from the 6 drilling sites. Each data point represents the average of 1–2 samples collected at each sampling radii indicated. Data point size is as explained in Fig. 3 except here the smallest box represents only one sample analyzed at the sampling radii indicated. Dotted surface Ba overlays are from Fig. 3(A–C). Inset G summarizes the surface Zn and Cu data as described in Fig. 3 for the Ba data. A dot indicates no significant ($p < 0.05$) loading or correlation. No subsurface samples were analyzed for Zn or Cu.

the trends in sediment Zn and Cu concentrations, respectively, surrounding the drilling sites studied. No similar Cd plot was prepared since many samples at several sites had Cd concentrations below the detection limit of flameless AAS (< 0.01 ppm). The elevations of these metals in the sediments were greater, i.e. ≤ 2–10 times, than could be accounted for by the level of trace metal contaminants in the discharged barite alone. Much of the observed elevations are apparently caused by the metals present in other drill mud additives or other types of discharges from the drilling platform.

Of the 3 metals, Zn showed the most consistent gradients which occurred at the High Island and Vermilion 321 sites. At both sites, Zn was significantly correlated with both distance and Ba (Spearman $p < 0.01$). At all sites elevated sediment Zn levels (Ba/Zn = 20–150) were much greater than could be accounted for by Zn present in natural barite (Ba/Zn > 310).

Copper showed no consistent trend at any site except small elevations within 125 m of the High Island site (Fig. 6(D–F)). The elevations in sediment Cu levels (Ba/Cu = 1400) were much greater than could be explained by Cu normally present in barite ore (Ba/Cu $\geq$ 9400, Kramer *et al.*, 1980).

Cadmium appeared to be elevated near the High Island and Vermilion 321 sites. At each site, 3 samples showed high concentrations within 125 m of the drilling site. These elevations were all more than 10 times control levels. Cadmium was significantly correlated with Ba ($p < 0.01$) at the Vermilion 321 site. The similarity of the Ba/Cd ratios observed in the sediments (44 000) and in natural barite ore ($\geq$ 71 000, Kramer *et al.*, 1980), indicates that Cd present in the discharged barite could account for a considerable fraction of the elevated sediment Cd levels observed.

4. CONCLUSIONS

1. The Ba mass balance suggests that retention of discharged Ba in near-field sediments is low, < 1% for shallow water sites and < 12% for deep water sites.
2. Significant elevations in sediment Hg (4–7 fold) and Zn (5–10 fold) were observed, especially at deep water development and production sites.
3. Little or no significant elevations in other sediment trace metals were observed including Cd, Cr, Cu and Pb.

REFERENCES

BOOTHE, P.N. & PRESLEY, B.J. (1985). Distribution and behavior of drilling fluids and cuttings around Gulf of Mexico drilling sites. Final Report to the American Petroleum Institute, Washington, DC, 140 pp.

GETTLESON, D.A. (1980). Effects of oil and gas drilling operations on the marine environment. In *Marine Environmental Pollution. I. Hydrocarbons*, ed. R.A. Geyer. Elsevier, New York, pp. 371–412.

GETTLESON, D. A. & LAIRD. (1980). In *Symposium/Research on the environmental fate and effects of driling fluids and cuttings*, Proceedings Volume II. Lake Buena Vista, Florida, 21–24 January 1980.

HATCH, W.R. & OTT, W.L. (1968). Determination of sub-microgram quantities of mercury by atomic absorption spectrophotometry. *Anal. Chem.*, **40**, 2085–87.

KRAMER, J.R., GRUNDY, H.D. & HAMMER L.G. (1980). Occurrence and solubility of trace metals in barite for ocean drilling operations. In *Symposium /Research on the environmental fate and effects of drilling fluids and cuttings.* Proceedings Volume II. Lake Buena Vista, Florida, 21–24 January 1980, pp. 789–98.

MENZIE, C.A. (1982). The environmental implications of offshore oil and gas activities. *Environ. Sci. Technol.,* **16**(8), 454A–72A.

NATIONAL RESEARCH COUNCIL (1983). *Drilling Discharges in the Marine Environment.* National Academy Press, 180 pp.

2

The Fate of Oil-based Drilling Muds at Two Artificial Island Sites in the Beaufort Sea

P. ERICKSON, B. FOWLER and D. J. THOMAS

Seakem Oceanography Ltd, 2045 Mills Road, Sidney, British Columbia, Canada V8L3SI

ABSTRACT

The fate of low aromatic content base oil (Vista ODC), discharged from two artificial island exploratory well sites in the Beaufort Sea drilled by ESSO Resources Canada Limited, was studied over a 2-year period. Oiled cuttings were discharged during the winter at both locations. At Minuk I-53 a sacrificial beach island in 14 m of water, base oil could not be detected in surface sediments immediately after drilling outside of an extensive area of grounded ice rubble surrounding the island. Vista oil, however, was widely dispersed in surface sediments at the end of the following open water season as delineated by a distinctive signature of low molecular weight isoprenoids in the GC trace. Most of the oil was dispersed around the south side of the island from the discharge point on the west side and thence in an easterly direction coinciding with the direction of net current flow in the area. Base oil could be reliably quantified more than 700 m to the east of the island. The maximum concentration of Vista ODC (defined in terms of the sum of 10 low molecular weight isoprenoid peaks in the GC trace of Vista) in all samples collected was more than 4000 µg/g to the southeast of the island centre. Dispersion was more restricted at Kaubvik I-43, a caisson-retained island. Oiled cuttings were dispersed in an east-west orientation from the discharge point on the south side of the island with quantifiable base oil detectable more than 400 m to the west. The maximum concentration of Vista observed was 2000 µg/g. Dispersal of oil-based muds appears to be enhanced by the presence of grounded ice rubble near a sacrificial island.

Grounded ice restricts dispersion in the winter but can carry oiled cuttings long distances from the island after ice break-up. The fate of oiled cuttings discharged from a caisson-retained island in deeper water appears to be governed mainly by currents and in this respect is more typical of dispersion from a conventional drilling platform in more temperate seas.

1. INTRODUCTION

Oil-based drilling muds (OBM) offer several advantages over water-based muds. The use of OBM reduces or eliminates borehole problems, reduces risk and results in considerable time and cost savings in the drilling of an offshore well, particularly where hole stability is a problem (Johancsik & Grieve, 1987a). The use and disposal of OBM in the offshore regions of Canada has been restricted, however, due to environmental concerns regarding the toxicity of the oils used in OBM formulations. The toxicity of OBM is thought to be primarily a function of its aromatic content, particularly the low molecular weight aromatics and higher molecular weight polycyclic aromatics. As a result, highly refined mineral oils with a low aromatic content (usually <1%) are now used almost exclusively in offshore drilling to produce a more environmentally acceptable formulation.

OBM have been used extensively in the North Sea and the experience there is the source of most of the available information on the environmental effects of OBM in the marine environment (Davies *et al.*, 1984; also Davies *et al.* in these proceedings). In Canada's offshore, the fate of low aromatic OBM has been studied at two offshore well sites near Sable Island (Yunker & Drinnan, 1987). In the Beaufort Sea, ESSO Resources Canada Limited has used OBM at two shallow water well sites and discharged treated cuttings to the sea from one of these on an experimental basis. No definitive guidelines, however, have been developed for the ongoing use and disposal of OBM in Canadian waters. Because of the many attractive features of OBM from a drilling standpoint, a need exists to determine the fate and environmental effects of the discharge of OBM to the Beaufort Sea. While experience in the North Sea and the two wells on the east coast offers much relevant information, there are several aspects of drilling in the Beaufort which limit the use of data collected in more temperate regions: (1) water depths are generally less than 30 m in the areas where major oil and gas discoveries have been made; (2) the region is ice covered for up to 9 months of the year; (3) the bottom is subject to ice scour; (4) artificial islands and berm-supported caissons are often used as drilling platforms. In addition, biological productivity is lower than in more temperate seas and the nearshore arctic marine habitat is thought to be more sensitive to the effects of discharged wastes and site disruption.

This study was carried out to address the fate of oil-contaminated cuttings discharged from two well sites in the Beaufort Sea based on measurements of oil in surrounding surficial sediments on the completion of drilling and following one open water season. The two wells, Minuk I-53 and Kaubvik I-43 were drilled by ESSO Resources Canada Limited in the winters of 1985–6 and 1986–7, respectively. Minuk I-53 was drilled from a sacrificial beach island in 14 m of water and Kaubvik I-43 drilled from a caisson supported by a subsea berm in 19 m of water. Both locations are near the outer edge of the average maximum extent of landfast ice in the southern Beaufort Sea (Fig. 1). The water depths and island types are representative of most offshore wells drilled in the Beaufort. The Kaubvik well was to be drilled in the summer of 1985 but was delayed until the fall/early winter of 1986/7. The Minuk island and rig were damaged by a storm in September 1985, delaying spudding of that well until the early winter of 1985. Drilling was completed in April 1986. As a result of these changes in drilling schedule, only two sampling trips were made to the Kaubvik site; predrilling and at the beginning of the open water season following well completion.

Cross-sectional views of a typical sacrificial beach island and an island formed by placing a caisson on a subsea berm as used at Minuk and Kaubvik, respectively, are shown in Fig. 2. Discharge of treated cuttings from a sacrificial beach island occurs into the shallow water of the upper island slope. Although discharge of cuttings from a caisson-retained island is usually into deeper water than that around a sacrificial beach island, most of the cuttings in either case will tend to accumulate on only one side of the structure and cannot be dispersed in any direction freely. Dispersion from a sacrificial beach island will be very dependent on wave action in open water as well as ice scour. Once abandoned, both sacrificial beach and caisson-retained islands erode quickly (EBA Engineering Consultants Ltd, 1984). Because of the greater depth, wave action will be less of a factor and local currents more important in dispersal of oiled cuttings around a caisson-retained island.

Approximately 740 m³ of treated drill cuttings were discharged at Minuk between November, 1985 and April 1986 and 643 m³ of cuttings discharged at Kaubvik in November and December 1986. A total of 278

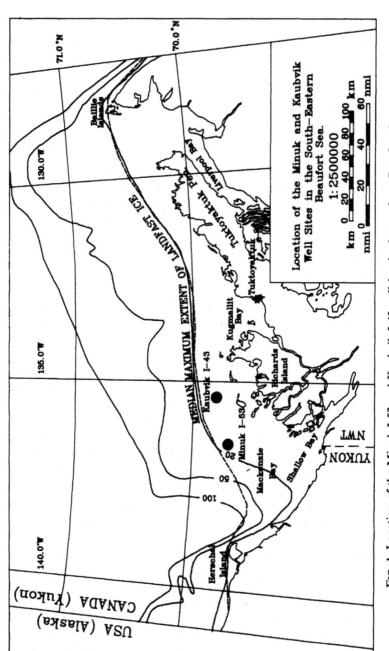

FIG. 1. Location of the Minuk I-53 and Kaubvik I-43 wellsites in the southern Beaufort Sea.

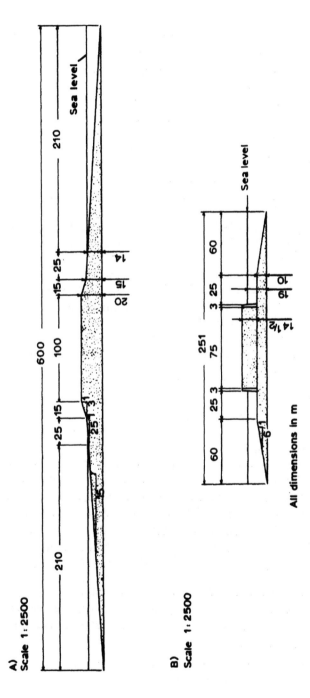

FIG. 2. Cross-sectional profiles of the Minuk I-53 and Kaubvik I-43 artificial islands.

TABLE 1
Characteristics of Vista ODC and Escaid 90 base oils
(from Johancsik & Grieve, 1987b)

Product	Viscosity (cSt)	Pour point (°C)	Density (kg/litre)	Aromatics (%)	100% WSF LC_{50} toxicity
Vista ODC	1·8	−59	0·805	0·2	Non-toxic
Escaid 90	1·24	−50	0·792	0·5	Non-toxic

tonnes of oil were discharged along with 1249 tonnes of barite at Minuk. At Kaubvik, a total of 143 tonnes of base oil were discharged with 332 tonnes of barite. Overall oil retention at Minuk and Kaubvik was 13.6 and 14.9 g/100 g dry solids, respectively. Vista ODC* was used at Minuk and Vista ODC supplemented with smaller amounts of a second base oil, Escaid 90, were used at Kaubvik. Both Vista ODC and Escaid 90 are paraffinic, naphthenic oils with low (<1%) aromatic contents. A summary of their characteristics is given in Table 1.

Vista ODC was also the base oil used in the study of the fate of OBM discharged at two well sites on the east coast offshore Nova Scotia (Yunker & Drinnan, 1987).

2. METHODS

2.1. Sampling Design
Samples were collected on a stratified random basis. The maximum expected radius of detectable input was set at 1 km from the island centre for Minuk and 2 km for Kaubvik. The area within this radius was divided into 5 subareas on the basis of the likelihood of a detectable impact. The strata are indicated in Fig. 3. As it was believed that the maximum impact would occur on the same side of the islands as the point of discharge, most of the sampling effort was allocated to this area. Strata 1, 2 and 3 were selected on the same side of the island/berm as the discharge pipe at distances of 0–300, 300–500 and 500–1000(2000) m respectively in a 120° arc centred at the discharge point. It was expected that the area encompassed by these would experience the greatest input with most of

*Editors' Note: Vista ODC base oil is chemically similar to the Conoco base oil referred to elsewhere in these proceedings.

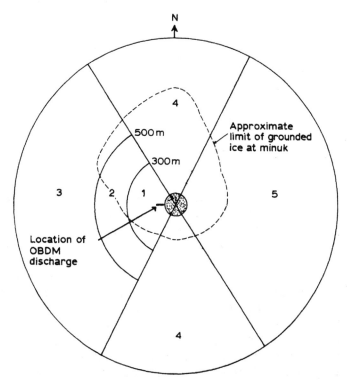

FIG. 3. Sampling strata and outline of the ice rubble boundary at the Minuk island, Winter 1986.

the base oil deposited within 300 m of the discharge point based on the east coast and North Sea experience (Yunker & Drinnan, 1987; Davies *et al.*, 1984). Stratum 4 was assigned as the area immediately adjacent to strata 1, 2 and 3. Stratum 5, the area on the opposite side of the island, was assumed to be the least likely area to show a measurable effect. Numbers of sampling locations were allocated equally among areas so that the density of samples was greatest in strata 1, 2 and 3. In post-drilling sample collections, additional samples were taken in a systematic fashion approximately 200 m from the islands at intervals of 30–45°. The winter sampling at Minuk was modified further because of a large area of grounded ice rubble on the west and northwest sides of the island. No samples could be taken within this area which covered much of strata 1, 2, 3 and 4 (Fig. 3).

TABLE 2

Stratum No.	No. of samples	
	Minuk	Kaubvik
1	5	5
2	6	6
3	4	4
4	8	8
5	3	3
Reference	3	3

Sampling sites were chosen randomly within each stratum by dividing them into substrata (10° arcs, 50 m concentric circles), assigning numbers to the substrata and randomly selecting the numbers. The procedure of randomly selecting the sampling locations within the strata was performed before each sampling time. Additional samples were taken from locations randomly chosen within two reference zones (zones beyond the outer radius), one for each drill site. The reference zones at each site were 500 m diameter circles centred 750 m outside the study area to the east in water of similar depth and with a relatively flat bottom (no borrow pits or trenches). The number of samples taken from each stratum plus the number of reference samples are given in Table 2 (based on a maximum of 29 sites per location). Additional samples (randomly selected from each stratum) were taken, but not analysed.

A strategy involving the preparation of composite samples was employed to reduce the sample variance. Three replicate grabs were taken at each sampling site and 3 subsamples, taken from the top 1 cm of sediment of each grab, combined. From the results of Piotrowicz *et al.* (1981) the composite samples should be an unbiased estimate of the mean of the three replicates. Analysis of a composite sample would thus substantially reduce the sampling variance for little additional effort and cost. The variance of replicate grabs within sites was not measured because it is not used in the statistical analysis.

2.2. Sampling
The sampling schedule in this study was revised several times as a result of alterations in drilling schedules. The Kaubvik I-43 well, originally planned for the summer of 1985 was not spudded until the fall of 1986 just prior to freeze-up. The Minuk island and drilling rig suffered severe

TABLE 3

Sampling trip	Dates	Description
1	29 August–4 September	Predrilling sampling at Kaubvik and Minuk collection at Kaubvik and Minuk
2	29 September–11 October 1985	Repeat of predrilling sampling at Minuk
3	18–27 April 1986	Postdrilling sampling through the ice at Minuk
4	20–22 September 1986	Follow-up sampling at Minuk after one open water season
5	1–4 August 1987	Postdrilling sampling at Kaubvik

damage in a storm in September 1985 which delayed spudding of that well until after freeze-up in 1985. It is estimated that approximately 2300 tonnes of barite and 338 000 litres of diesel fuel were lost around the island as a result of the damage to the island and storage tanks. Loss of this material as well as erosion and large scale alterations to the island necessitated repeat background sampling. Table 3 gives a summary of the sampling actually completed.

Samples from each site were assigned a series number corresponding to the sampling trip and the island location. Samples from the first trip taken at Minuk are termed series 1; background samples from Kaubvik are series 2; repeat background samples at Minuk from the second trip are series 3; post-drilling samples at Minuk on trip 3 are series 4; open water follow-up samples from Minuk are series 5 and post-drilling samples from Kaubvik collected on the fifth sampling trip are series 6. These terms (series 1–6) are used in subsequent discussions of the data.

Surface samples were collected with a modified Ponar grab with a 0.06 m² bite. The grab sampler was deployed with an aluminium hand winch and davit welded to the deck of the support vessels. Three separate grabs were taken at each sampling site and subsamples from each combined into a single 250 ml hydrocarbon clean glass storage jar. The sampler was equipped with hinged stainless steel doors allowing access to the surface of the sample after the grab had been retrieved. The upper 2 cm of sediment were skimmed from the surface of the grab with a cleaned, stainless steel scoop. In April, 1986, the grab sampler was deployed through 60 cm holes in the ice made with a portable hot water ice melter. In September 1986, seven sample locations 200 m or less from the Minuk island centre were not accessible by boat and were sampled by diver. Diver-collected samples were taken by scooping the top 2 cm of sediment into a 500 ml sampling jar from three locations within 10 m of the target site. The sample jar was recapped before returning to the surface. Core sampling was attempted at Minuk on the fourth sampling trip close to the island in an attempt to see whether any drilling muds were buried and detectable in deeper portions of the sediment. Coring was unsuccessful as the sediments were exclusively sand. Three gravity cores were taken on the final sampling trip at Kaubvik from just beyond the berm. Corers consisted of a 10×50 cm acrylic core tube with stainless steel cutting head on a lead weighted, aluminium core body. The corer was lowered to within 3 m of the bottom and then allowed to free fall. Cores were successfully collected at three locations 400 m and 90°, 180° and 270° from the island centre. Cores were extruded with a plunger covered with cleaned aluminium foil onto a surface covered with aluminium foil. The cores were sectioned into subsamples 2–6 cm in length then transferred to glass storage jars as for grab samples. All sediment samples were stored cool until returned to the laboratory where they were homogenized and then frozen until analysis (1–3 months). In April 1986 (Series 4) samples were frozen immediately after collection and not thawed until analysed.

Samples for Ba and particle size analysis were taken from each grab or core sample after samples were taken for hydrocarbon analysis. Approximately 50 g of wet sediment from the upper few cm was skimmed off with a scoop and placed in a plastic 'Whirl-Pak' bag. Samples were stored as for hydrocarbons.

2.2.1. Positioning

Several different methods of positioning were used. Supply vessels used to sample on trips 1, 2 and 4 were equipped with a Syledis SR3 positioning system with a HP 9836 computer and display. The positioning system recorded the location of the sampling davit at the time of grab impact and positions are believed accurate to within 10 m. On the final sampling trip, positioning was via radar range and bearing to the caisson centre. Positioning accuracy under these conditions is believed to be within 50 m close to the caisson and to within 200 m at 2 km from the caisson. Positions in April 1986 (trip 3) were surveyed by triangulation from reference points on the island with range determined to within 5 m. Sampling was complicated on trips 1, 2 and 4 because of the fast drift rate of the vessel.

In all cases, the vessel was positioned so as to drift over the station. When within a 50 m radius of the target area, the grab was deployed. On the final sampling trip, the *Arctic Nanook* was able to maintain its position over the target area through the use of bow thrusters.

2.2.2. Sediment Traps

Sediment traps were deployed at each site prior to the start of drilling. Traps were deployed at three sites at each island on the same side as the discharge pipe and off the island or berm slope. Each mooring consisted of two 10-cm diameter × 70-cm long PVC cylinders with a glass collection cup at the bottom suspended approximately 3–4 m above the sediment. Only one trap mooring from the Kaubvik site was successfully recovered.

2.2.3. Current Meters

Aanderaa RCM-4 current meters were deployed at Minuk during drilling at locations approximately 1 km due north and south of the island to obtain a current record during OBM discharge. One meter was deployed at each site. Both meters were deployed initially on trip 1 and were to have been recovered prior to freeze-up. On 10 October 1985 the instruments were checked as a result of the storm that had damaged the island. The north meter was damaged (rotor missing; vane bent) and no usable data obtained. Another meter was deployed at this location. The south meter was operating properly, the tape replaced and the meter redeployed. The south meter was at a depth of 14.6 m for the initial period and 13 m over winter. Recovery of the meters was attempted by diver 20–22 April 1986. The north meter was not found and is presumed to have been dragged off position by ice. The south meter was recovered on 21 April. The south meter had been moved about 75 m off location and was buried in the mud with only the vane and vinyl floats showing. This had evidently occurred shortly after deployment as only about a 2 week record was obtained.

2.3. Hydrocarbon Analyses

The method used for the determination of hydrocarbons in sediments and drill cuttings is an adaptation of the method of Cretney *et al.* (1980) as used previously for the determination of low toxicity base oil in cuttings and marine organisms (Hutcheson *et al.*, 1987). As Vista oil is relatively volatile (boiling range 188–270°C, corresponding to nC10–nC15) special consideration was given to retain the volatile components during work-up, primarily by the use of Kuderna-Danish solvent evaporators for solvent removal.

Sediments were stored frozen and prior to analysis were allowed to thaw at 40°C overnight, thoroughly mixed and subsampled for hydrocarbon analysis and dry weight determination.

A sediment subsample (20–30 g) was weighed into a 250 ml round bottom flask with 40 ml methanol, 4 g 50% aqueous potassium hydroxide and aliquots of internal standards for the nonpolar hydrocarbons (*n*-decane-d22, 5-alpha-androstane and *n*-hexatriacontane and aromatic compounds perdeuterated naphthalene, anthracene, pyrene, chrysene, benzo(k)fluoranthene, and perylene). The flask was initially stirred then the sample refluxed for 1 h. Water (40 ml) was added to hydrolyse esters and the sample refluxed for an additional 30 min. The cooled digest was serially extracted by shaking with pentane (3 × 60 ml) and the extracts decanted through a glass fibre filter (47 mm Whatman GF/C in a Millipore holder) into a 250-ml separatory funnel. The combined extracts were washed by back extraction with pre-extracted water (3 × 60 ml), dried over anhydrous sodium sulphate and taken down to 500 μl in a Kuderna-Danish concentrator. The concentrated extract was loaded onto a silica gel column (10 ml bed volume, 13 cm × 1·0 cm, 5% water deactivated, topped with a 1 cm layer of anhydrous sodium sulphate). Non-polar hydrocarbons were eluted with 25 ml pentane and the aromatics with 40 ml dichloromethane, and the fractions taken down to 100 μl by Kuderna-Danish for GC/FID and GC/MS/SIM analysis.

An additional subsample (5 g) was weighed into a 50 ml beaker and dried at 80°C for 24 h for dry weight determination.

Nonpolar hydrocarbons were analysed using Hewlett-Packard 5830A/5840A gas chromatographs with flame ionisation detectors (FID). Peak areas were measured using the HP 5830/40 integrator and compounds were quantified relative to the primary internal standard 5-alpha-androstane. Perdeuterated *n*-C10 was used as a volatilisation check.

An even *n*-alkane calibration standard was run at least once a day and relative response factors calculated relative to 5-alpha-androstane.

Aromatic fractions were analysed on a Finnigan 9500/3200 gas chromatograph-mass spectrometer (GC/MS), with a Finnigan 6100 data system. PAH and masses monitored were naphthalene (128), C1–C4 alkylated naphthalenes (142, 156, 170, 184), fluorene (166), dibenzothiophene (184), C1–C2 alkylated dibenzenothiophenes (198 and 212), phenanthrene (178) anthracene (178), C1–C4 alkylated phenanthrene/anthracene (192, 206, 220 and 234), dibenzanthracene (228), chrysene (228), benzofluoranthenes (b, k, j and a) (252), benzo(e)pyrene (252), benzo(a)pyrene (252), and perylene (252).

2.3.1. Quality Control and Quality Assurance

A comprehensive quality control program was followed to maintain and document accuracy and precision for hydrocarbon analyses. Procedures included the following:

(1) Procedural blanks run initially and regularly throughout the sample suite.
(2) 10% of the samples analysed as blind replicates.
(3) GC response calibration standards run at the beginning and end of each working day. The relative response is required to be within prescribed limits before analyses can proceed.
(4) Frequent duplicate GC injections to monitor instrumental variation.
(5) Mass spectrometer resolution and response to a fragmentation standard required to meet US EPA criteria (Eichelberger *et al.*, 1975).

The precision of nonpolar hydrocarbon analyses based on the pooled variance of 9 samples analysed blind in duplicate (expressed as the relative standard deviation) was 21% for total nonpolar hydrocarbons; 19% for the sum of the 10 Vista marker isoprenoids (Section 2.4); and 22% for the total *n*-alkanes.

The accuracy of PAH analyses was monitored by the analysis of a certified reference material (US NBS SRM No. 1647) consisting of an acetonitrile solution of several PAH of interest at a certified concentration. Results for most compounds were lower than certified levels (8% low for benz(a)anthracene and chrysene to 40% for benz(a)pyrene). Results for fluorene were within 5% of certified values whereas anthracene results were 24% high. The precision of total PAH analyses based on six blind replicate analyses was 11% (relative standard deviation).

2.4. Determination of Vista Base Oil Detection Limit

Vista ODC is a narrow boiling range raffinate which gives a distinctive complex GC signature within the nC10–nC16 range. Under the capillary GC conditions used in this study, over 150 components are resolved including approximately 15–20 prominent resolved peaks over an envelope of an unresolved complex mixture (Fig. 4). The major resolved components appear to be isoprenoid hydrocarbons, naphthenes and the normal alkanes from nC11 to nC14. The normal alkanes are not a prominent set in the GC trace. At lower levels, the distinctive features of the Vista GC pattern are lost leaving the isoprenoids as the Vista markers. Several of these same isoprenoid hydrocarbons are prominent in the Beaufort Sea surficial

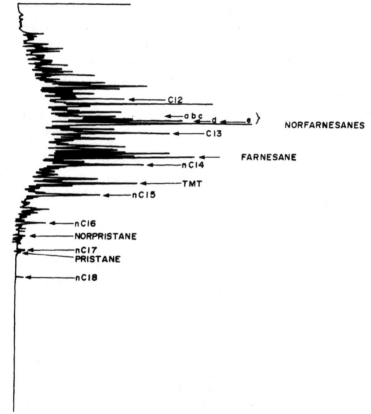

FIG. 4. Gas chromatogram of the aliphatic fraction Vista ODC base oil.

sediments, and the detection limit of Vista in these sediments is raised. As discussed in Section 3 the hydrocarbon levels in the study area sediments are a function of sediment particle size. Total alkanes range from an average of 9200 ng/g for predominantly (>98·5%) clay/silt typical of area surficial sediments to 310 ng/g for predominantly (>98·5%) sand used for island and berm construction. Isoprenoid concentrations vary in a similar manner with particle size. Clearly, the Vista detection limit will also be a function of particle size and will be highest, i.e. the GC method least sensitive, for predominantly clay/silt surficial sediments and lowest, i.e. most sensitive, for the sand/gravel used for island construction.

The Vista detection limit was determined for the instrument, then in samples directly in a Minimum Detectable Vista Oil (MDVO) experiment by spiking a typical study area sediment and a sediment extract. The instrumental detection limit was determined for the GC using a series of Vista standards (0.9–500 mg/ml), 1.0 ml of each spiked with the working internal standard and analysed by GC using the standard conditions. Calibration data were plotted for six combinations of distinctive peaks, and the total of the ten conspicuous peaks, assigned generally as isoprenoid hydrocarbons, were selected as the Vista indicators. With the standard GC conditions, the sum of the selected isoprenoid peaks can be converted to Vista using the expression:

$$(\text{Vista}) = (\text{sum selected isoprenoids}) \times 0 \cdot 019 \ 53 + 13 \cdot 77$$

where the isoprenoid values are entered in nanograms and the resulting Vista values are given in micrograms. The absolute instrumental detection limit for Vista was found to be 20 ng Vista injected which is equivalent to $0 \cdot 2$ mg/g for a 10 g sample, analysed using standard conditions.

The detection limit in study area sediments was determined directly by a Minimum Detectable Vista Oil (MDVO) experiment. A typical area sediment (Minuk predrilling sample 3-07, collected October 1985 300 m from Minuk island centre, $52 \cdot 2\%$ clay and silt) was homogenised by stirring and split into five subsamples. Four subsamples were spiked with 1.0 ml aliquots of Vista standards to give Vista dry weight concentrations of $24 \cdot 57$, $4 \cdot 95$, $2 \cdot 47$ and $0 \cdot 495$ mg/g and these analysed by the standard procedure. The characteristic Vista pattern is masked by the background hydrocarbons at lower concentrations so that Vista could not be distinguished from the background at 5 ppm and below.

Using the Vista indicators found in the background samples, a detection limit was calculated based on the mean level and variance of these components for a given particle size range. Vista was taken to be the sum of the 10 target low molecular weight isoprenoids that exceeded the 95% confidence limit for a given particle size as predicted from background samples (Section 3.1). On this basis the detection limit ranged from 15 ppm (Vista) in 100% sand to 130 ppm Vista in 100% clay/silt sediments (Fig. 5). This was a conservative definition for background sediments with a high clay + silt content as a distinctive Vista pattern could be seen in some samples which did not have detectable Vista by the above definition. The relationship of the isoprenoids, PAH and barium to particle size is discussed in Section 3.2.

Escaid 90 was used to supplement stocks of Vista at Kaubvik. Although GC traces of the two oils are qualitatively similar, there are significant

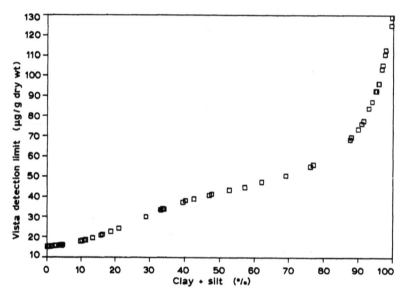

FIG. 5. Detection limit for Vista ODC as a function of sediment particle size.

differences in composition (Fig. 6). Escaid 90 is a more volatile base oil with major components eluting from 900 to 1300 (1100 maximum) on the Kovats scale (Kovats & Keulemans, 1964). By comparison, Vista major components elute from Kovats 900 to 1600 (1200 maximum). Normal alkanes nC10, nC11, and nC12 comprise about 20% of the total peak area of Escaid, but less than 5% in Vista. The norfarnesane group (at 1270 Kovats) are minor components of Escaid and the higher isoprenoids are present in negligible quantities. Consequently, the selected isoprenoids used as a set of Vista markers are not sensitive to the presence of Escaid. No attempt was made to quantify Escaid. Qualitatively its presence in samples containing base oil at Kaubvik could be detected on the basis of the magnitude of the nC10–nC12 peaks.

2.5. Barium Analyses

Sediment samples were prepared for analysis by first homogenizing, subsampling, drying to constant weight, grinding in an agate mortar and sieving through a 63 mm (230) mesh sieve. Samples (0·5 g) were digested by fusion with $Li_2B_4O_7$ in $LiNO_3$ with dissolution of the melt in nitric acid. Digests were analysed by flame AAS using a N_2O–C_2H_2 flame. This method was chosen on the basis of its excellent detection limit and high

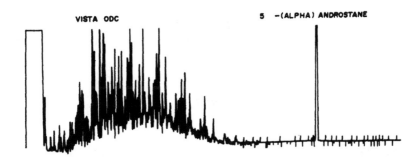

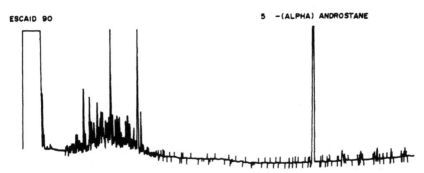

FIG. 6. Comparison of gas chromatograms of Vista ODC and Escaid 90 base oils.

degree of precision. Barium sulphate generally resists dissolution by the standard acid digests employed for many elements. As a result these methods suffer from poor sensitivity and incomplete recoveries.

Precision of blind replicate analyses of the same sample was 3% (relative standard deviation). Blind analyses of reference materials (NRC MESS-1; BCSS-1) gave results 16% high in series 2 and within 3% for series 3, 4, 5 and 6.

2.6. Particle Size Analysis

Sediment grain size analysis was done by the hydrometer and sieve method using seven sieves (No. 10, 20, 30, 60, 100, 200 and 230 standard mesh sizes). The analysis provided percentage gravel, sand, silt and clay (silt defined as particles from 2 to 63.5 μm and clay less than 2 μm). A number of blind replicate samples were analysed with each suite to provide an indication of precision. Results for replicate samples were all within 3% of the original results for the 3 size classes (percentage sand, silt and clay).

3. RESULTS

3.1. Sediment Grain Size

Sacrificial beach islands and subsea berms are constructed from sand which has a low metal and hydrocarbon content. The background surface sediments in the area of both the Minuk and Kaubvik well sites were found to be a mixture of roughly 50% silt-sized and 50% clay-sized particles. As a result, there was a gradation from 100% sand on the island or berm to 0% sand at locations removed from any influence of island or berm construction. Triangular plots of grain size characteristics of sediment samples from both well sites prior to drilling (Series 2 and 3, Fig. 7) clearly indicate the two end members with a straight addition of sand from the island/berm to background sediments giving rise to the variable sand content sediment observed at many of the sites.

3.2. Analysis of the Grain Size Relationship for Barium and Hydrocarbons

There is a positive relationship in Beaufort Sea sediments as elsewhere between the percentage fine-grained sediment and metal and hydrocarbon concentrations (Hoff & Thomas, 1986). As a result of the wide range in sediment particle size, it was necessary to examine the relationship between the percentage fine-grained sediment and Ba and base oil (defined in Section 2.4) concentrations in predrilling samples. The sum of the clay + silt was used as a measure of the fine grained sediment (no difference was found if clay or clay + silt was used; therefore the sum was chosen).

The following nonlinear (univariate) regression model was used to describe the predrilling relationship for both Ba and Vista:

$$Y = b_o^i + b_1^i x = b_2^i x^2 + \ldots + b_m^i x^m + e$$

where for a given series i (2 or 3), Y is the $\log_{10}$ transform of the concentration of either Ba or Vista, x is the particle size (percentage clay + silt) of the sediment, the bs are the unknown coefficients to be estimated and e is the random, or unexplained, part of Y. The constant b_o^i represents the overall (background) level for series i, and $b_1^i, b_2^i, \ldots, b_m^i$ measure the particle size effect for that series. Other than particle size, no other variables or factors were considered important in determining the concentration of

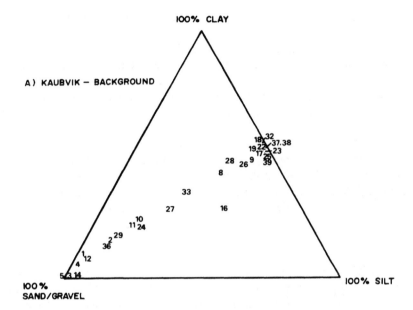

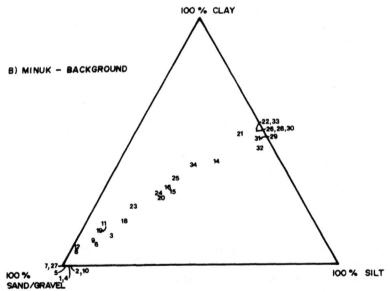

FIG. 7. Triangular plots of sediment grain size for samples collected in background sampling of (A) Kaubvik I-43 and (B) Minuk I-53.

barium or the sum of the 10 low molecular weight isoprenoids used as a measure of Vista.

The results of fitting a separate equation to each of series 2 and 3, are summarised below. In each case, the coefficients b_0, b_1,...,b_m were estimated using unweighted least squares. The degree of the polynomial m was determined by plotting the data, the fitted equation, and the residuals; and by examining R_2 = coefficient of determination and the associated F statistic.

The fitted equations for series 2 and 3 were used to construct approximate 95% prediction intervals for the corresponding postdrilling samples, assuming that drilling has no impact and that there are no other interim changes in the particle size relationship. Series 2 was used to predict values for series 6; series 3 was used to predict the series 4 and 5 values. In general, the prediction interval is of the form:

$$Y(x^*) = y(x^*) + 1.96 \, \text{s.e.}(y(x))$$

where $y(x^*)$ is the fitted value of y that corresponds to a postdrilling sample sediment with particle size $x = x^*$ (i.e. calculated by substituting $x = x^*$ into the fitted regression equation) and s.e. $(y(x^*))$ is the associated standard error when $y(x^*)$ is used to predict a future value of Y. Postdrilling levels that exceed the upper limit of the corresponding prediction interval suggest that there is a statistically significant drilling impact at the given sample site.

3.2.1. Barium

There does not appear to be any obvious relationship between Ba and particle size for either of series 2 or 3. The mean ($\pm$ the standard deviation) Ba concentration at Kaubvik (series 2) and Minuk (series 3) were 1122 ± 196 ppm and 812 ± 310 ppm, respectively. A 95% prediction interval for postdrilling samples is shown for Minuk and Kaubvik in Fig. 8. Most of the postdrilling samples at Kaubvik (series 6) are consistent with the background Ba levels whereas most of series 4 and 5 samples appear to be significantly elevated at Minuk.

3.2.2. Vista ODC

The sum of the 10 target low molecular weight isoprenoids used to indicate and quantify the presence of Vista ODC in sediments (Section 2.4) shows a strong positive relationship with particle size at both sites. A third order polynomial was used to fit log(Vista) as a function of clay + silt. A 95% prediction interval for postdrilling samples is shown for Minuk and

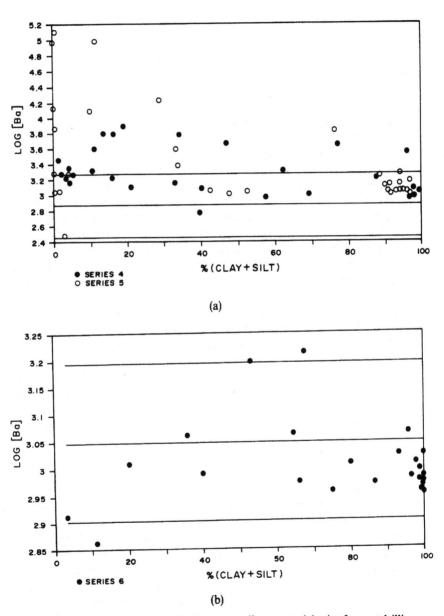

FIG. 8. Log$_{10}$ barium concentrations versus sediment particle size for postdrilling samplings at Minuk (a) and Kaubvik (b) (95% confidence limits based on background samples are shown as solid lines).

Kaubvik in Fig. 9. All the postdrilling samples at Minuk in series 4 (immediately after drilling during ice cover) are within the predicted levels whereas many of the samples from series 5 and 6 are significantly elevated over background levels.

3.2.3. PAH

A randomly selected subset (approximately one third) of all samples were analysed by GC/MS for a suite of 25 selected PAH indicated in Section 2.3. The total PAH concentration shows a similar relationship to particle size as found for the nonpolar hydrocarbons and selected isoprenoids. Total PAH (nonlog transformed) plotted against particle size for postdrilling samples is shown in Fig. 10. These plots indicated the PAH concentrations were not significantly different after drilling. Samples 6-04 and 5-26 were the only postdrilling samples that were significantly above background. These samples also had high levels of base oil. None of the other samples with quantifiable levels of Vista, however, had elevated levels of PAH. The composition of PAH through all series was remarkably consistent as indicated by the maxima in alkyl homologue distributions (AHD) and parent compound distribution (PCD, Lake *et al.*, 1979). The two samples with elevated PAH were qualitatively different from the rest, having higher phenanthrene with respect to perylene. This suggests a different source of higher aromatic content material at these locations not directly related to dispersion of oiled cuttings. It was concluded therefore, that there was no significant aromatic input from the drilling as expected on the basis of the low aromatic content of the base oil. An increase in the concentration of aromatics in the sediments would be expected only if production wastes contained appreciable amounts of formation oils (Yunker & Drinnan, 1987).

3.3. Currents

Currents as measured 1 km south of Minuk in the September–October period had a general northeasterly trend. For the period in November for which a current record was obtained, the net flow was initially almost due east. Current speeds were generally less than 20 cm/s in September and October with maximum speeds of over 40 cm/s occurring during periods of strong winds (speeds were between 40 and 50 cm/s for the 3 day period during the storm which damaged the Minuk island in September). Current speeds of over 30 cm/s were recorded on 10–11 November 1985 but decreased to less than 15 cm/s for the remainder of the 8 day record of the winter deployment.

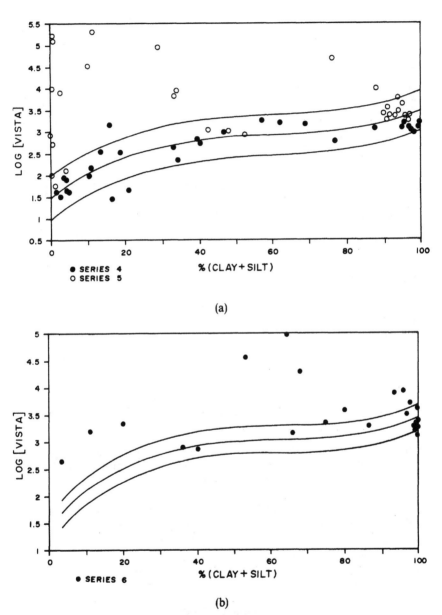

(a)

(b)

FIG. 9. Log$_{10}$ Vista concentrations versus sediment particle size for postdrilling samples at Kaubvik (b) and Minuk (a) (95% confidence limits for background samples are shown as solid lines).

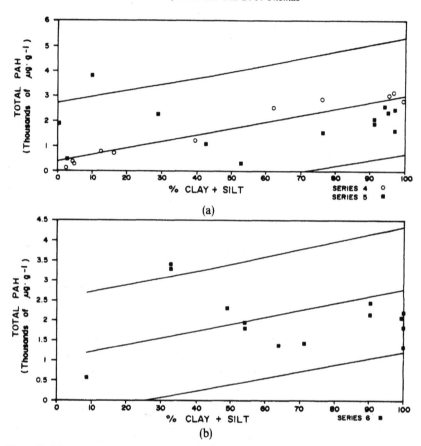

FIG. 10. Total PAH versus sediment particle size for postdrilling samples at Minuk (a) and Kaubvik (b) (95% confidence limits for background samples are shown as solid lines).

No direct current measurements were made at Kaubvik during the drilling period. However, current meters had been deployed the previous summer (27 July–7 October 1986) at depths of 6 and 19 m in the vicinity of Kaubvik (MacLaren Plansearch Limited, 1987). During this period there was a net residual flow of 4·6 cm/s to the northeast.

4. DISCUSSION

The results for each set of postdrilling samples are discussed below in terms of the areal extent of dispersal of base oil to surface sediments from

OBM discharged from each well site, potential mechanisms for dispersal and the weathering of base oils over the period of the study.

4.1. Dispersion of OBM

4.1.1. Post-drilling Samples at Minuk (series 4)

There was no base oil detectable in any samples collected outside the ice rubble field. A single sample to the west, well beyond the ice rubble field did have elevated levels of the 10 isoprenoids, but examination of the GC trace of this sample did not indicate a distinctive Vista pattern of peaks. The rubble field was not symmetrical and was most extensive to the west and northwest of the island (Fig. 3). Ice accumulates where there are bottom obstructions that impede its movement so that the extension of the ice rubble field to the northwest reflects shallower depths in that region as a result of the migration of the island after the storm in September 1985 as well as the presence of some lost material and equipment. The edge of the rubble field was close to the island on the east and south sides however, so that it was possible to take some samples in that area that were within 200 m of the discharge line from the island. The results suggest that all waste OBM was contained within the perimeter of grounded ice rubble.

Barium concentrations in most samples close to the island were elevated. A contour plot of Ba concentrations (Fig. 11) indicates that the excess Ba originated from the island and, in the absence of base oil data, point to the discharge of OBM during the winter as the most likely source. As no trace of Vista ODC was found in any samples with excess Ba, however, another source must be involved. Two other possible sources which would not have an association with Vista were: (1) water-based drilling muds used in the first few weeks of the drilling program, and (2) barite lost from the island as a result of storm damage to the island the previous fall. Because water-based muds would have been discharged early in the winter before an extensive ice rubble field developed, barium could have been more widely dispersed. However, the quantity of barium from this source was small compared to the large quantity of barite (2300 tonnes) lost from damaged storage tanks during salvage and clean-up operations after the storm damage in September 1985. The barite was released during salvage on the west side of the island (the same side as the discharge line) at about the same time as series 3 samples were being collected. It is likely that series 3 samples were not elevated because the barite had not been widely dispersed when samples were taken. Although barite was visibly present

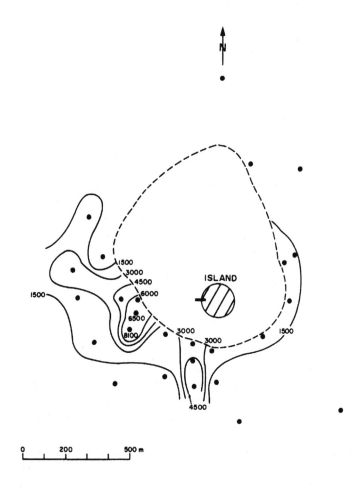

FIG. 11. Contour plot of excess barium in surface sediments at Minuk I-53 in April 1986 within 2 weeks of drilling completion. ●, Sample locations; — — —, approximate extent of ice rubble field.

in the water, it was not possible to sample close to the storage tanks for logistic reasons. Given the large quantities of barite lost, this source would appear to be the most likely explanation for elevated sediment Ba observed in the postdrilling samples.

4.1.2. Samples from Minuk at the End of the Following Open Water Season (series 5)

Whereas all the OBM appears to have been contained within the rubble field in winter, widespread dispersion of Vista oil occurred after break-up. A contour plot of equivalent Vista oil concentrations (see Section 2.4) shown in Fig. 12, indicates dispersal of oil around the south side of the island and then in an easterly direction. Because of the position of the sampling locations, it is not possible to define the outer limits of Vista dispersal. The maximum concentration of Vista found was in a sample to the southeast of the island, about 350 m and on the opposite side of the island from the discharge point. Concentrations of low molecular weight isoprenoids were more than 100 times background values. Extrapolation of the contours suggests that Vista would be quantifiable more than 1 km from the island to the east. Some samples that did not have levels of low molecular weight isoprenoids in excess of the 95% confidence level for the particle size of the sample, did have detectable levels of Vista on the basis of the distinctive pattern of peaks in the GC trace. A signature of Vista was noted in the GC trace of a sample taken 950 m at 260° from the island centre, indicating dispersal of small quantities almost 1 km to the west of the island as well.

The direction of movement of most of the dispersed oil is in agreement with the net residual flow recorded by the current meter 1 km south of the island (Section 3.3). Dispersal of Vista oil was, however, far more extensive than might be assumed given the apparent confinement of oil before ice break-up, the short open water season and available data on maximum currents. On the east coast, near Sable Island, for instance, Yunker & Drinnan (1987) found that dispersal of base oil was more confined at a well site in similar water depths but with a much more energetic ocean environment. Although a signature of base oil was noted in GC traces of some sediments as far as 1 km away from the site, concentrations were not elevated more than 10 times background beyond 200 m from the well centre (sediments in that area were 100% sand making detection limits of base oil the same close to the well as at distant locations).

The reason for the dispersal of large quantities of OBM over a large

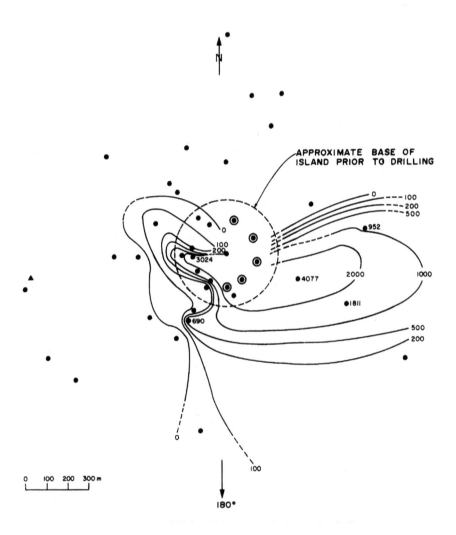

FIG. 12. Contour plot of Vista ODC concentrations in surface sediments at Minuk I-53 in September 1986 one open water season after drilling was completed. ▲, Vista present, not quantified; ●, sample locations; ◉, samples collected but not analyzed.

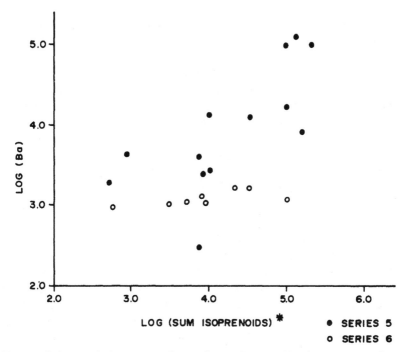

FIG. 13. Relationship between surface sediment Vista and barium concentrations in follow-up samples from Minuk and postdrilling samples from Kaubvik. (*Vista = sum of selected isoprenoids (0·019 53) + 13·77, as given in Section 3.2.)

area and greater distances at Minuk may be related to ice movements. While the grounded ice rubble on the island slope confined the OBM in winter, it also provided a potential mechanism for dispersal after ice break-up. As the ice around the island melts and breaks-up in early summer, the grounded ice rubble eventually floats free and drifts off in the direction of the prevailing currents and wind direction. As it drifts away, some of this ice may scour the bottom and if oiled cuttings are present, may carry them some distance from the island. In addition, some cuttings may actually be held in the ice nearest the discharge point and be released as the ice melts.

There was a direct correlation between Ba concentrations and samples with Vista oil present in series 5 (Fig. 13) suggesting that a similar mechanism is responsible for Ba dispersal. Maximum Ba concentrations (100 000 ppm) were an order of magnitude higher than maximum levels observed in series 4 immediately after drilling and two orders of magnitude above background.

4.1.3. Post-drilling Samples from Kaubvik (series 6)

Sampling at Kaubvik occurred approximately 7 months after drilling was completed (early January, 1987) and about 3–4 weeks after ice break-up. Samples were collected just prior to dismantling of the caisson so that there was a lot of ship activity in the area. A contour plot of excess low molecular weight isoprenoids shows clearly the presence of Vista in the immediate vicinity of the discharge point and dispersal mainly to the east as observed at Minuk with some cuttings dispersed up to 400 m to the west (Fig. 14). There was a very steep gradient in base oil concentrations in sediments directly to the south of the caisson.

On the basis of the normal alkanes nC10, C11 and C12, the presence of Escaid 90 could be inferred in GC traces of many of the samples with detectable amounts of Vista. Because Escaid was not measured, concentrations of base oil in these samples will have been underestimated. These samples are noted in the contour plot with an asterisk. The sediment trap provided a time integrated estimate for the relative contribution of the two base oils. Escaid 90 is present in this sample, but at much lower concentrations than Vista in agreement with the relative abundance of the two oils used in drilling mud (ESSO Resources, personal communication).

The areal extent of dispersion was more restricted than observed at Minuk after one open water season. OBM was discharged from Kaubvik in the early winter and was not confined by an extensive build-up of ice rubble. Ice scouring and wave action will not, therefore, have been factors in the dispersion of oiled cuttings at the time samples were taken. The observed distribution of base oil at Kaubvik, therefore, can be assumed to be primarily the result of currents and is in agreement with the observed direction of net residual currents observed in the area the previous summer (Section 3.3).

There was only a very weak relationship between Ba and base oil concentrations in sediments at Kaubvik (Fig. 13). Barium concentrations were much lower than at Minuk with the highest concentration of 1600 ppm being only slightly above the upper 95% confidence interval predicted from background sampling. Samples at Minuk with similar levels of Vista had Ba concentrations up to 50 times higher. Some of the difference may reflect the greater quantities of barite either lost or discharged with drilling wastes at Minuk (approximately 10 times as much Ba was discharged/lost). The weaker correlation between Ba and

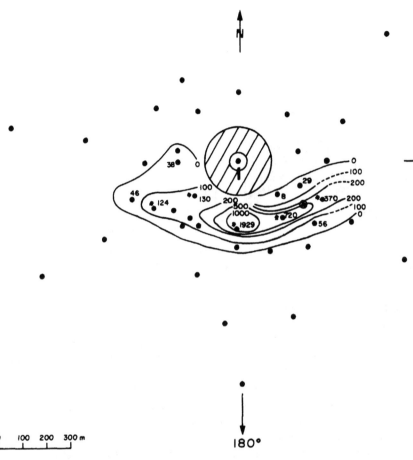

FIG. 14. Contour plot of Vista ODC concentrations in surface sediments at Kaubvik I-43, August 1–4, 1987, 7 months after completion of drilling. ▨, Approximate area of berm; ●, sample locations; ═, approximate discharge line position; *, Escaid 90 present; ▲, sediment trap.

Vista at Kaubvik may also reflect a difference in the principal mechanism for dispersal. Moving ice would be able to transport higher density material such as barite greater distances and in larger quantities than ocean currents. The observed Ba/Vista relationship at Kaubvik may therefore indicate a differentiation between base oil and Ba on the basis of density, whereas the strong correlation at Minuk is an indication that ice movement was the dominant factor in dispersal at that site.

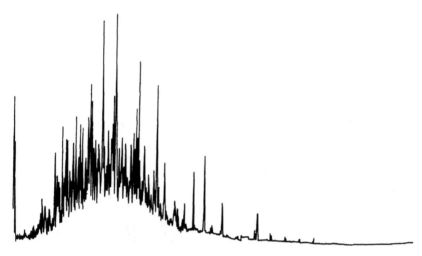

Fig. 15. Gas chromatogram of postdrilling sediment sample containing Vista ODC base oil at Minuk I-53.

4.2. Weathering of Base Oils

Because a time series of samples was not collected from sites where base oils were detected, no direct information is available for the rate of loss of oil from the sediment. Dissolution and evaporation will gradually decrease sediment oil concentrations. Yunker & Drinnan (1987) noted an almost 100 fold decrease in the concentration of Vista oil in the cuttings pile at the West Venture site (16 m water depth) over a 3-month period. The cold water temperatures and less energetic environment in the Beaufort will likely result in a slower loss than observed on the east coast.

Quantitative changes in Vista ODC were observed, however, at both sites. Differences were a function of distance from the discharge point and concentration. There were also differences between sites. The Vista in use at Minuk had slightly more volatile components than in the standard Vista originally supplied and used for calibration. However, there were negligible changes in the concentrations of the selected isoprenoids.

At Minuk in series 5 samples, substantial changes can be seen in the original Vista on the basis of GC traces of samples containing distinctive base oil patterns (Fig. 15). The changes can be summarised as a decrease in complexity with 10 conspicuous peaks emerging from the original complex Vista pattern. The unresolved envelope also decreased and there is a shift in the centre of the envelope from around nC13 to nC14 or higher. Included in the group of remaining conspicuous peaks are most

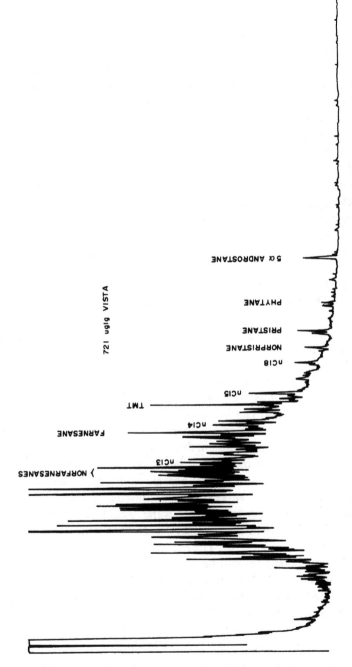

FIG. 16. Gas chromatogram of postdrilling sediment sample containing Vista ODC base oil at Kaubvik I-43.

of the selected 'Vista' isoprenoid marker peaks which will artificially improve the sensitivity of the method for weathered or aged base oil. The calibration factor used to convert from total selected isoprenoid to Vista was based on undegraded Vista, so that the reported 'Vista' values for weathered samples will be higher than the sum of the residual peaks would indicate.

Similar changes are apparent at Kaubvik (Fig. 16), but the inclusion of a different base oil, Escaid 90, makes direct comparison between the sites difficult.

At both sites, weathering is most pronounced in samples farthest from the discharge point presumably as a result of the greater distance travelled and hence longer water column residence time.

The material recovered from the sediment trap, deployed for 11 months 250 m at an angle of 120° from the Kaubvik caisson centre, showed a substantial quantity of Vista and Escaid 90. Qualitatively, the trap material was unchanged base oil, and similar to the Vista used at Minuk. A series of high molecular weight *n*-alkanes with a pronounced odd-even predominance is also apparent in this sample showing input from either suspended particulate matter from the Mackenzie River or resuspended bottom sediment.

4.3. Relationship of the Data to other Beaufort Sea Locations

The data for the Minuk I-53 and Kaubvik I-43 well sites indicates the extent of dispersal that can be expected in these water depths at other Beaufort Sea locations. However, other discharge scenarios and different sea and ice conditions may result in different dispersal patterns. Discharge of oiled cuttings during the open water season would mean that oiled cuttings would enter a more typical oceanic environment and be subject to the immediate effects of waves and wave induced currents. The discharge of OBM from a drillship into deeper water would be similar to other areas such as the North Sea or offshore Nova Scotia and dispersion governed primarily by currents. The Minuk results suggest that the extent and direction of dispersal will be most difficult to predict in shallower areas within the landfast and transitional ice zones where moving ice in the winter and during break-up may scour the sea-bed and transport oiled cuttings large distances.

ACKNOWLEDGEMENTS

Logistic support for all sampling was provided by Esso Resources Canada Limited. Supply vessels, the *Arctic Nanabush*, *J. Mattson* (trips 1, 2 and

4) and the *Arctic Nanook* (trip 5) were used for open water sampling and helicopter support and a polar bear monitor, snowmobile time and accommodation at Minuk provided for through the ice sampling.

The project was funded under contract by the Environmental Studies Research Funds. Dr Rainer Engelhardt was the Scientific Adviser for the project.

REFERENCES

CRETNEY, W. J., WONG, C.S., CHRISTENSEN, P.A., MCINTYRE, B.W. & FOWLER, B.R. (1980). Quantification of polycyclic aromatic hydrocarbons in marine environmental samples. In *Hydrocarbons and Halogenated Hydrocarbons in the Aquatic Environment*, Environmental Science Research Series, ed. B.K. Afghan and D. McKay. Plenum Press, New York.

DAVIES, J.M., ADDY, J.M., BLACKMAN, R.A., BLANCHARD, J.R., FERBRACHE, J.E., MOORE, D.C., SOMERVILLE, H.J., WHITEHEAD, A. & WILKINSON, T. (1984). Environmental effects of the use of oil-based drilling muds in the North Sea. *Mar. Polut. Bull.*, **15**, 363–70.

EBA ENGINEERING CONSULTANTS LTD (1984). Abandonment of offshore artificial islands in the Beaufort Sea. A report for the Environmental Protection Service, Environment Canada, Yellowknife, N.W.T., 120 pp. + Appendices.

EICHELBURGER, J.W., HARRIS, L.E. & BUDDE, W.L. (1975). Reference compound to calibrate ion abundance measurements in gas chromatography–mass spectrometry. *Anal Chem.*, **47**, 995–1000.

ERICKSON, P.E., THOMAS, D., PETT, R. & DE LANGE BOOM, B. (1983). Issungnak Oceanographic Survey. Part A: Oceanographic Properties. A report prepared for Esso Resources Canada Limited, Gulf Canada Resources Inc., and Dome Petroleum Limited by Arctic Laboratories Limited, 194 pp.

GREEN, R.H. (1979). *Sampling Design and Statistical Methods for Environmental Biologists*. John Wiley, New York, 257 pp.

HOFF, J.T. & THOMAS, D.J. (1986). A compilation and statistical analysis of high quality Beaufort Sea sediment data with recommendations for future data collections. A report prepared for the Environmental Protection Service, Yellowknife, N.W.T., by Arctic Laboratories Limited, Sidney, B.C., 118 pp.

HUTCHESON, M.S., STEWART, P.L., ODENSE, R. & FOWLER, B. (1987). The effects of mineral oiled drill cuttings in marine invertebrates: Lethality, hydrocarbon accumulation/depuration, burrowing success and molting. In *Pollution and Physiology of Marine Organisms*, ed. A. Calabrese, F.T. Thurberg, W.B. Vernburg & F.J. Vernberg, University of South Carolina Press, South Carolina, pp. 9–46.

JOHANCSIK, C.A. & GRIEVE, W.R. (1987a). Oil-based mud reduces borehole problems. *Oil & Gas Journal*, April 27, 46–58.

JOHANCSIK, C.A. & GRIEVE, W.R. (1987b). Solids control evaluated during oil-based-mud drilling. *Oil & Gas Journal*, May 4, 42–5.

KOVATS, E. & KEULEMANS, A.I.M. (1964). The Kovats retention index system. *Anal. Chem.*. **36**. 31A–41A.

LAKE, J.L., NORWOOD, C., DIMOCK, C. & BOWDEN, R. (1979). Origins of polycyclic aromatic hydrocarbons in estuarine sediments. *Geochim. Cosmochim. Acta*, **43**, 1847–54.

MACLAREN PLANSEARCH LIMITED (1987). Final report: 1986 Beaufort Sea Oceanographic Program Current Meter and Wave Data Analysis. An unpublished report for ESSO Resources Canada Limited, Calgary, Alberta.

PIOTROWICZ, S.R., HOGAN, C.A., SHORE, R. & PSZENNY, A.A. (1981). Variability in the distribution of weak acid leachable Cd, Cr, Cu, Fe, Ni, Pb, and Zn in the sediments of the Georges Bank, Gulf of Maine region. *Environ. Sci Technol.*, **15**, 1067–72.

THOMAS, D.J., GREENE, G.D., DUVAL, W.S., MILNE, K.C. & HUTCHESON, M.S. (1983). Offshore oil and gas production waste characteristics, treatment methods, biological effects and their applications to Canadian Regions. Environmental Protection Service, Ottawa, 365 pp.

YUNKER, M.B. & DRINNAN, R.W. (1987). Dispersion and fate of oil from oil-based drilling muds at West Venture C-62 and South Des Barres 0-76, Sable Island, Nova Scotia. Environmental Studies Revolving Funds, Report No. 060. Ottawa.

3

The Environmental Effect of Oil-based Mud Drilling in the North Sea*

J. M. DAVIES

Department of Agriculture and Fisheries for Scotland, Marine Laboratory, Victoria Road, PO Box 101, Torry, Aberdeen AB9 8DS, Scotland, UK

D. R. BEDBOROUGH

Department of Energy, London, UK

R. A. A. BLACKMAN

Ministry of Agriculture, Fisheries and Food, Burnham-on-Crouch, UK

J. M. ADDY

Britoil Plc, Aberdeen, Scotland, UK

J. F. APPELBEE

Conoco UK Ltd, London, UK

W. C. GROGAN

Marathon Oil UK Ltd, Aberdeen, Scotland, UK

J. G. PARKER

Shell Exploration and Production, Aberdeen, Scotland, UK

and

A. WHITEHEAD

BP Petroleum Development, Aberdeen, Scotland, UK

*This paper has been produced by the UKOOA/UK Government Working Group set up to establish and monitor the environmental impact of offshore oil development in the UK.

ABSTRACT

Drilling activity in the North Sea has increased from a level of 150 wells drilled in 1978 to a peak level of 290 wells in 1985. With the collapse of world oil prices, approximately 190 offshore wells were drilled in 1987. By 1984, 70–75% of all wells drilled on the UK Continental Shelf used oil-based mud (OBM) in at least one section. This percentage has remained relatively constant despite the reduction in overall drilling activity.

Offshore drilling operations are now a significant source of oil discharged to the North Sea. The Quality Status Report produced for the 'Second International Conference on the Protection of the North Sea' has estimated an input of some 71 000–150 000 t of oil to the North Sea in 1986, of which some 29 000 t, was derived from the offshore industry.

In the North Sea the majority of oil contaminated cuttings are generated and discharged outside the territorial waters of littoral states. Agreement on how to control these, and other discharges from the offshore industry, is reached within the Paris Commission which deals with discharges from land-based sources (offshore installations are considered to be land based within the Paris Convention) into the seas covered by the Convention.

In 1985 the Paris Commission Working Group on Oil Pollution met and considered the environmental data available in the North Sea on the impact of oil-based drilling muds and compiled a list of 'agreed facts' which described the scale of contamination from oil-based drilling muds and the severity and extent of the biological effects of this practice. They also identified gaps in our knowledge of the effects of discharging residual drilling mud on cuttings.

The objective of this paper is to examine the most recent (post-1983) environmental survey data around UK oil fields in relation to the Paris Commission 'agreed facts' and to reach a conclusion as to whether the 'agreed facts' still hold. A further objective of the paper is to fill in some of the gaps in knowledge identified in the original set of 'agreed facts'.

1. INTRODUCTION

Offshore oil and gas exploration and production in the North Sea is relatively recent. The first commercial quantities of gas were discovered on the UK Continental Shelf (UKCS) in 1965 and the first oil in 1969. Drilling activity has varied from year to year but the last 10 years has seen a steady increase from 150 wells drilled in 1978 to a peak level of 290 wells in both 1984 and 1985. The collapse of the price of oil in 1986 has had an adverse effect and only 190 wells were drilled in 1986. The largest reduction in drilling activity has occurred in the 'oil province', found in the deeper waters of the northern North Sea, with less effect in the 'gas province' found in the shallower waters of the southern North Sea (Fig. 1). In 1987 activity picked up in the last quarter of the year and a total of 258 wells were drilled.

In the mid- to late-1970s the use of oil-based mud (OBM) began to increase. By 1984 70–75% of all wells drilled on the UKCS used OBM in at least one section. It is probable that this percentage will remain constant, but overall drilling activity will be determined by the price of oil.

Drilling operations are now a significant source of oil discharged to the North Sea. The Quality Status Report produced for the 'Second International Conference on the Protection of the North Sea' has estimated an input of between 71 000 and 150 000 t of oil to the North Sea in 1986 of which some 29 000 t was derived from the offshore industry. Table 1 shows the inputs of oil from the UK offshore industry into the North Sea and it can be seen that between 1981 and 1986 around 90% is in the form of oil adsorbed onto cuttings.

The use of diesel as a base oil in drilling mud decreased after 1983 when low toxicity base oils came into regular use. Since 1984 no diesel based muds have been used on the UKCS. Moreover, the last 3 years have seen a steady reduction in levels of oil on cuttings from an average of 190 g of oil per kg of dry cuttings in 1984 to around 150 g/kg in 1987, with far fewer wells now having levels of over 200 g/kg. With current mud systems and state of the art shale shakers, levels of 120–130 g/kg are achievable. (Development of oil and gas resources in the UK, 1987.) Claims are made that cuttings cleaning equipment can achieve levels of 50–100 g/kg but it is accepted that reliability has been poor. However, new mud formulations which contain less base oil may lead to reduced levels of oil on cuttings.

In the North Sea the majority of oil-contaminated cuttings are generated and discharged outside the territorial waters of littoral states. Agreement on how to control these and other discharges from the offshore industry, is reached within the Paris Commission which deals with discharges from land-based sources into the seas covered by the Convention (offshore installations are considered to be land-based within the Paris Convention).

In 1985 the Paris Commission Working Group on Oil Pollution considered the environmental data available on the impact of OBM and compiled a list of 'agreed facts' which described the scale of contamination

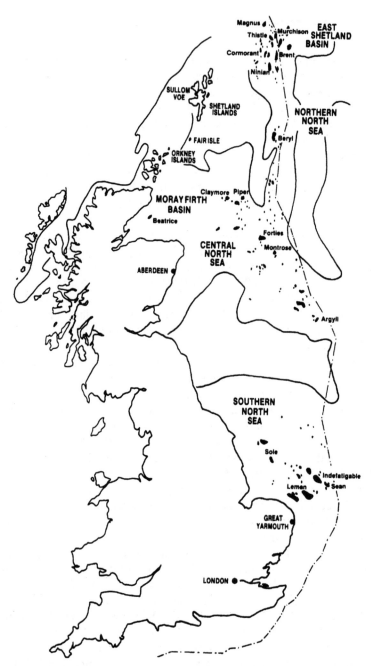

FIG. 1. The areas of oil and gas activity in the UK sector of the North Sea.

TABLE 1
Oil discharged by the UK offshore industry (tonnes per annum)

	1981	1982	1983	1984	1985	1986
Oil spills	104	162	186	130	310	540[a]
Produced water	525	927	1 700	1 430	2 150	2 710
Oil on cuttings	5 800	8 600	14 500	19 800	20 200	13 000
Oil on cuttings as % of total	(90)	(89)	(88)	(93)	(89)	(80)
Total	6 429	9 689	16 386	21 360	22 660	16 250

[a] For comparative purposes the spill of 2 500 t from a fractured pipeline has not been included.
Source: Development of Oil and Gas Resource of UK, 1987.

from OBM and the severity and extent of the biological effects of this practice. They also identified gaps in our knowledge of the effects of discharging residual drilling mud on cuttings (Paris Commission Working Group on Oil Pollution, 1985). The main environmental points are summarised below:

1. Discharges of cuttings from water- or oil-based drilling can have an adverse effect on the seabed biological community. Beneath and in the immediate vicinity of the platform, this is due mainly to physical burial of the natural sediment. However, the extent of the biological effect of OBM cuttings from multiple-well drilling is substantially greater than that with water-based muds. Information is needed as to the relative biological effects of drilling a single well with OBM as against water-based muds.

2. Despite the scale of inputs in all fields studied, the major deleterious biological effects were confined within the 500 m safety zone and associated primarily with burial under the mound of cuttings on the seabed. Seabed recovery in this zone is likely to be a long process.

3. Surrounding the area of major impact is a transition zone in which lesser biological effects are detected as community parameters return to normal, generally within 200–1000 m. The shape and extent of this zone is variable, and is largely determined by the current regime and the scope of the drilling operation. With greater currents and more extensive drilling, this delineation may be extended 2000 m in the direction of greatest water movement.

4. Elevated hydrocarbon concentrations attributable to OBM were observed beyond the areas of biological effects. These elevated hydrocarbon concentrations have been measured out to as far as 4000 m in the direction of the prevailing current.

The objective of this paper is to examine the most recent (post-1985) environmental survey data around UK oil fields in relation to the 'agreed facts' and to reach a conclusion as to whether these facts still hold. A further objective of the paper is to fill in some of the gaps in knowledge identified in the original set of agreed facts.

2. RECENT ENVIRONMENTAL EVIDENCE FROM MAJOR PRODUCTION SITES

Monitoring of the effects of OBM discharges has continued since 1983 around existing UK installations, new installations which have come into production since then and at single well exploration sites. In 1983 the concept of four zones of effect was established (Davies *et al.*, 1984). These zones were defined in terms of the hydrocarbon chemistry and the detectable changes in the benthic animal populations.

Figure 2 shows the relationship between sediment hydrocarbon concentrations and distance from platforms using data collected prior to and after 1983. These data include several surveys at a number of sites carried out between 1978 and 1987, i.e. during the time diesel-based muds were used (up to 1983) and beyond. The data obtained since 1983 indicate that hydrocarbon concentration gradients around platforms remain very similar to those found earlier. However, some of the new fields monitored after 1983 are in the southern North Sea, for example the Forbes and Gordon fields, and the coarser sediments in these areas have naturally lower total organic and hydrocarbon concentrations.

Figure 3 shows the relationship between macrobenthic species diversity, as expressed by the Shannon–Weiner index (H_s), and distance from the platform. This includes data from 1978 to 1987. The data are shown for purposes of comparison, as a ratio to the background (3000–8000 m diversity values) since diversity indices vary naturally with sediment type. Data obtained since 1983 conform to the same pattern as that prior to 1983, with the major deleterious biological effects being confined within the 500 m safety zone.

In 1985 the effects of OBMs had not been so clearly defined in the UK

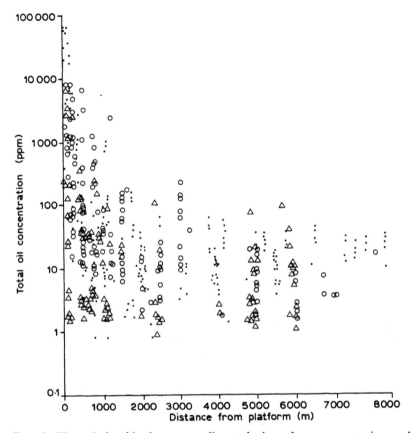

FIG. 2. The relationship between sediment hydrocarbon concentration and distance from production site. ●, Pre-1983; ○, 1983–1985; △, since 1985.

Sector of the southern North Sea, where depths are shallower, current regimes stronger, sediment characteristics more variable and a smaller number of wells are drilled from any one installation. We now have data for a southern North Sea field, Vulcan, which indicate that the same pattern of effects operates within this area. Vulcan, first surveyed in March 1986 to determine the baseline conditions, was surveyed in July 1987 following the completion of eight wells drilled with low toxicity OBMs. Table 2 summarises the data obtained from both surveys and Fig. 4 shows the sediment hydrocarbon concentrations and the macrobenthic species diversity (expressed as H_s) related to distance from the platform. This

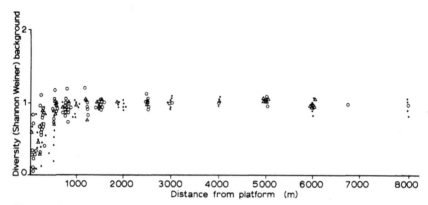

FIG. 3. The relationship between macrobenthic diversity, shown as a ratio to the background diversity, and distance from production site. ●, Pre-1983; ○, 1983–1985; △, since 1985.

figure also shows the range of values encountered during the baseline survey for total sediment hydrocarbon concentration and diversity.

Hydrocarbon contamination from the drilling cuttings discharge is marked at the 200 m site with low levels of contamination at 500 m and extremely low levels from 800 to 1200 m, possibly also at 2500 m. At the 5000 m site hydrocarbon values were at background for the area (the range for 15 Vulcan baseline survey sites being 0·3–3·6 μg/g with a mean of 1·24 μg/g).

Although a high value for sediment hydrocarbon concentration was recorded at 200 m, the depression of the diversity index (H_s) was not marked. A more detailed investigation of the species present does, however, show the effects of the cuttings discharge. Analysis of the top five species present and curves produced by the Sanders rarefaction method indicated that a normal population was present at the 1200 m site. The fauna within this area is naturally more sparse, less diverse and more variable than in the northern North Sea and some of the differences at the intermediate stations may be accounted for in this way (Fig. 4). This spatial and temporal variability is due to physical factors such as sediment grain size and mobile bottom sediments.

Thus, the cumulative results of data collected to date, in the UK Sector, including the southern North Sea, has not led to any changes in the concept of zones of effect. Surrounding the area of major biological impact is a transition zone in which lesser biological effects are detected as

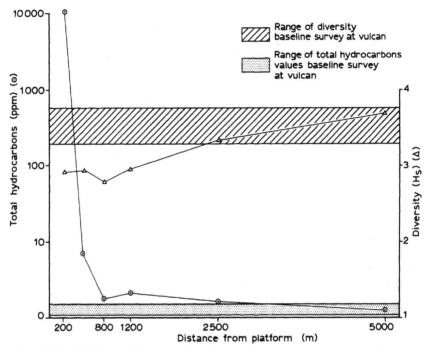

FIG. 4. Total hydrocarbons and diversity (H_s) of the macrobenthos within the sediments related to distance from the Vulcan platform, 1987 survey. The range of baseline survey data is also shown.

community parameters return to normal, generally within 1000 m. The shape and extent of this biological transition zone is variable, and is largely determined by the current regime and the scope of the drilling operation. With greater currents and more extensive drilling, it may be extended 2000 m in the direction of greatest water movement. Elevated hydrocarbon concentrations attributable to OBM were observed beyond the areas of biological effects. These elevated hydrocarbon concentrations have been measured out to as far as 4000 m in the direction of the prevailing current.

3. ENVIRONMENTAL SURVEYS AROUND SINGLE WELL SITES

Environmental effects resulting from the discharge of oil-contaminated cuttings at single well drilling sites in the UK sector have not previously

TABLE 2

Species abundance and diversity of the macrobenthic and sediment hydrocarbon levels, as measured by infra-red spectroscopy at six transect stations sampled in the Vulcan field for 1986 baseline and 1987 monitoring survey

	Year	Distance from platform (m)					
		200	500	800	1 200	2 500	5 000
Number of species	1986	—	29	38	34	33	28
	1987	17	17	21	29	32	20
Number of individuals	1986	—	516	516	416	413	152
	1987	167	117	236	298	446	271
H_s	1986	—	3·29	3·44	3·75	3·43	3·69
	1987	2·90	2·93	2·78	2·96	3·34	3·69
Total hydrocarbons by IR (expressed as Brent Crude Oil equivalents, μg/g dry wt sediment)	1986 (mean of 2 values)	—	1·25	1·5	0·85	0·6	0·6
	1987 (mean of 3 values)	11 347	8·5	2·6	3·4	2·2	1·1

been evaluated. Several surveys have therefore been carried out recently in the UK sector of the southern and central North Sea in order to assess the extent of hydrocarbon contamination of sediments and associated biological effects. These surveys are reviewed on a case history basis before summarising the pattern that is indicated at single well sites.

3.1. Recently Drilled Sites

Seabed surveys were conducted in 1986 before and after drilling two exploration wells in Block 16/28 in the central North Sea. The two wells, 16/28H and 16/28I, were located approximately 10 km apart in a water depth of approximately 100 m. Both wells had the 17½ in sections drilled using a low toxicity OBM and well 16/28H also had the 12¼ in section drilled using OBM. Pre-drilling surveys indicated that both sites were physically and chemically similar with a minor degree of hydrocarbon contamination typical of North Sea background conditions in this area.

Post-drilling surveys were carried out immediately upon completion of the drilling programme. Figure 5 shows the concentrations of hydro-carbons in sediments before and after the drilling discharges occurred.

Minor contamination extended out to 800 m at site I but concentrations

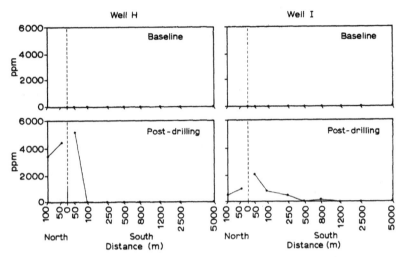

FIG. 5. The concentration of hydrocarbons in the sediments before and after drilling two single wells on block 16/2B in 1986.

were not greatly elevated above background beyond 500 m. High levels of contamination up to 100 × background were restricted to distances within 250 m of the discharge at site I and to within 100 m at site H.

Close to the discharge source, sediment hydrocarbon concentrations appeared similar to those found at multi-well sites, probably due to a significant incorporation of contaminated cuttings in the sediments. Hydrocarbon levels dropped off rapidly to not much greater than 10–100 × background within 500–1000 m of the discharge locations at sites 16/28 H and I.

3.2. Historical Exploration Sites

In 1987 a comprehensive investigation of three historical single well sites in the central North Sea was carried out to assess the longer-term pattern of seabed sediment hydrocarbon contamination and biological recovery at exploration drilling sites. Details of the wells studied are shown in Table 3. Ten stations were sampled at each well site on a transect along the residual current. Sample site locations are also described in Table 3.

Figure 6(a) shows the aliphatic and aromatic sediment hydrocarbon concentrations (μg/g) as determined by gravimetric analysis of sediment extracts. Elevated hydrocarbon concentrations extended to 800 m at the diesel site and within 750 m of the site where a smaller quantity of low

TABLE 3

Details of the well type and location for single well sites examined

Well location	Spud date	Mud type	Sections drilled	Estimated weight of oil discharged on cuttings
14/11	April 1980	Water-based mud	All	—
16/27	December 1981	Low toxicity oil-based mud	$12\frac{1}{4}''$ and $8\frac{1}{2}''$	85
21/12	December 1982	Diesel oil-based mud	$17\frac{1}{2}''$ and $8\frac{1}{2}''$ and $6''$	181

Sample locations were at 50 m north and 50, 100, 200, 500, 800, 1 200, 2 500 and 5 000 m south at the well site. A control station was located at 6 000 m east.

toxicity OBM cuttings was discharged. No significant contamination was detected at the well site where water-based mud was used.

Figure 6(b) summarises the results of the macrofaunal biological analysis. The species diversity at each station was determined. At the site where water-based mud had been used there was no observable effect on the benthos. At both sites where OBM was used a marked reduction in species diversity occurred between 50 and 500 m. At both well sites the normal diverse benthic community typical of the area was evident at the 500 m S station although the well site using low toxicity base oil appeared to have a well established benthic community at 200 m.

Although further work is being carried out on these sites, these preliminary findings have indicated that although oil-contaminated cuttings are still very much in evidence 5–6 years following a single well drilling programme, the biological effects were confined to within a zone of 750 m or less (Cranmer, 1988).

4. SEABED RECOVERY ON CESSATION OF DRILLING

At the time of the first reviews of the effects of OBM drilling discharges in the North Sea there were no data available from locations which had completed their drilling programme. It was recognised that biological and chemical data were needed from such sites in order to assess the long-term implications of the discharges and describe the response of the seabed macrobenthos to curtailment or cessation of OBM cuttings discharges. Seabed monitoring by offshore operators at fields approaching and past

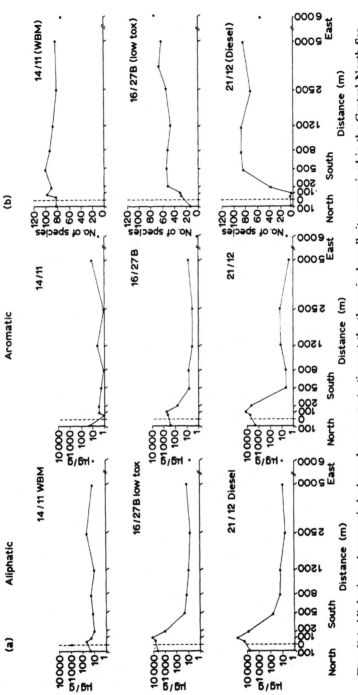

FIG. 6(a). Aliphatic and aromatic hydrocarbon concentrations at the three single well sites examined in the Central North Sea. (b) The number of macrofaunal species in the sediments around the three single well sites examined in the Central North Sea.

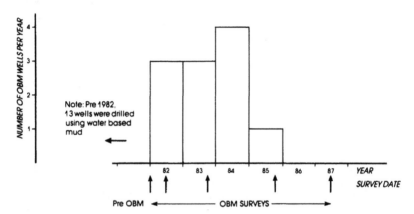

FIG. 7. History of oil-based mud drilling in the Beatrice oilfield and timing of seabed monitoring surveys.

completion of their development drilling programme has been carried out and data are now becoming available to describe seabed recovery and to test the predictions made in an earlier paper (Davies *et al.*, 1984).

4.1. Beatrice

The Beatrice field lies 12 miles from the coast in the Moray Firth and is the first nearshore oilfield to be developed in the North Sea. The main development drilling programme was carried out between 1978 and 1986 (Fig. 7) involving completion of 24 wells, 11 of which were drilled using low toxicity OBM. A series of five seabed monitoring surveys has been carried out since February 1982, just prior to first use of OBM through to May 1987, 20 months after the last well was completed. The results of the first three surveys (up to 1983: five OBM wells drilled) were summarised in Addy *et al.* (1984). It was concluded that the zones of seabed biological effect were consistent with other investigations and appeared to be largely caused by organic enrichment, except in the cuttings pile beneath the platform where physical smothering was responsible for impoverishment of the fauna. However, Kingston (1987) has interpreted the same data in another way, suggesting that they demonstrate a direct toxic effect rather than organic enrichment.

Between the 1983 and 1985 surveys a further six wells had been completed using OBM and an increase in the area of seabed biological effect was detected (Hobbs, 1985). The most recent survey in May 1987 (Cranmer, 1987) described a general decrease in sediment normal alkane

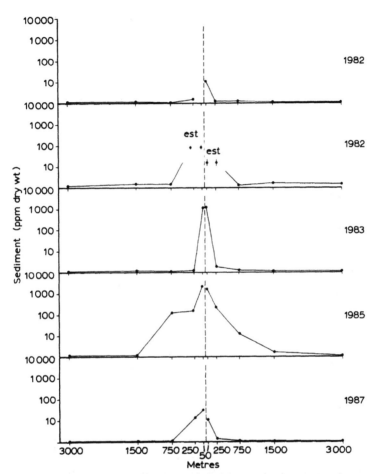

FIG. 8. Concentration of *n*-alkanes in ppm dry wt of sediment on the NNE-SSW transect through the Beatrice field 1982–7.

concentrations attributed to both biodegradation and probable sediment transport (Fig. 8). Benthic macrofauna had generally responded to the improvement in seabed conditions since cessation of drilling. Although the 'Zones of Effect' previously identified were still discernible, the condition of the benthic communities at many stations had improved. This was particularly noticeable in the 'transition' zone surrounding the cuttings pile. These trends are illustrated in Fig. 9 which shows data from the NNE to SSW transect through the field—this being the major axis

J. M. Davies et al.

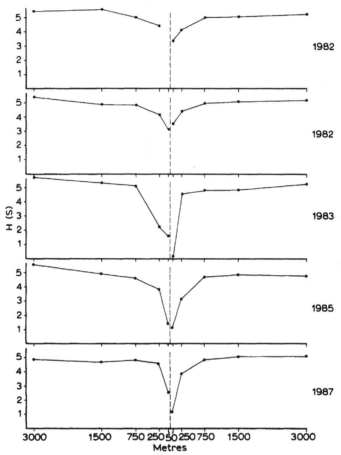

FIG. 9. Species diversity measured by Shannon–Weiner index (H_s) on the NNE-SSW transect through the Beatrice field 1982–7.

of water movement and so also the axis on which the most extensive biological effects can be detected. The macrobenthic species diversity close to the platform has increased since drilling ceased, particularly in the transition zone surrounding the cuttings pile (100 and 750 m from the platform).

The reduction in *n*-alkane concentrations between 1982 and 1987 is marked and gas chromatography supports the conclusion that biodegradation is taking place. Some mass transport due to sediment disturbance

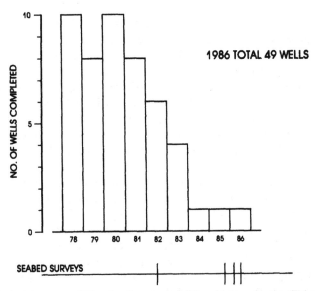

FIG. 10. A summary of the development drilling history in the Thistle oilfield (1978–86). The timing of seabed biological surveys is also shown.

from seabed currents and storm action in the relatively shallow water (50 m), is also probable. In the 20 months between completion of the last well and the 1987 survey, there had been dramatic reductions in hydrocarbon concentrations close to the platform, typically of at least one order of magnitude. In addition to this general decrease, in both aliphatic and aromatic hydrocarbon concentrations, Cranmer (1987) noted from the gravimetric data that there appeared to be a relatively greater reduction in aliphatics, possibly suggesting more rapid degradation of this fraction.

4.2. Thistle
Development drilling in the Thistle oilfield commenced in 1977 and was completed in 1986 during which period 49 wells were drilled, 31 of which involved the use of OBM in some sections. Between late 1979 and 1983 diesel-based muds were used. The final wells up to 1986 were drilled using low toxicity OBM (Fig. 10). Four seabed monitoring surveys were carried out starting at the height of the diesel-based mud drilling in 1982, through to 1986, by which time drilling had virtually ceased. The results of this monitoring were summarised by Hannam *et al.* (1987). A comprehensive

survey in 1982 during the period of diesel OBM drilling showed that the large volumes of material discharged from the platform altered sediment characteristics out as far as 3000 m. There was a 200 m zone around the platform where high levels of fresh diesel and continuous smothering made sediments inhospitable to macrofaunal species. Throughout the series of surveys there had been fluctuations in sediment hydrocarbon concentration around the platforms. However, the general biological trend since 1982 has been towards increasing diversity and since 1982 the boundary of the 'transition zone' has moved closer to the platform (Fig. 11).

The surveys carried out towards the completion of the development drilling programme showed significant environmental improvements. The clearest signs of recovery were in the physical characteristics of the sediments and macrobenthos, with steady improvements following the decrease in drilling input. Hydrocarbon data showed a less clear picture, with substantial variation in concentrations recorded especially between 1985 and January 1986. A number of stations showed marked increases over this period. The new wells drilled between the two surveys were not sufficient to explain the recorded increases in terms of new input. Hannam *et al.* (1987) concluded that the increases were most likely due to redistribution of the old and heavily contaminated accumulated pile of cuttings in combination with new inputs. Despite the increases in hydrocarbon concentration in 1986 the material appeared degraded in character (Hannam *et al.*, 1987) and had no major adverse influence on the biological recovery process. In June 1986 the hydrocarbon levels had again fallen, accompanied by further improvements to the macrobenthos.

4.3. Murchison

Four pre-operational seabed surveys in the Murchison field, carried out between September 1978 and August 1980, were followed up by two surveys during and after the development drilling phase. Development drilling commenced in June 1980. At the time of the September 1982 survey 20 wells had been completed using diesel OBM in the lower sections. A further six wells were drilled up to April 1984, two using diesel OBM and four using low toxicity OBM. The 1985 survey was carried out 16 months after the cessation of drilling. The result of the preoperational or baseline surveys are summarised in Appelbee & Mair (1981) and Grigson & Grogan (1981).

Details of the 1982 and 1985 monitoring surveys are described in Mair *et al.* (1987). Table 4 summarises data collected during these surveys.

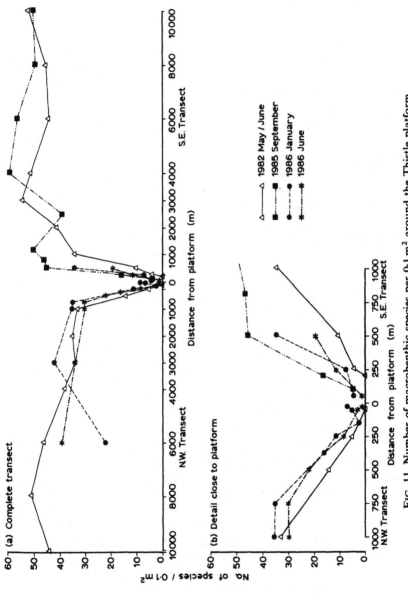

FIG. 11. Number of macrobenthic species per $0 \cdot 1 \, \text{m}^2$ around the Thistle platform.

Both monitoring surveys indicated a reduction in species abundance and diversity close to the platform with an inverse relationship to hydrocarbon concentration. There was a general increase in both monitoring surveys in species abundance and diversity with increasing distance from the platform; with most values returning to background levels at approximately 1000 m in 1982 and 2000 m in 1985 (see Figs 12 (a) and (b)). Comparing changes at stations between years, it is apparent that species abundance and diversity had improved markedly at the two inner stations in 1985, only 16 months after cessation of drilling, and that hydrocarbon levels had significantly decreased. At the 500 m and 1000 m sites the species diversity had decreased, reflecting the slight elevation of hydrocarbons at these sites in 1985.

The general improvement in biological conditions parallels a major reduction in seabed hydrocarbon concentration close to the platform, with almost an order of magnitude reduction at the 100 m and 250 m sites. This reduction is believed to be largely due to biodegradation, although sediment transport is probably also involved as suggested by the increase in hydrocarbon concentration measured at the 500 m and 1000 m sites.

Data, such as summarised above from fields close to the end or past completion of the development drilling programme, provide the first possibility for assessment of recovery processes and the long-term implications of OBM drilling discharges. Future monitoring will permit more detailed interpretation, particularly of the long-term fate of hydrocarbons. However, at this stage a number of preliminary conclusions may be drawn in relation to earlier predictions about seabed recovery.

The above studies suggest the early onset of macrobenthic recovery particularly in the inner zone once drilling has ceased. In response to reducing (and not necessarily ceased) organic input, the macrobenthic species diversity has recovered to near normal levels at some previously depleted stations around Beatrice, Thistle and Murchison. Subtle species compositional changes are still discernible along the gradient of organic enrichment, but generally the biological improvement appears to be rapid. As expected, the macrobenthos in the most heavily contaminated area very close to the platform remains highly modified and impoverished. However, despite the continuing high concentration of hydrocarbons associated with the cuttings pile itself, there are signs of increasing species diversity. The transition zone may increase in area (e.g. Murchison) but still does not extend beyond 2000 m from the platform.

Substantial decreases in seabed hydrocarbon concentrations have been

TABLE 4

Species abundance, diversity and evenness measures of the macrobenthos at the five transect stations sampled in 1982 and 1985 and at reference stations in the 1980 and 1985 surveys of the Murchison oilfield. Sediment hydrocarbon levels, as measured by infra-red spectroscopy, are also given

		Transect stations SE or Murchison platform distance from platform (m)					Range of values from reference stations (>4000 m from platform in 1982 and 1985 surveys) and stations in 1980 baseline survey
		100/150	250	500	1 000	2 000	
Number of species (S) per 0·5 m²	1980						127–159 (6 stations)
	1982	20	48	138	134	154	133–171 (5 stations)
	1985	71	106	110	109	159	
Number of individuals (N) per 0·5 m²	1980						1740–2114 (6 stations)
	1982	1911	2975	2695	2621	2675	1810–2645 (5 stations)
	1985	4468	1629	2698	2284	3143	
Shannon–Wiener index (H_s)	1980						5·23–5·63 (6 stations)
	1982	1·33	1·76	5·08	5·31	5·60	5·38–5·88 (5 stations)
	1985	3·14	4·50	4·32	4·69	5·26	
Pielou Evenness (J)	1980						0·73–0·80 (6 stations)
	1982	0·31	0·32	0·71	0·75	0·77	0·75–0·83 (5 stations)
	1985	0·51	0·67	0·64	0·69	0·72	

(continued)

TABLE 4—contd.

			Transect stations SE or Murchison platform distance from platform (m)					Range of values from reference stations (>4000 m from platform in 1982 and 1985 surveys) and stations in 1980 baseline survey
			100/150	250	500	1 000	2 000	
Hydrocarbon level as measured by infra-red analysis (expressed as µg Brent Crude oil equivalents/g dry sediment, sample replicates I and II)	—I	1980					2·7	0·8–2·7 (21 stations) mean—1·7
	—I	1982	2 293·0	1 050·0	11·5	8·1	5·5	2·5–6·5 (3 stations)
	—II		4 387·0	535·0	10·6	11·1	5·5	mean—4·2
	—I	1985	291·4	231·0	42·2	34·5	2·3	1·7–6·0 (6 stations)
	—II		677·5	91·0	98·9	89·6	4·9	mean—3·4

ªThe results exclude one anomaly, a high value of 5·6 µg/g from site 1.

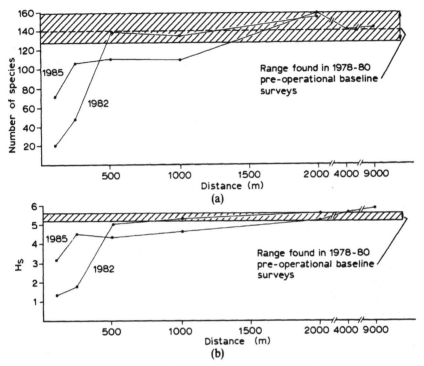

FIG 12. (a) Variation in number of macrobenthic species in sediments around the Murchison field. (b) Variation in macrofaunal species diversity measured by Shannon–Weiner index (H_s) in sediments around the Murchison field.

found on curtailment of drilling discharge, although the picture can be confused by fluctuations (for example at Murchison and Thistle) which may be due to natural sediment and deliberate cuttings redistribution, for example, to permit inspection of the platform structure at the seabed.

5. OIL TAINT IN FISH FLESH

When it became evident that oil contamination around platforms could be detected beyond the 500 m 'no fishing' zone the question of fish becoming tainted with oil was considered. During 1985 and 1986 fish were caught from three areas in the North Sea close to oil and gas exploration and production platforms, which had drilled many wells using OBM, and also from areas outside the influence of drilling activity as reference (control) samples. These fish were then tasted by a trained panel

TABLE 5
The taste panel results for fish caught around oil platforms

Distance (km)	Species	No. of fish	No. of assessments	No. detecting taint	No. of fish with half or more assessors detecting taint
3·7–9·3	Cod	10	100	1	0
0·93	Cod	4	36	1	0
0·93	Haddock	3	27	0	0
1·40	Haddock	10	80	2	0
0·93	Haddock	10	80	0	0
		37	323	4	0
0·40	Cod	2	14	0	0
0·40	Haddock	1	7	0	0
0·40	Tusk	1	7	0	0
		4	28	0	0
0·55–0·86	Dab	30	250	36	2
1·00–1·90	Dab	10	90	3	0

to determine the presence of any oily taint in the flesh. A fish was deemed to be oil tainted if more than half the panel detected an oily taint.

Haddock (*Melanogrammus aeglefinus*) and cod (*Gadus morhua*) were caught by trawls from four areas in the vicinity (1–2 km) of an oil producing platform in the East Shetland Basin and the same species caught by long line 400 m from a producing platform to the south of the Basin. The taste panel results (Table 5) showed no oil taint in the fish caught by long line or trawl.

Flatfish live in close contact with the sediments and are thus likely to be more vulnerable to oil in sediments. Therefore, dabs (*Limanda limanda*) were harvested by trawl from 550–860 m and 1000–1850 m from an oil-producing platform in the Moray Firth and compared to reference fish trawled 20 km from the platform (McGill *et al.*, 1987). The taste panel data (Table 5) showed that none of the fish were oil-tainted but that in some of the fish caught close to the platform some assessors did detect oily taint.

In order to determine whether the oily taint detected in the flatfish harvested from close (500–800 m) to an oil platform was related to the presence of petrogenic hydrocarbons in the flesh, four fish, identified as being possibly tainted, were analysed for *n*-alkanes and PAH, as were

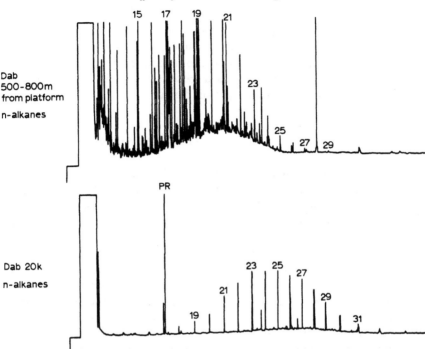

FIG. 13. Gas chromatographic analysis of the *n*-alkane fraction from the muscle of dabs, collected at 500–800 m and over 20 km from platform. (Reprinted by kind permission from McGill *et al.* (1987).)

several control samples. All the samples from the 500–800 m zone, except one, exhibited *n*-alkane profiles which are characterised by two unresolved complex mixtures (UCM) (Fig. 13). The UCM from carbon numbers C14–C27 typifies an input from a petrogenic source similar to that found for sediments in the same area and contrasted markedly with those of the reference samples. The second UCM, eluting before C14, may be associated with the presence of base oils used in low toxicity muds.

In a study to examine possible tainting of flatfish from the southern North Sea (Parker *et al.*, in preparation), samples of plaice (*Pleuronectes platessa*), dab (*Limanda limanda*) and sole (*Solea solea*) were collected from within 500 m of selected gas production platforms. The platform locations were chosen to represent different drilling histories. The reference site was located in an exploration block where exploration wells had been drilled before 1970 using only water-based muds. Initially, the taste panel did not differentiate between petroleum-derived and other taints. In the first series of experiments approximately one third of the fish samples were classed as having a detectable taint, the intensity of which was 'slight', on the scale used. Similar taints were found in both reference and platform sites. When selected fish were assessed specifically for 'oily' taint, only one fish was considered to have a slight oil flavour.

6. DISCUSSION AND CONCLUSIONS

This paper has considered the immediate and longer-term effects that can be measured around single well sites, at platforms during development drilling, and in the phase after drilling has ceased. There has been much recent discussion on the scale of effect of drilling mud cuttings in an overall North Sea context, and whilst it is difficult to produce a precise figure for the area involved, it was felt important to produce a 'best estimate' for the UK sector of the North Sea. The evidence discussed in this paper suggests that the major deleterious benthic effects are confined to an inner zone within 500 m of development drilling operations and within 250 m of single well sites. Beyond this, more subtle biological changes are seen in the transition zone generally extending to around 1000 m from development sites and 500 m from single well operations. In terms of contamination, the area of seabed involved is greater. The evidence from single well sites suggest that contamination along the axis of the prevailing current extends out to 1000 m but for development drilling operations may extend out to 4000 m. Recently concern has been expressed that in areas such as the East Shetland Basin, where oil production is highly concentrated, low level hydrocarbon contamination may occur over a larger area of the seabed. This is being investigated.

Based on 40 development and 380 single well sites drilled with OBM in the UK sector of the North Sea, the estimated areas of seabed affected by major deleterious effects on the benthos, subtle biological changes and hydrocarbon contamination are shown in Table 6. The major biological impact under and close to the cuttings piles totals 106 km^2 which represents about 0·04% of the UK North Sea. Adding on the area in which subtle biological effects can be detected the area involved is about 400 km^2 or 0·13% of the total UK North Sea. If the term 'effects' is defined to include the outer area in which OBM hydrocarbon contamination can be measured, but without accompanying biological effects, then the total area is approximately 1602 km^2 or 0·55% of the area of the UK sector of the North Sea.

TABLE 6
The calculated area of seabed affected by oil based drilling muds

Zone	Number of locations	Size/shape of zone	(km^2)	Cumulative areas % of UK North Sea	(km^2)	% of UK North Sea
Major biological effects	40 development sites	500 m radius	31·4	0·01	106	0·04
	380 single well sites	250 m radius	74·6	0·03		
Subtle biological changes	40 development sites	2 000 × 1 000 m ellipse	251·4	0·09	400	0·13
	380 single well sites	1 000 × 500 m ellipse	149·3	0·05		
OBM hydrocarbons present	40 development sites	4 000 × 8 000 m ellipse	1 005	0·35	1 602	0·55
	380 single well sites	1 000 × 2 000 m ellipse	597	0·2		

The results of further monitoring around UK platforms and single wells since 1983 largely support the 'agreed facts' drawn up by the Paris Commission Working Group on Oil Pollution in 1985, and have provided no evidence which would require major alteration, except that several gaps in our knowledge, then identified, have now been filled. The biological effects of drilling a single well with oil-based mud have been shown to be similar in nature, but much more restricted in extent, than at multi-well OBM sites.

The Paris Commission 'agreed facts' also drew attention to two other gaps in our knowledge, namely; 'That the effects of any degraded or partially degraded oil components are not known' and 'That there is little information on the sub-lethal effects that the cuttings might have in the marine environment.'

Whilst neither of these has been studied directly in the field, the monitoring around UK platforms where drilling discharges have ceased indicates that recovery and recolonisation of the transition zone begins within 1–2 years, accompanied by degradation of the oil hydrocarbons, and that the biological transition zone begins to move inwards, despite some outward redistribution of oil-contaminated materials. This suggests that the deleterious effects of oil on benthic organisms beyond the immediate zone, where smothering and organic enrichment dominate, result from the toxic effects of the freshly discharged oil, but that once this oil has been sufficiently degraded in the surface sediments, recolonisation can proceed. This is supported by experimental evidence (Blackman et al., 1986; Leaver et al., 1987). The 'degraded or partially degraded oil components' may therefore cause subtle, long-term effects, but these do not appear to hinder recolonisation.

One sublethal effect which has been examined is that of taint. No oily taint was found in round fish caught 400–2000 m around major oil production platforms or in flatfish harvested from outside 1000 m of an oil producing platform. It is apparent, however, that for flatfish caught closer to platforms, some assessors detected an oil taint, indicating a tendency towards taints by petroleum hydrocarbons in flatfish caught up to 850 m from platforms. However, the method of cooking the fish for taste panel assessment would ensure that any taint present was maximised, and yet the intensity of the taint detected was barely above the threshold limit for trained and experienced assessors. Thus, it is considered unlikely that any taint would be detected by untrained persons and that the typical consumer eating these fish, cooked and presented in a more usual manner, would not detect any taint. Tainting of flatfish has been reported around Danish inshore platforms discharging oil-based muds (DOE, 1987) (PARCOM GOP INFO 11/9 E, 1987) but it is not clear whether this was specifically tainting by oil, or a general 'off-flavour' due to other contaminants or natural sedimentary materials.

It is difficult to compare the results from these UK monitoring studies around oil platforms conducted since 1983 with experience elsewhere of the effects from the discharge of residual oil-based mud on cuttings. The large reported literature from the North American continent refers exclusively to the use of water-based muds with only occasional, minor addition of refined oil (Neff, 1987). A paper by Reiersen *et al.*, in these proceedings presents results of monitoring surveys around installations in the Norwegian sector of the North Sea.

At what rate will an affected area recover? Monitoring will continue at selected platform sites to chart the rate of recovery of affected areas. This information will be necessary in order to predict what is likely to happen, and what further controls might be needed, if the development of new fields, in perhaps more environmentally sensitive areas, is considered. While it is predicted with some confidence that recovery of the outer zone of effect will be comparatively rapid, the long-term fate of the innermost zone beneath and in the immediate vicinity of the platform remains to be determined. What will be the recovery mechanism here? What measures might best be taken to hasten its recovery? With platform decommissioning at the end of a field's productive life already being considered for the North Sea, the potential for long-term recovery of the cuttings pile and the mechanisms by which it most rapidly occurs will require study.

REFERENCES

ADDY, J.M., HARTLEY, J.P. & TIBBETS, P.J.C. (1984). Ecological effects of low toxicity oil-based mud drilling in the Beatrice oilfield. *Mar. Pollut. Bull.*, **15**, (12), 429–36.

APPELBEE, J.F. & MAIR, J.M. (1981). Hutton–Murchison fields, environmental baseline study, volume II—macrofaunal assessment August 1980 survey. Report by the Institute of Offshore Engineering for Conoco (UK) Ltd.

BAMBER, R.N. (1984) The benthos of a marine fly-ash dumping ground. *J. Mar. Biol. Ass. UK*, **64**, 211–26.

BLACKMAN, R.A.A., FILEMAN, T.W., LAW, R.J. & THAIN, J.E. (1985). The effects of oil-based drill-muds in sediments on the settlement and development of biota in a 200-day tank test. ICES CM 1985/E:23, 8pp, mimeo.

BLACKMAN, R.A.A., LAW, R.J. & THAIN, J.E. (1986). The effects of new oil-based drill-muds in sediments on the settlement and development of biota in an improved tank test. ICES CM 1986/E:13, 8pp, mimeo.

CRANMER, G. (1987). Environmental survey of the benthic sediments around the Beatrice AD/AP platform complex 1987. Report by Aberdeen University Marine Studies Ltd, for Britoil plc.

CRANMER, G. (1988). Environmental surveys of sediments around three exploration well sites in the North Sea. AUMS Ltd, Report for UKOOA.

DAVIES, J.M., ADDY, J.M. BLACKMAN, R.A.A., BLANCHARD, J.R., FERBRACHE, J.E., MOORE, D.C., SOMERVILLE, H.J., WHITEHEAD, A. & WILKINSON, T. (1984). Environmental effects of the use of oil-based drilling muds in the North Sea. *Mar. Pollut. Bull.*, **15**(10), 363–70.

DICKS, B. BAKKE, T. & DIXON, I.M.T. (1986/87). Oil exploration and production: impact on the North Sea. *Oil Chem. Pollut.*, **3**, 289–306.

DIXON, I.M.T. (1987). Experimental application of oil-based muds and cuttings to sea bed sediments. In *Fate and Effects of Oil in Marine Ecosystems*, ed J. Kuper & W.J. Van den Brink. Martinus Nijhoff, Dordrecht, 1987.

DOE. Quality status of the North Sea. Second International Conference on the Protection of the North Sea, September 1987. Department of the Environment.

EISMA, D. (1987). The North Sea: an overview. *Phil. Trans. R. Soc. Lond.*, **B316**, 461–85.

GRIGSON, S. & GROGAN, W. (1981). Hutton–Murchison fields, environmental baseline study, volume I — hydrocarbons heavy metal appraisal August 1980 survey. Report by the Institute of Offshore Engineering for Conoco (UK) Ltd.

HANNAM, M.D., ADDY, J.M. & DICKS, B. (1987). Ecological monitoring of drill cuttings discharges to the seabed in the Thistle oil field. TNO Conference, Amsterdam, February 1987.

HOBBS, G. (1985). Environmental survey of seabed samples near the Beatrice AD and AP platforms in August 1985 associated with the use of low toxicity oil-based drilling fluids. Report by Field Studies Council, Oil Pollution Research Unit for Britoil plc.

KINGSTON, P.S. (1987). Field effects of platform discharges on benthic macrofauna. *Phil Trans. R. Soc. Lond.*, **B316**, 545–65.

LEAVER, M.J., MURISON, D.J., DAVIES, J.M. & RAFAELLI, D. (1987). Experimental studies of the effects of drilling discharges. *Phil. Trans. R. Soc. Lond.*, **B316**, 625–40.

MAIR, J.McD., MATHIESON, I. & APPELBEE, J.F. (1987). Offshore macrobenthic recovery in the Murchison field following the termination of drill cuttings discharge. *Mar. Pollut. Bull.*, **18**(12), 628–34.

McGILL, A.S., MACKIE, P.R., HOWGATE, P. & McHENERY, J.G. (1987). The flavour and assessment of dabs (*Limanda limanda*) caught in the vicinity of the Beatrice Oil Platform.

NEFF, J.M. (1987). Biological effects of drilling fluids, drill cuttings and produced waters. In *Long Term Environmental Effects of Offshore Oil and Gas Development*, ed. D.D. Boesch, & N.N. Rabalais. Elsevier Applied Science, London and New York, pp. 469–538.

PAYNE, J.F., FANCEY, L., KICENIUK, J., WILLIAMS, U., OSBORNE, J. & RAHIMTULA, A. (1985). Mixed function oxygenases as biological monitors around petroleum hydrocarbon development sites: potential for induction by diesel and other drilling mud base oils containing reduced levels of polycyclic aromatic hydrocarbons. *Marine Environ. Res.*, **17**, 328–32.

TIBBETTS, P. (1985). A study of the aliphatic hydrocarbons in benthic sediments around the Beatrice A platform in 1985. Report by M-Scan Ltd, for Britoil plc.

4

Monitoring in the Vicinity of Oil and Gas Platforms; Results from the Norwegian Sector of the North Sea and Recommended Methods for Forthcoming Surveillance*

L.-O. REIERSEN

The Norwegian State Pollution Control Authority, PB 8100 Dep, 0032, Oslo 1, Norway

JOHN S. GRAY

Department of Biology, Section of Marine Zoology and Chemistry, University of Oslo, PB 1064, Blindern, 0316 Oslo 3, Norway

KARSTEN H. PALMORK

Marine Research Institute, PB 1870, 5011 Nordnes, Bergen, Norway

and

ROLF LANGE

Cooperating Marine Scientists AS, Billingstadsletta 19, 1362 Billingstad, Norway

*Editors' Note: This paper has aroused considerable controversy among peer reviewers and the various groups with an interest in the topic. In particular, the controversy focuses on the authors' interpretation of the data which they contend suggests a significant extension of the impact zone around production platforms in the Norwegian sector. Peer reviewers have raised concern over comparability of the data sets from 1984 and 1986 because of changes in sampling equipment, accuracy of station locations, appropriateness of statistical analysis techniques, differences in hydrocarbon extraction techniques, and interpretation of cause-and-effect between diversity indices shifts and sediment hydrocarbon levels.

While the reviewers have expressed concern that the paper does not include enough data to support the authors' conclusions regarding extension of the zone of effects, the Review Committee feels that there is substantive information that is valuable regarding the North Sea and that the paper is important in that it

ABSTRACT

Studies of environmental effects of oil activities in the Norwegian sector of the North Sea show that significant contamination from oil and trace metals are found around all installations using Oil-based mud (OBM). Contamination is greatest along the direction of the primary current axis and effects are found to distances beyond those previously reported. At one field significant contamination was observed as far out as 7000 and 12 000 m and a change in fauna was found out to 5000 m. Indications are that this is an effect due to discharges of oil-contaminated cuttings. There is a clear relationship between the amounts of oil-contaminated cuttings discharged and the area of seabed affected. The depth to which sediment is contaminated varies, but 2000 m from one field the contaminated layer was more than 9 cm.

Despite the change from diesel-based muds to low aromatic muds (low toxicity muds) there has been no concomitant improvement in the benthic communities studied. There are, however, signs that following cessation of drilling activities degradation of oil has occurred and the benthic communities show signs of recovery. The degree of degradation seems to be correlated to the concentration of hydrocarbons in the sediment.

In the vicinity of platforms using OBM the base oil discharged with the cuttings is found accumulated in fish livers of near bottom feeders but not in pelagic fish.

In the Norwegian surveys there were large variations from field to field and from year to year both within and between surveys. This was shown to be largely due to methodological differences and it was felt necessary therefore, to introduce a standardised monitoring programme, which is presented here.

---

summarises a considerable amount of work which is otherwise unpublished in the open literature concerning the effects of drilling discharges in the Norwegian North Sea.

Although the controversy has not been resolved by the authors in the paper, we decided to proceed with publishing the paper since the conclusions arising from the authors' interpretations may have important regulatory repercussions in the Norwegian sector of the North Sea. It is important that the reader be aware of the technical differences of opinion. To seek further information, the readers are encouraged to refer to the current scientific literature and contact the authors and the Exploration and Production Forum (25/28 Burlington Street, London, W1X 1LB, UK).

1. INTRODUCTION

The amount of oil bound to drill cuttings discharged in the Norwegian sector is much lower than that of the British sector (Table 1). In 1985 3331 t of oil were discharged in the Norwegian sector together with cuttings compared with 20 230 t in the British sector (Anon., 1988a), (in 1986 drilling activity was reduced). Yet the amount of oil entering the North Sea from oil and gas production is less than 20% of the total (Anon., 1987a).

Nations with exploration rights in the North Sea require that the effects of exploration for and production of oil and gas have to be monitored. For the British sector a review of the distribution of and environmental effects of Oil-based Muds (OBM) was produced in 1984 by Davies *et al.* The findings showed that 250 m from platforms typically hydrocarbon levels were between 1000 and 10 000 times background, but that background levels occurred 2000–3000 m from the platform. Elevated hydrocarbon levels were found to follow the predominant current pattern and were detected beyond the areas showing biological effects. The major deleterious biological effects were found within 500 m of platforms and the limit of effects to 1000 m.

Recently Bedborough *et al.* (1987) have reviewed the inputs of hydrocarbons to the North Sea from oil and gas developments and Kingston (1987) has reviewed effects of platform discharges on macrofauna, largely in the British sector.

No reviews are available for the Norwegian sector, yet environmental studies date back to 1973 for the Ekofisk field and now cover 15 different fields (Table 2). This paper reports the major findings of the monitoring programmes and discusses the approaches and methods used as the results obtained varied greatly from survey to survey. These findings suggested

TABLE 1

Discharges of oil bound to drill cuttings (t) in the North Sea. Reported to the Paris commission (Anon., 1988a).

	1981	1982	1983	1984	1985	1986
Norway	400	950	1471	3466	3331	2030
UK	6900	10600	14000	19800	20230	13000
North Sea total[a]	7426	12732	17749	24959	25757	19680

[a] Includes discharges in Denmark and Holland.

TABLE 2

Monitoring of the marine environment in the vicinity of oil and gas platforms in the Norwegian sector of the North Sea. (B = biological monitoring, C = chemical monitoring, F = fish monitoring)

Field	1973	1975	1977	1978	1979	1980	1981	1982	1983	1984	1985	1986	1987
Ekofisk	B and C	B and C	B and C	B and C	B and C	B	B and C	B and C	C	B and C	C and F	—	B, C and F
Statfjord				C	C	F and C	C	B and C	F	B and C	C	B, C and F	C and F
Frigg				C	C	—	—	C	—	—	—	B, C and F	—
Heimdal										B and C	—	B, C and F	—
Valhall						C	C	C	C	C	B, C and F	C and F	C and F
Hod								C	C	C	C	C	C
Traena B								B and C		B and C			
Halten B											B and C		
Gullfaks										B and C		B and C	C and F
Oseberg										B and C		B, C and F	C and F
Odin										C	B, C and F		
Ula										B and C			C and F
Gyda													B, C and F
Veslefrikk													B and C
Tommeliten													B and C

that better standardisation of techniques and quality control was necessary. Suggestions as to how these may be achieved are presented.

1.1. Norwegian Regulations on Monitoring around Oil and Gas Platforms

The primary aim of monitoring around oil and gas platforms in the Norwegian sector is to ensure that no unnecessary damage is caused to the marine environment. To this end biological and chemical monitoring must be done at regular intervals and be done in such a way that it is possible to show changes that occur in the marine environment in both space and time. The monitoring programme must be designed so that it is possible to ascertain whether or not the required discharge limits and level of cleaning are working satisfactorily or whether they should be changed.

A baseline study is required a minimum of 1 year before the start of drilling of production holes and the plan must be developed with advice from relevant experts and approved by the State Pollution Control Authority (SFT). The investigation itself must be done by an independent company approved by SFT. The primary requirement for the baseline survey is a detailed study of background hydrocarbon and selected heavy metal concentrations in bottom sediments. When the platform is in production the breakdown of hydrocarbons must be followed using such indicators as changes in single components or changes in the n-C_{17}/pristane and n-C_{18}/phytane ratio.

Biological monitoring should be based on macrofaunal communities so that changes in species composition or/and diversity can be followed. In addition, when in production, bacterial breakdown of organic material, e.g. drilling muds should be followed.

For oil fields it has been required that chemical monitoring is conducted annually and biological monitoring every 3 years. For gas fields chemical monitoring is required every third year and biological monitoring every sixth year. Environmental studies should be done in the period April–July in order to give comparability to the data sets obtained. Similarly it is required that the methods used must be comparable so that it is possible to follow changes in chemical and biological properties over time.

2. MONITORING STUDIES OF DISCHARGES IN THE NORWEGIAN NORTH SEA

Figure 1 shows the location of the Norwegian oil and gas fields and Table 3 lists the monitoring programmes that have been conducted in the Norwegian sector and on which this paper is based.

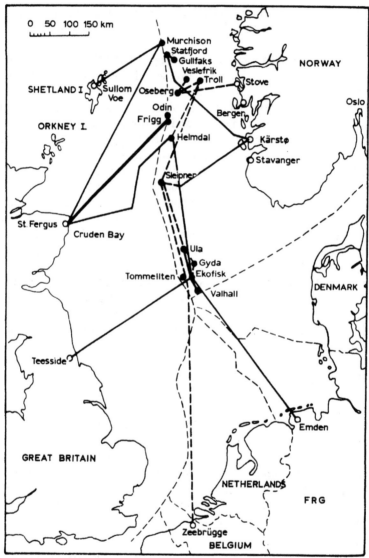

FIG. 1. Norwegian fields and pipelines in the North Sea. ●, areas in production or under development; ——, existing pipelines; ——, pipelines sanctioned developed.

TABLE 3

Sources of variability in data from monitoring surveys in the Norwegian sector of the North Sea

Monitoring source	Chemical		Biological
	Hydrocarbon	Metals	
Planning	+	+	+
Sampling	+ + +	+ + +	+ + +
Treatment	+ +	+ +	+ + +
Storage	+	+	0
Extraction	+ +	+	+ +
Concentration and isolation	+	+	−
Identification	+ +	0	+ +
Data Analysis	+ +	+	+
Reporting	+	+	+

+ + + = major source, + + = moderate source, + = minor source, 0 = no problem, − = not relevant

2.1. Methodological Variations

Great variations in both chemical and biological properties were observed both over time and between fields. However, detailed analyses of the data obtained showed that the variation did not necessarily reflect field to field differences or changes over time, but was largely due to methodological variation.

Table 3 lists the major sources of variability encountered in evaluating and comparing monitoring surveys from different fields and years. The major problem was undoubtedly differences in sampling methods. Whilst attempts were made to obtain quantitative samples in all surveys (both for hydrocarbon, heavy metal and biological analyses), the size and type of sampler used varied enormously, from a small corer taking $0.025\,m^2$ to a large boxcorer taking $0.25\,m^2$. For the chemical analyses it is imperative that the samples come from undisturbed surface samples, yet this was rarely controlled. Thus much of the variability reported between surveys (see also Davies *et al.*, 1984, Fig. 1) can be attributed to sampling variability; this is also true for the biological studies.

2.1.1. Hydrocarbon Analyses

In the analyses of Total Hydrocarbon Concentration (THC) most surveys used the IOC standard method (1982) but the estimated procedure used to determine total hydrocarbons varied greatly with two surveys using n-C_{13}–n-C_{23}, four surveys n-C_{10}–n-C_{30}, two surveys n-C_{10}–n-C_{32} and three surveys n-C_{10}–n-C_{40}. Three other methods for estimating hydro-carbons were also used so that again it is extremely difficult to compare results either from field to field or from year to year. As an example of the difficulties, some surveys reported hydrocarbon content as mg/kg or μg/kg sediment on a wet weight basis whereas others used a dry weight basis.

2.1.2. Analyses of Heavy Metals and Major Elements

Most variability was felt to be due to differences in sampling methods and post-sample treatment, for example, whether or nor they were homogenised, standardised with the use of reference samples as controls and in whether or not values were reported for total concentrations or only for acid-extracted fractions. Unfortunately, there is no international reference standard for sediment containing a high amount of barium as barite. Development of such a standard which could be used in comparative tests is eminently desirable.

2.1.3. Biological Analyses

In the biological analyses again sampling methods were the greatest source of variation. Many reports did not give the volume of sample obtained on each occasion and quite often the size of grab used, number of replicates and total area sampled were inadequate for biological sampling. Once sediment samples are obtained the fauna must be extracted by elutriation or washing. It is extremely important that elutriation or washing are done carefully as damage to the fauna and loss of organisms can easily result. Few reports even mentioned how elutriation or washing was done and none gave any quality control measures for this aspect.

At the most extreme for the same field in two different years the number of taxa varied from 20 (66 grab samples) the first year to over 400 (140 grab samples) the next year! Clearly the first surveys used quite inadequate methods.

A further source of error for the biological sampling was in the identification of the fauna. Analysis of the species lists showed that some surveys had used quite inappropriate taxonomic literature and identifications were wrong. If biological surveys are to be used as reference material it is essential that good taxonomy be done by the different companies involved. This can be achieved by constant access to the relevant literature (which by no means all consultancy firms have) and a reference collection identified by acknowledged experts for the fauna of the geographical area studied. Whilst there was no standard suite of statistical analyses used, in general most surveys used adequate techniques.

2.2. Monitoring Results from the Norwegian Sector of the North Sea

Due to the variations mentioned above field results for individual fields will be presented rather than compiled composite data for all fields.

2.2.1. Water-based Mud

At fields where predominantly Water-based Muds (WBS) were used (Odin, Heimdal and Frigg) minor contamination can be found 100–1000 m from the platform, traceable either as raised levels of heavy metals and/or hydrocarbons, or as small changes in the benthic fauna (Anon, 1986*a*, 1987*b*, *c*).

2.2.2. Oil-based Mud

Table 4 shows comparative data from different fields of the amounts of oil and cuttings discharged. The Statfjord field has clearly the greatest amount of oil discharged. Whereas at Statfjord A drilling activity was reduced in 1986 the activity was still high at Statfjord C. The activity is reflected in the tonnage of discharged cuttings and hydrocarbons. It should be noted that at Valhall, Statfjord A and B prior to 1983–4 drilling was done using diesel-based muds.

Figure 2 shows total hydrocarbon data for the three Statfjord fields for July 1986 showing that the predominant current direction has a major influence on the spread of the hydrocarbons. The area that is polluted with values up to 10 000 times background levels is within 250 m of the platform. It is clear that stations in all directions out to 1000 m were affected and along the predominant current direction out to 5000 m (Anon., 1987*d*). To get a better estimate of the area contaminated the outermost stations in the 1987 survey was changed from 5000 to 7000 m along the predominant current and from 1000 to 2000 m in the three other directions. The results from the 1987 survey showed that all stations were contaminated. Figure 3 shows the result from Statfjord A. Even the reference stations 12 000 m downstream from Statfjord B had 22·5 ppm THC (Anon., 1988*b*). These results indicate that a large area around the Statfjord field is contaminated. Statfjord is the most polluted of the Norwegian fields and the area impacted by hydrocarbons has spread over time as drilling activity and use of OBM has increased. This is best illustrated looking at Statfjord C where the amounts of drilling mud discharged has increased from 304 t in 1984 to 8839 t in 1986 (Table 4),

L.-O. Reiersen et al.

TABLE 4
Discharges of oil and cuttings from production platforms in the Norwegian sector of the North Sea

Platform (operator)	Year	No. of wells using		Discharge	
		Diesel mud	Low-aromatic mud	Oil (t)	Cuttings (t)
Ekofisk	83	0	3	65	—
(Phillips)	84	0	3	53	697
	85	0	1	11	206
	86	0	0	0	0
Statfjord A	83	0	5	384	—
(Mobil/	84	0	5	1·078	5·804
Statoil)	85	0	6	781	5·203
	86	0	1	196	1·199
Statfjord B	83	7	0	644	—
(Mobil/	84	4	2	967	6·256
Statoil)	85	0	7	909	6·810
	86	0	5	305	3·166
Statfjord C	84	0	1	51	304
(Mobil/	85	0	11	1·188	9·704
Statoil)	86	0	11	876	8·839
Valhall	83	4	3	183	—
(Amoco)	84	0	5	351	4·707
	85	0	6	353	4·309
	86	0	7	134	1·854
Oseberg	85	0	1	61	521
(Hydro)	86	0	6	416	3·776

and the concomitant increase in contaminated area (Table 5). Similar data are available from the other platforms where OBM have been used, e.g. Valhall, Oseberg, Ekofisk and Gulfaks (Anon., 1985*b*, 1986*b*, 1987*e*, *f*, 1988*c*) and the general conclusion that can be drawn is that the most hydrocarbon impacted areas are within 250–1000 m of the platform and elevated levels are found to 3000–5000 m. There seems to be a direct relationship between the size of the impacted area and the amount of discharged cuttings contaminated with OBM, but this is modified by the water current in the area and the depth of discharge.

There are indications that resuspension and transport of the sedimented

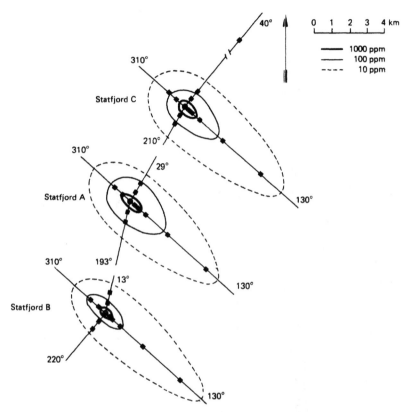

FIG. 2. The approximate extent of hydrocarbon contamination in bottom sediment around Statfjord field in 1986 (from Anon., 1987*d*). ■, sampling stations.

material away from the area occur near the Statfjord A and B platforms (Anon., 1987*d*).

The depth of the contaminated layer was measured at some fields and seems to vary with distance from platform and the type of bottom sediment. At 2000 m from the Valhall platform the contaminated layer is more than 9 cm deep (Anon., 1987*e*), indicating a mixing of hydrocarbons into the sediment, probably due to both biological activity and chemical/physical processes.

2.2.3. Heavy Metals and Major Elements

Among the major elements monitored barium is of direct interest as it is a component of both water- and oil-based drilling muds. Around the

L.-O. Reiersen et al.

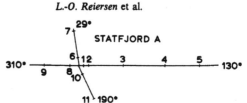

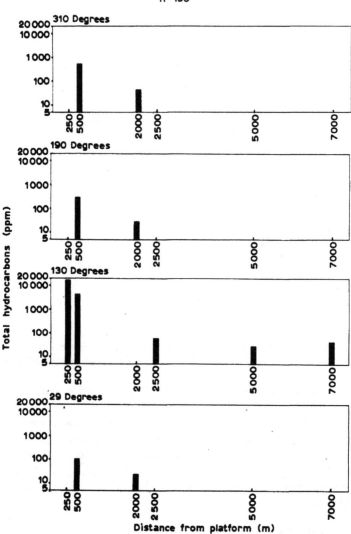

FIG. 3. The total hydrocarbon concentration (THC) in ppm at different stations along transects from Statfjord A in 1987 (from Anon., 1988b).

TABLE 5
The total concentration of hydrocarbons (THC) in top sediment sampled at different distances from Statfjord A and C. The values are in mg/kg (ppm) dry weight and are presented in Anon. (1985a) and Anon. (1988b).

Stations		Year					
		1987[a]	1986[a]	1985[b]	1984[c]	1982[a]	1981[a]
Statfjord A							
500 m	130°	4 220·0	1 180·0	3 333·0	2 317·0	206·0	669·0
2 000 m	130°	—	—	—	21·0	12·8	15·8
2 500 m	130°	58·9	45·8	31·4	23·0	—	—
5 000 m	130°	36·8	18·7	18·3	10·0	4·2	5·9
Statfjord C							
250 m	130°	16 600·0	3 310·0	997·0	4·0	—	—
500 m	130°	752·0	2 340·0	1 216·0	5·0	2·0	—
1 000 m	130°	—	471·0	121·0	5·0	2·9	—
2 500 m	130°	88·6	41·5	—	5·0	—	—

[a] THC analysed by GC.
[b] THC analysed by gravimetry on the non-polar fraction.
[c] THC analysed by IR.

Statfjord field barium data show a distribution pattern almost identical to that for total hydrocarbons (Fig. 2). Levels of barium can therefore be expected to be correlated to hydrocarbon levels if the primary source of the hydrocarbon contamination is from drilling muds. Figure 4 shows the relationship between barium concentration and THC for (a) Statfjord fields A, B and C and (b) Valhall and Hod fields. In both cases over 80% of the variance can be explained by the regression and the two regression equations were not significantly different.

Heavy metals such as lead, chromium, copper, strontium and cadmium show increased concentrations close to the platforms, but normally drop to background levels within 1000 m. The heavy metals probably originate from operational discharges, e.g. the barite used in the drilling muds.

2.2.4. Effects on Benthic Ecology

The general view of effects of hydrocarbon discharges on biological systems is that areas affected are restricted to well within those showing elevated hydrocarbon levels and usually within 500 m of platforms (Davies *et al.*, 1984; Kingston, 1987). For the fields listed in Table 2 all except the Statfjord field show effects on the biological systems limited to within 1000 m of the platform. Findings at the Statfjord field were however,

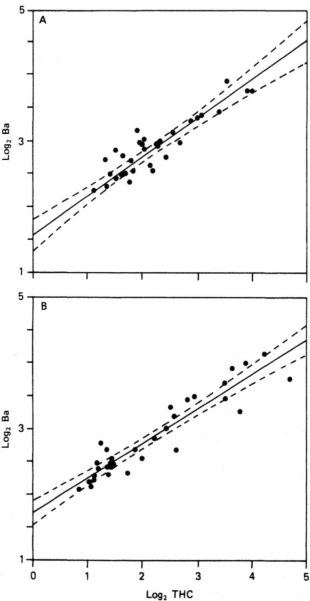

FIG. 4. The correlation between total hydrocarbon concentration (THC) and barium (Ba) in the sediment: A, Statfjord field; B, Valhall and Hod fields. – – –. 95% C.I.

TABLE 6

Mean values of diversity (H_s base log 2) at the Statfjord field. Significance tested by t test. $n =$ the number of samples

Year	THC level (ppm)	n	$H_s \pm SD$	p values of t test
1984	Background ($X = 9.68$)	17	5.51 ± 0.19	
				0.065 (5.51 versus 5.30)
1986	10–99	14	5.30 ± 0.39	
				0.04 (5.30 versus 4.76)
1986	100–1 000	14	4.76 ± 0.51	

different. Slightly different methods were used in the surveys done in 1984 and in 1986 which affected the total area sampled and the number of species obtained there being fewer in 1986. However, diversity values at the 'control sites' 2500 m and 5000 m from the platforms at Statfjord A and B were almost identical in the 1984 and 1986 surveys ($H_s = 5.45 \pm 0.36$ and 5.56 ± 0.24, respectively). Thus diversity, which integrates numbers of species and the distribution of individuals amongst species, is here unaffected by the reduced number of species found as a consequence of changed sampling methods. At Statfjord C the species found at 2500 m and 5000 m were closely similar in 1984 and 1986 thus comparisons of diversity between the two years are valid. From the survey in 1984 background diversity is known to be $H_s = 5.51 \pm 0.19$, $n = 17$ (Table 6).

Table 7 shows changes in diversity and THC levels over time for the 130° transect at the Statfjord field. Whilst there has been no significant change at Statfjord A and B there is a statistically highly significant reduction in diversity ($p = 0.001$ paired t test) over all stations at Statfjord C between 1984 and 1986. For stations at 2500 and 5000 m at Statfjord C in 1986 the mean diversity is 4.53 ± 0.16 compared with the 1984 'background' data for the Statfjord field (mean 5.51 ± 0.19, Table 6). For another comparison the outer stations at 2500 and 5000 m for Statfjord A and B ('control sites'), were 5.45 ± 0.36 in 1984 and 5.56 ± 0.24 in 1986. Thus using the conventional test of ± 2 s.d. there is a significant reduction in diversity out to 5000 m at Statfjord C in 1986 compared to the 1984 data or control sites for 1986. Figure 5 shows that only the Statfjord C field in 1986 shows effects out to 5000 m whereas at Statfjord A and B platforms diversity is reduced out to 1000 m. Thus at this, the most impacted field within the Norwegian sector, lowered diversity is found within 1000 m of the platform, but more important significantly lowered diversity can be found out to 5000 m.

TABLE 7

Comparison of (H_s base log 2) along the 130° transect at Statfjord field in 1984 and 1986. Data from Anon. (1985a) and Anon. (1987d). Significance tested by paired t test

Platform	Distance (m)	THC (ppm)		H_s (base log 2)		Significance H_s 1984/1986 t test
		1984	1986	1984	1986	
A	250	5748	7960	1·36	0·96	
	500	2317	1180	2·29	2·74	
	1000	183	940	5·18	3·86	$p = 0.20$
	2500	23	46	5·56	5·29	
	5000	10	19	5·82	5·43	
B	250	670	9550	3·34	1·53	
	500	51	194	4·96	4·12	
	1000	134	113	5·39	5·15	$p = 0.50$
	2500	13	28	5·46	5·70	
	5000	8	14	4·97	5·82	
C	250	4	3310	5·63	0·29	
	500	5	2340	5·45	0·55	
	1000	5	471	5·52	4·38	$p = 0·001$
	2500	5	52	5·63	4·64	
	5000	3	24	5·50	4·41	

THC analysed by GC in 1986 and IR in 1984.

That this decrease in diversity is more than just a random event can be supported by other biological data. At the 5000 m station at Statfjord C in 1986 dominance was higher than in 1984 yet four of the top five dominants were identical. In addition plots of individuals per species against number of species showed that changes in community structure had occurred in 1986 compared with 1984. We believe therefore, that the lower diversity in 1986 reflects significant changes in community structure rather than random effects. The comparable THC levels in the sediment for Statfjord C have increased from <6 ppm in 1984 to 52 at 2500 m and 24 at 5000 m in 1986. Thus there are indications that the significant reduction in diversity is due to increased hydrocarbon contamination.

Figure 6 shows a plot of macrofaunal diversity against THC for the combined Statfjord fields for 1986. The figure shows that diversity is clearly reduced above THC levels of 100 ppm. A t test comparing the

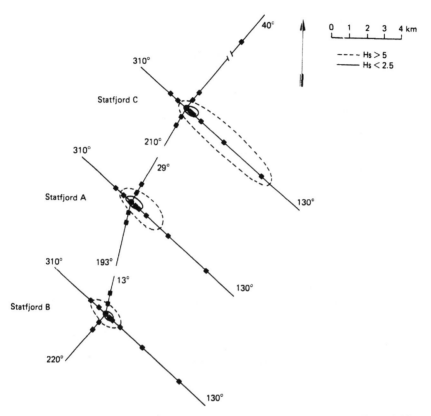

FIG. 5. The approximate extent of macrofaunal diversity (H_s) at Statfjord field 1986. ■, sampling stations.

diversity of stations having THC values between 100 and 1000 ppm ($n = 14$) and those between 10 and 99 ppm in 1986 showed a probability of rejecting the H_0 of no difference of $p = 0.04$. A similar test comparing background sites in 1984 ($n = 17$) with data for stations having THC levels between 10 and 99 ppm in 1986 ($n = 14$) showed a probability of rejecting the H_0 of no difference of $p = 0.06$ (Table 6). These data suggest that reductions in diversity occur between THC levels of 10 and 99 ppm.

2.2.5. Degradation and Recovery
Improvements with time have been recorded at the Oseberg field in 1986 (Anon., 1987*f*), after the use of OBM in one bore hole in 1984. Breakdown

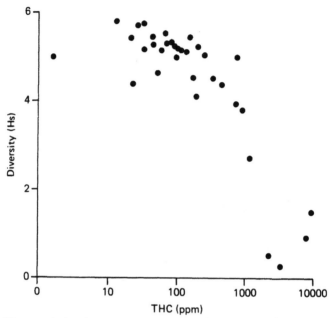

FIG. 6. The correlation between the total concentration of hydrocarbon (THC) in the sediment and the diversity (H_s) of macrobenthic fauna, Statfjord field 1986. — —, regression lines.

products of the oil have been recorded and the fauna shows signs of recovery with the organic-enrichment indicator species *Capitella capitata* increasing in abundance from 4 individuals in 1984 to 65 in 1985. This was accompanied by an increase in the total number of species.

There are also chemical indicators of biodegradation; the results are, however, ambiguous. Observations at the Statfjord field indicate highest degradation rate close to Statfjord A, i.e. where the content of hydrocarbons is highest, but no significant trends at Statfjord B and C (Anon., 1987*d*). At Valhall and Hod fields the results indicate an inverse correlation between degradation activity and the oil content in the sediment, and no sign of degradation was observed at the station heavily polluted by diesel 5 years ago (Anon., 1987*e*). For further information regarding degradation see Grahl-Nielsen *et al.*, this volume.

From the Ekofisk field a clear reduction in hydrocarbon content over the last 3 years 100 m from the platform (43 700 ppm in 1984 to 440 ppm in 1987) was observed. This is assumed to be due to biodegradation and mixing into deeper layers (Anon., 1988*d*).

2.2.6. Hydrocarbons Accumulated in Fish

The accumulation of hydrocarbons in fish has been analysed the last few years. Although the total number of fish analysed is small ($n = 70$) and there is large individual to individual variation, some general conclusions can be drawn (Table 8). In the vicinity of platforms using OBM, fish living close to the bottom (e.g. cod) have accumulated the base oil, but no accumulation was observed in saithe which is a more pelagic fish. In the vicinity of platforms using WBM no accumulation was observed in bottom fish (haddock). From the Danish sector oil-tainted fish (plaice) have been caught 20–45 km from platforms drilled (9–18 wells) using 'low-toxicity' OBM (Randløv & Poulsen, 1986). These results show that accumulation of base oils in fish caught in the vicinity of platforms might be a problem for the fishing industry and should be studied further.

3. REGULATORY INTERPRETATIONS

3.1. Contamination of Hydrocarbons

In the Norwegian sector hydrocarbon concentrations in sediments are remarkably similar to those reported by Davies *et al.* (1984) for the British sector of the North Sea. Total hydrocarbon levels were up to 10 000 times background near platforms. Background levels in the North Sea (Dogger Bank) are 0·4–2·0 ppm (dry weight) and off the west coast of Britain 1·1–2·0 ppm (Law & Fileman, 1985). From the monitoring performed in the Norwegian sector comparable figures for background levels are reported to be 2–10 ppm (dry weight). The extension of the area of seabed with values over background has increased to 5000 m or more at the Statfjord, Valhall and Oseberg fields (Anon., 1987*d*, *e*, *f*, 1988*b*), probably reflecting the increasing use of OBM in recent years (Table 4). The total area contaminated by hydrocarbons on the Norwegian sector of the North Sea is estimated to be at least 200 km^2.

3.2. Effects on Benthic Ecology

The biological data in general showed gross effects located within a 1 km zone where diversity, number of species and dominance patterns were all altered. Davies *et al.* (1984), Kingston (1987) and Addy (1987) record closely similar effects in the British sector. The shape of the area affected

TABLE 8

The concentration of hydrocarbons in liver (mg/kg dry weight) from fish caught in the vicinity of platforms. X is the mean value and n the number of fish

Field	Species	THC	NPD	Decalines	Benzenes	Cyclohexanes
Valhall[a] 1987 OBM	Cod	1890.0	2.35	65.0	27.0	26.0
		950.0	3.23	12.0	11.0	5.6
		580.0	1.47	1.0	0.26	0.11
		510.0	0.44	1.8	0.50	1.1
		710.0	1.54	2.0	1.6	1.3
	X±SD	928±563	1.81±1.05	16.4±27.6	8.1±11.5	6.2±11.1
Ref. (Gyda)		660.0	0.097	2.1	0.5	0.9
		430.0	0.058	1.6	0.2	0.4
		670.0	0.060	1.4	0.14	0.5
		480.0	0.067	2.0	0.4	1.1
	X±SD	560±123	0.07±0.018	1.8±0.3	0.31±0.2	0.7±0.3
Frigg[b] 1986	Haddock n = 5					
WBM Ref.	X±SD n = 5	750±250	0.02±0.02	—	—	—
15 000 m Oseberg[c] 1987	X±SD Saithe n = 5	1000±0.0	0.03±0.03	—	—	—
OBM Ref.	X±SD n = 5	265±22.0	0.12±0.05	—	—	—
3500 m	X±SD	260±14.0	0.17±0.14	—	—	—

[a] Anon. (1988e).
[b] Anon. (1987c).
[c] Anon. (1988f).
THC: total hydrocarbon concentration.
NPD: naphthalene, phenanthrene/anthracene and dibenzothiophene.
OBM: Oil-based mud is used.
WBM: Water-based mud is used.

was in response to the dominant current direction showing greatest effect along the major current axis (Fig. 2). As far removed as 5000 m from the Statfjord C there was a significant reduction in diversity from 1984 to 1986 (Table 7). To date no one has reported effects of oil in macrofaunal diversity as far as 5000 m from a platform. The generally accepted view is that 1000 m is the maximal range of effects on the macrofauna (see Davies *et al.*, 1984; Kingston, 1987).

Different grabs were used in the two surveys; a large Van Veen ($0.1\,m^2$) in 1984 and a smaller Shipek ($0.04\,m^2$) in 1986. The change in grab size unfortunately opens the way for alternative explanations of the differences between years. It is to us quite unacceptable that scope for alternatives can occur in surveys which are primarily designed to find out whether or not deleterious trends occur.

The reduced fauna found at 5000 m from Statfjord C could possibly be due to natural biological events such as patchiness in the fauna whereby a highly diverse patch was sampled in 1984 and a less diverse patch in 1986. We are, however, not inclined to this view as diversity at the control sites at Statfjord A and B and along other axes than that of the dominant current at Statfjord C were closely similar in 1984 and 1986. Furthermore, there is no evidence of an increase in pollution-tolerant opportunist species. The fauna at the 2500 m and 5000 m stations 'closely resembles that found in 1984', although total species number was reduced from 147 to 85 (Anon., 1987*d*). Four of the five most dominant species were identical in both surveys.

Diversity integrates the number of species and distribution of individuals among species. Although the number of species was less in the 1986 survey there was a comparable reduction in individuals (due to small sample size) at the control sites (Statfjord A and B) so that diversity values remained similar. Dominance increased from 1984 to 1986 at 2500 and 5000 m (Statfjord C) and this is one indication of worsening conditions which could account for the reduction in diversity. The lack of change in the diversity of controls from 1984 to 1986 argues against there having been any natural influence (e.g. climatic) on the fauna between years. It seems to us most parsimonious therefore, to suggest that there were reductions in diversity at 5000 m in response to the discharge of oil contaminated cuttings from Statfjord C (Table 4).

Environmental control authorities have a responsibility to react to the first signals that significant deleterious effects have been demonstrated. At the Statfjord field we have the problem with changed methodology from one survey to the next. The surveys should, however, be of such a quality that authorities can rely on the results obtained and not give scope for the operator to doubt the validity of the data. Although based on few samples, changes have been observed at Statfjord C in 1986 at 5000 m from the platform which are statistically significant and suggest deteriorating conditions due to hydrocarbon discharge. It is clearly necessary to follow up this finding and perhaps alter the use of OBM and the discharge permits. It will also be of importance to perform a new biological monitoring at Statfjord to establish the effect of the continuing high drilling and discharge activity at Statfjord C in 1987.

3.3. Limit of Response

The other significant finding from this survey is that at the Statfjord field reductions in diversity begin at total hydrocarbon concentrations between 10 and 100 ppm (Fig. 6 and Table 6). In their review Davies *et al.* (1984) stated that there was evidence of a linear relationship between oil concentration and decreasing diversity at concentrations below 100 ppm. Our findings strengthen this statement and suggest that reductions in diversity probably begin at concentrations as low as 25 ppm (5000 m from Statfjord C). However, the same mean THC concentration was also measured 2500–5000 m from Statfjord A and B, but no effect was observed on diversity. At these stations the THC chromatograms showed that the hydrocarbons originated from the diesel discharged prior to 1984 at Statfjord A and B (Anon., 1987*d*). At Statfjord C only low aromatic mud has been used.

This result indicates that one should not only focus on the exact THC value measured when looking at limit of response, but also take into account the type of oil (diesel or low-aromatic) and oil components (e.g. naphthalenes), the degree of weathering and degradation, of the sedimented oil, other mud chemicals (e.g. emulsifiers) and natural conditions such as the oxygen supply in the top sediment layer.

3.4. Diesel-based Mud or Low-aromatic Mud

Whether or not the trend to use low-aromatic drilling muds has led to an improvement in the environment is a key question. In the North Sea there has been a trend since 1983 away from the use of diesel-based drilling muds to 'low toxicity' muds (Table 4). Despite this trend there seems not to have been a comparable improvement in the benthic fauna, for example at Statfjord A and B where diesel-based muds were used from the start of drilling in 1978 and 1981 until 1983 and 1984 respectively (Table 4). If the change in benthic fauna detected at 5000 m at Statfjord

C is due to effects of oil, it does not appear that so-called 'low-toxic' drilling muds do less damage to the benthic environment than diesel-based muds. This finding will also have a bearing on the selection of toxicity tests to be used in giving approval for drilling muds. So far only the water soluble fraction of the 'whole' drilling mud has been tested and done on non-sediment living animals. No tests have been performed on the actual 'receiver' of the oil-contaminated cuttings, the animals living in the top sediment.

3.5. Standardised Procedure for Monitoring Surveys

Whilst the major effects of oil activities around platforms are similar in both British and Norwegian sectors the main characteristic of both chemical and biological data sets is their variability. Davies *et al.* (1984) noted that direct comparability between sets are difficult due to different analytical approaches. Kingston (1987) reports from underwater video surveys that drill cuttings are extremely patchily distributed around platforms. This fact probably plays a large part in explaining the great variability in levels of contamination found around platforms. Nothwithstanding this finding, from our detailed analyses of the data from the Norwegian sector it is clear that another important source of variability is in sampling rather than due to differences in analytical methods.

What is clear from the above is that it is essential to standardise procedures so that comparative results can be obtained enabling analyses of field to field and year to year variations to be followed. A guide-line for the monitoring of the vicinity of platforms in the North Sea has now been produced by a collaboration between authorities and scientists in UK, the Netherlands, Denmark, Sweden and Norway. Table 9 shows the main point in the guide-line: monitoring frequencies, sampling strategies, parameters to be measured, analytical methods, etc. The guide-line will be available from summer 1988.

It is also essential that quality control measures are added to the requirements so that it is possible for the authorities to rapidly assess the quality of the data obtained. Table 10 lists a suggested protocol for quality control for the monitoring surveys.

The company performing the analytical work on heavy metals and hydrocarbons should participate in intercalibration exercises. If for some reason the sampling or analytical methods used at one field are changed, intercalibration must be done.

If this guide-line can be followed by the members of the Paris

TABLE 9
Recommended monitoring programme for platforms in the North Sea
(C = chemistry, B = biology)

Baseline survey		Monitoring programme	
		Gas fields	
Frequency	B and C once before drilling production well	Drilling phase	C every year
			C and B every 3rd year
		Production phase	C and B every 6th year
		Oil fields	
		Drilling phase	C every year
			C and B every 3rd year
		Production phase	C and B every 3rd year
Stations	A grid system including future ref. station and radial transects	16 stations in 4 radial transects. Ref. station: 10 000 m from platform, same depth and sediment	

Sampling	*Chemical monitoring*	*Biological monitoring*
Time of the year		April–June
Equipment	corer	grab or corer (0·1 m²)
Replicates	3	5
Total area sampled	—	0·5 m²
Penetration depth: mud	15 cm	15 cm
sand	5 cm	5 cm
Parameters to be measured		
Sediment parameter		
Grain size		×
Dry weight	×	×
Redox potential (E_h)		×
Total organic content (TOC)		×
Major elements/heavy metals (Ba, Cd, Cu, Pb, Zn, Hg, Fe, by EAAS)	×	×
Total content of hydrocarbons (THC) by GC	×	×
Selected aromatics (NPD) and bicyclic aliphatics by GCMS	×	×
Biodegradation	×	×
Biological parameters		
Macrobenthic fauna as number of individuals and species, diversity, etc.		×

Special programmes if decided:
  Musselwatch, fish eggs and larvae, plankton.
  Hydrocarbon content and selected aromatics (NPD) in fish and shellfish.

TABLE 10
Protocol for quality control

Prior to survey

The consultant company must document its:

Expertise and equipment for sampling, analysis and statistical treatment of the data

Taxonomic literature and reference collection

Earlier experience of similar work

The company should participate in an intercalibration system for the analysis of heavy metal and hydrocarbons

Survey

A detailed protocol must be written from planning to reporting

The volume of the sediment samples is to be measured

The number of species at the reference station should be > 100, and the number of individuals per sample should be several hundred.

Commission then we can achieve comparability of data from oil field to oil field and over time so that any changes both detrimental or advantageous can be accurately recorded. Although this is an extremely detailed protocol it is felt necessary to specify in such detail, as clearly the methodology in use today only in a few cases allows us to validly compare samples.

One way that effective rationalisation can be achieved is for surveys of neighbouring fields to be done at the same time by one company rather than the present free-for-all. This would achieve both cheaper monitoring and more consistent sampling, which would allow the control authorities to be better able to make comparisons between fields and to be more able to distinguish natural from oil-induced changes.

ACKNOWLEDGEMENTS

We extend grateful thanks to Statoil for financial support in the preparation of this paper. The authors alone are responsible for the views expressed.

REFERENCES

ADDY, J.M. (1987). Environmental monitoring of the Beatrice oilfield development. *Phil. Trans. R. Soc. Lond.*, **316B**, 655–68.

ANON. (1985a). Statfjord environmental survey, June 1984. Institute of Offshore Engineering, Heriot-Watt University. IOE/84/251, 203 pp.

ANON. (1985b). Valhall and Hod fields 1985 environmental survey. Part I: Chemical analyses of sediments and fish. Senter for Industriforskning. Report No. 850532–1, 45 pp.

ANON. (1986a). Odin–1985 Environmental Survey. Senter for Industriforskning og Norwegian Institute for Water Research. Report No. 850810–1, 65 pp.

ANON. (1986b). 1985 Ekofisk field hydrocarbon survey. Senter for Industriforskning. Report No. 850811, 40 pp.

ANON. (1987a). Summary Report. International conference on environmental protection of the North Sea. Water Research Centre, 18 pp.

ANON. (1987b). Heimdal platform environmental survey, 1986. 1. Physical, chemical and biological characterization of the surface sediments. 2. Aromatic hydrocarbons in fish. Oceanographic Center of Cooperating Marine Scientists. Report No. SF59.0045.00/02/86, 208 pp.

ANON. (1987c). Frigg environmental survey, 1986. 1. Physical, chemical and biological characterization of the surface sediments. 2. Aromatic hydrocarbons in fish. Oceanographic Center of Cooperating Marine Scientists. Report No. SF59.0045.00/01/86, 217 pp.

ANON. (1987d). Statfjord environmental survey, July 1986. Metals, hydrocarbons and macrobenthic fauna. A/S Miljøplan, Report No. P86–077.

ANON. (1987e). Valhall and Hod fields environmental survey 1986. Senter for Industriforskning. Report No. 860705–1, 51 pp.

ANON. (1987f). Oseberg environmental survey, 1986. 1. Physical; chemical and biological characterization of the surface sediments: 2. Hydrocarbons in fish tissues. Oceanographic Center og Norsk Institutt for Vannforskning. Report No. 02.0785.00/01/87, 572 pp.

ANON. (1988a). Discharges from offshore exploration and exploitation installations in 1986. GOP 122/2-E, Paris Commission, 10 pp.

ANON. (1988b). Miljøundersøkelser på Statfjordfeltet 1987. IKU Sintef group of A.S. George. Report No. 02.0840.00/01/87, 138 pp.

ANON. (1988c). Miljøundersøkelser på Gullfaksfeltet 1987. IKU Sintef group og A.S. George. Report No. 02.0841.00/01/87, 97 pp.

ANON. (1988c). Ekofisk and Eldfisk environmental surveys 1987. Senter for Industriforskning. Report No. 870709–01, 29 pp.

ANON. (1988d). Valhall and Hod environmental surveys 1987. Senter for Industriforskning. Report No. 870511, 30 pp.

ANON. (1988e). Oseberg environmental survey 1987. 1. Physical and chemical characterisation of the surface sediments. 2. Hydrocarbons in fish tissues. IKU Sintef group. Report No. 02. 0837.00/02/87, 212 pp.

BEDBOROUGH, D.R., BLACKMAN, R.A.A. & LAW, R.J. (1987). A survey of inputs to the North Sea resulting from oil and gas developments. *Phil. Trans. R. Soc. Lond. B.*, **316**, 495–509.

DAVIES, J.M. ADDY, J.M., BLACKMAN, R., BLANCHARD, J.R., MOORE, D.C. SOMERVILLE, H.J., WHITEHEAD, A. & WILKINSON, T. (1984). Environmental effects of oil based mud cuttings. *Mar. Pollut. Bull.*, **15**, 363–70.

INTERGOVERNMENTAL OCEANOGRAPHIC COMMISSION (1982). The determination of petroleum hydrocarbons in sediments. Manuals and Guides, No. 11, UNESCO.

KINGSTON, P.F. (1987). Field effects of platform discharges on benthic macrofauna. *Phil. Trans. R. Soc. Lond. B.*, **316**, 545–65.

LAW, R.J. & FILEMAN, T.W. (1985). The distribution of hydrocarbons in surficial sediments from the central North Sea. *Mar. Pollut. Bull.*, **16**, 335–7.

RANDLØV, A. & POULSEN, E. (1986). Environmental impact of low-toxic oil-based drilling mud. Taint in fish and possibilities of reduction of the impact. COWIconsult, Mærsk olie og gas a.s., 65 pp.

5

Patterns of Oil-based Drilling Fluid Utilization and Disposal of Associated Wastes on the Canadian Offshore Frontier Lands

PAUL G. CHÉNARD,* F. RAINER ENGELHARDT, JEAN BLANE‡

and

DUNCAN HARDIE

Canada Oil and Gas Lands Administration, Environmental Protection Branch, 355 River Road, Ottawa, Ontario, Canada K1A 0E4

ABSTRACT

This paper presents a summary of quantitative data on the use of oil-based mud on the Canadian offshore frontier lands. The information is used to test two simple empirical models useful in prediction of environmental impacts. The first model, designed to estimate the volume and mass of oily cuttings produced at a well site, shows close agreement with results based on actual caliper logging data. The second model, to predict the oil content of cuttings as a function of geographical location and several common downhole parameters, failed to demonstrate significant correlations.

A survey of mud losses indicates that usually 60% of the oil-based drilling fluids consumed during drilling are discharged to the environment, mostly but not exclusively on cuttings. Weighting agents are by far the most abundantly discharged mud additive. Canadian government guidelines have fostered the use of mineral or paraffinic base oils low in aromatic compounds instead of diesel as well as low toxicity additives, thereby minimizing the potential for environmental impact.

* Present address: Energy, Mines and Resources Canada, Office of Environmental Affairs, 580 Booth Street, Ottawa, Ontario, Canada K1A 0E4.
‡ Present Address: University of Waterloo, Faculty of Environmental Studies, Waterloo, Ontario, Canada.

1. INTRODUCTION

The spudding of the Alma F-67 well on the Scotian Shelf in December 1983 marked the introduction of oil-based drilling mud (OBM) in the Canadian offshore frontier lands. From then to the time of this writing, 28 additional wells have been spudded or approved (Table 1). Although

TABLE 1

Wells drilled with OBM on the Canadian frontier lands from 1983 to 1987

Year[a]	Well	Area[b]	Operator	Mud oil
1983	Alma F-67	SS	Shell	Conoco ODC
1984	South Desbarres O-76	SS	Shell	Conoco ODC
1984	West Venture C-62	SS	Mobil	Conoco ODC
1984	Glenelg E-58	SS	Shell	Conoco ODC
1984	Glenelg E58A	SS	Shell	Conoco ODC
1984	Glenelg H-38	SS	Shell	Shell DF
1985	Alma K-85	SS	Shell	Conoco ODC
1985	Peskowesk A-99	SS	Shell	Vista ODC
1985	Kegeshook G-67	SS	Shell	Vista ODC
1985	Merigomish C-52	SS	Shell	Vista ODC
1985	North Triumph G-43	SS	Shell	Shell DF
1985	Baccalieu I-78	GB	Esso	Esso Sarnia A
1985	Adgo G-24	A	Esso	Vista ODC
1985	Minuk I-53	A	Esso	Vista ODC; Esso DMO-75
1985	Nipterk L-19A	A	Esso	Esso DMO-75
1986	Glenelg N-49	SS	Shell	Shell Sol DMS
1986	North Triumph B-52	SS	Shell	Shell Sol DMS
1986	Panuke B-90	SS	Shell	Shell Sol DMS
1986	West Chebucto K-20	SS	Husky Bow Valley	Shell Sol DMS
1986	Tantallon M-41	SS	Shell	Shell Sol DMS
1986	Kaubvik I-43	A	Esso	Vista ODC; Escaid 90
1986	Unak L-28	A	Shell	Shell Sol DMS
1987	Whycocomagh N-90	SS	Canterra	Shell Sol DMS
1987	Springdale J-29[c]	GB	Husky Bow Valley	Shell Sol DMS
1987	Whiterose E-09	GB	Husky Bow Valley	Shell Sol DMS
1987	Bonne Bay C-73	GB	Husky Bow Valley	Shell Sol DMS
1987	Whiterose C-19[c]	GB	Husky Bow Valley	Shell Sol DMS
1987	Fortune D-85[c]	GB	Husky Bow Valley	Shell Sol DMS
1987	Terra Nova E-09	GB	Petro-Canada	Shell Sol DMS

[a] Year spudded.
[b] A = Arctic; GB = Grand Banks; SS = Scotian Shelf.
[c] Approved but not spudded as of December 1987.

TABLE 2
Drilling activity on the Canadian frontier lands from 1983 to 1987

Year	Total no. wells spudded	Total no. wells spudded offshore	OBM wells				
			Scotian Shelf	Grand Banks	Arctic	Total	Percentage of offshore wells
1983	95	33	1	0	0	1	3
1984	120	44	5	0	0	5	11
1985	99	37	5	1	3	9	24
1986	60	23	5	0	2	7	30
1987	48[a]	12[a]	1	6	0	7	58
Total	422	149	17	7	5	29	19

[a] Estimate includes 3 wells on the Grand Banks that have been approved but are not spudded as of 87/12/08.

there has been a marked decrease in drilling since the peak year of 1984, the incidence of scheduled use of OBM in offshore wells has increased every year (Table 2). During the period 1983–7, 19% of all offshore wells spudded on Canadian frontier lands were drilled with OBM for 30–90% of their total depth (based on operators' records submitted to the Canada Oil and Gas Lands Administration (COGLA)).

The Canadian government decided early on to regulate the use of OBM because of a general concern over the discharge of oily cuttings into the environment, and because diesel was frequently the base oil used elsewhere (i.e. North Sea, Gulf of Mexico) and has known toxicity to marine organisms and humans. COGLA issued provisional guidelines in 1983, prepared with input from industry and other involved government departments. These guidelines were subsequently amended in November 1985 to their present formal version (COGLA/CNOPB 1986). A guideline section recommending base oils of low toxicity and of low aromatic hydrocarbon content, in conjunction with the development of several mineral or paraffinic substitutes for diesel by the industry, led to the situation that diesel oil has not been used in OBM on the Canadian frontier lands. In addition, the guidelines have allowed COGLA to obtain from the offshore operators quantitative data on mud composition, oil content of cuttings, and losses of mud during field activities. These data are required to help the regulator address possible concerns about the environmental impact of OBM use and oily cuttings discharges around well sites.

This paper presents an up-to-date synthesis of the use of OBM in the

Canadian offshore, and provides an analysis of these data with the objective of providing the environmental regulator with the means to calculate in advance a reliable estimate of the amount of oily cuttings, and thus of mud oil and other ingredients, that would be discharged at a well site on the Canadian offshore frontier lands.

2. MATERIALS AND METHODS

The first step in the data analysis was to estimate the theoretical volume of cuttings discharged at each well, assuming perfect borehole geometry and equivalence between cuttings volume and borehole volume. In some cases, hole diameters and lengths were directly given in the drilling bit programs supplied in advance by the operators. In the other cases where bit programs were not found in the records, diameters and interval lengths were derived from casing programs according to standard engineering conversions. The accuracy of the theoretical volume can be evaluated in each well by comparing this result with the actual volume measured by caliper logging, which integrates the hole volume at regular depth intervals. The total actual borehole volume is the sum of the interval volumes provided on the logging curves that are included by an operator in a final well report submitted to COGLA after the well is plugged and abandoned. The volumes and masses of dry cuttings produced at wells drilled with OBM on the frontier lands were subsequently tabulated for the period 1983 to 1987 (Table 3). A simple regression plot of actual versus theoretical cuttings volumes for each well provided an estimate of the accuracy of predicted volumes (Fig. 1).

In order to express the environmental discharge data in mass units, the cuttings volumes were converted to masses using the formation densities measured by the FD logs that are included in the final report for each well. The cuttings masses for each well were then plotted against their corresponding volumes (Fig. 2).

In order to estimate the amount of mud oil that would be discharged on cuttings, the mass of cuttings was combined with the oil content measured by the retort method on cuttings from various depths in the well. These results are reported by the operators in their final well reports, as required by the OBM guidelines.

The FD and compensated neutron log (CNL) curves included in the final well reports were examined in an attempt to relate oil content on cuttings with the parameters that Strong (1984) identified as possible

TABLE 3
Volumes and masses of cuttings produced at wells drilled with OBM on
the Canadian Frontier Lands from 1983 to 1987

Well cuttings dry	Area[a]	Cuttings volume		Cuttings mass (t)
		Caliper log method[b] (m³)	Nearest casing method[c] (m³)	
Alma F-67	SS	152	135	343
South Desbarres 0-76	SS	504	493	1 060
West Venture C-62	SS	115·1	116·2	285
Glenelg E-58	SS	196	200	367
Glenelg E-58A	SS	84	76	186
Glenelg H-38	SS	212·1	212·1	459
Alma K-85	SS	172	181	327
Peskowesk A-99	SS	215·2	204·1	499
Kegeshook G-67	SS	181·2	176	387
Merigomish C-52	SS	218·2	204·5	469
North Triumph G-43	SS	276·9	268	611
Glenelg N-49	SS	236	219·3	497
North Triumph B-52	SS	204·7	187	441
Panuke B-90	SS	140·7	135·9	429
West Chebucto K-20	SS	175	175·7	395
Tantallon M-41	SS	244·8	216·6	490
Whycocomagh N-90	SS	195·1	161·8	387
Kaubvik I-43	A	294·4	296·5	652
Adgo G-24	A	126·9	123·4	221
Minuk I-53	A	292·5	187·6	610
Nipterk L-19A	A	108	81	219
Unak L-28	A	163·3	153·9	357
Baccalieu I-78	GB	163·8	161	387

[a] A = Arctic; GB = Grand Banks; SS = Scotian Shelf.
[b] Actual hole volume obtained by summing the caliper log readings.
[c] Estimated hole volume obtained from bit or casing programs submitted by the operators.

determinants of oil adsorption on cuttings. The logging data provided enough information on five of these parameters: hole cross-section (caliper readings), formation density, formation porosity, depth and lithology. All the corresponding measurements from each well were compared against oil content for simple linear regression analysis, using Lotus 1-2-3 spreadsheet software on an IBM microcomputer. The significance of each regression was estimated by comparing its correlation coefficient (r) with the critical values for r given by Rohlf & Sokal (1969) at the 0·05 level of significance. The lithological parameter was given an integer number

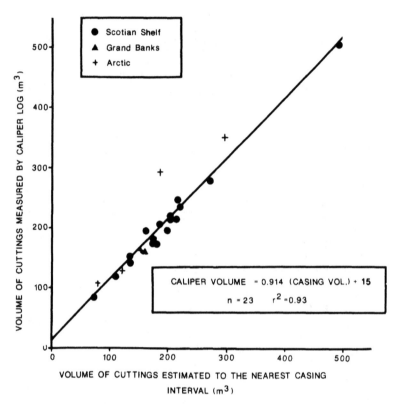

FIG. 1. Linear regression plot of cuttings volume measured by caliper log on cuttings volume estimated from bit and casing programs. Data are for wells drilled with oil-based mud on the Canadian frontier lands from 1983 to 1987.

corresponding to shale, limestone, sandstone and siltstone. Regressions were performed for individual wells and for data pooled by geographic region. Three distinct 'geological provinces' were considered: Scotian Shelf, Grand Banks, and Arctic (the combination of Mackenzie Delta, Beaufort Sea and Arctic Islands).

Data on the drilling muds, including identity of OBM ingredients and quantities used, were also provided by the operators as requested by the OBM guidelines.

3. RESULTS AND DISCUSSIONS

There is a significant correlation between cuttings volumes estimated to the nearest casing interval and the volumes given by caliper log measure-

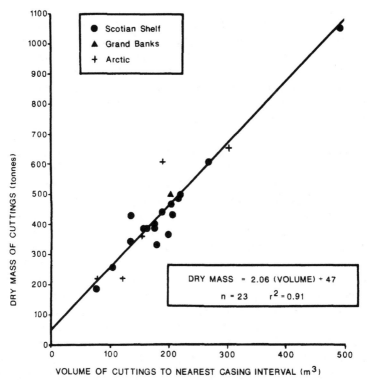

FIG. 2. Linear regression plot of cuttings dry mass calculated from logging measurements on cuttings volume estimated from bit and casing programs. Data are for wells drilled with oil-based mud on the Canadian frontier lands from 1983 to 1987.

ments (Fig. 1). This confirms a previous preliminary conclusion that an excellent estimate of the volume of cuttings from an interval drilled with OBM can be obtained by assuming perfect borehole geometry (Chénard *et al.*, 1986).

Significant correlation is also found between estimated cuttings volumes and dry masses derived from logging information (Fig. 2). The slope of this regression line indicates a mean density of 2·06 for dry cuttings, which is very close to 2·02 found previously (Chénard *et al.*, 1986). This estimated density can be used to accurately predict the mass of oily cuttings discharged after drilling an interval of known length with OBM, and save the analyst from performing the laborious task of reading the logging curves for each well.

Forecasting the amount of mud oil discharged requires an accurate prediction of the oil content of cuttings. But Tables 4 and 5 do not show a simple pattern as for volume and mass. For instance, it was assumed that oil content of cuttings from Arctic wells would be significantly greater than from East Coast wells because of the smaller grain size associated with the silty nature of Beaufort Sea and Mackenzie Delta Sediments (B. Young, 1987, personal communication). Although the data in Table 4 suggest mean oil content to be different in both regions, a *t*-test (Stanley, 1963) between the overall means for the Scotian Shelf and Arctic regions does not support a significant difference ($t = 0.313$, $p > 0.50$). Similarly, comparisons with the one Grand Banks well shows it to resemble the Arctic data which further points to a lack of geographical pattern.

Simple linear regressions of oil content against five different downhole parameters derived from logging data yielded regression coefficients (r^2) that were low (Table 5). Nevertheless, many of the corresponding correlation coefficients (r) are significant at a 0.05 level of probability when compared with the critical values of r given by Rohlf & Sokal (1969). It is important to note, however, that the difference between the actual and critical values of r, and hence the strength of the significance, decrease as the number of observations increases (compare Grand Banks with Scotian Shelf in Table 5). The data were pooled by geographic area for the purposes of this discussion, but the same lack of consistent correlation patterns could also be seen when comparing regressions performed on individual wells. The number of significant correlation coefficients varied from none at all (e.g. West Chebucto K-20) to four (e.g. Baccalieu I-78). No single parameter was always significantly correlated with the oil content of cuttings.

Results from a study conducted by Esso at the Kaubvik I-43 well site suggest that the oil content is inversely related to particle size of cuttings but only up to some limit (Fig. 3). Oil retention also seems to increase with depth, perhaps because cuttings tend to be finer, but the regressions of oil content against depth (Table 5) do not firmly support the trend in Fig. 3. There is still considerable uncertainty about the processes that determine particle size distribution at the drill bit and about the interactions between cuttings and OBM while the solids are being transported to the surface. We cannot even be reasonably certain that the size of the cuttings sampled at the surface is the same as the size occurring at the drill bit when the oil is first adsorbed (G. Morrell, 1987, COGLA, personal communication). Absolute comparisons between sets of measurements may not be possible but relative comparisons might be (Høiland et al.,

TABLE 4
Cuttings production and oil content from wells drilled on the Canadian offshore frontier lands between 1983 and 1987

Area	Well	Oil on cuttings (g per 100 g dry wt)				Cuttings dry mass (t)	Oil discharged on cuttings (t)
		Mean ± SD	Min.	Max.	n		
SS	Alma F-67	11·1 g ± 3·5	5·0	23·0	74	343	38
SS	South Desbarres O-76	11·7 ± 3·0	6·0	19·0	135	1060	124
SS	West Venture C-62	13·6 ± 4·8	4·7	29·7	89	285	39
SS	Glenelg E-58	13·7 ± 2·7	8·0	18·0	52	367	50
SS	Glenelg E-58A	15·4 ± 1·7	11·3	18·0	28	186	29
SS	Glenelg H-38	N/A	N/A	N/A	N/A	459	N/A
SS	Alma K-85	8·6 ± 2·4	2·9	14·2	49	327	28
SS	Peskowesk A-99	N/A	N/A	N/A	N/A	499	N/A
SS	Kegeshook G-67	14·7 ± 4·1	7·0	20·8	46	387	57
SS	Merigomish C-52	N/A	N/A	N/A	N/A	469	N/A
SS	North Triumph G-43	13·7 ± 3·7	6·3	21·8	55	611	84
SS	Glenelg N-49	10·6 ± 4·0	4·4	16·4	34	497	53
SS	North Triumph B-52	11·8 ± 4·9	4·0	19·9	40	441	52
SS	Panuke B-90	N/A	N/A	N/A	N/A	429	N/A
SS	West Chebucto K-20	9·5 ± 2·7	3·8	16·3	49	395	38
SS	Tantallon M-41	9·6 ± 4·1	3·0	26·5	55	490	47
SS	Whycocomagh N-90	N/I	N/I	N/I	N/I	387	N/I
SS	Area Summary (mean ± SD)	12·1 ± 4·2	2·9	29·7	706	449 ± 179	53 ± 27
A	Adgo G-24	24·2 ± 11·1	9·9	67·6	54	221	53
A	Minuk I-53	14·2 ± 3·7	7·7	23·7	104	610	87
A	Nipterk L-19A	27·3 ± 5·1	8·0	38·1	87	219	60
A	Unak L-28	16·3 ± 15·0	0·2	93·3	40	357	58
A	Kaubvik I-43	N/A	N/A	N/A	N/A	652	N/A
A	Area Summary (mean ± SD)	20·2 ± 10·2	0·2	93·3	287	412 ± 186	65 ± 15
GB	Baccalieu I-78	20·9 ± 10·0	4·9	38·8	62	387	81
	Summary of all areas (mean ± SD)	14·5 ± 5·2	0·2	93·3	994	(mean) 441 ± 182	(mean) 58 ± 24

Abbreviations: A = Arctic, GB = Grand Banks, N/A = data summary not available, N/I = data not included due to insufficient sample size, SD = standard deviation, SS = Scotian Shelf.

Paul G. Chénard et al.

TABLE 5

Regression coefficients (r^2) and corresponding correlation coefficients (r) obtained from simple linear regressions of oil content on cuttings (g per 100 g dry weight) against various downhole parameters measured in wells drilled on the Canadian offshore frontier lands (n = number of observations; r^* = significant r at 0·05 level for $n-2$ degrees of freedom and 1 independent variable, according to Rohlf & Sokal (1969); N/A = data not available)

Parameter	Result	Scotian Shelf	Arctic	Grand Banks
Hole size (caliper reading)	Wells	13	4	1
	n	678	270	62
	r^2	0·008	0·072	0·705
	r	0·089	0·268	0·840
	r^*	0·075	0·119	0·250
Formation density	Wells	13	4	1
	n	678	270	62
	r^2	0·027	0·067	0·277
	r	0·164	0·259	0·526
	r^*	0·075	0·119	0·250
Depth in well	Wells	13	4	1
	n	705	270	62
	r^2	0·016	0·000	0·675
	r	0·126	0·000	0·822
	r^*	0·073	0·119	0·250
Lithology	Wells	11	3	N/A
	n	616	167	N/A
	r^2	0·006	0·039	N/A
	r	0·077	0·197	N/A
	r^*	0·079	0·150	
Formation porosity	Wells	13	2	1
	n	678	176	62
	r^2	0·026	0·000	0·266
	r	0·161	0·000	0·516
	r^*	0·075	0·146	0·250

1986). It is interesting to note that the overall mean oil content measured for the Canadian offshore frontier lands (14·5 g per 100 g dry weight; Table 4) is quite similar to the values reported by Davies *et al.* (1984) for the North Sea (6–17 with a mean close to 15 g/100 g), indicating that the phenomenon might be ultimately governed by a mechanism that remains to be expressed empirically. This deserves further study.

Drilling operators also quantify the various losses of OBM so that environmental managers can acquire some reliable estimates on the amounts of drilling mud and oily cuttings that may have been discharged to the local or regional environment. In this study, the total mud loss

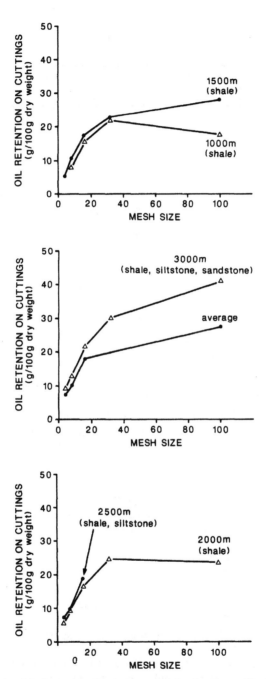

FIG. 3. Relationship between oil retention and size fractions of cuttings brought up from various depths of the Kaubvik I-43 well.

Paul G. Chénard et al.

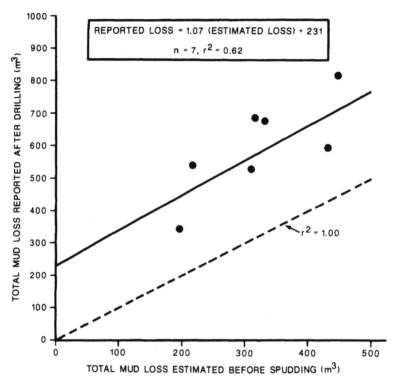

FIG. 4. Comparison of estimated and measured losses of OBM at wells drilled on the Scotian Shelf between 1985 and 1987.

estimated by the operators was always lower than what was actually measured but the amount of discrepancy is not constant (Fig. 4). Mud losses can be categorized under 5 pathways:

1. loss on discharged cuttings;
2. other losses from the solids control system (i.e. centrifuge overflow, etc.);
3. accidental discharges (i.e. round tripping, mud changeover, etc.);
4. seepage into formations; and
5. mud left in the hole prior to plugging for suspension or abandonment.

The first three pathways eventually lead to the environment while the last two do not under normal circumstances. Eventual losses to the environment outweigh losses to the well by approximate ratios between

TABLE 6
Fate of oil-based mud lost at several well sites on the Canadian offshore frontier lands

Area	Well	Percentage lost	
		To environment[a]	To the well[b]
SS	Kegeshook G-67	51·2	48·8
SS	North Triumph G-43	67·2	32·8
SS	Glenelg N-49	53·3	46·7
SS	North Triumph B-52	48·9	51·1
SS	West Chebucto K-20	84·2	15·8
SS	Tantallon M-41	53·5	46·5
SS	Whycocomagh N-90	58·1	41·9
SS	Area Mean	59·5	40·5
A	Adgo G-24	61·6	38·4
A	Unak L-28	23·7	76·3
GB	Baccalieu I-78	35·1	64·9

[a] Sum of losses on cuttings, solids control effluent, accidental dumping and unaccounted losses.
[b] Sum of losses into formation and mud left in the hole prior to abandonment of the well.

3:1 and 1:1 in 8 of the 10 wells for which we have enough data (Table 6), although estimates from the Whycocomagh N-90 should be treated with caution because only 6 measurements of oil on cuttings were made (W.H. Sawotin, 1987, Canterra Energy Ltd, personal communication). On the other hand, the wells Unak L-28 (ratio 0·3:1) and Baccalieu I-78 (ratio 0·5:1) are known to have lost more mud to the well than to the environment.

In addition to the base oil, Canadian operators have reported the use of any number of 37 different additives in their OBM formulations (Table 7). However, only six of these compounds were used in 50% or more of all wells drilled: barite 92%, lime (81%), calcium chloride (81%), Primul (77%), Technivis HT (56%), and Technimul (50%). Since cuttings were analyzed only for total content, there are no data on the actual proportions of mud additives on cuttings. However, there is probably no reason why the additives would not occur on the cuttings in the same proportion as in the mud. This assumption is reasonable because the drilling fluid is formulated to remain stable and homogeneous under downhole temperatures and pressures and is not likely to segregate from the base oil through cuttings treatment. Rough estimates of the quantities discharged of these

TABLE 7

Additives and base oils used in oil-based muds for wells drilled on the Canadian offshore frontier lands from 1983 to 1987. Information compiled from regulatory submissions to COGLA[a]

Additive	Scotian Shelf (16 wells)		Arctic (4 wells)		Grand Banks (6 wells)[b]	
	n	Total amount	n	Total amount	n	Total amount
Barite	14	5451 MT	4	2773 MT	6	2996 MT
Biomul	4	41·7 m³				
Calcium carbonate			1	56 MT		
Calcium chloride	11	293·97 MT	4	103·11 MT	6	243·01 MT
Calcium oxide			1	19·40 MT	1	0·4 m³
Driltreat						
Duratone	1	7·26 MT	2	33·41 MT		
Duratone HT			1	7·76 MT	1	2·95 MT
Envirokleen	1	1·14 m³				
Envirovis	1	77·01 MT				
Envirospot			1	0·41 m³		
EZ Mul			1	36·87 MT		
EZ Mul NT	4	113·71 MT	2	44·19 MT	1	7·13 MT
EZ Spot			1	0·59 MT		
Geltone II	3	63·29 MT	3	67·18 MT	1	3·83 MT
Hematite	1	3·21 MT				
Hot lime			2	45·43 MT		
Invermul			2	93·0 MT		
Invermul NT	4	597·3 MT	1	14·71 MT	1	22·3 MT
Lime	13	703·90 MT	2	26·79 MT	6	90·89 MT
Mica Fine			1	5·1 MT		
Primul	13	136·29 MT	2	8·16 MT	5	194·23 MT
Quicklime	4	35·25 MT				

	n		n		n	
Sawdust	2	0·90 MT	1	1·56 MT		
Soda ash	6	481 MT	1	0·09 MT		
Sodium chloride	1	2·80 m³				
Technikleen						
Technimul	6	49·72 MT	2	8·92 MT	5	32·21 MT
Technisurf	1	0·20 m³	2	1·85 m³		
Technitrol	2	14·97 MT	4	60·86 MT		
Technivis			1	9·7 MT		
Technivis HT	9	96·68 MT	1	0·45 MT	5	754·91 MT
Technivis R	2	40·6 MT				
Mud oils						
Conoco ODC/Vista ODC	8	1 546·6 m³	3	2 788·9 m³		
Escaid 90			1	20 m³		
Esso DMO-75			2	590·0 m³		
Esso Sarnia A					1	314·9 m³
Shell DF	1	520·7 m³				
Shell Sol DMS	6	1 860·0 m³	5	2 300·8 m³		

[a]Note: There are similarities among additives (for example: calcium oxide and hot lime and quicklime, Duratone and Duratone HT, EZ Mul and EZ Mul NT, Invermul and Invermul NT), and among product lines (for example: Driltreat through Geltone II and Technikleen through Technivis R).

[b]Includes amounts estimated for wells approved or still drilling in December 1987.

MT = metric tonnes; *n* = number of wells using the compound.

TABLE 8

Estimated discharges of the six most frequently used OBM additives for drilling on the Canadian offshore frontier lands

Additive	Consumption[a] (MT)	Wells[a] (n)	Discharge per well[b] (MT)	Acute 96-hour LC-50[c]	
Barite	11 220	24	280·5	100 000	ppm
Calcium chloride	640·1	22	17·5	8 000	mg/litre
Lime	821·6	21	23·5	10 000	ppm
Primul	338·7	20	10·2	240 000	μl/litre
Technimul	90·9	13	4·2	N/A	
Technivis HT	852·0	15	34·1	135 000	mg/litre

[a] Includes Grand Banks wells approved but not yet spudded in December 1987, in which case consumption is estimated by operators in regulatory submissions.
[b] Assumes that 60% of mud consumed is lost to the environment (Table 6).
[c] For the three-spined stickleback *Gasterosteus aculeatus*.
MT = metric tonnes; N/A = data not available; ppm = parts per million.

six compounds can thus be calculated (Table 8), further assuming from Table 6 that 60% of the mud consumed is actually lost to the environment in most cases.

The domination of weighting agents in discharges can still be clearly seen in Table 9 which presents total discharges of the four main functional groups of OBM additives.

The low toxicities of the most common additives (Table 8) would lead us to suppose that an acute aquatic contamination risk is slight at the various well sites, in addition to high dilution and cuttings dispersal rates. Chemical monitoring of oily cuttings piles at the West Venture C-62 and the South Desbarres 0–76 well sites on the Scotian Shelf suggests that cuttings disappear quickly in a shallow water location (in this case, 16 m) but persist longer in deeper water (70 m), probably due to wave action at the shallow location (Yunker *et al.*, 1986). Water depths at the various offshore OBM well sites in Canada range from 16 m to over 1000 m, with most in the range of 30–100 m.

4. CONCLUSION

Decisions on any environmental concern require sound factual information for them to be effective and relevant. This was the underlying objective of the reporting requirements in the COGLA Guidelines on the Use of

TABLE 9
Estimated discharges of the four principal functional
categories of OBM additives used for drilling on the
Canadian offshore frontier lands

Function	Consumption[a] (MT)	Wells[a] (n)	Discharge per well[b] (MT)
Alkalinity and ph control[c]	1 619	26	37
Emulsifier[d]	1 359	26	31
Viscosifier[a]	1 037	26	24
Weighting agent[f]	11 223	26	259

[a] Includes Grand Banks wells approved but not yet spudded in December 1987, in which case consumption is estimated by operators in regulatory submissions.
[b] Assumes that 60% of the mud consumed is lost to the environment (Table 6).
[c] Includes calcium carbonate, calcium chloride, calcium oxide, hot lime, lime, quicklime and soda ash.
[d] Includes EZ Mul, EZ Mul NT, EZ Spot, Invermul, Invermul NT, Primul and Technimul.
[e] Includes Geltone II, Technivis, Technivis HT and Technivis R.
[f] Includes barite and hematite.

Oil-based Drilling Mud: to provide an adequate level of monitoring knowledge to ensure that the environmental consequences of drilling with oil-based mud could be understood or, better yet, anticipated. The regulation of oil and gas activities benefits from any predictive capability that derives from such detailed information. The information presented in this paper verifies the hypothesis that the key variable of oil absorption on cuttings cannot be predicted with any great accuracy using specific well characteristics. Furthermore, there does not seem to be any noticeable geographic pattern at least in the Canadian context. It appears also that the overall mean absorption of 14·5% (i.e. 14·5 g per 100 g dry weight of cuttings, Table 4) could be reasonably applied to all well sites. This value is quite similar to the concentrations reported elsewhere (North Sea: Davies *et al.*, 1984) which raises the possibility that some fundamental limiting factor may be at work.

REFERENCES

CHÉNARD, P.G. (1984). Composition of oil-based drilling muds. In Report of the Workshop on Environmental Considerations in the Offshore Use of Oil-

Based Drilling Muds, ed. G.D. Greene & F.R. Engelhardt. Ottawa, Canada, May 23–24, 1984, pp. 12–20. Environmental Protection Branch Technical Report No. 2, Canada Oil and Gas Lands Administration, May 1984, 125 pp. + App.

CHÉNARD, P.G., ENGELHARDT, F.R., DRINNAN, R.W. & YUNKER, M.B. (1986). Environmental Perspectives on the Marine Disposal of Oil-Based Drilling Muds. In M.L. Lewis (ed.) Northern Hydrocarbon Development — Environmental Problem Solving. *Proceedings of the Eighth Annual Meeting of the International Society of Petroleum Industry Biologists.* Banff, Alberta, Canada. 24–26 September 1985. pp. 79–87.

COGLA/CNOPB (1986). Drilling for Oil and Gas on Frontier Lands — Guidelines and Procedures. Appendix J: Guidelines for the Use of Oil-Based Drilling Muds. Canada Oil and Gas Lands Administration and Canada Newfoundland Offshore Petroleum Board, September 1986.

DAVIES, J.M., ADDY, J.M., BLACKMAN, R.A., BLANCHARD, J.R., FERBRACHE, J.E., MOORE, D.C., SOMERVILLE, H.J., WHITEHEAD, A. & WILKINSON, T. (1984). Environmental effects of the use of oil-based drilling muds in the North Sea. *Mar. Pollut. Bull.,* **15**(10), 363–70.

HØILAND, H., LJOSLAND, E., WOLD, R. & VEGGELAND, K. (1986). The nature of bonding of oil to drill cuttings. In *Proceedings of the International Conference on Oil-Based Drilling Fluids,* 24–26 February 1986, Trondheim, Norway, pp. 43–46. State Pollution Control Authority (SFT) and Statfjord Unit (Mobil Exploration Norway Inc), 136 pp.

ROHLF, F.J. & SOKAL, R.R. (1969). *Statistical Tables.* W.H. Freeman, San Francisco, 253 pp.

STANLEY, J. (1963). *The Essence of Biometry.* McGill University Press, Montreal, 147 pp.

STRONG, D.C. (1984). Cuttings cleaning, mud handling and disposal. In Report of the Workshop on Environmental Considerations in the Offshore Use of Oil-Based Drilling Muds, ed. F.R. Engelhardt and G.D. Greene. Ottawa, Canada, May 23–24, 1984, pp. 25–34. Environmental Protection Branch Technical Report No. 2, Canada Oil and Gas Lands Administration, May 1984, 125 pp. + App.

YUNKER, M.B., DRINNAN, R.W. & ENGELHARDT, F.R. (1986). A study of the behavior and fate of the hydrocarbons in sediments resulting from the use of oil-based drilling muds at two east coast rig sites. In Northern Hydrocarbon Development—Environmental Problem Solving. *Proceedings of the Eighth Annual Meeting of the International Society of Petroleum Industry Biologists,* ed. M.L. Lewis. Banff, Alberta, Canada. September 24–26, 1985, pp. 68–78.

6

Exposure of Deep Seagrass Beds off the West Coast of Florida to Discharged Drilling Effluents

M. JOHN THOMPSON, ALAN D. HART

Continental Shelf Associates, Inc., 759 Parkway Street,
Jupiter, Florida 33477, USA

and

CARY W. KERLIN

Sohio Petroleum Company, 9401 Southwest Freeway, Suite 1200,
Houston, Texas 77074, USA

ABSTRACT

During the summer and fall of 1985 Sohio Petroleum Company drilled an exploratory well off the west coast of Florida in what is commonly known as the Big Bend Area. Drilling-related impacts to the seagrass Halophila decipiens *were possible in this area, and Sohio had the options: (1) not to drill; (2) to drill with zero discharge; (3) to use other means of mitigation; or (4) to conduct a monitoring program in conjunction with drilling. After consultation with State and Federal regulatory agencies, Sohio elected to conduct a monitoring program investigating the relationships between drilling effluent discharges and* H. decipiens *growth.*

Twenty-one stations were established around the discharge site at distances ranging from 50 to 3700 m. The results from these stations were compared to those from a reference station located 9·6 km from the discharge site. Changes in barium concentrations and barium-to-iron ratios in surficial sediment samples were utilized as tracers of the settleable fraction of the drilling effluent discharges. Seagrass growth parameters (changes in leaf count and leaf biomass) were used as measures of impacts.

Reduced growth occurred where greater increases of the drilling effluents were observed ($p < 0·01$). Complete mortality of seagrass was observed within 300 m of the discharge site. At stations located greater than 300 m, *leaf counts and biomass estimates were reduced compared to the reference station. These results indicated that some aspect of the discharge of drilling effluents impacted the growth of the seagrass around the discharge site. Such impacts were likely ephemeral because recolonization of the affected area was observed on an October 1987 survey.*

1. INTRODUCTION

A recent exploratory drilling effort by Sohio Petroleum Company in the Big Bend Area of Florida presented an opportunity to study the effects of discharged drilling effluents on a major biotic component of the coastal ecosystem, seagrasses. The purpose of this study was to determine the impacts of drilling effluent exposure on the fringing seagrass species *Halophila decipiens*. If relationships between the standing crop of the seagrass and exposure to drilling effluents could be established, then the spatial impacts on the seagrass throughout the area around the drillsite could be determined by modeling the fates of the discharged materials.

The study area was located in the Gainesville OCS Area Block 707 (Fig. 1(a)) where the seafloor is flat, broken only by widely scattered low-relief rock outcrops. Water depth at the drillsite was approximately 21 m and depths over the study area varied less than 0·6 m. A thin veneer of medium-to-coarse quartz-carbonate sand overlay a very hard, aerially weathered, limestone bedrock. Sediment cover was patchy and the underlying bedrock was exposed in many places.

Two species of marine seagrasses, *H. decipiens* and *H. englemanni*, were known to inhabit the depth zone in which the study area occurred (Continental Shelf Associates, Inc. & Martel Laboratories, Inc., 1985). These species are considered to be fringing or pioneer species, adapted to grow in areas where environmental conditions are unsuitable for larger, climax species such as *Thalassia*, *Syringodium* and *Halodule*. The *Halophila* species are much smaller than these climax species and form part of an offshore seagrass-macroalgal community present in the Florida Big Bend Area in water depths from 10 to 25 m. *H. decipiens* occurs seasonally and, depending on the water depth and temperature, appears in the spring or early summer; reaches peak standing crop and flowers in the late summer or early fall; then dies back during winter (Continental Shelf Associates, Inc. & Martel Laboratories, Inc., 1985). A pre-drilling television survey of the Sohio drillsite in October of 1984 revealed that the seafloor supported extensive stands of *H. decipiens* (Continental Shelf Associates,

Inc., 1984). The study area around the drillsite was located near the seaward edge of the distribution of *H. decipiens*.

2. METHODS

To determine whether or not discharged drilling effluents affected the seagrass around the drillsite, the relationships between the presence of drilling effluents indicators and measures of the seagrass growth were evaluated. The presence of the settleable fraction of drilling effluents was determined by measuring the trace metals barium and iron in the fine-grain fraction ($< 63 \mu$m) of the surficial sediments. The change of the mean barium concentrations and the mean barium-to-iron ratios in the fine-grain fraction have been used effectively in many studies to evaluate exposure to drilling fluids (Continental Shelf Associates, Inc., 1986; Trefry et al., 1981; Trocine & Trefry, 1983).

Monitoring stations were established near the drillsite and at reference distances (> 9 km) for comparison. Three replicate surficial sediment samples were collected at each station for trace metal analysis, sediment traps were deployed to measure the depositional rate during drilling, and quantitative photographs were taken at six randomly-placed, repetitively-visited seagrass quadrats. Samples of seagrass were collected near each station for leaf biomass determinations.

Surficial sediment samples were collected by carefully sliding a flattened plastic coring device across the seafloor at a depth of 2 cm. Samples were frozen aboard ship for shipment to the laboratory. All samples were wet sieved through a 63-μm screen.

Deployed sediment trap arrays consisted of three cylindrical PVC traps attached to a steel frame. The mouths of the individual sediment traps were positioned 0·5 m above the seafloor to minimize resuspension effects. Trap diameter was 10·2 cm and the height was 50·8 cm, yielding an aspect ratio of 5·0 (Lorenzen et al., 1981). Once deployed, the sediment trap array became the center of the station. The three replicate sediment traps were recovered on each survey and frozen aboard ship. They were analyzed in the same manner as surficial sediments.

Trace metal concentrations were determined using Instrumental Neutron Activation Analysis. The fine material was freeze-dried and ground in a Al_2O_3 'Diamonite' mortar and pestle. Each finely ground sample was sealed in a plastic vial and subjected to a neutron flux (10^{-13} neutrons per cm^2/s) to activate the elements of interest (barium and iron) without further pre-treatment of the samples. Samples were irradiated for 14 h and allowed to 'cool' for 10 days. The cooling period permits isotopes of chlorine, sodium, and other background-producing elements to decay away, improving counting statistics for barium and iron. Samples were counted using a Ge (Li) detector coupled to a Nuclear Data Corporation Model 66 pulse height analyzer and computer data acquisition system. Concentrations in the sediment samples were obtained by comparing counts for each sample with counts from standard rock powders of accurately known amounts of barium and iron.

Photographic seagrass monitoring stations to be sampled repetitively were established by positioning six photoquadrat marker frames within a 6-m radius of a station center marker (the sediment trap array). The frames were positioned using distances and compass headings derived from random number tables. After being positioned, the photographic quadrats were permanently marked by driving iron stakes into the seafloor. The stakes were placed so that a specially designed photoframing device fitted precisely over the marked quadrat to ensure exactly the same area was photographed on each survey. A smaller frame, 23 × 43 cm, was attached to the base of the photoframer stand, and delineated the actual area quantitatively analyzed. Quadrats were photographed using a Nikonos III camera equipped with Subsea 100 watt-second strobes and 28 mm lens. Photographs were analyzed using a Vanguard Model S-13SP Motion Analyzer. All seagrass blades were counted and all suitably positioned seagrass blades were measured. Each photograph showed a certain number of seagrass blades that were twisted or folded in a manner making precise measurement impossible.

Representative leaves from each collected sample of seagrass were sorted into size categories. Each group of leaves was dried to constant weight in a Boekel oven at 60°C. The average biomass for a leaf within a particular size category was determined. Leaf biomass in each photograph was estimated by converting lengths to biomass using relationships developed by measuring and weighing the collected leaves. All biomass estimates are expressed in mg dry weight.

During the pre-drilling cruise, 24 May to 12 June 1985, eleven stations were established within 300 m of the drillsite (Fig. 1(b)). Only 11 could be established at that time because the growing season had just begun and the seagrasses were not sufficiently established to justify erecting more stations. Three reference stations were established at distances greater than 9 km from the drillsite (Fig. 1(c)), but two of these reference stations were later determined to be inappropriate for comparison with the

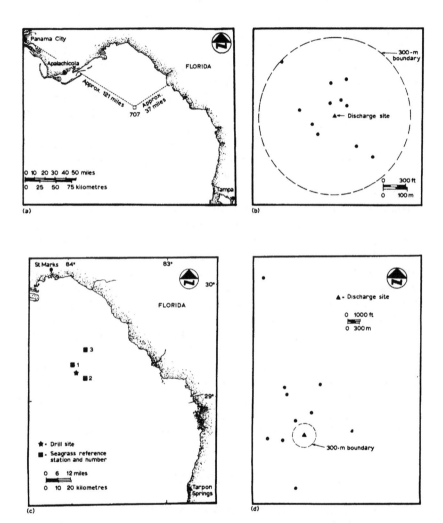

FIG. 1. (a) Location of Gainesville OCS Area Block 707 in the Eastern Gulf of Mexico, (b) locations of stations within 300 m of the discharge site, (c) reference station distances, (d) stations between 300 and 3700 m from the discharge site.

monitoring stations because the seagrass stands occurring at these locations were not monospecific. During the second survey, 10 additional stations were established beyond the 300-m radius (Fig. 1(d)).

Endeco Type 105 recording current meters were deployed at mid-water and near-bottom depths adjacent to the drilling area. Current speed and direction were recorded at 1-h intervals. Hydrographic profiling was conducted on each cruise using a Beckman RS5-3 portable salinometer. The instrument was calibrated across its temperature and conductivity range aboard ship prior to each use. Values of temperature and salinity were recorded at 3-m intervals. Calibration samples for field measurements were collected at the surface, at mid-depth, and near the bottom by divers. Salinity calibrations were performed using a Grundy precision salinometer in the laboratory. The mud engineers aboard the *Glomar High Island VI* maintained daily drilling mud discharge logs specifically designed for this program. Data recorded included type of solids control device, discharge rate, length of discharge, percent solids, and density of the discharge material. Samples of each bulk discharge were collected. Solids control devices and the muds pits were also periodically sampled. These hydrographic and discharge data were used in conjunction with the Offshore Operators Committee (OOC) mud discharge simulation model for drilling mud dispersion to hindcast the overall fate of the settleable discharges.

Immediately after Survey 2, Hurricane Elena passed directly over the study area (31 August to 3 September 1985). A follow up survey (3–6 October 1985) revealed that the study area had been severely disturbed during Hurricane Elena. Based on these observations, the study was discontinued and only the data collected on the first two surveys were analyzed. In addition, subsequent surveys were conducted in August 1986 and October 1987 to evaluate seagrass recovery in the study area.

3. RESULTS AND DISCUSSION

3.1. Drilling Effluent Indicators and Seagrass Growth

Prior to drilling, mean barium concentrations ranged from 48 to 204 ppm in the fine-grain fraction at the 11 stations located within 300 m of the drillsite. The mean concentration at the reference station was 89 ppm. During Cruise 2, after 2·5 months of drilling, the mean fine-grain concentrations within 300 m of the drillsite ranged from 22 000 to over 300 000 ppm. Mean concentrations at the 10 stations established during Cruise 2 beyond 300 m ranged from approximately 1000 to over

54 000 ppm. The mean Cruise 2 concentration at the reference station was 140 ppm.

Changes of the mean barium concentrations in the fine-grain fraction of the surficial sediments are presented in Fig. 2(a). Changes at the stations established during Cruise 2 were estimated using the mean Cruise 1 concentration (91 ppm). Greater changes of the mean concentrations were observed nearer to the drillsite. Substantive changes compared with the reference station were observed to a distance of 3·7 km.

Similar results were observed for the mean changes of the barium-to-iron ratios (Fig. 2(b)). Mean pre-drilling levels of the ratio ranged from 0·005 to 0·021. During Cruise 2, mean ratios ranged from 1·75 to 18·4 within 300 m of the drillsite. Immediately beyond 300 m, levels were greater than 2·0. At about 3·7 km, the mean level was 0·12; the mean Cruise 2 level at the reference station was 0·015.

Pre-drilling leaf counts at stations within 300 m of the discharge site ranged from 32 to 412 blades/m². The number of leaves at the reference station was 190/m². Pre-drilling leaf biomass at stations within 300 m of the discharge site ranged from 2 to 41 mg/m². Leaf biomass at the reference station was 16 mg/m².

No seagrass leaves and, correspondingly, no leaf biomass were observed in the photoquadrats within 300 m of the discharge point during Cruise 2. This survey occurred in August, the height of the growing season, when maximum standing crop would be expected. At the stations located between 300 and 3700 m, leaf counts ranged from 217 to 1040 blades/m², and leaf biomass estimates ranged from 6 to 411 mg/m². Cruise 2 levels of leaf counts and biomass at the reference station were 1668 leaves/m² and 561 mg/m², respectively.

The mean changes of leaf count and biomass with respect to the changes in the mean barium concentrations and with respect to the mean barium-to-iron ratios in the surficial sediments at the 21 stations sampled are presented in Figs 3(a) and 3(b). Changes at the stations established during Cruise 2 were estimated using mean Cruise 1 levels. Increases of leaf count and biomass — growth — were associated with lesser changes of barium; decreases in seagrass standing crop — mortality — were associated with greater changes of barium concentrations and barium-to-iron ratios. However, these relationships were not linear.

In order to standardize the seagrass growth parameters, changes of leaf number and biomass were converted to percentage change. Percentage change was calculated as the difference between the levels of each parameter in the respective photoquadrats during Cruises 1 and 2 divided

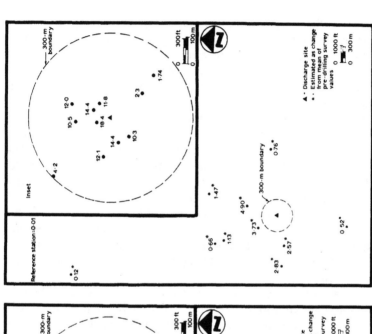

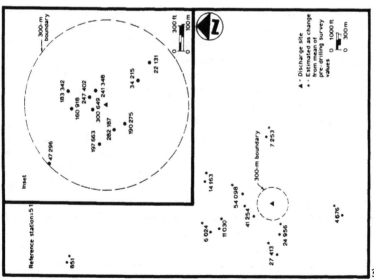

FIG. 2. Spatial distributions of changes of (a) mean barium concentrations (ppm) and (b) mean barium-to-iron ratios in the fine-grain fractions of the surficial sediment samples collected during Cruises 1 and 2.

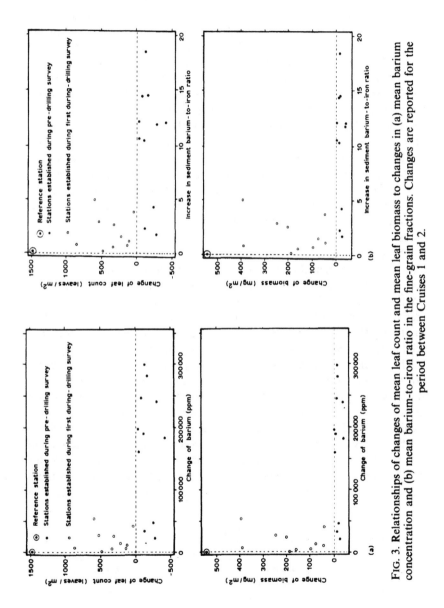

FIG. 3. Relationships of changes of mean leaf count and mean leaf biomass to changes in (a) mean barium concentration and (b) mean barium-to-iron ratio in the fine-grain fractions. Changes are reported for the period between Cruises 1 and 2.

by the Cruise 1 levels. The mean percentage change was determined for each station based on the six quadrats sampled. Percentage change was more indicative of the relative changes of the seagrass growth parameters than were raw changes. The spatial distribution of seagrass growth in terms of leaf count and leaf biomass are presented in Figs 4(a) and 4(b).

Within 300 m of the discharge site, the seagrass standing crop was eliminated entirely between the beginning of drilling (12 June 1985) and the second survey (7 August 1985). Farther from the discharge site, growth was inhibited, compared to that observed at the reference station. Growth observed at the monitoring station most distant from the discharge point (3·7 km northwest of the drill site), was 23% and 20% of that observed at the reference station in terms of leaf count and biomass, respectively. The magnitude of this reduction is, unfortunately, based on the results from a single reference station. Growth at this one reference station may, or may not, be entirely representative of growth over other unimpacted areas at this depth.

The relationships between net changes of the drilling effluent indicators and the percentage changes of the seagrass growth parameters were investigated using correlation analysis. Because these relationships were nonlinear, the net changes of the drilling effluent indicators were transformed using common logarithms. Use of this transformation was equivalent to stating that a proportional change in drilling effluent indicators was related to a linear response of the seagrass growth parameters (Fig. 5).

The results of the correlation analyses (Pearson's product moment) are given in Table 1. The correlations were consistently negative and statistically significant ($p < 0.01$). These results show that some properties of the drilling effluent discharges, as indicated by the tracers, were significantly related to reduction of seagrass growth. The nature of this relationship was elimination of standing crop within 300 m and reduction of growth farther from the discharge site. These correlations do not indicate what physical or chemical attributes of the discharged drilling effluents were responsible for the inhibition of seagrass growth. The

TABLE 1

	Increases in	
	Barium concentration (log transformation) (%)	*Barium-to-iron ratio (log transformation)* (%)
Leaf count	−0·84	−0·83
Leaf biomass	−0·78	−0·77

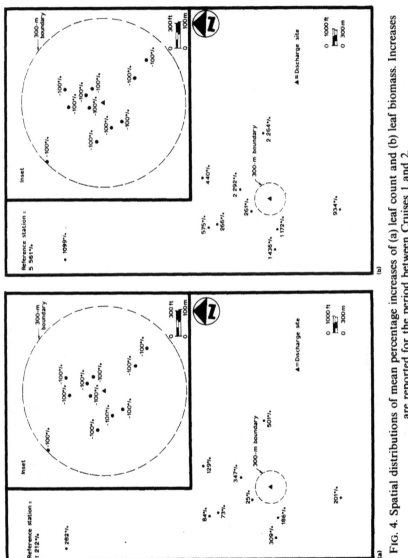

FIG. 4. Spatial distributions of mean percentage increases of (a) leaf count and (b) leaf biomass. Increases are reported for the period between Cruises 1 and 2.

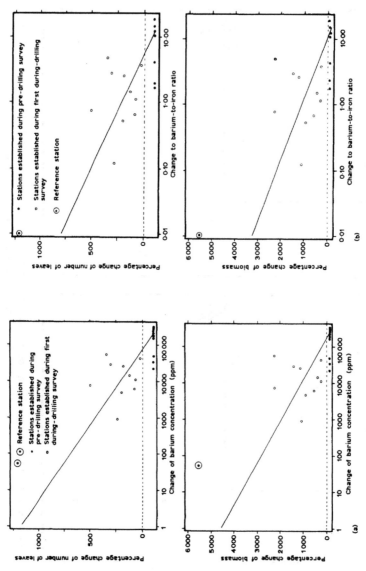

FIG. 5. Relationships of the mean percentage changes of the number of leaves and mean leaf biomass to changes of (a) mean barium concentrations and (b) mean barium-to-iron ratios in the fine-grain sediment fractions. Changes are reported for the period between Cruises 1 and 2.

correlations only indicate that there was an association between higher levels of drilling effluents tracers and lower seagrass growth.

The barium and barium/iron trace metal parameters are only indicators or tracers for the settleable fraction of the drilling effluents—the component of the drilling effluents measured in this program. Correlations of barium and the barium-to-iron ratio with seagrass growth or mortality do not indicate that these tracers were themselves responsible for differences in seagrass growth; rather it indicates these drilling effluent tracers and seagrass growth parameters may be covarying with one or more characteristics of the drilling effluents that are directly responsible for the observed relationships.

3.2. Spatial Impact Assessment

One of the objectives of the study was to estimate the spatial pattern of the biological effects. To accomplish this objective, it was necessary to: (1) hindcast the fate of the drilling effluent discharges; (2) determine a predictive relationship between biological impacts and the results of the fate modeling; and (3) determine the spatial distribution of the biological impacts using the results of the first and second steps.

To successfully hindcast the fates of the discharges, data from the monitoring program and previous surveys of the study area were used. These data included: (1) the discharge rates and quantities based on the discharge log maintained by the mud engineer during drilling; (2) settling velocities of discharged solids based on analysis of selected samples collected on board the drilling rig; (3) bathymetry of the study area reported by Continental Shelf Associates, Inc. (1984); (4) vertical hydrographic profiles of the water column collected on 29 May and 19 August 1985 near the drillsite; and (5) hourly current measurements recorded at two depths near the discharge site during the study period.

The fate of the high-volume, high-rate discharges was modeled using the Offshore Operators Committee (OOC) mud discharge model developed by Brandsma & Sauer (1983). Each of nine bulk discharges from the mud pits was modeled in this manner. A deterministic trajectory model was used to simulate the low-rate discharges from the various solids control devices. In this model, small discharge quantities were tracked from their discharge point to their deposition on the bottom. The initial trap depth and plume size for the trajectory model were determined using the OOC model. After these parameters were determined, movements of the solids were controlled by the hourly current estimates recorded as part of the

TABLE 2

| Deposition | Percentage change | |
	Barium concentration	Barium-to-iron ratio
Cumulative	0·81	0·77
Bulk	0·93	0·91

monitoring program, the settling velocities of the discharged solids, and diffusion.

Results of the fates modeling are presented in Fig. 6(a). A generally symmetrical pattern of deposition was determined by the modeling. The discharged materials were primarily deposited within 230 m of the discharge site. Deposition within this distance exceeded 350 g/m². Within 300 m of the discharge site, simulated depositions exceeded 100 g/m².

Correlations between simulated depositions from bulk and cumulative discharges and data collected in the field were statistically significant ($p < 0.05$). These correlations are shown in Table 2.

The correlation between the observed depositional rates of barium, based on the sediment trap samples, and the depositions hindcast for the bulk discharges was high ($r = 0.79$). Similarly, the depositional rates for total materials were correlated with the hindcast total cumulative deposition ($r = 0.65$); however, there was some discrepancy between the quantities deposited in the sediment traps and those estimated by the modeling (Fig. 6(b)). Results from the model consistently underestimated the deposition observed in the sediment traps. Correcting the modeling results for the discrepancy between the materials used during drilling and the estimated discharges did not completely explain the discrepancy of higher deposition. A possible source for this discrepancy is resuspension of material on the bottom into the traps (a process for which the modeling does not account). In addition, behavior of the particles may not have been precisely mimicked by the modeling due to variations of settling velocities of the particles and variations of the currents at the exact time of discharge. The results of the fates modeling effort generally appeared realistic, however, this study does not constitute a field validation of the model. This comparison of the simulated results with field data serves only to quantify confidence in the modeling results.

The second step in spatial impacts assessment was to determine relationships between the fate of the discharged material and seagrass growth. The major characteristics relating seagrass growth and discharges were changes of barium concentrations and barium-to-iron ratios in the sediments. Another means of relating discharges and growth was to determine the relationship of the simulated depositions to the measurements of changes in the seagrass at the specific stations. The functional relationship was determined using regression analysis. The dependent variable was seagrass growth, percentage change of leaf count in this case, and the independent variable was simulated cumulative deposition. The results of this analysis are presented in Fig. 6(c).

As a final step in the spatial impact assessment, the results of the first two steps were combined to contour the area of effect based on seagrass growth as indicated by blades/m² (Fig. 6(d)). This was done by determining the simulated deposition which would result in specific levels of seagrass growth based on the growth observed at the reference station. These impact levels were then contoured using the results of the fates modeling (Fig. 6(c)). Based on these results, seagrass in an area of $0.16 \, km^2$ around the discharge site was eliminated. An additional $0.12 \, km^2$ would have seagrass growth levels less than 10% of that observed at the reference station. An area of $4.0 \, km^2$ would exhibit growth less than 25%, compared to the reference station.

3.3. Seagrass Recovery

Selected stations were reoccupied during two surveys to the study area after conclusion of the monitoring program. During August 1986, one year after the program ended, two stations were sampled. These results are compared to results from the monitoring program in Table 3.

The August 1986 levels of drilling effluent tracers remained elevated compared to pre-drilling levels but were lower than those observed during

TABLE 3

Distance from discharge site (m)	Date	Barium concentration (ppm)	Barium to iron ratio	Number of leaves (m^{-2})	Estimated leaf biomass (mg/m^2)
65	June 1985	55	0.006	77	10.0
	August 1985	247 457	14.4	0	0.0
	August 1986	2 920	0.41	50	10.3
300	June 1985	72	0.008	243	20.3
	August 1985	48 628	4.28	0	0.0
	August 1986	1 112	0.17	189	41.1

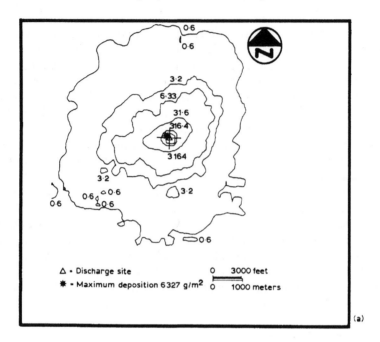

(a)

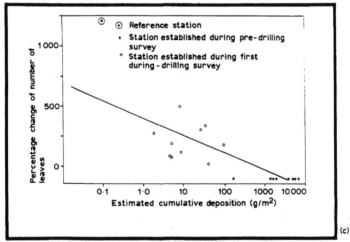

(c)

FIG. 6. Results of the fates modeling and impact assessment: (a) deposition of solids from fates modeling; (b) comparison of deposition estimated by fates modeling to deposition measured by sediment traps; (c) relationship between estimated cumulative deposition and percentage increase of number of leaves; (d) hindcast results of impact based on number of leaves.

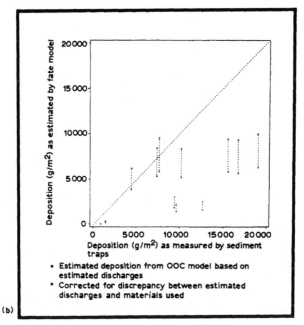

(b)

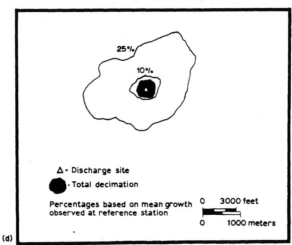

(d)

FIG. 6.—*contd.*

TABLE 4

Distance from discharge site (m)	Barium concentration (ppm)	Barium-to iron ratio	Number of leaves (m^{-2})	Estimated leaf biomass (mg/m^2)
65	3 650	0·34	845	487
85	1 290	0·19	858	487
85	1 240	0·17	798	404
150	810	0·14	540	199
3 700	130	0·03	1 415	770
>9 000	70	0·01	538	275

Survey 2. Estimated leaf biomass values in August 1986 were comparable to pre-drilling values, having increased from $0.0\,mg/m^2$ within a year; however, the August 1986 biomass levels were lower than would be expected because August is generally the peak of the growing season (the estimated leaf biomass at the reference station during August 1985 was $561\,mg/m^2$).

Another survey of the study area occurred in October 1987. Due to the length of time between the end of the program and this survey, the original quadrats were not surveyed. Seagrass and sediments were sampled near six of the original stations. These results are presented in Table 4.

Because the number of samples was small, correlations between the drilling effluent tracers and the seagrass standing crop estimates were examined using Spearman's coefficient of rank correlation (Conover, 1971), a nonparametric measure of association between two variables based on ranks. Correlations using barium concentrations and barium-to-iron ratios were equal because their ranks coincided. Correlations between barium concentrations and the two standing crop estimates were not statistically significant (number of leaves: $r_s = 0.37$, $p > 0.05$; estimated leaf biomass: $r_s = 0.26$, $p > 0.05$). These results indicated that there was no association between the seagrass standing crop and levels of the drilling effluent tracers two years after drilling ended.

Data from the two follow-up surveys (August 1986 and October 1987) indicated recovery of the seagrass had occurred. Passage of Hurricane Elena (31 August to 3 September 1985) destroyed virtually all the deep seagrass beds in this area of the west Florida continental shelf. It is impossible to determine whether the seagrass recovery observed in Block 707 was from drilling alone because the effects of drilling and the hurricane were confounded. The seagrass has recovered from both the hurricane and the drilling, but Hurricane Elena may have altered the rate of recovery from drilling.

4. SUMMARY AND CONCLUSIONS

Although the causative factors of the seagrass growth inhibition cannot be determined based on the results of this study, two factors are thought to have contributed, individually or in concert, to the observed impacts on the seagrass around the drillsite. Smothering by drilling muds and cuttings may have been important in the nearfield. Farther from the drillsite, increased turbidity in the water column probably reduced the light levels at the seafloor. As a result, seagrass productivity would decrease. Based on the recovery data, it appears such reductions in productivity would be ephemeral because recolonization of affected areas would occur when light levels returned to normal conditions. The results of this program are site- and time-specific, a limitation inherent in all field studies (National Research Council, 1983).

Successful application of these results to other areas will depend on the similarity of the environmental conditions, drilling regime, and physiological state of the seagrass. When used under appropriate conditions, however, the modeling approach provides a means of estimating impacts prior to drilling if the fate of discharged material can be reliably modeled. Review of the impact assessments developed by modeling can provide regulators with a means of evaluating environmental differences in given proposals.

This study serves as a first step in understanding the relationship between biological assemblages on the Florida continental shelf and the discharge of drilling effluents. In Gainesville OCS Area Block 707, discharged drilling effluents had a short-term, quantifiable effect on the seagrass in the vicinity of operations. Questions as to the specific agents responsible for the observed impacts and applicability of these results to other sites and species remain to be studied.

ACKNOWLEDGEMENTS

This research effort was funded entirely by the Sohio Petroleum Company of Lafayette, Louisiana. The authors wish to express their appreciation to Drs David A. Gettleson and Neal W. Phillips for their reviews and comments on preliminary drafts of this manuscript.

REFERENCES

BRANDSMA, M.G. & SAUER, T.C (1983). *Mud Discharge Model: Report and User's Guide*. Exxon Production Research Co., Houston, TX, 144 pp.

CONOVER, W.J. (1971). *Practical Nonparametric Statistics*. John Wiley, New York, NY, 462 pp.

CONTINENTAL SHELF ASSOCIATES, INC. (1984). Live-bottom survey of drillsite locations in Gainesville Area Block 707 off the northwest coast of Florida. A final report for Sohio Petroleum Company, Lafayette, LA.

CONTINENTAL SHELF ASSOCIATES, INC. (1986). Environmental monitoring program for exploratory well No. 1, Lease OCS-G 6281, East Breaks Area Block 166 near Applebaum Bank. A final report for Texaco USA, New Orleans, LA.

CONTINENTAL SHELF ASSOCIATES, INC. & MARTEL LABORATORIES, INC. (1985). Florida Big Bend Seagrass Habitat Study Narrative Report. A final report for the Minerals Management Service, Metairie, LA. Contract No. 14-12-0001-30188.

LORENZEN, C.J., SHUMAN, F.R. & BENNETT, J.T. (1981). *In situ* calibration of a sediment trap. *Limnol. Oceanogr.*, **26**, 580–5.

NATIONAL RESEARCH COUNCIL (1983). *Drilling Discharges in the Marine Environment*. National Academy Press, Washington, D.C., 180 pp.

TREFRY, J.H., TROCINE, R.P. & MEYER, D.B. (1981). Tracing the fate of petroleum drilling fluids in the northwest Gulf of Mexico. In *Conference Record: Oceans '81*, Boston, MA, pp. 732–6.

TROCINE, R.P. & TREFRY, J.H. (1983). Particulate metal traces of petroleum drilling mud dispersion in the marine environment. *Environ. Sci. Technol.*, **17**(9), 507–12.

PART II
Treatment Practices

7

Options for Treatment and Disposal of Oil-based Mud Cuttings in the Canadian Arctic

A. H. GILLAM

*CBR International Corp., P.O. Box 2010, 101–9865 West Saanich Road,
Sidney, British Columbia, Canada V8L 3S3*

R. W. DRINNAN

*Aquatic Science Consultants,
239 Chambers Place, R.R. No 3, Site B, C-6, Nanaimo,
British Columbia, Canada V9R 5K3*

S. R. H. DAVIES

*Institute of Offshore Engineering, Heriot-Watt University,
Research Park, Riccarton,
Edinburgh EH14 4AS, Scotland, UK*

and

F. R. ENGELHARDT

*Canada Oil and Gas Lands Administration,
355 River Road,
Ottawa, Ontario, Canada K1A 0E4*

ABSTRACT

*This paper reviews the treatment and disposal options, for use in the
Canadian Arctic, for cuttings contaminated with oil-based drilling mud
(OBM).*

*Routine solids control equipment, plus additional cuttings cleaning systems,
were considered and the consequences of offshore disposal, onshore disposal,
and incineration were evaluated for each. The emphasis of the paper is on*

presenting information to assist in decision-making by operators and regulatory agencies. Background information on the physical environment, the environmental effects of oil-contaminated cuttings and cuttings cleaning technology was reviewed and used as a basis for developing matrices to enable operators and regulators to evaluate the engineering, environmental, and financial consequences when choosing a treatment system and method of disposal.

With solids control plus cuttings cleaning systems, the offshore discharge of cuttings will result in a localized environmental impact. Coastal (shallow) Beaufort Sea regions have the greatest environmental sensitivity, with the deep waters of the high Arctic having the least sensitivity. Physical dispersion and biodegradation rates of discharged cuttings in the Arctic are unknown.

Onshore disposal of cuttings requires additional specialized handling equipment. The incomplete freezing of oil-based mud cuttings may restrict land spreading or burial. Land-based incineration and landfill disposal of cuttings would create the least environmental impact, although additional capital and operating costs would be incurred.

1. INTRODUCTION

A significant number of oil and gas finds have recently been announced by operators in the Mackenzie delta and high Arctic, and further delineation drilling and extended flow testing is taking place at the Amauligak site. There is increased interest by some operators to consider oil-based drilling muds (OBM), because of the lithology of the region and the logistics of drilling in the Arctic. The advantages of oil-based muds include improved drilling rates, especially in those areas where the drilling season is limited; increased lubricity, important when directional drilling is required; and improved borehole stabilization in the reactive clays and shales frequently encountered in the Arctic.

There are a number of environmental concerns related to the disposal of OBM or cuttings contaminated with oil, including short-term toxicity and longer term organic enrichment.

This paper is an updated summary of a report funded by the Canadian federal government's Environmental Studies Research Funds (ESRF). The original purpose of the study was to provide both operators and regulatory agencies with a series of options, summarized in tabular format, outlining the engineering, environmental and financial considerations to be reviewed when using OBMs. Drilling mud and cuttings treatment systems are reviewed and the environmental consequences of both land and marine disposal options for the cuttings are evaluated. Two types of base oils are considered—diesel (high-aromatic content) and lower-toxicity (<1% aromatics)—for a number of marine and land disposal areas.

2. ENVIRONMENTAL CONSIDERATIONS

2.1. Offshore Disposal

2.1.1. Physical Dispersion
The fate of drilling fluids and cuttings discharged into the marine environment is determined by a range of physical, chemical and biological processes which serve to disperse, change, or concentrate constituent materials. Discharged materials are distributed by the dispersive energy of the ocean at the disposal site which, in turn, is a function of wind, tide, waves and mean currents (National Academy of Sciences, 1983).

In the case of cuttings discharges, the relatively large particles will settle rapidly near the well site. Soluble and particulate fluid additives adhering to the cuttings will, to some extent, be washed off as the large particles settle. Ayers *et al.* (1980) found that over 90% of discharged water-based drilling fluid solids settled directly to the bottom, in a study carried out under calm sea conditions in the Gulf of Mexico. They found that the distance the cuttings travelled from the well site and the settlement time were primarily a function of current and water depth. This observation has been confirmed for oiled-cuttings discharges in the North Sea (Poley & Wilkinson, 1983).

In waters deeper than 20 m, or in areas that are ice-covered for large portions of the year, the likelihood of cuttings being significantly dispersed by wave action decreases. In the deeper-water drilling areas of the high Arctic, for example, the bottom sediments will rarely, if ever, be influenced by wave action.

2.1.2. Environmental Effects
(i) Toxicity. Initial concern over the use of oil-based muds (OBM) was related to the high acute toxicity of the base oil (diesel) that was used. This toxicity was thought to be a function of the concentration of aromatic hydrocarbons present. As a result, alternative, or lower toxicity, base oils were developed which were much lower in aromatics; typically 1–10%, compared with 20–30% for marine diesel.

Analysis of a range of the low toxicity base oils themselves (Blackman et al., 1983) showed some correlation between total naphthalene concentrations and short-term acute toxicity (96 h LC_{50}) to brown shrimp (Crangon crangon), although there were some exceptions. At present there is no standardized method for the determination of the aromatic content of base oils used in drilling fluids and, thus, it is difficult to relate the toxicity of base oils to their aromatic content. Furthermore, the wide range in toxicity of the lower toxicity base oils, compared to the uniformly low toxicity of drilling muds formulated using them (Blackman et al., 1983), raises the question of the suitability of short-term, acute toxicity testing for assessing the environmental effects of discharging these materials.

In Canada, a number of studies have investigated the effects of oil-based mud cuttings on fish. Addison et al. (1984) observed no toxic effects of low toxicity oil-based cuttings on winter flounder over a 24-day test period. Sub-lethal assays of these flounder, based on mixed-function oxidase (MFO) measurements, showed no induction either of MFO enzymes or of the cytochromes associated with them. Even in the 'worst-case' scenario of prolonged static exposures, no lethal or sublethal effects were noted in the winter flounder, a species which shows MFO induction in the presence of other hydrocarbons. There was little or no indication, even on injection of base oils, of MFO induction in several species of fish.

Also on the basis of MFO studies in fish, Payne et al. (1985) concluded that any potential for MFO induction by hydrocarbon-contaminated cuttings would likely be reduced by substitution of low aromatic base oils for diesel. They also noted that different classes of aromatic compounds appear to affect the fish differently, pointing out the difficulty of trying to deduce any sublethal effects, such as induction of MFO enzymes, on the basis of total aromatic content.

Hutcheson et al. (1984) assessed both acute (96 h) and chronic (32 day) lethal toxicities of oiled cuttings (low toxicity base oil) to a variety of marine benthic species. They concluded that toxic aromatic compounds seemed to play no role in the toxicity response. Additional studies investigating this relationship are currently being completed under the Canadian ESRF program.

The benthic fauna beneath and immediately adjacent to the drilling platform will be completely buried by discharges of both diesel and low toxicity OBM cuttings, but there is little evidence that the latter will have less environmental effect on the sea bed. The hydrocarbon load of low toxicity, oil-based mud cuttings results in significant organic enrichment of the sea bed (Addy et al., 1984). Within this organically-enriched area, hydrocarbon analyses point to active biodegradation, particularly of the *n*-alkane fraction. This degradation is accompanied by severe oxygen depletion in sediments within 250 m of the platform. However, although high hydrocarbon concentrations are clearly correlated to biological effects, this correlation does not necessarily imply a toxic effect because the faunal response is consistent with organic enrichment and not with direct toxicity.

Blackman *et al.* (1985), using an initial oil concentration of 1000 times the background total hydrocarbon content in North Sea field studies, found that there was a marked difference between the biota developing in tanks containing OBM and in a control tank, which received drilling mud solids only without any oil. There was a difference in effect between two drilling muds based on alternative oils of moderate and low aromatic hydrocarbon content, but a greater difference existed between these two muds and a diesel OBM. When the total surficial sediment oil concentrations decreased, biota were observed to develop even in the diesel OBM tank, even though the oil concentrations in the subsurface sediments remained high. Blackman *et al.* (1985) concluded that although it is difficult to extrapolate from experimental tanks to deep-water offshore oil fields, it is expected that surface recolonization will occur, even on sediments heavily oiled to a depth of several centimeters once the surficial deposits are sufficiently clean. They also concluded that recolonization will be promoted by the adoption of low toxicity base oils in preference to diesel, but that the recolonizing communities are likely to be different from those in an unaffected area.

(ii) Field studies. Most of the information on the impact of oil-based muds on the marine environment has been from the North Sea area. This has been reviewed by a joint UK government and industry working group (Davies *et al.*, 1984). They found that despite both differences in inputs from different platforms and the variations in seabed area at different locations, the seabed chemical and biological effects associated with these discharges could be summarized in terms of four zones each affected differently (Table 1).

The spread of cuttings has been found to be greatly influenced by particle size. Cuttings resulting from the use of an OBM are generally larger than when using a WBM and tend to fall, as agglomerations, more directly to the sea bed. A number of workers have noted that OBM cuttings are cohesive and undergo little resuspension and that cuttings with an oil content of greater than 4–5% by weight aggregate into large pellets which sink rapidly (100–700 m/h) (Blackman & Law, 1981). These

TABLE 1

Zones of effect of discharge of oil-based mud and cuttings[a]

(modified from Davies *et al.*, 1984)

Zone	Maximum extent within range (m)	Biology	Chemistry
I	0–500 (Usually <250)	Impoverished and highly modified benthic community (beneath and close to the platform the seabed can consist of cuttings with no benthic fauna)	Hydrocarbon (HC) levels high; sediments largely anaerobic; HC's 1 000 plus × background
II	200–2 000	Transition zone in benthic diversity and community structure	Hydrocarbon levels above background; HC's 10–700 × background
III	800–4 000	No benthic effects detected	Hydrocarbon levels return to background HC's 1–10 × background
IV	4 000	No benthic effects	No elevation of hydro-carbons

[a] This study considered diesel and first generation, i.e. high aromatic, low-toxicity base oils only.

authors have concluded that very little absorbed oil is stripped from the solids during settlement but that oil will leach slowly to the surrounding water after deposition.

Davies *et al.* (1984) found that the extent of biological effects was greater from OBM cuttings than from WBM cuttings beyond the area of physical smothering, e.g. Zones II and III (Table 1). These effects of oil-contaminated cuttings may result from organic enrichment of the sediment or toxicity of certain fractions, such as aromatic hydrocarbons, or both. This topic will remain an area of active interest until field studies from locations discharging only lower toxicity oil-based cuttings are reported.

(iii) Arctic environment. There are comparatively few data on the environmental effects of the discharge of OBM cuttings in the Arctic. A recent report (Yunker, 1986) of a study to determine the approximate rates and durations of release of oil from OBM cuttings concluded that OBM cuttings discharged into shallow Arctic water during open water will be strongly leached or weathered by wave action, particularly summer

storms. Oiled cuttings discharged through the ice after freeze up and/or into deep water can be expected to persist for extended periods of time and to show only a slow loss of oil.

2.2. Onshore Disposal

2.2.1. Introduction

Disposal of drill cuttings onshore is one alternative to offshore discharge. The implications arising from onshore disposal, primarily the location of landfill sumps, are discussed in conjunction with alternatives, including landspreading and dispersal in fresh water. There have been no reports of the disposal of OBM cuttings in landfill sumps. The following discussion is concerned mostly with sump disposal of WBM cuttings, although the information presented is also of relevance to the disposal of OBM cuttings onshore.

Water-based mud wastes from onshore exploratory wells in the Arctic regions have been effectively handled by the use of sumps and subsequent burial (Beak Consultants Ltd, 1974; Dames & Moore, 1974; Canadian Petroleum Association, 1977). Sump disposal involves the construction of a pit area, disposal of waste material and subsequent filling in of the sump with the original overburden material. This overburden in turn is often capped with gravel. Studies made during the past decade of abandoned Arctic wells show that when the proper construction and restoration criteria are met, drilling wastes can be effectively contained by sumps (French, 1978a, b, 1980; Smith & James, 1979, 1985). In general, the area affected by sump construction and backfill is of the order of 200–300 m².

Surveys of sumps made throughout the Arctic in 1976–7 indicated that both the location of the sump and the timing of construction and restoration are major concerns. A further consideration with OBM cuttings is that they may not freeze as completely as those using WBM, resulting in the potential for leaching from the sump. To date, there is insufficient information to determine the degree to which this leaching may occur.

2.2.2. Geographical Location

Substantial geophysical and climatic differences exist between the high-Arctic and low-Arctic regions. High-Arctic tundra is characterized by both well-vegetated lowlands or oases, e.g. Banks and Southwest Cameron Islands, and barren polar deserts, e.g. Prince Patrick and Ellef Ringnes Islands. Low-Arctic tundra is deltaic, and is typified by the Mackenzie delta. Differences between the two tundra (French, 1978*a*) include duration and magnitude of summer melt; vegetation cover; ground ice and degree of permafrost; underlying bedrock; and accessibility to aggregate material.

(i) High-Arctic. Within the high-Arctic, both polar deserts and well-vegetated areas exist. The well-vegetated lowlands are typically a surficial layer of beach materials covered by a thick organic mat, composed primarily of moss. Such areas are probably best suited for waste disposal by sumps (Smith & James, 1979). Problems that may arise from construction and restoration of sumps are outlined in Table 2.

These regions frequently are associated with rivers and high drainage. Therefore, sump location and construction must be sufficient to prevent undermining or erosion. Sumps should be of sufficient size initially to prevent overflow and to minimize terrain disturbance, if additional sumps are needed. Sumps also should be located away from any shoreline to prevent seawater thawing and long-term erosion.

Common difficulties associated with the restoration stage can also be mitigated provided that proper procedures are followed. The major problems are: (i) sump subsidence and collapse, which results either from the incorporation of snow and ice with the infill, from underground erosion, or from both; and (ii) mixing of wastes and infill material (volcano effect) arising from the weight of the overburden collapsing the surface of the sump. The latter occurs when the sump is infilled before the wastes are completely frozen. Sump subsidence can be prevented by careful restoration practices, including the elimination of snow or ice from the infill material, construction of upslope drainage lines, and doming of the sump with gravel caps. The volcano effect can be prevented by late-winter restoration and by covering the sump with gravel aggregate.

Damage to vegetation can result from burial, crushing or ripping and direct toxicity from any surface oil leaking from the sump. Burial often occurs when the sump is too full and overflows at the edges, or by dumping the infill material on vegetation during sump construction. It can be avoided by correct sump size and storing infill material away from vegetated areas. Burial will result in localized death of vegetation, although recolonization can occur by plants with aerial stems (Heginbottom, 1973; Babb & Bliss, 1974; Bohn, 1974; Smith & James, 1985). Crushing and ripping of vegetation is generally caused by vehicular traffic and can be corrected by restricting traffic movement to the winter period. Contamination of vegetation may result from improper construction (see Table 2) or from spills during filling.

TABLE 2
Summary of onshore disposal concerns for WBM

Problem	Cause	Solution
Non-containment	Sumps too small initially	Additional sump; truck wastes elsewhere; modify existing sump
Terrain disturbance	Initial or additional construction in summer; trucking fluids	Gravel dome; cease construction until winter
Volcano effect	Infilling when freezing incomplete	Domed gravel caps; winter infilling
Subsidence/collapse	Incorporation of snow/ice during infilling; water seepage	Pre-inspection in summer; careful infilling (no snow); drainage lines upslope
Lack of infill	Meltout resulting in a heat sink over sump	Domed gravel caps
Fluid leakage	Ice lenses in walls during meltout; sump too full	Proper sump size
Vegetation kills	Contamination from: (a) erosion (b) spring runoff (c) spills (d) overflow (too full, volcano)	(a) Planting of native grasses (b) Winter construction (c) Careful handling (d) Proper sump size
	Toxicity to plants unknown	
Burial	Overflow; location of infill material during storage	Proper sump size; locate away from vegetation zone
Crushing, ripping of plants	Vehicular movements during backfilling or transport	Winter movement only

References: French (1978*a*, *b*, 1980); Smith & James (1979, 1985).

Polar deserts of the high-Arctic are typically barren arid regions that lack vegetation, wildlife and ground ice. Many of the problems that might occur in wetter more-vegetated areas would be avoided in this region. On the other hand, sump construction and restoration costs would be higher because of the greater difficulty in blasting continuous permafrost and in back-filling frozen infill material. French (1978a) stated that land-spreading, or at least non-containment, of wastes may be a preferable alternative in these circumstances. This is, however, not currently considered by regulatory agencies as a viable option for disposal of OBM wastes in the Canadian Arctic.

(ii) Low-Arctic. Low-Arctic tundra is best exemplified by the Mackenzie delta: an ice-rich terrain covered by shrub vegetation (north of tree-line) and innumerable lakes and drainage channels. The climate is affected by its proximity to the Beaufort Sea, and above-freezing temperatures are reached in summer.

The low-Arctic delta region is the most sensitive area for waste disposal because of its greater biological importance, greater human habitation and annual thaw and flooding. The use of sumps for containment of drilling wastes has been recommended for this region (Smith & James, 1985). Similar guidelines apply for construction and restoration as discussed for the high-Arctic. Problems associated with non-containment of water and terrain disturbances would be mitigated in a manner similar to the vegetated areas of the high-Arctic (Table 2).

Additional problems could arise from the thawing of the upper layer of the tundra, which would result in a melt-out of ice crystals in the permafrost. This melting could cause subsidence or collapse of the sump, leakage of waste material, enlargement of the sump, and the formation of standing water bodies. Standing water acts as a heat sink, thawing the overburden layer of the sump and allowing the infill and waste material to mix (volcano effect). Some of these problems could be overcome by using gravel aggregate to dome the sump after restoration, although use of natural disposal sites, e.g. borrow pits or enclosed lakes without sensitive biota, may be an alternative (French, 1980; French & Smith, 1980).

Burial of oiled cuttings in the permafrost below the active zone may immobilize the material, particularly if the sump is properly constructed and includes a lining of impermeable material. However, no data are available to assess this option. The active zone is thought to cause movement of heavy metals into the ground and surface waters, and a similar problem may exist for some of the oil and other organic components of OBM, although this may also be a result of excess filling of the sump above the permafrost.

Environmental effects from onshore disposal of WBM cuttings to date are not well documented, other than those on vegetation. Table 3 outlines some of the possible environmental concerns, particularly in relation to birds and mammals, which could result from improper sump construction or restoration or from accidental spills of OBM cuttings. The most likely effect associated with OBM disposal would result from the coating of feathers or fur.

(iii) Local considerations The local placement of a sump is particularly important in the low-Arctic where the permafrost experiences spring thaw and annual flooding during which the potential for ecological damage is greatest. Poor location may lead to erosion and to subsequent subsidence or collapse of the sump, with the potential for leakage of wastes and for contamination. Erosion by spring melt-out and alluvial streams can be prevented by a pre-inspection of the site during summer whereas erosion by downslope drainage can be prevented by locating the sump in a shallow depression or, if that is not possible, by constructing drainage lines upslope.

2.2.3. Seasonal Concerns
To reduce the potential for problems, sumps should be constructed, filled and restored during the same winter, to minimize any thawing of the permafrost and to ensure the in-situ freezing of waste material. Terrain disturbances by vehicles are also minimized during the winter. In some areas, e.g. low-Arctic and Mackenzie delta, summer movement by heavy equipment is impossible.

Problems arising from subsidence, volcano effects, leakage resulting from degradation of sump walls and excessive terrain damage were reported from surveys that investigated the effects of drilling operations during a summer and over a two-winter season (French, 1978a; Smith & James, 1985). Restricting equipment movement to winter, capping the sump with a gravel dome, locating the sump in a continuous permafrost location (polar deserts) or constructing and restoring sumps over a single winter season would alleviate many of these concerns.

2.2.4. Alternatives to Sumps
While land-spreading for WBM wastes has been considered a practical method of disposal, there is concern that during summer thaws the base oil may migrate when OBM is involved, making this option unattractive.

TABLE 3

Possible effects on terrestrial and estuarine biota from onshore disposal of OBM drill cuttings

Species	Effect	Mitigation	Comments
Birds			
Waterfowl, shore birds	Loss of insulation; ingestion; mortality due to direct contact with oil on open water (from leakage); loss of insulation due to contamination; loss of prey species (contamination of mudflats); loss of habitat; loss of staging and breeding areas	Allow wastes to freeze completely before restoring; careful handling of wastes; prevent oil from reaching watercourse or marsh areas	Large numbers of ducks summer in Mackenzie delta and along Beaufort Sea coast; species undergo annual moult, making them more vulnerable due to flightlessness; Mackenzie delta and Yukon coastal plain are staging areas for geese during August–October
Mammals	Interruption of migration paths; interruption or impact on breeding, calving or denning areas	Coordinate disposal to minimize interaction during high use periods; protect large scale damage to vegetation	Actual drilling operations may have as much (or greater) impact as the land disposal options for OBM cuttings

Fish	Contamination of habitat and loss of prey species due to leakage to fresh water	Proper construction and operation of burial pit; restrict disposal to fresh waters to noncritical periods	Many fish species enter estuaries during spring (after breakup) and return to fresh water in late summer to spawn or to over-winter
Cultural	Loss of domestic and commercial income or recreation due to oil contamination of foxes and other fur bearers or to anadromous fish	Avoidance of high use areas; proper disposal management	

Discharge of OBM cuttings to a freshwater environment has not been studied to date although reports on this disposal method for other drilling wastes have been prepared (Beak Consultants Ltd, 1974; Hrudey *et al.*, 1976). The use of small lakes as borrow pits may prove suitable, providing there is sufficient protection to prevent groundwater contamination.

In areas where fish are resident, especially commercial and subsistence species, considerable environmental concern may exist. Movement and burial of the cuttings is likely to occur in rapidly flowing rivers with high sediment loadings, but there is no information to predict how quickly this may occur or what effect it might have.

3. EVALUATION OF TREATMENT OPTIONS FOR OILED CUTTINGS

When using an OBM, cuttings treatment systems may be considered to reduce the oil adhering to the cuttings, either to minimize the environmental risk, or to recover the oil itself for reuse, or both. All drilling fluids and cuttings generated during the drilling process pass through routine solids control equipment. Under some circumstances, solids control may be sufficient. However, if additional cleaning is required, a number of treatment systems can be utilized. Four general categories of cleaning systems are reviewed: spray wash; immersion wash; thermal systems; and stabilization (solidifier) systems.

3.1. Solids Control/Cuttings Cleaning Systems

3.1.1. Routine Solids Control

A solids-control system is installed with the prime purpose of separating drilled solids from the drilling mud and mud solids. The equipment used will depend, in part, on the nature of the geological formation being drilled. In regions where the cuttings are generally large and intact ($> 200 \, \mu m$) most can be removed with shale shakers alone. Some geological structures, however, result in higher amounts of fine-sized particles in the drilling fluids and additional procedures, including screening, hydrocyclones and centrifuges, are required to separate them.

In the Beaufort Sea/Mackenzie delta region, where much of the structure consists of poorly lithified sedimentary rock with reactive clays, the cuttings tend to break down when WBM is used. The particle size is typically about $5–50 \, \mu m$ in diameter, a size range not easily removed by screening devices. In fact, depending upon the well, from 40 to 90% of the solids generated using WBM will pass through the shale shaker screens to be handled by equipment downstream. Over half of this underflow material is removable only by centrifuging. This factor is one reason why OBM has been considered, since it results in much larger sized cuttings which can be more easily removed.

The first stage in the treatment of oiled cuttings is a solids-control system. High-efficiency shale shakers are often used to remove as much of the solids as possible before mechanical attrition can wear the particles down in size. A well-operated shale shaker, under favourable conditions, should be capable of removing around 90% (by weight) of the oiled-drill solids.

Desanders and desilters have not been used extensively in the wells drilled to date with OBM in the Arctic. The main reasons for this are both environmental and economical since the discard, or underflow, has a very high oil content. This is not desirable environmentally and in addition results in the loss of large amounts of valuable drilling fluid. To maintain acceptable fluids properties, centrifuges must be used to remove the fine particles the shale shakers are not able to discard. A primary centrifuge is run to recover barite and return it to the active mud system. A secondary centrifuge processes the liquid discarded by the first centrifuge, discarding solid waste and returning the salvaged liquid to the active mud system. Because of the very fine nature of the solids produced during Arctic drilling, a system such as that described above would generally be the most efficient.

Solids and any associated liquid wastes discarded from the shale shakers, desanders, desilters, mud cleaners and centrifuges may be treated in a number of ways. If no cuttings cleaning system is installed, the oily cuttings may be sluiced overboard with seawater down a disposal chute. If a cleaning system is used, waste from the solids control equipment would be routed to the cleaning units. In the latter case, routine solids control systems would be considered as the first stage in the overall treatment and disposal process.

3.1.2. Spray Wash Systems

With these systems, cuttings from all or part of the solids-control equipment are sluiced to a vibrating screen unit. As oversize cuttings travel along the screen, they are first sprayed with wash fluid (which may be either diesel or aqueous-based) and are then allowed to drain for the remainder of the screen.

Undersized cuttings fall through the screen mesh along with the wash fluid. In some cases the latter is treated in a desilting cyclone or in a centrifuge to remove some of the undersized material. Generally, the separated cuttings are then discharged down a caisson where any remaining free oil on the water surface may be removed by skimming.

Once the wash fluid becomes unacceptably contaminated with oil and fines it must be disposed of in some manner. It is claimed that diesel-based wash fluid could be reincorporated into the active mud system. Spent aqueous wash fluid is usually discharged overboard, often down the cuttings disposal caisson, to allow any free oil to be skimmed off the surface of the water and recovered.

3.1.3. Immersion Wash Systems

Cuttings are sluiced from the solids-control system to an agitated tank containing diesel or aqueous-based wash fluid. The resulting slurry is then pumped over vibrating screens. Oversized cuttings are discharged whereas undersized cuttings flow back to the agitated tank. Once the wash fluid becomes unacceptably contaminated with fines (and oil), it must be disposed of in a manner similar to that used for spray wash systems. Generally, there is no net gain in reducing total oil discharges to the environment.

More advanced systems include wash fluid treatment using cyclones or centrifuges. Diesel wash fluid, and any remaining solids are recycled to the wash tank. Once spent, the diesel is either recycled to the active mud system or pumped to a tank prior to disposal by burning in oil test burners.

Immersion wash systems can also incorporate multi-stage washes (diesel/seawater or diesel/seawater/seawater). After each stage, separated wash fluid is recycled. A three-phase decanting centrifuge is used at the final stage of the treatment system to remove dispersed oil in the seawater wash fluid and to recycle it to the first stage (diesel) wash tank, or to the active mud system, or to both.

Solvent extraction units are another group classified as immersion wash systems. Cuttings are treated in solvents, e.g. trichloroethane or liquid CO_2, which, having considerably greater oil-removal properties than aqueous detergents, give much improved cleaning performance. Because solvent costs are relatively high, it is logical that these systems should incorporate cleaning and recovery steps.

3.1.4. Thermal Systems

These systems involve distillation, combustion or a combination of the two to drive the oil and water from the cuttings.

In batch vacuum-distillation systems, cuttings are stored in a buffer storage tank. At the commencement of the cycle a batch is fed into the retort barrel of the unit through grinding blades. The ground cuttings are heated under vacuum which causes oil and water to distil off. Any fines generated are removed in a heated cyclone, after which the vapors are condensed and recycled to the active mud system. At the end of the distillation cycle, all solids are discharged overboard.

In a second type of distillation system, basically a two-stage process, cuttings pass through a surge tank and are transported down a heat tube by an internal auger. Some of the vapors distil off and are collected and ignited with air to heat the heat tube for final hydrocarbon recovery. These vapors are also recycled and ignited. The treated cuttings are then cooled with seawater and discharged from the bottom of the unit.

In vibrator bed systems, cuttings pass through a surge hopper and are conveyed across the drier bed, with heated air blown upward through the material. The hot air drives oil and water from the cuttings, and the gaseous stream then exits at the top of the drier along with some fine cuttings. The fine cuttings are removed by a cyclone and scrubber which also condenses the oil and water. These fluids are separated in an oil/water separator. Some of the oil may be used to power the air furnace; the remainder (it is claimed), may be recycled to the active mud system although this recycling will depend on the actual state of the recovered oil. All solids and the de-oiled water are discharged overboard.

A third option is incineration. Few tests on the incineration of cuttings alone have been carried out. Milburn (1984) reported on the results of open-pit burning of whole mud wastes, and concluded that rotary kiln incineration (or similar equipment) would be required for complete disposal, especially for cuttings. However, the equipment is large, complicated and expensive and cuttings from several sources would have to be processed in a central treatment facility to reduce costs.

National guidelines exist for air emissions from incinerators, which should be considered when dealing with the incineration of cuttings from oil-based muds. Emission limits are given for particulates, HCl and SO_2, but as yet not for unburnt hydrocarbons or heavy metals. The ash from incineration operations would be landfilled.

Dome/Canmar performed a series of incineration tests in 1985 on whole oil-based mud, formulated using both diesel and a low-toxicity base oil.

Four separate incinerators were tested: a TOPS (Technical Offshore Petroleum Services) burner; a Saacke (rotary cup) burner; a reciprocating kiln; and an air portable incinerator. The rates of throughput of mud ranged from 400 bbl/day (TOPS), through 350 bbl/day (Saacke), to 40 bbl/ day (reciprocating kiln and air portable incinerator). The reciprocating kiln was also used to burn oil mud cuttings. Although this proved to be labour-intensive, it did produce a very clean waste product.

3.1.5. Stabilization Systems
These systems stabilize the oil on cuttings rather than removing it. In the Leco system, a specially-treated quicklime (CaO) is used to produce a product suitable for use as a filler for road construction material. Following the Amoco Cadiz incident, oiled beach sands were stabilized with a similar system, using standard grades of industrial quicklime. Recently, there have been reports discussing a European quicklime stabilization process incorporating pulverized fly ash (PFA) and oiled beach sands.

3.2. Engineering Considerations
Cuttings treatment systems available up to late 1985 were evaluated using a relative rating scale (1–4) for each of the following categories:

— cleaning performance
— processing capacity
— size, weight and power requirements
— process supplies
— safety concerns
— personnel requirements
— vessel support requirements
— shorebase support requirements.

Descriptions of the rating system for each category follow. The values for each category assigned to the different ratings are presented in Table 4. Engineering considerations for each treatment system were considered to be independent of the base oil used. Table 5 is a summary of the ratings for each treatment system.

3.2.1. Cleaning Performance
Cleaning performance is based on the amount of oil that is retained on the waste material discharged from the treatment system. In general, thermal and solvent extraction systems should provide considerably better cleaning performance than aqueous or base-oil wash systems. A thermal

TABLE 4
Rating values for engineering and logistical factors

	1	2	3	4
Cleaning performance (% wet weight of residual oil)	<1	1–5	5–10	>10
Processing capacity (t/h)	1–5	5–10	10–20	>20
Size (m³)	<50	50–100	100–500	>500
Weight (t)	<25	25–50	50–100	>100
Power (kW)	<25	25–100	100–500	>500
Process supplies	Increasing need for specialized process supplies ⟶			
Health and safety	Increasing health and safety concerns ⟶			
Personnel	Increasing specialized personnel requirements ⟶			
Vessel and shorebase support	Increasing support requirements ⟶			
Capital costs (US $)	0·2–0·4 M	0·4–0·6 M	0·6–0·8 M	0·8 + M
Operating costs/day (US $)	0·5–1·5 K	1·5–3·0 K	3·0–4·5 K	4·5 + K

system would probably reduce residual oil on cuttings to less than 1% wet weight, whereas the oil retained on cuttings from some of the better wash systems would be about 10%.

3.2.2. Processing Capacity
Processing capacity refers to the amount of cuttings from the solids-control equipment that can be handled by a particular treatment system. The relatively poor cleaning performance of early systems was, in part, a function of the capacity of the system. One problem was that the peak cuttings flow rate expected from a well was often underestimated, partly because of faster rates of penetration (ROP) from the use of OBM. The processing capacities should be compared to the average hourly solids generation of 0·5–10 t for Arctic wells. The lower range is from smaller hole sections (198 mm) at a slow rate of penetration (ROP) (5–6 m/h),

TABLE 5
Rating values for cuttings treatment systems

Treatment system	Factors										
	Cleaning performance	Cap[a]	Size[c]	Weight[d]	Power requirements	Process equip. supplies	Safety	Manpower[e]	Vessel/ shorebase support	Capital costs	Operating costs
Solids control	4	4	1	1	1	1	2	2	1	1	1
Spray wash	4[b]	1-3	1	1	1-2	3	1-2	2	2-4	1*	1
Immersion wash	2-4	1-4	1-3	1-4	1-3	2-4	1-4	1-4	2-4	1-4*	1-3
Distillation	1	1	1-4	1-2	2-4	1	4	3-4	2-4	1-4*	3
Combustion	1	2	3-4	3	3	1	4	1-3	2-4	1-4	3
Stabilization	NA	1-4	1-4	1-2	1	3	1-3	2	2-4	1	3

[a] In the case where multiples of the principal treatment were configured together, e.g. tandem arrangements, the largest common configuration was rated for capacity.

[b] Caisson systems are thought to be ineffective for oil recovery. Much of the oil released from the cuttings, and separated out in the caisson, gets entrained in the caisson discharge. Thus, most of the oil removed from the cuttings is discharged into the sea.

[c] The system dimensions, plus any probable tank storage requirements, were obtained from reference material. Tank volumes were converted to m³ and were tallied with the system size. This total was then rated.

[d] Where possible, the net weight was used to rate a system.

[e] Systems rated by more than one number are claimed by the manufacturer to be automatic, requiring few personnel. However, the systems remain to be field tested and thus would require, initially, a dedicated crew of trouble-shooters. NA Not applicable.

whereas the upper range of solids generated would be from large-diameter holes (445 mm).

3.2.3. Size, Weight and Power Requirements
The importance of these factors varies depending on the size of the rig or platform on which the system is to be installed. Compact systems should be more attractive. Flexibility of an installation is important in terms of retrofitting the system to another rig or fitting the system into tight confines.

3.2.4. Process Supplies
Most wash systems require a continuous supply of chemicals for efficient operation. The stock of special chemicals that can be maintained at a drilling site may be limited by weight, space and logistics considerations. Where lack of space is a particular problem, it would be better to use a chemical that is also used elsewhere, e.g. low-toxicity base oil. Solvents such as trichloroethane and liquid CO_2 may impose a significant cost if losses are high.

3.2.5. Safety Concerns
Wash systems have the advantage over thermal systems in that they run 'cold' and hence operate with a lower risk of fire and explosion. Restrictions governing the location of such units on a drill unit should be less severe than for a thermal system. Some operator health risks also exist from toxic vapors from the use of thermal systems, as well as from wash systems using more exotic wash fluids such as trichloroethane. Other potential hazards considered include fire and explosion risks; contamination in the working environment; and temperatures that may be encountered.

3.2.6. Personnel Requirements
The need for full-time operators should be minimized on smaller drilling platforms because accommodation is likely to be limited. In this case, a system with a high degree of automation and low maintenance requirements would be preferable. Most of the thermal systems are claimed to fall into this category.

3.2.7. Vessel and Shorebase Support Requirements
This will range from the case where there is no requirement for vessel or shorebase support (a rating value of 1), to the case where a dedicated

vessel is required to transfer drilling supplies and dispose of waste materials and additional dedicated equipment and shorebase personnel are required (a rating value of 4).

3.2.8. Capital and Operating Costs

The various treatment systems have been grouped into four ranges with respect to capital costs. Data on many systems were not available and these have been marked with an asterisk in the matrix (Table 5). Judgements were made on the basis of complexity and comparisons with other systems, with respect to the level (1–4) into which they might fall.

Operating costs include dedicated personnel, process supplies, power and financing. They are ranked on the basis of total costs per operating day. There was little information on operating costs for most of the systems and, therefore, the ranking is based on the experience and judgement of the authors.

Both capital and on-structure operational costs were considered to be independent of the geographical location. However, the costs associated with onshore disposal options, particularly from marine-based drilling platforms, will be affected by the area of operations.

4. EVALUATION OF DISPOSAL OPTIONS FOR OILED CUTTINGS

In the following section, the environmental concerns associated with (1) the discharge of cuttings into marine waters, (2) disposal in landfills, and (3) disposal by incineration are assessed for each treatment option.

4.1. Offshore Disposal Options

4.1.1. Base Oil Type

Environmental effects may differ with the type of base oil used. Most of the literature on environmental effects of OBM refers to information related to diesel-based muds, with only the more recent reports evaluating low-toxicity formulations. Few data are available on oils in the 5–20% aromatic range. Because most of the available data and concerns relate to the aromatic content of the base oil, the environmental matrices consider low aromatic (<1% aromatics, Brandes IR method) and diesel

(> 20% aromatics, Brandes IR method). It is assumed that the environmental effects of other base oils will fall somewhere between these two extremes.

It must be emphasized that the concerns related to the differences in the effects of cuttings containing low aromatic base oils and diesel base oils are based on the short-term (96 h LC_{50}) toxicity of the base oils. Although it seems reasonable that there may be some difference between the environmental effects of the cuttings from the two types of base oils, there are three potential difficulties with this assumption. First, no field studies have been reported where it is possible to make a direct comparison of the effects of cuttings discharged using different types of base oil. Second, organic enrichment and smothering of the sediments by the oiled cuttings may override the 96 h LC_{50} toxicity effects in the environment and there may be little actual difference in field effects of the two types of muds (Addy *et al.*, 1984). Third, it is difficult to extrapolate the results of studies of multi-well production situations in the North Sea to single exploratory wells in Canada. Furthermore, whereas low toxicity base oils can vary considerably in their short-term (96 h) toxicity, the toxicity of the drilling muds formulated with them is quite similar (Blackman *et al.*, 1983).

4.1.2. Geographical Location

The environmental effects of oil-contaminated cuttings discharged to Arctic marine waters are a function of the ultimate fate of the cuttings themselves and the biological resources present. The dispersion/location of cuttings is influenced by wave-induced mixing, which in turn is a function of water depth, by ice-scouring which is also a function of water depth, by suspended sediment depositions, and by tidal currents.

In shallow water, and particularly along the Mackenzie delta, waves and ice will be major forces affecting sediment/cuttings movement especially during break-up, although it must be remembered that there is a seasonal component to this. For example, there is likely to be a minimal disturbance of bottom material during the winter period of permanent ice cover but the cuttings can be expected to be affected by ice movement and by wave action during the break-up and open-water period. It is assumed that the fate of cuttings discharged under the ice in the shallow Beaufort Sea area will be relatively similar to that of those discharged in greater than 20 m of water during the open-water season.

Mixing and movement of bottom sediment material and cuttings from wave-induced energies, residual tidal currents (which are typically very low in the Beaufort Sea) or from ice-scouring are expected to be minimal in depths greater than about 20 m. In the high-Arctic, the total ice cover in winter and the presence of a substantial number of icebergs in summer, will ensure that the bottom sediment material will only be influenced to any large degree in shallow areas (less than 20 m). Hence, cuttings would remain in the discharge area for long periods, except in areas of relatively strong currents.

The two geographical areas considered in this report are therefore based on the oceanographic and ice forces that control the movement of cuttings:

— shallow Beaufort Sea (< 20 m depth), and

— Beaufort Sea (> 20 m depth) and high-Arctic islands (low-current areas).

4.1.3. Rating Criteria

A number of treatment systems have been described which vary with respect to cleaning performance, engineering and logistical factors, and cost. For offshore or coastal drilling operations and marine disposal the above factors have to be balanced against the relative environmental concerns (ratings) for each treatment option.

The ratings criteria for a number of environmental considerations are presented in Tables 6–11 for a number of different geographical, seasonal and base oil scenarios. The rationale behind the evaluation of these environmental considerations is discussed below.

One of the difficulties in establishing these criteria is that the impact is not only a function of the amount of OBM adhering to the cuttings, but is also and perhaps to a greater degree a function of the volume of the cuttings discharged, which is in turn related to the size, depth and number of wells drilled.

In comparing treatment options, only the oil retention was considered, i.e. it was assumed that the area covered by the cuttings would be similar for all systems. The ratings also refer to the area immediately adjacent to the well-head, the zone that will show the greatest impact. It is important to recognize, however, that as one moves away from the well, the degree of impact diminishes and that for any one well all four levels of impact, that are used as ratings criteria, will be present.

The criteria are based, in part, on the information presented in Section 2 and, in particular, from data gathered in the North Sea. The ratings (1–4) used in Tables 6–11 are based on the following definitions.

TABLE 6
Environmental considerations

Treatment systems	Oil retention (%)[a]	Burial of benthos	Toxicity (96 h LC_{50})	Community changes	Tainting	Anoxic/ H_2S conditions	Impact on resource use/users	Formation of bottom pavement
Solids control	15–25	4	4	4	4	4	2	4
Spray wash	8–12	4	3–4	3–4	4	4	2	4
Immersion wash	<5–>20	3–4	2–4	2–4	2–4	2–4	2	3–4
Thermal	<1	3	2	2	2	2	2	3

[a] g oil/100 g oil and water wet effluent.
Region of activity: shallow water—winter
Disposal option: marine
Base oil used: diesel

TABLE 7
Environmental considerations

Treatment systems	Oil retention (%)[a]	Burial of benthos	Toxicity (96 h LC_{50})	Community changes	Tainting	Anoxic/ H_2S conditions	Impact on resource use/users	Formation of bottom pavement
				Issues				
Solids control	15–25	3–4	3–4	3–4	3–4	3–4	3–4	3–4
Spray wash	8–12	3–4	3–4	3–4	3–4	3–4	3–4	3–4
Immersion wash	<5–>20	2–4	2–4	2–4	2–4	2–4	2–4	2–4
Thermal	<1	2–3	2	2	2	2	2–3	2–3

[a] g oil/100 g oil and water wet effluent.
Region of activity: shallow water—summer
Disposal option: marine
Base oil used: diesel

TABLE 8
Environmental considerations

Treatment systems	Oil retention (%)[a]	Issues						
		Burial of benthos	Toxicity (96 h LC_{50})	Community changes	Tainting	Anoxic/ H_2S conditions	Impact on resource use/users	Formation of bottom pavement
Solids control	15–25	4	2	4	3–4	4	2	4
Spray wash	8–12	4	2	3–4	3–4	4	2	4
Immersion wash	<5–>20	3–4	1–2	2–4	1–4	2–4	2	3–4
Thermal	<1	3	1	2	1–2	3	2	3

[a] g oil/100 g oil and water wet effluent.
Region of activity: shallow water—winter
Disposal option: marine
Base oil used: low-toxicity

TABLE 9
Environmental considerations

Treatment systems	Oil retention (%)[a]	Burial of benthos	Toxicity (96 h LC_{50})	Community changes	Tainting	Anoxic/ H_2S conditions	Impact on resource use/users	Formation of bottom pavement
Solids control	15–25	3–4	2	3–4	2–3	3–4	3–4	3–4
Spray wash	8–12	3–4	2	3–4	2–3	3–4	3–4	3–4
Immersion wash	<5– >20	2–4	1–2	2–4	1–3	2–4	2–4	2–4
Thermal	<1	2–3	1	2	1	2	2–3	2–3

[a] g oil/100 g oil and water wet effluent.
Region of activity: shallow water—summer
Disposal option: marine
Base oil used: low-toxicity

TABLE 10
Environmental considerations

Treatment systems	Oil retention (%)[a]	Issues						
		Burial of benthos	Toxicity (96 h LC_{50})	Community changes	Tainting	Anoxic/H_2S conditions	Impact on resource use/users	Formation of bottom pavement
Solids control	15–25	4	4	4	4	4	1	4
Spray wash	8–12	4	4	4	4	4	1	4
Immersion wash	<5->20	3–4	2–4	2–4	2–4	2–4	1	3
Thermal	<1	3	2	2	2	2	1	3

[a] g oil/100 g oil and water wet effluent.
Region of activity: high-Arctic region (low current areas) Beaufort Sea (>20 m)
Disposal option: marine
Base oil used: diesel

TABLE 11
Environmental considerations

Treatment systems	Oil retention (%)[a]	Issues						
		Burial of benthos	Toxicity (96 h LC_{50})	Community changes	Tainting	Anoxic/ H_2S conditions	Impact on resource use/users	Formation of bottom pavement
Solids control	15–25	4	2	4	2–3	4	1	4
Spray wash	8–12	4	2	4	2–3	4	1	4
Immersion wash	<5–>20	3–4	1–2	2–4	1–3	2–4	1	3–4
Thermal	<1	3	1	2	1	2	1	3

[a] g oil/100 g oil and water wet effluent.

Region of activity: high-Arctic region (low current areas) Beaufort Sea (>20 m)

Disposal option:　marine

Base oil used:　low-toxicity

Rating scale	*Criteria*
1	Negligible impact—no noticeable change in biological community or elevated hydrocarbon concentrations in sediments; cuttings rapidly dissipated; no conflicts with other resources/resource users.
2	Minor impact—no noticeable biological change; sedimentary hydrocarbon concentrations 1–10 times background; cuttings persistent only for duration of well; little likelihood of resource/resource user conflict.
3	Moderate impact—noticeable change in biological community as a result either of direct toxicity or of organic enrichment; sedimentary hydrocarbon levels 10–1000 times background; cuttings persistent for one season only; presence of cuttings may conflict with resource use.
4	Major impact—impoverished and highly modified benthic community; hydrocarbon levels high, greater than 1000 times background; anaerobic sediments; cuttings persistent for more than 1 year; acutely toxic, e.g. within a few days; presence of cuttings piles likely to conflict with resource use such as fishing.

The criteria, or ratings, used to develop the environmental matrices were applied to the following issues or concerns: (1) burial of benthic community; (2) toxicity (96 h LC_{50} of the base oil only) to benthic, epibenthic or pelagic organisms; (3) community alterations, e.g. enrichment; (4) potential for tainting; (5) formation of anoxic/H_2S conditions; (6) impact on resource use/users; and (7) formation of bottom pavement.

The matrices which summarize the ratings evaluating the impact of cuttings discharged to a marine environment were determined for the following combination of geographical areas, seasons and base oil types:

— shallow-water areas—winter period (ice cover)—DBM
— shallow-water areas—winter period (ice cover)—LTM
— shallow-water areas—summer period/ice break-up—DBM
— shallow-water areas—summer period/ice break up—LTM
— Beaufort Sea (>20 m depth) and high-Arctic Region (low-current areas)—DBM

— Beaufort Sea (>20 m depth) and high-Arctic Region (low-current
   areas)—LTM.

DBM: Diesel oil-based mud; LTM: low toxicity oil-based mud.

Because the environmental factors must be determined on a site-specific
basis, the matrices were based on a comparison of the different treatment
systems, assuming a common environmental setting for each. The primary
consideration when comparing the different systems was the amount of
oil retained on the cuttings, which can vary between less than 1% to over
20%. A secondary consideration was the consistency of the waste material,
which can range between a viscous, oily semi-solid (wash systems) to a
fine, dry powder (combustion systems).

The following assumptions were also made when considering the effects
of cuttings in different geographical regions during different seasons:

   (a) It was assumed that the mixing of cuttings with or burial by natural
       sediments, and spreading of the cuttings pile by oceanographic
       processes, would reduce the overall impact by allowing a greater
       opportunity for biodegradation and by lowering the concentration
       of oil present in the sediments. This assumption is based on the
       fact that burial of sediments by cuttings and organic enrichment
       were the two major impacts that were identified in the review of
       the literature.
   (b) It was also assumed that the mixing process would be much more
       likely in water depths of less than 20 m, when there was no ice
       cover, compared to (i) areas in water depths of greater than 20 m,
       or (ii) any water depth during the period of ice cover. The 20 m
       contour was selected because, under most oceanographic conditions,
       little sediment movement from waves was expected at depths greater
       than this.
   (c) Toxicity and potential for tainting were assumed to be greater for
       cuttings from DBM compared to cuttings contaminated with LTM.
       It was also assumed that the potential for tainting would be higher
       under conditions where the cuttings pile would be more compact,
       for example during ice cover or in waters greater than 20 m in depth.
   (d) It was assumed that, in general, the nearshore Beaufort Sea was
       more important as a resource to user-groups compared to the
       deeper water regions of the Beaufort Sea and high-Arctic waters.

4.2. Onshore Disposal Options

Land disposal options include burial in sumps or spreading over the land.
The latter has been suggested for WBM and cuttings in the barren regions

of the high-Arctic. It is not likely to be considered for oil-based cuttings, however, because this material is less likely to freeze to the same degree and the oil presents a continuing contamination problem. This may be alleviated if the cuttings are combined with solidifiers, e.g. quicklime. The material may remain relatively inert, although no data from Arctic regions are available.

Sumps are more probable candidates for onshore disposal. If properly constructed and maintained, the environmental risk is considered slight. On the other hand, poorly constructed sumps, during summer and in an important waterfowl staging area, could result in considerable environmental impact. The lack of complete freezing of oil-based mud and cuttings, however, remains a problem with sump disposal.

The range of potential environmental impacts used to evaluate onshore disposal options of untreated cuttings, stabilized cuttings and incineration ash are based on the information presented in Tables 2 and 3. The ratings used are given below.

Rating scale	*Criteria*
1	Negligible—properly constructed burial site with drainage lines upslope and no leakages; located away from vegetation zones and from any water body with sensitive biota.
2	Minor—properly constructed site; no leakage to local drainage; no minor damage to aquatic habitat or to surrounding vegetation.
3	Moderate—poorly constructed site; small leakages to nonsensitive aquatic habitats; local damage to surrounding vegetation.
4	Major—improperly constructed site; oil released to major staging area and/or sensitive aquatic habitat; permanent damage to or alteration of tundra.

The environmental impacts for landfilled wastes are expected to be moderate or major (3–4) only for untreated cuttings in a poorly constructed landfill. All other combinations are considered to have negligible or minor environmental impact ratings (1–2).

4.3. Incineration Options
Thermal systems for incineration can be installed on board the drilling platform or placed at a convenient shore-based facility. The waste material from these systems is low in residual oil. At onshore drilling operations,

or if the waste material is transported to shore from offshore sites, the residual material left after incineration can be buried.

Thermal units, incinerators, or open-pit burning will generate air emissions in the form of particulates, partially combusted hydrocarbons, and NO_x and SO_x gases. The long-term effects of such emissions are considered to be negligible because the total quantity of waste is small. Local and short-term problems could occur if, for example, the units were located up-wind of a human settlement, in an area of poor circulation, e.g. temperature inversions in valleys, or near sensitive environmental habitats, e.g. nesting sites.

Precautions related to the timing of the use of the incinerator, to the location of burn sites and to operator safety conditions should mitigate against potential short-term environmental difficulties.

5. SUMMARY

Cuttings treatment systems are used to reduce the oil content of drilling wastes, either to recover the oil for reuse or in certain marine areas to enable the drilling wastes to be discharged offshore. High efficiency shale shakers with a dual centrifuge system have been suggested as an efficient, cost-effective treatment method for Canadian Arctic drilling.

5.1. Treatment and Disposal Options—Offshore Treatment with Marine Disposal

Routine solids-control equipment is widely used, reliable and uses proven technology. It has relatively low operating and capital costs, requiring minimal dedicated operation or maintenance personnel. With additional cuttings cleaning systems there are higher capital and operating costs with an increase in specialized personnel, chemicals, space and power requirements.

With solids control alone or with solids control plus a cleaning system, the discharge of cuttings will result in a localized environmental impact. Short-term toxicity concerns can be addressed by use of lower toxicity base oils. Environmental concerns are also related to the sensitivity of the area in which the cuttings are discharged, with coastal (shallow) Beaufort Sea regions having the greatest sensitivity and the deep waters of the high-Arctic the least sensitivity. The physical dispersion and biodegradation rates of discharged cuttings in the Arctic are not known.

5.2. Treatment and Disposal Options—Offshore Treatment with Land Disposal

Support vessels or barges to transport the cuttings to shore, specialized shore base and off-loading requirements, and on-board 'buffer' storage of cuttings will be required, all at additional cost. There would be reduced marine environmental concerns with offshore discharges limited to accidental spills. Environmental concerns would be with the proper preparation of the landfill site; some locations, e.g. low Arctic delta regions, may not be suitable. The incomplete freezing of OBM cuttings may restrict land spreading or burial. Estimates are that an area between 200 and $300\,m^2$ would be required to construct a shore-based sump for a single, exploratory well.

5.3. Treatment and Disposal Options—Onshore Treatment with Landfill Only

With routine solids-control equipment and on-site disposal, there would be no transportation or storage costs and minimal logistical constraints. Environmental concerns would be with the proper preparation of the landfill site (see above).

5.4. Treatment and Disposal Options—Onshore Treatment with Incineration and Landfill

The environmental impact of this onshore option would be the smallest, although some short-term air emissions would be generated. There would be additional storage and logistical requirements combined with increased capital and operating costs for the equipment.

ACKNOWLEDGEMENTS

Many people contributed to the original report and in reviewing the information. In particular we would like to thank Dr D Stone, Department of Indian Affairs and Northern Development (DIAND), Ottawa; and Mr E. Birchard, Esso Resources Ltd, Calgary, for their support.

REFERENCES

ADDISON, R.F., DOE, K. & EDWARDS, A. (1984). Effects of oil-based drilling mud cuttings on winter flounder (*Pseudopleuronectes americanus*): Absence of acute toxicity on mixed function oxidase induction. *Canadian Technical Report Fisheries and Aquatic Science*, 1307, 19 pp.

ADDY, J.M., HARTLEY, J.P. & TIBBETTS, P.J. (1984). Ecological effects of low toxicity oil-based mud drilling in the Beatrice Oilfield. *Mar. Pollut. Bull.*, **15**, 429–36.

AYERS, R.C. JR, SAUER, T.C. JR, STEUBNER, D.O. & MEEK, R.P. (1980). An environmental study to assess the effect of drilling fluids on water quality parameters during high rate, high volume discharges to the ocean. In *Proceedings of a Symposium on Research on Environmental Fate and Effects of Drilling Fluids and Cuttings*, Washington, D.C. Courtesy Associates, pp. 351–91.

BABB, T.A. & BLISS, L.C. (1974). Effects of physical disturbance on Arctic vegetation in the Queen Elizabeth Islands. *J. Appl. Ecol.*, **11**, 549–62.

BEAK CONSULTANTS LTD (1974). Disposal of waste drilling fluids in the Canadian Arctic. A report for Imperial Oil Ltd, Edmonton, Alberta, 170 pp.

BLACKMAN, R.A. & LAW, R.J. (1981). The oil content of discharged drill cuttings and its availability to benthos. ICES CM1981/E:23, 7 pp.

BLACKMAN, R.A., FILEMAN, T.W. & LAW, R.J. (1983). The toxicity of alternative base-oils and drill-muds for use in the North Sea. ICES CM1983/E:11, 20 pp.

BLACKMAN, R.A., FILEMAN, T.W., LAW, R.J. & THAIN, J.E. (1985). The effects of oil-based drill muds in sediments on the settlement and development of biota in a 200-day tank test. ICES CM1985/E:23, 18 pp.

BOHN, D. (1974). Long-term effects of mud sump toxicity on vegetation. Indian and Northern Affairs, Ottawa. June 7 Report.

CANADIAN PETROLEUM ASSOCIATION (1977). *Handbook of Environmental Guidelines for Drilling Sump Construction*. Environmental Conservation Committee, Canadian Petroleum Association, Calgary, 20 pp.

DAMES & MOORE, O. (1974). Report on containment and disposal of drilling fluids in the N.W.T. A report for the Arctic Petroleum Operators Association and the Government of Canada, Department of Indian Affairs and Northern Development, 82 pp.

DAVIES, J.M., ADDY, J.M., BLACKMAN, R.A., BLANCHARD, J.R., FERBRACHE, J.E., MOORE, D.C., SOMERVILLE, H.J., WHITEHEAD, A. & WILKINSON, T. (1984). Environmental effects of the use of oil-based drilling muds in the North Sea. *Mar. Pollut. Bull.*, **15**, 363–70.

FRENCH, H.M. (1978a). Terrain disturbances, Report No. 6 to Department of Indian and Northern Affairs, Ottawa, Ontario, 52 pp.

FRENCH, H.M. (1978b). Terrain and environmental problems of Canadian Arctic oil and gas exploration. *Musk-Ox*, **21**, 11–17.

FRENCH, H.M. (1980). Terrain, land use and waste drilling fluid disposal problems, Arctic Canada. *Arctic*, **33**, 794–806.

FRENCH, H.M. & SMITH, M.W. (1980). Sump Studies II: Geothermal disturbances adjacent to wells drilled in permafrost. Environmental Studies No. 14, Indian and Northern Affairs, Ottawa, 61 pp.

HEGINBOTTOM, J.A. (1973). Effects of surface disturbances upon permafrost. Report 73-16. Environ-Social Committee Northern Pipelines, Information Canada, 29 pp.

HRUDEY, S.E., MICHALCHUK, J. & MCMULLEN, J.D. (1976). A preliminary assessment of water pollution from abandoned oil and gas drilling sumps in the N.W.T. Report to the industry-government working group 'A' on the disposal of waste drilling fluids in the Canadian Arctic. Environmental Protection Service, Northwest Region, Environment Canada, 42 pp.

HUTCHESON, M.S., STEWART, P.L., ODENSE, R., FOWLER, B.F. & GREEN, D. (1984). Development of toxicity testing guidelines for oiled cuttings. Final report for the Environmental Protection Service, Environment Canada, prepared by Atlantic Oceanics Company Ltd and Seakem Oceanography Ltd.

MILBURN, D (1984). Use of diesel oil-based muds at Panarctic Oil Limited's Cisco M-22 wellsite, High Arctic Islands, N.W.T. Water Resources Division, Northern Affairs Program, Indian and Northern Affairs Canada, Yellowknife, N.W.T., 34 pp.

NATIONAL ACADEMY OF SCIENCES (1983). Drilling discharges in the marine environment. Report of the panel on assessment of fates and effects of drilling fluids and cuttings in the marine environment. National Academy Press, Washington, D.C.

PAYNE, J.F., FANCEY, L., KINENIUK, J., WILLIAMS, U., OSBORNE, J. & RAHINTULA, A. (1985). Mixed-function oxygenases as biological monitors around petroleum hydrocarbon development sites: Potential for induction by diesel and other drilling mud base oils containing reduced levels of polycyclic aromatic hydrocarbons. *Mar. Environ. Res.*, **17**, 328–32.

POLEY, J.P. & WILKINSON, T.G. (1983). Environmental impact of oil-base mud cuttings discharges—a North Sea perspective. IADC/SPE Drilling Conference, New Orleans, Louisiana, USA. IADC/SPE 11400: 335–339.

SMITH, D.W. & JAMES, T.D.W. (1979). Biological changes in permafrost terrain adjacent to high Arctic oil and gas well sites. Department of Indian Affairs and Northern Development: Report No. 16, 150 pp.

SMITH, D.S. & JAMES, T.D.W. (1985). Ecological changes adjacent to sumps at exploratory well sites in the Mackenzie Delta and Northern Yukon. Department of Indian Affairs and Northern Development: Report No. 35, 53 pp.

YUNKER, M.B. (1986). Long term potential for leaching of oil contaminated cuttings from the use of oil-based muds: The Arctic marine environment. Final report for Indian and Northern Affairs Canada, September 1986.

8

Environmental Analysis of a Saline Drilling Waste Disposal Site

KATHERINE E. O'LEARY*

Shell Canada Ltd, PO Box 2506, Calgary, Alberta, Canada T2P 3S6

MARIAN W. KEMBLOWSKI and GEORGE M. DEELEY

Shell Development Co., PO Box 1380, Houston, Texas 77251–1380, USA

ABSTRACT

A KCl drilling fluid was used for a large-scale drilling project located in the Peace River region of Alberta. The drilling waste liquids were disposed downhole. Site-specific conditions were examined to select a disposal method for the drilling waste solids that would result in the least environmental impact. Land disposal was rejected since the volume and salinity of the waste solids would have necessitated that an extensive area of land be disturbed in order to avoid soil salinization. Dilution-burial was the disposal method selected; the solids were mixed with the clay till, buried in clay till and capped with clay till. Three years have elapsed since wastes were first brought to this site and 1·5 years since disposal operations were completed. Environmental impact has been minimal. There should continue to be a minimal environmental impact as a result of the low moisture content of the buried waste/till mixture and the hydrogeological characteristics of the site.

1. INTRODUCTION—DRILLING WASTE DISPOSAL

Environmental impacts associated with the disposal of drilling wastes depend on the characteristics of the wastes and the disposal environment, together with the disposal method. This paper examines the environmental

* Present address: Apt 7, 170 Weber St. N, Waterloo, Ontario, Canada N2J 3H17.

197

impact associated with the dilution-burial method used for the saline drilling waste solids generated from a large-scale drilling project in northwestern Alberta.

2. BACKGROUND

The Peace River Oil Sands deposit is located in northwestern Alberta (Fig. 1). The Peace River In-Situ Project (PRISP) began in 1979 and demonstrated the technical feasibility of in situ recovery of bitumen from the Peace River Oil Sands using a pressure cycle steam drive process. The technical success of PRISP led to the development of the commercial Peace River Expansion Project (PREP). A potassium chloride (KCl) based drilling fluid was used to directionally drill 212 wells during Phase I of PREP.

The drilling pads were built on organic terrain (muskeg) and therefore an off-lease site was required for the drilling wastes. Drilling waste management activities conducted prior to and during the PREP drilling operations have been described by Davies and Wensley (1986).

Two disposal methods for the $22\,000\,m^3$ of generated drilling waste solids were considered. Mixing of the solids with native clay till and burial of this mixture (dilution-burial) has been Shell Canada Limited's (SCL) standard approach in areas where subsurface conditions are suitable. However, surface disposal was favored by the land management agency (Lloyd, 1985). A survey of the literature and an evaluation of the site-specific conditions was conducted to determine which method would have the least environmental impact in the PREP area.

3. REGIONAL DESCRIPTION OF PREP AREA

3.1. Climate

The area is characterized by long cold winters and short cool summers. The mean temperature of the warmest month is below 22°C and fewer than 4 months have mean temperatures exceeding 10°C (Borneuf, 1981). Mean annual precipitation is 455 m. Precipitation is heaviest in summer and spread evenly throughout the remainder of the year. Snow accounts for about 25% of the total precipitation. Potential evapotranspiration is in the order of 400 mm (Borneuf, 1981). Evaporation generally exceeds precipitation during the growing season.

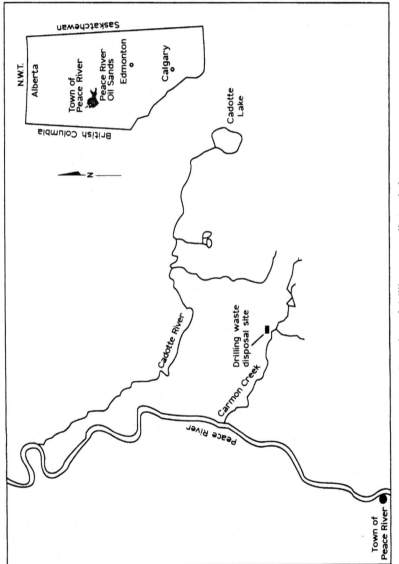

FIG. 1. Location of drilling waste disposal site.

3.2. Physiography

The overburden thickness in the development area is at least 100 m (Borneuf, 1981). The subsurface is a clayey till, underlain at 3–15 m by deposits of sand, gravel, cobbles and boulders which are, in turn, underlain by a much thicker, dense till. Topography is level to undulating.

The drainage network is mostly intermittent. A patchwork pattern of muskeg and better drained till deposits characterize the area. The terrain slopes to the northwest, towards the Peace and Cadotte Rivers. Carmon Creek flows west-northwest, to the Peace River. An un-named creek flows 15 km northward to join the Cadotte River, which in turn flows northwest to join the Peace River (Fig. 1).

Vegetation on well-drained sites ranges from deciduous (aspen and balsam poplar) through mixed-wood to coniferous (white spruce), Sphagnum mosses, black spruce, sedges and willows are common on sites with poor drainage and a willow-alder association occurs on wet sites over mineral soils.

4. DRILLING WASTE DISPOSAL

4.1. Literature Review

An American Petroleum Institute (API) field study concluded that saline drilling wastes were not amenable to surface disposal (Whitmore, 1981). Surface disposal and dilution-burial were used to close 81 saline drilling waste pits in Mississippi and Louisiana (Freeman & Deuel, 1986). Surface disposal was found to be generally unsuitable in upland areas due to the high potential for accidental damage to surface soils by the saline waste. Dilution-burial was preferred because it required the least land area, while eliminating adverse effects of salt on surface soils and plants. It was concluded that mixing of the wastes with native soil and burial of this mixture on-site was preferable to surface disposal in upland areas, provided that site characteristics minimize the risk of groundwater contamination.

The effects of drilling waste disposal pits on groundwater has been examined by laboratory, field and groundwater modelling studies. Laboratory studies indicate that the flux of aqueous leachate from a drilling waste pit should be small (Deeley, 1986). A nonsite-specific analysis that accounted for attenuating mechanisms, such as dispersion, predicted that chemical concentrations in leachate will decrease with increasing distance from the source (Kemblowski & Deeley, 1987). Groundwater modelling of specific sites predicted that concentrations of contaminants at receptor locations between 50 and 2000 m down gradient from the drilling waste pits would not exceed health-based water quality standards (ERT, 1987).

A field study found that contaminants that migrated from drilling waste pits resided in narrow shallow bands close to the point of origin (Henderson, 1982). Concentrations of contaminants in groundwater were reduced within several hundred feet of the source to levels which were acceptable with respect to human health and potability.

4.2. Implications of Drilling Waste Disposal at PREP

Site-specific characteristics in the PREP area were examined to assess the implications associated with surface disposal and dilution-burial of the KCl drilling waste solids.

4.2.1. Surface Disposal

The volume and salinity of the drilling waste, existing government guidelines, and site characteristics made surface disposal a difficult proposition at PREP. The drilling waste solids contained approximately 200 000 kg of chloride; therefore, adherence to the Alberta Energy Resources Conservation Board's 450 kg/ha chloride loading rate for surface application of drilling waste would have required approximately 444 ha of land. This degree of surface disturbance was contrary to SCL's objective of minimizing environmental impact.

The logistics of surface disposal were complicated by the fact that the closest suitable land area of this size was a partially cleared provincial grazing reserve 10 km away. The expense for transportation of the wastes to this site, site preparation, waste application and incorporation, and site restoration, was expected to be prohibitive.

4.2.2. Dilution-Burial

The primary concern with burial of the PREP saline drilling waste solids was the possibility of groundwater contamination. A hydrogeologic investigation of the proposed waste containment/disposal site was conducted to assess this concern.

5. HYDROGEOLOGICAL INVESTIGATION OF PROPOSED WASTE CONTAINMENT SITE

A reconnaissance of the area near the drilling operations identified a site (Fig. 1) that appeared to be suitable for containment and burial of drilling wastes. Hydrogeological conditions of this particular site were investigated.

Soil stratigraphy to a depth of 16 m was determined by drilling boreholes around and within the 8 ha site. The site is underlain by up to 5 m of weathered clay till, over a relatively homogeneous, very stiff, unweathered clay till.

Piezometers were installed (Fig. 2). Piezometer depths vary between 3 and 16 m below ground level. Hydraulic conductivity values were estimated by the time required for recovery of the water level in the piezometers after bailing, and by the falling head test in piezometers that were dry. Hydraulic conductivity values obtained from these tests were low, ranging between 1.6×10^{-7} and 2.5×10^{-11} m/s (Table 1). The higher values in this range reflect the presence of fractures. A shallow well pump-in technique outlined by Van Schilfgaarde (1978) was also conducted. This involved filling a test hole with KCl solution and maintaining the fluid depth until the flow into the soil became constant. The hydraulic conductivity value obtained from this technique was 1.03×10^{-8} m/s which is within the range of values obtained from the tests conducted in the piezometers.

The natural groundwater quality at this site is brackish, with calcium, magnesium and sulfate ions dominating. Concentrations of sulfate and total filterable residue (TFR) are in the vicinity of 2000 mg/litre and 3000 mg/litre, respectively. The water is therefore not considered suitable for domestic purposes because the maximum acceptable concentration for domestic consumption for both sulfate and TFR is 500 mg/litre (Health and Welfare Canada, 1978). Concentrations of chloride are below 10 mg/litre. Areal and vertical variations in TFR indicate a lack of hydraulic continuity within the till.

Based on the hydrogeological investigation, it was concluded that burial of the KCl drilling waste solids at this site would have a minimal impact on the groundwater.

6. CONTAINMENT AND DISPOSAL OPERATIONS

Drilling wastes were transported daily to the site beginning in January, 1985. Liquid wastes were contained in pits which were periodically treated to settle suspended solids. Treated liquids were either recycled into the active mud system or pumped into deep disposal wells. The solids were stored in shallow ramped pits which allowed drainage of entrained liquids (Davies & Wensley, 1986).

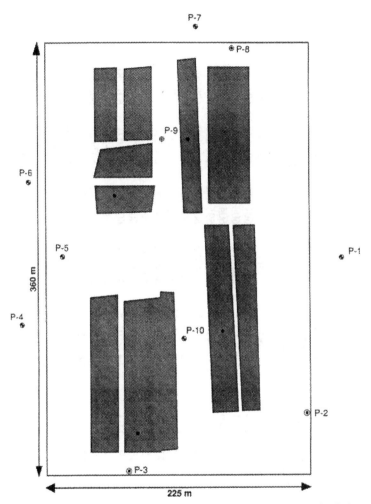

FIG. 2. Layout of drilling waste disposal site. Shaded area denotes buried waste (till mixture); ⊙, single piezometer; ⊗, double piezometer; ⊕, triple piezometer; ●, sample locations.

TABLE 1
Ground water monitoring

Piezometer no.	Sample interval (metres below ground level)	Hydraulic conductivity[a] (m/s)	Groundwater elevation (m.a.s.l.)			
			87-06-17	86-10-10	87-06-22	87-10-12
P 1A	5 – 8·5	$1·4 \times 10^{-10}$	604·930	604·767	Dry	Dry
P 1B	13 –16·1	NT	Dry	Dry	598·289	597·813
P 2	11·6–13·3	$8·8 \times 10^{-8}$	601·197	601·511	601·643	601·541
P 3	9·5–11·4	$5·2 \times 10^{-8}$	608·791	608·849	610·342	607·708
P 4A	7 – 8·6	NT	Dry	604·411	607·134	605·718
P 4B	13 –16·1	NT	Dry	595·942	600·234	597·657
P 5A	1·8– 3·1	$5·2 \times 10^{-8}$	610·582	609·720	610·140	609·803
P 5B	10 –11·5	$6·5 \times 10^{-8}$	600·118	601·003	602·463	610·119
P 6A	7 – 8·5	$4·2 \times 10^{-11}$	602·319	Dry	Dry	Dry
P 6B	13 –16·1	$2·6 \times 10^{-8}$	600·672	600·356	600·630	600·539
P 7A	5 – 8·7	$2·5 \times 10^{-11}$	602·632	602·542	602·172	602·180
P 7B	13 –16·2	$8·9 \times 10^{-8}$	600·870	600·582	600·730	600·617
P 8	9 –11·3	$7·4 \times 10^{-8}$	599·641	601·078	603·071	601·281
P 9A	1·5– 3·3	NT	611·302	610·514	610·240	610·155
P 9B	9·3–11·7	$6·2 \times 10^{-9}$	608·610	608·793	607·681	608·100
P 9C	13·8–15·6	$1·6 \times 10^{-7}$	601·078	600·794	600·788	600·661
P 10A	1·7– 3·1	$6·5 \times 10^{-8}$	611·420	610·870	611·026	610·684
P 10B	9·7–11·5	$2·2 \times 10^{-8}$	611·709	602·709	606·709	604·172

[a] Hydraulic conductivity values were calculated based on recovery of the water level with time after bailing, except for P1A, P6A and P7A, which were based on falling head tests. The geometric average of these values is 10^{-8} m/s.
NT = not tested.

Once the solids had been drained of free water, they were spread out on the surface, allowed to dry, and mixed with the clay till that had been excavated from the pits. Altogether $22\,000\,m^3$ of drilling waste solids were mixed with $42\,000\,m^3$ of clay till. After removal of the liquid drilling wastes, the waste/till mixture was pushed into the pits and compacted into $3\,m$ thick layers. A $2\,m$ thick clay cap was placed over the top to reduce infiltration. A total of $64\,000\,m^3$ waste/till mixture was disposed of at this site in the locations shown in Fig. 2.

The entire site was graded to enhance surface runoff and then revegetated with a grass legume mixture to control erosion. Restoration was completed in September, 1986. Figure 3 shows the site shortly after restoration. The cost to dispose of the solids and restore the site was \$443\,000 or \$20\,m^3$ of solids.

Auger samples at depth intervals of approximately $0.5\,m$ were taken through the waste/till mixture profile at four locations at the disposal site (Fig. 2). These samples were analyzed for moisture content, percentage saturation, chloride, electrical conductivity (EC) and pH (Table 2).

FIG. 3. Restored drilling waste disposal site.

TABLE 2
Chemical characteristics of drilling waste solids/clay till mixture[a]

	Moisture content (%)	Saturation (%)	Chloride (mg/litre)	EC (mmhos/cm)	pH
			(saturated paste extract)		
Mean	11·2	59·0	2 650	10·5	8·2
Standard deviation	4·3	4·7	1 950	5·3	1·0

[a] Auger samples at depth intervals of approximately 0·5 m were taken throughout the waste/till mixure profile at four locations at the disposal site (Fig. 2). The composition of the waste/till mixture as a whole is represented by the mean and standard deviation of the analytical results from these samples.

7. GROUNDWATER MONITORING

Three years have elapsed since drilling wastes were first brought to this site, and 1·5 years since disposal and restoration operations were completed.

Biannual sampling and analysis of the groundwater for potability and total organic carbon (TOC) has been conducted to detect changes in groundwater composition related to the drilling waste. The concentrations of chloride, sulfate, total filterable residue (TFR) and TOC found in those piezometers which have consistently yielded water are shown in Figs 4 and 5.

The concentrations shown in Fig. 4 are from piezometers located around the perimeter of the drilling waste disposal site. The water yielded from these piezometers continues to be characteristic of the background groundwater quality. Concentrations of chloride continue to be low, while concentrations of sulfate and TFR continue to exceed the maximum acceptable concentrations for drinking water. This consistency with background water quality demonstrates that contaminants have not migrated to the boundaries of the site.

Figure 5 shows analytical results from the two piezometer nests (P9 and P10) located within the drilling waste disposal site. The drilling waste does not appear to be a source of either sulfate or organic contaminants since sulfate and TOC concentrations in samples from these piezometers do not exceed background levels. Chloride and TFR concentrations exceed background levels, particularly in Piezometers 9A and 9B.

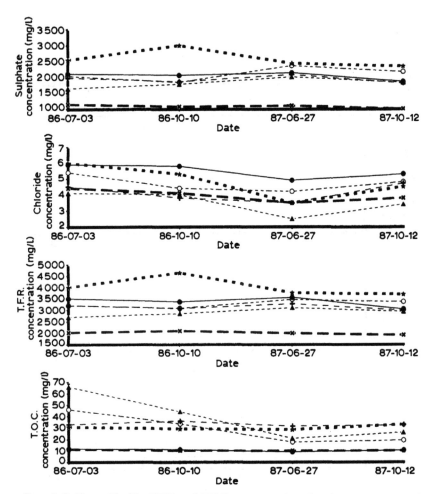

FIG. 4. Sulfate, chloride, TRF and TOC concentrations in piezometers around perimeter of drilling waste disposal site (86-07-03 to 87-10-12). —●—, P2; -+-, P3; —★—, P5A; —O—, P5B; — ×—, P6B; —▲—, P8.

The pits where the drilling waste solids/till mixture was buried were initially used for storage of the liquid drilling wastes. It is therefore likely that the elevated chloride and TFR concentrations in piezometers P9 and P10 are attributable to the time period when the liquids were contained

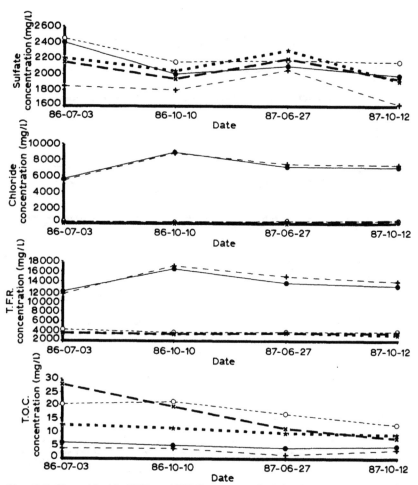

FIG. 5. Sulfate, chloride, TFR and TOC concentrations in piezometers within the drilling waste disposal site (86-07-03 to 87-10-12). —●—, P9A; – + –, P9B; —★—, P9C; —O—, P10A; — x —, P10B.

in the pits. There should be less generation and release of leachate in the post-closure period since the hydraulic head exerted by the liquids have been removed, the buried waste/till mixture is relatively dry, and infiltration is limited by the clay cap and by the contour of the site.

Continued groundwater monitoring will detect any movement of soluble constituents.

8. FATE OF LEACHATE FROM DRILLING WASTE PIT

Although it has been established that background quality limits ground-water usage, and existing site conditions effectively attenuate the transport of soluble constituents from waste to groundwater, the ultimate fate of chloride in the waste drilling pit was examined based on available information and reasonable assumptions for the location. It should be noted that the assumptions are, in all cases, very conservative in overestimating ultimate aqueous phase concentrations of chloride.

Hydrogeological studies have indicated that a shallow sand and gravel aquifer which underlies the disposal pit area discharges into the Cadotte River (Fig. 6) (Borneuf, 1981). In this exercise, it was therefore assumed that all pit leachate eventually discharges into the Cadotte River. A determination of leachate flow rate and Cadotte River flow rate provided a leachate dilution factor which was applied to the maximum leachate chloride concentration.

Groundwater table elevations of shallow wells at the site indicated some mounding (Table 1). Data on groundwater levels in the multi-depth wells, P9A (610·15 m), P9C (600·66 m), P10A (610·68 m) and P10B (604·17 m),

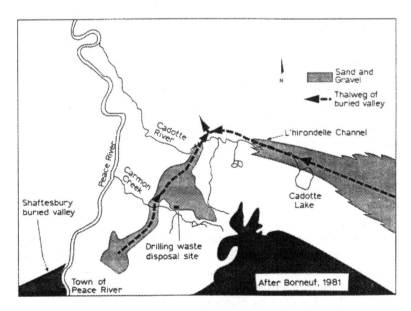

FIG. 6. Shallow aquifer flow toward Cadotte River.

provided an estimate of the vertical gradient. The maximum vertical gradient (i) (defined as the difference in potentiometric heads between two piezometers/the vertical distance between their screens), was estimated to be 0·770 for P9 and 0·775 for P10.

The geometric average of the hydraulic conductivity (K) was approximately 10^{-8} m/s (Table 1). Therefore, the maximum leachate flow rate (q_L) would be:

$$q_L = Ki = 10^{-8} \text{ m/s} \times 0.0772$$
$$= 7.72 \times 10^{-9} \text{ m/s}$$

The maximum total leachate flow rate (Q_L) from the 360 × 225 m pit area (A) would be:

$$Q_L = 7.72 \times 10^{-9} \text{ m/s} \times 360 \text{ m} \times 225 \text{ m}$$
$$= 6.25 \times 10^{-4} \text{ m}^3/\text{s}$$

If this leachate were diluted in the Cadotte River which has a conservatively estimated flow rate (Q_R) of 3·7 m³/s (based on flow rate typical for rivers in the area that are tributary to the Peace River), the dilution factor (F_D) would be:

$$F_D = Q_R/Q_L = (3.7 \text{ m}^3/\text{s})/(6.25 \times 10^{-4} \text{ m}^3/\text{s}) = 5920$$

Thus, the expected river concentration (C_R) of chloride from leachate based on a maximum leachate chloride concentration (C_L) of 8940 mg/litre (Fig. 5) should not exceed:

$$C_R = C_L/F_D = (8940 \text{ mg/litre})/5920 = 1.5 \text{ mg/litre}$$

This indicates the minimal effect that any leachate would have on the most likely receptor. Any recharge into the Peace River would, of course, result in even greater dilution.

9. SUMMARY

- 22 000 m³ of KCl drilling waste solids were mixed with 42 000 m³ of clay till. This mixture was buried in and then capped with clay till.
- Groundwater quality at the disposal site is naturally brackish and not suitable for domestic consumption.
- Soluble constituents from the drilling waste/clay till mixture have been contained within the 8 ha disposal site.
- There has been minimal environmental impact associated with this

disposal. There should continue to be a minimal environmental impact as a result of the low moisture content of the buried mixture and the hydrogeological characteristics of the site.

REFERENCES

BORNEUF, D. (1981). Hydrogeology of the Peace River Area, Alberta. Alberta Research Council, Report No. 81-2.

DAVIES, G.E. & WENSLEY, J.H. (1986). Waste management for a large scale drilling project. 37th Annual Technical Meeting of the Petroleum Society of CIM held in Calgary, June 8–11, 1986.

DEELEY, G.M. & WHITAKER, G.A. (1986). Hydraulic conductivity and leachability of waste drilling fluids. Presented at the Ninth Annual Energy-Sources Technology Conference and Exhibition, New Orleans, Louisiana, February 23–27, 1986, ed. J.B. Dicks, University of Tennessee Space Institute.

ERT (1987). Oil and gas industry exploration and production wastes. Prepared for API, July 1, 1987.

FREEMAN, B.D. & DEUEL, L.E. (1984). Guidelines for closing drilling waste fluids in wetland and upland areas. In *Proceedings of the Seventh Annual Energy-Sources Technology Conference and Exhibition*. American Society of Mechanical Engineers, New Orleans, Louisiana, February 12–16, 1984, pp. 47–110.

FREEMAN, B.D. & DEUEL, L.E. (1986). Closure of freshwater base drilling mud pits in wetland and upland areas. In *Proceedings of a National Conference on Drilling Muds*, May 29–30, 1986, ed. R.E. Kamat, Environmental and Ground Water Institute, Norman, Oklahoma, pp. 303–27.

HEALTH AND WELFARE CANADA (1978). *Guidelines for Canadian Drinking Water Quality.*

HENDERSON, G. (1982). Analysis of hydrologic and environmental effects of drilling mud pits and produced water impoundments, Vol. 1, Executive Summary and Report. Dames and Moore, Houston, Texas, October, 1982.

KEMBLOWSKI, M.W. & DEELEY, G.M. (1987). A ground water screening model for drilling fluid disposal pits. In *Proceedings of a National Conference on Drilling Muds*, May 28–29, 1987, ed. R.E. Kamat & L.W. Canter. Environmental and Ground Water Research Institute, Norman, Oklahoma, pp. 205–34.

LLOYD, D.A. (1985). Drilling waste disposal in Alberta. In *Proceedings of a National Conference on Disposal of Drilling Wastes*, held May 30–31, 1985, ed. D.M. Fairchild. Environmental and Ground Water Institute, Norman, Oklahoma, pp. 114–31.

VAN SCHILFGAARDE, J. (1978). Drainage for Agriculture Agronomy Series No. 17, American Society of Agronomy Inc.

WHITMORE, J.C. (1981). Water base drilling mud land spreading and use as a site reclamation medium. A study for the American Petroleum Institute, Forsgren-Perkins Engineering, PA.

9

Drilling Wastes Management and Closed-Loop Systems*

MANOHAR LAL and NEAL THURBER

Amoco Production Company, Research Center, 4502 East 41st Street, PO Box 3385, Tulsa, Oklahoma 74102, USA

ABSTRACT

This paper discusses the basic requirements for environmentally safe and cost-effective management of the drilling wastes problem. The key elements of an optimum drilling wastes management system are then developed from these basic requirements. These elements essentially consist of optimizing the drilling operation to reduce the quantity of drilling wastes, keeping the wastes nonhazardous, and then selecting an optimal wastes disposal method which is both economical and environmentally acceptable. Obviously, minimization and control of the wastes during the operation reduce the magnitude and complexity of the problem and make the task of wastes disposal easier and more economical. Substantial reductions in mud and disposal costs, trouble costs, and improvement in drilling rates are obtained by implementing optimum waste management practices. The paper discusses the development of the closed-loop systems approach and the steps required for successful implementation. These steps include the selection of waste disposal methods based on prevailing regulations in the area, identification and isolation of all potential waste sources, selection of drilling fluid and additives, and selection of proper closed-loop mud processing or solids-control system. A few simple results from an economic analysis model for solids control and closed-loop mud processing systems are briefly discussed. Finally, case histories are presented where in one case disposal and reclamation costs were decreased by 50% and in another drilling days were reduced by over 20% through better waste management. This confirms the importance of

* Editors' Note: Some supporting data are still proprietary to Amoco Production Company. Publication is planned in the near future. Interim availability requests should be addressed directly to authors.

various elements required for optimum waste management and benefits gained from implementation of the proposed closed-loop systems approach.

1. INTRODUCTION

Current regulatory trends indicate that different local, state and federal regulatory agencies are imposing different and more stringent regulations concerning the disposal of drilling wastes. For example, the Environmental Protection Agency (EPA) in the United States is re-evaluating the current exempt status of the exploration and production (E&P) wastes from hazardous waste laws. In Canada, the Energy Resources Conservation Board is re-examining the present guidelines for sump disposal and new guidelines may be forthcoming. In the United States, the recent EPA's interim report on this subject (US EPA, 1987) had generated a great deal of controversy and drawn a strong response from the American Petroleum Institute (API, 1987) and other interested parties. The US Environmental Protection Agency's addendum to its earlier draft report indicates a change in EPA's preliminary conclusions regarding wastes from oil and gas operations and shows improved understanding of the problem (EPA, 1987).

For a better appreciation and understanding of the problem, it is necessary to recall certain basic facts. The protection of the environment is important for everyone and not just for certain groups or government agencies. Next, we must realize that if we wish to recover and utilize the natural resources, the very actions required to do so are bound to affect the environment to some extent and generate some kind of 'wastes'. Finally, it is the responsibility of the industry exploiting the natural resources to ensure that the impact on environment is minimal and the wastes are properly recycled and/or disposed.

In the context of a drilling operation, it simply means that in order to produce the hydrocarbons, a hole has to be drilled in the ground. This operation temporarily affects the immediate environment and generates some drilling wastes. A responsible drilling operator/contractor must manage the drilling wastes such that the environmental impact is insignificant and the wastes are disposed of properly even in areas where no regulations are in effect. The fact that so many oil and gas wells have been drilled in the past 80 years with rather limited effect upon the environment indicates that this activity is not hazardous for the environment. At the same time, it is equally important for society, which needs this critical energy resource, to realize that the zero-risk option does not exist. It must therefore ensure that its government agencies do not impose unduly harsh, unworkable and expensive regulations. Such regulations usually do not add much to the protection of environment, but only make it extremely difficult for the oil industry to operate.

In other words, the oil industry and various local, state and federal regulatory agencies should not have an adversarial relationship. They need to work for the same goals, namely, availability of this critical energy resource at a reasonable price without significantly impacting the environment. It is in their common interest to replace the spirit of confrontation with cooperation and work together on necessary regulations. They must try to understand the magnitude and complexity of the technical problems and find mutually acceptable technical solutions to the problems standing in the way of achieving the desired goals. This paper discusses some of these technical problems and the proposed solutions.

In the next section, we introduce the basic requirements for environmentally safe and cost effective management of the drilling wastes problem. The key elements of an optimum drilling wastes management system are then developed from these requirements. This leads to the discussion of the closed-loop systems approach and the steps that are required for successful implementation of this approach. Finally, two case histories are briefly described to confirm that the proposed approach does indeed achieve the desired objectives. In the first case, disposal and reclamation costs decreased by 50% and, in the second case, the drilling days were reduced by more than 20% through better waste management practices.

2. BASIC REQUIREMENTS AND KEY ELEMENTS OF DRILLING WASTE MANAGEMENT

We should attempt to run our operations in closed loops and generate products/byproducts which are useful or easily disposable/recycled without adversely affecting the environment. Ideally, waste generated in any operation should be processed and recycled back into the environment in such a manner that it loses its identity and does not in any way affect the environment. In practice, waste must be processed and/or disposed of properly. In no case should it be simply thrown away or abandoned without first insuring that it poses no significant health and environmental risks.

Waste management must start from the beginning of an operation since control over the quantity and quality of waste throughout the operation simplifies the subsequent task of wastes disposal. Obviously, a good waste management system must meet the following basic requirements:

(1) Minimize the quantity of wastes during the operation.
(2) Control the quality to keep the wastes nonhazardous.
(3) Process/dispose of wastes in an environmentally-safe and cost-effective manner satisfying the regulations.

In drilling, the key elements of a good wastes management system thus consist of optimizing the drilling operation to reduce the quantity of drilling and associated wastes, keeping the wastes nonhazardous, and then selecting an optimal wastes disposal method which is both economical and environmentally acceptable. The three actions required for successful waste management, namely, minimizing, controlling and processing of drilling wastes need further discussion for better understanding of the wastes management systems.

As stated earlier, minimization and control of the wastes during the operation reduce the magnitude and complexity of the problem and make the task of wastes processing easier. It must be noted in this regard that the minimization of wastes includes not only the drilling fluids and well-bore cuttings, but also the associated wastes and water contacting the drilling operation from nature and other well site activities. In fact, any wastes generated or contacting the drilling operation become the operation's responsibility. The first logical step for successful waste management is therefore the identification and isolation of the potential wastes sources. The drilling wastes mainly consist of drill cuttings/solids, drilling fluids, and small amounts of associated wastes. The associated wastes include miscellaneous fluids such as completion, treatment, stimulation, packing and formation, and other miscellaneous wastes such as lubricants, solvents, paints, cement returns, sanitary, etc.

Various factors affecting the quantity and quality of wastes are geology, well geometry, drilling fluid and additives, solids removal equipment, water usage, and the problems encountered during the drilling operation. In general, these factors are interrelated. For example, the choice of drilling fluids and additives is related to the formations; solids removal equipment is dictated by both the formations or cuttings and the drilling fluid; the dilution and disposal volumes and problems such as differentially stuck pipe are directly related to the choice of solids-control equipment. The proper selection of controllable factors like the drilling fluids and its additives, closed-loop mud processing or solids control, and better water management practices can substantially reduce the quantity and control the quality of the drilling wastes. We should also isolate and not combine the wastes if it adversely affects the quality of the combined waste. For example, mixing brines with freshwater fluids in the same reserve pit affects the quality of the combined wastes and makes disposal more difficult.

The most important but often neglected element that has an extremely significant effect on the quantity of drilling wastes generated during the operation is the mud processing or solids control/removal/separation equipment. The reason for the neglect is often based on the upfront extra rental costs, etc., required for better and additional solids control equipment. However, not well understood or appreciated are the substantial reductions in the dilution and subsequent waste disposal volumes/costs as a direct result of more effective solids control. Recent studies with an economic analysis model (Lal, 1988), show that, in most cases, the savings in dilution and disposal costs alone more than offset the increased costs associated with additional solids-control equipment. The extra benefits from increased penetration rates, reduced trouble costs and minimal environmental impact also justify using proper closed-loop mud processing systems. Recent field studies, reported later in this paper, confirm the view that closed-loop solids-control systems are not only environmentally desirable but also cost-effective in most cases.

A closed-loop solids-control system is defined as one which processes the mud and separates such a percentage of dry (liquid 50% by volume or less) drill solids that there is no excess mud left after dilution which would have to go to the reserve pit. The economic model uses this definition and basic mass balance equations, etc. (Lal, 1988). It computes the dilution and disposal volumes and the corresponding gross and net savings for the closed-loop condition and also for different drill solids percentage removals. It is interesting to note in this regard that the closed-loop processing condition can be achieved with currently available equipment and technology.

It is also worthwhile to emphasize the importance of better water management practices since their impact is usually not well understood. Water that is used on site or contacts the operation from precipitation or run-off usually ends up in the reserve pits and adds to wastes. In order to appreciate the magnitude of the problem, average discharge to the reserve pit from various sources of water is listed in Table 1. The data was collected from twelve drilling operations in the Utah-Wyoming

TABLE 1
Common sources of water discharge to reserve pits

Source	Average discharge (bbls/day)
10″ Rainfall in 200′ × 200′ area	6 000 bbls
Location run-off/near surface aquifers	0–4 000
Water hoses	0–250
Jetting mud pits	0–300
Pump rod lubrication	0–200
Desanders or desilters	150–700
Shakers overs	0–50
Centrifuge solids slide	20–100
Water lubrication of centrifugal pumps	10–50

overthrust belt and from a rig in the Gulf Coast area.

Obviously, water from various sources that ends up in the reserve pit throughout a drilling operation can add up to a significant amount. A little planning to develop methods for diverting or handling run-off or precipitation, and recycling/reusage of water on well site can pay rich dividends. It can substantially reduce the quantity of wastes and the volume of water used in a drilling operation. This reduction accordingly decreases the water costs, waste disposal costs, and location preparation/reclamation costs as the excess water does not make a one-way trip to the reserve pit.

Having discussed the first two elements, let us now discuss the third and final element of a wastes management system. Several methods are available for processing and/or disposal of the drilling wastes. These methods include dewatering and backfilling, surface application/landfarming, downhole/annular disposal, fixation and/or solidification, centralized disposal sites and treatment facilities, incineration and microbiological treatments, etc. Most of these methods have already been described in detail in technical literature (Proc., 1985–86; Conner, 1986; Hanson & Jones, 1986; Hinds et al., 1986). Although factors such as geological and geographical settings influence the choice of a disposal method, prevailing regulations in the area usually dictate the final choice. Unfortunately, it is becoming increasingly difficult to predict and comply with changing and more restrictive legislation concerning the wastes disposal. Evidently, the problem of selecting an optimum disposal method needs to be discussed in the context of prevailing regulations.

The closed-loop systems approach requires that the drilling wastes generated by a drilling operation be disposed of at the drilling site, if possible, and not taken out of the loop to a different location. This approach is also desirable for the following reasons. The potential liability from use of third party disposal pits, where we have no control over the types of wastes being mixed from other sources, is great. Even if a company decides to operate its own central disposal site, one operation, which may discard contaminated fluids/solids, could contaminate the entire disposal pit and increase liability and disposal costs for the total volume of wastes. Furthermore, transportation of drilling wastes can subject us to additional environmental liability in the event of mishandling by the carrier or a transportation accident. Finally, on-site drill cuttings/fluids disposal is typically the most economical method of reclamation.

When reserve pits cannot be used, regulations governing the drilling operation must be thoroughly understood. Often there may be flexibility in the regulation which allows location discharge of treated drill solids and/or fluids meeting established discharge criteria. The fluid and solids wastes treatment and discharge or annular disposal of fluids, if allowed, can eliminate off-location wastes disposal. This may also be cost-effective in comparison to transporting and disposing the wastes at a centralized disposal facility.

If regulations in an area completely ban the on-site reserve pit and/or disposal of drilling wastes even after proper treatment, the solids and fluids have to be transported and disposed off-location. In such a case, as stated earlier, preference must be given to the construction of an approved off-location reserve pit in view of the potential liability from use of third-party disposal facilities. Also, on-location storage and handling of solids and fluids would require sufficient tankage to collect, hold and ready the drill solids and/or fluids for transportation. It was stated earlier that closed-loop mud processing or a solids-control system is an essential element of an optimum waste management system, regardless of the wastes disposal method. Assuming this element is in place for a strict 'zero-discharge' regulation, tankage would be required primarily to contain relatively dry drill solids. Solids discharge into tankage that is suitable for transportation is ideal from the viewpoint of efficiency and cost. Such an arrangement requires separated drill solids discharging directly or via a conveyor belt into a dump truck or bin which may be hauled-off location. Excess fluid from this operation should be collected and drained to a sump for reuse in the mud system. If it becomes necessary to dispose mud at the end of a hole section, or due to weighting-up, or any excess mud, it may be pumped into frac tanks on location. Then it can be either annularly disposed, if permitted, or loaded into a vacuum tank for off-location disposal.

Regarding the on-site processing and/or disposal methods, the following discussion gives the order of preference and a brief description of each method itself. In practice, even when all the elements required for optimal waste management are implemented, we are left with some drilling wastes that need to be processed and/or disposed. This waste would normally consist of separated dry drill cuttings/solids, a relatively small amount of drilling fluid, and very little miscellaneous wastes. When managed properly, drill cuttings and water-based drilling fluid should hardly need any further treatment to meet established criteria for on-site disposal. Even for oil-based drilling fluids, usually, the fluid can be reused at a different location and, depending on regulations, the oil-wet cuttings can be disposed using one of the methods discussed later.

Evaporation of water-based fluid wastes is usually the least expensive land method, and should be investigated first. This evaporation may occur naturally with time and an appropriate drying environment, or it may be hastened by heating and/or spraying the fluid into the atmosphere. In an ideal climate and with high-pressure sprayers, it is possible to spray up to 900 bbls per day according to field data collected in Wyoming by Thurber (1988). The regulatory concerns regarding evaporation are that the fluid be evaporated within a time limit; it should be confined within the reserve pit boundaries; and we should ensure that the fluid is indeed evaporating and not seeping into the surrounding ground or groundwater. In areas of high humidity and/or precipitation, or where reserve pit stability is questionable, or where regulations do not allow, evaporation is obviously not an acceptable choice.

The next method in order of preference is usually surface application of reserve pit fluids, which refers to discharging water-based fluids to the surface environment, either land or water. Surface application of fluids may involve direct pumping or spraying of fluid into the environment, or treating it before it is applied to the surface. Expenses for the surface application of fluids are discharge and/or fluid treatment costs. If fluids are kept clean and contaminant-free, as should be the case in a closed-loop operation, the only significant cost of fluid discharge is pumping. However, if the fluid becomes increasingly contaminated, treatment costs may escalate so that it is not economical to treat the fluids to surface discharge standards. Treatment of fluids to meet surface discharge standards can include pH adjustment, solids flocculation with polymers, activated carbon treatment, and use of a high-pressure reverse osmosis to meet discharge water standards.

Most regulatory agencies have established standards for fluid discharge, which should be known and applied to the drilling operation. For example, regulatory agencies usually need to know the chemistry of the fluid to be discharged. If on land, the agency may require soil samples before and after fluid discharge. If a regulatory agency is developing or modifying water quality standards for surface application, key concepts that need to be emphasized are the one-time application of treated fluids and the area of fluid application. Many water quality standards are based on continual discharge to the environment, unlike drilling where it is a one-time application. Surface application of fluids on a one-time basis significantly reduces the impact on the environment. Regarding the area of application, the greater the surface area over which the fluid is applied, the lesser the fluid's impact because of lower concentration. Regulatory agencies use different tests, such as those developed for agricultural irrigation or bioassay testing, to determine the discharge allowances.

Another unique feature of drilling that the regulatory agencies need to be made aware of is the permit timing necessary for surface discharge of fluids. Since a drilling operation goes on 24 h a day, it is important that agencies understand and address the need for a timely discharge process. Several states have already provided this necessary procedure. Finally, proper planning and implementation of the timely discharge process can substantially reduce disposal costs. For instance, if the surface hole fluids are fresh water and relatively clean, these fluids should be treated and discharged before they become contaminated by another drilling mud (e.g. salt-water based) from another section of the hole. This would also reduce the size of reserve pit required because total pit fluid volume does not have to be stored until completion of the drilling operation.

A downhole/annular disposal method is to be considered next, in order of preference, when evaporation and/or surface application methods cannot be used for fluid-waste disposal. Downhole disposal of pit fluids refers to pumping fluids downhole, either between the annulus of casing strings or through perforations, or into formations below a competent casing shoe. The regulations for downhole fluid disposal demand that the disposal fluids not contaminate fresh water aquifers. While fluid is commonly disposed downhole in approved Underground Injection Control (UIC) wells, permitting is time consuming. Typically, regulations for the downhole disposal of drilling operation fluids involve the following:

(1) Advance notification to the regulatory agency and timely approval process.
(2) Disposal of fluid only from a given drilling operation down that operation's wellbore. (Multiple drilling operations' fluids disposal down one wellbore regulated in US by EPA and the approval process is time-consuming.)
(3) Availability of competent downhole formation that will receive and contain the disposed fluid.
(4) Verification from logs, etc., that the fluid was properly placed.

While the downhole disposal of drilling fluids can be economical and dependable, the pumping operation must be properly engineered to prevent casing damage. The wellbore design, including casing/cementing designs, should be carefully reviewed. Maximum pumping pressures and formation fracture pressures should be known and not exceeded. If disposal operation occurs while a well is producing, production fluid temperatures should be monitored.

Regarding the disposal of solid wastes, as stated in the beginning of this discussion, closed-loop operation should result in dry drill solids which are usually harmless and can be spread on site without any further treatment. In general, depending on the condition and quality of solid waste or prevailing regulations, it should be possible to use one or more of the following methods to treat and/or dispose of the solids waste. These methods are soil farming of drill cuttings (Whitfill & Boyd, 1987), fixation and/or solidification of wastes (Conner, 1986), or properly burying the resulting wastes on location.

3. IMPLEMENTATION OF THE CLOSED-LOOP SYSTEMS APPROACH

Based on the preceding discussion, the following four main steps are required to achieve the goals of optimal wastes management through a closed-loop system. These steps are listed and discussed in the order in which they would have to be implemented in an actual drilling operation.

(1) Study the prevailing regulations in the area and select the most cost-effective method for fluid and drill solids disposal which is environmentally safe, processes and/or disposes of the wastes on-location, if possible, and satisfies the regulations.

(2) Identify and isolate all potential waste sources; in particular, divert, minimize, and reuse the water contacting the drilling operation from nature and other well-site activities.
(3) Select the drilling fluid and additives which minimize drill solids degradation and pose no risk to health or environment.
(4) Select, install and operate, if necessary, additional solids-control/removal equipment to implement a closed-loop mud processing system.

The first step in obtaining legal fluid and cuttings disposal methods can be, and usually is, time consuming. It involves interaction with regulatory agencies. It is, however, very important that regulations are available which clearly describe how fluid and drill solids may be disposed. The environmental coordinators from the oil companies should work with regulatory personnel and obtain methods for the safe and timely disposal of drill site fluids and solids. During the drilling operation, if drilling wastes are disposed of in accordance with regulations and previously agreed upon and approved methods, future environmental liability is minimized. When timely procedures are not available, recourse for disposal often results in expensive remedies. These remedies could be excavation of another pit or expensive intermediate trucking or barging of fluids/solids to another pit.

Another benefit of having disposal regulations outlined is that they provide a basis for designing the closed-loop system. For example, a zero-discharge regulation would necessitate use of the best solids-control equipment, including possibly a Mud Processor (flocculation/centrifuge process), and require an elaborate fluid/drill solids wastes disposal procedure. On the other hand, fluid discharge regulations may allow some latitude with equipment/operational design because the fluid can be properly treated on-location and discharged. In short, the design of a closed-loop system and, in particular, the selection of an appropriate disposal method are dictated by the regulations.

The second step concerning identification and isolation of waste sources and better water management practices has already been discussed in detail in the previous section. Isolation of various drilling wastes instead of dumping them in the same pit, and design measures, such as uphill trenches or dikes to keep run-off away from the reserve pit, indeed have significant impact on the quality and quantity of the wastes. The third step, the selection of drilling fluid and additives, impacts two areas of closed-loop systems. First, a properly chosen mud system can minimize drill solids degradation which leads to more efficient removal by the mechanical solids control equipment. Second, carefully chosen mud additives minimize potential environmental liability which generally means minimal fluid disposal concerns and costs.

The importance of the mud selection becomes apparent when reviewing past attempts at operating closed-loop systems. The less-than-optimum performance of many attempted systems, combined mud systems and solids-control equipment that were not effective in controlling fine drill solids build-up. Freshwater muds were often used to drill formations with high concentrations of reactive clays. As these clays rapidly degraded into ultra-fine solids, the solids-removal equipment became increasingly less efficient. Thus, attempts to operate closed-loop systems resulted in an undesirable increase in mud weight and viscosity.

There are essentially two methods of improving closed-loop system performance when reactive shale formations are drilled. One method is to use a Mud Processor, which combines flocculation and centrifuging processes, strips all solids from freshwater mud, and returns clear water for the dilution of the active mud system. The second method is to use an inhibitive polymer KCl or a mineral oil-based mud. Various other factors such as cost versus potential benefits like improved hole and temperature stability, etc., also influence the choice of a mud system. Depending on these factors, for example, it may be cost-effective and environmentally desirable to use water-based mud with a Mud Processor for the surface hole and mineral oil-based mud for the intermediate and production hole sections.

Regarding the mud additives, it is important to investigate and use substitute additives that have reduced potential for adverse environmental impact. Replacement additives can be directly substituted for most of the less desirable additives. For example, new polymer thinners like polysaccharide polymers can replace chrome lignosulfonate/sodium chromate for mud thinning, and organic phosphonates/sodium sulfite/ammonium bisulfite can replace chromate/dichromate for corrosion control. More research is needed to quantify the undesirable effects like toxicity, etc., of each mud additive in a drilling fluid.

The fourth step concerning the selection and operation of adequate solids control/removal equipment is the most important final step for closed-loop operation. It is obviously not possible to recommend a single design for a closed-loop mud processing or solids-control system since the drilling conditions and regulations in an area would affect the economics, design, and implementation of such a system. For example, wellbore geometry, penetration rate, lithology, and mud properties determine the flow rate, drill cutting concentration, particle-size distribution, and the tendency for drill solids degradation. These interacting variables in turn largely govern the choice of solids removal equipment and its arrangement for closed-loop operation. For example, drilling nonreactive formations at high rates would require more emphasis on using enough fine screen shakers while slow drilling in reactive formations requires emphasis on centrifuging and/or flocculation techniques. It is therefore difficult to quantitatively predict the precise suite of drill solids removal equipment required for a closed-loop mud system.

Some progress has been made recently in developing a performance analysis model (Lal, 1988), which analyzes the performance of a given solids control system in terms of the percentage drill solids it can remove, assuming certain particle-size distribution. This model when coupled with the economic analysis model, mentioned earlier, can assist in the design of a closed-loop system. The economic and performance analysis models are not discussed further because of the space limitation. We close this discussion with one final comment regarding the misuse of the term 'closed-loop system'. Packaged systems which provide solids-control equipment and tankage rigged-up and ready-to-use are sometimes called closed-loop systems. A closed-loop system should use linear-motion high-performance shakers (Lal & Hoberock, 1985), high-performance hydrocyclones (Young, 1987), high-*g* and large-volume centrifuges (Thurber, 1988), and proper system arrangements. A single service company does not yet provide such a package.

Finally, let us discuss briefly a few case histories which confirm the importance of various elements required for optimum waste management and the benefits gained from their implementation. The first case concerns drilling wastes disposal regulations and the benefits from technical interaction with the regulatory agencies. In a certain district before 1982, evaporation or hauling fluids to a disposal pit were the only two legal and practical methods of fluid disposal. However, evaporation was slow, and third party pits were expensive and risked potential liability. When pit fluid volumes built up due to drilling activity, safe and timely methods of fluid disposal were not available. By involving the Oil and Gas Commission and the Department of Environmental Quality with drilling operations, a mutual understanding of environmental and operational goals was accomplished. In this favorable working environment, new regulations emerged, such as legal surface application of treated pit fluids and legal downhole disposal of pit fluids, with adequate safeguards to protect the environment. In the same district, improved solids control and better water management practices substantially reduced the fluid disposal volumes which, coupled with improved regulations, *decreased* the disposal and reclamation costs by almost 50%. Before 1983, a reserve pit reclamation in this district on the average was costing $240 000–$507 000, depending on waste salinity. The implementation of closed-loop technology decreased the corresponding reclamation costs to $109 000 and $257 000, respectively.

A more recent case in 1986 concerns a nine-well infill drilling program, two vertical and seven directional holes drilled to approximately 1310 m TVD, within city limits in another district. Location constraints required minimal or no reserve pit with wastes hauled for off-location disposal. Previous success in reducing dilution and disposal costs by greater use of drill solids removal equipment suggested application of closed-loop technology. Two closed-loop mud processing system designs were considered and implemented; one used only mechanical solids removal equipment mainly consisting of a shaker and banks of 4- and 2-inch hydrocyclones used as concentrators with their wet underflow fed to a high-performance centrifuge; the other used mechanical equipment with flocculation/centrifuge process (Mud Processor). To evaluate the performance of the closed-loop systems, a number of mud samples were collected at various depths at the feed, discard and effluent streams for all pieces of equipment on seven of the nine wells. Solids control on a typical well in 1983 consisted of only a rig shaker and dilution. In 1984–5, the equipment was upgraded to include banks of 4- and 2-inch hydrocyclones and a small centrifuge to process the underflow from 2-inch cones. The benefits realized through these improvements and favorable projections from the economic analysis study led to the implementation of closed-loop technology in 1986.

Table 2 shows the comparative economics for these three groups of wells and the drill solids removal efficiency achieved in each case. Comparing 1984–5 results with 1986, the average dilution volume for the surface hole decreased from 1350 bbls to 900 bbls, and the solids removal efficiency improved from 46% to 68% resulting in a gross savings of $1500. In the production hole, the dilution volume decreased from 2300 bbls to 1100 bbls, and the solids removal efficiency improved from 67% to 80% with a corresponding gross savings of $6000. Taking into account the additional cost of $1000 for improved solids control, the net savings were $6500 which are substantial for such a shallow well. It is interesting to note that the ideal solids removal efficiency for a closed-

TABLE 2
Comparative economics for a typical well

	1983	1984–5	1986	Closed-loop condition
Surface hole removal efficiency	15%	46%	68%	81%
Production hole removal efficiency	20%	67%	80%	89%
Surface hole mud and disposal costs	$10 200	$7 800	$6 300	$4 500
Production hole mud and disposal costs	$25 600	$14 300	$8 300	$4 800
Total costs	$35 800	$22 100	$14 600	$9 300

loop system, as predicted by the economic analysis model, was not realized. However, the savings over 1984–5 are significant despite a small increase in removal efficiency. A reduction in lost circulation and differential sticking problems was also realized, which may have contributed to an average 3 days per well decrease in rig days.

In conclusion, the closed-loop system approach to management of the drilling wastes appears to be very promising. As evident from the discussion and field studies, this approach is not only environmentally desirable but also cost-effective in most cases.

REFERENCES

API (1987). Oil and gas industry exploration and production wastes. Document No. 471-01-09, July 1987.

CONNER, J.R. (1986). Fixation and solidification of wastes. *Chem. Engng* (Nov. 1986) 79–85.

EPA (1987). EPA details findings on mud, water waste regs. Oil & Gas J. Oct. 5.

HANSON, P.M. & JONES, F.V. (1986). Mud disposal—an industry perspective. *Drilling* (May 1986) 16–21.

HINDS, A.A., DONOVAN, D.M., LOWELL, J.L. & LIAO, A. (1986). Treatment reclamation and disposal options for drilling muds and cuttings. IADC/SPE 14798 paper presented at the 1986 IADC/SPE Drilling Conference, Dallas, TX, Feb. 10–12, 1986.

LAL, M. (1988). Economic and performance analysis model for solids control, submitted for presentation at the 1988 SPE Conference, Houston, Texas, Oct. 2–5, 1988.

LAL, M. & HOBEROCK, L.L. (1985). Solids conveyance dynamics and shaker performance. SPE 14389 paper presented at the 1985 SPE Conference, Las Vegas, Nevada, Sept. 22–25, 1985.

PROCEEDINGS OF THE NATIONAL CONFERENCE ON DISPOSAL OF DRILLING WASTES (1985–86). University of Oklahoma, Norman, Oklahoma.

THURBER, N.E. (1988). Decanting centrifuge performance study. M.S. Petroleum Engineering Thesis, University of Tulsa, April 1988.

US EPA (1987). Interim report on wastes from the exploration, development, production of crude oil, natural gas and geothermal energy.

WHITFILL, D.L. & BOYD, P.A. (1987). Soil farming of oil mud drill cuttings. SPE/IADC 16099 paper presented at the 1987 SPE/IADC Drilling Conference, New Orleans, Louisiana, March 15–18, 1987.

YOUNG, G.A. (1987). An experimental investigation of the performance of a 3-in hydrocyclone. SPE/IADC paper 16175 presented at the 1987 SPE/IADC Drilling Conference, New Orleans, Louisiana, March 15–18, 1987.

10

Treatment of Drilling Waste: a Dutch Case History

A. G. VELTKAMP and J. W. JURG

*Environmental Department, Nederlandse Aardolie Maatschappij BV,
P.O. Box 28000, 9400 HH Assen, The Netherlands*

ABSTRACT

Disposal of waste originating from onshore drilling activities in the Netherlands, and in particular of the fluid part containing untreated surplus drilling mud, has been a problem for many years. Due to a change in views regarding environmental quality over the last decade, the formerly accepted practice of disposing liquid waste in a brackish estuarine area is now not considered desirable.

In the absence of alternative disposal routes for the waste, possible improvement of its quality has been investigated in order to attain a more environmentally acceptable end product. Several pilot tests employing physical and chemical treatment have been carried out to reduce the total amount of solids.

In this paper, details about these tests, the results, and the implementation of the results are described.

1. INTRODUCTION

1.1. Historical Background

Onshore drilling activities in the Netherlands generate 60 000 t of drilled cuttings and 300 000 m³ of liquid waste each year. Prior to 1970, disposal of the solid waste occurred at public and private dumping sites. In addition, the solid waste was used to fill ditches and canals.

Since 1970 the disposal of cuttings has been concentrated at a few sites, where disposal is controlled by the authorities. Here disposal occurs according to the land-fill principle, followed by coverage in such a

229

way that leaching to groundwater and subsoil, as a consequence of precipitation, is prevented as much as possible.

The liquid waste is transported by boat to an assigned disposal site in a brackish estuarine area, being in fact the only salty environment where liquids with a high chloride concentration can be disposed. Thirty per cent of this waste consists of surplus drilling mud which is no longer suitable for use. The rest is rain water and water used to clean the sites: both fluids may be contaminated with spilled mud.

In 1984, objections were raised against the disposal of the fluid waste stream in the estuary. These objections concerned the ecological consequences of the high load and the high content of suspended particles and solids, reducing the transparency in the water column (i.e. increased turbidity).

1.2. Approach

Improvement of the quality of the fluid waste stream was intended to be attained along the whole pathway from generation to dumping. Therefore, an attempt was made to minimize the amounts of mud in the waste by optimization of the mud circulation system at the drill-site (Fig. 1).

The following steps were taken:

—Introduction of other mud types and adjustment of the drilling programs; this contributed immediately to a reduction of the amount of wastes to be disposed.
—Building of a Central Mud Plant for storage and reconditioning of mud, and treatment and preparation of workover, packer, well killing and gravel pack fluids (Fig. 2).
—Improved housekeeping at the drilling site.

By providing the sites with an impermeable surface layer and tailored drains and drain pits, dispersion of spilled mud became minimal. Spilled mud is pumped back into the mud system and contaminated rain and waste water are, after simple treatment, used for washing and preparation of new muds.

The remaining quantity to be disposed of amounted to approximately $200\,000\,m^3$ a year. This waste was roughly characterized by:

—an average density of $1·15\,kg/litre$;
—a suspended solid content of 20%;
—chloride concentration of $40\,g/litre$;
—COD of $20\,g/litre$.

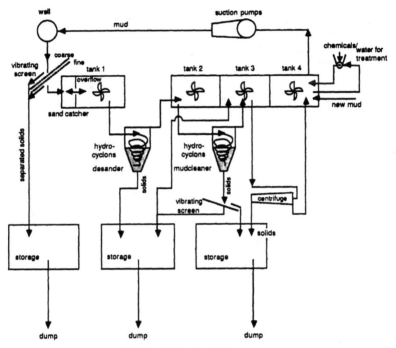

FIG. 1. Mud circulating system.

—In addition, drilling fluid components and chemical additives can be present. A treatment method had to be found for this waste at short notice so that the effluent would fulfil the environmental requirements for disposal in the estuary. Therefore, a series of field tests were initiated to get an insight into the possibilities of achieving this aim. In the meantime an effort had to be made to find a definitive solution for disposal.

2. FIELD TESTS

2.1. Aims and Scope
In preparing the tests several objectives were formulated. These objectives were as follows.

2.1.1. The Treatment System Effluent
By the time the tests were ready to commence, quality requirements for the waste in relation to the receiving environment were not sufficiently

A. G. Veltkamp and J. W. Jurg

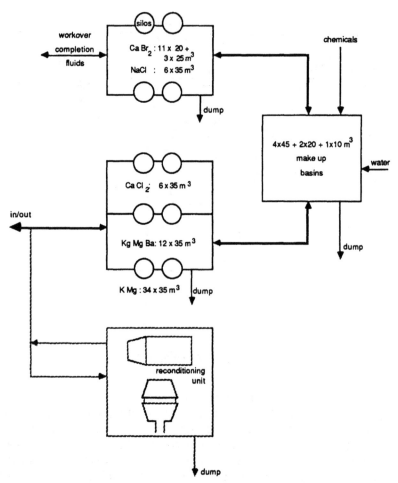

FIG. 2. Central mud plant. Total storage: $2325\,m^3$. Total make up: $230\,m^3$.

set. Therefore, the tests were aimed at an improvement of the waste's quality using an economically and technically feasible method. Special attention had to be paid to the suspended solid content of the water treated. In addition, with regard to the sludge produced, dryness and stability of the sediment (filter cake) would be relevant in view of transportation and dumping requirements.

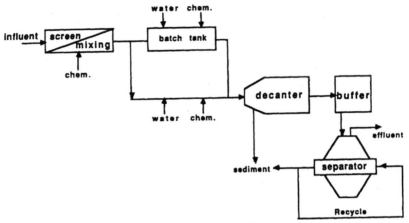

FIG. 3. Diagram showing the working principle of the centrifuging unit.

2.1.2. *The Treatment System Itself*
The test should provide sufficient data to make a system comparison possible with regard to reliability, performance, ease of operation and economics.

2.2. Techniques Tested
Based on the prerequisites that (1) the tests should be restricted to existing physical-chemical treatment systems, and (2) the treatment units should be suitable both at a central plant and at each separate drill site, market and literature research were carried out. Finally, it was decided that the following tests should be carried out:

1. centrifugal separation,
2. pressure filtration,
3. pressure belt filtration.

The *centrifugal separation unit* (Fig. 3) consisted of an Alfa Laval NX418 decanter (centrifugal accelerations 1000–2200 g; rotational speed 1500–3000 rpm) and an Alfa Laval QX210 disc stack separator (up to 5000 g; 0–6000 rpm).

The efficiency of the decanter was studied by varying machine parameters (e.g. speed, depth of liquid ring, etc.) and operating parameters (e.g. feed flow, type of polyelectrolyte, etc.).

The separator was operated at a constant speed of 6000 rpm. The flow

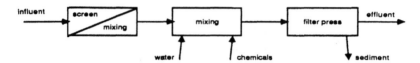

FIG. 4. Diagram showing the working principle of the pressure filtration unit.

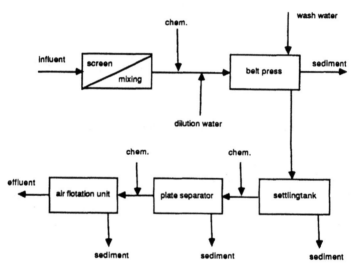

FIG. 5. Diagram showing the working principle of the pressure belt filtration unit.

rate of the whole unit was adjusted to the dewatering efficiency of the separator and consequently the necessity of recirculating the extracted sediment.

The *pressure filtration unit* (Fig. 4) consisted of a Rittershaus & Blechner chamber filter press (total filter cloth area $120\,m^2$, cloth weight $330\,g/m^2$ and an air flow through cloth 0–$50\,litres/dm^2$). As this technique is an intermittent process, it is very easy to carry out many tests within a short time.

The *pressure belt filtration unit* (Fig. 5) consisted of an Andritz S7-2200 filter belt press, a Serpac 3CIDH lamella separator (settling area $30\,m^2$, capacity $50\,m^3/h$ and load per unit area $1\cdot6\,m^3/m^2\,h$) and a Serflo 2L flotation system (flotation area $10\,m^2$, capacity $50\,m^3/h$ and load per unit area $5\,m^3/m^2\,h$).

2.3. Test Conditions
Many combinations of test conditions were used including:

—a variety of technical parameters;
—fluids to be treated: common mud types and brines, and mixtures thereof, with and without dilution, as well as the fluid normally disposed into the estuary;
—a range (combinations of different concentrations) of chemical reagents to improve the conditioning, i.e. polyelectrolytic flocculants (anionic and cationic polymers), inorganic coagulants, and slaked lime (for neutralization, precipitation and carbonate removal).

3. RESULTS

3.1. Test Conditions
The tests were conducted under the following conditions:

—All systems were tested under practical conditions in the open field over a 1–3 month period.
—The operation was conducted entirely by the service-contractor of the system concerned. NAM took care of the influent and the processing of effluent and solids.
—Treatment criteria, chemicals to be used and other parameters were established through consultation between NAM and the contractor.
—The data of separate day trials were averaged over periods of approximately 2 weeks. An indication of the conditions under which the filter belt press was used is supplied in a model test sheet in Table 1.
—In a number of tests, particle size distributions were measured in order to determine the possible use of the effluent for the production of new mud. In Fig. 6 the separation is shown of a treated fresh water bentonite mud and in Fig. 7 that of a Gypsum mud.

Because of the large number of trials with different types of waste water from drilling and workover activities, the summarized results can only be given as arithmetic means. Table 2 must therefore be regarded as indicative only. The net throughput of the systems tested varied from 5 to 15 m^3/h.

3.2. Reagents
The types of flocculants and coagulants which were used are highly dependent on the chemical composition of the mud which is to be treated.

TABLE 1
Belt filter press trial conditions

Test sheet pressure belt filtration unit	
Treatment of a mixture of mud and clean-up water	
Mud throughput ($d = 1{\cdot}19\,\text{t/m}^3$)[a]	$6{\cdot}2\,\text{m}^3/\text{h}$
Dry solids flow rate of mud and salt	$1{\cdot}7\,\text{t/h}$
Dry solids content of mud	23 wt %
Coagulant consumption of filter belt press (35 wt % $FeCl_3$)	22 litres/h
Flocculant consumption of filter belt press (0·2 wt %)	$1{\cdot}8\,\text{kg/h}$
Throughput of conditioned mud ($d = 1{\cdot}17\,\text{t/m}^3$)	$7{\cdot}7\,\text{m}^3/\text{h}$
Dry solids content of filter cake	$51{\cdot}2\,\text{wt}\,\%$
Coagulant consumption for water post-conditioning (poly $AlCl_3/FeCl_3$)	14 litres/h
Flocculant consumption for water post-conditioning ($d = 1{\cdot}05\,\text{t/m}^3$)	$0{\cdot}3\,\text{kg/h}$
Net quantity of filtrate water ($d = 1{\cdot}05\,\text{t/m}^3$)[b]	$8{\cdot}5\,\text{m}^3/\text{h}$
COD reduction	80–90%
Mineral oil reduction	97%
Suspended components in filtrate	2–3 g/litres
Reduction of chloride content	50–60%
pH of filtrate water	5–6
Reduction of settleable components	80–90%

[a]The above data are averages of 9 measuring days (with a total of 63 pressing hours). The worst and the best daily averages were 3·1 and $8{\cdot}4\,\text{m}^3/\text{h}$ respectively.
[b]The filtrate water flow rate could have been higher because of the large capacity of the water post-conditioning installation. During the tests, however, the emphasis was put fully on the dewatering of the mud with the filter belt press.

In conducting the tests the following substances were used:

—anionic and cationic polymers;
—$Ca(OH)_2$;
—$FeClSO_4$;
—$FeCl_3$;
—$Al_2(SO_4)_3$.

The quantities of flocculant and coagulant needed are mainly determined by the solid content of the influent. If pre-separation by means of settlement and/or centrifuging is carried out chemical consumption could be reduced by 20–30%.

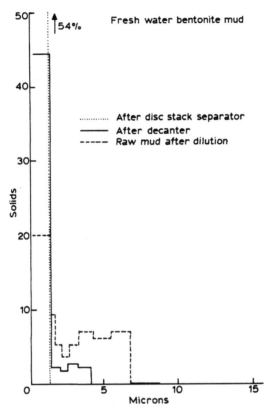

FIG. 6. Particle size distributions before and after treatment.

4. ASSESSMENT AND EVALUATION

4.1. Assessment
The three systems tested each have their own advantages and disadvantages as well as their own specific uses.

4.1.1. Qualitative Aspects
As to the quality of the effluent the following is observed:

—All three systems produce a clear effluent. The reduction of solids amounted to more than 80%.
—Influents with a density of ± 1.2 kg/litre produce the highest removal efficiency in all three systems; influents with a higher density can be diluted with the effluent.

A. G. Veltkamp and J. W. Jurg

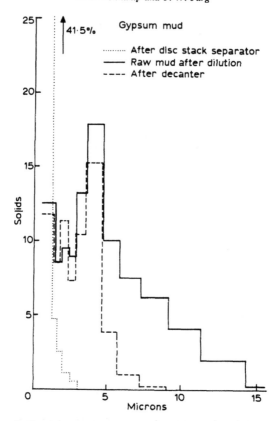

FIG. 7. Particle size distributions before and after treatment.

TABLE 2
Summary of trial tests

Parameters	Systems	Centrifuges		Chamber filter press		Filter belt press	
		In	Out	In	Out	In	Out
COD	g/litre	20	3	25	4	16	4
Suspended solids (>64 μm)	g/litre	100	0·2	100	0·04	100	2
Settleable solids (Imhoff)	ml/litre	300	0/20	300	0·1	300	10/15
Chloride content	g/litre	15	10	20	15	80	20

—The spadability (dryness) of the solids separated showed differences; the dryness varied from moderately spadable (with the centrifuge system) to excellently spadable (with the chamber filter press), with dry solids contents from 40 to 65%.

4.1.2. Operational Aspects

As to sensitivity of the process, it was observed that this sensitivity was the highest with the three-stage purification carried out with the filter belt press. The chamber filter press was the least sensitive to changes in the quality of the influent.

Of all systems, it can be said that the process was least sensitive when a mix of mud was treated. The different types of mud themselves probably provide the destabilization of particles, i.e. the removal of resistance to flocculation. The degree of process sensitivity clearly influences the number of operating staff and the amount of training needed.

4.2. Evaluation

Based on its ease of operation, good qualitative performance and economy, it was decided to install a chamber filter press at a drill site as a follow-up test for treatment of all liquid waste. In this way the effluent could be reused for making up new mud and wash water. Thus, use of ground and surface water can be reduced.

The follow-up test, which was carried out on a single well drill site in the winter of 1986, largely produced the same results as the previous field trial, both in qualitative and in operational aspects. Reuse of the effluent as basic water for new mud, however, was minimal because of the high levels of calcium chloride and pH. These high levels were caused by the use of the filtering reagent $Ca(OH)_2$ (30 litres/m^3, 30% solution) and the coagulant $FeCl_3$ (3–5 litres/m^3, 50% solution). As a consequence of the exchange of muds and/or heavy rainfall there were peak quantities of waste water at certain times. The collection and processing of this turned out to be difficult to realize, both operationally and because of the amount of space needed on the drill site. Consequently, several hundred m^3 of waste water still had to be discharged untreated.

5. CURRENT SITUATION

After termination of the trials described in the previous section, discussions were held with the appropriate authorities on the practical application of the results obtained. It was decided that mud-containing waste streams from all NAM drill sites in Holland should be treated with a chamber filter press in a centrally located plant. After settlement and homogenization in two basins of about $1500\,m^3$, the mix is to be treated by the press while $Ca(OH)_2$ and $FeCl_3$ are added. The throughput of waste streams to be treated at present amounts to $\pm 10\,m^3/h$, the solids content of the effluent being below 0·1%. The effluent is now so clear that discharge does not reduce the receiving water transparency of the estuary.

Based on these results permission has now been obtained to continue dumping in the estuary for the time being. A licence has also been granted to discharge the effluent into a public biological sewage purification plant.

11

Feasibility of Land Application as a Waste Management Practice for Disposal of Residual Diesel Invert-based Muds and Cuttings in the Foothills of Alberta

J. ASHWORTH

Norwest Soil Research Ltd, 9938 67th Avenue, Edmonton, Alberta, Canada T6E 0P5

R. P. SCROGGINS

Oil, Gas and Energy Division, Industrial Programs Branch, Conservation and Protection, Environment Canada, Ottawa, Ontario, Canada K1A 0H3

and

D. McCOY

Environmental Affairs, Canterra Energy Ltd, P.O. Box 1051, Calgary, Alberta, Canada T2P 3H5

ABSTRACT

The disposal of spent diesel invert mud residues and cuttings (DIMR) from wells drilled for oil or natural gas is an issue of concern to environmentalists as well as to producers of energy, partly because of the well-known bio-toxicity (Wolfe, 1977. Fate and effects of petroleum hydrocarbons in marine organisms and ecosystems. Proc. Symp., Seattle, Wa, Pergamon Press, Oxford) of many of the various hydrocarbons in DIMR.

In a land-treatment field demonstration (commenced in June 1986), DIMR from a recently completed well site with ca. 11% total hydrocarbons and 0·8% aromatic hydrocarbons, was roto-tilled into soil near the lease on test plots at the rate of 150, 300 or 450 m³/ha on June 3, 1986. These rates include that required to dispose of a typical volume of DIMR on a 1 ha drilling lease. Soil amendments include manure, charcoal and spent carbon granules; a partial split-plot design allowed the effects of cultivation and/or fertilizer nitrogen (N) to be evaluated as well. The DIMR used contained very high levels of salts (predominantly calcium chloride) so that electrical conductivity (EC) of soil given the highest rate of DIMR was initially near 30 mS/cm. Little or no growth of a mix of sown grasses occurred in 1986 on any plot at any rate of DIMR.

However, analysis of soil samples taken at regular intervals until October 1986 showed that the EC of treated soil fell rapidly, especially where N had been applied. The decline in EC was probably due to heavy rain in July and September 1986, which helped leach soluble salts from the soil profile.

Summer rainfall was again high in 1987 and growth was good on all plots given DIMR at the lowest rate, but poor if N had been withheld. At the two higher rates of DIMR, growth was reasonably good only on plots given a mulch of manure in addition to N. Cultivation after mulching tended to be detrimental, despite re-seeding.

In contrast to the results of a previous greenhouse/laboratory study on DIMR (Peake, Visser & Danielson, 1986. Biodegradation and analysis of diesel invert mud residues, Phase I. A report prepared for the Canadian Petroleum Association and Environment Canada. Kananaskis Centre for Environmental Research, University of Calgary), activated charcoal or carbon granules had a detrimental effect on growth in the field. Growth in 1987 was best on plots that showed the most rapid decline in residual total hydrocarbon levels during 1986 and 1987. Concentrations of target polyaromatic hydrocarbons (PAHs) were apparently unrelated to growth. All target PAH concentrations declined rapidly in 1986 leaving a small residue which degraded more slowly in 1987.

It is impossible to draw general, reliable conclusions from one specific trial which is not yet complete. Results to date suggest: (1) Optimum rates of DMIR application are in the 150 m³/ha range; (2) at rates of 300 m³/ha or higher, it appears that the length of time and level of management required to achieve reclamation may be both economically and environmentally limiting. At least one more field season is required before definitive conclusions can be drawn.

1. INTRODUCTION

1.1. Scope of the Investigation

The use of diesel invert muds in drilling operations for oil and natural gas in Alberta has greatly increased in recent years. Although quantities of DIMR vary with both well depth and efficiency of residue recovery, several hundred cubic metres is a typical volume for sour gas wells drilled in the Alberta foothills.

This quantity of residue must be dealt with during the drilling lease restoration phase. Incineration or solidification disposal options for DIMR are not economically attractive and technical problems have arisen with these approaches. To industry, the most cost-effective disposal option for residues is usually to spread them over the lease and re-vegetate the site. However, land application of DIMR has adverse environmental implications, because of the well-known inability of plants to grow on soil that is badly contaminated with oil (Plice, 1948). Also chemical analysis shows that toxic and highly refractory compounds such as PAHs are often present in large concentrations in DIMR (Peake *et al.*, 1986).

1.2. Research Undertaking
The Environmental Research Advisory Council (ERAC) of the Canadian Petroleum Association (CPA) and the Environmental Protection Service (EPS) of Environment Canada agreed to co-fund a two phase, 5 year research program to assess the feasibility of land-application of DIMR on drilling leases in an environmentally acceptable manner.

These considerations were the subject of two years of laboratory and greenhouse work (Phase 1, carried out by staff at the Kananaskis Centre for Environmental Studies, University of Calgary) followed by a field demonstration, begun in 1986 (Phase 2) which is the responsibility of staff at Norwest Soil Research Ltd, Edmonton. Phase 2 is the main subject of this report and only a brief summary of the Phase 1 work is given below.

1.3. Laboratory and Greenhouse Work
DIMR samples, from two completed sour gas well sites near Bragg Creek (in the Rocky Mountain foothills, near Calgary), with approximately 1% aromatic hydrocarbon content (c. 11% total hydrocarbon content) were mixed (at four rates of up to 40 g per 100 g of soil) with topsoil from near the two sites. Degradation in moist soil at 23°C was monitored by measuring carbon dioxide evolution and extractable hydrocarbons. Up to 63% of the aliphatic and aromatic hydrocarbons in 30% DIMR/soil mixtures had disappeared after 8 months, the rate being dependent on soil type and the amount of nitrogen (N) fertilizer supplied.

Germination and growth of grasses was strongly inhibited by DIMR but had returned almost to normal 5 weeks after application. Alfalfa was

more sensitive and did not develop past the cotyledon stage until at least
14 weeks after DIMR application.

Growth was greatly improved when activated charcoal was applied at
heavy rates to the DIMR/soil mixtures. Spent carbon granules (from
filters which purify monoethanolamine used to refine natural gas) also
improved growth in these greenhouse trials.

2. 1986 FIELD STUDIES

2.1. Materials and Methods

2.1.1. Site Characteristics
The layout of the field experiment (Fig. 1) was a compromise between the
requirements of a demonstration trial (as specified in the Terms of
Reference, ERAC 84 04 12) and those of testing various amendments in a
partial factorial design.

The experimental site (Sarcee site) was a cleared area (previously aspen
woodland) in the foothills of the Rocky Mountains at an elevation of
about 1500 m. The soil at the site was a clay loam, low in N and P, with
pH 7·5 and 4% organic matter.

2.1.2. Treatments
The diesel invert cuttings used were obtained from a newly completed
sour gas well on a nearby lease. Six individual subsamples of the bulk of

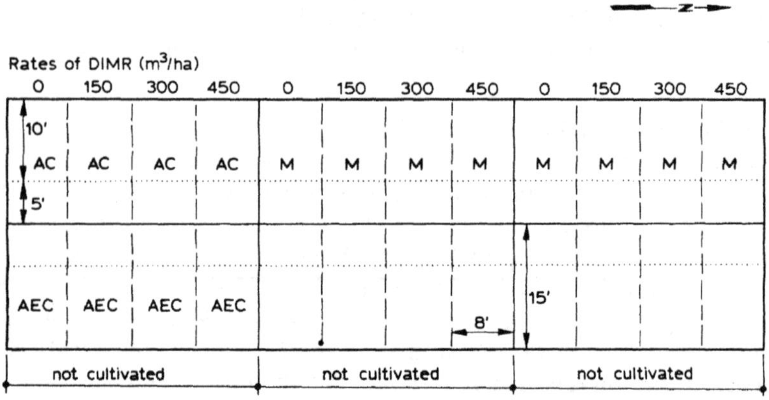

FIG. 1. Layout of field experiment. AC, activated charcoal; AEC, Alberta Ethylene
Carbon; M, manure (these treatments applied to an area 15′ × 8′). The 10′ wide
central strip received no fertilizer.

the material were obtained before application by augering the pit after mixing with a backhoe. The total volume of material available was estimated at $300 \, m^3$.

Heavy snow in May delayed the start of operations. On June 3, 1986 the DIMR was spread and incorporated mechanically on plots measuring $9 \cdot 1 \times 2 \cdot 4 \, m$ using a small front-end loader/rototiller at 150, 300 or $450 \, m^3$/ha. These rates include the rate required to dispose of a typical volume of DIMR within a 1 ha lease.

The experiment also tested the effects of the following amendments: activated charcoal ($0 \cdot 7$ cm layer, incorporated, approximately 25 t/ha), spent carbon granules ($0 \cdot 2$ cm layer, incorporated, approximately 10 t/ha), and well-rotted manure at approximately 45 t/ha either incorporated or left on the soil surface (Fig. 1). Spent carbon granules, used in filters as an absorbent for degradation products in a monoethanolamine sour gas refining process, were supplied by Alberta Gas Ethylene Ltd. In addition, fertilizer supplying 240 kg N, 44 kg P, 41 kg K and 18 kg S per ha was added to part of the area occupied by the above treatments.

One week after applying the treatments the site was sown to a mixture of perennial ryegrass, creeping red fescue, slender wheatgrass and oats. Seed was broadcast at 84 kg/ha and lightly raked in.

2.1.3 Sampling

Soil samples were taken immediately after application and four more sets of samples were taken on June 23, July 31, August 28 and October 23, 1986, using an Oakfield core sampler. On these occasions, up to 12 cores were taken to 10 cm depth, the cores combined, mixed and a representative sample taken. Sampling to depths below 10 cm was badly hampered by small rocks. Where possible, cores were retrieved and combined but the objective of 12 cores was impractical and most subsoil samples consisted of 4–5 cores only. Analysis of cuttings and soil for inorganic parameters and for methylene chloride (total) extractable hydrocarbons was done at Norwest Labs. Individual hydrocarbon constituents were identified and quantified at Enviro-Test in Edmonton.

2.1.4. Analyses

Analyses for pH, EC, major cations and anions and sodium adsorption ratio were done on the filtrate from a saturated paste of the soil sample and water. Soil fertility parameters (available nutrients, organic carbon) were measured by routine procedures available from Norwest Labs.

Metals in the invert cuttings were analyzed by two methods: (a) by nitric/perchloric acid digestion followed by atomic absorption (AA) or inductively coupled plasma (ICP) spectrometry; (b) by similar analysis of a leachate obtained by tumbling a composite sample (40 g) of cuttings (obtained by combining roughly equal portions of the six individual samples from the pit) in deionized water (800 ml) for 24 h and filtering under suction through a micropore filter (0·45 μm).

In preliminary work, six sub-samples of the invert cuttings used were analyzed for target hydrocarbons. The heterocyclic nitrogen compounds (HNCs) suggested as target compounds, which were quinoline, carbazole and dibenzo(a,h)acridine, were not detected. Five groups of PAHs were detected and chosen as target compounds for the field study: naphthalenes, phenanthrenes, biphenyls, fluorenes and dibenzothiophenes. No benzo (a) pyrene was detected.

It should be borne in mind that the hydrocarbon component of invert cuttings is typically composed of only 5–6% aromatics and over 90% saturated hydrocarbons. The PAH compounds were selected as targets because of their known toxicity (Wolfe, 1977) and their measurable presence in DIMR.

The detected PAHs were measured in sets of soil samples taken on four dates (June 23, July 31, August 28 and October 23, 1986) from the plot given the middle rate of invert application (300 m^3/ha).

Samples were mixed with granular sodium sulphate and Soxhlet extracted with dichloromethane (DCM) for 16 h. Some samples were spiked with a series of deuterated PAHs and one deuterated HNC to determine method efficiencies.

The DCM extract base-neutral fraction from the spiked samples was further extracted with 6N sulphuric acid to separate basic compounds (HNCs) from neutral compounds. The DCM fraction (neutral) was taken to near dryness, dissolved in 2 ml of hexane and quantitatively added to an 18 g alumina column. The separation efficiency of each freshly treated batch of alumina was checked by adding to the alumina a standard mixture of diesel and selected PAHs. If separation of aliphatics from the PAHs was achieved the alumina was considered adequate for the samples.

The alumina column was eluted with 45 ml of pentane to isolate the nonpolar saturated hydrocarbons (Fr. 1) then by 150 ml of benzene to isolate the aromatics (Fr. 2) followed by 150 ml of DCM to isolate the polar compounds (Fr. 3). All three neutral fractions were reduced to a known volume and analyzed by GC/FID. The aromatic fraction was further characterized by GC/MS. The acidic aqueous portion (base compounds) was adjusted to pH 10 with 6N NaOH and extracted with DCM to isolate the basic (HNC) compounds. The DCM was reduced to a known volume and analyzed by GC/MS.

2.1.5. Later Field Operations in 1986

The original intention was to cultivate a section of the experimental area several times during the 1986 growing season. However, due to heavy rain no cultivation was possible in July. Preparations for cultivation at the end of August when the site was drier were again thwarted by further rain which persisted well into September. Finally the appropriate section was rototilled in late September (under very moist soil conditions). The effects on degradation of this cultivation may not be marked because the entire experiment was rototilled again soon afterwards (on October 18). This last cultivation was in preparation for re-seeding the site on October 15, 1986 with a seed mixture containing one part each of perennial ryegrass, tall wheatgrass, alkali grass, creeping red fescue and winter wheat.

All top growth on the control plots was cut and removed before rototilling. Ammonium nitrate (34-0-0) supplying 210 kg N/ha was applied (not to the central strip of the experiment) and manure at 40 t/ha was either incorporated or left on the surface as before (Fig. 1). Seed was broadcast at 84 kg/ha and lightly raked in.

2.2. Results and Discussion

2.2.1. Invert Cuttings

Six individual samples of DIMR were found to contain, on the average, 11·4% hydrocarbons on an as-received basis. The bulk density of the fresh material was approximately 1·75 g/cm^3.

Over 90% of the hydrocarbons were nonaromatic; about 6% were aromatic; only about 1% were found to be polar hydrocarbons. This analysis indicates that the material was similar to the Quirk Creek DIMR described by Peake *et al.* (1986). Concentrations of both total and extractable metals (results not shown) indicated no contamination by metals.

2.2.2. Analysis of Soil Samples

Amounts of methylene chloride-soluble, i.e. 'total' hydrocarbons in surface soil samples (0–10 cm) are shown in Fig. 2. Hand-drawn curves have been fitted to indicate the change in average recovery with time. The three curves start out at 1·8, 3·6 or 5·3% on June 3, calculated values which are

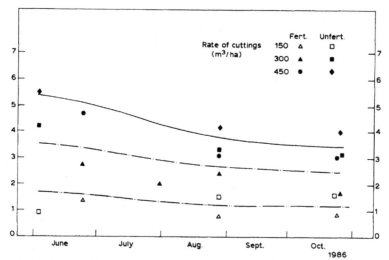

FIG. 2. Methylene chloride-extractable hydrocarbons in DIMR-treated soil during 1986 growing season.

consistent with the volumes and bulk density of cuttings applied, assuming a 12 cm depth of mixture (soil plus cuttings) at all three rates of application. Samples from fertilized portions of the experiment gave consistently lower results for total hydrocarbons than were obtained for samples from unfertilized portions. This finding is discussed further below.

Analysis of a preliminary set of soil samples taken from each rate of DIMR application on June 3 showed that electrical conductivity (EC) was much too high for plant growth. Analysis of DIMR and of one soil sample for major ions indicated that the salinity was due to calcium chloride.

Electrical conductivity (EC) fell rapidly during the 1986 growing season (Fig. 3). Hand-drawn curves have again been fitted to the points. At the end of October all plots, except the unfertilized portion of the plot given the top rate of DIMR, had EC values at or below 6 mS/cm, which should not limit growth severely.

On fertilized areas, conductivity was consistently lower than on unfertilized areas. The explanation for this may be that fertilizer application assisted in the rapid degradation of an appreciable component of the hydrocarbons in the invert cuttings, thereby improving soil permeability and increasing leaching losses of chloride. In turn, lower EC values on the fertilized portion of the experiment could be expected to improve rates

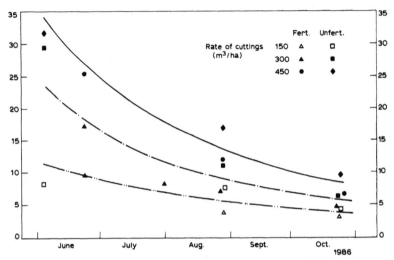

FIG. 3. Electrical conductivity (dS/m) of DIMR-treated soil during 1986 growing season.

of degradation of hydrocarbons by microorganisms (since high chloride concentrations have a sterilizing effect).

2.2.3. *Degradation of Target PAH Compounds and Total Hydrocarbons*

The concentrations of all the target PAH compounds fell rapidly during the 1986 growing season (see Figs 4–9). For example, dimethyl naphthalenes fell from about 170 ppm to 60, 20 and 10 mg/kg in samples collected in June, July, August and October 1986 respectively, from the plots given the middle rate of DIMR.

At this same rate of DIMR, amounts of dichloromethane-extractable hydrocarbons recovered were initially 3–4% of the surface layer of material (0–10 cm) on a dry weight basis and remained in the 2–3% range in the final set of samples. Similar persistence was observed at the other rates of application.

There was no apparent relationship between molecular weight of the target PAHs and their rate of degradation. Also, the target compounds have a range of molecular weights which approximately spans that of the bulk of the hydrocarbons in the material. For these reasons, losses of the target PAHs as a result of volatilization are not considered responsible for their rapid rate of disappearance.

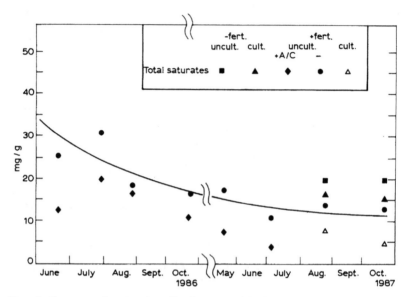

FIG. 4. Concentration (mg/g soil) of saturated hydrocarbons during 1986 and 1987 growing seasons.

2.2.4. *Availability and Degradation of Organic Compounds*

Unsubstituted PAHs can be expected to degrade fairly rapidly in soils because (a) they are not strongly sorbed on soil organic matter and (b) they lack substituent groups conferring resistance to microbial attack (Briggs, 1976). A measure of sorption on organic matter is the octanol/water partition coefficient which for simple PAHs is of the order of 10^{-4}. This indicates that roughly 1% of these PAHs is in the soil solution, and thereby available for degradation, compared with less than 0·01% for hydrocarbons with a partition coefficient of 10^{-8}. Long-chain aliphatic compounds can be expected to have very low octanol/water partition coefficients and will therefore be only gradually available for degradation.

2.2.5. *Leaching Losses*

Partition to organic matter is still very strong in absolute terms (99% or more) and losses of PAHs due to leaching were probably very small. This view is supported by the low recoveries of PAHs in soil 10–20 cm below the surface even after heavy rain. For the total hydrocarbons *and* the target PAHs, recoveries in subsoil averaged about 20% of the amounts found in the 0–10 cm layer. Most of this material was probably present

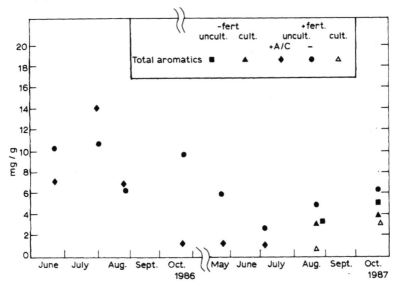

FIG. 5. Concentration (mg/g soil) of aromatic hydrocarbons during 1986 and 1987 growing seasons.

below 10 cm from the time of application, due to incorporation below that depth.

2.2.6. Effects of Invert Cuttings on Plant Growth in 1986

Some sparse growth occurred at the 10% rate of application and was noticeably better where fertilizer had been applied. At the middle rate, no growth occurred except where fertilizer had been used in combination with a manure application that had been left on the soil surface. There was no growth on any area given invert cuttings at the top rate.

No ameliorating effect of either form of carbon treatment was observed at any rate of invert application. The poor growth was the cause of the decision to cultivate and re-seed the entire site in October.

3. 1987 FIELD STUDIES

3.1. Materials and Methods

3.1.1. Field Work and Observations

In June 1987 the part of the plot area earmarked for cultivation was again rototilled. All appropriate plots received N and P at 220 and 35 kg/ha

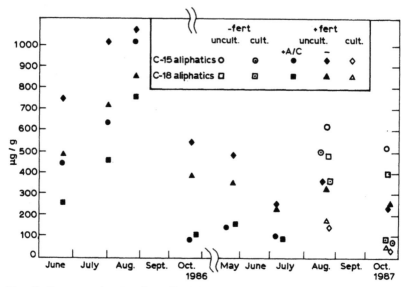

FIG. 6. Concentration (mg/kg soil) of long-chain aliphatic hydrocarbons during 1986 and 1987 growing seasons.

respectively. All plots were re-seeded with the same seed mixture as was used originally, mixed 50/50 with timothy seed, since volunteer timothy had been seen in 1986. All plots were lightly raked. There was no growth on the plots at this stage and soil conditions were very dry.

All appropriate plots again received 220 kg N/ha (as 34-0-0) in mid-August. Soil conditions were then very moist and some weeds were noticeable as well as grass on some plots. To kill broadleaved weeds an application of chlorsulfuron was made at 50 g/ha.

Soil samples were taken as previously described, in May, July, August and October.

Dry matter production (Table 1) was measured by clipping a 2 ft strip across the entire plot (8 ft wide) on October 22 and weighing the bagged, dried material.

3.1.2. Organic Analysis

Target PAHs were measured in soil samples taken from fertilized and unfertilized plots from July 1987 onwards. In addition, soil samples stored frozen since sampling in 1986 were analyzed for total aliphatic and aromatic hydrocarbons (see Figs 4 and 5) because of the correlation

TABLE 1
Dry matter yields 1987 (g/m^2) and %N in dried produce

Treatment		Amount of invert applied (m^3/ha)							
		0		150		300		450	
		+F	−F	+F	−F	+F	−F	+F	−F
None	Yield	270	80	100	16	—	—	—	—
	%N	2·0	0·9	1·6	0·8				
Activated charcoal		450	120	220	—	37	—	—	—
		2·2	1·0	1·7		2.0			
Carbon granules		a	a	140	—	—	—	—	—
				1·9					
Cultivation only		a	a	220	—	61	—	13	—
				2·3		1·7		1·9	
Manure only		a	a	180	8	130	—	190	—
				2·1	0·9	1·6		1·5	
Manure + Cult.		340	70	200	17	41	—	37	—
		1·2	1·0	2·2	0·8	1·7		2·2	

^{a}Not measured, but approx. 350 or 90 g/m^2 with or without added fertilizer (+F or −F) respectively. —: less than 4 g/m^2.

observed between growth and total hydrocarbon content rather than target PAH content.

3.2. 1987 Results and Discussion

3.2.1. Dry Matter Yields
Except on control plots, yields in 1986 were all negligible. In 1987, all plots treated with the low rate of DIMR and fertilizer had more than 80% coverage and good growth (even with no further treatment). However, at the two higher rates of invert treatment, only plots given fertilizer and a surface layer of manure (not cultivated in) had good coverage (75%) and growth. Yields are approximate because growth was not uniform, especially on DIMR-treated plots.

A noticeable feature of the manured-but-uncultivated plots was that the surface cover trapped dead leaves and other organic debris, thereby adding to the existing thin veneer of material. This discontinuous layer has two desirable properties; it is uncontaminated, and it acts as a mulch, conserving soil moisture. In July 1987, on disturbing loose portions of the surface layer, healthy seedlings could be seen in damp patches underneath.

Seed heads in the dried produce from each plot were examined in the

TABLE 2
Oil content (%) October 1986 and 1987 in top 10 cm of soil

Treatment	Amount of invert applied (m³/ha)							
	0		150		300		450	
	+F	−F	+F	−F	+F	−F	+F	−F
			1·8[a]		3·6[a]		5·3[a]	
(Oct. 1986)	—	—	2·0	1·4	2·1	3·6	3·5	4·9
None (Oct. 1987)	—	—	1·2	1·5	2·5	3·0	2·7	3·6
	—	—	0·4	2·0	1·6	3·3	3·1	4·3
Activated charcoal	—	—	0·5	1·7	1·6	3·4	2·7	4·4
	—	—	1·0	1·2	2·8	3·4	5·2	5·9
Carbon granules	—	—	0·9	1·1	1·9	3·3	4·0	5·3
	—	—	1·1	1·9	0·2	3·3	3·2	3·4
Cultivation only	—	—	0·7	0·9	0·8	1·8	2·0	3·0
	—	—	0·8	2·3	0·9	2·8	1·2	1·9
Manure only	—	—	0·6	1·3	0·7	2·1	0·7	3·1
	—	—	0·1	1·3	2·2	3·0	2·5	3·5
Manure + Cult.	—	—	0·4	0·9	1·3	1·8	1·5	2·2

—: background value (less than 0·1%).
[a]Approx. initial value, June 1986.

laboratory. Ryegrass was most frequently the dominant species present (about 3:1 over the next most prominent grass, timothy). Oats dominated on some plots; they are unlikely to survive well in a second season, but their value consists in providing ground cover in the current season and a residue of biomass to support subsequent growth of permanent forage. Alkaligrass and (rarely) slender wheatgrass were the only other grass species identified; winter wheat did very badly.

3.2.2. Oil Content of DIMR-treated Soil
In general, the total hydrocarbon content of topsoil fell significantly in 1987, especially on plots given the two higher rates of DIMR (Table 2). However, when oil content had fallen to around 1%, any further decline was apparently very slow.

The results of subsoil (10–20 cm) analysis (Table 3) show clearly that significant quantities of hydrocarbons moved down the soil profile on plots treated with heavy rates of invert cuttings. Adding fertilizer greatly ameliorated the problem. Downward movement was very noticeable where a carbon or charcoal amendment was made, consistent with slow degradation on these plots.

TABLE 3
Subsoil (10–20 cm) oil content (%) October 1986 and 1987

Treatment	Amount of invert applied (m³/ha)							
	0		150		300		450	
	+F	−F	+F	−F	+F	−F	+F	−F
(Oct. 1986)	—	—	0·2	0·1	0·1	0·2	0·5	0·7
None (Oct. 1987)	—	—	L	L	0·2	1·2	0·6	1·4
	—	—	L	0·1	0·2	0·3	0·3	0·2
Activated charcoal	—	—	NS	0·1	L	0·5	0·8	1·1
	—	—	0·1	0·1	0·3	0·4	0·8	1·1
Carbon granules	—	—	L	0·1	0·2	0·4	1·4	1·9
	—	—	0·1	0·4	0·2	0·5	0·3	1·6
Cultivation only	—	—	NS	0·2	L	L	0·2	1·1
	—	—	0·1	0·3	L	0·1	L	0·7
Manure only	—	—	NS	0·1	NS	1·0	NS	1·3
	—	—	0·5	0·2	0·3	NS	0·2	0·8
Manure + Cult.	—	—	NS	0·2	NS	0·1	NS	1·2

—: background level (less than 0·05%).
L = less than 0·1%
NS = no sample obtained because sampling difficult below 10 cm.

3.2.3. Target PAHs and other Organics in DIMR-treated Soil

Concentrations of target PAHs in soil given the middle rate of DIMR (Figs 7–9) tended to decline further from values observed in samples taken in October 1986. A discontinuity was observed in results for substituted naphthalenes (see Fig. 9) which had not been resolved at the time of submitting this article.

The ratio of total saturates to total aromatics in DIMR-treated soil (see Figs 4 and 5) was much lower (3:1) than in the original DIMR (10:1) from the start of the demonstration, but did not narrow much further as time went on. Possibly many saturated hydrocarbons of suitable chain length become rapidly incorporated into the soil biomass and cannot be extracted.

Substantial amounts of long-chain aliphatic hydrocarbons were detected (Fig. 6); in October 1987, a roughly 20-fold difference was observed between concentrations on cultivated, fertilized plots and uncultivated, unfertilized plots.

3.2.4. Salinity of DIMR-treated Soil

There was a very marked effect of fertilizer on topsoil salinity in the first season, but only a slight effect in the second season. No damaging salinity

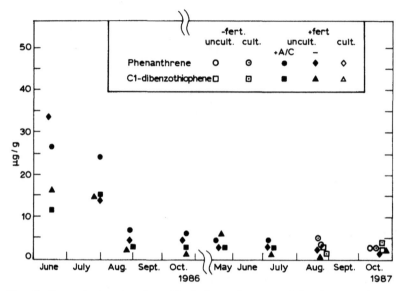

FIG. 7. Concentration (mg/kg soil) of phenanthrenes and C1-dibenzothiophenes during 1986 and 1987 growing seasons.

was present anywhere in the experiment from May 1987 onwards (Table 4). Subsoil salinity was below 4 mS/cm on all plots by October 1987 (Table 5).

4. CONCLUSIONS

The most successful soil management strategy for revegetation after DIMR application at this one site was broadcasting nitrogen fertilizer at generous rates and covering the surface with a mulch of well-rotted farmyard manure. After rototilling at the end of the first season, this treatment was repeated, with no subsequent cultivation other than lightly working the seed mixture into the loose surface material.

No rehabilitation treatment was completely effective, even after two growing seasons, at rates of DIMR greater than 150 m^3/ha. Activated charcoal or carbon granule amendments were detrimental.

The most successful grass species employed were perennial ryegrass and timothy. Oats functioned well as a companion crop.

The concentrations in soil of the target PAHs studied in this work were not good indicators of whether or not revegetation could be achieved.

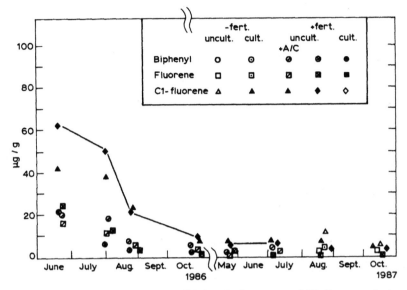

FIG. 8. Concentration (mg/kg soil) of biphenyls, fluorenes and C1-fluorenes during 1986 and 1987 growing seasons.

TABLE 4
Electrical conductivity (mS/cm) in topsoil (0–10 cm) in October 1986 and 1987

Treatment		Amount of invert applied (m³/ha)						
		0		150		300		450
		+F	−F	+F	−F	+F	−F	+F	−F
				12ᵃ		23ᵃ		34ᵃ	
None	(Oct. 1986)	—	—	2·9	4·3	3·9	6·2	3·7	9·1
	(Oct. 1987)	—	—	1·6	1·7	2·0	2·5	2·3	2·4
		—	—	4·6	6·5	5·5	8·4	10·3	10·3
Activated charcoal		—	—	0·9	1·9	2·2	2·4	3·5ᵇ	2·9
		—	—	2·5	4·1	4·5	7·0	5·1	12·0
Carbon granules		—	—	1·3	2·0	2·2	2·7	2·1	2·8
		—	—	2·5	4·4	4·5	6·0	6·8	11·9
Cultivation only		—	—	1·8	1·6	1·5	1·9	2·0ᵇ	1·6
		—	—	2·9	4·0	3·7ᵇ	3·3	4·4	7·5
Manure only		—	—	1·2	1·5	1·6	1·8	1·6	2·2
		—	—	3·0	3·5	6·1	6·8	7·1ᵇ	6·9
Manure + Cult.		—	—	1·2	1·7	1·7	1·7	2·4ᵇ	1·7

—: background value (0·5 mS approx.).
ᵃApprox. initial value, June 1986.
ᵇTopsoil EC was higher where fertilizer had been used on these plots.

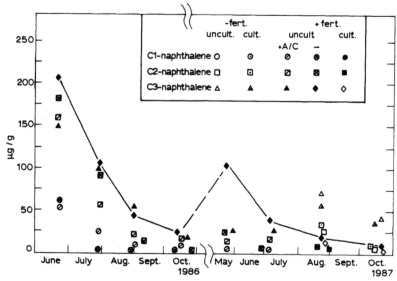

FIG. 9. Concentration (mg/kg soil) of substituted naphthalenes during 1986 and 1987 growing seasons.

TABLE 5
Subsoil conductivity (10–20 cm) EC in mS/cm, Oct. 1986 and 1987

Treatment	Amount of invert applied (m³/ha)							
	0		150		300		450	
	+F	−F	+F	−F	+F	−F	+F	−F
None (1986)	—	—	3·0	4·5	4·0	5·9	3·7	10·0
(1987)	—	—	1·8	2·4	2·1	2·6	2·3	3·2
Activated charcoal	—	—	2·0	1·8	4·6	5·6	7·7	9·3
	—	—	NS	1·5	2·2	2·5	4·1	2·8
Carbon granules	—	—	2·4	3·3	4·1	5·9	6·6	11·2
	—	—	1·0	1·7	2·0	2·6	1·8	4·0
Cultivation only	—	—	3·1	5·2	5·7	6·8	5·5	9·2
	—	—	NS	1·9	1·4	2·1	2·2	1·7
Manure only	—	—	3·0	4·6	5·5	4·2	5·2	6·8
	—	—	NS	2·2	NS	2·3	NS	2·0
Manure + Cult.	—	—	3·4	4·0	6·3	7·2	5·3	6·2
	—	—	NS	1·7	NS	2·1	NS	1·9

NS = no sample taken. Where there was any appreciable plant growth, soil cores were very hard to obtain below 10 cm.
—: background EC (0·5 mS approx.).

Total hydrocarbon concentration was a better indicator; when it fell below 1%, revegetation was possible given adequate N and moisture.

There was little or no leaching of target PAHs into subsoil even though the site was moderately well drained and received heavy rain in two consecutive growing seasons. Many DIMRs may contain large concentrations of soluble salts which will leach into subsoil and groundwater. The environmental hazard associated with leaching is considered small but is one of the factors to consider in deciding the feasibility of land application of DIMR.

It is impossible to draw general, reliable conclusions from one specific trial which is not yet complete. Results to date suggest: (1) Optimum rates of DIMR application are in the 150 m^3/ha range; (2) at rates of 300 m^3/ha or higher, it appears that the length of time and level of management required to achieve reclamation may be both economically and environmentally limiting. At least one more field season is required before definitive conclusions can be drawn.

ACKNOWLEDGEMENTS

We thank R. D. Glasrud for field work, Cheryl Engler for seed identification and D. Erickson (Enviro-Test Labs, Edmonton) for organic analyses.

REFERENCES

BRIGGS, G.G. (1976). Degradation in soils. *Proc. British Crop Protection Council Symposium on Persistence of Insecticides and Herbicides*, pp.41–54.

PEAKE, E., VISSER, S. & DANIELSON, R.M. (1986). Biodegradation and Analysis of Diesel Invert Mud Residues, Phase I. A Report prepared for the Canadian Petroleum Association and Environment Canada. 84 pp. Kananaskis Centre for Environmental Research, University of Calgary.

PLICE, M.J. (1948). Some effects of crude petroleum on soil fertility. *Soil Science Society of America Proceedings*, 13, 413–16.

WOLFE, D.A. (Ed.) (1977). Fate and effects of petroleum hydrocarbons in marine organisms and ecosystems. *Proc. Symposium held in Seattle, Wa.* Pergamon Press, Oxford, 478pp.

12

The AGIP Experience in Treatment and Disposal of Wastes on Deep Drilling Sites

G. Dossena*a*, G. P. Carta*b*, V. Crico*c* and F. Vallorani*c*

*a*AGIP Geodynamics and Environmental Studies Department, *b*AGIP
Environmental Studies Department, *c*AGIP Drilling Department, San
Donato Milanese, P.O. Box 12069, Milano, Italy

ABSTRACT

*Italian regulations governing the disposal of industrial wastes require the
use of an approved site, e.g. sanitary landfill, and prohibit any sludge farming
techniques. The liquid effluents may be discharged into surface water bodies
provided that these effluents meet the acceptability standards imposed by
the present Italian legislation. Facing such stringent requirements, the
Italian Oil Operator (AGIP) has selected an effective on-line method for
treatment and disposal of wastes.*

*In deep drilling operations, the method eliminates massive quantities of
wastes left behind upon completion of the well. The concept adopted by
AGIP allows minimum waste pit size, and maximizes final restoration of
the location. The present paper deals with the major aspects of AGIP
methodology for treatment of drilling wastes.*

1. LEGAL CONTEXT

Disposal of wastes and effluents in Italy are regulated by recently passed
environmental legislation. With regard to the disposal of industrial solid
waste and sludges in particular, the Presidential Decree No. 915/82 rules
that:

—authorization for each type of waste disposal must be obtained;
—wastes may not be abandoned and no form of land farming of sludge
  is allowed;
—industrial waste may not be dumped into municipal landfills; and

Parameter	Limit (mg/litre, exc. pH)
pH	5·5–9·5
Suspended solids	80
BOD	40
COD	160
Mineral oils and grease	5
Surfactants	2
Chloride	1 200
Sulphate	1 000
Ammonia	15
Lead	0·2
Mercury	0·005
Copper	0·1
Arsenic	0·5
Selenium	0·03
Cadmium	0·02
Chromium III	2
Chromium VI	0·2
Barium	20

FIG. 1. Some meaningful limits for effluent discharge in surface water bodies.

—waste dumped into a landfill must be of suitable consistency, i.e. without free water.

The disposal of waste water is subject to Law No. 319/76, the main points of which, with regard to discharge into surface waters, are the following:

—prior authorization of discharge;
—compliance of effluent waters with quality standard limits (Fig. 1); and,
—banning the use of dilution water to reach water quality.

Water quality standards are National Standards and they are independent of the characteristics of the receiving water body.

2. AVAILABILITY OF WASTE DISPOSAL SITES

As the law governing waste disposal is relatively new at present, the only landfills available in sufficient numbers are those for inert wastes (10–20 US$/t). In theory, drilling wastes may not be dumped in these sites, even if suitably compacted. In practice, however, local authorities allow the dumping of industrial wastes that meet a standard leaching test (IRSA, 1985).

This means that wastes may be dumped if the concentration of heavy metals in the leachate is within the limits shown in Fig. 1.

3. CHOICES FOR THE MANAGEMENT OF DRILLING WASTES

Faced with large amounts of waste (20 000–40 000 m³) produced while drilling one deep well (*c.* 5000 m) with the rig on site for over 200 days, it became evident that the traditional methods, collecting all wastes in a single, high capacity basin, were inadequate.

The particular problems were:

—finding, especially in Italy, a sufficient surface area to create large enough basins;
—having to hand back the area in its original condition;
—restoring the site in a reasonably short time;
—handling the wastes efficiently; and
—finding treatments for the disposal of this type of heterogeneous waste.

It became clear that the existing technology for treating wastes would no longer be sufficient in view of the recent legislation. This might have meant developing entirely new systems and more sophisticated equipment. But before embarking on this route, we decided to have a look at the crux of the matter, that is, the wastes themselves and how they were formed.

It was thought that the properties of wastes could be controlled or directed in such a way as to allow their disposal using already existing and tested knowledge. We adopted the following method:

—confinement and separation of wastes by type;
—reduction of the quantity of wastes produced;
—periodic treatment and disposal of wastes during drilling; and
—use of purified chemicals to reduce the toxicity of wastes.

This approach led to the development of standard methods and equipment for treatment and disposal of every type of waste allowing:

—use of much smaller basins for each type of waste;
—reduction of the maximum volume of all collected wastes to 2000 m³;
—leaving the site free of wastes and ready for immediate restoration; and

—optimization of costs for disposal of drilling wastes in compliance with the law.

4. DRILLING LOCATION LAYOUT

In order to meet the above objectives, it was necessary to review the regulations for location layout through defining the optimum number of waste basins required, as well as their position and size.

Figure 2 shows the traditional layout of drilling waste basins for a deep drilling site. Figure 3 shows standard layout for waste basins as used by AGIP.

The various structures, numbered progressively in Fig. 3, have the following functions:

(1) preliminary collection of waste drilling fluids (WDF);
(2) final lagooning of WDF awaiting treatment. The basin has a capacity of about 800–900 m^3;
(3) storage of any acid and/or oily fluid as well as any completion fluid with a high salt content;
(4, 5) collection, clarification and recycling of waste waters;
(6) collection and on-line consolidation of cuttings;
(7) storage of consolidated cuttings; and
(8) lagooning of depurated water before discharge.

5. TECHNIQUES TO REDUCE THE QUANTITY OF WASTES TO BE DISPOSED

Reduction in total drilling waste volume is mainly a result of improved mechanical solids-control equipment. Storage of different fluids in specific basins allowed further reductions in the quantity of wastes as well as its reuse. The best example is the reuse of waste water for drilling operations such as making up or diluting drilling fluids, washing, etc., thus avoiding the use of additional water. Table 1 gives some quantities of the waste produced by the new method of drilling waste management.

6. TREATMENT AND DISPOSAL OF DRILLING WASTES

Systematic application of the methods and criteria described above has led to five main types of waste on drilling sites, each having predefined

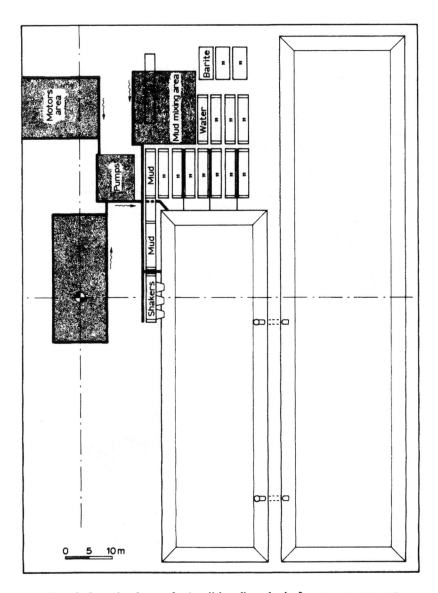

FIG. 2. Location layout for 'traditional' method of waste management.

G. Dossena et al.

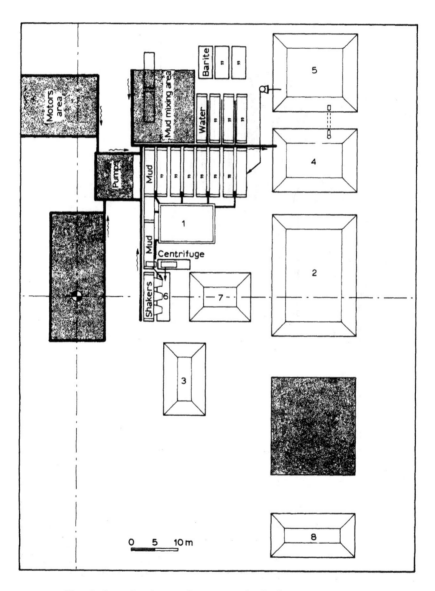

FIG. 3. Location layout for new method of waste management.

TABLE 1
Statistical data on wastes production and average disposal costs

Type of waste	Production $(m^3/drilled\ m)$	Average disposal cost $(US\$/m^3)$
Drilling fluids	1·30	42·9
Cuttings	0·36	27·8
Waters	0·15	13·5
Contingency fluids	0·015	158·7

characteristics and being reasonably consistent. They consist of:

—waste drilling fluids;
—cuttings;
—slightly polluted water;
—contingency fluids, such as oily and acid fluids; and
—civil wastes.

A suitable disposal system can be identified for each of these.

6.1. Waste Drilling Fluids
Waste drilling fluids are periodically treated and disposed of by using a mobile unit. The treatment process involves chemical treatment followed by mechanical dewatering. Dewatering is achieved by centrifuges or filter presses.

The aqueous phase is then subjected to neutralization and powdered activated carbon treatment before being discharged into surface waters near the site. The solid phase is sent to a suitable disposal site. Such treatment reduces waste volumes and disposal costs. Also, suitable geotechnical characteristics of solid waste are obtained. Disposal of the solid waste produced by dewatering may be difficult because of the already mentioned lack of sites. Also, the possibilities of using solid wastes from dewatering as additives in brick and tile manufacturing, were investigated. This is an important issue in Italy where the brick and tile industries are widespread. The research regarding the technical feasibility of such a proposed disposal method was undertaken by a specialized institute (Fabbri *et al.*, 1982).

6.2. Cuttings
Cuttings, along with wastes from the mud-cleaners and centrifuges, are consolidated by cementing binders. These materials present minor

problems in terms of disposal as they are suitable infills for worked-out quarries that are common all over the country.

6.3. Slightly Polluted Water (Waste Water)

These waters are usually reused by the rig. A double water circuit is required on the rig because this waste water is usually quite turbid (500–700 mg/litre) and therefore unsuitable for use in the electric generators and dynamic brakes to which only limpid water is recycled. Reuse of waste water results in an average reduction of 50% in water consumption of a rig (40 m^3/day against 80 m^3/day).

6.4. Waste Contingency Fluids

Waste contingency fluids, acid and oily spots, the most polluting materials on a drilling site, have to be carefully recovered and segregated because even small quantities can contaminate other types of waste rendering treatment procedures ineffective. In practice, these fluids are separated as soon as they return to the surface and are sent to basin 3 (Fig. 3) via a metal trough. No on-site treatment is provided for these materials; they are promptly transported to specialized companies for disposal.

6.5. Civil Wastes

A collection system, consisting of a network of waste bins, has also been set up for 'civil wastes' (bottles, cans, old work gloves, etc.). The objective in this case is not so much to collect such waste, but to avoid it finding its way into the lagooning basins where it could damage or slow down treatment plants. Figure 4 shows the various phases in the management, treatment and disposal of all drilling wastes.

7. CONCLUSIONS

The described method has been developed mainly to allow the disposal of drilling wastes according to the new Italian environmental regulations and the availability of suitable landfills. The improvement in waste management led to 50% reduction in the quantity of waste to be treated. In spite of the complexity of the treatments, the disposal costs have been significantly reduced, though the exact figures are still not available.

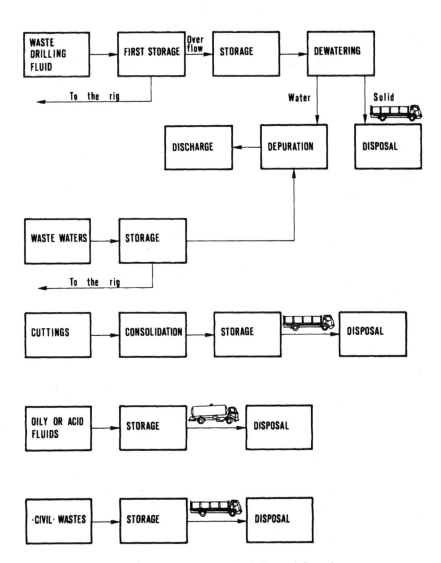

FIG. 4. Wastes treatment and disposal flow-chart.

G. Dossena et al.

REFERENCES

IRSA (1985). Metodi analitici per i fanghi—App. 2—*Quad. Ist. Ric. Acq.*, 64.
FABBRI, B., VENTURI, I. & CARTA, G.P. (1982). Smaltimento di fanghi di perforazione di pozzi per ricerca di idrocarburi mediante introduzione in impasti da laterizi, *Refrattari e Laterizi*, **82**, 171–5.

13

Treatment of Salt Contaminated Reserve Pit Material and Drilled Cuttings

BETH C. KNOL

Shell Western E&P Inc., Grandview Plaza, Traverse City, Michigan 49684, USA

ABSTRACT

A number of wells in the United States are now drilled without reserve pits in areas with highly permeable soils. Removal of salt contaminated drilling solids and cuttings from existing reserve pits has been required in environmentally sensitive areas. As an option to hauling these salt saturated materials to a state approved disposal site, a process was developed to reduce the salt concentration to a minimal level. The proposed process involves adding fresh water to the contaminated material at a centralized plant to dissolve crystallized salt and to reduce the salt concentration in the liquid phase. The salty water is then extracted and pumped to a disposal well. The washing process is repeated until the residual salt concentration is acceptable. The desalinized, dewatered material can be utilized for construction purposes or disposed in unclassified disposal sites. The concern of the regulatory agency over residual aromatic hydrocarbons in the processed solids has delayed implementation of this project.

1. DESIGN CRITERIA SUMMARY

The following criteria were considered in the development of the desalinization process. Some of these constraints are described in greater detail later in the text.

1. The amount of salt retained with the processed cuttings must be controlled by varying either the concentration of salt in the liquid phase of the cuttings or the liquid volume retained with the cuttings

271

or both. The volume of salt retained equals the volume of liquid in the cuttings multiplied by the concentration of salt in the liquid. Because the kinds of salt vary depending on the formation from which the cuttings came, a chloride ion concentration of 1500 ppm or less in the liquid phase was set as an arbitrary target.
2. The processing plant must be capable of handling a variety of aggregate mixtures, i.e. coarse, very fine or all salt.
3. The quality of the saline water extracted from the cuttings must be such that a salt water disposal well would not be plugged by suspended solids remaining in the liquid.
4. The quantity of liquid effluent and size of solids retained with the effluent must be within the range of disposal wells.
5. The processing equipment must be readily available with proven performance in field applications.
6. The cost of the plant and its operation must be less than the cost of hauling and disposing the material in a classified landfill and/or the plant must have some other utility.

2. PROCESS DESCRIPTION

After a cursory look at a number of pieces of equipment, two combinations appeared to be viable alternatives: aggregate washing equipment combined with a clarifier or aggregate washing equipment with mud plant and solids control equipment. The first option was capable of processing cuttings but it had some drawbacks. Principally, the clarifier was less efficient in handling fine materials because the exiting solids have a higher water content than the same material exiting from solids-control equipment. This means that less salt is disposed during a cycle and consequently more recycling is needed to reach the desired salt concentration. When the fresh water consumed per pound of salt removed was compared, the clarifier did not perform as well as the solids-control equipment. In addition, although the clarifier is a standard piece of equipment in the sewage and aggregate industries, oil field personnel are not familiar with it. Solids-control equipment, on the other hand, is utilized universally for drilling wells. The remainder of this paper will concentrate on the method of processing which appears optimum at this time: aggregate washing equipment combined with solids-control equipment. The overall process is schematically represented in Fig. 1 and is described below.

Prior to processing, salt contaminated material is stored on a curbed

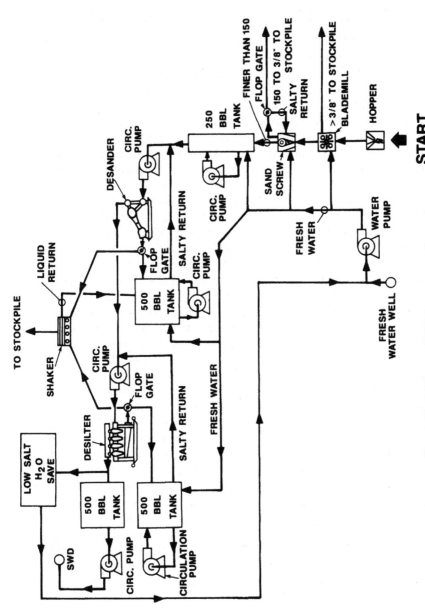

Fig. 1. Shell Western E&P Inc. process flow scheme

asphalt pad. Once enough material is accumulated to begin processing, the aggregate is placed on a conveyor which transfers the material to the blademill. The blademill is basically a horizontal tub which encloses a shaft on which are mounted screw flights and paddles. The purpose of the blademill is to precondition the material for separation. Water is introduced to uniformly saturate the material while the paddles scrub and abrade the material so that lumps are broken up. Because of this mechanical action, clays and silts can be separated more readily. The broken up material then drops from the blademill onto a screen where water jets spray the cuttings as they pass through the screen; particles larger than 0·0095 m are rejected and transferred to the stockpile via a conveyor. The rationale for not recycling materials larger than this size is that they were not part of the drilled aggregate† but part of the native soil used to stabilize reserve pits prior to closure. Hence, they are only surface wet with salty water. The fresh water spray over the screen after the blademill is sufficient to remove this salt. After passing through the screen, the material finer than 0·0095 m is fed to the sand screw via chutes.

The sand screw consists of a base tub where the material enters and an inclined tub which contains a continuous screw flight (an auger). The purpose of this machine is to separate sand size particles (150 mesh to 0·009 5 m) from finer material. This is accomplished by the tumbling action of the rotating continuous screw and the introduction of a rising current of fresh water into the base tub of the sand screw. Fine materials are carried upward by the flow of water over a weir into a 250 barrel mixing tank. The coarse material on the other hand settles to the bottom of the sand screw tub where it falls onto the continuously rotating screw. This screw carries the water saturated 150 mesh to 0·009 5 m size material upward within the inclined tub. As the material is carried upward, the water within the material is squeezed out by the rotation of the screw against the curved base of the tub. The exiting material has been dewatered from saturation to a water content of 18–25% (water weight/solids weight) depending on the material. The salinity of the water that is squeezed from the coarse material is measured continuously by a conductivity sensor. This gauge controls the position of the flop gate which determines which of two conveyors the dewatered material will land on. If the salinity is acceptable, the dewatered material is sent to the 150 mesh to 0·0095 m stockpile. If the salt concentration is too high, the dewatered material is returned to the sand screw by means of a transfer conveyor. The dilution/

†The cutoff size of 0·009 5 m was based on field data specific to Michigan. Screen size may have to be adjusted according to the area of use.

extraction process is repeated until an acceptable level of salt concentration is reached. During this recycling process, the feed from the hopper to the blademill is stopped until the previous batch of material exits the plant.

Cuttings finer than the 150 mesh (100 μm) have been carried to the 250 barrel mixing tank by the upward current in the sand screw tub. More water is added to this mixture until the percentage of suspended solids within the fluid is 5–7% by volume, which is an operational constraint of the desander and desilter. The volume of solids can be determined automatically with gauges typically used in municipal sewage plants. To prevent the solids from settling to the bottom of the tank, the fluid is continuously mixed by hydraulic agitators run by the circulation pump. Once the proper consistency is obtained, the fluid in the mixing tank is pumped to the desander.

Both the desander and the desilter consists of a series of tapered cones. Particles are separated from the liquid phase by means of centrifugal force as the fluid swirls inside the cone. Solids exit through the tapered end of the cone while the liquid comes off the top. The size of particle which can be removed is a function of the cone dimensions and the rate at which the fluid is pumped to the cones. Larger cones remove larger particles and have higher throughput per cone. Cones sized to remove sand particles and silt size particles are called desanders and desilters respectively. For this study, a two cone (12″) desander and a twenty cone (4″) desilter were assumed. Both require a fluid throughput of $3\cdot78\,\text{m}^3/\text{min}$ (1000 gpm).

The desander will remove all particles coarser than 80 μm and a portion of the particles finer than this size. Most of the fluid and all particles smaller than those described are discharged to the desilter. The salinity of the exiting fluid is measured by the conductivity sensor. If the salinity is acceptable, the flop gate beneath the desander directs the material to a shaker where any free moisture can drain through the shaker screen and be returned to the desander's storage tank. The drained cuttings from the shaker fall onto a conveyor which carries them to the stockpile. Salty cuttings, on the other hand, fall into the 500 barrel tank beneath the desander. Since the exiting material from the desander has a 35% water content or less, fresh water must be added to meet the suspended solids constraint and to dilute the salt concentration. Once the proper density has been obtained, the fluid is pumped through the desander again until an acceptable level of salinity is reached.

The fluid and particles outside of the range of the desander are pumped to the desilter which nominally removes all particles down to around 40 μm and some of the particles down to 12 μm. Material from the desilter goes to the same shaker as the desander if the salinity is acceptable. If not, the material is discharged to the tank beneath the desilter for recycling where fresh water must be added to meet the suspended solids constraint and to dilute the salt. Fluid from the desilter is discharged to a salty water surge tank. Depending on the characteristics of the disposal well and the amount of solids carried over with the wash water, it may be necessary to flocculate or to filter the wash water prior to disposal in a well. If the exiting water has low salinity (5000 ppm or less), the flow can be diverted to a storage tank for reuse.

3. LOADING CONDITIONS

The proposed process is designed to treat salt contaminated aggregate which comes from two sources: excavated (old) reserve pits and closed system drilling. Reserve pit material is a mixture of drilled cuttings (natural mineral aggregate which comprises the well bore), salt water-based drilling mud and native soil which was used to solidify the pit material prior to burial. A reserve pit typically contains about 900 m^3 of this mixture including the native soil.

In areas where reserve pits are not permitted, the drilled cuttings are contained in tanks. The material is similar to reserve pit material except that only a small amount of local soil is added to the tanks for stabilization prior to hauling. The total volume per well is about 150–310 m^3 depending on well depth and other factors.

For the purpose of this study, it was assumed that a truckload (nominally 7·6 m^3) of material delivered for processing could consist entirely of all coarse, i.e. greater than 150 mesh, or all fine aggregate rather than a mixture of the two. This is the worst loading condition for the plant because one piece of equipment (sand screw, desander or desilter) must process the entire volume of the material. This is not an unreasonable assumption for reserve pits since they originally functioned as settling basins for the drilled cuttings. Hence, layers of all coarse or all fine materials might be encountered. For the desilter and desander, the fine material is the worst loading condition, whereas all coarse material is the most stringent loading condition for the sand screw. To assess the effectiveness of each piece of equipment, its performance was evaluated assuming these extreme loading conditions. The liquid phase of the aggregate is assumed to be saturated with NaCl. Table 1 presents the initial and exiting material conditions. In addition, the time and fresh

TABLE 1
Processing data for sand screw and desander[a]

A. Full volume to sand screw—all coarse material

	Initial condition (N)	Exiting condition (N)
	10 570 salt	16·90 salt
	119 130 solids	119 130 solids
	30 020 H_2O	30 020 H_2O

Pass No.	Fresh H_2O (m^3)	Salt concentration[b] liquid phase (ppm)	Salt left (N)	Elapsed time (min)
1	4·775	69 000	2 122	5
2	4·775	14 100	422	10
3	4·775	2 800	84·5	15
4	4·775	600	16·9	20
	19·100			

B. Full volume to desander or desilter—all fine material

	Initial condition (N)	Exiting material (N)
	15 060 salt	32·02 salt
	85 400 solids	85 400 solids
	42 700 H_2O^c	29 890 H_2O^c

Pass No.	Fresh H_2O (m^3)	Salt concentration[b] liquid phase (ppm)	Salt left (N)	Elapsed time (min)
1	61·32	23 100	694	18·3
2	63·03	1 070	32·02	36·6
	124·35			

[a]Data based on one truckload of material or approximately 7·6 m³.
[b]Salt used for analysis is NaCl; multiply values by 0·063 to obtain Cl ion concentration.
[c]Water content of unprocessed material is 50% based on field data; processed material has a reduced water content of 35%.

water requirements are also included. A mixture of coarse and fine materials would result in a reduction in total batch time.

It should be noted that the values obtained in this table are based on conservative assumptions with respect to the equipment itself. For example, the density of the fluid to the desander and the desilter was limited to 5% suspended solids by volume and an overall density of 9·5 lb/gal. This equipment is typically capable of handling 18 pound mud when weighted with barite. Performance data on the sand screw indicate that the moisture content of coarse material which has been through the sand screw is 18% and no more than 25% for materials on the fine end of its range. A moisture content of 25% was assumed for this study. This means that for every pass through the sand screw a larger percentage of dissolved salt is retained with the solids which in turn necessitates more cycles for desalinization of the solids.

4. PLANT PROCESSING CAPACITY

In this study, a sand screw capable of moving 179 t/h of aggregate was chosen. The two cone desander and twenty cone desilter selected process 3·78 m³/min of fluid. This corresponds to about 12–14 m³/h of aggregate. It should be noted that all these pieces of equipment are available in a wide range of capacities. The key is to select units that are compatible and that are appropriate for the expected size distribution of the drilled cuttings. In addition, fresh water availability and disposal capacity must also be considered. The subsequent paragraph exemplifies these points.

As shown in Table 1, it takes approximately 20 min to process one 10 yard batch of coarse material through the sand screw and 37 min to process fine material. This is the fastest that material can be processed. Besides equipment limitations, it is also necessary to consider salt water disposal and fresh water supply requirements in determining realistic daily volumes. In addition, if fresh water cannot meet processing demands, it may be necessary to retard the processing rate or limit the number of hours of processing per day and provide fresh water storage tanks. It is important to note that the desander and desilter must operate at their design rate of 3·78 m³/min (1000 gpm) to function properly. If the capacity of the salt water disposal well is below this value, then it is necessary to include surge tanks. Table 2 presents various operating conditions from the least constrained to the most constrained with the corresponding volume processed.

TABLE 2
Daily processing capabilities under various operating constraints

		Fine material	Coarse material
Case I	Equipment limit per day		
	Volume (m^3)	319	573
	Cycle time (min)	37	20
	Processing time (h)	24	24
	Fresh water required (m^3)	2 387	1 755
Case II	Saltwater disposal limited to 8000 BPD		
	Volume (m^3)	162	419
	Cycle time (min)	37	20
	Processing time (h)	12·2	17·4
	Fresh water required (m^3)	1 210	1 270
	Salt storage required (m^3)	573	350
	Salt surge volume (m^3)	45·2	23·5
Case III	Hauling limited to 10 h		
	Volume (m^3)	132	241
	Cycle time (min)	37	20
	Processing time (h)	10	10
	Fresh water required (m^3)	1 005	731
	Fresh storage required (m^3)	587	427
	Salt storage required (m^3)	476	201
	Salt surge volume (m^3)	45·2	23·5
Case IV	Fresh water limited to 4000 BPD		
	Volume (m^3)	86	210
	Cycle time (min)	37	20
	Processing time (h)	6·5	10
	Fresh water required (m^3)	645	636
	Fresh storage required (m^3)	375	368
	Salt storage required (m^3)	304	201
	Salt surge volume (m^3)	45·2	23·5
Case V	Salt water disposal—(0·844 m^3/min)		
	Volume (m^3)	71	178
	Cycle time (min)	69	28
	Processing time (h)	10	10
	Fresh water required (m^3)	531	541
	Fresh water storage (m^3)	265	273
	Salt storage required (m^3)	0	0
	Salt surge volume (m^3)	45·2	23·5

NOTE:1. Each case includes the limitations of all previous cases.
2. Storage requirements are total needed. No reduction has been taken for the 200 m^3 available from the desander, desilter, and mixing tank.
3. Recycling of fresh water is assumed.
4. The incoming material is assumed to be saturated with NaCl.

To ameliorate both fresh and salt water limitations, recycling of low salinity water is desirable. Effluent from the third and fourth pass of the sand screw and the second pass of the desander and desilter can be used for the initial washing of salt saturated cuttings.

5. PROCESS QUALITY CONTROL

To insure that no aggregate material exits the plant until it is 'desalinized', the salinity of the material must be continuously monitored; sporadic checking is not sufficient. As mentioned earlier, conductivity sensors can be used to control the process because electrical conductivity is proportional to salinity and can be measured with great accuracy. In addition, the change in specific conductance is directly proportional to the change in molarity in the range of interest, i.e. very low salt concentrations. Table 3 presents specific data for solutions of sodium chloride.

Figure 2 presents schematically how conductivity sensors can be used to control the desalinization process. A mixture of saline water and aggregate enters the separation device such as a sand screw which mechanically separates the mixture into aggregate and saline water. The conductivity (salinity) of the fluid is measured by the sensor and transmitted to the controller. The controller 'determines' if the conductivity is higher

TABLE 3
Conductivity sensor and specific
conductance data
(Resolution to 1×10^{-9} Mho
available; process range 0·225 Mho
to 0·001 Mho)

Example NaCl

Molarity	Cl (ppm)	Specific conductance Mho/cm
0·01	354	0·0010
0·05	1 770	0·0050
0·10	3 545	0·0097
1·00	35 450	0·0778
2·00	70 950	0·1350
5·326	188 940	0·2250

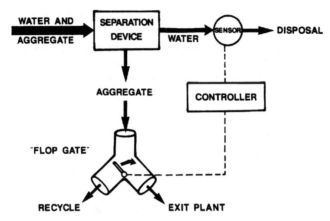

FIG. 2. Method for controlling recycling.

than the desired conductivity or salinity which has been preset by the operator. If it is, the flop gate remains closed and the material is recycled through the separation device again. If the salinity is less than or equal to the desired value, the controller opens the flop gate which causes the material to exit the plant.

The instrumentation system proposed should be designed 'fail safe'. A 'fail safe' system is one in which failure of a sensor or controller results in the aggregate being recycled through the plant until the problem is fixed. For example, the controller first checks to see if the sensor is producing a minimum signal level. If not, the flop gate is left in the recycle position. The normal position of the flop gate is such that the material recycles; mechanical power must be exerted to move the flop gate to the 'exit' position. Therefore, if the controller malfunctions the material continues to be recycled. In summary, the salinity of the aggregate can be continuously monitored with reliable instrumentation with a high degree of accuracy and a fail safe design.

6. LAND APPLICATION

The desalinated, dewatered aggregate can be used for construction purposes or disposed of in an unclassified landfill. If utilized for construction, the Department of Natural Resources of the State of Michigan has proposed that the total chloride applied be limited to $0.574 \, \text{N/m}^2$ of chloride (0.012 pounds/ft^2) per year. This application rate results in a concentration of salt in the leachate which meets drinking water standards when dilution due to precipitation is considered. The application rate is expressed in terms of chloride ions in consideration of the fact that other salts besides NaCl are also encountered. Assuming a maximum level of 1500 ppm of chloride and the predicted moisture content of the processed material, the following application rate results:

Fine Material—desander and desilter—336 m³/acre
Coarse Material—sand screw—433 m³/acre

Although the proposed application rate is expressed on a square metre basis, from a practical standpoint the area covered by one truckload of

SIZE: 300'x300'≅ 2.07 ACRES

MATERIAL REQ'D: COARSE ONLY

ALLOWABLE APPLICATION: 567yds³/ACRE TOTAL

VOLUME PERMITTED: 2.07x567= 1173cy

TRUCK SIZE: 40cy/LOAD

MAXIMUM LOADS: 1173cy÷40= 29LOADS

$$\textbf{AREA SPACING:} \quad \frac{\textbf{AREA}}{\textbf{LOAD}} = \frac{\textbf{300'x300'}}{\textbf{29}} = \frac{\textbf{3100ft}^2}{\textbf{LOAD}}$$

$$\textbf{LINEAR SPACING:} \quad \sqrt{\textbf{3100ft}^2} \cong \textbf{55ft}$$

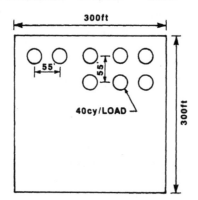

FIG. 3. Example—drilling location.

material is the smallest area over which can be controlled. Shell Western has proposed to control the spacing of dumped loads so that the average application rate meets the required limit. Figure 3 gives an example of how this can be done.

7. STATUS

Unfortunately, this project has not been implemented to date. The delay is due to a disagreement about the level of benzene which is tolerable. The source of the benzene is the reservoir rock which is drilled through, small amounts from pipe 'dope' used on the drillstring and hydrocarbons in the drilling mud which is made from produced water that contains trace quantities of benzene. Michigan's Department of Natural Resources has proposed a limit of 20 parts per billion of benzene on a dry soil weight basis in the draft operating permit. Data are currently being developed to demonstrate that this level is unreasonably stringent. Volatilization, absorption by the fresh water during the desalinization process, soil adsorption, and biodegradation all contribute to reducing very low initial levels even lower.

On the bright side, a large percentage of the equipment needed to process drilled cuttings is also suitable for storing and circulating drilling mud. Shell Western has found that it is highly profitable to store and reuse drilling fluids from one well to the next instead of paying for disposal and then making the mud from 'scratch' at the next well site. Consequently, a decision was made in May of 1987 to install all equipment that could be used both for solids processing and mud recycling. In addition, a large asphalt storage pad has been built for containing salty cuttings. So, when the benzene problem is resolved, a large percentage of the project will already be in place and functioning.

14

Disposal of Salt Water Drilling Mud

Surjit S. Nagra

Research Department, Esso Resources Canada Ltd, 339 50th Avenue SE, Calgary, Alberta, Canada T2G 2B3

and

Ron R. MacDonald

Esso Resources Canada Ltd, 237 Fourth Avenue SW, Calgary, Alberta, Canada T2P 0H6

ABSTRACT

Saturated salt water mud is commonly used to drill through salt zones. Potassium chloride salt-based muds are also used to drill through water sensitive shales. The major environmental concerns when disposing of salt water mud are surface soil and ground water contamination from soluble salts. This makes it necessary to reduce salt from mud to acceptable levels before disposal. Initial attempts using a fresh water extraction technique to remove salts from drilling mud sumps failed to reduce the salt levels below 15 000 ppm even after 3 or 4 extractions. Various other techniques such as filtration, centrifugation and chemical precipitation were also considered but did not appear to be viable or cost effective. Re-examination of the fresh water extraction technique in the laboratory continued to show that it was possible to reduce the salt content of the mud to acceptable levels by thorough and efficient mixing of the mud with the fresh water. A modified technique was tried in the field using a submersible pump unit to thoroughly mix the sump fluids with the added fresh water. The chloride levels were reduced to around 4000 ppm from 98 000 ppm after 3–4 extractions with fresh water. Based on these results, regulatory approval was obtained for a trial on-lease spreading of sump solids and downhole disposal of the liquid.

285

1. INTRODUCTION

Drilling mud is usually used in drilling wells. The major functions of the drilling mud are to remove and carry drill cuttings to the surface, control subsurface pressures, lubricate and cool the bit, provide bore hole stability and provide maximum information about the formation being penetrated (Moore, 1974; Gray *et al.*, 1979). The different types of drilling fluids include fresh water, salt water, oil base, polyemulsion, air and foam, etc. The selection of mud type depends on the cost of the system, associated drilling expenses, requirements for evaluating well characteristics, and anticipated problems related to the drilling formation to be encountered.

Once the drilling is completed, the waste mud and the drilled solids are dumped in a pit or sump. The sump fluids and solids must then be disposed of. Considerable attention has been focused in recent years on potential environmental impacts from these drilling fluid wastes (Miller & Honarrar, 1975; Nesbitt & Sanders, 1981; Hinds *et al.*, 1986). The major concerns expressed in the disposal of saline drilling fluids are surface soil and ground water contamination from soluble salts and trace metals contained in the waste mud (Moseley, 1983). However, there is a lack of information about environmentally acceptable limits to the disposal of salt water muds.

Esso Resources Canada Limited has several produced water disposal wells drilled into the Cambrian formation at Cold Lake. During the course of drilling, two salt zones are penetrated, which require a saturated salt-base mud to avoid extreme erosion of the salt zone. Because of the saline nature of the resulting drilling fluid wastes (approximately 150 000 ppm NaCl), the Alberta Forestry Services (AFS) and Energy Resources Conservation Board (ERCB) continued to express their concerns about the disposal methods.

In 1980, Esso's Research Department investigated the removal of salt from the waste mud by a fresh water extraction technique. This consisted of mixing the salty mud with fresh water in a one to one ratio (1:1) several times and decanting the supernatant liquid each time. Two field trials of this method took place in 1981. Approximately 70% reduction in the salt content was obtained after two flushings but little was achieved by a third flushing. The chloride content of the mud was still around 15 000 ppm or higher. It was speculated that further chlorides could not be removed by this technique, as these chlorides could possibly be chemically bound to the clays in the mud. Further laboratory tests confirmed that extraction was a viable method to reduce the salt concentration of the mud, provided the mud was mixed thoroughly with the water. When this was done, over 98% of the salt was removed from the mud. Efficient mixing of the mud with the water was identified as the key to success. The laboratory and field results are presented in this paper.

2. LABORATORY EXPERIMENTAL PROCEDURES

Samples of the salt water mud from four different sumps from disposal wells in the Cold Lake area were brought to the laboratory. The fresh water extraction technique was tried on mud samples obtained from a number of sumps under various conditions of dilution and mixing.

Five hundred millilitres (500 ml) of the mud were poured into a graduated cylinder. A known amount of fresh water was added to the cylinder. The cylinder was inverted and shaken a few times by hand in order to mix the water with the mud. This resulted in a homogeneous mixture. The solids were then allowed to settle. The water from the top layer was decanted from the mud. The decanted liquid was analyzed for Na^+, K^+, Ca^{2+}, Cl^- and SO_4^{2-} ions. The metal ions Na^+, K^+ and Ca^{2+} were analyzed by the ICP (Inductively Coupled Plasma) technique. The Cl^- and SO_4^{2-} were determined quantitatively by volumetric and gravimetric techniques. A known weight of the mud was taken from the bottom layer. It was thoroughly washed with water a number of times until no more chlorides were eluted. This was then diluted to a fixed volume and analyzed for Na^+, K^+, Ca^{2+}, Cl^- and SO_4^{2-} ions. The experiments were also repeated when the mixing of the water with the mud was poor.

3. LABORATORY RESULTS AND DISCUSSION

3.1. Drilling Muds for Cambrian Wells

Figure 1 is a schematic of the Cambrian Cold Lake wells showing various mud systems employed during drilling operations. The top 650 m of the wells were drilled with a fresh water–bentonite system mud. The salt zone interval was then drilled with the saturated NaCl–bentonite mud system. Once the drilling of each interval was completed, the drilling fluids were dumped into a common sump. From the volumes of each drilling mud system used, the mass balance indicated that the concentration of NaCl in the fluids in the sump would be around 155 000 ppm. This was found

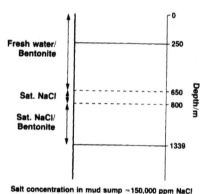

FIG. 1. Mud systems used to drill Cambrian salt water disposal wells at Cold Lake.

to be in good agreement with a laboratory determined salinity of the sump fluids of 165 000 ppm NaCl.

3.2. Disposal Process

The approach to the salt water mud disposal problem was to significantly reduce the salinity of the mud. A process by which the salt could be effectively removed from the mud solids before disposal had to be established. Since the salt is readily soluble in water, a means for thorough mixing of the sump contents with added fresh water and an efficient method of separation of the liquids from the solids were required. The recovered saline water could then be disposed of in a disposal well and the solids which contained a lower level of salt could then either be buried, land farmed or reused. The schematic of this process is shown in Fig. 2. A total volume of 1500–2000 m^3 of the drilling mud effluents had to be disposed.

3.3. Separation of Liquids from Solids

The options considered for the separation of liquid effluents from the solids were as follows.

3.3.1. Centrifugation

A 50 ml sample of drilling mud waste was centrifuged in a high speed centrifuge. Liquids were decanted from the solids. After drying the centrifuged solids in the oven at 100 °C to constant weight, it was found

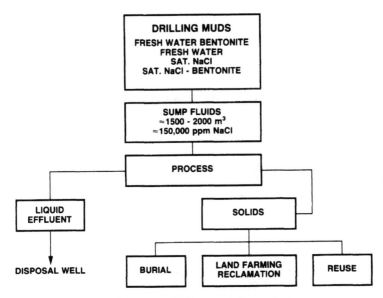

FIG. 2. Salt water drilling mud disposal scheme.

that 40% of the mass of the solid was water and because the water retained the salt, a considerable amount of salt was left behind in the solids. Therefore, this did not appear to be a viable method of separation of salt from the drilling sump fluids.

3.3.2. Pressure Filtration
Filter presses are used in the mining industry for dewatering the mineral concentrates. Dewatering by a filter press depends on the salt water being displaced by fresh water forced into the mud cake under pressure. Mud filtration tests on sump effluents were carried out at the Outokumpu Equipment Laboratory in Mississauga, Ontario, using a Larox PF 0·1 H2 lab pressure filter with a filter cloth of Tamfelt 71-2155 polypropylene. The test results indicated that a very long time was required to form a solid cake. The filtration rate of the mud was very low during the pumping cycle even under a 700 kPa pumping pressure. Increasing the pressure to 1600 kPa had little effect on the filtration rate. Even after washing the cake with water followed by pressure filtration, the NaCl content in the mud cake remained higher than 20 000 ppm. The salt and moisture levels remaining in the filter cake were unacceptable. Thus pressure filtration

was found to be an impractical field method of dewatering bentonite salt water drilling mud.

3.3.3. Evaporation
This process was considered but did not appear to be worthwhile since after evaporation of the water from the sump fluids, the salt is left behind with the solids.

3.3.4. Chemical Precipitation
This process would be highly uneconomic and impractical since the only chlorides which are insoluble in water are silver, mercury and lead which are highly toxic.

3.3.5. Fresh Water Extraction
This technique appeared to be the most practical and economical of all the options considered. The principle of the method is that salt readily dissolves in water. By repeated extractions with fresh water the extracted saline water from the mud could be disposed of down hole. In this way it was considered possible to remove most of the salt from the sump effluents.

3.4. Fresh Water Extraction with Efficient Mixing (Laboratory Results)
In a typical experiment, 500 ml of the waste drilling mud were thoroughly mixed with 500 ml of fresh water in a measuring cylinder. About 400 ml of the water separated from the mixture after settling for about 2 h. The moisture content of the settled mud was determined to be about 75% by weight. The decanted liquid and mud were subsequently analyzed for various ions. The laboratory results of the fresh water extraction technique with various amounts of water added to the mud and the number of extractions are shown in Fig. 3. Initially, the liquid decanted from the mud (without dilution) contained 95 390 ppm Cl^- and the mud contained 74 500 ppm Cl^- by weight.

The results shown in Fig. 3 demonstrate that the fresh water extraction technique is capable of substantially reducing the salt content of the sump fluids. The number of repeated extractions required to reduce the salt content depends upon the amount of fresh water added to the sump fluids and ultimate concentration of salt desired. The key to success is thorough mixing of the mud with the added fresh water.

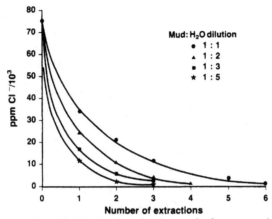

FIG. 3. Concentration of chloride ions in the mud after successive extractions with water.

3.5. Fresh Water Extraction with Poor Mixing

Experiments were also conducted in which a known amount of the mud in a test tube was diluted with water in a one to one (1:1) and one to two (1:2) ratio but was not mixed thoroughly. The water was added on top of the mud and allowed to mix poorly by slow agitation. The process was repeated 5 times; samples of the mud phase were collected from the top, middle and bottom and analyzed for Na^+, K^+, Ca^{2+}, Cl^- and SO_4^{2-} ions. The results of these experiments are plotted in Fig. 4 along with the

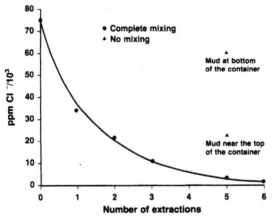

FIG. 4. Concentration of chloride ions in the mud after successive extractions with water in 1:1 dilution.

results of experiments where the mixing was thorough. It shows that when the mixing was very poor, the mud at the bottom had Cl^- concentration of 65 000 ppm even after 5 extractions, in comparison to the Cl^- of 74 500 ppm in the original mud sample. Even the mud at the interface of the liquid had a Cl^- concentration of 25 000 ppm. These experiments clearly demonstrated the importance of efficiently mixing the water with the mud using the fresh water extraction technique.

4. FIELD TEST RESULTS OF THE FRESH WATER EXTRACTION TECHNIQUE WITH EFFICIENT MIXING AT COLD LAKE

After successful laboratory experiments, a field test was conducted. Initial attempts at mixing the sump contents and fresh water with a backhoe were found to be unsuccessful. The chloride content of the mud could not be brought any lower than 15 000 ppm even after a number of repeated extractions. A second trial was made using a high speed submersible centrifugal pump shown in Fig. 5. The pump was able to suck the mud from the bottom and thoroughly agitate the sump contents. The mud would come out from the top in the form of a fountain as shown in Fig. 6. The submersible pump was suspended into the sump from a crane which could move the pump around from one end of the sump to another. In this way the contents of the sump were efficiently mixed with the added fresh water. In some tests flocculants were added to the sump contents but were later found not to be necessary. After mixing, the solids were allowed to settle while the water from the top was pumped into tank trucks and hauled to a disposal site. It was then pumped down disposal wells after going through sand filters.

This fresh water extraction technique was repeated a number of times until the chloride content was brought down to acceptable levels. The field results from four different sumps are reported in Table 1 together with the number of one to one (1:1) extractions employed. The reason for employing only (1:1) mud to fresh water ratio in the field was to minimize the amount of fresh water trucked to the sump site. The data reported in Table 1 suggest that in most cases the chloride content was brought down to less than 5000 ppm chlorides.

At this reduced concentration, the sump solids were approved to be mixed with the lease surface soil and spread over the well site lease. This approval was granted on a trial basis and the run-off from the lease is

FIG. 5. Centrifugal submersible pump unit used for efficient mixing of the mud solids with the added fresh water.

being checked for salt leaching. Additionally, surface accumulations of water have not been drained off but were collected for disposal in a disposal well. To-date, off lease salt contamination from this process on the four sumps has not occurred.

5. CONCLUSIONS

Salt water drilling mud was surface disposed in the field after reducing salt to acceptable levels by repeated extractions with fresh water. The

FIG. 6. Mud coming out in the form of a fountain during mixing with the
submersible pump in a drilling sump.

TABLE 1

Field results showing reduction of salt content of fluids
of drilling mud sumps at Cold Lake by fresh water
extraction using a submersible pump for enhanced mixing

Sump	Number of 1:1 extractions	ppm $Cl^-/10^3$	
		Before	After
A	4	98	5·1
B	4	47	3·8
C	3	25	4·0
D	3	21	3·7

number of extractions required depended on the ratio of the mud to the
added fresh water and the concentration of the salt in the mud. This
technique was found to be an efficient and cost effective method of disposal
of salt contaminated muds.

The key to this process was the efficient mixing of the added water with
the sump solids. A submersible centrifugal pump worked well for mixing
the sump contents with the added water.

The salt water after separation was disposed in a disposal well and the solids were approved to be mixed with the lease surface soil and spread over the well site lease.

Regulatory approval was granted on a trial basis and the run-off from the lease is being checked for salt leaching. To-date, off lease salt contamination from this process on the four sumps has not occurred.

ACKNOWLEDGEMENTS

The authors would like to thank Esso Resources Canada Limited for permission to publish this work. Thanks are also due to Messrs D McCabe and R. J. Obrigewitsch for assistance in executing the field trials.

REFERENCES

GRAY, G.R., DARLEY, H.C.H. & ROGERS, W.F. (1979). *Composition and Properties of Oil Well Drilling Fluids*. Gulf Publishing Co., Houston, Texas.

HINDS, A.A., DONOVAN, D.M., LOWELL, J.L. & LIAO, A. (1986). Treatment reclamation and disposal options for drilling muds and cuttings. Paper 14798, presented at the IADC/SPE Drilling Conference, Dallas, Texas, February 10–12.

MILLER, R.W. & HONARRAR, S. (1975). Effect of drilling fluid component mixtures on plants and soils. Environmental Aspects of Chemical Use in Well Drilling Operations, EPA-560/1-75-004 pp.125–143.

MOORE, P.L. (1974). *Drilling Practice Manual*. Penwell Publishing Co., Tulsa, Oklahoma.

MOSELEY, H.R. (1983). Summary and analysis of API onshore drilling mud and produced water environmental studies. API Bulletin, D19.

NESBITT, L.E. & SANDERS, J.A. (1981). Drilling fluid disposal, *J. Petrol. Technol.*, December. 2377–81.

15

Physical/Chemical Fate of Organic and Inorganic Constituents within Waste Freshwater Drilling Fluids

GEORGE M. DEELEY

Shell Development Company, P.O. Box 1380, Houston, Texas 77251-1380, USA

ABSTRACT

A recent study by the American Petroleum Institute has provided an extensive characterization of drilling wastes collected in the field. A number of these wastes were subjected to column leachability tests to determine the hydraulic conductivity to be expected in a waste disposal pit. Hydraulic conductivities were found to decrease rapidly to values of 10^{-8}–10^{-7} cm/s as a result of settling in the columns, confirming results from previous studies.

The effluent from these glass columns were analyzed for a wide range of organic and inorganic constituents. These data indicated that chemicals were attenuated within the drilling fluids, with the exception of the more soluble elements such as sodium and chloride. The maximum effluent concentration of most constituents was less than 10% of their initial total concentration in the solid portion of the waste drilling fluid.

The results of this and previous studies reflect the likelihood of similar behavior occurring in a disposal pit, with low permeabilities and constituent attenuation decreasing the flux of soluble compounds to the subsurface. Groundwater modeling, based in part on these results, predicts that there is very limited potential for significant contamination from drilling fluid disposal pits.

1. INTRODUCTION

1.1. Objectives
The primary objectives of this study were to:

(a) investigate the hydraulic conductivity (permeability) of waste drilling fluids as they settled in glass columns in the laboratory;

(b) determine concentrations of chemical constituents in the column filtrate;

(c) compare effluent constituent concentrations with initial total waste drilling fluid constituent concentrations as a measure of attenuation within waste drilling fluid.

The results of this investigation have been incorporated into mathematical models to predict the environmental fate of drilling fluid waste constituents.

1.2. Sealing Capability of Waste Drilling Fluid

One of the capabilities of water/bentonite clay drilling fluids is to prevent their loss into formations surrounding a bore-hole. Drilling fluids are designed to form a low permeability filter cake on the inside surface of a bore-hole to prevent this loss. This property of bentonite clay fluids indicates that waste drilling fluids may form a low permeability filter cake on the bottom of an earthen pit being used during drilling operations. This would be expected because bentonite clay, the major constituent of drilling fluids, has also been used extensively to seal canals, ponds and reservoirs (Kays, 1986).

This formation of a low permeability filter cake has been described by considering a total volume of a stable suspension of solids filtered against a permeable substrate (Gray & Darley, 1980). Larsen (1938) found that the volume of filtrate was proportional to the square root of filtering time when a drilling fluid was filtered through paper at a constant temperature and pressure. A plot of cumulative filtrate volume versus square root of time results in a straight line which generally intercepts the *y*-axis above zero. This zero error is commonly called the mud spurt, and is largely caused by the tendency of finer mud particles to pass through a formation (or filter paper) until its pores become plugged. With most drilling fluids the zero error is small and plugging occurs almost instantly. This process has been confirmed for freshwater drilling fluid waste (Deeley & Whitaker, 1986).

Hydraulic conductivity provides a measure of water flux from a pit and can be used in conjunction with aqueous phase constituent concentrations to determine the flux of an individual chemical from a pit. Laboratory studies have been performed on drilling fluid wastes collected in the field to determine the hydraulic conductivity to be expected in waste drilling fluid disposal pits (Deeley & Whitaker, 1986). Hydraulic conductivities were found to decrease rapidly, after addition of waste drilling fluid to glass columns, to final values of 10^{-7}–10^{-6} cm/s depending upon mud type and form of cuttings present. This decrease in hydraulic conductivity was most likely due to continued settling and compression of the filter cake with time.

1.3. Inorganic Constituents

Inorganic constituents commonly found in waste drilling fluids include arsenic, barium, cadmium, chloride, chromium, lead, sodium, sulfate, and zinc (Deeley, 1984). The transport and fate of these constituents in subsurface environments may involve several processes (sorption, cation/anion exchange, chemical precipitation and particulate transport) and are influenced by several variables (constituent type, soil type/properties, oxidation-reduction potential, pH and temperature).

Henderson (1982) found that the migration of components from freshwater drilling fluid disposal pits was insignificant, based on the observation that contaminated locations were found in relatively narrow, shallow bands close to the point of origin. A study by Murphy & Kehew (1984) examined salt-based drilling fluid disposal pits ranging in age from 2 to 23 years. Concentrations of compounds of interest were found to decrease with distance from the buried drilling fluid. The reduction was attributed to attenuation, mixing, and dispersion processes within the soils. The restriction of pollutants, including chloride, within the local area of these older disposal pits is a strong indication of the efficacy of pits as a final repository for waste drilling fluid at a properly managed site. Literature on barium, a major component of drilling fluids, was reviewed in detail by Crawley *et al.* (1987). It was concluded that barium is not a threat to water resources.

Canter *et al.* (1984) and Reimers *et al.* (1986) compared average solid and liquid phase concentrations of major elements present in drilling fluid disposal pits and showed that more than 99% of pit concentrations of the elements of concern were retained in the sediments. In a study of 125 reserve pits dating from 1979 to 1986, Leuterman *et al.* (1987) determined that water soluble heavy metals (Cr, Pb, Zn, Mn) in reserve pits were generally low to nondetectable, or if found in the total analysis were usually too tightly bound to clays or organics to exceed limitations as determined by the EP Toxicity Leachate Test. It should be noted that when the EP Toxicity Test has been applied to waste drilling fluids, resulting concentrations of constituents of interest were generally less than regulatory limits (Deeley *et al.*, 1987; Canter *et al.*, 1984).

The influence of pH and ionic strength of a leaching solution on the availability of arsenic, barium, chromium and lead in three different drilling fluid wastes was examined with samples obtained from disposal pits and equilibrated in the laboratory under controlled conditions (Deeley & Canter, 1986). In no case was there a substantial release to the aqueous phase with decreasing pH or ionic strength even though pH was decreased to less than 5 for each of the wastes studied. The lower pH values are unlikely to occur in the environment because drilling fluid wastes themselves were shown to have a large neutralizing capacity. This indicates that extraction tests based on rigorous, low pH conditions, such as the Extraction Procedure (EP) Toxicity Test or the Toxicity Characteristic Leaching Procedure (TCLP) (USEPA, 1986), do not simulate the behavior of metals in waste drilling fluids in the field.

Stohs (1986) used a geochemical flow simulator based on local thermodynamic equilibrium to analyze the pollution potential of metals present in drilling wastes. Based on this model, arsenic, barium, cadmium, chromium (III) and lead are not a threat to groundwater under the prevailing chemical environment in a freshwater drilling fluid waste disposal pit.

The more soluble but less threatening constituents such as chloride, sodium and sulfate tend to be released with aqueous leachate from a pit. However, the total flux of these materials leaving a pit under prevailing low hydraulic conductivity conditions is very small from a freshwater drilling fluid disposal pit (Deeley, 1986).

1.4. Organic Constituents

Organic compounds present in drilling waste could originate from the formations being drilled or from release of petroleum products such as diesel fuel into a poorly managed pit. Lower molecular weight organic compounds, such as benzene, are generally found at less than 1 mg/litre. Heavier organic compounds, such as naphthalene, may occur at levels up to 10 mg/litre but tend to be tightly bound within the bentonite clay/soil matrix of the waste (Reimers *et al.*, 1986; Deeley, 1986). Bentonite has been shown to efficiently adsorb organic compounds (Lakatos, 1986).

Leachability tests have been performed on freshwater drilling fluid wastes collected in the field to determine the attenuation of organic compounds to be expected in a waste disposal pit (Deeley, 1986). Benzene, naphthalene and phenol were chosen as test compounds because they cover a wide range of chemical properties and have been identified as potential pollutants in waste drilling fluids. Effluent concentration decreases were greater than 95% for each compound when compared with the initial total concentration of each component within the solid portion of the drilling fluid waste. This attenuation was likely the result of the organics sorbing onto naturally occurring insoluble soil humic substances and clay surfaces in the waste drilling fluid.

Biodegradation is another process potentially influencing the fate of the low level organics found in or being released from waste drilling fluid pits. Recent investigations indicate that microorganisms are present in and may transform many of the organic pollutants that enter soils and subsurface environments (Wilson & McNabb, 1983). Ehrlich *et al.* (1982) provide a case history of biodegradation of phenols in groundwater by anaerobic bacteria at a coal tar facility. They collected soil cores and confirmed the presence of methane generating bacteria, measured methane in groundwater samples, and demonstrated that phenol concentration decreased more rapidly than conservative tracers. Wilson (1986) obtained similar results for degradation of alkylbenzenes from a gasoline spill. Chiang *et al.* (1986) studied rates of benzene, toluene and xylene biodegradation in an aquifer contaminated by a refinery flare pit. They found that the biodegradation reactions can be described as a first order rate process.

1.5. Fate and Transport Modeling

It is clear from these studies that both organic and inorganic compounds within a drilling fluid waste may be acted upon by attenuating mechanisms. Therefore, these processes must be considered when determining proper handling and disposal routes for freshwater drilling fluid wastes. The net effect of these mechanisms can be assessed by incorporating them within mathematical models which describe the bulk flow, attenuation and degradation of these compounds within the soil/groundwater matrix at a given site.

Kemblowski and Deeley (1987) developed a screening model for predicting maximum steady-state concentrations of chemicals at some distance from a pit under a given set of conditions at a predetermined probability level. Under the wide range of conditions examined, the concentration of a constituent at a point 100 feet downgradient from a pit would have an 80% likelihood of falling between 0·001 and 29·4% of the initial leachate concentration, depending upon the specified conditions of flow and biodegradation rate.

The present study expanded upon previous investigations by examining many inorganic and organic constituents in the leachate generated by waste drilling fluids in glass columns. In addition to leachate composition, flow data were obtained in the form of hydraulic conductivity measurements.

2. MATERIALS AND METHODS

2.1. Drilling Fluid Wastes

The American Petroleum Institute (1987) conducted a field sampling and analysis study to characterize exploration and production wastes. As part of this study drilling fluid waste consisting of both liquid and solid phase were collected from 18 sites. Drilling fluid pits were sampled from a boat or shore (where boat access was not possible). Liquids were sampled using weighted bottles or bottom-filling devices at the surfaces of pits. Solids were sampled using a pipe driven through the entire depth of the pit. Each final sample was a composite of individual samples from 30 to 60 areas of a pit to represent solids in the pit. Each sample was analyzed for 125 organic constituents, 29 inorganic constituents, 15 conventional parameters, and 2 RCRA* characteristics.

Seven waste drilling fluid samples were selected for column experiments. These samples were collected in the field concurrent with waste characterization sampling. All samples were stored in glass containers with no headspace under refrigeration (4 °C) until used in experiments.

Samples were not given any special preparation or treatment prior to use other than mixing to provide a homogeneous material. The samples were selected because they represent a variety of mud types with a wide range of analyzed organic and inorganic constituents.

2.2. Soils

Site soil physical characteristics were analyzed, but these soils were not used for the column study because of the possibility of collecting soils influenced by some aspect of drilling activity. Therefore, soils were collected from pristine areas, mixed and analyzed. These soils were then matched as closely as possible to the soils present at a site. These soils were placed under their respective waste in column experiments to simulate site soils. A complete chemical analysis was performed on these soils along with TCLP and control column leach tests to allow for correction of any soil influence on the column experiments where actual waste was leached.

---

*Editors' Note: Resource Conservation and Recovery Act—enacted as US Public Law 94–580 in 1976 as an amendment to the Solid Waste Disposal Act (SWDA). The RCRA has been amended several times subsequently.

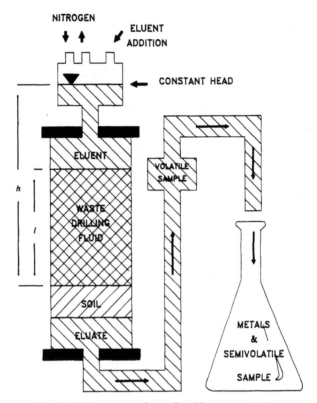

FIG. 1. Laboratory column leaching apparatus.

3.2. Procedures

The laboratory apparatus used in the leaching study consisted of glass columns containing the waste drilling fluid being studied, soil representative of the site where waste was collected, and washed sand through which the leachate passed (Fig. 1). A pressure head provided by a sodium chloride/deionized water solution was maintained at a constant level. The sodium chloride solution concentration was at a strength representative of sodium chloride concentration in the waste liquid. No other ions were represented in the leaching solution. A nitrogen atmosphere was maintained in the column headspace to discourage oxidative effects.

In summary, a known mass (M) of selected mud was added to each column. A subsample of each mud was taken, while filling each column, to determine percentage solids. The flow initially was relatively high due

to the mud spurt, and the water head was allowed to fall during the time
that fluid was settling to permit the water level to reach the top of the
mud cake. This minimized the amount of dissolved constituents from the
waste dispersing into the leaching liquid. There was undoubtedly some
transfer of soluble constituents across the mud cake/water interface.
However, this would be minimal because of the slow rate of diffusion to
be expected within the mud cake.

The filtrate was allowed to flow from the bottom of the column by
gravity through the sample collection apparatus. The leachate was
collected at timed intervals (t). Measurements were taken of the volume
of leachate collected (V), hydraulic head (h), and mud cake thickness (l).
The collected samples were analyzed for a complete set of major organic
and inorganic parameters.

The pore volume (PV) of the mud cake was determined at each time
interval from the particle density of waste fluid solids (d_p), mud cake
thickness (l), cross-sectional area (A) and assuming the density of water
(d_w) = 1 g/cm^3. The total mass (M) and percentage solids by weight (S)
of the added mud was known, therefore,

$$\text{Pore Volume (PV)} = \text{Total Volume} - \text{Particle Volume}$$

$$\text{PV} = Al - \frac{\text{MS}}{d_p}$$

An effective pore volume is assumed to equal $0\!\cdot\!8 \times \text{PV}$. This is based on
an estimated 20% of the total water in the column being trapped within
the soil structure and not involved in actual flow through the column.

The hydraulic conductivity (K) and permeability (k) were determined
using measured parameters according to Black *et al.* (1965):

$$k = \frac{\mu}{d_w g} K = \frac{\mu}{d_w g} \frac{Vl}{Aht}$$

where: k = intrinsic permeability with water (cm^2)
K = hydraulic conductivity (cm/s)
V = volume of leachate in time t (cm^3)
h = hydraulic head (cm)
A = cross-sectional area of mud cake (cm^2)
t = time interval (s)
μ = viscosity of water ($0\!\cdot\!08904$ g/cm s)
d_w = density of water (1 g/cm^3)

g = acceleration due to gravity ($980 \cdot 66 \, cm/s^2$)
l = mud cake thickness (cm).

3. RESULTS AND DISCUSSION

3.1. Waste Characterization

Of the parameters analyzed, only 35 out of 125 organics and 28 of the 29 inorganics were detected in at least one liquid or solid pit sample. No samples tested exceeded reactivity limits for reactive sulfide or cyanide. One liquid sample associated with a lime-based drilling mud exceeded the pH ion value corrosivity limit. One liquid sample contained $0 \cdot 11$ mg/litre benzene which exceeded a proposed TCLP regulatory limit of $0 \cdot 07$ mg/litre. The sampling report for that site indicated there was a floating layer of oil on the pit which was likely the source of benzene. Solids from one pit and the associated liquid phase contained $5 \cdot 1$ mg/litre and $23 \cdot 0$ mg/litre of pentachlorophenol (PCP), respectively. This exceeded a proposed regulatory limit of $3 \cdot 6$ mg/litre. However, it should be recognized that PCP is not a representative constituent of normal drilling wastes. The site report indicated PCP was introduced into this site as a downhole biocide for which alternatives exist. None of the metals associated with drilling solids or liquids exceeded any proposed TCLP regulatory limit, based on total or TCLP analyses. The number of positive analyses per site and number of exceedances of proposed regulatory limits are summarized in Table 1.

3.2. Hydraulic Conductivity

Physical parameters were measured and calculated routinely during column studies. Measurements continued for up to 250 days depending on the behavior of the columns. Of the original nine columns, five developed leaks as a result of column edge effects or inappropriate leaching fluid effects and were discontinued. Any future studies should use a leaching solution designed to maintain the original sodium absorption ratio (SAR) within the columns. The use of only sodium chloride in this study changed the SAR of some of the columns, resulting in a disruption of clay structure. The leaching solutions for columns 1, 2, 6 and 7 did not significantly affect this process.

The hydraulic conductivities of the remaining columns decreased to average levels of from 10^{-8} to 10^{-7} cm/s depending upon mud type, form

TABLE 1

Number of positive results and exceedances of proposed limits for compounds
detected in waste drilling fluid liquids and solids. Based on total or TCLP analyses,
as appropriate

Parameter	Drilling waste solid		Drilling waste liquid	
	Times analyzed per 18 sites	*Exceedance of proposed limits*	*Times analyzed per 18 sites*	*Exceedances of proposed limits*
1,1-Dichloroethane	1		0	
2-Chlorophenol	0		1	
Acenaphthene	4		1	
Acetone	4		6	
Aluminum	18		15	
Anthracene	3		1	
Antimony	2		1	
Arsenic	18	0	13	0
Barium	18	0	18	0
Benzene	7	0	3	1
Beryllium	11		5	
Biphenyl	3		2	
Boron	18		18	
Cadmium	3	0	3	0
Chloride	18		18	
Chromium	18	0	11	0
Cobalt	18		6	
Dibenzothiophene	1		2	
Ethylbenzene	18		3	
Fluorene	12		10	
Lead	18	0	10	0
Mercury	4	0	3	0
Methylene chloride	0	0	1	0
Molybdenum	15		14	
Naphthalene	14		10	
Nickel	18		10	
Pentachlorophenol	1	1	1	1
Phenanthrene	18		13	
Phenol	1	0	3	0
Potassium	18		18	
Pyrene	2		6	
Selenium	1	0	4	0
Silver	0	0	2	0
Sodium	18		18	
Strontium	18		18	
Styrene	1		1	
Tin	2		0	

TABLE 1—*contd.*

Parameter	Drilling waste solid		Drilling waste liquid	
	Times analyzed per 18 sites	*Exceedance of proposed limits*	*Times analyzed per 18 sites*	*Exceedances of proposed limits*
Titanium	18		14	
Toluene	18	0	8	0
Trichloroethene	2	0	0	0
Vanadium	18		13	
Zinc	18		17	
α-Terpineol	0		1	
p-Cymene	2		0	
trans-1,2-Dichloroethene	4		0	

TABLE 2

Average hydraulic conductivities and aqueous phase flux from waste drilling fluids in glass columns

Column number	Hydraulic conductivity (cm/s)	Flux (ml/min cm²)
1	$7·8 \times 10^{-7}$	$4·7 \times 10^{-4}$
2	$6·4 \times 10^{-8}$	$4·2 \times 10^{-5}$
6	$6·0 \times 10^{-8}$	$2·8 \times 10^{-4}$
7	$6·5 \times 10^{-7}$	$3·4 \times 10^{-4}$

of cuttings present, and duration of experiment (Table 2). A hydraulic conductivity of 10^{-7} cm/s ($2·1 \times 10^{-3}$ gal/day/ft²) will result in a flow of 13 gal/day from a 5000 ft² pit containing 25 000 ft³ of waste drilling fluid with a 5 ft thick solid layer overlain by 1 ft of liquid. This attests to the potential for waste drilling fluids to be self-sealing when added to disposal pits.

3.3. Chemical Parameters

Column experiment results indicate that organic and inorganic constituents were attenuated within the waste drilling fluids. With the exception of the more soluble constituents such as sodium, chloride, molybdenum, vanadium, and p-cresol, the compounds examined were significantly attenuated: the maximum leachate concentration being less than 10% of the total concentration in the waste drilling fluid (Table 3). The attenuation was computed by comparing the maximum measured leachate concen-

TABLE 3
Average percentage attenuation of some compounds within waste drilling fluid columns

Component	Average attenuation %
TOC	98·0
Ethyl benzene	98·7
p-Cresol	64·4
Naphthalene	99·9
Toluene	98·7
Arsenic	97·1
Barium	99·8
Beryllium	99·0
Chromium	95·0
Cobalt	99·8
Copper	95·5
Lead	99·8
Molybdenum	85·3
Nickel	96·5
Potassium	95·3
Sodium	49·7
Strontium	95·8
Vanadium	74·3

tration (mg/litre) with the initial total concentration measured in the solid portion of the waste (mg/kg) which was a worst case analysis because maximum leachate concentration was a temporary peak and effluent concentrations were generally lower.

Some results from column 1 where 6·2 pore volumes of liquid passed through the mud cake during the course of the experiment (231 days) are shown in Fig. 2. Total organic carbon decreased sharply after 1 pore volume of liquid flushed the column while pH remained relatively constant throughout the experiment (Fig. 2(a)). Throughout the experiment a total of 3% of the initial concentration of TOC in the waste was leached. Benzene and toluene decreased initially followed by an increase in the last data point (Fig. 2(b)). This could have been caused by a heterogeneity in the waste or a contaminated sample. Benzene was not detected in the initial total waste analysis, so no removal percentage can be determined. Of the initial total waste level of toluene, 11% leached from the column. Xylenol was rapidly flushed from the waste while cresol was retarded with

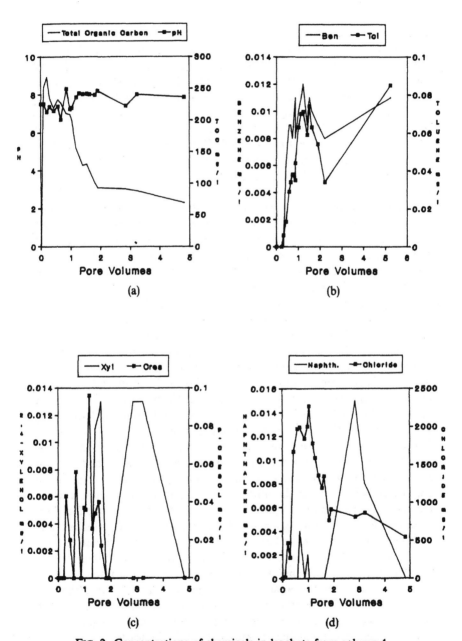

FIG. 2. Concentrations of chemicals in leachate from column 1.

a peak effluent concentration at 3 pore volumes (Fig. 2(c)). No xylenol
was found in the initial waste analysis and 35% of the cresol leached
during the experiment. The chloride concentration rapidly decreased while
the much less soluble naphthalene was retarded in the waste (Fig. 2(d)).
Chloride was not retarded (100% leached) while only 0·04% of the initial
naphthalene leached during the experiment. With the exception of total
organic carbon and chloride, all effluent concentrations were very low.
These results reflect the sorptive capability of waste freshwater drilling
fluids for nonionic organic compounds such as benzene and naphthalene,
while the more hydrophilic compounds such as xylenol and cresol more
readily leach.

Concentrations of other inorganic constituents found in the column
effluents are shown in Fig. 3. All of these constituents were found at low
concentrations relative to the total amount in the waste or associated
soils. Columns 2, 6 and 7 had such a low water flux that only about 2
mud-cake pore volumes of liquid had been released by the end of the
experiments which lasted from 30 to 254 days. Thus it was not possible
to monitor the complete behavior of some of the slowly eluting compounds
in these columns. Arsenic increased in the column 1 effluent up to 5 pore
volumes and then decreased, it was still increasing in the remaining
columns (Fig. 3(a)). Barium, chromium, molybdenum and nickel behaved
similarly with an initial concentration peak by 2 pore volumes followed
by decreasing concentrations (Fig. 3(b–e)). The total barium and chromium
leached during the experiments were less than 1% of the total initial waste
concentrations, while nickel was 6·7% leached. Potassium and strontium
were more soluble and concentrations decreased rapidly (Fig. 3(f, g)).
However, only 1·3% of the initial potassium and 5% of the initial
strontium leached during the course of the experiment. Vanadium was
still increasing in concentration in columns 2 and 6 at experiment
completion with 20% of the initial waste concentration having leached.
However it appeared that concentrations were starting to level off in these
columns (Fig. 3(h)) although a steady-state had not been reached and
additional time would have been required to verify the behavior of these
components. Columns 2 and 6 were duplicate columns and their similar
behavior attests to the repeatable nature of these column tests.

In summary, chemical analysis of column leachate indicated that the
majority of soluble material generally left the waste after flushing with
two mud cake pore volumes of leaching solution. While a major portion
of most components remained attenuated within the waste solids, only to
be released at very low levels over a long time period. Organic materials

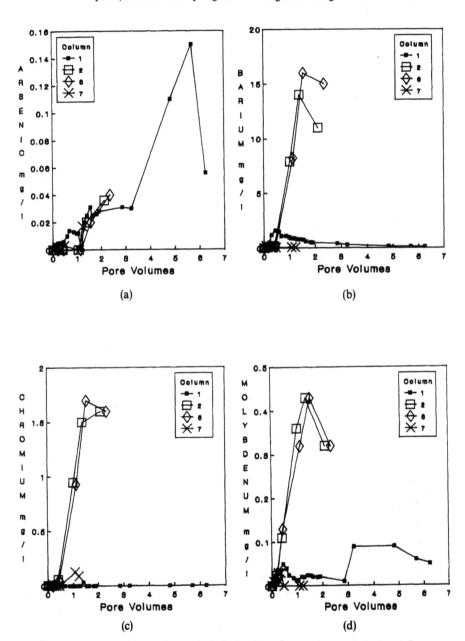

FIG. 3. Concentrations of chemicals in leachate from columns 1, 2, 6 and 7.

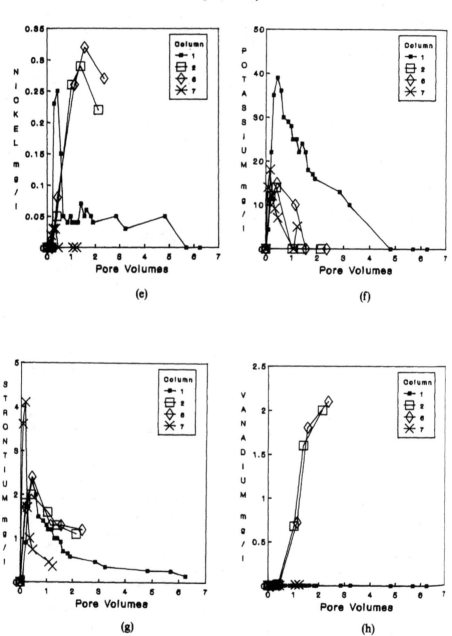

FIG. 3.—*contd.*

were sorbed within the waste solids and inorganics remained as insoluble precipitates or also sorbed to the waste solids. The rate of release depended upon the flux of water from the mud. Therefore, although the sealing of pits by waste muds does not stop the flow of soluble constituents from a pit, the low permeability decreases the rate of release to very low levels.

4. CONCLUSIONS

The characterization of both liquid and solid portions of drilling waste fluids reflected the absence or nonavailability to aqueous solution of hazardous chemicals in the samples analyzed. This was substantiated through Toxic Characteristic Leaching Procedure (TCLP) testing of these same wastes. The TCLP being a worst case analysis since it subjects the wastes to extreme conditions not experienced in the real world.

The hydraulic conductivity of waste drilling fluids, added to glass columns, decreased rapidly to low values depending upon the type of waste fluid and cuttings present. Hydraulic conductivities in the range of 10^{-8}–10^{-7} cm/s were obtained with the drilling fluid wastes studied.

The results of this and previous studies reflect the likelihood of similar behavior occurring in a disposal pit, with low permeabilities and constituent attenuation decreasing the flux of soluble compounds to the subsurface.

Groundwater modeling, based in part on these results, predicted that there is a very limited potential for contamination of groundwater from drilling fluid disposal pits (American Petroleum Institute, 1987). Concentrations at receptor locations did not exceed health-based water quality standards in either deterministic modeling of specific sites, qualitative evaluation of specific sites, or Monte Carlo simulation of representative sites.

ACKNOWLEDGEMENTS

The author thanks the American Petroleum Institute (API) for permission to present this paper. I gratefully acknowledge the API's Production Waste Issue Group, ERT, Rocky Mountain Analytical Laboratory, and Deuel & Zahray Laboratories for their efforts in performing various aspects of this study.

REFERENCES

AMERICAN PETROLEUM INSTITUTE (1987). Oil and gas industry exploration and production wastes. Document No. 471-01-09, prepared by ERT, Houston, Texas, July.

BLACK, C.A., EVANS, D.D., WHITE, J.L., ENSMINGER, L. E. & CLARK, F. E. (1965). Methods of soil analysis: Part 1—Physical and mineralogical methods. American Society of Agronomy Inc., Soil Science Society of America Inc., Madison, Wisconsin.

CANTER, L.W. *et al.* (1984). *Environmental Implications of Off-Site Drilling Pits in Oklahoma.* Environmental and Ground Water Institute, University of Oklahoma, Norman, Oklahoma.

CHIANG, C.Y., KLEIN, C.L., SALANITRO, J.P. & WISNIEWSKI, H.L. (1986). Data Analysis and Computer Modelling of the Benzene Plume in an Aquifer Beneath a Gas Plant. *Proceedings of the NWWA/API Conference on Petroleum Hydrocarbons and Organic Chemicals in Ground Water—Prevention, Detection and Restoration,* NWWA, Dublin, Ohio, pp. 157–78.

CRAWLEY, W., ARTIOLA, J. & REHAGE, J.A. (1987). Barium containing oilfield drilling wastes: effects on land disposal. *Proceedings of the Third National Conference on Drilling Muds,* University of Oklahoma, Norman, Oklahoma, May 28–29, pp. 235–59.

DEELEY, G.M. (1984). Chemical speciation and flyash stabilization of arsenic, barium, chromium, and lead in drilling fluid waste, Ph.D. Dissertation, University of Oklahoma, Norman, Oklahoma, 272 pages.

DEELEY, G.M. (1986). Attenuation of chemicals within waste freshwater drilling fluids. *Proceedings of the Second National Conference on Drilling Muds,* University of Oklahoma, Norman, Oklahoma, May 29–30, pp. 365–94.

DEELEY, G.M. & CANTER, L.W. (1986). Distribution of Heavy Metals in Waste Drilling Fluids under Conditions of Changing pH. *J. Environ. Qual.* 17(2), 108–12.

DEELEY, G.M. & WHITAKER, G.A. (1986). Hydraulic Conductivity and Leachability of Waste Drilling Fluids. *Proceedings of the Ninth Annual Energy-Sources Technology Conference and Exhibition,* PD-Vol. 5, American Society of Mechanical Engineers, New Orleans, Louisiana, February 23–27, pp. 243–54.

DEELEY, G.M., CANTER, L.W. & LAGUROS, J.G. (1987). Stabilization of drilling fluid waste with fly ash. *Proceedings of the Materials Research Society Symposium,* Vol. 86, p. 77.

EHRLICH, G.G., GOERLITZ, D.R., GODSY, E.M. & HULT, M. F. (1982). Degradation of phenolic contaminants in ground water by anaerobic bacteria: St Louis Park, Minnesota. *Groundwater,* 10(6) 703–22.

GRAY, G.R. & DARLEY, H.C.H. (1980). *Composition and Properties of Oil Well Drilling Fluids,* 4th edn. Gulf Publishing Company, Houston, Texas.

HENDERSON, G. (1982). Analysis of hydraulic and environmental effects of drilling mud pits and produced water impoundments. Vol. 1, Executive Summary and Report. Dames and Moore, Houston, Texas, October.

KAYS, W.B. (1986). *Construction of Linings for Reservoirs, Tanks, and Pollution Control Facilities,* 2nd edn. John Wiley, New York.

KEMBLOWSKI, M.W. & DEELEY, G.M. (1987). A ground water screening model for drilling fluid disposal pits. *Proceedings of the Third National Conference on Drilling Muds*, University of Oklahoma, Norman, Oklahoma, May 28–29, pp. 205–34.

LAKATOS, G. (1986). Use of natural bentonite in an oil-polluted wastewater treatment system (Tifo, Hungary). *Proceedings of the 2nd International Conference on Environmental Contamination*, Amsterdam, September, pp. 256–8.

LARSEN, D.H. (1938). Determining the filtration characteristics of drilling muds. *Petroleum Engineering*, September, 42–8.

LEUTERMAN, A.J.J., JONES, F.V., CHANDLER, J.E. & MOFFITT, C.M. (1987). Drilling Fluids and Reserve Pit Toxicity. *Proceedings of the Third National Conference on Drilling Muds*, University of Oklahoma, Norman, Oklahoma, May 28–29, pp. 160–77.

MURPHY, E.C. & KEHEW, A.E. (1984). The effect of oil and gas well drilling fluids on shallow groundwater in Western North Dakota. Report of Investigation No. 82, North Dakota Geological Survey, Fargo, North Dakota.

REIMERS, R.S., DEKERNION, P.S. & ABDELGHANI, A.A. (1986). Long-term assessment of non-hazardous oilfield waste pits. *Biocycle*, August, 44–9.

STOHS, M. (1986). A study of metal ion migration in soils from drilling mud pit discharges, M.S. Thesis, University of Texas, Austin, Texas, May, 205 pp.

USEPA (1986). Hazardous waste management system; identification and listing of hazardous waste; notification requirements; reportable quantity adjustments; proposed rule. *Federal Register*, **51** (114) June 13, 21 648–93.

WILSON, B.H. (1986). Biological fate of hydrocarbons at an aviation gasoline spill site. *Proceedings of the NWWA/API Conference on Petroleum Hydrocarbons and Organic Chemicals in Ground Water — Prevention, Detection, and Restoration*, NWWA, Dublin, Ohio, pp. 78–92.

WILSON, J.T. & McNABB, J.F. (1983). Biological transformation of organic pollutants in groundwater. *EOS*, **64**(33), August 16, 505–7.

16

Impact of Heavy Oil Drilling Wastes from Alberta, Canada on Soils and Plants

T. M. MACYK, F. I. NIKIFORUK and S. A. ABBOUD

Terrain Sciences Department, Alberta Research Council, PO Box 8330, Postal Stn F, Edmonton, Alberta, Canada T6H 5X2

ABSTRACT

A greenhouse pot trial was undertaken to assess the effect on soils and plants of different application rates of drilling wastes resulting from freshwater, KCl and NaCl mud systems used in the Cold Lake Region of Alberta, Canada. The wastes were added to a mixture of the upper 25 cm of a Gray Luvisolic soil from the area based on chloride content and applied at the equivalent rate of 200, 400, 600, 800 and 1200 kg Cl/ha of soil. A rate of 2400 kg Cl/ha was added to the KCl and NaCl treatments. Additional waste rates added to assess liming effect were based on the acid neutralizing capacities of the wastes. Each treatment was replicated three times and brome grass planted. Moisture content was maintained at or near field capacity by watering on a pot weight basis. All treatment yields were greater than the control soil yield except for the 2400 kg/ha chloride rate of KCl waste. There was no statistically significant increase or decrease in yield with increased application rate. A significant increase in soil pH resulted from freshwater and NaCl waste addition. Significant increases in EC and SAR occurred with increased concentration of the three wastes added. Elemental uptake by the brome grass varied relative to the different wastes. Calcium and potassium content increased favourably for all additions of KCl and NaCl wastes. Boron levels in the brome grass improved with the addition of freshwater waste. Chloride content was increased significantly with excessive concentrations occurring at the highest application rates of NaCl and KCl wastes.

1. INTRODUCTION

Three major types of drilling muds and variations thereof are used for oil and gas exploration in Alberta. These include the freshwater gel bentonite, the salt water systems which use sodium and potassium and the oil invert mud.

The wastes produced from drilling operations contain many complex organic and inorganic compounds that are added at various stages of the drilling process. The materials that start out as a drilling mud before going down hole are usually quite altered once they enter a waste pit. The materials found in the waste pit can also be changed chemically and physically by other products that are purposely or inadvertently added to the pit.

Review of the literature indicates that effects of disposal of drilling wastes on plants and soils varies with the type of waste, rate of application and plant species (Miller & Honarvar, 1975; Peseran, 1977; Nelson et al., 1980; Miller & Peseran, 1980).

This paper presents the results of a greenhouse pot experiment to assist in better understanding the effect of different types and application rates of wastes on soils and plants. The pot trial experiment was part of an overall project which included a thorough physical and chemical analysis of freshwater, KCl and NaCl wastes found in the Cold Lake Region (Macyk et al., 1987) and a literature review pertinent to an evaluation of the potential for migration of drilling waste components in earthen materials (Pauls et al., 1987).

2. MATERIALS AND METHODS

2.1. Trial Treatments

The soil used in the trial was comprised of the upper 25 cm (Ap, Ae and upper portion of AB horizons) of an Orthic Gray Luvisol commonly occurring in the study area. The rates of waste material added were based on chloride content and applied at the equivalent of 200, 400, 600, 800 and 1200 kg Cl/ha of soil. The chloride content of the drilling wastes and pot soil were as given in Table 1. The lowest waste application rate of 200 kg/ha required 113·6, 6·9 and 1·95 g of freshwater, KCl and NaCl

TABLE 1

Waste type	Chloride content (mg/kg)
Freshwater	903
KCl	14 900
NaCl	52 700
Pot soil	7·51

waste respectively per 1 kg of soil/waste mixture. This resulted in a ratio of 1:9, 1:145 and 1:513 of freshwater, KCl and NaCl:soil respectively indicating the considerably large volume of freshwater waste required for the pot mixtures.

A rate of 2400 kg Cl/ha was added to the KCl and NaCl treatments when it was apparent that germination in the previously established mixtures was very good.

The liming effect of the waste materials was also considered in assessing the rate of drilling waste application to soils. To increase the pH of the pot soil from its actual value of 6·35 to a pH of 7 necessitated 12·5, 15 and 30 g waste/kg of soil for the freshwater, NaCl and KCl waste respectively. It should be noted that the 200 kg/ha rate of chloride which is equivalent to 113·6 g freshwater waste per 1 kg waste/soil mixture exceeded the rate required to bring the pot soil mixture to a pH of 7 by almost 10 fold.

For these reasons additional waste rates were added to the study based on liming effect. The additional rates based on their equivalent chloride rate were 20, 40 and 80 kg/ha of chloride for freshwater waste and 1520 kg/ha of chloride for the NaCl waste.

Each treatment had three replicates. Three control (soil only) pots had plants and six did not. The label on each pot described the pot contents as described in Table 2. Examples are:

1K2—Rep 1, KCl waste,	200 kg/ha of Cl
2FW4—Rep 2, Freshwater waste,	400 kg/ha of Cl
2FWL3—Rep 2, Freshwater waste, lime rate = 80 kg/ha of Cl	

Samples of each treatment were submitted to the Agricultural Soil and Feed Testing Laboratory for available nutrient analysis.

2.2. Trial Establishment

The greenhouse experiment was conducted in the University of Alberta Phytotron. The trial was initiated February 13 and the harvesting completed May 19, 1987.

The pots used were 12·7 cm in diameter and accommodated 1 kg of soil and drilling waste mixture. The pots were lined with plastic bags and kept in saucers throughout the trial. The drilling waste and pot trial soil were weighed, mixed and retained in the same plastic bags throughout the experiment.

Prior to planting, the individual pots were watered to field capacity and fertilized on the basis of results from the available nutrient analysis.

320   *T. M. Macyk, F. I. Nikiforuk and S. A. Abboud*

TABLE 2

Drilling waste rate (g waste/kg soil) treatments based on chloride content and liming potential

Pot label	Waste	Rep.	Rate (kg Cl/ha soil)	Waste (g)	Soil (g)	Waste and soil (g)
1FWL1	Freshwater	1	20	12	988	1 000
2FWL1	Freshwater	2	20	12	988	1 000
3FWL1	Freshwater	3	20	12	988	1 000
1FWL2	Freshwater	1	40	23	977	1 000
2FWL2	Freshwater	2	40	23	977	1 000
3FWL2	Freshwater	3	40	23	977	1 000
1FWL3	Freshwater	1	80	45	955	1 000
2FWL3	Freshwater	2	80	45	955	1 000
3FWL3	Freshwater	3	80	45	955	1 000
1FW2	Freshwater	1	200	114	886	1 000
2FW2	Freshwater	2	200	114	886	1 000
3FW2	Freshwater	3	200	114	886	1 000
1FW4	Freshwater	1	400	227	773	1 000
2FW4	Freshwater	2	400	227	773	1 000
3FW4	Freshwater	3	400	227	773	1 000
1FW6	Freshwater	1	600	341	659	1 000
2FW6	Freshwater	2	600	341	659	1 000
3FW6	Freshwater	3	600	341	659	1 000
1FW8	Freshwater	1	800	455	545	1 000
2FW8	Freshwater	2	800	455	545	1 000
3FW8	Freshwater	3	800	455	545	1 000
1FW12	Freshwater	1	1 200	682	318	1 000
2FW12	Freshwater	2	1 200	682	318	1 000
3FW12	Freshwater	3	1 200	682	318	1 000
1K2	KCl	1	200	7	993	1 000
2K2	KCl	2	200	7	993	1 000
3K2	KCl	3	200	7	993	1 000
1K4	KCl	1	400	14	986	1 000
2K4	KCl	2	400	14	986	1 000
3K4	KCl	3	400	14	986	1 000
1K6	KCl	1	600	20	980	1 000
2K6	KCl	2	600	20	980	1 000
3K6	KCl	3	600	20	980	1 000
1K8	KCl	1	800	27	973	1 000
2K8	KCl	2	800	27	973	1 000
3K8	KCl	3	800	27	973	1 000
1K12	KCl	1	1 200	41	959	1 000
2K12	KCl	2	1 200	41	959	1 000
3K12	KCl	3	1 200	41	959	1 000
1K24	KCl	1	2 400	83	917	1 000

TABLE 2—*contd.*

Pot label	Waste	Rep.	Rate (kg Cl/ha soil)	Waste (g)	Soil (g)	Waste and soil (g)
2K24	KCl	2	2 400	83	917	1 000
3K24	KCl	3	2 400	83	917	1 000
1NaL	NaCl	1	1 520	15	985	1 000
2NaL	NaCl	2	1 520	15	985	1 000
3NaL	NaCl	3	1 520	15	985	1 000
1Na2	NaCl	1	200	2	998	1 000
2Na2	NaCl	2	200	2	998	1 000
3Na2	NaCl	3	200	2	998	1 000
1Na4	NaCl	1	400	4	996	1 000
2Na4	NaCl	2	400	4	996	1 000
3Na4	NaCl	3	400	4	996	1 000
1Na6	NaCl	1	600	6	994	1 000
2Na6	NaCl	2	600	6	994	1 000
3Na6	NaCl	3	600	6	994	1 000
1Na8	NaCl	1	800	8	992	1 000
2Na8	NaCl	2	800	8	992	1 000
3Na8	NaCl	3	800	8	992	1 000
1Na12	NaCl	1	1 200	12	988	1 000
2Na12	NaCl	2	1 200	12	988	1 000
3Na12	NaCl	3	1 200	12	988	1 000
1Na24	NaCl	1	2 400	23	977	1 000
2Na24	NaCl	2	2 400	23	977	1 000
3Na24	NaCl	3	2 400	23	977	1 000
1BP	Blank	1	0	0	1 000	1 000
2BP	Blank	2	0	0	1 000	1 000
3BP	Blank	3	0	0	1 000	1 000
1B	Blank	1	0	0	1 000	1 000
2B	Blank	2	0	0	1 000	1 000
3B	Blank	3	0	0	1 000	1 000
4B	Blank	4	0	0	1 000	1 000
5B	Blank	5	0	0	1 000	1 000
6B	Blank	6	0	0	1 000	1 000

The field capacity moisture constants for the freshwater, KCl and NaCl wastes and the soil material were 41, 37, 41 and 13% respectively. The amount of water added to the pots varied according to the ratio of waste: soil and the respective field capacity moisture values. Fertilization involved the addition of 5 ml of a solution (7·35 g NH_4NO_3 and 5·13 g $NaHPO_4$ in 500 ml solution) to the water added to bring the mixtures to field capacity.

Approximately 75 (0·4 g) brome grass seeds were planted randomly and covered with 1 cm of soil/waste mixture. The pots were placed in a random order and then rerandomized every 2 weeks.

2.3. Trial Maintenance
The greenhouse compartment temperature was maintained between 20 and 25 °C during the day and between 15 and 20 °C at night. A day length of 16–18 h was achieved by the use of overhead lighting. Moisture content was maintained at or near field capacity by watering on a pot weight basis twice a week. As the plants matured and outdoor daylight hours and temperature increased, watering took place three times a week.

2.4. Growth Measurement
Observations were recorded once a week and photographs of each waste series were taken once every 2 weeks. At 2-week intervals the height of the grass was measured by placing a ruler at the soil surface and reading the average height of the plants.

2.5. Harvest Methods
The pots were harvested May 4, 1987 except for the 2400 kg/ha chloride treatments which were harvested May 19. Before harvesting pictures were taken of each pot and notes were taken on plant characteristics including height and coloration.

The plants were harvested 1 cm above the surface of the growth medium to minimize potential contamination. Fresh weight of each treatment was determined as soon as possible after clipping.

The plants were washed in a dilute metal free detergent solution, rinsed three times in distilled water, dried at 105 °C for 24 h and the dry weights recorded.

Following the harvest, the waste/soil mixture was gently removed from the roots, bagged and labeled for drying.

2.6. Methods of Plant Analysis
The grass samples were digested with a concentrated HNO_3–$HClO_4$ acid mixture in a teflon bomb heated in a CEM microwave digestion unit and the solution concentration of Al, Fe, Zn, Mn, Ca, Mg, Na, K, Sr, P, Ba, Mo, B, S, Si and As measured by ICP-AES, and Cd and Pb by graphite furnace atomic absorption. Chloride content was determined by the sodium nitrate extraction procedure of Gaines *et al.* (1984).

2.7. Methods of Soil and Waste/Soil Mixture Analysis

The pH was measured in a paste (Doughty, 1941) and in a 2:1 slurry of 0·01 M $CaCl_2$ (Peech, 1965). Total carbon content was measured with a LECO CR12 carbon analyser (Leco Corporation, 1979), $CaCO_3$ equivalent by acid dissolution (Bascomb, 1961) and acid neutralizing capacity by addition of 0·5 M HCl and back titration with 0·25 M NaOH (methods 1·004 and 1·005 (AOAC, 1981)). Saturated pastes were prepared according to the USDA Soil Salinity Laboratory method (USDA, 1954; Rhoades, 1982), extracted and the extracts filtered through a 0·45 μm filter and analysed for pH, electrical conductivity, alkalinity, chloride, and for soluble salts (Na, K, Ca, Al, Cr, Fe, V, Ti, Cd, Cu, Pb, Zn, Mn, Mg, Li, Sr, B, Ba, P, S, Mo, Ni, Se, As, Co, Si) using an ARL model 34000 simultaneous Inductively Coupled Plasma Atomic Emission Spectrometer. Cation exchange capacity (CEC) and extractable cations were determined by extraction with a normal (1 M at pH 7·0) ammonium acetate solution (Holmgren *et al.*, 1977), where NH_4 ions were determined by a Tecator Kjeltec Auto 1030 Analyser distillation and titration unit and the extractable ions by the ICP-AES. The particle size analysis was done using a simplified hydrometer method (Gee & Bauder, 1979).

DTPA-NH_4HCO_3 extractable elements (Fe, Cd, Cu, Pb, Zn, Mn, Ca, Mg, Na, K, B, P, Mo, Ni, Se) were determined by the method of Soltanpour and Workman (1981). Total elemental analysis was done by digestion in a CEM microwave digestion system. The procedure used included ashing the material overnight in a 425 °C muffle furnace, digestion in a teflon bomb in the microwave oven with 1·5 ml HNO_3, 4·5 ml HCl and 10 ml HF for 10 min at 100% power, 20 min at 50% power and 10 min at 100% power. The digested solutions were transferred and made up with saturated H_3BO_3 to 50 ml, and the metal concentrations measured using ICP-AES.

3. RESULTS AND DISCUSSION

3.1. Height Growth Results

A total of 82 days elapsed from the time of planting to harvest. Germination was relatively uniform despite the fact that some crusting occurred in most of the pots. A comparison of means of the cumulative growth are presented in Table 3.

Height growth was enhanced by the addition of all three wastes. Plant height in the pot soil or control was less than that which occurred in the

TABLE 3

Comparison of the means of the cumulative height growth of brome grass

Waste	Cl rate (kg/ha)	N	Week					
			3	5	7	8	10	12
Freshwater	0	3	3·33a[a]	7·00a	12·0a	12·7bc	13·3abc	13·3bc
Freshwater	20	3	3·67a	8·67a	14·3a	15·3ab	15·3ab	15·3ab
Freshwater	40	3	3·67a	9·33a	15·0a	15·0ab	15·3ab	15·3ab
Freshwater	80	3	3·00a	8·67a	14·3a	15·0ab	15·3ab	14·7ab
Freshwater	200	3	3·67a	8·00a	14·7a	16·0a	16·0a	16·0a
Freshwater	400	3	3·33a	8·67a	14·0a	15·0ab	15·0ab	15·0ab
Freshwater	600	3	3·00a	7·67a	13·3a	13·7abc	14·0abc	14·3ab
Freshwater	800	3	2·33a	6·67a	11·3a	12·7bc	13·0bc	13·0bc
Freshwater	1 200	3	2·33a	6·67a	10·3a	11·0c	11·3c	11·3c
KCl	0	3	3·33a	7·00a	12·0ab	12·7a	13·3a	13·3a
KCl	200	3	3·00a	8·67a	14·0a	14·3a	14·7a	14·3a
KCl	400	3	3·00a	7·33a	13·0a	14·0a	14·0a	14·0a
KCl	600	3	3·67a	7·67a	12·0ab	14·7a	14·7a	14·0a
KCl	800	3	3·67a	6·33a	12·3ab	13·7a	13·3a	13·7a
KCl	1 200	3	2·67a	6·67a	11·3ab	13·7a	14·0a	14·0a
KCl	2 400	3	3·00a	6·00a	8·67b	13·7a	14·0a	15·0a
NaCl	0	3	3·33ab	7·00b	12·0a	12·7a	13·3a	13·3b
NaCl	200	3	3·33ab	7·67ab	14·0a	14·3a	15·7a	15·7ab
NaCl	400	3	3·33ab	7·00b	14·3a	14·7a	15·3a	14·7ab
NaCl	600	3	3·33ab	7·67ab	14·0a	14·7a	14·7a	14·7ab
NaCl	800	3	2·67ab	7·67ab	14·3a	15·0a	15·7a	15·3ab
NaCl	1 200	3	2·67ab	6·33b	12·7a	14·7a	15·3a	14·7ab
NaCl	1 520	3	1·67b	7·67ab	13·3a	14·0a	14·0a	13·3b
NaCl	2 400	3	4·67a	11·3a	13·0a	14·3a	16·0a	17·3ab

[a]Treatment means in any one column not followed by a common letter are significantly different at 0·01 probability by Tukey's Studentized Range (HSD) Test.

pots containing the highest concentrations of waste mixed with the same soil. Statistical analysis indicated virtually no significant differences in height growth relative to concentration for each of the three different waste types.

3.2. Yield Measurements

The fresh and dry weights of the brome grass were determined for each of the treatments and are presented in Table 4. Table 4 presents a comparison of the mean weights of the grass shoots. Except for the

TABLE 4
Comparisons of the mean weights of the brome grass
shoots

Waste	Cl rate (kg/ha)	N	Fresh grass (g)	Dry grass (g)
Freshwater	0	3	4·50ab[a]	1·10a
Freshwater	20	3	5·25a	2·03a
Freshwater	40	3	4·74ab	1·98a
Freshwater	80	3	3·54b	1·64a
Freshwater	200	3	5·57a	1·65a
Freshwater	400	3	5·72a	1·47a
Freshwater	600	3	5·25a	1·28a
Freshwater	800	3	4·74ab	1·22a
Freshwater	1 200	3	3·54b	1·03a
KCl	0	3	4·50a	1·10ab
KCl	200	3	5·15a	1·71a
KCl	400	3	4·94a	1·52ab
KCl	600	3	5·41a	1·55ab
KCl	800	3	4·85a	1·34ab
KCl	1 200	3	4·81a	1·37ab
KCl	2 400	3	4·11a	0·97b
NaCl	0	3	4·50a	1·10b
NaCl	200	3	5·32ab	1·92a
NaCl	400	3	5·37ab	1·75a
NaCl	600	3	5·45a	1·82a
NaCl	800	3	5·65a	1·92a
NaCl	1 200	3	5·31ab	1·70a
NaCl	1 520	3	5·10ab	1·65ab
NaCl	2 400	3	5·75a	1·62ab

[a]Treatment means in any one column not followed by a common letter are significantly different at 0·01 probability by Tukey's Studentized Range (HSD) Test.

2400 kg/ha chloride rate of KCl waste, all treatment yields were greater than the control soil yield. The greatest yields occurred at the lowest waste application rates and decreased with increased rates of waste application. There was no statistically significant increase or decrease in grass weight (yield) with increased waste application rate. This result could be explained by the fact that the controlled environment of the greenhouse may have masked some impacts by not creating any stresses. For example, moisture levels were maintained at or very near field capacity.

3.3. Pot Trial Soil and Waste Characteristics

The physical and chemical properties of the soil used in the pot trial experiment are presented in Table 5. Five samples were analysed from the bulk soil sample obtained and mean values determined therefrom.

Data for samples obtained from the bulk waste samples are also presented. The bulk waste samples were composites of material from all digs and all depths from each of the waste sample site locations (Macyk et al., 1987).

3.4. Chemical Properties of the Waste/Soil Mixtures

The waste/soil mixtures contained in each of the pots were analysed following removal of the plant root and shoot material. A comparison of mean values is presented in Table 6.

For the freshwater waste/soil mixtures a significant increase in $CaCO_3$ equivalent and pH resulted from waste addition. The increase in total organic carbon was likely due to organic carbon from the waste such as oils and greases.

For the potassium chloride waste/soil mixtures an apparent increase in $CaCO_3$ equivalent and pH was evident, however the differences were not statistically significant. Total organic carbon was significantly increased for the 2400 kg/ha rate only.

For the sodium chloride waste/soil mixtures pH increased significantly for the higher rates due to waste addition.

3.5. Saturated Paste Extract Data for the Waste/Soil Mixtures

A comparison of mean values relative to the saturated paste extract data for the pot trial waste/soil mixtures is presented in Table 7. The data indicated significant increases in pH, EC, SAR and chloride content with increasing concentrations for each of the three wastes. The EC levels could be limiting to plant growth at the higher application rates utilized. Miller & Pesaran (1980) suggested that many plants exhibit severe growth reductions when the soil-paste extract conductivity reaches about 8–10 dS/m. The Alberta Soils Advisory Committee (1977) suggested that moderate limitations to plant growth occur at EC levels of 4–8 dS/m and that severe limitations occur at levels greater than 8 dS/m. Using these criteria severe limitations might possibly occur in the 1200 kg Cl/ha application rate for freshwater, and at the 2400 kg Cl/ha application rate of potassium chloride waste.

The SAR values also suggest potential limitations with increasing concentration of waste application. The Alberta Soils Advisory Committee

TABLE 5
Chemical and physical properties of the pot trial soil and waste

Wastes	Exchangeable and total cations (meq/100g)					CaCO₃ (kg/kg)	pH		Acid neut. cap.	Total C (%)	Oil (%)	Field water (%)	Particle size (%)		
	Na	K	Ca	Mg	TEC		H₂O	CaCl₂					Sand	Silt	Clay
Freshwater	14·34	1·34	84·11	5·75	22·9	0·057	1·4	11·6	12·08	3·230	0·93	127·55	46	38	16
KCl	16·97	31·14	39·62	1·82	20·1	0·063	9·3	9·6	8·62	6·310	3·05	82·11	42	36	22
NaCl	78·65	0·77	88·15	8·30	17·3	0·115	10·3	10·3	18·31	2·480	0·14	87·98	51	33	16
Pot soil	0·04	0·41	5·15	1·68	9·6	0·001	5·4	5·3		0·541	0·01	15·1	64	19	17

Saturation extract data

Wastes	Sat'n (kg/kg)	EC (dS/m)	SAR	pH	Cl	Alk.	Na	K	Ca	S	Al	Cr	Fe	V
								(mg/litre)						
Freshwater	0·885	7·66	39	11·08	1 020·0	360·0	1 874·1	65·3	174·5	1 365·0	1·11	0·156	<0·001	0·069
KCl	0·703	56·80	41	8·91	21 200·0	89·0	5 564·1	14 464·8	1 367·5	1 965·0	4·20	<0·002	<0·001	<0·003
NaCl	0·630	137·85	259	10·05	83 700·0	110·5	50 679·1	338·8	2 897·5	1 367·5	6·64	<0·002	<0·001	<0·003
Pot soil	0·353	0·19	1	6·33	21·3	58·6	13·4	14·2	20·1	7·0	4·82	0·021	18·605	0·059

(continued)

TABLE 5—contd.

Wastes	Ti	Cd	Cu	Pb	Zn	Mn	Mg (mg/litre)	Li	Sr	B	Ba	P	Mo	Ni
Fresh water	0·004	<0·001	0·046	<0·010	0·196	<0·001	0·032	0·014	0·63	0·20	0·03	0·159	1·587	<0·003
KCl	<0·001	<0·001	<0·001	<0·010	0·463	0·057	11·2	0·285	9·44	0·30	0·37	0·029	1·256	<0·003
NaCl	<0·001	<0·001	<0·001	<0·010	<0·002	<0·001	<0·010	<0·004	4·82	0·08	0·26	<0·010	0·124	<0·003
Pot soil	1·188	0·003	0·116	0·045	0·189	0·19	5·693	0·015	0·11	0·07	0·10	0·623	0·010	0·023

Wastes	Se	As (mg/litre)	Co	Si
Freshwater	0·062	0·117	<0·001	21·51
KCl	<0·030	0·011	<0·001	5·15
NaCl	<0·030	<0·003	<0·001	5·55
Pot soil	0·065	0·065	0·008	20·90

Total elemental analysis data

Wastes	Mn	Na	K	Ca	Mg	Cu	Zn (µg/g)	Cd	P	Mo	Al	Fe	Ti
Freshwater	186·0	5 410	7 550	26 880	4 700	39·2	47·4	20·8	338	29·2	26 320	12 130	1 190
KCl	372·0	10 950	33 620	40 560	9 190	32·2	209·0	6·9	729	36·8	57 640	25 320	3 115
NaCl	260·0	41 820	11 810	70 200	14 880	16·2	77·0	<4·3	401	<21·4	37 760	15 600	1 930
Pot soil	198·5	6 890	11 890	4 250	2 690	7·8	12·9	<3·2	246	<16·0	24 320	8 760	1 660

Wastes	V	Cr	Ba	Ni	Sr	Li	Zr
				(μg/g)			
Freshwater	66·0	47·8	431	35·2	111·0	60·5	106·5
KCl	131·0	136·5	749	37·6	240·0	91·0	133·5
NaCl	50·5	62·5	430	18·2	192·5	59·5	98·5
Pot soil	32·6	32·4	311	19·2	35·0	19·0	128·5

DTPA extract data

Wastes	Fe	Cd	Cu	Pb	Zn	Mn	Ca	Mg	Na	K	B	P	Mo
								(μg/g)					
Freshwater	260·0	0·2	10·6	18·7	25·0	20·5	100	159	1 730·0	247·0	3·6	4·8	1·7
KCl	212·5	0·2	8·0	10·7	56·0	15·8	148	93	2 400·0	7 985·0	2·2	2·5	1·2
NaCl	180·0	<0·1	6·0	12·0	21·5	12·4	89	315	18 000·0	87·0	1·9	5·2	0·2
Pot soil	55·0	<0·1	0·6	0·4	0·4	2·4	415	182	7·6	78·5	<0·1	5·9	<0·2

Wastes	Ni	Se
		(μg/g)
Freshwater	1·9	1·0
KCl	2·2	<0·8
NaCl	0·8	<0·8
Pot soil	0·2	<0·8

TABLE 6
Comparison of the chemical properties (mean values) of the waste/soil mixtures

Waste	Cl rate (kg/ha)	N	CaCO₃ equiv.	pH		TC	TIC	TOC
				H₂O	CaCl₂	(%)		
Freshwater	0	3	0·32fᵃ	5·73e	5·40g	0·59d	0·04f	0·55e
Freshwater	20	3	0·05f	7·03d	6·93f	0·65d	0·01f	0·64e
Freshwater	40	3	0·09f	7·50c	7·20e	0·62d	0·01f	0·61e
Freshwater	80	3	0·43f	7·57c	7·47d	0·67d	0·05f	0·62e
Freshwater	200	3	1·00e	7·83bc	7·80c	0·84d	0·12e	0·72de
Freshwater	400	3	1·80d	8·03ab	7·97bc	1·19c	0·22d	0·97cd
Freshwater	600	3	2·91c	8·17ab	8·17ab	1·49c	0·35c	1·14bc
Freshwater	800	3	3·93b	8·20a	8·27a	1·74b	0·47b	1·27b
Freshwater	1 200	3	5·68a	8·33a	8·40a	2·50a	0·68a	1·82a
KCl	0	3	0·32ab	5·73a	5·40f	0·59b	0·04ab	0·55b
KCl	200	3	0·00b	6·50a	6·40e	0·66b	0·00b	0·66b
KCl	400	3	0·15ab	7·00a	6·57de	0·66b	0·02ab	0·64b
KCl	600	3	0·15ab	5·97a	6·87cd	0·65b	0·02ab	0·64b
KCl	800	3	0·29ab	7·20a	7·13bc	0·73b	0·04ab	0·69b
KCl	1 200	3	0·35ab	7·40a	7·37b	0·75b	0·04ab	0·71b
KCl	2 400	3	0·52a	7·57a	7·77a	1·04a	0·06a	0·98a
NaCl	0	3	0·32a	5·73d	5·40g	0·59a	0·04a	0·55a
NaCl	200	3	0·06ab	6·13cd	5·93f	0·65a	0·01ab	0·64a
NaCl	400	3	0·01b	6·23cd	6·00ed	0·68a	0·01b	0·68a
NaCl	600	3	0·06ab	6·43bcd	6·33de	0·66a	0·01ab	0·65a
NaCl	800	3	0·09ab	6·93abc	6·63cd	0·62a	0·01ab	0·61a
NaCl	1 200	3	0·13ab	7·03abc	6·90bc	0·66a	0·02ab	0·65a
NaCl	1 520	3	0·15ab	7·27ab	7·23ab	0·70a	0·02ab	0·68a
NaCl	2 400	3	0·31a	7·47a	7·33a	0·77a	0·04a	0·73a

ᵃTreatment means in any one column not followed by a common letter are significantly different at 0·01 probability by Tukey's Studentized Range (HSD) Test.

(1977) presented the SAR criteria given in Table 8. Using these criteria it is apparent that SAR would be most limiting in the freshwater waste/soil mixtures and least limiting in the KCl waste/soil mixtures. Application rates greater than 400 kg Cl/ha of freshwater waste have the potential to result in severe limitations relative to SAR levels. Severe limitations could result in NaCl waste/soil mixtures at the 2400 kg Cl/ha application rate.

Significant increases in saturation percentage for the freshwater waste/soil mixtures can be attributed to the large amount of waste added in each pot mixture (Table 1).

The decrease in iron and aluminum compared to the control was

TABLE 7
Comparison of the saturated paste extract data (mean values) for the waste/soil mixtures

Waste	Cl rate (kg/ha)	N	pH	Satn (kg/kg)	EC (dS/m)	SAR	Cl	Alk.	NO₃	Al	Fe	V	Ti	Cu	Zn
										(mg/litre)					
Freshwater	0	3	6·50c[a]	0·343ef	0·16h	1·02h	7·95e	78·7d	0·09a	3·05a	19·3a	0·05a	0·99a	0·04a	0·18bcd
Freshwater	20	3	7·30b	0·337f	0·77g	1·69gh	9·05e	272b	0·07a	0·50b	0·09b	0·03a	0·01b	0·01b	0·13cd
Freshwater	40	3	7·57ab	0·330f	1·06g	2·58fg	13·3e	285ab	0·09a	0·61b	0·06b	0·03a	0·01b	0·01b	0·12d
Freshwater	80	3	7·70ab	0·378de	1·53f	3·68f	28·6de	277b	0·10a	0·85b	0·08b	0·03a	0·01b	0·01b	0·13d
Freshwater	200	3	7·92ab	0·390d	2·91e	7·08e	102d	274b	0·12a	1·32ab	0·07b	0·03a	0·01b	0·01b	0·16bcd
Freshwater	400	3	7·97ab	0·440c	4·65d	10·8d	260c	315a	0·10a	1·90ab	0·05b	0·03a	0·01b	0·02ab	0·19bcd
Freshwater	600	3	8·05a	0·478c	6·00c	14·7c	451b	290ab	0·11a	2·06ab	0·04b	0·03a	0·01b	0·02ab	0·21abc
Freshwater	800	3	8·05a	0·521b	6·92b	18·1b	520b	286ab	0·11a	2·09ab	0·03b	0·03a	0·01b	0·02ab	0·22ab
Freshwater	1 200	3	8·24a	0·608a	8·08a	22·7a	645a	210c	0·11a	2·10ab	0·01b	0·03a	0·01b	0·02b	0·28a
KCl	0	3	6·50b	0·343a	0·16f	1·02f	7·95d	78·7c	0·09a	3·05a	19·3a	0·05a	0·99a	0·04a	0·18a
KCl	200	3	6·57b	0·340a	0·96ef	1·81e	209d	138b	0·12a	0·66b	0·18b	0·03b	0·01b	0·01b	0·18a
KCl	400	3	6·95ab	0·343a	1·86de	2·13de	702c	173ab	0·09a	1·00ab	0·08b	0·03b	0·01b	0·01b	0·11a
KCl	600	3	7·03ab	0·343a	2·56cd	2·50cd	672c	180a	0·13a	1·26ab	0·04b	0·03b	0·01b	0·01b	0·09a
KCl	800	3	7·13ab	0·350a	3·43c	2·91c	966bc	186a	0·10a	1·63ab	0·02b	0·03b	0·01b	0·01b	0·11a
KCl	1 200	3	7·33ab	0·355a	4·88b	3·64b	1320b	180a	0·15a	2·26ab	0·02b	0·03b	0·01b	0·01b	0·10a
KCl	2 400	3	7·47a	0·372a	10·5a	6·53a	3130a	174ab	0·35a	3·70ab	0·02b	0·03b	0·02b	0·01b	0·16a
NaCl	0	3	6·50ab	0·343ab	0·16e	1·02e	7·95e	78·7c	0·09b	3·05a	19·3a	0·05a	0·99a	0·04a	0·18a
NaCl	200	3	6·30b	0·350ab	0·57e	2·82de	126de	99·2c	0·11ab	0·62b	1·11b	0·03a	0·09b	0·02b	0·21a
NaCl	400	3	6·45ab	0·338b	1·36de	4·19cd	363de	141bc	0·17a	0·48b	0·06b	0·03a	0·01b	0·01b	0·19a
NaCl	600	3	6·57ab	0·345ab	2·07cde	5·16cd	559cde	164abc	0·13ab	0·68b	0·04b	0·03a	0·01b	0·01b	0·21a
NaCl	800	3	6·90ab	0·333b	3·24bcd	6·61bc	897bcd	186ab	0·17a	0·98ab	0·03b	0·03a	0·01b	0·01b	0·20a
NaCl	1 200	3	7·03ab	0·353ab	4·30bc	8·21b	1260bc	204ab	0·12ab	1·36ab	0·03b	0·03a	0·01b	0·01b	0·24a
NaCl	1 520	3	7·13ab	0·371a	4·81b	8·94b	1410b	209ab	0·13ab	1·46ab	0·04b	0·03a	0·01b	0·01b	0·22a
NaCl	2 400	3	7·28a	0·332b	7·66a	13·7a	2490a	241a	0·09b	2·11ab	0·07b	0·03a	0·01b	0·01b	0·21a

[a]Treatment means in any one column not followed by a common letter are significantly different at 0·01 probability by Tukey's Studentized Range (HSD) Test.

(continued)

TABLE 7—contd.

Waste	Cl rate (kg/ha)	Mn	Ca	Mg	Na	K	Sr	B	Ba	P	Mo	As	S	Si
								(mg/litre)						
Freshwater	0	0·30b[a]	18·9e	5·29f	19·5g	5·87f	0·08e	0·06d	0·10a	0·57a	0·05e	0·06d	3·75h	19·1a
Freshwater	20	0·38a	128d	22·2e	78·3fg	4·69f	0·24d	0·06d	0·09a	0·22d	0·05e	0·17ab	67·8gh	5·08c
Freshwater	40	0·30b	150cd	25·0de	130fg	4·72f	0·29d	0·07d	0·08a	0·23d	0·05e	0·18a	132fg	5·97c
Freshwater	80	0·19b	201c	32·1cd	214f	6·30ef	0·35d	0·13cd	0·12a	0·26cd	0·05e	0·14c	239f	8·11c
Freshwater	200	0·05c	320b	43·7b	511e	10·1e	0·58c	0·24c	0·12a	0·35b	0·18e	0·15bc	537e	10·7bc
Freshwater	400	0·05c	460a	53·6a	922d	17·9d	0·85b	0·45b	0·13a	0·34bc	0·37d	0·17ab	885d	14·9ab
Freshwater	600	0·04c	500a	55·4a	1300c	27·5c	1·02a	0·56ab	0·10a	0·25d	0·58c	0·15bc	1130c	17·5a
Freshwater	800	0·04c	502a	50·1ab	1590b	35·6b	1·12a	0·62a	0·09a	0·23d	0·80b	0·14c	1310b	19·1a
Freshwater	1 200	0·02c	488a	34·3c	1930a	49·6a	1·16a	0·68a	0·35a	0·13e	1·13a	0·09d	1520a	20·3a
KCl	200	0·30d	18·9g	5·29g	19·5e	5·87b	0·08g	0·06b	0·10d	0·57a	0·05a	0·06ab	3·75f	19·1a
KCl	200	1·34a	111f	26·2f	81·6de	19·1b	0·26f	0·04b	0·16d	0·19bc	0·05a	0·05b	32·8ef	5·43b
KCl	400	1·02ab	224e	46·4e	134d	44·8b	0·54e	0·04b	0·25c	0·17c	0·05a	0·06ab	63·1de	3·38b
KCl	600	0·75bc	307d	60·2d	183cd	70·3b	0·77d	0·04b	0·29bc	0·16c	0·05a	0·06ab	85·1cd	2·56b
KCl	800	0·55cd	417c	75·9c	247c	107b	1·06c	0·04b	0·32bc	0·17c	0·05a	0·06ab	121c	2·09b
KCl	1 200	0·38cd	590b	96·0b	363b	208b	1·55b	0·04b	0·34b	0·20bc	0·05a	0·06ab	176b	2·07b
KCl	2 400	0·49cd	1050a	143a	854a	915a	2·76a	0·09a	0·50a	0·26b	0·05a	0·07a	400a	3·84b
NaCl	0	0·30a	18·9f	5·29e	19·5d	5·87cd	0·08f	0·06a	0·10e	0·57a	0·05a	0·06bcd	3·75d	19·1a
NaCl	200	0·95a	44·8ef	11·4e	818d	5·04d	0·10f	0·04b	0·09e	0·20c	0·05a	0·03d	4·76d	7·56b
NaCl	400	1·65a	104ef	23·0de	181cd	6·38cd	0·22ef	0·04b	0·15de	0·18c	0·05a	0·05cd	2·6cd	5·22b
NaCl	600	1·94a	158de	39·2cd	280bcd	7·37bcd	0·32de	0·04b	0·20cd	0·18c	0·05a	0·08bcd	21·6bcd	4·48b
NaCl	800	1·55a	232cd	54·5bc	432bcd	8·79bcd	0·46cd	0·04b	0·25bc	0·19c	0·05a	0·09abc	37·3bcd	3·48b
NaCl	1 200	0·96a	323bc	69·9ab	627bc	10·6abc	0·63bc	0·04b	0·28bc	0·20c	0·05a	0·10ab	53·0abc	3·01b
NaCl	1 520	0·89a	360b	74·4ab	715b	12·1ab	0·69b	0·04b	0·31ab	0·19c	0·05a	0·10ab	62·4ab	2·93b
NaCl	2 400	1·61a	502a	90·0a	1280a	13·76a	0·93a	0·04b	0·36a	0·39b	0·05a	0·13a	90·8a	4·28b

[a]Treatment means in any one column not followed by a common letter are significantly different at 0·01 probability by Tukey's Studentized Range (HSD) Test.

TABLE 8

	SAR value
No limitations	< 4
Slight limitations	4–8
Moderate limitations	8–12
Severe limitations	> 12

probably due to precipitation resulting from lower solubility as pH increased. The increases in boron, calcium, magnesium and potassium were not harmful to the brome grass and were probably beneficial. Barium and strontium levels were not high enough (in view of calcium and magnesium concentrations) to be harmful to the crop.

3.6. DTPA Extract Data for the Waste/Soil Mixtures
A comparison of mean values relative to the DTPA extract data for the pot trial waste/soil mixtures is presented in Table 9.

The literature indicates that DTPA soil extractions are correlated with plant uptake, i.e. as DTPA extractable elemental content increases plant uptake of that element generally increases.

The most significant change as application rate increased occurred in the freshwater waste/soil mixtures. Iron, copper, lead and zinc increased and calcium decreased relative to the control or unamended soil material. Increases in sodium and potassium were due to additions of the waste materials.

3.7. Total Elemental Content of the Brome Grass
A comparison of mean values relative to the total elemental content of the brome grass shoots is presented in Table 10. For interpretation purposes the values from Table 10 were compared to values reported in the literature (Chapman, 1966). The values reported in Chapman (1966) are for a very wide range of plants and in the case of only two elements were values available specifically for brome grass. For the remaining elements the reported values represent a mean of values for 'grasses' including bluegrass, ryegrass and timothy.

Calcium levels in the tissue decreased with increasing application rate of freshwater waste whereas the reverse was true for the KCl and NaCl waste mixtures. Chapman (1966) reported a low range value of 0·64% (6400 ppm) and an intermediate range value of 1·04–1·75% (10 400 to 17 500 ppm) for ryegrass. Values in Table 10 range from low to intermediate levels for calcium.

TABLE 9
Comparison of the DTPA extract data (mean values) for the waste/soil mixtures

Waste	Cl rate (kg/ha)	N	Fe	Cu	Pb	Zn	Mn	Ca	Ma	Na	K	B	P	Mo	Ni
									(µg/g)						
Freshwater	0	3	51·0e[a]	0·43g	0·30e	0·43e	4·13c	442a	146a	22·4f	85·0e	0·10d	7·00ef	0·20c	0·60c
Freshwater	20	3	39·5e	0·47fg	0·47e	0·63de	2·80d	450a	127ab	58·2f	64·0f	0·10d	8·87de	0·20c	0·80bc
Freshwater	40	3	40·6e	0·53fg	0·77e	1·07de	2·67d	451a	110b	85·7f	59·2f	0·10d	10·6dc	0·20c	1·10ab
Freshwater	80	3	47·3e	0·73f	1·47e	1·40de	3·07d	438a	134ab	163f	71·7ef	0·10d	11·7bc	0·20c	1·17a
Freshwater	200	3	75·3d	1·27e	3·30d	2·10d	4·37c	396a	138a	369e	87·3e	0·30d	14·5a	0·13c	0·50c
Freshwater	400	3	111c	2·17d	6·47c	4·33c	5·90b	368a	148a	720d	121d	0·63c	13·1ab	0·20c	0·73c
Freshwater	600	3	129bc	2·73c	8·67b	5·10bc	6·67b	346d	146a	1090c	157c	0·90b	10·8cd	0·40b	0·67c
Freshwater	800	3	135b	3·33b	9·77b	6·10b	6·80b	328e	145a	1500b	206b	1·10b	8·90de	0·50b	0·70c
Freshwater	1 200	3	171a	4·53a	12·6a	10·0a	8·30a	273f	134ab	2178a	289a	1·63a	6·03f	0·87a	0·70c
KCl	0	3	51·0a	0·43c	0·30c	0·43c	4·13bc	442a	146a	22·4d	85·0c	0·10a	7·00c	0·20a	0·60a
KCl	200	3	42·2ab	0·43c	0·33c	0·47d	4·40b	430ab	140a	56·2cd	125c	0·10a	7·60bc	0·20a	0·20a
KCl	400	3	38·9b	0·47bc	0·43c	0·70cd	3·23cd	427ab	131a	72·3cd	187bc	0·10a	9·03bc	0·20a	0·20a
KCl	600	3	35·3b	0·47bc	0·40c	0·83cd	2·90d	421ab	135a	99·0bcd	244bc	0·10a	8·57bc	0·20a	0·20a
KCl	800	3	36·7b	0·57bc	0·50c	1·13bc	2·93d	407b	138a	136bc	319bc	0·10a	8·93bc	0·20a	0·20a
KCl	1 200	3	39·1b	0·67b	0·80b	1·63b	3·27cd	425ab	129a	182b	459b	0·10a	9·90b	0·20a	0·27a
KCl	2 400	3	48·9a	0·97a	1·77a	4·23a	5·70a	417ab	111a	414a	1290a	0·17a	12·7a	0·20a	0·73a
NaCl	0	3	51·0a	0·43a	0·30c	0·43ab	4·13a	442a	146a	22·4e	85·0a	0·10a	7·00b	0·20a	0·60a
NaCl	200	3	52·0a	0·43a	0·47abc	0·47abc	6·20a	414b	124ab	69·8e	63·7b	0·10a	8·10b	0·20a	0·47ab
NaCl	400	3	46·6ab	0·43a	0·30c	0·30c	5·83a	435ab	134ab	126de	65·5ab	0·10a	8·07b	0·20a	0·20b
NaCl	600	3	44·2bc	0·43a	0·30c	0·30c	5·23a	435ab	136ab	180de	64·7b	0·10a	8·27b	0·20a	0·23b
NaCl	800	3	41·7bc	0·40a	0·30c	0·30c	4·50a	418b	134ab	241cd	63·5b	0·10a	8·83b	0·20a	0·20b
NaCl	1 200	3	38·5c	0·50a	0·37bc	0·37bc	3·90a	415b	153a	378bc	74·3ab	0·10a	9·23b	0·20a	0·20b
NaCl	1 520	3	41·8bc	0·53a	0·57a	0·57a	4·03a	418b	149a	422b	73·7ab	0·10a	9·53b	0·20a	0·33ab
NaCl	2 400	3	44·4bc	0·50a	0·50ab	0·50ab	5·53a	425ab	114b	629a	60·0b	0·17a	12·5a	0·20a	0·20b

Treatment means in any one column not followed by a common letter are significantly different at 0·01 probability by Tukey's Studentized Range (HSD) Test.

TABLE 10

Comparison of the total elemental content (mean values) of the brome grass

Waste	Cl rate (kg/ha)	N	Al	Fe	Zn	Mn	Ca	Mg	Na	K	Sr	P	Ba	Mo
								(μg/g)						
Freshwater	0	3	26·6a[a]	51·8a	20·7a	235a	6290a	2790ab	167d	27300a	13·3a	1610a	48·7a	4·43a
Freshwater	20	3	42·3a	204a	10·2bc	67·1d	6780a	3190a	190d	30600a	14·6a	1400ab	39·4ab	10·5a
Freshwater	40	3	26·5a	53·0a	8·84c	52·3d	6650a	3000ab	257d	28400a	13·3a	1420ab	33·1bc	4·42a
Freshwater	80	3	35·8a	89·6a	11·8bc	73·7d	6270a	2840ab	268d	31500a	13·3a	1460ab	25·0cd	4·26a
Freshwater	200	3	32·9a	103a	15·7abc	121bc	6080a	2680ab	446cd	31100a	12·8a	1330abc	18·5de	4·26a
Freshwater	400	3	17·6a	27·8a	13·9abc	130bc	5040ab	2110bc	681c	26900a	10·9a	1040bc	13·2de	8·07a
Freshwater	600	3	24·8a	89·2a	17·5abc	154b	5350ab	2190bc	1455b	31100a	13·1a	1190abc	13·1de	8·74a
Freshwater	800	3	24·1a	48·3a	16·9abc	166b	5100ab	2120bc	1830a	32900a	14·1a	1150abc	14·1de	12·7a
Freshwater	1 200	3	14·7a	46·4a	13·8abc	97·8cd	4080b	1700c	2880a	31500a	13·8a	871c	8·87e	15·8a
KCl	0	3	26·6ab	51·8a	20·7ab	235a	6290d	2790a	167b	27300a	13·3c	1610a	48·7a	4·43a
KCl	200	3	20·8b	35·7a	14·8b	140d	6040d	2390abc	119b	37200a	13·4c	1210bc	38·6ab	4·46a
KCl	400	3	23·5b	39·2a	18·5b	142d	6900d	2550ab	120b	29100a	16·9bc	1110bc	41·4ab	5·18a
KCl	600	3	26·1b	37·8a	17·5b	156cd	7250cd	2520ab	136b	30400a	17·5bc	1050c	32·7abc	4·37a
KCl	800	3	27·7ab	40·8a	20·4ab	175c	8310bc	2560ab	188b	31900a	20·4ab	1170bc	30·6bcd	4·67a
KCl	1 200	3	30·2ab	37·3a	18·7b	182bc	8710ab	2340bc	267b	28400a	21·6ab	1050c	21·6cd	4·32a
KCl	2 400	3	62·6a	88·4a	28·4a	208ab	9710a	2000c	1580a	36700a	24·8a	1410ab	14·6d	5·76a
NaCl	0	3	26·6ab	51·8a	20·7a	235a	6290b	2790a	167a	27300a	13·3a	1610ab	48·7a	4·43a
NaCl	200	3	19·6b	32·0a	14·5ab	182ab	5510b	2480a	150a	25000a	12·4a	1320bc	40·0ab	4·38a
NaCl	400	3	23·1ab	37·5a	14·4ab	170ab	6000b	2620a	192a	28600a	11·5a	1370bc	38·9b	4·33a
NaCl	600	3	24·3ab	35·7a	12·9ab	162ab	6410b	2760a	259a	28300a	12·9a	1310bc	34·4bc	4·30a
NaCl	800	3	20·6b	35·5a	13·3ab	139b	6520b	2680a	195a	26700a	13·3a	1280bc	31·1bc	4·44a
NaCl	1 200	3	31·7ab	54·8a	9·526b	135b	6800b	2620a	420a	27300a	13·2a	1200c	26·5c	4·41a
NaCl	1 520	3	37·2a	42·9a	11·4ab	143b	7320ab	2660a	494a	27700a	12·9a	1270bc	25·7c	4·29a
NaCl	2 400	3	37·8a	63·3a	15·8ab	219ab	8920a	2800a	1205a	31900a	15·9a	1820a	24·8c	6·07a

[a] Treatment means in any one column not followed by a common letter are significantly different at 0·01 probability by Tukey's Studentized Range (HSD) Test.

TABLE 10—contd.

Waste	Cl rate (kg/ha)	B	S	Si	As (μg/g)	Pb	Cd	Cl
Freshwater	0	19·2c[a]	694b	691b	23·6a	0·31a	0·41ab	2660d
Freshwater	20	17·6c	1560a	589b	23·3ab	0·62a	0·45a	3590cd
Freshwater	40	24·3bc	1620a	685b	22·1abc	0·22a	0·26ab	6060b
Freshwater	80	20·7c	1830a	925ab	17·6abcd	0·20a	0·20b	7100ab
Freshwater	200	24·2bc	1770a	955ab	17·0abcd	0·28a	0·20b	9130a
Freshwater	400	32·1abc	1540a	1170a	16·0abcd	0·20a	0·20b	6620b
Freshwater	600	49·5ab	1820a	1170a	8·71bcd	0·29a	0·20b	6520b
Freshwater	800	52·2a	1940a	1100a	5·60d	0·28a	0·20b	5850bc
Freshwater	1 200	39·1abc	2210a	1140a	7·99cd	0·20a	0·21b	5450bc
KCl	0	19·2a	694d	691a	23·6a	0·31b	0·41b	2660c
KCl	200	11·9a	1270c	616ab	13·4b	0·20b	0·22b	9970b
KCl	400	13·3a	1460bc	537b	15·5ab	0·20b	0·20b	10 500b
KCl	600	13·1a	1540bc	532b	11·0b	0·20b	0·22b	11 500b
KCl	800	13·2a	1710b	545b	11·7b	0·20b	0·20b	11 100b
KCl	1 200	16·1a	1670b	531b	14·3ab	0·23b	0·20b	12 600b
KCl	2 400	15·9a	2130a	559b	16·0ab	2·35a	1·22a	22 700a
NaCl	0	19·2a	694d	691ab	23·6a	0·31a	0·41ab	2660c
NaCl	200	13·0a	983cd	640ab	13·9a	0·23a	0·23b	9750b
NaCl	400	12·9a	1260bc	748a	14·4a	0·20a	0·20b	11 400b
NaCl	600	11·5a	1240bc	669ab	17·4a	0·31a	0·21b	11 700b
NaCl	800	10·3a	1280bc	619ab	16·3a	0·20a	0·24b	11 700b
NaCl	1 200	13·1a	1570b	612b	15·4a	0·20a	0·21b	12 300b
NaCl	1 520	14·2a	1700b	682ab	14·3a	0·20a	0·23b	11 900b
NaCl	2 400	10·6a	2411a	645ab	16·6a	0·91a	0·81a	18 200a

[a]Treatment means in any one column not followed by a common letter are significantly different at 0·01 probability by Tukey's Studentized Range (HSD) Test.

Sodium and sulphur levels increased with increasing application rate of the three waste types. Sodium levels reached the upper end of the normal range scale for treatments of 600 kg Cl/ha and greater for the freshwater waste and the 2400 kg Cl/ha rate of KCl and NaCl wastes.

Boron levels increased with increasing application rate of freshwater waste whereas the reverse was true for the other waste types. The boron levels in Table 10 are comparable to the intermediate range of 15–80 ppm B in grasses reported by Chapman (1966).

Chloride levels increased with increasing application rates of potassium and sodium chloride wastes (Fig. 1). For both wastes the 2400 kg Cl/ha application rate resulted in significantly increased chloride content in the

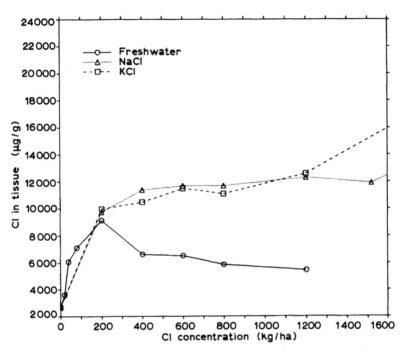

FIG. 1. Chloride content of brome grass relative to different waste application rates.

plant tissue (Table 10). No such clear trend was apparent for the freshwater waste material. Chloride values in Table 10 ranged from 2600 to 22 700 ppm. Chapman (1966) reported an intermediate range of 7000 ppm and a high range of 8700–15 400 ppm for chloride in grasses. Using these values as a guideline suggests that the 2400 kg Cl/ha rate of potassium and sodium chloride wastes has resulted in relatively high chloride content in the brome grass.

Zinc levels were generally lower in the amended soils than in the control soil. The levels reported in Table 10 are comparable to the intermediate range of 14–48 ppm reported by Chapman (1966).

Lead values were elevated in the 2400 kg Cl/ha application rates for sodium and potassium chloride wastes and significantly different for the latter. Chapman (1966) reported an intermediate range mean value of 0·9–1·4 ppm lead in the edible portion of 13 crops. The lead values in Table 10 varied from 0·20 to 2·35 with most values in the 0·20–0·30 range.

4. SUMMARY AND CONCLUSIONS

The results obtained indicated both positive and negative impacts resulting from different waste application rates.

4.1. Impact on Soil

4.1.1. Positive Impacts
The lime content of the wastes can be considered a positive characteristic. The increase in $CaCO_3$ equivalent and pH values to optimum levels for specific soils is useful, however application rates need to be controlled so that excessive pH values do not result. This would be of greater concern with the freshwater than the KCl and NaCl wastes.

4.1.2. Negative Impacts
In assessing the soil/waste mixture analytical data it was apparent that EC, SAR and chloride values were the only factors considered potentially limiting to plant growth.

4.2. Impact on Plants

4.2.1. Positive Impacts
Calcium and potassium status in the brome grass increased favorably for almost all additions of KCl and NaCl wastes to the soil. Boron status in the brome grass was improved by the addition of freshwater waste.

4.2.2. Negative Impacts
Chloride content in the brome grass was increased significantly as a result of addition of the waste material with excessive concentrations occurring at the 2400 kg Cl/ha application rates of KCl and NaCl wastes.

ACKNOWLEDGEMENTS

We would like to recognize the financial support of the Heritage Savings Trust Fund through the Reclamation Research Technical Advisory Committee (RRTAC) of the Land Conservation and Reclamation Council of the Alberta Government. The advice and assistance of the Oil and Gas Reclamation Research Program Technical Committee is gratefully acknowledged.

REFERENCES

ALBERTA SOILS ADVISORY COMMITTEE (1977). Soil quality criteria for agriculture. Report printed by Agriculture Canada.

AOAC (1981). Official methods of analysis, 14th edn. Association of Official Analytical Chemists, Arlington, VA.

BASCOMB, C.L. (1961). A calcimeter for routine use on soil samples. *Chemistry and Industry*, Part II, 1826–7.

CHAPMAN, H.D. (ed.) (1966). Diagnostic criteria for plants and soils. Department of Soils and Plant Nutrition, University of California Citrus Research Center and Agricultural Experiment Station, Riverside, California.

DOUGHTY, J.L. (1941). The advantages of a soil paste for routine pH determination. *Soil Science*, **22**, 135–8.

GAINES, T.P., PARKER, M.B. & GASCHO, G.J. (1984). Automated determination of chlorides in soil and plant tissue by sodium nitrate extraction. *Agronomy Journal*, **76**, 371–4.

GEE, G.W. & BAUDER, J.W. (1979). Particle size analysis by hydrometer: a simplified method for routine textural analysis and a sensitivity test of measurement parameters. *Soil Science Society of America Journal*, **43**, 1004–7.

HOLMGREN, G.G.S., JUVE, R.L. & GESCHWENDER, R.C. (1977). A mechanically controlled variable rate leaching device. *Soil Science Society of America Journal*, **32**, 568–70.

LECO CORPORATION (1979). CR-12 carbon system 781–600. Instrument Manual 200-195.

MAAS, E.V. (1986). Physiological response of plants to chloride. In *Chloride and Crop Production*, ed. T.L. Jackson. Papers of an American Society of Agronomy Annual Meeting (November 1984), published by the Potash and Phosphate Institute (August 1986).

MACYK, T.M., ABBOUD, S.A. & NIKIFORUK, F.I. (1987). Oil and gas reclamation research program: drilling mud disposal: sampling and detailed characterization. Volume I: Report, Volume II: Appendices. Terrain Sciences Department, Alberta Research Council. Unpublished report prepared for the Land Conservation and Reclamation Council, Reclamation Research Technical Advisory Committee, Alberta Environment.

MILLER, R.W. & HONARVAR, S. (1975). Effect of drilling fluid components and mixtures on plants and soils. *Conference Proceedings: The Environmental Aspects of Chemical Use in Well Drilling Operations.* EPA 560/1-75-004, pp. 125–43.

MILLER, R.W. & PESERAN, P. (1980). Effects of drilling fluids on soils and plants: II. Complete drilling fluid mixtures. *Journal of Environment Quality*, **9**, 552–6.

NELSON, D.W., LIU, S. & SOMMERS, L. (1980). Plant uptake of toxic metals present in drilling fluids. In *Symposium, Research on Environmental Fate and Effects of Drilling Fluids and Cuttings*, January 21–24, Lake Buena Vista, FL, Vol. 1, pp. 114–38.

PAULS, D.R., MORAN, S.R. & MACYK, T.M. (1987). Oil and gas reclamation research program: drilling mud disposal: review of literature related to clay liners for sump disposal of drilling waste. Terrain Sciences Department, Alberta

Research Council. Unpublished report prepared for the Land Conservation and Reclamation Council, Reclamation Research Technical Advisory Committee, Alberta Environment.

PEECH, M. (1965). Hydrogen-ion activity. In *Methods of Soil Analysis, Part 2*, ed. C.A. Black *et al. Agronomy*, **9**, 914–26. American Society of Agronomy, Inc., Madison, Wisconsin.

PESERAN, P. (1977). Effect of drilling fluid components and mixtures on plants and soils. M.S. thesis, Utah State University. API sponsored research project.

RHOADES, J.D. (1982). Soluble salts. In *Methods of Soil Analysis, Part 2*, ed. A. L. Page *et al. Agronomy*, **9**, 167–79. American Society of Agronomy, Inc., Madison, Wisconsin.

SOLTANPOUR, P.N. & WORKMAN, S.M. (1981). Soil-testing methods used at Colorado State University soil testing laboratory. Colorado State University.

USDA (1954). Diagnosis and improvement of saline and sodic soils. Agriculture Handbook 60, United States Department of Agriculture.

PART III
Land Disposal and Effects

17

API Survey Results on 1985 Onshore Wastes Volumes and Disposal Practices within the US Petroleum Extraction Industry

B. D. Freeman

Shell Oil Company, PO Box 2463, Houston, Texas 77001, USA

and

P. G. Wakim

American Petroleum Institute, 1220 L Street, N.W. Washington, D.C. 20005, USA

ABSTRACT

The Production Waste Issue Group Subcommittee of API directed a survey in 1986 to estimate waste volumes and to identify disposal practices within the United States (US) Petroleum Extraction Industry (PEI). A nationwide onshore survey of this type has never been conducted. This survey was precipitated by the 1980 Resource Conservation and Recovery Act amendments which required EPA to collect this information.

The paper first addresses the compilation and statistical analysis of survey data. Secondly, the estimated onshore exploration and production (E&P) waste volumes are presented for 1985. The nationwide wastes volumes are broken down into drilling fluids, produced water and associated waste classifications. Thirdly, the waste disposal practices for each waste classification are identified. Drilling pit closure, produced water disposal and associated waste disposal techniques are reported.

Finally, the potential environmental impact of the waste disposal practices is evaluated. Eighty-seven percent of drilling waste liquids are hauled off site, evaporated or injected underground. Sixty-eight percent of buried drilling waste solids are generated from freshwater base mud systems. The remaining 32% of buried drilling waste solids are generated from salt and oil base mud systems of which 86% are buried in lined pits. Ninety-five percent of drilling waste solids landspread on site are generated from freshwater base mud systems. As regards produced waters, 92% are injected underground. Sixty-two percent of produced waters are injected in enhanced recovery projects and are not a waste. Associated wastes represent about 3% of the total drilling fluid and associated wastes volume. The majority of associated wastes is hauled offsite to licensed disposal facilities.

Overall, the PEI disposes of drilling muds, produced water and associated wastes utilizing environmentally sound techniques.

1. INTRODUCTION

The Petroleum Extraction Industry (PEI) plays a vital role in the United States economy. It contributes a significant part of the gross national product as well as providing a natural resource that is essential to the national security of this country. The PEI, however, does generate a large volume of waste in onshore drilling and producing operations. This industry originated in the mid-part of the 18th century, but as time has proceeded, the waste disposal techniques utilized by the PEI have changed to provide adequate environmental protection. Presently, state and federal agencies regulate onshore disposal practices for drilling fluids, produced water, and other wastes generated by the PEI.

In 1976, Congress passed the Resource Conservation and Recovery Act (RCRA) to regulate the disposal of solid wastes. In 1978, EPA proposed regulations under Subtitle C of the Act. The regulations addressing the PEI were onerous and expensive. The PEI did not believe the regulations were justified. Congress did not either, and in 1980, amended the RCRA to temporarily exempt exploration and production (E&P) wastes from Subtitle C. Congress did require that EPA conduct a study to determine if onshore E&P wastes should be regulated under Subtitle C, and based upon the study results, make a recommendation to Congress as to how the wastes should be regulated. The EPA did not conduct the study as required, and in December 1985, an Alaskan environmental group filed a suit demanding the study be conducted as required under the Act. The court required EPA to publish a final report by August 31, 1987, but later granted an extension to December 31, 1987. As a result, the American Petroleum Institute (API) determined that a parallel study (API, 1987) must be conducted along with EPA to assure that a truly representative study resulted. This is because, if EPA determined that wastes should be regulated under Subtitle C, PEI would have to spend approximately $44 billion the first year of regulatory implementation and $5 billion thereafter.

One of the seven tasks required under Section 8000(m) of RCRA was for EPA to determine how onshore E&P wastes were generated, the volume of wastes generated and the industry practices used to dispose of these wastes. This paper discusses how the API survey was conducted, the results of the survey and an evaluation of the waste generation and disposal practices.

2. WASTE GENERATION AND DISPOSAL PRACTICES

The Production Waste Issue Group subcommittee of API determined that the best method to gather this information was to conduct a survey of PEI using a written questionnaire (API, 1987, pp. 215–27). The questionnaire would solicit operators for onshore drilling and associated wastes volumes generated in their operations, along with the disposal practice used for each waste. Produced water volumes would be collected from state agencies, along with a supplemental survey of companies to substantiate state agency data.

2.1. Drilling Fluids
Drilling muds are formulated from bentonite clay, water and a variety of organic and inorganic additives. Organic additives include petroleum products and compounds altered or man-made. Inorganic additives consist of alkaline earth and metal salts employed to alter rheological properties of the bentonite clay. Depending upon the geological strata being explored, drilling fluids utilized can be oil, salt, polymer or freshwater based muds. Their potential to damage soil or groundwater environments decreases in that same order.

The PEI normally uses earthen pits for storing drilling wastes during the drilling of oil and gas wells. Drilling wastes are defined as a 'heterogeneous mixture of drilling muds, borehole cuttings, additives and various wastes generated at drilling sites' (Freeman & Deuel, 1984). In environmentally sensitive areas, pits are usually lined with a synthetic material or wastes are stored in steel tanks to prevent the leaching of contaminants to soil and groundwater environments. However, earthen pits used in most geographical areas have minimal groundwater impacts. Column studies show that a filter cake from waste muds is quickly formed on the bottom of pits greatly minimizing the leaching of constituents to groundwater (Deeley, 1986).

The closure of an earthen drilling pit is normally conducted using a two-step method. First, the liquids are removed from the pit for disposal. The liquids can be discharged under a National Pollutant Discharge Elimination System (NPDES) permit, injected underground between surface and intermediate casing, evaporated into the atmosphere or hauled off site to a licensed disposal facility. Secondly, the solids can be mixed with soil in the pit and buried in place; removed from the pit and landspread on site; removed from the pit, solidified and buried on site; or removed from the pit and hauled to a licensed disposal facility. The pit closure method chosen by the operator, however, is usually dependent upon State regulatory requirements. For drilling rigs using closed systems, the drilling wastes are normally hauled off site to a licensed disposal facility. There can be exceptions whereby closed system wastes can be managed on site with state agency and landowner permission.

2.2. Produced Water
Produced water is fluid produced in association with the production of oil and gas wells. Nearly all produced water is saline (greater than 10 000 mg/litre total dissolved solids). In most cases, however, this water is not disposed as a waste but is used to enhance the recovery of hydrocarbons in secondary recovery projects.

Produced water can be disposed of or used in several ways. Ninety-two percent of waters produced onshore are injected underground for enhanced recovery projects or for disposal. The remaining 8% are disposed or used in various other ways. As regards disposal, the waste waters can be discharged under NPDES permit to waters of the United States, stored and evaporated in lined earthen pits or stored, evaporated and percolated in earthen pits. Regarding use, the produced waters can be discharged to adjacent property for beneficial use (livestock, crops, etc.) or spread on roads for dust control or prevention of frozen road surfaces. The disposal or use of produced waters is heavily regulated by state agencies.

2.3. Associated Wastes
The remaining wastes generated in the PEI are labeled 'associated wastes' under the 1980 RCRA amendments. These wastes include sludges, contaminated soil, well treatment fluids, inhibitors, tank bottoms, produced sand and a host of other small volume materials intrinsically derived from the drilling, production and treatment of oil and gas. These wastes comprise approximately 3% of the total drilling and associated wastes volume generated by industry. The wastes are normally found in

tanks, earthen pits, vessels, sumps and flow lines (API, 1987, pp. 151–67). Wastes can also be generated during spills of materials on the ground and in the water.

Associated wastes are disposed of in a number of ways. Wastes are recycled, spread on roads, incinerated, injected underground, stored in earthen pits, landspread, buried, evaporated, discharged or transported off site to a licensed facility. The volumes of these wastes at well sites and facilities are generally small and are transported to off-site disposal facilities or disposed of on site. Most of the wastes are regulated by state agencies and require responsible disposal.

3. COMPILATION AND STATISTICAL ANALYSIS OF DATA BASE

In order to collect information on onshore E&P wastes, a survey of a cross-section of the PEI was conducted beginning in July 1986. This production waste survey was divided into several parts: Drilling Wastes, Associated and Other Wastes, and Produced Waters. The first part was designed to determine the source, volumes and disposal practices of drilling fluids for all wells drilled in 1985. The second part was designed to estimate the volume of other wastes associated with exploration, development and production of oil and gas resources. The third part was used to estimate volumes of produced water.

This section discusses some of the statistical issues that were associated with the survey. It shows the extent of the sample coverage and presents a brief description of the statistical analysis. These topics are discussed in more detail in the API Waste Survey (API, 1987, pp. 17–50).

3.1. Statistical Issues
The validity of the estimates depends upon the sample being representative of the population and providing a sufficient number of observations to identify and display the patterns and trends inherent in the larger population.

The instructions accompanying the questionnaire asked each operator to submit a stratified sample of 10% of the wells drilled in 1985 or a minimum of ten wells if it was larger. The stratification was based upon areas of drilling activity with some weight given to including states with lower drilling activity in order to provide some representation in sparse areas. Stratified samples provide a gain in precision and hence improve the quality of the estimates made from the survey.

The issue of data quality assurance was approached in several stages. First, the original survey forms were reviewed by two experienced drilling engineers prior to computer entry. Second, for each numeric item of the questionnaire, the extreme values were reviewed for feasibility by a team of experienced persons. A third stage of review was conducted when the data were plotted; obvious outliers were identified and investigated.

3.2. Sample Coverage

For the first part of the survey (Drilling Wastes), the sample contained 659 wells or about 1% of all the wells drilled in 1985. These wells were classified into four depth categories in order to (1) analyze the sample coverage by depth class and (2) produce estimates by state and depth class. The depth categories were defined as follows: 0–3750 ft; 3751–7500 ft; 7501–15 000 ft; and over 15 000 ft. The four depth classes were represented by 196, 192, 246 and 25 wells, respectively.

For the second part of the survey (Associated and Other Wastes), 209 operators responded, representing 25 different companies. However, several companies were excluded because their crude oil production by state was not known. Consequently, the number of operators that were actually used in the analysis decreased to 141.

For the third part of the survey (Produced Waters), 186 questionnaire forms were received of which 16 had to be excluded for missing key information. The remaining 170 forms represented 14 different companies in 23 states and accounted for 51% of the total onshore US production of crude oil.

3.3. Statistical Analysis

The complexity of the estimation procedures used for the first part of the survey ranges from a simple tabulation of the sample data to multi-step model-based inference. For the key items, the sample was used to obtain the relationship between the waste volume discharged from a given well and its depth. These equations were then applied to all the wells in the API Dynamic Well Data System (or DWDS data base) which contained about 96% of all the wells completed in 1985. Finally, these estimates were slightly inflated to account for the missing 4% of the wells.

Because of the unclear statistical relationship between the volume of associated wastes and the volume of crude oil production, no direct estimation was performed for the second part of the survey. Instead, a tabulation of the sample results was produced.

The third part was used to estimate volumes of produced water. These estimates were based on the water to oil ratios calculated from the sample for each state and each well depth class.

4. SURVEY RESULTS

The 1985 API survey estimates US drilling fluid and produced water volumes at approximately 21·2 billion barrels. The drilling fluid and produced water volumes estimated by API are reasonably accurate. The associated waste volumes, however, were not estimated on a nationwide basis. Instead, a tabulation of the survey volume results are shown.

Survey results show disposal practices used to handle drilling and production wastes have not changed substantially over the past decade. More importantly, this survey shows that drilling and production wastes disposal should have little impact on soil and groundwater environments.

4.1. Drilling Fluid Wastes

For 1985, API estimated there were 361 million barrels of drilling fluids generated throughout the United States (Table 1). This waste volume was generated as a result of the PEI drilling approximately 70 000 onshore wells. The typical pit stores approximately 5200 barrels of drilling wastes. Liquids and solids make up approximately 90% and 10% of this volume, respectively (Table 1). Drilling waste volumes with disposal methods are tabulated in Table 2.

TABLE 1
US Petroleum Extraction Industry 1985 onshore drilling waste fluid generation

	Barrels (M)	Percentage
Liquids	323 662	89
Muds and cuttings	35 674	10
Other solids	2 073	1
Total	361 409	100

4.1.1. Drilling Waste Liquid Disposal
There were approximately 294 million barrels of free liquids stored in drilling pits (Table 2). Operators can use a combination of disposal techniques at a specific site to dispose of liquids. For example, an operator

TABLE 2
US Petroleum Extraction Industry 1985 onshore drilling waste fluid disposal

Disposal method	Barrels ($\times 10^6$)	Percentage
Liquids		
Discharge	36 972	13
Evaporation	106 431	36
Underground injection	48 255	16
Off-site disposal	102 584	35
Sub-total	294 242	100
Solids with residual liquids		
Burial	42 303	63
Landspread	24 386	36
Solidification	478	1
Sub-total	67 167	100
Total	361 409	

may use evaporation, underground injection and off-site disposal as methods to dispose of liquids at a particular site. The API survey of pit closure techniques for all survey wells shows operators use evaporation 52%, off-site disposal 38%, underground injection 18% and discharge to water or land environments 7% of the time to dispose of liquids (API, 1987, pp. 119–20). With regard to liquid disposal techniques, discharging liquids to the land or water environments is the less reliable disposal technique for preventing environmental damage. Evaporation, underground injection and off-site disposal will have little or no environmental impacts.

The PEI discharged 37 million barrels of liquids into the on-site land and water environments under the Clean Water Act (CWA) (Table 2). However, 36 million barrels of this total were generated from freshwater and oil-based mud systems (Table 3). Most of the liquids stored in the pit were excess water, recycled water or rainwater. The remaining 0·7 million barrels of salty liquids were spread on roads (Table 3).

The PEI injected approximately 48 million barrels of pit fluids into the surface/intermediate casing annulus of the drilled well or into Class II injection wells (Table 2). This practice is regulated under most, if not all, state or federal UIC programs.

Approximately 103 million barrels of pit liquids were hauled to off-site commercial facilities which are heavily regulated by state agencies. Lastly,

TABLE 3
1985 Onshore drilling waste fluid disposal (Barrels ($\times 10^6$))

	Freshwater/ polymer/ other	Saltwater		Oil	
		No pit liner	*With pit liner*	*No pit liner*	*With pit liner*
Liquids					
Evaporation	62 650	2 764	35 468	3 170	2 379
Discharge	32 591	0	685	312	3 384
Underground injection	41 496	271	1 099	5 390	0
Off-site disposal	69 202	445	23 508	6 263	3 166
Sub-total	205 939	3 480	60 760	15 135	8 929
Solids with residual liquids					
Burial	28 669	476	10 946	1480	732
Landspread	23 060	0	847	479	0
Solidification	384	0	94	0	0
Sub-total	52 113	476	11 887	1 959	732
Total	258 052	3 956	72 647	17 094	9 661

106 million barrels of pit liquids were evaporated (Table 2).

In summary, this evaluation shows that 87% of pit liquids are evaporated, injected or hauled off site, and not disposed into the on site land or surface water environments. The remaining 13% of pit liquids are discharged under the CWA into the land or surface water environments of which 98% is essentially freshwater and 2% is saltwater placed on roads for dust and ice suppression. This survey evaluation shows that pit liquids disposal in the PEI will have little impact on the on-site land and surface water environments.

4.1.2. Drilling Waste Solid Disposal

In 1985, there were approximately 67 million barrels of solids and residual liquids left in drilling pits prior to closure (Table 2). Operators can use a combination of pit closure techniques at drilling sites to dispose of solids. Operators used burial 90%, landspreading 10% and off-site disposal 4% of the time (API, 1987, pp. 119–20).

Forty-two million barrels of solids were buried on site in 1985 (Table 2). Approximately 29 million barrels of these solids were generated from freshwater base mud systems (Table 3). These buried freshwater wastes will have minimal impacts on soil and groundwater environments. This is because salts and organic constituent concentrations in freshwater muds are normally low. Moreover, heavy metal concentrations are low enough to assure that migration to groundwater in excessive concentrations is greatly retarded due to the pH and adsorption properties of soils (Freeman & Deuel, 1984). Approximately 14 million barrels of buried solids are generated from salt- and oil-based mud systems (Table 3). However, 86% of salt- and oil-based muds are buried in synthetic lined pits which prevent the leaching of constituents to soil and groundwater environments (Table 3). This leaves approximately 1·5 million barrels of oil-based solids and 0·5 million barrels of saltwater-based solids buried in pits without synthetic liners throughout the US.

As regards potential leaching of these buried solids, API study conclusions show: 'The results of the deterministic modeling and of the Monte Carlo study indicate that there is very limited potential for contamination of groundwater from reserve pits. Concentrations at receptor locations did not exceed health-based water quality standards (Table 7.2) in either deterministic modeling of specific sites, qualitative evaluation of specific sites, or Monte Carlo simulation of representative sites (API, 1987, p. 16). Buried drilling waste impacts on groundwater are very limited. Moreover, the inherent sealing properties of drilling muds and the attenuative properties of soils will greatly reduce or eliminate any of these potential effects (Deeley, 1986).

Approximately 24 million barrels of waste solids were landspread on site at drilling locations. Where this practice was used, 95% of the waste solids landspread are freshwater-based solids (Table 3). Salt- or oil-based solids are not routinely landspread because these wastes require professional planning and management intensive operations. However, operators did landspread 1·3 million barrels of salty and oily waste solids throughout the US (Table 3). The environmental impact created by the landspreading of freshwater waste solids is definitely minimal due to the low levels of salts and organics. Studies conducted by Freeman & Deuel (1984, 1986) show that landspreading of freshwater waste solids at on-site drilling locations is a sound disposal practice and will result in minimal environmental impacts. The landspreading of salty or oil-based drilling waste solids is a more difficult task and requires management intensive operations.

Solidification is not a widespread disposal practice used by PEI. Only 0·5 million barrels of waste solids are disposed using this technique (Table 2). When conducted properly, burial of solidified wastes minimizes the migration of constituents to the soil and groundwater environments.

This is because the constituents are encapsulated in the solids material preventing leachates from transporting constituents to groundwater. It is questionable, however, that solidification can be utilized effectively on salt- and oil-based mud solids because of interferences with the chemical solidification process.

In summary, this evaluation shows that 68% of buried drilling waste solids are generated from freshwater-based mud systems (Table 3). These freshwater solids will have little impact on groundwater after burial due to low constituent concentrations and soil attenuation mechanisms. Thirty-two percent of buried drilling solids are generated from saltwater- and oil-based mud systems of which 86% are buried on site in synthetic lined pits. The remaining 14% of these salty and oily drilling waste solids are buried in unlined earthen pits. Whenever drilling waste solids are landspread, 95% are generated from freshwater-based mud systems. However, based on modeling studies, the 1987 API study indicates there is a very limited potential for groundwater contamination from the burial or landspreading of reserve pit waste solids (API, 1987, p. 16).

4.2. Produced Water

In 1985, the PEI produced 21 billion barrels of saltwater from hydrocarbon bearing reservoirs (Table 4). The estimates were made from a supplemental survey of 14 companies that produced 51% of the total US crude oil production. The survey showed that 92% of produced waters are injected into Class II wells regulated under the Safe Drinking Water Act (SDWA).

Sixty-two percent of injected waters are used in enhanced recovery projects and are not recognized by the PEI as a waste (Table 4). Produced

TABLE 4
US Petroleum Extraction Industry 1985 onshore produced water volumes supplemental survey

	Barrels ($\times 10^6$)	Percentage
Underground injection		
Saltwater disposal	6 191 851	30
Enhanced recovery	12 878 909	62
NPDES discharge	1 180 350	6
Other		
Percolation ponds, evaporation pits, POTWs, offsite disposal, etc.	633 607	2
Total	20 884 715	100

waters are injected into reservoirs to maintain pressure or to transport hydrocarbons to the surface in secondary recovery projects. Thirty percent of produced waters are injected for disposal (Table 4). These waters are normally injected into subsurface reservoirs not productive of hydrocarbons. Six percent of produced waters are discharged under the CWA. Most of these discharges are to adjacent land surfaces that could lead to Waters of the United States. The remaining 2% of produced waters are disposed in various ways. For example, percolation ponds are used to dispose of produced waters in geographical areas where groundwater is not usable for domestic or industrial use; produced waters are hauled to off-site commercial facilities which are licensed and regulated under state regulations; and earthen storage pits are permitted by state or federal agencies for evaporation of produced waters.

The 1985 survey demonstrates that the disposal of produced waters as regulated by state and federal agencies should have minimal impacts on the land and surface water environments.

4.3. Associated Wastes

There were approximately 6 million barrels of associated and other wastes reported in Part II of the API Survey, which represents operators producing 51% of the total 1985 US onshore crude oil production (API, 1987, pp. 151–67). This volume was tabulated directly from the survey questionnaires and was not extrapolated to a national volume.

Generally, most associated wastes are generated in production and workover activities, not drilling operations. These wastes are listed in Table 5. The disposal practices utilized by operators for each waste vary considerably (Table 6). For example, tank bottoms can be recycled, spread on roads, landspread, or buried.

The results of this survey show that 48% of wastes are disposed of at off-site disposal facilities, 21% spread on roads or land, 5% recycled, 4% buried on site, 1% stored in pits, 1% injected underground, and 1% evaporated (Table 6). Upon review of the survey data, it appears that most wastes are disposed of using environmentally sound techniques. For example 92% of solvents, 65% of tank bottoms, 76% of waste lubricating oils, and 67% of workover fluids are either recycled or hauled to an off-site disposal facility (API, 1987, pp. 151–67). The remainder is spread on roads, injected underground or stored in on-site pits. Nineteen percent of associated wastes are disposed of using techniques not described in the survey, but these techniques have now been identified and will be reported later to the EPA. Therefore, the 1985 survey demonstrates that disposal

TABLE 5

US Petroleum Extraction Industry 1985 associated waste types generated onshore

Associated wastes	Volume percentage
1. Tank bottoms, separator sludge, pig trap solids	10
2. Lubricating and hydraulic oils	1
3. Oily debris, filters, contaminated soil	12
4. Emulsions	2
5. Produced sand	11
6. Spent iron sponge	1
7. Dehydration wastes (glycol, amine, etc.)	4
8. Workover, swabbing, unloading, completion fluids	34
9. Used solvents, cleaners	2
10. Cooling water, engine and other water	8
11. Solid wastes, drums, insulation, etc.	1
12. Other production liquid wastes	14
	100

TABLE 6

US Petroleum Extraction Industry 1985 associated waste disposal techniques utilized onshore

Disposal technique	Volume percentage
Recycle	5
Roadspread	16
Landspread	5
Injection	1
Evaporation	1
Pit storage	1
Burial	4
Off-site disposal	48
Not defined	19
	100

of associated wastes as regulated by state and federal agencies should produce minimal impacts on the land and surface water environments.

5. CONCLUSIONS

The 1985 API survey results produced a number of interesting facts and conclusions.

1. Overall, the PEI disposes of drilling muds, produced water and associated wastes utilizing environmentally sound techniques. Disposal of wastes is heavily regulated by state and federal agencies.
2. In 1985, there were 361 million barrels of drilling wastes generated by the PEI.
   A. There were 294 million barrels of drilling liquids. Eighty-seven percent of liquids are evaporated, injected underground or hauled off site to disposal facilities. Thirteen percent of liquids are discharged off site of which 98% are essentially freshwater- or oil-based being discharged under the CWA, and 2% are saline and spread on roads.
   B. There were 67 million barrels of drilling solids with residual liquids. Sixty-eight percent of buried drilling solids are generated from freshwater-based mud systems with some polymers added. Buried freshwater waste solids result in minimal environmental impacts. Thirty-two percent of buried drilling wastes are generated from salt- and oil-based mud systems of which 86% are buried in synthetic lined pits. The remaining 14% of salt- and oil-based solids are buried in unlined earthen pits. Ninety-five percent of waste solids landspread are generated from freshwater-based mud systems. Based upon 1987 API modeling studies, there is a very limited potential for groundwater contamination resulting from the burial or landspreading of reserve pit waste solids (API, 1987, p. 16).
3. In 1985, there were approximately 21 billion barrels of water produced by PEI. Ninety-two percent of produced waters are injected underground and 6% are discharged off site under the CWA. The remaining 2% of produced waters are disposed of using various disposal techniques (percolation, roadspread, etc.), which are regulated by state and federal agencies. Sixty-two percent of produced waters are injected in enhanced recovery projects and are not

recognized by the PEI as waste. The use or disposal of produced water by PEI results in minimal environmental impacts.

4. There were 6 million barrels of associated wastes tabulated in the API 1985 survey. This documented volume does not represent a nationwide estimate. Eighty-one percent of associated wastes reported are disposed of using environmentally sound techniques. Nineteen percent of associated wastes disposal techniques are not identified in this paper, but will be described later in a report to EPA.

ACKNOWLEDGEMENTS

The authors wish to thank the American Petroleum Institute for permission to present the paper. We also gratefully acknowledge the API Production Waste Issue Group for their invaluable help and guidance in completing this project.

REFERENCES

AMERICAN PETROLEUM INSTITUTE (1987). Oil and Gas Industry Exploration and Production Wastes. Document No. 471-01-09, prepared by ERT, July 1987.

DEELEY, G. H. (1986). Attenuation of chemicals within waste freshwater drilling fluids. Presented at the *Proceedings of a National Conference on Drilling Muds*, sponsored by the University of Oklahoma Environmental and Groundwater Institute, May 29–30, 1986.

FREEMAN, B. D. & DEUEL, L. E. (1984). Guidelines for closing drilling waste fluid pits in wetland and upland areas. Presented at the Seventh Annual Energy Sources Technology Conference and Exhibition, sponsored by The Petroleum Division-ASME, New Orleans, Louisiana, February 12–16, 1984.

FREEMAN, B. D. & DEUEL, L. E. (1986). Closure of freshwater base drilling mud pits in wetland and upland areas. Presented at the *Proceedings of a National Conference on Drilling Muds*, sponsored by the University of Oklahoma Environmental and Groundwater Institute, May 29–30, 1986.

18

The Analytical Methods Utilized and Results from the Analyses of Field Collected Drilling Wastes*

Susan L. DeNagy† and William A. Telliard

United States Environmental Protection Agency, Industrial Technology Division (ITD) (WH-552), 401 M St. SW, Washington, D.C. 20460, USA

ABSTRACT

The United States Environmental Protection Agency (EPA) conducted a nationwide field sampling and analysis study to characterize wastes associated with the oil and gas extraction industry. One of the primary wastes examined were drilling wastes. Of most interest was the identification and quantification of organic analytes in these wastes. This paper presents the analytes selected for study, the analytical methods utilized, and the results produced.

Two types of samples of drilling wastes were collected and analyzed: supernates and sludges from reserve pits. In addition, approximately one half of the sludge samples were leached using EPA's Toxicity Characteristic Leaching Procedure (TCLP). Samples and the TCLP leachates were tested for organic and inorganic analytes on the 'Industrial Technology Division (ITD) List of Analytes' using Revision C of EPA Methods 1624 and 1625 and other EPA methods. Organic compounds frequently detected were aliphatic and aromatic hydrocarbons, and organic acids.

*Editors' note: As stated in the conclusions and discussion section of this paper, the evaluation of the data presented in this report is continuing, as well as of additional data on the topic gathered by the American Petroleum Institute (API). Further information on the API data set, and the concerns regarding these data, can be obtained from EPA, by referencing: 'API comments on EPA's 'Report to congress: management of wastes from the exploration, development and production of crude oil, natural gas, and gas thermal energy', 53 FR 81 (Jan. 4, 1988), Docket No. F-88-OGRA-FFFFF. Submitted to EPA March 15, 1988'.

†Deceased 20 September 1987. This paper is dedicated to her memory.

1. INTRODUCTION

This paper presents the findings of the field sampling and analysis project conducted on wastes associated with exploration, development, and production activities of crude oil and natural gas (the 'Project'). A summary of the analytical data is presented. The Project was designed to develop information about wastes from four types of sites: drill sites, production sites, centralized pits and centralized treatment facilities. Primarily, EPA sampled drilling wastes and produced water. EPA also sampled tank bottoms and several commingled oil and gas extraction industry wastes disposed of via centralized pits or centralized treatment facilities. In total, 101 samples were collected; from this total, 42 were classified as sludges and 59 were classified as liquids. Classification of samples was determined by sampling location and/or the condition of the sample (e.g. solids content of the sample).

Nearly all samples were analyzed for 229 organic compounds, 68 metals, and 22 conventional analytes (e.g. ammonia, chloride, TSS and BOD5). In addition, approximately half of the sludge samples were leached using EPA's Toxicity Characteristic Leaching Procedure (TCLP). The leachate extracts from this procedure were tested for a subset of organics and metals. On selected samples, such as those collected from centralized pits and centralized treatment facilities, EPA analyzed for 136 chlorinated dioxins and furans, and 79 pesticides and herbicides.

Analytical results and quality assurance data were reported by laboratories to EPA on magnetic tape or disk and in hard copy form. After data validation and quality assurance, all data were entered into the IBM mainframe at EPA's National Computer Center (NCC). Summary statistics were computed on both the analytical and quality assurance results.

The data have been weighted by production volumes and, therefore, represent EPA's best estimate of average, nationwide pollutant concentrations in this industry. These data show that the highest frequencies of occurrences and highest average concentrations are for pollutants normally associated with oil and gas exploration, based on the technical literature and other EPA studies.

2. REGULATORY BACKGROUND

The United States Environmental Protection Agency (EPA) is required to regulate the oil and gas extraction industry under several major environmental statutes. These statutes include the Clean Water Act (CWA) (i.e. appropriate effluent limitations guidelines), the Safe Drinking Water Act (SDWA) (i.e. the Underground Injection Control (UIC) program), and the Resource Conservation and Recovery Act (RCRA) (i.e. the regulatory determination under Section 3001(b)(2)(B), which uses information from the study under Section 8002(m)).

The Technical Report on which this paper is based fulfills an obligation in the settlement agreement of Alaska Center for the Environment *et al.* v. EPA (Civil Action No. A89-471 (D. Alaska).

3. SAMPLING AND ANALYSES

3.1. Project Objectives
The objectives of the Project were to identify and quantify waste constituents, to document site-specific waste sources and volumes, and to aid in the documentation of nationwide and specific regional industry practices. These objectives were met by collecting technical data through the literature and industry sources and by collecting and analyzing samples of waste.

Specific objectives were to:

— Provide data to be included in a report to Congress on wastes associated with oil and gas exploration, development, and production activities as required by RCRA.
— Provide nationwide data on sources and volumes of oil and gas wastes.
— Provide information on the complexity and diversity of the wastes generated by the industry, current disposal practices, and ultimate treatment.
— Identify characteristics and constituents of the waste streams and estimate variability of these waste streams.
— Provide data that can be used in the design of a larger, more comprehensive sample survey of the industry.

3.2. Sampling Strategy and Sampling

3.2.1. Target Population
The sampling program was designed to develop information about wastes from four types of sites: drill sites, production sites, centralized pits and centralized treatment facilities. The total number of sample sites was 49, distributed as shown in Table 1.

Susan L. DeNagy and William A. Telliard

TABLE 1
Distribution of types of sample sites

Type of site	Number of sites sampled
Drill sites	19
Production sites	
Produced water	21
Tank bottom sludges	2
Centralized pits	4
Centralized treatment	3
Total	49

Eighty-six percent of the sites sampled were active drill or production sites. This allocation reflects EPA's interest in developing information regarding the most prevalent waste sources within the industry. Fourteen percent of the sites sampled were centralized waste storage or treatment facilities. This allocation addresses EPA's need to develop information about the types of processes, characteristics of commingled waste, and the characteristics of treated commingled wastes.

3.2.2. Selection of Sample Sites
A sample frame was constructed by geographically defining eleven zones (or strata). These zones were primarily developed by grouping common geological formations and operating practices, using state borders as boundaries between zones. Figure 1 shows the zones used for site selection. Zone 1 and Zone 3 were excluded from the sample frame because there were essentially no oil and gas exploration, production, or development activity in those states. Thus, sample sites were distributed through nine zones.

The site selection process was designed to handle two types of sites: (1) sites randomly selected, and (2) sites specifically selected on the basis of judgement. Randomly selected sites were distributed uniformly across the zones to ensure coverage. Specifically selected sites were used to supplement equal distribution of sites, to examine particular practices of interest, or to replace randomly selected sites which could not be sampled.

3.2.3. Sampling
Sampling was conducted from June through September 1986. Trip reports were written for each site sampled. Trip reports include site identification,

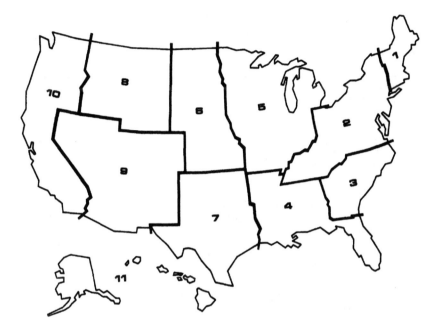

FIG. 1. Sample frame showing eleven geographical zones.

status during sampling, description, plot plan, sampling information, and other technical information. Details are included in the Technical Report.

3.3. Selection of Analytes

The analytes selected for testing were those on various regulatory lists or those specific to characterization of this industry. The regulatory lists from which the pollutants were taken are the:

— Priority Pollutant List [NRDC v. Train, 8 ERC 2120 (DDC1976)].
— Priority Pollutant Appendix C List (ibid.).
— RCRA 40 CFR Part 261 Appendix VIII List [50 FR 1999].
— Michigan List [49 FR 49793].
— Superfund Hazardous Substances List.
— Analytes to be added to the RCRA Appendix VIII List (EPA memo from Robert April to Marcia Williams, 20 Dec 85).
— Paragraph 4(c) List (NRDC v. Train, op. cit.: high priority compounds detected in wastewaters).

TABLE 2
Summary of analytes

Analyte type and analytical technique	Number of analytes
Organics	
Volatiles by GCMS	
Isotope dilution	32
Reverse search	23
Semi-volatiles by GCMS	
Isotope dilution	82
Reverse search	94
Dioxins and furans by GCMS	136
Pesticides by GC	
Electron capture detector (ECD)	39
Flame photometric detector (FPD)	35
Herbicides by GC/ECD	3
Total organics	444
Metals	
Atomic absorption	6
Calibrated Inductively Coupled Plasma (ICP)	21
ICP screening	41
Total metals	68
Conventionals by wet chemistry	19
RCRA (Corrosivity, Ignitability, Reactivity)	3
Total number of analytes	534

— ITD List (analytes specific to ITD programs).

The complete list of analytes tested for in this Project is given in Appendix F to the Technical Report. That list covers all conventional pollutants (e.g. BOD5 and TSS) and metals; but it does not list every individual organic compound. Specifically, classes of the tetra- through octachlorodibenzo-*p*-dioxins and furans (136) and classes of polychlorinated biphenyls (210) are not listed. The analytical methods employed in the Project for groups of analytes are summarized in Table 2. This table is divided into three categories: organic analytes are specific chemical compounds based on carbon chemistry; metals are chemical elements (substances that cannot be divided without altering their physical and chemical properties); conventional pollutants are, for the most part, chemical substances that have been historically used to assess the performance of water treatment plants (drinking water and sewage) and for assessing water quality. Other analytes included in the conventional category are those that test for substances as a group, rather than as a single chemical compound.

3.4. Analyses

Samples were tested by laboratories using EPA analytical methods. Nearly all of these methods are either approved wastewater methods (CWA section 304(h)), proposed or approved methods for testing wastes (RCRA SW-846, TCLP), or EPA Comprehensive Environmental Response, Compensation and Liability Act (CERCLA; Superfund) Contract Laboratory Program (CLP) methods. The exceptions are methods for pH using paper, for chlorine using the Coastal kit (Coastal Chemical Co. Style 13 107) and for the Oil and Grease test which employs the IMCO Retort (Imco Services Model No. R2100), for which no EPA or Standard Method is available. The IMCO Retort is the oil and gas extraction industry operator's method for determining the oil content of drilling fluids. The methods are summarized in Table 3.

Some methods were extended to cover an extended list of analytes. For example, Methods 1624 and 1625 were extended to cover the organic compounds on the RCRA Appendix VIII and IX Lists (50 FR 1999 and 41 FR 26 639, respectively), where it was demonstrated feasible to test for a given analyte. ITD tested these analytes using Methods 1624 and 1625 prior to sample analyses and therefore knew the accuracy and precision that could be expected for these compounds prior to application of these methods (Narrative for SAS 106; Narrative for SAS 109).

4. QUALITY ASSURANCE

Quality assurance and quality control (QA/QC) were integral parts of the Project. The details of this QA/QC for all aspects of the Project are given in the Technical Report and are outside of the scope of this paper. However, for the analytical portions of the Project, data quality objectives were established, quality control limits for recovery and precision were required by the analytical methods or were imposed by QA external to the methods, and a statement of data quality, in terms of a confidence limit, was developed for each analyte where precision and recovery data were available, and for surrogate analytes where it was not possible to test every analyte (chlorinated dioxins and furans, for example).

TABLE 3
List of analytes, matrices, fractions and analytical methods

(1) Analysis category	(2) Matrix	(3) Fraction	(4) Analysis technique	(5) Method	(6) Modification
Organics	Water	Pesticide	GCEC	1618	
			GCFPD	1618	
		Herbicide	GCEC	1618	
		Volatiles	GCMS	1624C	
		B/N	GCMS	1625C	
		Acid	GCMS	1625C	
		Diox/furan	HRGCLRMS	613M	C14 to C18
			HRGCLRMS	8280	High res. MS
	Sludge	Pesticide	GCEC	1618	
			GCFPD	1618	
		Herbicide	GCEC	1618	
		Volatiles	GCMS	1624C	
		B/N	GCMS	1625C	
		Diox/furan	HRGCHRMS	8280M	High res. MS
Org/TCLP			——Same as for waters——		
Metals	Water	Mercury	CVAA	245.5	
		furnace	FURNAA		
			Sb	204.2	
			As	206.2	
			Se	270.2	
			Ag	272.2	
			Tl	279.2	
		ICP	ICP	200.7M + 42	Element screen
	Sludge	Mercury	CVAA	245.5M CLP	
		furnace	FURNAA	3020	
		Sb	204.2M CLP		
			As	206.2M CLP	
			Se	270.2M CLP	
			Ag	272.2M CLP	
			Tl	279.2M CLP	
		ICP	ICP	200.7M + 42	Element screen + HCl if nec.

TABLE 3—*contd.*

(1) Analysis category	(2) Matrix	(3) Fraction	(4) Analysis technique	(5) Method	(6) Modification
Met/TCLP			——Same as for waters——		
Conventional	Water	Ammonia	Electrode	350.3	
		BOD5	Probe	405.1	
		Chloride	Ion chrom.	300.0	
		Chlorine	Color.		Field test
		COD	Color.	410.4M	Saline
		Cyanide	Distill.	335.2	
		Fluoride	SPADNS	340.1	
		Nitrate/ nitrite	Ion chrom.	300.0	
		pH (field)	Paper.	—	
		(lab)	Electrode	150.1	
		Oil and gr.	Grav.	413.1	
		Residue	Grav.–TDS	160.1	
			–TSS	160.2	
			–Tot.	160.3	
		Specific conduct.	Wheatstone bridge	120.1	
Conventional	Water	Sulfide	Titr.	376.2	
		TOC	Combust.	415.1	
		TVO	TOC	415.1M	Purge and trap
	Sludge	Ammonia	Electrode	350.3	
		BOD5	Probe	405.1	
		Chloride	Ion chrom.	300.0	
		COD	Color.	410.4M	Saline
		Cyanide	Distill.	335.2	
		Fluoride	SPADNS	340.1	
		Nitrate/ nitrite	Ion chrom.	300.0	
		pH (field)	Paper	—	
		(lab)	Electrode	150.1	
		Oil and gr.	Grav.	413.1	
		Oil and gr.	Retort	—	
		Residue	Grav.–Tot.	160.3	
		Sulfide	Color.	376.2	
		TOC	Combust.	9060	
		TVO	TOC	415.1M	Purge and trap
Haz. waste	Sludge	Ignit.		1010	
		Corrosiv.		1110	
		React.		SW-846	

(*continued*)

TABLE 3—contd.

Notes:
(1) Analysis category—general category into which analytes can be classified.
   Organics—carbon based chemical compounds. The list of these compounds can be found on the ITD/RCRA List of Analytes.
   Organics/TCLP—organic compounds leached from sludge and sediment using the RCRA Toxicity Characteristic Leaching Procedure.
   Metals—elements found on the ITD/RCRA List of Analytes.
   Metals/TCLP—metals leached from sludge and sediment using the TCLP.
   Conventional—conventional wastewater chemistry analytes.
   Hazardous waste characteristics—RCRA analytes which determine if a waste is hazardous. The TCLP has been proposed to replace the current EP Toxicity procedure.
(2) Matrix—the nature of the sample.
   Water—produced water, runoff water, or other sample which is nearly all water.
   Sludge—tank bottoms, drilling muds, or other sample which contains a significant quantity of solids (normally greater than 1%).
(3) Fraction—a means of further categorizing the sample for purposes of analysis.
   Volatile—volatile organic compounds analyzed by GCMS.
   B/N—base/neutral organic compounds analyzed by GCMS.
   Acid—organic acids analyzed by GCMS.
   Diox/furan—chlorinated dibenzo-p-dioxins and chlorinated dibenzofurans.
   TVOC—total volatile organic carbon.
   Furnace—metals analyzed by furnace atomic absorption spectrometry.
   ICP—metals analyzed by inductively coupled spectrometry.
   BOD5—biochemical oxygen demand.
   COD—chemical oxygen demand.
   Oil and gr.—oil and grease.
   Specific conduct.—specific conductivity.
   TOC—total organic carbon.
   Ignit.—hazardous waste characteristic of ignitability.
   Corrosiv.—hazardous waste characteristic of corrosivity.
   React.—hazardous waste characteristic of reactivity.
(4) Analysis technique
   GCEC—gas chromatography combined with an electron capture detector.
   GCFPD—gas chromatography combined with a flame photometric detector.
   GCMS—gas chromatography combined with a mass spectrometer detector.
   HRGCLRMS—high resolution gas chromatography combined with low resolution mass spectrometry.
   HRGCHRMS—high resolution gas chromatography combined with high resolution mass spectrometry.

TABLE 3—*contd.*

TOC—total organic carbon analyzer.
CVAA—cold vapor atomic absorption spectrometry.
FURNAA—furnace atomic absorption spectrometry.
Ion chrom.—ion chromatography.
Color.—colorimetric.
Titr.—titrimetric.
Distill.—distillation.
SPADNS—distillation followed by calorimetric.
Grav.—gravimetric.
TDS—total dissolved solids.
TSS—total suspended solids.
Tot.—total solids.
Retort—the platform operator's method of determining the oil content of a sample from a well.
(5) Method—the EPA method number. Water methods are three-digit numbers (some include a decimal). ITD methods are 1618, 1624C and 1625C. Office of Solid Waste SW-846 methods are all other four-digit numbers.
(6) Gives a modification to the method.
C14 to C18—Samples were screened for all tetra- through octa-isomers of chlorinated dibenzo-*p*-dioxin and dibenzo-furan.
High res. MS—high resolution mass spectrometer was used in place of low resolution instrument to gain specificity.
+ 42 element screen—search of a specific ICP wavelength for 42 metals in addition to the 27 determined by calibration and search.
CLP—method modified for application to solids by the Superfund Contract Laboratory Program.
Purge and trap—volatiles are purged from water or sludge.
+ HCl if nec.—hydrochloric acid added to aid in digestion of organic sludges if necessary.
Field test—test performed in at the site.
Saline–Hach method 8000.

5. RESULTS

5.1. Statistical Analysis

Summary statistics were computed for each sampling category, including number of samples analyzed, number with each compound detected, and the minimum, mean and maximum concentrations of the analyte for the sampling category. Tables 4 and 5 show the results of analyses of samples from centralized pits and drilling pits, respectively. Centralized pits are pits to which drilling wastes are hauled from remote sites. In Tables 4

TABLE 4

Results of analyses of field collected drilling wastes—centralized pits (analytes detected in one or more samples only) (site = central pit; location = central pit; phase = liquid; extract = direct)

Compound	Compound name	N	Detected Prop.	Detect	Minimum	Wtd. Mean	Maximum	Units
1-001	Total volatile organic carbon	4	4	1·000	0·3	0·6	0·8	mg/liter
1-002	Biochemical Oxygen Demand	4	4	1·000	11·0	385·2	630·0	mg/liter
1-003	Chloride	4	4	1·000	1 620·0	11 530·0	31 000·0	mg/liter
1-004	Chemical Oxygen Demand	4	4	1·000	2 280·0	7 257·5	19 500·0	mg/liter
1-005	Nitrate/nitrite	4	4	1·000	1·1	4·3	6·8	mg/liter
1-006	Hydrogen ion	4	4	1·000	5·7	7·5	8·5	pH
1-007	Oil and grease	4	4	1·000	8·0	37·2	111·0	mg/liter
1-009	Residue, non-filterable	4	4	1·000	192·0	4 246·0	16 000·0	mg/liter
1-010	Residue, filterable	4	4	1·000	2 600·0	17 150·0	26 000·0	mg/liter
1-011	Specific conductivity	4	4	1·000	8 000·0	42 750·0	80 000·0	µMho/cm
1-012	Total organic carbon	4	4	1·000	27·0	481·5	996·0	mg/liter
1-013	Ignitability	4	4	1·000	200·0	200·0	200·0	degF
1-014	Corrosivity	4	4	1·000	5·7	7·5	8·5	pH
65850	Benzoic acid	4	2	0·500	<DL	623·8	2 390·2	µg/liter
67641	2-Propanone	4	3	0·750	<DL	2 381·5	5 624·2	µg/liter
71432	Benzene	4	2	0·500	<DL	73·1	216·7	µg/liter
75092	Methylene chloride	4	4	1·000	108·2	261 848·0	1 011 774·9	µg/liter
75150	Carbon disulfide	4	1	0·250	<DL	3·4	13·6	µg/liter
78933	2-Butanone	4	1	0·250	<DL	12·5	50·1	µg/liter
84742	Di-n-butyl phthalate	4	1	0·250	<DL	390·4	1 561·8	µg/liter
85018	Phenanthrene	4	2	0·500	<DL	24·7	74·7	µg/liter
86737	Fluorene	4	2	0·500	<DL	9·4	26·9	µg/liter
91203	Naphthalene	4	3	0·750	<DL	38·5	94·2	µg/liter
91575	2-Methylnaphthalene	4	3	0·750	<DL	68·6	128·5	µg/liter
92524	Biphenyl	4	2	0·500	<DL	11·7	26·1	µg/liter
95487	o-Cresol	4	2	0·500	<DL	23·2	69·7	µg/liter

CAS	Compound						
98862	Acetophenone	15·1 µg/liter	3·7	<DL	0·250	1	4
99876	p-Cymene	13·6 µg/liter	9·3	<DL	0·750	3	4
100414	Ethylbenzene	21·0 µg/liter	5·2	<DL	0·250	1	4
100516	Benzyl alcohol	341·2 µg/liter	85·3	<DL	0·250	1	4
105679	2,4-Dimethylphenol	44·9 µg/liter	11·2	<DL	0·250	1	4
106445	p-Cresol	364·9 µg/liter	125·5	<DL	0·500	2	4
108101	4-Methyl-2-pentanone	25·8 µg/liter	11·4	<DL	0·500	2	4
108883	Toluene	21 159·9 µg/liter	5 339·1	<DL	0·500	2	4
108952	Phenol	1 171·3 µg/liter	292·8	<DL	0·250	1	4
112403	n-Dodecane	1 547·7 µg/liter	815·1	38·6	1·000	4	4
112958	n-Eicosane	2 421·3 µg/liter	740·8	22·7	1·000	4	4
117817	bis(2-Ethylhexyl) phthalate	2 231·0 µg/liter	577·6	<DL	0·500	2	4
124185	n-Decane	832·8 µg/liter	416·7	<DL	0·750	3	4
129000	Pyrene	11·5 µg/liter	2·8	<DL	0·250	1	4
132649	Dibenzofuran	13·7 µg/liter	3·4	<DL	0·250	1	4
132650	Dibenzothiophene	19·1 µg/liter	7·4	<DL	0·500	2	4
142621	Hexanoic acid	566·9 µg/liter	141·7	<DL	0·250	1	4
544763	n-Hexadecane	3 859·8 µg/liter	1 297·2	37·3	1·000	4	4
593453	n-Octadecane	3 575·3 µg/liter	1 014·1	20·8	1·000	4	4
629594	n-Tetradecane	2 405·4 µg/liter	711·2	31·3	1·000	4	4
629970	n-Docosane	1 472·7 µg/liter	444·1	17·8	1·000	4	4
630013	n-Hexacosane	179·6 µg/liter	90·6	15·5	1·000	4	4
630024	n-Octacosane	113·6 µg/liter	47·0	<DL	0·750	3	4
638686	n-Triacontane	236·0 µg/liter	64·0	<DL	0·500	2	4
646311	n-Tetracosane	220·5 µg/liter	125·6	17·6	1·000	4	4
832699	1-Methylphenanthrene	69·1 µg/liter	17·2	<DL	0·250	1	4
1730376	1-Methylfluorene	11·8 µg/liter	2·9	<DL	0·250	1	4
7429905	Aluminum	289 000·0 µg/liter	78 310·5	822·0	1·000	4	4

(continued)

TABLE 4—contd.

Compound	Compound name	N	Detected Prop.	Detect	Minimum	Wtd. Mean	Maximum	Units
7439896	Iron	4	4	1·000	4 120·0	296 100·0	1 130 000·0	µg/liter
7439921	Lead	4	2	0·500	<DL	721·2	2 460·0	µg/liter
7439932	Lithium	4	3	0·750	<DL	2 728·7	5 730·0	µg/liter
7439954	Magnesium	4	4	1·000	59 600·0	280 975·0	842 000·0	µg/liter
7439965	Manganese	4	4	1·000	127·0	4 186·7	12 400·0	µg/liter
7439987	Molybdenum	4	2	0·500	<DL	123·2	292·0	µg/liter
7440020	Nickel	4	3	0·750	<DL	331·7	1 220·0	µg/liter
7440097	Potassium	4	3	0·750	<DL	1 385 000·0	4 010 000·0	µg/liter
7440213	Silicon	4	4	1·000	19 800·0	42 400·0	99 800·0	µg/liter
7440224	Silver	4	2	0·500	<DL	0·8	2·2	µg/liter
7440235	Sodium	4	4	1·000	604 000·0	7 413 500·0	15 200 000·0	µg/liter
7440245	Strontium	4	4	1·000	6590·0	39 397·5	58 400·0	µg/liter
7440315	Tin	4	4	1·000	117·0	422·2	756·0	µg/liter
7440326	Titanium	4	3	0·750	<DL	213·2	390·0	µg/liter
7440382	Arsenic	4	2	0·500	<DL	690·7	2 720·0	µg/liter
7440393	Barium	4	4	1·000	1060·0	3 297·5	7 660·0	µg/liter
7440417	Beryllium	4	1	0·250	<DL	11·7	47·0	µg/liter
7440425	Boron	4	4	1·000	3 020·0	15 532·5	39 700·0	µg/liter
7440439	Cadmium	4	2	0·500	<DL	49·2	184·0	µg/liter
7440451	Cerium	4	2	0·500	<DL	192·2	537·0	µg/liter
7440473	Chromium	4	2	0·500	<DL	998·0	3 490·0	µg/liter
7440484	Cobalt	4	1	0·250	<DL	124·5	498·0	µg/liter
7440508	Copper	4	2	0·500	<DL	188·2	610·0	µg/liter
7440622	Vanadium	4	2	0·500	<DL	308·5	944·0	µg/liter
7440655	Yttrium	4	1	0·250	<DL	94·5	378·0	µg/liter
7440666	Zinc	4	4	1·000	50·0	2 232·7	4 720·0	µg/liter

Site	Analyte	n	frac	min	mean	max	Units
7440702	Calcium	4	1·000	109 000·0	1 125 750·0	3 340 000·0	µg/liter
7664417	Ammonia	4	1·000	1·0	6·6	12·0	mg/liter
7704349	Sulfur	4	1·000	13 800·0	161 700·0	362 000·0	µg/liter
7723140	Phosphorus (black, white, red, yellow)	2	0·500	<DL	2 271·2	8 200·0	µg/liter
16984488	Fluoride	4	1·000	1·7	2·1	3·2	mg/liter

(Site = central pit; location = central pit; phase = solid; extract = direct)

Site	Analyte	n	frac	min	mean	max	Units
1-001	Total volatile organic carbon	3	1·000	17·0	87·3	220·0	mg/kg
1-002	Biochemical Oxygen Demand	3	1·000	44·0	4 308·0	10 400·0	mg/kg
1-003	Chloride	3	1·000	5 140·0	34 013·3	77 500·0	mg/kg
1-004	Chemical Oxygen Demand	3	1·000	6 150·0	30 250·0	71 500·0	mg/kg
1-005	Nitrate/nitrite	3	1·000	3·1	20·7	30·0	mg/kg
1-006	Hydrogen ion	3	1·000	7·2	8·0	9·2	pH
1-007	Oil and grease	3	1·000	3 750·0	34 683·3	81 300·0	mg/kg
1-008	Residue, total	3	1·000	14·4	44·6	67·4	%
1-012	Total organic carbon	3	1·000	10 500·0	24 366·6	49 000·0	mg/kg
1-013	Ignitability	3	1·000	200·0	200·0	200·0	degF
1-014	Corrosivity	3	1·000	7·2	8·0	9·2	pH
1-015	Reactivity	2	0·667	<DL	0·6	1·0	S(−/+)
1-016	Oil and grease (retort)	3	1·000	10 800·0	42 500·0	67 900·0	mg/kg
1-331	Tetrachlorodibenzo-p-dioxins	2	0·500	<DL	4·2	8·5	ng/kg
67641	2-Propanone	3	1·000	353·6	2 771·4	7 218·3	µg/kg
71432	Benzene	3	0·333	<DL	5 190·5	15 571·7	µg/kg
75092	Methylene chloride	3	0·333	<DL	42·0	126·2	µg/kg
78933	2-Butanone	3	0·333	<DL	72·9	218·7	µg/kg
84742	Di-n-butyl phthalate	3	0·333	<DL	163·7	491·3	µg/kg
85018	Phenanthrene	3	0·333	<DL	7 856·8	23 570·5	µg/kg
91203	Naphthalene	3	0·333	<DL	10 645·2	31 935·8	µg/kg
91576	2-Methylnaphthalene	3	0·667	<DL	63·4	168·9	µg/kg
92524	Biphenyl	3	0·333	<DL	5 868·2	17 604·7	µg/kg

(continued)

TABLE 4—contd.

Compound	Compound name	N	Detected	Prop. Detect	Minimum	Wtd. Mean	Maximum	Units
99876	p-Cymene	3	1	0·333	<DL	3 700·9	11 102·9	µg/kg
100414	Ethylbenzene	3	2	0·667	<DL	6 486·3	19 089·1	µg/kg
106445	p-Cresol	3	1	0·333	<DL	6·1	18·3	µg/kg
108101	4-Methyl-2-pentanone	3	2	0·667	<DL	17·6	30·4	µg/kg
108883	Toluene	3	1	0·333	<DL	28 755·7	86 267·3	µg/kg
112403	n-Dodecane	3	3	1·000	5 332·5	148 156·8	425 362·9	µg/kg
112958	n-Eicosane	3	3	1·000	6 788·4	83 810·1	236 351·1	µg/kg
117817	bis(2-Ethylhexyl) phthalate	3	2	0·667	<DL	8 870·8	22 131·1	µg/kg
124185	n-Decane	3	3	1·000	2 417·2	102 979·7	291 914·1	µg/kg
127184	Tetrachlorethene	3	1	0·333	<DL	6·3	18·9	µg/kg
544763	n-Hexadecane	3	3	1·000	12 313·3	152 957·1	431 419·4	µg/kg
593453	n-Octadecane	3	3	1·000	3 761·4	148 494·2	430 702·9	µg/kg
629594	n-Tetradecane	3	3	1·000	3 351·4	166 648·5	486 199·4	µg/kg
629970	n-Docosane	3	3	1·000	2 430·9	66 645·0	194 617·6	µg/kg
630013	n-Hexacosane	3	3	1·000	1 258·3	34 872·2	101 505·8	µg/kg
630024	n-Octacosane	3	2	0·667	<DL	18 029·4	53 604·1	µg/kg
638685	n-Triacontane	3	2	0·667	<DL	15 064·7	43 514·1	µg/kg
646311	n-Tetracosane	3	3	1·000	1 423·8	36 785·3	104 378·8	µg/kg
832699	1-Methylphenanthrene	3	2	0·667	<DL	21·4	53·2	µg/kg
1730376	1-Methylfluorene	3	1	0·333	<DL	3·5	10·7	µg/kg
1746016	2,3,7,8-Tetrachlorodibenzo-p-dioxin	2	1	0·500	<DL	9·8	19·7	ng/kg
3268879	Octachlorodibenzo-p-dioxins	2	2	1·000	23·5	198·9	374·3	ng/kg
7429905	Aluminum	3	3	1·000	4 280·0	14 726·6	21 800·0	mg/kg
7439896	Iron	3	3	1·000	11 900·0	24 033·3	36 900·0	mg/kg
7439921	Lead	3	3	1·000	15·0	131·3	233·0	mg/kg
7439954	Magnesium	3	3	1·000	5 730·0	8 293·3	12 400·0	mg/kg

CAS	Analyte						
7439965	Manganese	3	3	1·000	157·0	394·6	666·0 mg/kg
7439987	Molybdenum	3	1	0·333	<DL	5·3	16·0 mg/kg
7440020	Nickel	3	2	0·667	<DL	18·6	49·0 mg/kg
7440097	Potassium	3	1	0·333	<DL	3 600·0	10 800·0 mg/kg
7440213	Silicon	3	2	0·667	<DL	982·0	2 340·0 mg/kg
7440224	Silver	3	1	0·333	<DL	0·0	0·2 mg/kg
7440235	Sodium	3	3	1·000	2 990·0	21 196·6	46 200·0 mg/kg
7440246	Strontium	3	2	0·667	<DL	97·6	162·0 mg/kg
7440315	Tin	3	2	0·667	<DL	17·3	37·0 mg/kg
7440326	Titanium	3	3	1·000	80·0	190·0	316·0 mg/kg
7440382	Arsenic	3	1	0·333	<DL	3·6	11·0 mg/kg
7440393	Barium	3	3	1·000	150·0	4643·3	8 710·0 mg/kg
7440428	Boron	3	2	0·667	<DL	72·3	110·0 mg/kg
7440439	Cadmium	3	2	0·667	<DL	3·3	7·0 mg/kg
7440473	Chromium	3	3	1·000	21·0	63·0	139·0 mg/kg
7440508	Copper	3	3	1·000	13·0	27·0	40·0 mg/kg
7440622	Vanadium	3	3	1·000	38·0	57·3	85·0 mg/kg
7440655	Yttrium	3	1	0·333	<DL	4·3	13·0 mg/kg
7440666	Zinc	3	3	1·000	71·0	171·0	255·0 mg/kg
7440702	Calcium	3	3	1·000	34 900·0	38 800·0	45 500·0 mg/kg
7664417	Ammonia	3	2	0·667	<DL	9·3	27·0 mg/kg
7704349	Sulfur	3	3	1·000	1 380·0	4 156·6	7 480·0 mg/kg
7723140	Phosphorus (black, white, red, yellow)	3	2	0·667	<DL	512·0	910·0 mg/kg
16984488	Fluoride	3	3	1·000	190·0	516·6	690·0 mg/kg
18496258	Sulfide	3	2	0·667	<DL	1 760·0	4 790·0 mg/kg

(Site = central pit; location = central pit; phase = solid; extract = TCLP)

| 67641 | 2-Propanone | 3 | 2 | 0·667 | <DL | 77·3 | 159·5 µg/liter |

(continued)

375

TABLE 4—contd.

Compound	Compound name	N	Detected	Prop. Detect	Minimum	Wtd. Mean	Maximum	Units
71432	Benzene	3	2	0·667	<DL	81·4	229·5	µg/liter
75092	Methylene chloride	3	2	0·667	<DL	50·5	140·1	µg/liter
75150	Carbon disulfide	3	1	0·333	<DL	17·5	52·6	µg/liter
79005	1,1,2-Trichlorethane	3	1	0·333	<DL	23·8	71·5	µg/liter
91203	Naphthalene	3	1	0·333	<DL	43·1	129·5	µg/liter
91576	2-Methylnaphthalene	3	3	1·000	10·3	49·1	116·0	µg/liter
92524	Biphenyl	3	1	0·333	<DL	6·8	20·4	µg/liter
100414	Ethylbenzene	3	2	0·667	<DL	42·4	99·8	µg/liter
107062	1,2-Dichloroethane	3	1	0·333	<DL	32·3	96·9	µg/liter
108101	4-Methyl-2-pentanone	3	2	0·667	<DL	31·9	58·8	µg/liter
108583	Toluene	3	2	0·667	<DL	495·3	1 331·4	µg/liter
108952	Phenol	3	1	0·333	<DL	5·1	15·4	µg/liter
112403	n-Dodecane	3	1	0·333	<DL	9·2	27·8	µg/liter
112958	n-Eicosane	3	1	0·333	<DL	9·5	28·5	µg/liter
117817	bis(2-Ethylhexyl)phthalate	3	1	0·333	<DL	40·4	121·2	µg/liter
544763	n-Hexadecane	3	1	0·333	<DL	15·4	46·2	µg/liter
593453	n-Octadecane	3	1	0·333	<DL	9·5	28·5	µg/liter
629594	n-Tetradecane	3	1	0·333	<DL	10·6	31·9	µg/liter
629970	n-Docosane	3	1	0·333	<DL	4·2	12·7	µg/liter
7429905	Aluminum	3	3	1·000	476·0	4 008·6	9 310·0	µg/liter
7439896	Iron	3	3	1·000	24 700·0	42 066·6	64 800·0	µg/liter
7439921	Lead	3	1	0·333	<DL	71·0	213·0	µg/liter
7439954	Magnesium	3	3	1·000	25 400·0	37 166·6	45 800·0	µg/liter
7439965	Manganese	3	3	1·000	3 000·0	3 976·6	4 940·0	µg/liter
7439987	Molybdenum	3	2	0·667	<DL	76·6	118·0	µg/liter
7440020	Nickel	3	1	0·333	<DL	16·0	48·0	µg/liter

7440097	Potassium	3	1	0·333	<DL	72 666·6	218 000·0	µg/liter
7440213	Silicon	3	2	0·667	<DL	5 423·3	9 380·0	µg/liter
7440235	Sodium	3	3	1·000	119 000·0	896 333·3	2 380 000·0	µg/liter
7440246	Strontium	3	2	0·667	<DL	2050·0	3610·0	µg/liter
7440315	Tin	3	3	1·000	181·0	317·3	444·0	µg/liter
7440326	Titanium	3	1	0·333	<DL	43·3	130·0	µg/"liter
7440382	Arsenic	3	1	0·333	<DL	8·0	24·0	µg/liter
7440393	Barium	3	3	1·000	1 980·0	3 603·3	4580·0	µg/liter
7440428	Boron	3	3	·1·000	211·0	1 060·3	1530·0	µg/liter
7440473	Chromium	3	1	0·333	<DL	63·0	189·0	µg/liter
7440505	Copper	3	1	0·333	<DL	11·3	34·0	µg/liter
7440622	Vanadium	3	1	0·333	<DL	57·3	172·0	µg/liter
7440666	Zinc	3	3	1·000	73·0	963·0	2080·0	µg/liter
7440702	Calcium	3	3	1·000	445 000·0	719 666·6	999 000·0	µg/liter
7704349	Sulfur	3	3	1·000	2 580·0	10 603·3	24 800·0	µg/liter
7723140	Phosphorus (black, white, red, yellow)	3	2	0·667	<DL	246·6	532·0	µg/liter

Site: A description of type of site sampled
  Central treatment (centralized treatment facility)
  Central pit (centralized pit)
  Drilling
  Production
Location: A description of the point at which the sample was taken at the site
  Influent
  Effluent
  Tank
  Central pit (commercial facility)

(continued)

TABLE 4—contd.

Mid-point (of production)
End-point (of production)
Tank bottom
Pit (drilling)
Drilling mud
Phase: Refers to the physical state of the sample and in the case of pits is an indicator of whether the sample was a supernatant or sludge
  Liquid (supernatant)
  Solid (sludge)
Extract: Refers to the analytical procedure used to characterize the waste. The samples were analyzed by either constituent analysis (direct) or the TCLP
  Direct
  TCLP

Compound: CAS compound number
Compound name: Compound name
N: The number of samples analyzed for the compound
Detected: The number of samples in which the compound was detected
Prop. Detect: Estimated population proportion of samples with detectable amount of compound, calculated as the volume-weighted proportion of detection, as the mean of a 0/1 indicator of compound detection in each sample
Minimum: The minimum concentration reported, or <DL if one or more samples had no detectable amount of the compound
Mean: The volume-weighted mean concentration
Maximum: The maximum concentration reported
Units: Concentration units
pH: Refers to the negative base ten logarithm of the hydrogen ion concentration

378

TABLE 5

Results of analyses of field collected drilling wastes—drill pits (analytes detected in one or more samples only)

(Site = drilling, location = pit, phase = liquid, extract = direct)

Compound	Compound name	N	Detected	Prop. Detect	Minimum	Wtd. Mean	Maximum	Units
1-001	Total volatile organic carbon	17	16	0·876	<DL	5·5	13·0	mg/liter
1-002	Biochemical Oxygen Demand	17	17	1·000	5·0	493·2	12 000·0	mg/liter
1-003	Chloride	17	17	1·000	198·0	6 686·7	47 000·0	mg/liter
1-004	Chemical Oxygen Demand	17	17	1·000	130·0	5 086·2	85 000·0	mg/liter
1-005	Nitrate/nitrite	17	14	0·975	<DL	5·7	85·0	mg/liter
1-006	Hydrogen ion	17	17	1·000	6·5	7·7	12·7	pH
1-007	Oil and grease	17	17	1·000	3·0	482·1	18 300·0	mg/liter
1-008	Residue, total	1	1	1·000	41·0	41·0	41·0	%
1-009	Residue, non-filterable	16	16	1·000	224·0	188 386·0	107 000·0	mg/liter
1-010	Residue, filterable	16	15	0·999	<DL	38 241·8	178 000·0	mg/liter
1-011	Specific conductivity	17	17	1·000	1 150·0	88 146·9	400 000·0	µmho/cm
1-012	Total organic carbon	17	17	1·000	15·0	449·9	75 500·0	mg/liter
1-013	Ignitability	17	17	1·000	200·0	200·0	200·0	degF
1-014	Corrosivity	17	17	1·000	6·5	7·7	12·7	pH
1-015	Reactivity	17	3	0·171	<DL	0·1	1·0	S(−/+)
57125	Cyanides (soluble salts and complex)	17	4	0·640	<DL	49·1	80·0	µg/liter
58902	2,3,4,6-Tetrachlorophenol	15	1	0·063	<DL	72·0	1 136·2	µg/liter
65850	Benzoic acid	15	3	0·074	<DL	12·1	185·0	µg/liter
67641	2-Propanone	16	12	0·891	<DL	1 333·0	5 330·2	µg/liter
71432	Benzene	16	1	0·055	<DL	0·9	16·5	µg/liter
75092	Methylene chloride	16	12	0·574	<DL	8 158·1	124 850·9	µg/liter
75150	Carbon disulfide	16	2	0·009	<DL	0·1	22·4	µg/liter
78591	Isophorone	15	1	0·411	<DL	4·1	10·1	µg/liter
78933	2-Butanone	16	6	0·076	<DL	7·2	1 602·5	µg/liter

(continued)

TABLE 5—contd.

Compound	Compound name	N	Detected	Prop. Detect	Minimum	Wtd. Mean	Maximum	Units
79005	1,1,2-Trichloroethane	16	1	0·000	<DL	0·0	138·2	μg/liter
85018	Phenanthrene	15	3	0·507	<DL	208·1	5990·2	μg/liter
86737	Fluorene	15	2	0·411	<DL	10·8	1464·7	μg/liter
87865	Pentachlorophenol	15	1	0·063	<DL	3826·3	60342·3	μg/liter
91203	Naphthalene	15	5	0·512	<DL	43·8	1357·5	μg/liter
91576	2-Methylnaphthalene	15	6	0·516	<DL	80·9	8725·0	μg/liter
91941	3,3'-Dichlorobenzidine	15	1	0·096	<DL	7·1	74·9	μg/liter
92524	Biphenyl	15	1	0·411	<DL	25·4	61·8	μg/liter
98862	Acetophenone	15	2	0·007	<DL	0·1	24·4	μg/liter
99876	p-Cymene	15	3	0·508	<DL	16·4	42·9	μg/liter
100414	Ethylbenzene	16	2	0·005	<DL	0·1	227·8	μg/liter
105679	2,4-Dimethylphenol	15	1	0·004	<DL	0·0	10·7	μg/liter
107028	2-Propenal	16	1	0·055	<DL	2·7	50·5	μg/liter
108101	4-Methyl-2-pentanone	15	6	0·559	<DL	26·3	1963·2	μg/liter
108883	Toluene	16	6	0·779	<DL	186·6	442·4	μg/liter
108907	Chlorobenzene	16	1	0·001	<DL	0·0	34·2	μg/liter
108952	Phenol	15	3	0·067	<DL	3·8	315·1	μg/liter
112403	n-Dodecane	15	7	0·516	<DL	2345·6	40654·2	μg/liter
112958	n-Eicosane	15	11	0·594	<DL	4094·2	89800·9	μg/liter
117817	Bis(2-ethylhexyl) phthalate	15	5	0·110	<DL	2·1	1419·9	μg/liter
117840	Di-n-octyl phthalate	15	1	0·000	<DL	0·0	1347·9	μg/liter
123911	1,4-Dioxane	16	1	0·003	<DL	0·6	189·0	μg/liter
124185	n-Decane	15	6	0·516	<DL	764·2	20824·2	μg/liter
132649	Dibenzofuran	15	1	0·001	<DL	0·0	21·2	μg/liter
142621	Hexanoic acid	15	2	0·159	<DL	8·8	74·9	μg/liter
544763	n-Hexadecane	15	10	0·834	<DL	4733·2	170361·9	μg/liter

CAS	Compound						
591786	2-Hexanone	16	1	0·003	<DL	0·0	14·4	μg/liter
593453	n-Octadecane	15	9	0·586	<DL	2510·6	41719·7	μg/liter
615225	2-(Methylthiol) benzothiazole	15	2	0·004	<DL	0·1	90·2	μg/liter
629594	n-Tetradecane	15	8	0·527	<DL	2658·6	83588·5	μg/liter
629970	n-Docosane	15	5	0·520	<DL	533·3	25898·3	μg/liter
630013	n-Hexacosane	15	2	0·507	<DL	3062·4	7246·7	μg/liter
630024	n-Octacosane	15	2	0·507	<DL	83·6	442·9	μg/liter
638686	n-Triacontane	15	3	0·108	<DL	90·2	9609·4	μg/liter
646311	n-Tetracosane	15	5	0·520	<DL	892·5	8795·3	μg/liter
832699	1-Methylphenanthrene	15	2	0·507	<DL	246·3	2009·7	μg/liter
7429905	Aluminum	17	16	0·998	<DL	202058·6	3760000·0	μg/liter
7439896	Iron	17	16	0·998	<DL	366030·3	6310000·0	μg/liter
7439910	Lanthanum	17	1	0·000	<DL	0·0	1240·0	μg/liter
7439921	Lead	17	12	0·364	<DL	693·7	52000·0	μg/liter
7439932	Lithium	17	7	0·323	<DL	532·3	9000·0	μg/liter
7439954	Magnesium	17	17	1·000	692·0	178942·7	3400000	μg/liter
7439965	Manganese	17	17	1·000	108·0	5228·2	117000·0	μg/liter
7439976	Mercury	17	3	0·162	<DL	0·4	6·0	μg/liter
7439987	Molybdenum	17	13	0·406	<DL	123·1	577·0	μg/liter
7440020	Nickel	17	15	0·997	<DL	306·5	6090·0	μg/liter
7440097	Potassium	17	15	0·998	<DL	2280009·9	3980000·0	μg/liter
7440213	Silicon	17	16	0·974	<DL	69350·8	1020000·0	μg/liter
7440224	Silver	17	7	0·205	0·4	0·4	19·0	μg/liter
7440235	Sodium	17	17	1·000	185000·0	13142056·6	59100000·0	μg/liter
7440246	Strontium	17	17	1·000	1110·0	24300·4	592000·0	μg/liter
7440291	Thorium	17	1	0·000	<DL	0·0	1630·0	μg/liter
7440315	Tin	17	12	0·404	<DL	295·1	2270·0	μg/liter
7440326	Titanium	17	12	0·871	<DL	550·6	32400·0	μg/liter

(continued)

TABLE 5—contd.

Compound	Compound name	N	Detected	Prop. Detect	Minimum	Wtd. Mean	Maximum	Units
7440360	Antimony	17	1	0·003	<DL	0·2	68·0	μg/liter
7440382	Arsenic	17	6	0·251	<DL	43·7	222·0	μg/liter
7440393	Barium	17	17	1·000	192·0	42 414·2	18 700 000·0	μg/liter
7440417	Beryllium	17	9	0·384	<DL	9·7	36·0	μg/liter
7440428	Boron	17	16	0·998	<DL	16 929·4	49 900·0	μg/liter
7440439	Cadmium	17	13	0·406	<DL	58·3	2 230·0	μg/liter
7440451	Cerium	17	3	0·329	<DL	402·8	1 640·0	μg/liter
7440473	Chromium	17	14	0·406	<DL	1 355·8	327 000·0	μg/liter
7440484	Cobalt	17	9	0·384	<DL	123·6	504·0	μg/liter
7440508	Copper	17	13	0·404	<DL	363·7	35 800·0	μg/liter
7440622	Vanadium	17	11	0·402	<DL	284·7	1 730·0	μg/liter
7440655	Yttrium	17	10	0·363	<DL	122·7	18 000·0	μg/liter
7440666	Zinc	17	17	1·000	30·0	1 539·9	100 000·0	μg/liter
7440699	Bismuth	17	1	0·002	<DL	1·7	1 030·0	μg/liter
7440702	Calcium	17	17	1·000	174 000·0	1 059 199·3	24 000 000·0	μg/liter
7553562	Iodine	17	1	0·002	<DL	2·2	1 350·0	μg/liter
7664417	Ammonia	17	17	1·000	0·1	3·4	46·0	mg/liter
7704349	Sulfur	17	15	0·976	<DL	230 393·4	5 850 000·0	μg/liter
7723140	Phosphorus (black, white, red, yellow)	17	13	0·995	<DL	5 682·9	19 300·0	μg/liter
16984488	Fluoride	17	17	1·000	0·8	10·1	310·0	mg/liter
18496258	Sulfide	17	3	0·171	<DL	0·0	350·0	mg/liter

(Site = drilling; location = pit; phase = solid; extract = direct)

Compound	Compound name	N	Detected	Prop. Detect	Minimum	Wtd. Mean	Maximum	Units
1-001	Total volatile organic carbon	20	18	0·995	<DL	15·4	120·0	mg/kg
1-002	Biochemical Oxygen Demand	20	20	1·000	4·0	2 852·3	18 400·0	mg/kg
1-003	Chloride	20	20	1·000	571·0	38 976·2	135 000·0	mg/kg
1-004	Chemical Oxygen Demand	20	20	1·000	1 580·0	110 457·5	906 000·0	mg/kg
1-005	Nitrate/nitrite	20	14	0·964	<DL	76·8	611·0	mg/kg

CAS	Compound	n	detects	freq	min	mean	max	Units
1-006	Hydrogen ion	20	20	1·000	6·8	9·0	12·8	pH
1-007	Oil and grease	20	20	1·000	798·0	32 742·0	282 000·0	mg/kg
1-008	Residue, total	21	21	1·000	14·8	32·9	69·5	%
1-012	Total organic carbon	20	20	1·000	8 330·0	34 336·6	114 000·0	mg/kg
1-013	Ignitability	20	20	1·000	200·0	200·0	200·0	degF
1-014	Corrosivity	20	20	1·000	6·8	9·0	12·8	pH
1-015	Reactivity	20	5	0·467	<DL	0·4	1·0	S(−/+)
1-016	Oil and grease (retort)	20	20	1·000	1 950·0	24 134·3	220 000·0	mg/kg
1-331	Tetrachlorodibenzo-p-dioxins	4	2	0·894	<DL	6·3	7·2	ng/kg
65850	Benzoic acid	19	1	0·005	<DL	0·1	30·2	µg/kg
67641	2-Propanone	18	13	0·877	<DL	24 087·2	32 647·9	µg/kg
71432	Benzene	18	7	0·732	<DL	96·0	1 420·5	µg/kg
71556	1,1,1-Trichloroethane	18	1	0·105	<DL	1·6	16·0	µg/kg
75092	Methylene chloride	18	5	0·242	<DL	276·9	2 279·6	µg/kg
75150	Carbon disulfide	18	2	0·574	<DL	11·8	20·5	µg/kg
78933	2-Butanone	18	6	0·843	<DL	14 300·6	44 695·9	µg/kg
79005	1,1,2-Trichloroethane	18	1	0·105	<DL	6·3	60·3	µg/kg
79345	1,1,2,2-Tetrachloroethane	18	1	0·000	<DL	0·0	84·1	µg/kg
85018	Phenanthrene	19	1	0·005	<DL	14·0	2 571·3	µg/kg
86737	Fluorene	19	1	0·005	<DL	5·6	1 038·5	µg/kg
87865	Pentachlorophenol	19	1	0·019	<DL	5 112·5	274 320·7	µg/kg
91203	Naphthalene	19	7	0·160	<DL	47 123·5	446 266·6	µg/kg
91576	2-Methylnaphthalene	19	9	0·662	<DL	12·0	1 120·0	µg/kg
92524	Biphenyl	19	2	0·006	<DL	6·5	1 187·2	µg/kg
99876	p-Cymene	19	3	0·106	<DL	13 852·9	131 327·9	µg/kg
100414	Ethylbenzene	18	7	0·723	<DL	1 450·7	12 241·1	µg/kg
100425	Styrene	19	1	0·005	<DL	359·4	65 802·6	µg/kg
108101	4-Methyl-2-pentanone	19	4	0·034	<DL	0·9	32·4	µg/kg

(continued)

TABLE 5—contd.

Compound	Compound name	N	Detected	Prop. Detect	Minimum	Wtd. Mean	Maximum	Units
108883	Toluene	18	9	0·848	<DL	1 209·9	6 292·3	µg/kg
108907	Chlorobenzene	18	1	0·032	<DL	0·8	25·3	µg/kg
108952	Phenol	19	1	0·005	<DL	45·3	8 309·4	µg/kg
112403	n-Dodecane	19	13	0·802	<DL	245 821·2	2 272 826·6	µg/kg
112958	n-Eicosane	19	14	0·927	<DL	154 434·4	1 397 026·6	µg/kg
117817	bis(2-ethylhexyl) phthalate	19	9	0·816	<DL	3 440·4	8 522·3	µg/kg
124185	n-Decane	19	9	0·634	<DL	1 645·1	19 386·4	µg/kg
127184	Tetrachlorethene	18	1	0·000	<DL	0·0	73·7	µg/kg
132650	Dibenzothiophene	19	1	0·005	<DL	4·1	768·4	µg/kg
544763	n-Hexadecane	19	16	0·932	<DL	593 974·6	5 500 293·3	µg/kg
593453	n-Octadecane	19	15	0·932	<DL	154 639·0	1 409 446·6	µg/kg
629594	n-Tetradecane	19	14	0·806	<DL	164 291·8	1 516 013·3	µg/kg
629970	n-Docosane	19	10	0·911	<DL	81 886·8	745 479·9	µg/kg
630013	n-Hexacosane	19	8	0·882	<DL	55 001·2	444 672·6	µg/kg
630024	n-Octacosane	19	6	0·816	<DL	29 678·5	189 633·3	µg/kg
638686	n-Triacontane	19	6	0·855	<DL	30 040·5	55 331·9	µg/kg
646311	n-Tetracosane	19	9	0·877	<DL	63 493·6	530 537·9	µg/kg
832699	1-Methylphenanthrene	19	4	0·042	<DL	0·9	25·8	µg/kg
1746015	2,3,7,8-Tetrachlorodibenzo-p-dioxin	4	2	0·894	<DL	9·2	13·6	ng/kg
7429905	Aluminum	21	21	1·000	2320·0	9682·9	21 300·0	mg/kg
7439896	Iron	21	21	1·000	3090·0	23 576·7	56 800·0	mg/kg
7439921	Lead	21	12	0·440	<DL	79·5	446·0	mg/kg
7439954	Magnesium	21	21	1·000	1420·0	6 538·3	16 400·0	mg/kg
7439965	Manganese	21	21	1·000	117·0	446·7	940·0	mg/kg
7439976	Mercury	21	3	0·044	<DL	0·0	2·1	mg/kg
7439987	Molybdenum	21	2	0·039	<DL	0·6	16·0	mg/kg

384

7440020	Nickel	21	0·865	15	<DL	25·3	61·0 mg/kg
7440097	Potassium	21	0·169	10	<DL	3047·0	173 000·0 mg/kg
7440213	Silicon	21	0·426	15	<DL	1729·4	14 600·0 mg/kg
7440224	Silver	21	0·059	6	<DL	0·0	9·1 mg/kg
7440235	Sodium	21	1·000	21	580·0	27 098·3	67 400·0 mg/kg
7440245	Strontium	21	0·437	8	<DL	155·7	1090·0 mg/kg
7440315	Tin	21	0·243	8	<DL	13·2	101·0 mg/kg
7440326	Titanium	21	1·000	21	17·0	143·9	554·0 mg/kg
7440382	Arsenic	21	0·339	11	<DL	4·1	29·0 mg/kg
7440393	Barium	21	1·000	21	30·0	8 921·1	56 200·0 mg/kg
7440417	Beryllium	21	0·020	3	<DL	0·0	3·0 mg/kg
7440428	Boron	21	0·703	18	<DL	75·2	290·0 mg/kg
7440439	Cadmium	21	0·882	13	<DL	4·0	14·0 mg/kg
7440473	Chromium	21	0·928	20	<DL	33·7	368·0 mg/kg
7440484	Cobalt	21	0·010	5	<DL	0·1	17·0 mg/kg
7440508	Copper	21	0·995	18	<DL	35·0	82·0 mg/kg
7440622	Vanadium	21	0·481	15	<DL	12·0	74·0 mg/kg
7440655	Yttrium	21	0·017	2	<DL	0·1	12·0 mg/kg
7440666	Zinc	21	1·000	21	22·0	139·3	823·0 mg/kg
7440702	Calcium	21	1·000	21	9360·0	40 147·3	144 000·0 mg/kg
7664417	Ammonia	20	1·000	18	<DL	40·2	4700·0 mg/kg
7704349	Sulfur	21	1·000	21	1170·0	6970·4	19 900·0 mg/kg
7723140	Phosphorus (black, white, red, yellow)	21	0·216	15	<DL	84·4	3090·0 mg/kg
16984488	Fluoride	20	1·000	20	5·4	475·5	1800·0 mg/kg
18496258	Sulfide	20	0·435	4	<DL	703·8	1680·0 mg/kg

(continued)

TABLE 5—contd.

Compound	Compound name	N	Detected	Prop. Detect	Minimum	Wtd. Mean	Maximum	Units
	(Site = drilling; location = pit; phase = solid, extract = TCLP)							
59507	4-Chloro-3-methylphenol	15	1	0·000	<DL	0·0	31·9	µg/liter
65850	Benzoic acid	15	2	0·097	<DL	11·0	132·6	µg/liter
67641	1-Propanone	18	12	0·170	<DL	54·6	12 005·9	µg/liter
71432	Benzene	18	1	0·000	<DL	0·0	19·0	µg/liter
75092	Methylene chloride	18	14	0·964	<DL	26·2	60·4	µg/liter
84662	Diethyl phthalate	18	1	0·007	<DL	2·2	308·7	µg/liter
84742	Di-n-butyl phthalate	18	1	0·005	<DL	0·0	20·1	µg/liter
87875	Pentachlorophenol	15	1	0·061	<DL	990·6	16 221·1	µg/liter
91203	Naphthalene	18	5	0·150	<DL	217·9	2061·4	µg/liter
91576	2-Methylnaphthalene	18	9	0·768	<DL	39·9	285·5	µg/liter
100414	Ethylbenzene	18	3	0·150	<DL	2·5	96·4	µg/liter
108101	4-Methyl-2-pentanone	18	12	0·860	<DL	17·4	52·7	µg/liter
108883	Toluene	18	8	0·762	<DL	12·8	99·7	µg/liter
108952	Phenol	18	1	0·011	<DL	1·2	116·3	µg/liter
117817	Bis(2-ethylhexyl) phthalate	18	8	0·216	<DL	5·4	106·1	µg/liter
124185	n-Decane	18	2	0·043	<DL	21·7	672·8	µg/liter
142621	Hexanoic acid	15	2	0·381	<DL	31·7	87·5	µg/liter
593453	n-Octadecane	18	1	0·105	<DL	83·4	790·6	µg/liter
615225	2-(Methylthiolbenzothiazole	18	1	0·002	<DL	0·0	33·0	µg/liter
629594	n-Tetradecane	18	1	0·105	<DL	103·1	978·1	µg/liter
630013	n-Hexacosane	18	1	0·120	<DL	8·1	67·6	µg/liter
638685	n-Triacontane	18	1	0·105	<DL	11·0	104·4	µg/liter
646311	n-Tetracosane	18	1	0·105	<DL	67·9	644·3	µg/liter
7429905	Aluminum	21	11	0·817	<DL	1670·9	12 300·0	µg/liter
7439896	Iron	21	20	0·986	<DL	7111·2	56 200·0	µg/liter
7439921	Lead	21	8	0·807	<DL	582·0	830·0	µg/liter

386

7439954	Magnesium	21	21	1·000	6700·0	24876·9	95500·0	µg/liter
7439965	Manganese	21	21	1·000	1130·0	2751·5	7060·0	µg/liter
7439976	Mercury	20	1	0·072	<DL	0·2	3·6	µg/liter
7439987	Molybdenum	21	2	0·237	<DL	24·8	105·0	µg/liter
7440020	Nickel	21	11	0·647	<DL	91·1	181·0	µg/liter
7440097	Potassium	21	10	0·169	<DL	17023·4	171000·0	µg/liter
7440213	Silicon	21	14	0·413	<DL	25836·9	119000·0	µg/liter
7440224	Silver	21	1	0·004	<DL	0·0	21·0	µg/liter
7440235	Sodium	21	21	1·000	25300·0	1065826·8	7350000·0	µg/liter
7440246	Strontium	21	8	0·437	<DL	3695·5	23600·0	µg/liter
7440315	Tin	21	13	0·964	<DL	358·2	531·0	µg/liter
7440326	Titanium	21	1	0·072	<DL	4·0	57·0	µg/liter
7440393	Barium	21	21	1·000	312·0	1494·0	3870·0	µg/liter
7440417	Beryllium	21	1	0·003	<DL	0·0	7·0	µg/liter
7440428	Boron	21	16	0·822	<DL	716·1	4010·0	µg/liter
7440439	Cadmium	21	10	0·425	<DL	23·2	363·0	µg/liter
7440473	Chromium	21	7	0·577	<DL	76·6	1680·0	µg/liter
7440484	Cobalt	21	1	0·417	<DL	642·8	1540·0	µg/liter
7440508	Copper	21	6	0·654	<DL	50·9	241·0	µg/liter
7440655	Yttrium	21	2	0·080	<DL	4·0	64·0	µg/liter
7440666	Zinc	21	21	1·000	70·0	5782·9	11100·0	µg/liter
7440702	Calcium	20	20	1·000	224000·0	1133899·6	2220000·0	µg/liter
7704349	Sulfur	21	21	1·000	3590·0	34414·6	122000·0	µg/liter
7723140	Phosphorus (black, white, red, yellow)	21	3	0·011	<DL	2·6	553·0	µg/liter

Site: A description of type of site sampled
Central treatment (centralized treatment facility)
Central pit (centralized pit)
Drilling
Production

(continued)

TABLE 5—contd.

Location: A description of the point at which the sample was taken at the site
    Influent
    Effluent
    Tank
    Central pit (commercial facility)
    Mid-point (of production)
    End-point (of production)
    Tank bottom
    Pit (drilling)
    Drilling mud

Phase: Refers to the physical state of the sample and in the case of pits is an indicator of whether the sample was a supernatant or sludge
    Liquid (supernatant)
    Solid (sludge)

Extract: Refers to the analytical procedure used to characterize the waste. The samples were analyzed by either constituent analysis (direct) or the TCLP
    Direct
    TCLP

Compound: CAS compound number
Compound name: Compound name
N: The number of samples analyzed for the compound
Detected: The number of samples in which the compound was detected
Prop. Detect.: Estimated population proportion of samples with detectable amount of compound, calculated as the volume-weighted proportion of detection, as the mean of a 0/1 indicator of compound detection in each sample
Minimum: The minimum concentration reported, or '<DL' if one or more samples had no detectable amount of the compound
Mean: The volume-weighted mean concentration
Maximum: The maximum concentration reported
Units: Concentration units
pH: Refers to the negative base ten logarithm of the hydrogen ion concentration

388

and 5, results are shown for the sample phase (liquid; solid) and the sample extraction technique (direct; TCLP). These results are given as 'Site/Location/Phase/Extract' data because each of these variables has been treated separately.

5.2. Calculation of Volume-Weighted Mean Concentration Estimates

In order to produce estimated national mean concentration estimates which are compatible with estimated waste volume figures and that could be appropriately multiplied by waste volumes to obtain pollutant loadings, the mean concentrations reported are weighted by the appropriate waste volume for the site and combined between zones by the following process, for each pollutant, site type, sampling location, and sample phase and extract (liquid, solid, or TCLP):

(1) Analyses of multiple samples at the same sample point (if more than one were taken and analyzed) at the same site are averaged to produce $\bar{C}_a$, the estimated average concentration of the pollutant for the site.

(2) Weighted mean concentrations for each zone are calculated as:

$$\bar{C}_z = \frac{\sum f_v \bar{C}_a}{\sum f_v} \tag{1}$$

where $\bar{C}_a$ are the average concentrations of the pollutant at each site, f_v are the associated waste volumes, and the summation is taken over the number of sites randomly sampled in the zone. (The number of random samples analyzed per zone differs for each analytical method.) This calculation is based on the assumption that each state in a zone has the same average total production as other sites in the same zone, and hence that the weighted selection of state, followed by a random selection of sites based on the list from each state, leads to an equal chance of selecting each site within the zone as a random sample site.

If other estimates of the total number of sites in a state and zone were available, an appropriate adjustment to the weighting would be used by the ratio of the actual (or more accurately estimated) number of sites in a state and zone to the original probabilities used to select the states.

(3) Zone estimates based on random samples are combined with the directed samples based on the total zone waste volume and the total volume at the directed sites by

$$\bar{C}_z' = \left[\left(F_z - \sum f_d\right)\bar{C}_z + \left(\sum f_d \bar{C}_d\right)\right] F_z \tag{2}$$

where $\bar{C}'_z$ is the combined estimate, F_z are the total zone waste volumes (independently estimated, as described below), and f_d and $\bar{C}_d$ are the waste volume and average concentration at each directed site in the zone.

(4) National estimates are produced by weighting by the zone volumes

$$\bar{C} = \frac{\sum F_z \bar{C}'_z}{\sum F_z} \qquad (3)$$

5.3. Waste Volumes

The drilling waste volumes associated with each site were calculated based on engineering inspection and information from site operators, based on the current volumes in the pit at time of visit; the amounts of previously removed material from the pit; and the percentage of drilling completed. As no appropriate measure of waste volume was available for other site types (central pit, central treatment), and those sites were all directed sites, a simple arithmetic mean of the reported concentrations is calculated for those sites.

The zone and national total waste volume amounts were estimated from independent information sources. Produced water volumes and total drilling waste volumes by state were estimated for the 5 years 1981–5. The 5-year average waste volume for each state was totalled to form the total zone average waste volume used in the calculation of the national concentration estimates.

Total drilling waste values for each state were computed as the sum of the 5-year average drilling waste estimates for each state. The proportion of liquid and solid drilling waste was estimated for each zone based on the (volume-weighted) average proportion of liquid and solid waste observed at the sites visited in the zone. This proportion was applied to the total drilling waste value to produce estimated zone liquid and solid drilling waste values for use in the concentration weighting.

5.4. Detection Limits

In computing summary statistics, those samples with concentrations reported less than the detection limit must be accounted for. A number of different statistical approaches are possible to the handling of detection limit values in calculating summary statistics. For purposes of calculating summary statistics for this paper, samples which showed no detectable level of a compound were assigned the concentration value of zero.

Statistics produced by this method give conservative lower bounds to the effect of the detection limit samples. Further analysis of ways to most appropriately account for the detection limit samples may be of interest in further work with these data.

5.5. Frequency of Occurrence of Pollutants

Frequency of occurrence was measured by determining the number of times a given analyte was detected relative to the number of times it was tested for, regardless of concentration. Pollutants occurring frequently are more likely to be indigenous to this industry than those occurring seldom. However, any frequency information, including the data presented here, must be qualified by noting that the pooling of data to gain a sufficient statistical sample size may mix results that should not be mixed, and conclusions drawn are therefore indicators rather than hard facts.

5.5.1. Frequently Occurring Organic Analytes

Organic analytes that occurred frequently and that were expected in samples collected from this industry (based on data from studies of offshore platforms and from the technical literature) were aliphatic hydrocarbons, aromatic hydrocarbons, and organic acids.

(1) Straight chain hydrocarbons, C12 through C30 (the carbon range tested), are the predominant organic compounds detected in liquid and solid samples. The most frequently occurring hydrocarbon is *n*-hexadecane, detected in approximately 85% of the solid samples and 65% of the liquid samples. The most frequently occurring hydrocarbon in TCLP extracts is *n*-decane, detected in approximately 15% of the extracts.

(2) Aromatic hydrocarbons, including single ring and polynuclear compounds, were detected in solid samples, liquid samples and TCLP extracts. Benzene, toluene and ethyl benzene are present in approximately half of the liquid and solid samples and approximately one-third of the TCLP extracts. Polynuclear aromatics detected ranged from naphthalene and substituted naphthalenes through four ring systems. Biphenyl was the only linked ring hydrocarbon tested for and it was detected. The terpenoid hydrocarbon *p*-cymene was also detected. Naphthalene occurred with a frequency of approximately 40% in solid samples, 60% in the liquid samples and 30% in the TCLP extracts. 2-methylnaphthalene occurred in a greater proportion of the TCLP extracts than in the direct analysis of the liquid or solid samples. This higher proportion of occurrence may indicate that this compound may be more readily leached from samples by the TCLP. Alternative explanations may be that a lower

detection limit was achieved using the TCLP than for direct analysis of the wastes, because of the interference from other substances in the direct analysis, or that it is formed in the leaching process, although this latter process is unlikely, based on the chemistry involved.

(3) The organic acids phenol, 2,4-dimethylphenol, *o*-cresol, *p*-cresol, benzoic acid and hexanoic acid were detected at a frequency of approximately 20–45% (depending on the analyte) in liquid samples, but at a much lower frequency (less than 5%) in solid samples and TCLP extracts.

(4) The heterocyclic ring compounds dibenzofuran and dibenzothiophene were detected in a few samples.

(5) Ketones, including acetone (2-propanone), methyl ethyl ketone (MEK) (2-butanone), methyl isobutyl ketone (MIBK) (4-methyl-2-pentanone) and acetophenone were detected, although acetone is likely attributable to sampling or bottle cleaning contamination.

(6) The high levels of methylene dichloride detected on analysis are probably attributable to laboratory contamination.

(7) Bis(2-ethylhexyl) phthalate was detected in approximately half of the samples; other phthalates were detected in approximately 5% of the samples.

6. CONCLUSIONS AND DISCUSSION

Certain of the specific objectives stated at the outset of this report were met, i.e. wastes from the production of oil and gas in the United States were characterized and nationwide estimates of concentrations were made.

EPA is now in the process of evaluating and interpreting the analytical data presented in this report; therefore, this report draws no conclusions or inferences from the data compiled at this point. Interpretations and findings derived from the data will be contained in future reports that will be available for public comment.

Additional field sampling data may be collected by the Agency in the future if EPA determines that more field work is appropriate and necessary.

EPA has received similar data from the American Petroleum Institute's (API's) concurrent field sampling program. The API data, along with any other data submitted to the Agency will be evaluated by EPA to determine appropriateness for inclusion in further Agency reports. If EPA decides to include such data, they will be available for public comment when the reports are published.

ACKNOWLEDGEMENTS

The authors acknowledge the assistance of D. R. Rushneck in the writing of this paper, and of Barrett P. Eynon for statistical analyses.

REFERENCES

TECHNICAL REPORT (1987). Exploration, development, and production of crude oil and natural gas, field sampling and analysis results. United States Environmental Protection Agency Publication 530-SW-87-005, Office of Solid Waste and Emergency Response, 401 M St SW, Washington, D.C. 20460, USA.

NARRATIVE FOR SAS 106 (1986). Development of an isotope dilution GC/MS method for hot purge and trap volatile analysis. S-CUBED Division of Maxwell Laboratories (July 1986). Available from: USEPA Sample Control Center, PO Box 1407, Alexandria, VA 22313, USA.

NARRATIVE FOR SAS 109 (1986). Analysis of extractable organic pollutant standards by isotope dilution GC/MS. S-CUBED Division of Maxwell Laboratories, (July 1986). Available from: USEPA Sample Control Center, PO Box 1407, Alexandria, VA 22313, USA.

19

Effects of Spent Freshwater Gel Chem Drilling Mud on Cultivated Land Near Lloydminster, Saskatchewan

M. J. Lesky[a], R. P. Staniland[b] and R. J. Warren[a]

[a]Husky Oil Operations Ltd, 707, 8th Avenue S.W., Calgary, Alberta,
Canada T2P 3G7
[b]Considered Environments Ltd, 190 Whitefield Drive N.E., Calgary,
Alberta, Canada T1Y 5J2

ABSTRACT

Disposal of spent freshwater drilling muds on agricultural land is an acceptable practice subject to government guidelines and approvals. Husky Oil Operations Ltd has utilized this method for disposing of freshwater gel chem muds used at the Aberfeldy Steamflood Pilot near Lloydminster, Saskatchewan, and anticipates using the same muds and disposal procedures for the proposed Aberfeldy Steamflood Commercial Development.

In 1985, a program was implemented to evaluate the effect of disposing spent freshwater gel drilling mud on cultivated lands. Two non-saline, non-sodic sites adjacent to the Pilot were chosen for application of drilling mud at rates of $0\,m^3/ha$, $40\,m^3/ha$ and $80\,m^3/ha$.

The results indicated no negative effect on agricultural productivity as a result of applications of up to twice the rates presently recommended by Saskatchewan Energy and Mines (SEM).

The effect on soil chemistry reflected the amount of drilling mud applied and duration since application. Sodium ion concentration and the sodium adsorption ratio were subject to a minor, but statistically significant, increase. Sulphate and chloride ion concentrations also increased. However, 12 months following application no significant difference was observed. No effect was measured with respect to pH, electrical conductivity, calcium, magnesium and potassium.

395

1. INTRODUCTION

Disposal of spent freshwater drilling muds on cultivated land is an acceptable, regulated activity in Saskatchewan. Regulatory guidelines limit the total dissolved solids (4000 mg/litre), chloride (1000 mg/litre) and sulphate (2000 mg/litre) concentrations, pH (5·5–8·5), and the maximum rate of any single application (100 bbl/acre or 40 m³/ha). The operator must also ensure that the mud is controlled on the site without excessive ponding or runoff.

Since 1979, Husky Oil Operations Ltd (Husky) has operated the Aberfeldy Steamflood Pilot approximately 14 km east of Lloydminster, Saskatchewan. Freshwater drilling wastes generated during development of this project have been disposed of on cultivated land in accordance with government guidelines and with the approval of local landowners.

In 1985, Husky proposed the development of a commercial steamflood project in the area. In anticipation of generating approximately 3000 m³ of spent freshwater gel chem drilling mud annually, Husky initiated this study to determine the effects of the drilling mud applications upon the chemical characteristics and agricultural productivity of local soils used for mud disposal. The study results will enable Husky to predict and avoid potentially harmful effects of single and multiple applications of spent drilling mud.

The drilling mud used is basically a suspension of bentonite in fresh water. Small amounts of bentonite extenders, thinners and fluid loss control agents are used as required during drilling. Commonly used additives are polyacrylates, lignite, lignosulphonates and carboxymethyl cellulose. Dissolved solids in the source water and formation fluids also contribute to the ultimate composition of the spent mud.

The potential effects of the landspreading of drilling muds include increased soil salinity (soluble salt content) and sodicity (relative sodium component), and heavy metal concentration as well as the introduction of organic additives and their byproducts. In an evaluation of the potential effects of drilling wastes on crops, Abouguendia et al. (1985) concluded from laboratory data that increased salinity and sodicity constituted the greatest hazard. Heavy metal and organic compound concentration present a negligible hazard in common freshwater mud systems. For the muds generated by the Aberfeldy Steamflood Pilot and Commercial Development, soil salinity, sodicity and crop productivity were established as the parameters of concern.

TABLE 1
Selected chemical characteristics of 0–15 cm soil horizons at
Site 19 and Site 27

Parameter	Site 19 mean (SD[a])	Site 27 mean (SD[a])
pH	6·1 ± 0·2	5·3 ± 0·1
Conductivity (mS/cm)	0·9 ± 0·2	1·4 ± 0·3
Soluble sodium (mg/litre)	11·9 ± 1·1	17 ± 2·6
Extractable sodium (mg/litre)	16·7 ± 2·9	21·1 ± 3·1
Sodium adsorption ratio[b]	0·27 ± 0·05	0·36 ± 0·03
Chloride (mg/litre)	4·7 ± 0·6	3·1 ± 0·5
Sulphate (mg/litre)	38·3 ± 3·3	26·3 ± 3·0

[a]Mean and standard deviation of twelve pretreatment samples.
[b]$SAR = [Na/sqrt((Ca + Mg)/2)]$ all ions as milliequivalents/litre.

2. STUDY AREA

Two experimental sites were chosen which represent soil and physiographic conditions common in the area near the Aberfeldy Steamflood Pilot and the commercial development. The sites were not saline nor sodic; saline and sodic soils were not observed near enough to the facilities to be included in the study. The water table at both sites was greater than 2·2 m, as measured in April, 1985. Based on a rain gauge located at Site 27, the precipitation between April and October 1985 was 332·4 mm. The 1986 precipitation during this period at the Lloydminster airport was 395·6 mm. The 30 year long-term average was 341·5 mm.

Site 19 represents a black chernozemic silt loam to silty clay loam soil common immediately around the Pilot. The site is generally undulating with imperfect drainage in the low areas. Chemical characteristics of the 0–15 cm horizon of Site 19 are presented in Table 1.

Site 27 represents a black chernozemic loam to sandy loam soil common north and east of the Pilot. The site surface is flat, sloping slightly to the southeast and is well drained. Chemical characteristics of the 0–15 cm horizon of Site 27 are presented in Table 1.

3. METHODS AND MATERIALS

Site 19 and Site 27 were divided into nine plots (minimum 2000 m^2) and the treatments were randomly assigned. Figure 1 presents the site and

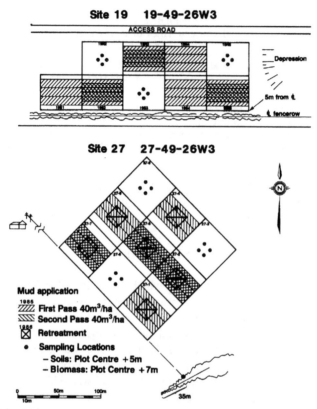

FIG. 1. Aberfeldy drilling mud disposal study: 1985 as-built site configurations.

treatment configurations. The fate and effect of drilling mud applications of $0\,m^3/ha$, $40\,m^3/ha$ and $80\,m^3/ha$ were evaluated.

In 1985, spent drilling mud was placed in a sump and treated using sulphuric acid to adjust the pH, according to standard local practice. After mixing and an equilibration period, the mud was pumped into vacuum trucks and spread with a rotating disc. The vehicle was capable of an application rate of $40\,m^3/ha$ on each pass. Two passes were required to achieve the $80\,m^3/ha$ rate. The mud being applied was sampled three times during the application procedure to determine its consistency within and between treatments and sites.

Both sites were treated in late April 1985 but due to reduced drilling activity in 1986, insufficient mud was available from the Lloydminster area. Spent drilling mud therefore was transported from the Pikes Peak Cyclic Steam Pilot area to enable retreatment of a reduced area centred within each plot at Site 27 in late April 1986 (Fig. 1). Prior to application, the material was treated to adjust pH, as above, but required some additional water to allow homogenization and pumping. The mud was sampled four times during application.

Sites were lightly cultivated after treatment and seeded by the farmer utilizing standard procedures and materials. Site 19 was seeded to canola. Site 27 was seeded to barley.

Soils were sampled at depths of 0–15 cm, 15–30 cm and 30–50 cm. Samples were collected at established distances and directions from plot centres (Fig. 1). Four samples were bulked by depth for each plot. Soils were sampled from each horizon, unless otherwise indicated, prior to treatment in April 1985 (time 0), 2 months later (time 2; 0–15 cm only) and at harvest (time 4). Only Site 27 was sampled in 1986, prior to retreatment in April (time 12) and at harvest time (time 16). Both sites were again sampled during 1987, in April (time 24) and August (time 28).

In August 1985, biomass was sampled at Site 19 using a 50 × 110 cm quadrat and cutting enclosed plants 1 cm above ground. Four biomass samples were bulked for each plot. This biomass procedure was revised at Site 27 where swathing had occurred. Two 50 cm wide sections were cut and bulked from each of these plots after ensuring swath width was consistent and the contributing area was entirely within the plot treatment. Assessments of germination success (stem density and physiologic stage), weed and pest occurrence and vigour were made during the 1985 growing season.

3.1. Laboratory Analyses
Samples of the spent drilling mud, as applied, were analyzed by the Husky Oil Laboratory in Lloydminster. Water from the mud was isolated using a Baroid mud press and passed through a 0·45 micron filter prior to analysis. Samples were analyzed for pH (meter), conductivity (meter), cations and trace metals (atomic absorption spectrophotometer), alkalinity (acid titration), and anions (dionex ion chromatograph).

Soils were analyzed at the Saskatchewan Soil Testing Laboratory in Saskatoon for pH, conductivity, soluble and extractable sodium, calcium, magnesium, potassium, chloride, sulphate and SAR. Procedures used were according to McKeague (1981). Biomass samples were oven dried (30 °C) to equilibrium and weighed before rehydration could occur.

3.2. Statistical Methods

Soil chemistry data from the 0–15 cm horizon was generated for each plot, resulting in three replicates of each treatment at each site. Samples from the 15–30 cm horizon were bulked by treatment resulting in no replication. The 30–50 cm samples were stored for future analysis. Biomass data was used in the form of the average of samples per plot, providing three replicates per treatment.

The repeated use of designated soil sampling locations enabled a paired sample analysis. Statistics were based on change in parameter levels. The resulting soils data for the 0–15 cm horizon were subject to a single factor analysis of variance at time 4, times 12 and 16 (Site 27 only), time 24 and time 28, to determine whether any treatment differed significantly. The Newman–Kuells Multiple Range procedure (Zar, 1974) was used to identify which treatments are different. The absolute values of the biomass data were also subject to a single factor analysis of variance. The F statistic generated increases as a function of the increased degree of difference between treatments and decreasing variability between replicates within treatments. In this case, the F calculated from the data must exceed the critical value F 0·05 (1), 2, 6 = 5·14 before a difference is accepted.

Simple linear regression equations were calculated based on the response of parameters to treatment rate 4 months after mud application. The calculated y intercept reflects the trend in the control, while the slope coefficient represents the treatment effect. A t statistic calculated for the slope tests the difference from zero, i.e. no effect. The regressions are simplified to allow contrast with the theoretical loading based on mud and soil characteristics.

4. DRILLING MUD CHARACTERIZATION AND THEORETICAL LOADING

Chemical data on the spent drilling muds as applied during the 1985 and 1986 treatment are presented in Table 2.

Mud from both years showed reasonably consistent levels of dissolved solids with the exception of sulphate and bicarbonate. The 1985 drilling muds showed elevated sulphate levels resulting from pH adjustment with H_2SO_4. The 1986 mud was only half as concentrated in most ions, probably as a result of the different source and dilution to enable pumping after extended tank storage.

Simple theoretical loadings for selected parameters are summarized in

TABLE 2

Chemistry of the freshwater drilling mud prior to application in 1985 and 1986

Parameter		1985 3 Samples			1986 4 Samples		
		Range	Mean	(SD[c])	Range	Mean	(SD[c])
pH		7.5–8.0	7.8		8.3–8.5	8.4	
EC	(mS/cm)[a]	4.8–5.4	5.2		3.3–3.7	3.4	
TDS	(mg/litre)[b]	3800–4800	4440	(87)	2300–2800	2670	(45)
Na	(mg/litre)	1064–1277	1166		560–670	635	
Ca	(mg/litre)	283–480	364		139–612	179	
Mg	(mg/litre)	11–44	26		18–24	20	
K	(mg/litre)	29–37	34		21–26	24	
Cl	(mg/litre)	432–520	484	(38)	200–268	236	(26)
SO$_4$	(mg/litre)	2000–2600	2333	(249)	500–1200	944	(307)
HCO$_3$	(mg/litre)	122–190	162	(29)	388–768	514	(150)
Al	(mg/litre)	N.D.–0.6	0.20		0.10–0.21	0.13	
V	(mg/litre)	0.1–4.8	2.40		0.10–0.18	0.13	
Cr	(mg/litre)	0.21–0.24	0.22		0.10–0.38	0.23	
Mn	(mg/litre)	N.D.–0.41	0.20		0–0.16	0.10	
Fe	(mg/litre)	0.14–0.18	0.16		0.13–0.29	0.23	
Co	(mg/litre)	0–10	0.10		0–10	0.10	
N	(mg/litre)	0.05–0.16	0.09		0.03–0.13	0.08	
Cu	(mg/litre)	0.03–0.04	0.03		N.D.–0.02	0.01	
Zn	(mg/litre)	0.014–0.101	0.05		0.01–0.018	0.016	
Ba	(mg/litre)	1.6–3.0	2.40		0.41–0.57	0.49	
Pb	(mg/litre)	0.36–0.38	0.37		0.20–0.40	0.28	

[a] Electrical conductivity in milliSiemens per centimetre.
[b] Total Dissolved Solids.
[c] Standard Deviation.

TABLE 3
Theoretical loading of sodium, chloride and sulphate on the
0–15 cm soil horizon

Parameter	Loading[a] 1 m³/ha mud (mg/litre)	Loading 40 m³/ha mud (mg/litre)	Loading 80 m³/ha mud (mg/litre)
Calculated for the 1985 mud:			
Sodium	1·20	48	96
Chloride	0·50	20	40
Sulphate	2·40	96	192
Calculated for the 1986 mud:			
Sodium	0·65	26	52
Chloride	0·24	10	19
Sulphate	0·97	39	78

[a]Loading indicates the concentration gain expected for a
0–15 cm soil depth before deeper leaching occurs.

Table 3. Loadings were calculated for a 15 cm soil column, assuming the average 85% water content in the mud and soil saturation percentage of 55. Application of 1 m³/ha is equivalent to 0·008 5 cm of solution contributed to an 8·25 cm potential water column. This results in a mud contribution factor of 0·001 03. Therefore, the estimated loading is equal to this factor (0·001 03) multiplied by the concentration in the mud and the application rate in m³/ha.

5. RESULTS AND DISCUSSION

Soluble sodium, extractable sodium, sodium adsorption ratio (SAR) and sulphate responded significantly (at 95% confidence) to different rates of mud application. Figures 2(a–e) illustrate the mean changes in parameter level during a 28 month period as well as indicating which treatments are significantly different. In some cases, the 'within treatment' variability reduced the significance of the difference, but the relative response remains consistent. The following discussion highlights several of the salient results.

Conductivity, pH and the remaining cations did not respond to treatment in a measurable or consistent manner and will not be addressed further. Table 4 summarizes the significant simple regression equations calculated from the response in parameters between time 0 and time 4.

TABLE 4

Simple linear regression analysis for select chemical parameters at Site 19 and Site 27

Parameter	Site	Regression	t	R^2 (adj)
Soluble sodium	19	$y = 0{\cdot}22 + 0{\cdot}575x$	3·76	62%
(mg/litre)	27	$y = -2{\cdot}72 + 0{\cdot}287x$	5·14	76%
Extractable sodium	19	$y = 5{\cdot}0 + 0{\cdot}875x$	4·11	66%
(ppm)	27	$y = -3{\cdot}33 + 0{\cdot}417x$	3·21	54%
Sodium adsorption ratio	19	$y = -0{\cdot}044 + 0{\cdot}010x$	3·66	61%
(no units)	27	$y = -0{\cdot}056 + 0{\cdot}0075x$	5·56	79%
	27 (time 12–16)	$y = -0{\cdot}289 + 0{\cdot}010x$	3·56	60%
Chloride	19	$y = 9{\cdot}22 + 0{\cdot}592x$	6·21	82%
(mg/litre)	27	$y = 4{\cdot}94 + 0{\cdot}287x$	6·90	85%
Sulphate	19	$y = 24{\cdot}1 + 3{\cdot}23x$	5·79	80%
(mg/litre)	27	$y = -4{\cdot}5 + 2{\cdot}3x$	3·40	57%

Simple linear regression equations: Significant at alpha = 0·05

$y \ = a + Bx$

y = change in parameter from time 0 to time 4 (or 12 to 16 if indicated)

x = treatment rate (m^3/ha)

a = y at $x = 0$; reflects site changes unrelated to treatment

B = slope of regression line; represents the magnitude of effect for each cubic metre per hectare of mud applied

t = student's t statistic; here t is greater than 2·36 (t 0·05 (2), 7)

R^2 = coefficient of determination; indicates percentage of variability explained by regression

The slope of the regression line indicates the observed loading for comparison with theoretical loadings summarized in Table 3. The observed values reflect the total input less losses due to leaching, mineralization and plant uptake.

The treatments had no discernible effect on germination or vigour, and had no statistically significant effect on productivity. Anticipated and actual loading of salts did not approach those necessary to bring total soil concentrations near levels recognized to limit plant growth.

5.1. Soluble and Extractable Sodium

Within the 0–15 cm horizon, extractable sodium (Fig. 2b) levels increased with treatment rate, exhibiting observed loading factors at time 4 of 0·417 (Site 27) and 0·875 (Site 19) after a theoretical loading of 1·20. Retreatment at Site 27 exhibited a similar but less significant response.

At Site 27, the 0–15 cm horizon showed evidence of reduced loading with leaching of soluble sodium to the 15–30 cm depth by time 4 and again at time 16, 4 months after retreatment. Site 19 results in the 15–30 cm horizon were not responsive to treatment; probably a response to the seasonally restricted and variable drainage conditions.

At Site 27, soluble sodium levels decreased over time to levels not significantly different from the control within 12–24 months illustrating effective leaching (Fig 2a). Extractable sodium at that site remained relatively stable after the retreatment, a result consistent with the lower mobility of the adsorbed component. Site 19 responded similarly to Site 27 in the 0–15 cm horizon but with a slower rate of leaching, especially evident at time 4.

5.2. Sodium Adsorption Ratio (SAR)

The observed increases in sodium levels (soluble and extractable) result in replacement of other cations by sodium on the soil exchange complex. SAR (Fig. 2c), which measures the relative activity of sodium, calcium and magnesium, showed significant increases in the 0–15 cm horizon proportional to application rate. The magnitude of the increase was relatively consistent at both sites, and for retreatment at Site 27. A 40 m^3/ha rate of application raised the SAR approximately 0·4 units, and the 80 m^3/ha rate raised the SAR by approximately 0·8 units. On Site 27, the 15–30 cm horizon reflected the effect in the 0–15 cm soil horizon.

The decrease in SAR levels within the 0–15 cm horizon was slow, with Site 27 exhibiting greater tendency to decrease than Site 19. This reflects shorter residency of soluble sodium in the Site 27 profile. SAR levels resulting from the 80 m^3/ha application rate approached the SAR level of the 40 m^3/ha application rate by time 28. The net effect after 1 or 2 years was significant but may be as dependent upon site textural and chemical characteristics and soluble sodium residency time as upon treatment rate.

5.3. Chloride

Chloride represented the parameter of highest solubility and mobility. Within the 0–15 cm horizon, chloride levels (Fig. 2d) increased with treatment rate, exhibiting observed loading factors of 0·287 (Site 27) and 0·592 (Site 19) after a theoretical loading of 0·50. Retreatment on Site 27 showed no net effect, probably resulting from the more dilute mud, and high rainfall and leaching that reduced levels for all treatments.

At Site 27, chlorides leached by time 4 in the 0–15 cm horizon, resulting

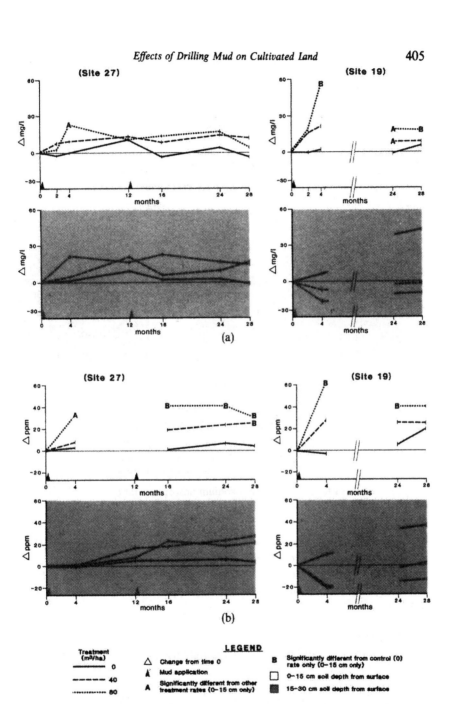

FIG. 2. Changes in (a) soluble sodium and (b) extractable sodium.

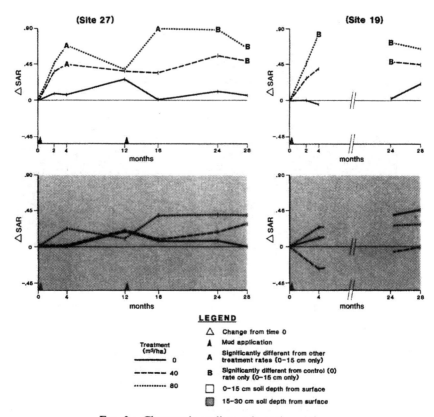

FIG. 2c. Changes in sodium adsorption ratio.

in a relatively high loading in the 15–30 cm horizon. On Site 19, the treatment effect was inconsistent at this depth.

On Site 27, chloride levels at 0–15 cm approached those of the control within 1 year of treatment. On Site 19, they approached the control within 2 years.

5.4. Sulphate

Sulphates are generally less soluble than chlorides and are the dominant anion in local soils.

Sulphate levels in the 0–15 cm horizon (Fig. 2e) increased with treatment rate exhibiting observed loading factors of 2·7 (Site 27) and 3·2 (Site 19) compared to predicted loading of 2·4. Site 27 showed little leaching to the

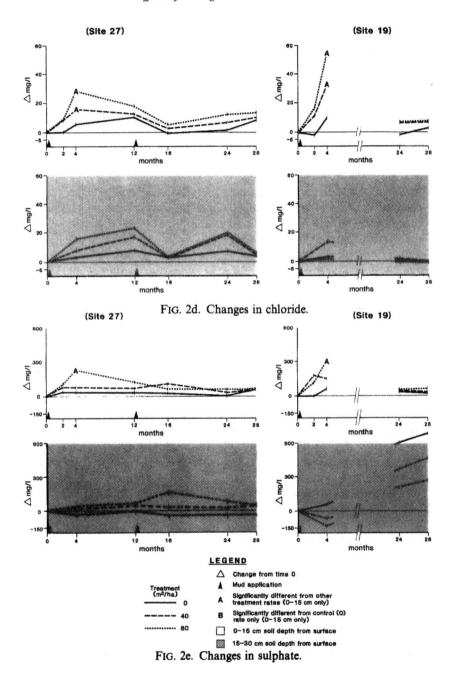

FIG. 2d. Changes in chloride.

FIG. 2e. Changes in sulphate.

15–30 cm horizon by time 4. Four months after retreatment, however, movement of sulphate to depth was obvious. Site 19 showed evidence of enrichment of the 0–15 cm horizon, illustrating the potential influence of a shallow water table.

Sulphates in the 0–15 cm horizon decreased to control levels within 1 year on Site 27, and within 2 years on Site 19.

5.5. Germination and Vigour

June, 1985 Observations
Site 19: The canola germinated erratically, but variability within plots appeared as great as between plots. Seedlings varied from cotyledon to five leaf. Density was highly variable within plots (e.g. 80–200 plants/m^2). Weeds were common throughout the site.
Site 27: Barley germination was consistent across the site with densities from 240 to 290 stems/m^2 and plants were consistently 0·25 m high. No weeds were present, a result of the landowner's effective control practices.

Mid July, 1985 Observations
Site 19: The canola was in flower with variations in growth associated with topography; poorly drained depressions showing relatively poor growth. Weed infestations were becoming obvious, with distributions unrelated to plot locations.
Site 27: The barley was in head, 0·8 m tall, and presented only slight symptoms of drought. The only distinguishable variation in condition was green pigment intensity with distribution unrelated to plot location. The minor variation was possibly due to slight inconsistencies in the anhydrous ammonia application conducted the previous fall.

Late August, 1985 Observations
Site 19: Rape beetle and weed infestations (particularly wild oats, green foxtail and thistles) severely affected canola vigour and productivity. Distributions of these infestations reflected agricultural practice rather than the experimental treatment.
Site 27: The barley showed consistent growth and full maturity. No weeds or pests were observed.

5.6. Biomass
The data was subjected to a single factor analysis of variance to determine whether any treatment differed significantly (at 95% confidence or

TABLE 5
Single factor analysis of variance and biomass production at Site 19 and Site 27

Site 19: Harvest Data August 19, 1985—Canola

Treatment	Replicate (plot)	Biomass (g/m²)
0 m³/ha	19-S-3	230
	19-N-2	284
	19-N-5	221
	Mean	245
40 m³/ha	19-S-1	205
	19-S-4	330
	19-N-4	248
	Mean	261
80 m³/ha	19-S-2	572
	19-S-5	266
	19-N-3	366
	Mean	401

(F calc.) 2·26 is less than 5·14 (F 0·05 (1), 2, 6); therefore, no significant difference.

Site 27: Harvest Data August 27, 1985—Barley

Treatment	Replicate (plot)	Biomass (g/m²)
0 m³/ha	27-3	320
	27-6	322
	27-9	522
	Mean	388
40 m³/ha	27-1	428
	27-4	424
	27-8	486
	Mean	446
80 m³/ha	27-2	474
	27-5	447
	27-7	373
	Mean	431

(F calc.) 0·33 is less than 5·14 (F 0·05 (1), 2, 6): therefore, no significant difference.

alpha = 0·05) from the others (Table 5). No significant difference was observed: a result consistent with the low natural soil salinity and limited expected contribution to salinity, metals and organic compounds made by the mud application.

6. CONCLUSIONS

Application of spent freshwater gel chem drilling mud generated from the Aberfeldy Steamflood Pilot or the proposed commercial development will not have a long term effect on crop productivity or local soils. The non-saline, non-sodic soils characteristic of the area can absorb multiple applications of these muds in excess of $40 \, m^3/ha$.

Application of spent drilling mud at $40 \, m^3/ha$ and $80 \, m^3/ha$ had no significant effect on biomass and no observed effect on germination or vigour. Conductivity, pH and divalent cations also remained essentially unaffected.

Soluble sodium, sulphate and chloride levels increased significantly with mud applications of $80 \, m^3/ha$ at Site 19 and Site 27. However, within 12 months no significant difference was observed. In the well drained soils at Site 27, soluble sodium, sulphate and chloride rapidly leached and elevated levels were evident in the 15–30 cm horizon. After 1 year, the maximum increase (0–15 cm horizon) in soluble sodium was less than 20 mg/litre; sulphate was less than 150 mg/litre; and chloride was less than 20 mg/litre. Extractable sodium and SAR levels remained slightly elevated, exceeding pretreatment levels by less than 30 mg/litre and 0·8 units, respectively.

Well drained soils characterized by Site 27 are capable of receiving $160 \, m^3/ha$ of spent freshwater gel chem drilling mud over a 2-year period without adversely affecting soil chemistry. Imperfectly drained soils characterized by Site 19 are capable of receiving $80 \, m^3/ha$ of freshwater mud over the same period. Progressive buildup of extractable sodium, however, may affect crop productivity if insufficient time is allowed, or drainage conditions are inadequate, for leaching of salts.

REFERENCES

ABOUGUENDIA, Z., BASCHAK, L. & GODWIN, R. (1985). Potential environmental effects of drilling wastes on crops and soils in Saskatchewan. Saskatchewan Energy and Mines.

McKEAGUE, J.A. (Ed.) (1981). Manual on soil sampling and methods of analysis. Canadian Society of Soil Science.

ZAR, JERROLD H. (1974). *Biostatistical Analysis*. Prentice Hall, New Jersey.

PART IV
Marine Fate and Effects

20

Acute Toxicity of Two Generic Drilling Fluids and Six Additives, Alone and Combined, to Mysids (*Mysidopsis bahia*)*

PATRICK R. PARRISH, JOHN M. MACAULEY

US Environmental Protection Agency, Environmental Research Laboratory, Gulf Breeze, Florida 32561-3999, USA

and

RICHARD M. MONTGOMERY

Technical Resources, Inc., Environmental Research Laboratory, Gulf Breeze, Florida 32561-3999, USA

ABSTRACT

Toxicity tests were conducted with two laboratory-prepared generic drilling fluids (muds) and six commonly used drilling fluid additives to determine their toxicity, alone and combined, to mysids (Mysidopsis bahia). In 25 tests, the acute toxicity of combinations of one, two or three of the drilling fluid additives mixed with either drilling fluid was less than the toxicity predicted from the empirical 96-h LC_{50}s for drilling fluid additive(s) and/or drilling fluid alone; the observed 96-h LC_{50}s of the mixtures were from 1·3 to 23·6 times the values predicted from the presumption of additive toxicity. Based on the drilling fluid additives and drilling fluids tested, a conservative estimate of the acute toxicity of mixtures of drilling fluid additives and drilling fluids would be derived if the toxicity of drilling fluid additive(s) and drilling fluid were separately determined and additive toxicity presumed.

*Contribution No. 617, Environmental Research Laboratory, Gulf Breeze, Florida, USA.

1. INTRODUCTION

The relationship between the toxicity of a substance tested alone and its toxicity when tested in combination with other substances has been investigated for many years. Alabaster & Lloyd (1982) critically reviewed the literature on the effects of mixtures of toxicants on fishes and other aquatic organisms and cited research conducted as long as 55 years ago (Southgate, 1932).

The acute toxicity of components and additives in drilling fluids (muds) has been investigated recently. Sprague & Logan (1979) determined 'separate and joint toxicity' of substances used in drilling fluids to rainbow trout (*Salmo gairdneri*). Carls & Rice (1984) used larvae of Pacific Ocean shrimp (*Pandalus hypsinotus*, *P. danae*, and *Eualus suckleyi*) and crabs (*Paralithodes camtschatica*, *Chionoecetes bairdi*, and *Cancer magister*) to determine the toxicity of common drilling fluid components; the contribution of the toxicity of an individual component to the overall toxicity of a mixture was calculated from an additive model. Conklin and Rao (1984) determined the toxicity of a mineral oil and diesel oil in a used and laboratory-prepared drilling fluid to grass shrimp (*Palaemonetes intermedius*) but did not discuss additive toxicity.

Our purpose was to test two laboratory-prepared generic drilling fluids (US EPA, 1983) and six commonly used drilling fluid additives to determine their acute toxicity to mysids (*Mysidopsis bahia*), a representative saltwater crustacean, when tested alone and in combination. We intended to determine if the toxicity of drilling fluid(s) and drilling fluid additive(s) changed when they were combined and, as a result, be able to classify the toxicity as being less than additive, additive, or more than additive. Additive toxicity, as we use the term, was defined by Alabaster & Lloyd (1982) as 'the toxicity of a mixture of poisons which is equal to that expected from a simple summation of the known toxicities of the individual poisons involved (i.e. algebraic summation of effects)'.

The study design was based on practical considerations of the amount and combinations of drilling fluid additives and drilling fluids that might be used in the US for oil and gas exploration. No drilling fluid additive was tested in excess of the manufacturer's recommended maximum concentration, and no drilling fluid additive was mixed with a drilling fluid in which it was a basic component. Chrome lignosulfonate, for example, was not tested in Generic Fluid No. 8 because the formula for that drilling fluid calls for chrome lignosulfonate as a basic component (US EPA, 1983).

2. MATERIALS AND METHODS

2.1. Test Materials

The drilling fluids tested were laboratory-prepared generic fluids purchased from the American Petroleum Institute, Washington, DC 20005, USA; formulation was according to US EPA (1983). The drilling fluid additives were technical grade and were provided by the Petroleum Equipment Suppliers Association, Houston, TX 77024, USA, except as noted (Table 1).

Sodium lauryl sulfate (dodecyl sodium sulfate) was used as a reference toxicant. The chemical used was Sigma Chemical Company No. L-5750, Lot 42F-0039, approximately 95% pure.

2.2. Test Methods

Methods for the static acute toxicity tests followed the principles outlined by Parrish (1985). Five concentrations and a control were maintained under controlled light and temperature. Water quality was measured at 24-h intervals and the number of live mysids was determined after 96 h. For techniques specific to drilling fluids toxicity testing, methods followed those of Duke *et al.* (1984), with the following exceptions:

(1) Each drilling fluid was tested as received, that is, without dilution, pH adjustment, or aeration; and
(2) only 20 mysids, instead of 60, were tested per litre of control seawater or test mixture and there was no replication of treatments.

Test temperature was $20 \pm 2°C$; salinity was $20 \pm 2°/_{00}$.

Concentrations of drilling fluids, drilling fluid additives and the reference toxicant were prepared by adding amounts calculated to produce test concentrations directly to seawater. Volumes (μl or ml) of the liquid test materials (drilling fluids, isothiazolin mixture and mineral oil) were added to seawater and weighted amounts (mg) of three dry test materials (chrome lignosulfonate, paraformaldehyde and polycarboxylic acid salt) were added to seawater. For both zinc oxide and the reference toxicant, a stock solution was prepared by adding a weighed amount to deionized water; volumes (μl) of each stock solution calculated to produce test concentrations were then added to seawater.

Range-finding tests were conducted to determine the approximate median-lethal concentration of the drilling fluids and each drilling fluid additive; definitive tests provided concentration-lethality data. Concentrations for tests with mixtures were based on the 96-h LC_{50} of each drilling fluid and drilling fluid additive.

TABLE 1

Description of the six drilling fluid additives (listed alphabetically) and two laboratory-prepared, generic drilling fluids (muds) used in acute toxicity tests at US EPA Environment Research Laboratory, Gulf Breeze, FL, to determine toxicity to mysids (*Mysidopsis bahia*) when tested alone and in mixtures.

Test material	Use in drilling fluids	Maximum recommended concentration in a drilling fluid	
		(g/litre)	(Amount per barrel[2])
Chrome lignosulfonate	Deflocculant, gel strength reducer, temperature stabilizer, and fluid-loss control agent	34·2	12 pounds
Isothiazolin mixture	Bactericide	0·331	0·014 gallons
Mineral oil; Mentor 28[a]	Lubricating agent	100	4·2 gallons
Paraformaldehyde	Bactericide	1·4	0·5 pounds
Polycarboxylic acid salt	Deflocculant and thinner	2·9	1 pound
Zinc oxide	Hydrogen sulfide scavenger	14·3	5 pounds
Generic Drilling Fluid No. 1, seawater/ freshwater/potassium/polymer mud	—	—	—
Generic Drilling Fluid No. 8, lignosulfonate freshwater mud	—	—	—

[a]Provided by IMCO Services, Houston, TX.
[b]Standard drilling industry units. 1 barrel = 42 US gallons; 1 US gallon = 3·785 litres; and 1 pound = 453·6 g.

Test concentrations were not measured; results are reported as nominal (intended) concentrations in seawater.

2.3. Statistical Analyses

Mortality data from the drilling fluids, drilling fluid additives and the reference toxicant tests were used to calculate a 96-h LC_{50} (the concentration of each test material that was lethal to 50% of the test animals after 96 h of exposure) and 95% confidence limits according to the binomial method (Stephan, 1977). Because control mortality was $\leq 10\%$ in all tests, no correction was necessary for computation of LC_{50}s and confidence limits.

To calculate the additive toxicity of (1) each different drilling fluid additive in a drilling fluid or (2) combinations of the drilling fluid additives in a drilling fluid, the LC_{50}s of the drilling fluid additive(s) and drilling fluid had to be converted to similar units, g/l. The specific gravity of each drilling fluid additive and the density of each drilling fluid were determined and the exposure concentrations and LC_{50}s were calculated for the mixture.

Predicted LC_{50}s for one drilling fluid additive in a drilling fluid, presuming additive toxicity, were calculated by using a simplified approach based on the equations of Sprague & Logan (1979) and Petrazzuolo (1980). In our approach, all calculations were in terms of LC_{50}s which were expressed in similar units, g/l. The equation used was:

$$x/LC_{50_a} + (1 - x)/LC_{50_m} = 1/LC_{50_{ma}} \tag{1}$$

where:

x = fraction of the total weight of the test material due to the additive;

LC_{50_a} = LC_{50} of the additive;

$1 - x$ = fraction of the total weight of the test material due to the drilling fluid;

LC_{50_m} = LC_{50} of the unadulterated drilling fluid; and

$LC_{50_{ma}}$ = LC_{50} of the final mixture (drilling fluid plus drilling fluid additive).

To predict the LC_{50} when more than one drilling fluid additive was used, eqn (1) was expanded to add the fraction of that additive in the mixture, divided by its LC_{50} value. That is,

$$x_1/LC_{50_{a_1}} + x_2/LC_{50_{a_2}} \cdots + \frac{(1 - (x_1 + x_2 \ldots))}{LC_{50_m}} = 1/LC_{50_{ma_1a_2}} \cdots \tag{2}$$

where:

x_1 = fraction of the total weight of the test material volume due to additive 1;

$LC_{50_{a_1}}$ = LC_{50} of drilling fluid additive 1;

x_2 = fraction of the total weight of the test material due to drilling fluid additive 2; and

$LC_{50_{a_2}}$ = LC_{50} of drilling fluid additive 2.

3. RESULTS

3.1. Toxicity of Drilling Fluids Additives Tested Alone

The response of mysids to the two generic drilling fluids was as anticipated, based on earlier tests with Generic No 1 and Generic No 8 (Duke *et al.*, 1984). Because the effect of storage conditions (time, temperature, etc.) on the toxicity of drilling fluids is not well understood, we tested both drilling fluids near the beginning and end of the testing period. The toxicity of both varied by a factor of approximately 2, based on the high:low LC_{50} ratio. Similar variability in drilling fluids tested was reported by Duke *et al.* (1984) and Parrish & Duke (in press). For the calculations of additive toxicity (see below), the average of the LC_{50}s obtained at the beginning and end of the testing period was used.

The acute toxicity of the six drilling fluid additives, as expressed by 96-h LC_{50}s, was from 1·2 mg/l for zinc oxide to 1612 mg/l for chrome lignosulfonate (Table 2). The acute toxicity of isothiazolin mixture and paraformaldehyde in the low parts-per-million range was not surprising, nor was the toxicity of the mineral oil in the same range. However, the toxicity of zinc oxide was not anticipated. The Material Safety Data Sheet that accompanied the shipment indicated that the material was low in toxicity. When the toxicity data were communicated to the supplier, a second sample was provided but it was as toxic as the first sample.

3.2. Toxicity of Drilling Fluid Additives in Drilling Fluids

The toxicity of all six drilling fluid additives, tested singly in Generic No. 1 and Generic No. 8, was less than additive (Table 3). The greatest difference between the predicted and observed LC_{50}s was for zinc oxide in Generic No. 8 (23·6 times additivity); the least difference, mineral oil in Generic No. 8 (1·4 times additivity).

When two drilling fluid additives were tested in the drilling fluids, the

TABLE 2

Acute toxicity of six drilling fluid additives and two laboratory-prepared generic drilling fluids (muds), tested at the
beginning and end of the testing program, to mysids
(*Mysidopsis bahia*). The static toxicity tests were conducted
at US EPA Environmental Research Laboratory, Gulf
Breeze, FL. Test concentrations were not measured. The
LC_{50}s and 95% confidence limits were calculated by the
binomial method; units are in mg/litre

Test material	96-h LC_{50}	Confidence limits
Chrome lignosulfonate	1 612·3	1 000·0–2 000·0
Isothiazolin mixture	200·4	160·0–320·0
Paraformaldehyde	30·9	20·0–40·0
Polycarboxylic acid salt	353·6	250·0–500·0
Mineral oil	9·8	8·4–16·8
Zinc oxide	1·2	0·5–2·0
Generic No. 1, beginning	6 469	5 250–8 750
Generic No. 1, end	3 911	1 750–5 250
Generic No. 8, beginning	135 100	117 000–156 000
Generic No. 8, end	65 888	48 800–97 500

greatest difference between predicted and observed LC_{50}s was for mineral
oil and paraformaldehyde in Generic No. 8 (15·3 times additivity); the
least difference, chrome lignosulfonate and mineral oil in Generic No. 1
(1·3 times additivity).

Finally, the LC_{50}s of combinations of three different drilling fluid
additives in each generic drilling fluid were most different for isothiazolin
mixture, polycarboxylic acid salt, and zinc oxide in Generic No. 8 (8·2
times additivity); the least difference, chrome lignosulfonate, isothiazolin
mixture, and polycarboxylic acid salt or zinc oxide in Generic No. 1 (2·0
times additivity).

3.3. Reference Toxicant Tests

The results of nine tests with sodium lauryl sulfate and mysids from the
same populations used for the additives and drilling fluids testing indicated
that the culture system provided animals that were in good condition for
testing. The response of mysids over the 10-month testing period (October
1985 to July 1986) was consistent with our experience and the literature
(Roberts *et al.*, 1982); 96-h LC_{50}s were from 4·3 to 9·2 mg/l.

TABLE 3

Acute toxicity of combinations of six drilling fluid additives mixed with two laboratory-prepared generic drilling fluids (muds) to mysids (*Mysidopsis bahia*). The static toxicity tests were conducted at the US EPA Environmental Research Laboratory, Gulf Breeze, FL. Test concentrations were not measured. The additives are listed alphabetically. The LC_{50}s and confidence limits were calculated by the binomial method; units are g/litre

Test	Additive(s)	Predicted 96-h LC_{50}	Observed 96-h LC_{50} and confidence limits	Ratio of observed LC_{50} to predicted LC_{50}
One additive in Generic No. 1	Chrome lignosulfonate	2 871	4 707 (2 700–11 000)	1·6
	Isothiazolin mixture	2 425	4 362 (3 520–7 030)	1·8
	Mineral oil	2 660	5 352 (3 510–7 000)	2·0
	Paraformaldehyde	1 798	4 002 (1 770–7 080)	2·2
	Polycarboxylic acid salt	2 161	5 515 (3 900–7 800)	2·6
	Zinc oxide	1 444	2 671 (1 750–7 000)	1·8
Two additives in Generic No. 1	Chrome lignosulfonate and mineral oil	2 385	3 122 (1 380–5 510)	1·3
	Isothiazolin mixture and mineral oil	1 871	4 732 (3 520–7 480)	2·5
	Mineral oil and paraformaldehyde	1 497	2 662 (1 770–3 550)	1·8
	Mineral oil and polycarboxylic acid	1 772	5 527 (3 910–7 820)	3·1
	Mineral oil and zinc oxide	1 126	2 177 (878–3 510)	1·9

Three additives in Generic No. 1	Chrome lignosulfonate, isothiazolin mixture, and polycarboxylic acid	1 693	3 304 (2 410–4 820)	2·0
	Chrome lignosulfonate, isothiazolin mixture, and zinc oxide	1 665	3 396 (2 260–4 520)	2·0
	Paraformaldehyde, polycarboxylic acid salt, and zinc oxide	1 155	4 620 (3 830–7 670)	4·0
One additive in Generic No. 8	Isothiazolin mixture	40 550	103 533 (58 500–234 000)	2·6
	Mineral oil	49 610	70 161 (29 500–117 000)	1·4
	Paraformaldehyde	38 267	79 184 (58 600–117 000)	2·1
	Polycarboxylic acid salt	45 304	92 108 (58 800–118 000)	2·0
	Zinc oxide	5 660	122 689 (58 600–234 000)	23·6
Two additives in Generic No. 8	Isothiazolin mixture and mineral oil	28 686	58 581 (39 000–78 000)	2·0
	Mineral oil and paraformaldehyde	6 522	99 547 (19 500–156 000)	15·3
	Mineral oil and polycarboxylic acid	23 715	55 511 (39 300–78 500)	2·3
	Mineral oil and zinc oxide	5 353	25 955 (19 500–39 100)	4·8
Three additives in Generic No. 8	Isothiazolin mixture, polycarboxylic salt, and zinc oxide	7 093	57 789 (39 200–157 000)	8·2
	Paraformaldehyde, polycarboxylic acid salt, and zinc oxide	3 783	29 912 (0–157 000)	7·9

4. DISCUSSION

It stands to reason that chemicals added to drilling fluids would be less toxic than when present in seawater alone because of the physiochemical nature of drilling fluids. Comprised mainly of clays and clay-minerals, drilling fluids have the capacity to adsorb toxicants and make them less bioavailable. The results of our tests and the work of others support this concept.

Sprague & Logan (1979) tested several additives separately and mixed with a laboratory-prepared ('simulated') drilling fluid. They also tested the simulated drilling fluid and three used drilling fluids, after having determined the toxicity of their components. For the used fluids, lethality to rainbow trout was 0·71, 1·2 and 0·83 times the values predicted from the toxicity of the components, assuming additivity. The LC_{50} of the simulated fluid was 1·5 times the value predicted from its components, making its toxicity 'significantly less-than-additive'. When the seven most toxic additives tested (one of which was paraformaldehyde) were added to the simulated drilling fluid, toxicity was near additive or less than additive. 'Approximately additive action' was found in 12 of the 21 toxicity tests when single additives were added to the simulated fluid, and in 9 of 21 tests, antagonism was 'strikingly demonstrated'. Thus, the result of our tests agree with the findings of Sprague and Logan even though they tested a freshwater vertebrate and we tested a saltwater invertebrate.

The work of Conklin & Rao (1984) showed that a No. 2 fuel oil (diesel) and a mineral oil were made less toxic when added to a used drilling fluid or a laboratory-prepared drilling fluid. Our analysis of their data, utilizing eqn (1) described above, reflects the less-than-additive toxicity of all the mixtures tested but one (Table 4).

5. SUMMARY

The acute toxicity of six common drilling fluid additives to mysids was less when they were mixed with two generic drilling fluids than when tested alone. The degree of toxicity reduction ranged from 1·4 to 23·6 times the values predicted from a presumption of additivity.

Based on the additives and drilling fluids tested and on data in the literature, a conservative estimate of the acute toxicity of mixtures of additives in drilling fluids would be obtained in most instances if the toxicity of each additive and drilling fluid were determined alone and if

TABLE 4

Predicted 96-h LC_{50}s and the relationship of predicted and observed LC_{50}s from tests conducted with grass shrimp (*Palaemonetes intermedius*) by Conklin & Rao (1984). Calculations were by the binomial method; units are mg/litre

Test material	Predicted LC_{50}	Observed LC_{50}[a]	Observed LC_{50}/ predicted LC_{50}
P7[b] plus No. 2 fuel oil	111	177/184	1·6/1·7
P7 plus mineral oil	860	558/631	0·6/0·7
NBS[c] plus No. 2 fuel oil	52	114/116	2·2/2·2
NBS plus mineral oil	408	778/715	1·9/1·8

[a]The LC_{50} was determined for each mixture without treatment and after being treated to simulate drilling conditions ('hot-rolled').
[b]A used, lightly treated lignosulfonate drilling fluid from the Gulf of Mexico.
[c]A reference drilling fluid prepared by the National Bureau of Standards.

additive toxicity were assumed. Such an approach might be used to estimate the acute toxicity of a drilling fluid system in advance of its use so that the least potentially harmful system could be selected. A confounding factor under use conditions, however, might be the input of petroleum hydrocarbons from oil-bearing strata; the presence of petroleum could increase the toxicity of the drilling fluid to crustaceans. Further, such an approach would not provide an estimate of chronic toxicity or potential adverse environmental impact.

ACKNOWLEDGEMENTS

The technical assistance of S. Friedman and L. G. Smith is appreciated, and we thank the Petroleum Equipment Suppliers Association and IMCO Services for providing chemicals for testing.

REFERENCES

ALABASTER, J.S. & LLOYD, R. (Eds) (1982). Mixtures of toxicants. In *Water Quality for Fish*, 2nd edn, Butterworths, London, pp. 253–314.

CARLS, M.G. & RICE, S.C. (1984). Toxic contributions of specific drilling mud components to larval shrimp and crabs. *Mar. Environ. Res.* **12**, 45–62.

CONKLIN, P.J. & RAO, K.R. (1984). Comparative toxicity of offshore and oil-added drilling muds to larvae of the grass shrimp (*Palaemonetes intermedius*). *Arch. Environ. Contam. Toxicol.* **13**, 685–90.

DUKE, T.W., PARRISH, P.R., MONTGOMERY, R.M., MACAULEY, S.D., MACAULEY, J.M. & CRIPE, G.M. (1984). Acute Toxicity of Eight Laboratory-prepared Generic Drilling Fluids. EPA-600/3-84-067. US Environmental Protection Agency, Gulf Breeze, FL, 34 pp.

PARRISH, P.R. (1985). Acute toxicity tests. In *Fundamentals of Aquatic Toxicology*, ed. G.M. Rand and S.R. Petrocelli. Hemisphere Publishing Corp., Washington, D.C., pp. 31–57.

PARRISH, P.R. & DUKE, T.W. (in press). Variability of the acute toxicity of drilling fluids to mysids (*Mysidopsis bahia*). In *Proceedings of the Symposium on Chemical and Biological Characterization of Municipal Sludges, Sediment, Dredge Spoils, and Drilling Muds*. American Society for Testing and Materials, Philadelphia, PA.

PETRAZZUOLO, G. (1980). Guidance for technology-based permit options for offshore oil and gas discharges of drilling fluids and cuttings in areas of biological concern. Draft report submitted to US Environmental Protection Agency, Office of Water Enforcement and Permits, Washington, D.C. 20460.

ROBERTS, M.H. Jr., WARINNER, J.E., TSAI, C., WRIGHT, D. & CRONIN, L. E. (1982). Comparison of estuarine species sensitivities to three toxicants. *Arch. Environ. Contam. Toxicol.*, **11**, 681–92.

SOUTHGATE, B.A. (1932). The toxicity of mixtures of poisons. *O.J. Pharmacol.*, **5**, 639–48.

SPRAGUE, J.B. & LOGAN, W.J. (1979). Separate and joint toxicity to rainbow trout of substances used in drilling fluids for oil exploration. *Environ. Pollut.* **19**, 271–81.

STEPHAN, C. E. (1977). Methods for calculating an LC_{50}. In *Aquatic Toxicity and Hazard Evaluation*, ASTM STP 634, ed. F.L. Mayer & J.L. Hamelink. American Society for Testing and Materials, Philadelphia, PA, pp. 65–84.

US ENVIRONMENTAL PROTECTION AGENCY. (1983). Issuance of Final General NPDES Permits for Oil and Gas Operations on the Outer Continental Shelf (OCS) of Alaska; Norton Sound and Beaufort Sea. Federal Register Vol. 48, No. 236, December 7, 1983, pp. 54881–54897.

21

Bioaccumulation of Polycyclic Aromatic Hydrocarbons in Flounder (*Pseudopleuronectes americanus*) Exposed to Oil-Based Drill Cuttings

JERRY F. PAYNE, J. KICENIUK, U. WILLIAMS, L. FANCEY, W. MELVIN

Department of Fisheries and Oceans, PO Box 5667, St John's, Newfoundland, Canada A1C 5X1

and

RICHARD ADDISON

Department of Fisheries and Oceans, Bedford Institute of Oceanography, PO Box 1006, Dartmouth, Nova Scotia, Canada B2Y 4A2

ABSTRACT

Within the past few years considerable concern has arisen over the importance of polycyclic aromatic hydrocarbons (PAH) as sources of carcinogens and mutagens in the aquatic environment. This concern also frequently arises in discussions of offshore oil and gas development even though the potential for contamination, both in terms of concentrations of PAH in the environment and size of area of any negative impact, is quite small. This study addressed concerns related to the potential for bioaccumulation and contamination of fish stocks from the disposal of oil-based drill cuttings which often contain residual levels of PAH. Only low concentrations of PAH were found in the livers of flounder exposed to neat drill cuttings for a month. If bioaccumulation is low in fish exposed to neat cuttings, it is reasonable to suggest that oil-based drill cuttings present little potential for contamination of fish stocks over any significant geographical area. Further, given the results obtained on aromatic hydrocarbon levels in control flounder, it is suggested that from the viewpoint of 'comparative risk', PAH contamination in the inshore environment should present more of a real management concern.

1. INTRODUCTION

The use of oil-based drilling fluids has been discouraged in hydrocarbon exploration and production in the marine environment in the United States and Canada but these drilling fluids are presently being used to a considerable degree in the United Kingdom and Norwegian sectors of the North Sea. Field studies in the North Sea have demonstrated only localized impacts around individual drilling sites including those sites where 'toxic' diesel oils were employed as base oils in drilling fluids (Davies et al., 1984; Matheson et al., 1986). Similar observations have been made in studies of field plots in the United Kingdom experimentally contaminated with oil-based muds and drill cuttings (Dixon, 1987). We have also shown that base oils have a lower mixed-function oxygenase (MFO) enzyme induction potential than diesel oil and display a low toxicity potential (Addison et al., 1984; Payne et al., 1985) including in the sensitive chick-embryo bioassay (Payne et al., 1987a). Futhermore, as noted by Payne et al. (1987a), base oils are enriched in aliphatics compared to aromatics, and aliphatics can generally be expected to have a lower overall cellular toxicity potential. Yet the question of disposal of cuttings contaminated with oil from drilling fluids remains somewhat controversial (Engelhardt et al., 1983; Hutcheson et al., 1984; Bakke et al., 1986; Barchard et al., 1987) including in Canada where oil-based fluids have been permitted for exploratory wells in the Arctic and off the East Coast.

Although fish have relatively high levels of enzyme systems for the detoxification of xenobiotics such as hydrocarbons (Bend & James, 1978; Stegeman, 1981; Payne et al., 1987b) a major question surrounding the disposal of oil-based muds is the potential for hydrocarbon bioaccumulation. Because of mutagenesis and carcinogenesis concerns (Payne et al., 1979; Baumann et al., 1982; Black, 1983; Malins et al., 1985), the polycyclic aromatic hydrocarbon component often comes in for special attention in considerations of offshore oil and gas development (Thomas et al., 1983). There is little information available on the bioaccumulation of polycyclic aromatic hydrocarbons (PAH) in fish exposed to hydrocarbon contaminated sediments (McCain et al., 1978; Stein et al., 1987) and we have investigated the potential for bioaccumulation in a benthic fish species exposed to drill cuttings containing different levels of aromatic hydrocarbons. Results point to minimal potential for bioaccumulation of PAH in fish stocks from drill cuttings around petroleum development sites.

2. METHODS AND MATERIALS

Winter flounder (*Pseudopleuronectes americanus*) were exposed for 1 month (30 days) to uncontaminated sediment and three different types of drill cuttings from exploration wells in the Arctic (Minuk, Nipterk) and off the East Coast (Conoco). Cuttings were added to aquaria (300 litres) to give a bottom layer 1 cm deep and the aquaria were supplied with air and running seawater (4–5 litres/min) at seasonally ambient conditions of temperature and photoperiod. The number of animals used in the experiment were as follows: Nipterk, 10; Conoco, 14; Minuk, 12; and Control, 16. The experiment was carried out at the Bedford Institute of Oceanography (BIO), which receives water from the Bedford Basin. A sample of flounder ($N = 8$) collected in putatively more pristine Newfoundland waters (referred to as wild) and held in the Marine Sciences Research Laboratory (MSRL) in Logy Bay for approximately a month was also analyzed for comparison. No information is available from BIO but previous studies have indicated that the levels of PAH in water supplies at the MSRL are very low or essentially non-detectable (1 litre samples analyzed) (Payne *et al.*, 1983; Kiceniuk & Khan, 1987).

Liver hydrocarbons were extracted with total lipids according to the procedure of Williams *et al.* (1985) modified from Bligh and Dyer (1959). One gram of liver was homogenized in an Omni Mixer homogenizer in methanol (10 × sample volume) for 1 min followed by chloroform (20 × sample volume) for 2 min. The homogenate was then filtered through Whatman No. 1 filter paper. The chloroform and methanol extracts were combined and the residue was transferred to a 125 ml Erlenmeyer flask. The residue was re-homogenized for 3 min in chloroform-methanol (2:1), filtered and the extract was pooled with the first extract. Potassium chloride (0·88%) was added to the total extract (0·25 volume of total extract) which was allowed to settle for 10 min before the supernatant was discarded. The extracts were further washed with methanol:water (1:1, 0·25 volume of total extract), with the top layer being removed following each wash. Five grams of anhydrous sodium sulphate was added to the final extract and this was permitted to stand for 30 min. At this point the extract was filtered through glass wool and evaporated under reduced pressure (37°C). The extracts were dried in a desiccator and taken up in HPLC grade hexane. They were then filtered through a Millex Sr 0·5 μm PTFE filter before analysis.

For hydrocarbon analysis of sediments, 10 g (wet) were extracted with 3 × 50 ml of methylene chloride and the extracts were stored for 30 min over 10 g of anhydrous sodium sulfate to remove any excess moisture. The extracts were then concentrated to 2·0 ml in hexane by a Kuderna–Danish apparatus and filtered through PTFE filters prior to analysis. Hydrocarbon analysis of liver extracts and sediments was carried out by high pressure liquid chromatography (HPLC) with the aid of a Waters Energy Analysis Column; this gives a good approximation of the relative abundance of different classes of PAH (both parent and alkylated 2–5 ring aromatics). Extracts were separated on a 4·6 mm × 25 mm Nucleosil NH_2 column with hexane (2 ml per min) as the mobile phase. Two, three, four and five-ring aromatics were calculated as naphthalene, phenanthrene, fluorene, pyrene and benzo(a)pyrene equivalents. Components were monitored at a wavelength of 254 nm and the practical limits of detection for naphthalene, fluorene, phenanthrene, pyrene and benzo[a]pyrene were 30, 10, 5, 30 and 30 ng respectively.

In the normal phase, the NH_2 column separates PAH based on the number of condensed aromatic rings—and to some extent stereochemistry (Wise *et al.*, 1977). The addition of short chain alkyl groups to PAH results in small changes in retention on NH_2 columns. This and similar NH_2 columns have been shown to be quite effective for carrying out class analysis of petroleum and similar complex mixtures (Wise *et al.*, 1977; Miller, 1982; Sonnefeld *et al.*, 1982; Payne *et al.*, 1985; Grizzle & Sablotny, 1986). Using Venezuelan crude oil as reference material, we have demonstrated a concordance between the levels of PAH determined by GC-MS (13 mg/g) and HPLC (16 mg/g). (The GC-MS analysis was carried out by Sea Chem Oceanography, Sidney, British Columbia.)

An estimate of PAH recovery from fish tissues was obtained by spiking triplicate samples of liver (predetermined to have a very low basal level of hydrocarbons) with 8·9, 2·5, 2·5, 5·0 and 5·0 μm of naphthalene, fluorene, phenanthrene, pyrene and benzo[a]pyrene respectively. Recoveries were: naphthalene, $33 \pm 13\%$; fluorene, $60 \pm 18\%$; phenanthrene, $50 \pm 23\%$; pyrene $44 \pm 19\%$ and benzo[a]pyrene $62 \pm 2·1\%$. Boehm & Hirtzer (1982) carried out PAH determinations on fish from the Atlantic and Gulf but did not report recoveries for their particular method of extraction. We are not aware of any relevant data for comparison with our method but studies of various food products have reported recoveries for PAH in the 40–100% and 30–50% range (Touminen *et al.*, 1988).

Sediments were washed free of oil with methylene chloride, air dried and analyzed for particle size using 35, 120 and 325 μm sieves.

Effects of treatment were analyzed by one-way ANOVA and means were compared by Duncan's multiple range test.

3. RESULTS

The concentrations of different classes of PAH found in control flounder versus flounder exposed to drill cuttings containing different concentrations of aromatic hydrocarbons are given in Figs 1 and 2. Hydrocarbon

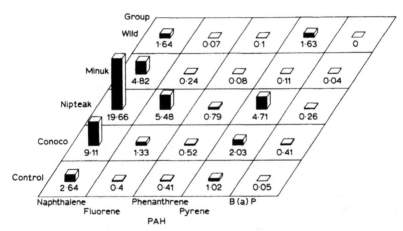

FIG. 1. Mean concentrations of PAH in flounder liver expressed on a wet weight basis (μg/g).

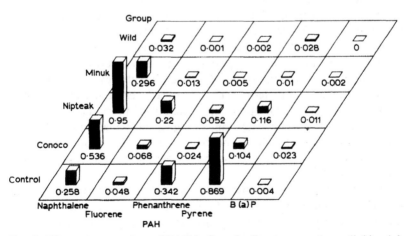

FIG. 2. Mean concentrations of PAH in flounder liver expressed on a lipid weight basis (mg/g).

TABLE 1

Statistical comparison for the various groups. Groups with the same bar are not significantly different. N, Nipterk; O, Conoco; M, Minuk; C, control; L, wild (from Newfoundland).

	Lipid Weight					Liver Weight				
Naphthalene	N	O	M	C	L	N	O	M	C	L
Fluorene	N	M	L	O	C	N	L	M	O	C
Phenanthrene	N	C	O	M	L	N	C	L	O	M
Pyrene	O	N	C	L	M	O	L	N	C	M
Benzo[a]pyrene	N	O	M	C	L	N	O	M	C	L

content is expressed in terms of wet weight of liver in Fig. 1 and liver lipid in Fig. 2. Duncan's multiple range statistical comparisons for the same are given in Table 1. The concentrations of different classes of hydrocarbons found in neat cuttings is presented in Fig. 3 while data on sediment particle size is presented in Fig. 4.

Three observations of importance stand out in the results. (1) Relatively low concentrations of PAH are accumulated in flounder liver. The highest level of accumulation was noted for naphthalenes but even here the magnitude of difference over the controls was less than 5 fold for Nipterk cuttings which contained the highest level of this particular class of hydrocarbons. (2) There was no obvious relationship between hydrocarbon concentrations found in cuttings and liver. For instance, the concentrations of naphthalenes found in fish exposed to the Conoco cuttings were similar to those found in fish exposed to the Nipterk cuttings even though the concentration of naphthalenes in the latter was over a hundred fold higher (~ 20 ppm versus ~ 3000 ppm). (3) Probably of most interest was the observation that the concentrations of higher molecular weight PAH (3–5 ringed compounds) in fish exposed to the various cuttings were little different from the levels found in control flounder from both Nova Scotia and Newfoundland.

Also of interest was the observation on sediment particle size. Minuk and Conoco cuttings appeared to be highly particulate in nature in comparison with the Nipterk cuttings. However, analyses of particle size showed all three to be remarkably similar, indicating greater caking and congealing of the fine sediments in Minuk and Conoco cuttings.

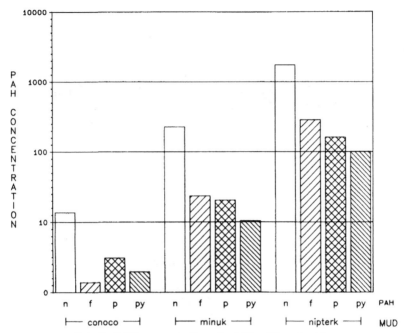

FIG. 3. PAH concentrations in drill cuttings (μg/g).

4. DISCUSSION

Flounder exposed to three different types of neat cuttings for a month accumulated only modest levels of PAH. Since the exposure simulated 'under-the-rig' conditions, due to dilution, it is reasonable to suggest that the potential for contamination of fish stocks by drill cuttings some distance away cannot be considered a problem of any major dimension at oil-development sites offshore. Overall, the results obtained in this study with drill cuttings are consonant with observations on active detoxification enzyme systems in fish, and the rapid turnover rate of individual PAH (Niimi & Palazzo, 1986).

However, the relatively high levels of naphthalenes found in flounder exposed to the Conoco cuttings—which actually contained very low levels of naphthalenes—is of interest. This cannot be attributed to differences in sediment grain size since the cuttings from the three sources were remarkably similar. Conoco cuttings are rich in alkylated cyclohexanes and is it possible that the naphthalenes detected are actually cyclohexane

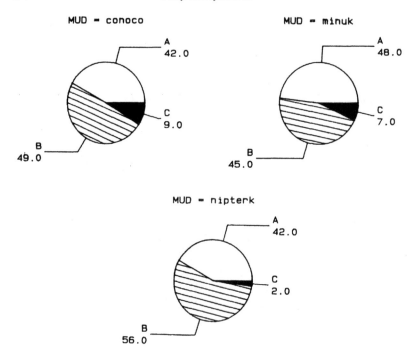

FIG. 4. Sediment grain size expressed on a percentage weight basis. A, mesh size of 35 μm; B, 120; C, 325.

compounds in which a degree of aromaticity has arisen through metabolism? Such a phenomenon of aromatic conversion would obviously be of scientific interest. In this regard it is worth noting that aromatization of dodecylcyclohexane has been recently reported in rainbow trout (*Salmo gairdneri*) (Craved i & Tulliez, 1987).

From the observations made on PAH levels in control fish, it is likely that even within a few hundred (or thousand) meters off disposal sites, the potential for PAH contamination from cutting piles would be minimal, probably not unlike the potential for contamination (from sewage, combustion, shipping and urban runoff sources) in inshore waters—a vast geographical area by comparison. Indeed from the viewpoint of 'comparative risks' in the marine environment, if a mutagenesis/carcinogenesis potential for PAH is to be entertained, it would seem more reasonable to focus on the latter. Earlier preliminary studies by Zitko (1975) as well as the more recent studies by Boehm and Hirtzer (1982) point to the importance of obtaining a better understanding of PAH levels in fish in order to answer concerns about possible effects on fish and human health.

5. CONCLUSION

It would appear that oil-based drill cuttings present little potential for contamination of fish stocks by PAH, and from the viewpoint of real 'comparative risk', if any, the PAH burden commonly found in fish in inshore waters should present more of a real concern for scientific and management interests.

ACKNOWLEDGEMENTS

The cooperation of Evan Birchard, Esso Resources (Calgary) and Jim Osborne, Environment Canada (St John's) in providing the drill cuttings is greatly appreciated. The study was supported in part through funds from the Program on Energy Research and Development, Government of Canada.

REFERENCES

ADDISON, R.F., DOE, K. & EDWARDS, A. (1984). Effects of oil based drilling mud cuttings on winter flounder (*Pseudopleuronectes americanus*): absence of acute toxicity or mixed function oxidase induction, Can. Tech. Rep. Fish. Aquat. Sci., 1307, 19 pp.

ADDY, J.M., HARTLEY, J.P. & TIBBETS, P.J.C. (1984). Ecological effect of low toxicity oil-based drilling mud in the Beatrice oilfield. *Mar. Pollut. Bull.*, **15**(12), 429–36.

BAKKE, T., GREEN, N.W., NAES, K. & PEDERSEN, A. (1986). Drill cuttings on the sea bed: field experiment on benthic recolonization and chemical changes in response to various types and amount of cuttings. Report of the Norwegian State Pollution Control Authority/Stratfjord Unit Joint Research Project on oil based drilling fluids: Cleaning and Environmental effects of oil contaminated drill cuttings, Feb., 1986, Trondheim, Norway.

BARCHARD, W.E., DOE, K.G., MAHON, S.D., MOORES, R.B., OSBORNE, J.M. & PARKER, W.R. (1987). Environmental implications of release of oil-based drilling fluids and oily cuttings into waters of the Canadian Northwest Atlantic. EPS-5-AR-87-1, Environment Canada, Dartmouth, Nova Scotia, 73 pp.

BAUMANN, P.C., SMITH, W.D. & RIBICK (1982). Hepatic tumor rates and polynuclear aromatic hydrocarbon levels in two populations of brown bullheads (*Ictalurus nebulosus*). In *Polynuclear Aromatic Hydrocarbons: Physical and Biological Chemistry*, ed. M.J. Cooke, A.J. Dennis & G.L. Fisher. Batelle Press, Columbus, Ohio, pp. 93–102.

BEND, J.R. & JAMES, M.O. (1978). Xenobiotic metabolism in marine and freshwater species. In *Biochemical and Biophysical Perspectives in Marine Biology*, ed. D.C. Malins & J.R. Sargent. Academic Press, New York, pp. 126–88.

BLACK, J.J. (1983). Field and laboratory studies of environmental carcinogenesis in Niagara River Fish. *J. Great Lakes Res.*, **9**, 326–34.

BLACKMAN, R.A.A., FILEMAN, T.W. & LAW, R.J. (1983). The toxicity of alternative base-oils and drill-muds for use in the North Sea. International Council for the Exploration of the Sea (ICES) CM 1983/E:11. Marine Environmental Quality Committee, pp. 1–7.

BLIGH, E.G. & DYER, W.J. (1959). A rapid method of total lipid extraction and purification. *Can. J. Biochem. Physiol.*, **37**, 911–17.

BOEHM, P.D. & HIRTZER, P. (1982). Gulf and Atlantic survey for selected organic pollutants in finfish. NOAA Technical Memorandum, NMFS-F/NEC-13, Woods Hole, 111 pp.

CRAVEDI, J.P. & TULLIEZ, J. (1987). Urinary metabolites of dodecylcyclohexane in *Salmo gairdneri*: evidence of aromatization and taurine conjugation in trout. *Xenobiotica*, **17**(9), 1103–11.

DAVIES, J.M., ADDY, J.M., BLACKMAN, R.A., BLANCHARD, J.R., FERBRACHE, J.E., MOORE, D.C., SOMERVILLE, H.J., WHITEHEAD, A. & WILKINSON, T. (1984). Environmental effects of the use of oil-based drilling muds in the North Sea. *Mar. Pollut. Bull.*, **15**, 363–70.

DIXON, I.M.T. (1987). Experimental application of oil-based muds and cuttings to seabed sediments. In *Fate and Effects of Oil in Marine Ecosystems*, ed. J. Kuiper & W.J. Van den Brink. Martinus Nijhoff Publishers, Dordrecht, pp. 133–50.

ENGELHARDT, F.R., HALL, H., PATERSON, R.J. & STRONG, D.C. (1983). Oil-based drilling muds in the North Sea: A perspective. Environmental Protection Branch Technical Report No. 3, Canada Oil and Gas Lands Administration, Energy Mines and Resources Canada and Indian and Northern Affairs Canada.

GRIZZLE, P.L. & SABLOTNY, D.M. (1986). Automated liquid chromatographic compound class group-type separation of crude oils and bitumens using chemically bonded aminosilane. *Anal. Chem.*, **58**, 2389–96.

HUTCHESON, M.S., STEWART, P.L., ODENSE, R. & FOWLER, B.F. (1984). Development of toxicity testing guidelines for oiled cuttings—final report, for Environmental Protection Service, Environment Canada, September, 1984. Atlantic Oceanics Company Ltd, Dartmouth, Nova Scotia, Canada.

KICENIUK, J.W. & KHAN, R.A. (1987). Effect of petroleum hydrocarbons on *Gadus morhua*, following chronic exposure. *Can. J. Zool.*, **65**(3), 490–4.

MALINS, D.C., KRAHN, M.M., MYERS, M.S., RHODES, L.D., BROWN, D.W., KRONE, C.A., McCAIN, B.B. & CHAN, S.-L. (1985). Toxic chemicals in sediments and biota from a creosote polluted harbor: relationships with hepatic neoplasms and other hepatic lesions in English sole, *Parophrys vetulus*. *Carcinogenesis*, **6**, 1463–70.

MATHESON, I., KINGSTON, P.F., JOHNSTON, C.S. & GIBSON, M.J. (1986). Statfjord field environmental study. Report of the Norwegian State Pollution Control Authority/Statfjord Unit Joint Research Project on Oil Based Drilling Fluids: Cleaning and Environmental Effects of Oil Contaminated Drill Cuttings, Feb. 1986, Trondheim, Norway.

McCAIN, B.B., HODGINS, H.O., GRONLUND, W.D., HAWKES, J.W., BROWN, D.W., MYERS, M.S. & VANDERMEULEN, J.H. (1978). Bioavailability of crude oil from experimentally oiled sediments to English sole (*Parophrys vetulus*), and pathological consequences. *J. Fish. Res. Board Can.*, **35**, 657–64.

MILLER, R. (1982). Hydrocarbon class fractionation with bonded-phase liquid chromatography. *Anal. Chem.*, **54**, 1742–6.

NIIMI, A.J. & PALAZZO, V. (1986). Biological half-lives of eight polycyclic aromatic hydrocarbons in rainbow trout. *Water Res.*, **20**, 503–7.

PAYNE, J.F., MALONEY, R. & RAHIMTULA, A. (1979). Are petroleum hydrocarbons an important source of mutagens in the marine environment? Oil Spill Conference, American Petroleum Institute Publication No. 553–536.

PAYNE, J.F., KICENIUK, J., MISRA, R., FLETCHER, G. & THOMPSON, R. (1983). Sublethal effects of petroleum hydrocarbons on adult American lobsters (*Homarus americanus*). *Can. J. Fish. Aquat. Sci.*, **40**(6), 705–17.

PAYNE, J.F., FANCEY, L., KICENIUK, J., WILLIAMS, U., OSBORNE, J. & RAHIMTULA, A. (1985). Mixed-function oxygenases as biological monitors around petroleum hydrocarbon development sites: potential for induction by diesel and other drilling mud base oils containing reduced levels of polycyclic aromatic hydrocarbons. *Mar. Environ. Res.*, **17**, 228–332.

PAYNE, J.F., FANCEY, L.L., KICENIUK, J., WILLIAMS, U., RAHIMTULA, A., KHAN, S. & OSBORNE, J. (1987a). Oil based drilling fluids: Are they an environmental risk? In *Fate and Effects of Oil in Marine Ecosystems*, ed. J. Kuiper & W.J. Van den Brink. Martinus Nijhoff Publishers, Dordrecht, pp. 279–89.

PAYNE, J.F., FANCEY, L.L., RAHIMTULA, A.D. & PORTER, E.L. (1987b). Review and perspective on the use of mixed-function oxgenase enzymes in biological monitoring. *Comp. Biochem. Physiol.*, **86C**(2), 233–45.

SONNEFELD, W.J., ZOLLER, W.H., MAY, W.E. & WISE, S.A. (1982). On-line multidimensional liquid chromatographic determination of polynuclear aromatic hydrocarbons in complex samples. *Anal. Chem.*, **54**, 723–7.

STEGEMAN, J.J. (1981). Polynuclear aromatic hydrocarbons and their metabolism in the marine environment. In *Polycyclic Hydrocarbons and Cancer*, Vol. 3, ed. H.V. Gelboin & P.D.P. Ts'o. Academic Press, New York.

STEIN, J.E., HOM, T., CASILLAS, E., FRIEDMAN, A. & VARANASI, U. (1987). Simultaneous exposure of English sole to sediment-associated xenobiotics: Part 2—Chronic exposure to an urban estuarine sediment with an added 3-H-benzo[a]pyrene and 14-C-polychlorinated biphenyl. *Mar. Environ. Res.*, **22**, 123–49.

THOMAS, D.J., GREEN, G.D., DUVAL, W.S., MILNE, K.C. & HUTCHESON, M.S. (1983). Offshore oil and gas production waste characteristics, treatment methods, biological effects and their applications to Canadian regions. Final Report for Water Pollution Control Directorate, Environmental Protection Service, Environment Canada, Ottawa.

THORESEN, K.J. & HINDS, A.A. (1983). A review of the environmental acceptability and the toxicity of diesel oil substitutes in drilling fluid systems. Report of the

IADC/SPE 1983 Drilling Conference, New Orleans, Louisiana, Feb., 1983.

TOUMINEN, J.P., PYYSALO, H.S. & SAURI, M. (1988). Cereal products as a source of polycyclic aromatic hydrocarbons. *J. Agric. Food Chem.*, **36**(1), 118–20.

WILLIAMS, U.P., KICENIUK, J.W. & BOTTA, J.R. (1985). Polycyclic aromatic hydrocarbon accumulation and sensory evaluation of lobsters (*Homarus americanus*) exposed to diesel oil at Arnold's Cove, Newfoundland. *Can. Tech. Rep. Fish. Aquat. Sci.*, 1402, 13 pp.

WISE, S.A., CHESTER, S.N., HERTZ, H.S., HILPERT, L.R. & MAY, W.E. (1977). Chemically bonded aminosilane stationary phase for the high-performance liquid chromatographic separation of polynuclear aromatic compounds. *Anal. Chem.*, **49**(14), 2306–10.

ZITKO, V. (1975). Aromatic hydrocarbons in aquatic fauna. *Bull. Environ. Contam. Toxicol.*, **14**, 621–31.

22

Bioaccumulation, Food Chain Transfer, and Biological Effects of Barium and Chromium from Drilling Muds by Flounder, *Pseudopleuronectes americanus*, and Lobster, *Homarus americanus*

JERRY M. NEFF[a], RONALD J. BRETELER[b] and R. SCOTT CARR[a]

[a]*Battelle Ocean Sciences, 397 Washington Street, Duxbury, Massachusetts 02332, USA*

[b]*Springborn Life Sciences, Inc., 790 Main Street, Wareham, Massachusetts 02571, USA*

ABSTRACT

A major concern about the routine discharge of water-based drilling muds to the ocean is that marine animals may accumulate metals from the drilling muds to concentrations that are harmful to themselves or to their consumers, including man. We performed long-term experiments to evaluate this concern. Juvenile lobsters and winter flounder were exposed for up to 99 days to natural marine sediments containing the settleable fraction of a used water-based drilling fluid, or to clean natural sediments alone. Test animals were fed either contaminated or uncontaminated food. The contaminated food consisted of sand worms, Neanthes virens, that had been exposed to sediments containing the settleable fraction of the drilling mud. The contaminated food contained barium at concentrations about 3 times background and chromium up to 1·6 times background. Flounder and lobsters accumulated small amounts of barium but not chromium from the contaminated sediments. Flounder, but not lobsters, exposed to contaminated sediment and food simultaneously accumulated slightly more barium than those exposed to contaminated sediments alone. Neither species accumulated chromium from contaminated food. Thus, biomagnification of barium and chromium from drilling fluid settleable fraction is unlikely to occur. The stress of chronic exposure to drilling mud was not sufficient to significantly decrease growth rate and concentration of nutrient reserves in the animals.

However, lobsters were moderately stressed by exposure to contaminated sediments as evidenced by higher mortality than among lobsters exposed for 99 days to clean sediments.

1. INTRODUCTION

During the drilling of offshore exploration and production wells, a variety of liquid, solid and gaseous wastes are produced on the drilling rig or platform. Some of these wastes may be discharged to the ocean, including cooling water from machinery, deck drainage and produced water following treatment in an oil/water separator, domestic sewage following primary treatment and chlorine addition, drill cuttings and water-based drilling muds. Drill cuttings and drilling muds are quantitatively the most important wastes discharged during exploration and development and have generated the greatest concern about possible adverse environmental impacts of offshore exploration and development.

There is concern that drill cuttings and drilling muds could be acutely toxic or produce deleterious sublethal responses in sensitive marine species, could alter the physical characteristics of benthic habitats by accumulating on the bottom, or could contaminate marine organisms with potentially toxic metals and organics (National Academy of Sciences, 1983). The objectives of the present investigation were to determine whether winter flounder and lobster can bioaccumulate barium and chromium derived from sediments and food contaminated with drilling muds. We also investigated the sublethal effects resulting from chronic exposure to the settleable fraction of drilling muds in sediments and food.

2. MATERIALS AND METHODS

Two experiments were performed. Juvenile winter flounder, *Pseudopleuronectes americanus* (experiment 1) and juvenile lobsters, *Homarus americanus* (experiment 2) were exposed for up to 98–99 days to clean sediments or sediments contaminated with the settleable fraction of drilling mud. Fish and lobsters were fed either uncontaminated sand worms (*Neanthes virens*) or worms contaminated with drilling mud. A fifth group of flounder or lobsters was exposed to clean natural sediment and fed worms that had been exposed to drilling mud and then allowed to depurate.

2.1. Exposure System: Flounder

Two large Fiberglas tanks were divided into five compartments, each with an area of $1\,m^2$. The compartments were separated by wooden frames covered with 4-mm mesh polyethylene screen. Each tank compartment contained a 4-cm layer of either clean sediment or a mixture of sediment and drilling mud. Water level was maintained at 25 cm with a standpipe. The five treatments were as follows:

1. Uncontaminated sediment; flounder fed uncontaminated worms (control) (US/UF).
2. Uncontaminated sediment; flounder fed contaminated worms (US/CF).
3. Uncontaminated sediment; flounder fed depurated worms (US/DF).
4. Sediment/drilling mud mixture; flounder fed contaminated worms (CS/CF).
5. Sediment/drilling mixture; flounder fed uncontaminated worms (CS/UF).

Clean natural seawater was supplied continuously from Duxbury Bay to the exposure tanks. Periodic analyses of the seawater has revealed that concentrations of several metals are in the range of typical clean estuarine waters. The water passed through a pressure sand filter before delivery to the tank. To ensure that the oxygen concentrations in the water remained above 80% saturation, high-capacity pumps were used to recirculate seawater in the tanks.

2.2. Exposure System: Lobster

Each lobster was isolated in a 1-liter polyethylene freezer container. The cover of each container had a 7-cm hole covered with 4-mm mesh plastic screening. Each container contained a 4-cm layer of sediment or sediment/drilling mud mixture. The containers, each containing a lobster, were submerged in a Fiberglas tank supplied with flowing natural seawater from Duxbury Bay. To enhance exchange of water in the small enclosures, each container was aerated with filtered compressed air. The five treatment groups were the same as those for flounder.

Seawater was passed through a pressure sand filter and a head tank equipped with a cooling unit before being pumped into the exposure tank. Seawater was introduced from the head tank to the exposure tanks through vinyl tubing.

2.3. Sediment/Drilling Mud Mixtures

The drilling mud was obtained from a platform in the Gulf of Mexico. It was a heavily treated lignosulfonate freshwater mud with a density of 16·8 lb/gal (2·0 g/ml). It contained 1630 g/liter barite, 42·2 g/liter chrome lignosulfonate, and 2·0 g/liter sodium chromate. Just prior to collection of the mud sample, the mud circulation system was shut off. The sample was collected from the mud pit just below the return line and, therefore, should be representative of the mud being circulated. The well depth at the time of sample collection was 3407 m.

The drilling mud was stored at 4°C until used. To prepare the settleable fraction of drilling mud, about 65 liters of drilling mud was added to a 1·2 m deep circular tank with a capacity of 208 liters. Flowing natural seawater was supplied to the bottom of the tank at a rate of about 1 liter/min and was allowed to overflow at the top. This allowed the soluble and fine particulate fractions to be flushed away while the settleable fraction remained in the tank. After about 1 h, seawater flow was stopped, and suspended particles in the water column were allowed to settle overnight. The aqueous phase was decanted and discarded, leaving a semisolid settleable fraction. This procedure was designed to simulate roughly the fractionation of drilling mud that takes place as it settles through the water column following discharge from an offshore platform.

The settleable fraction contained 33·4% water and had a barium concentration of 29·9 wt %. The settleable fraction was stored at 4°C until used.

Clean natural fine sand sediments were obtained from the intertidal zone of Duxbury Bay, Massachusetts. The sediment was sieved through 5-mm mesh hardware cloth to remove coarse particles and larger macroinfauna.

To obtain a drilling mud/sediment mixture containing approximately 9000 mg/kg barium, 0·556 liters of the homogenized, settleable fraction was mixed with 15·444 liters of clean, well-mixed sediment in an epoxy-coated cement mixture for 1 h.

The drilling mud/sediment mixture was equilibrated with flowing seawater for 1 month to establish a steady-state leaching rate of drilling mud ingredients to the ambient seawater. After the drilling mud/sediment mixture was added to exposure containers, the mixture was again equilibrated with flowing seawater for 2 weeks before experimental animals were added. This process was intended to reestablish interstitial water equilibrium and redox potential gradients.

2.4. Prey Organisms

Live sand worms, *Neanthes virens*, obtained from Maine Bait Corp., Newcastle, Maine, were chopped into small pieces (< 1 cm) with a carbon steel blade and fed in excess to flounder and lobsters. Three types of worms were used in the experiment. (1) Uncontaminated worms were maintained in Fiberglas tanks containing clean sediment and flowing seawater. (2) Contaminated worms were prepared by exposing clean worms to a drilling mud/sediment mixture similar to that described above for 3–7 days. Previous experiments had shown that worms reach a steady-state metal body burden after exposure for about 2 days to drilling mud-contaminated sediments (Neff *et al.*, 1985). Worms were retrieved from contaminated sediments by sieving through a 4-mm mesh screen. They were chopped into small pieces and fed immediately to test animals. (3) Depurated worms were obtained by placing groups of the drilling mud-contaminated worms in flowing seawater without sediments for at least 2 days. The worms were chopped into small pieces and fed immediately to the flounder and lobsters. Subsamples of each batch of worms were collected and frozen for later analysis of barium and chromium.

2.5. Experimental Animals

Juvenile winter flounder, *Pseudopleuronectes americanus*, were collected by otter trawl from Duxbury Bay, Massachusetts in June. They were acclimated to laboratory conditions for 1 month before the start of the exposure. During this time, they were fed clean, chopped sand worms. Each fish was labeled with a unique brand using a branding iron chilled in liquid nitrogen. Fish were allowed to recover for 2 weeks following branding. There were no mortalities or evidence of abnormal behavior among branded fish.

Fourth- and fifth-stage juvenile lobsters, *Homarus americanus*, 2–4 cm in length, were obtained from Sea Plantations, Inc., Salem, Massachusetts, in July. Lobsters were separated and acclimated for 2 weeks in individual test containers identical to the exposure containers. Mortality during acclimation was less than 10%. During acclimation, lobsters were fed clean chopped sand worms.

2.6. Exposure of Flounder

Initial total length and weight of each flounder were measured immediately before the fish were placed in the exposure tanks. Eighteen fish were assigned randomly to each exposure tank.

The flounder were fed six times per week until day 57. Previous experience had shown that, as the fish grow they can be fed larger meals at less frequent intervals. Thus, after day 56, feeding was reduced to every other day. Equal quantities of freshly chopped sand worms were fed to each group of fish in the different exposure compartments. The amount of food supplied was increased as the fish grew and total consumption increased. At each feeding, any uneaten food from the previous feeding was removed.

Daily observations were made of the general health of the fish, general appearance and condition of the sediments (in particular, the presence of black anaerobic patches), water quality parameters (temperature, salinity and dissolved oxygen), and general functioning of the exposure system. Fish were not subsampled during the experiment. All flounder were retrieved on day 98, weighed, measured, dissected and frozen immediately. Replicate sediment core samples were collected for barium and chromium analysis at the beginning, mid-point and end of the exposure period.

2.7. Exposure of Lobsters

Thirty-nine lobsters were used in each of the five treatments. Each lobster was weighed and placed in its exposure container. Each container had a unique identification number so that measurements could be made of each individual lobster. Three exposure tanks were used: one for containers with the drilling mud/sediment mixture, one for containers with clean sediment and contaminated food, and a third for containers with clean sediment and clean food (controls).

All lobsters were fed fresh chopped sand worms 6 days per week for the first 57 days of exposure. Subsequently, they were fed every other day.

Lobsters were monitored at the time of feeding. Observations were made of mortalities, presence of molts, condition of sediments and extent of feeding. Throughout the experiment, water quality parameters were measured twice weekly, and tanks were rinsed frequently with seawater to remove accumulations of silt and to prevent clogging of the screen covers of the test containers.

Twelve lobsters were sampled from each treatment after 56 days. The remaining lobsters were retrieved at the termination of the experiment on day 99. Lobsters were weighed, dissected and frozen immediately after retrieval from the exposure containers.

2.8. Chemical Analyses

Whole flounder samples were partially thawed and then dissected using a Teflon spatula with a sharpened edge. One or two filets of somatic muscle tissue, free of skin and bone, were obtained from each fish. A carbon-steel razor blade was used to dissect the abdomen (tail) of lobsters from the rest of the body. Teflon forceps were used to carefully remove the abdominal muscle from the exoskeleton. The muscle tissue of each flounder or lobster was placed in a glass liquid scintillation vial, freeze dried, and then homogenized with a glass pestle.

The vials containing freeze-dried homogenized muscle tissue were weighed to the nearest 0·1 mg to obtain a dry tissue weight. The samples were then charred for several hours at 250°C and ashed overnight at 450°C in a muffle furnace. The ash was dissolved with gentle heating in a mixture of 2 ml concentrated HNO_3 and 0·6 ml concentrated HCl. The samples then were diluted with deionized water, the amount of dilution depending on the expected analyte concentrations. Sufficient 0·1% KCl solution was added to bring the potassium concentration of the solution to 1000 ppm in order to suppress barium ionization. Final sample volume was estimated by weight.

Sediment samples were prepared for analyses of chromium by the method of Plumb (1981). A 10-g sample of freeze-dried sediment was digested with 10 ml of a 1:3 mixture of concentrated HNO_3/HCl in a 25-ml Erlenmeyer flask equipped with a Pyrex reflux cap. The flask was heated on a hot plate until the sample was almost dry. The sample was cooled, another 10 ml of the digestion mixture was added, and the sample was heated to near dryness again. The cooled sample was diluted with digestion mixture for analysis. This digestion procedure is designed to extract the 'total' metals from the sediment samples.

Concentrations of barium in tissue samples and chromium in tissue and sediment samples were determined with a Perkin Elmer Model 703 atomic absorption spectrophotometer equipped with an HGA 2000 graphite furnace and an AS-40 auto-sampler and printer. Instrument conditions were adjusted to obtain maximum recoveries from the different sample matrices. These conditions are described by Neff *et al.* (1985).

Barium in sediment samples was analyzed by instrumental neutron activation analysis at Battelle Pacific Northwest Laboratory in Richland, Washington.

2.9. Biological Measurements

Increases in length and weight, hepatosomatic index and liver glycogen concentration were measured in winter flounder as general indices of health status. Weight and the concentration of glycogen in digestive gland of lobsters were measured for the same purpose.

Liver or hepatopancreatic tissue samples were frozen in liquid nitrogen immediately upon dissection from the freshly killed animal to prevent enzymatic hydrolysis of glycogen. Tissue glycogen concentration was measured by the method of Carr & Neff (1984). The technique involved quantitative hydrolysis of glycogen to glucose with the enzyme amyloglucosidase, and analysis of glucose in the sample with the Beckman Glucose Analyzer II. The free glucose present in the tissue homogenate before enzymatic hydrolysis was also determined.

The total length of each branded winter flounder was measured to the nearest millimeter at the beginning and end of the exposure period. Length increase was then calculated as the difference in the two measurements or the final length/initial length ratio (relative growth).

Winter flounder and lobsters were blotted dry and weighed to the nearest 10 mg on a Mettler analytical balance or toploading balance. Weights were measured at the start of the exposures, at the midpoint of exposure (lobsters), and at the end of the exposure period (flounder and lobsters). Weight increments were calculated as the difference in values for successive measurements for each individual or the final weight/initial weight ratio (relative weight gain).

The hepatosomatic index of flounder was calculated at the termination of the experiment. Each fish was blotted dry and weighed to the nearest 10 mg. An incision was made in the abdominal cavity, and the entire liver was carefully dissected out and weighed. The ratio of liver wet weight to whole-body wet weight was calculated.

2.10. Data Analysis

Data analysis was performed on a VAX/11-780 computer using the BMDP (Dixon, 1981) and SPSS (Nie *et al.*, 1975) statistical packages. Both parametric and nonparametric procedures were employed in the data analysis. Parametric procedures were preferred when the data satisfied the necessary assumptions (e.g. homogeneity of variances and, to a lesser extent, normality of distributions). The former assumption was examined primarily by Cochran's C test. If the data did not allow use of parametric procedures, an attempt was made to meet the necessary requirements by using a log-transformation. If the data still did not satisfy requirements for a parametric test, the corresponding nonparametric procedure was used.

The basic parametric procedure used to compare two sets of data was the Student's *t*-test. One-way and factorial analysis of variance (ANOVA) was also performed on the data. If the one-way ANOVA was significant, a Student–Newman–Keuls (SNK) multiple range test was used to define which means were responsible for the observed differences. Two- and three-way factorial ANOVA were also used to simultaneously examine the effects of more than one treatment.

Several nonparametric procedures also were applied to the data. These included the Mann–Whitney U-test, the Kruskal–Wallis test, the nonparametric multiple comparisons test of Zar (1974), and Spearman's test.

3. RESULTS

3.1. Winter Flounder

The mean temperature of the flowing seawater during Experiment 1 was 17·4°C (SD = 3·5; $n = 27$). The temperature ranged from a high of 22·5°C in August to a low of 10°C at the end of the experiment in early November.

Throughout the experiment, salinity ranged between 30 and 32 ppt. Oxygen concentration in the water remained at saturation for the duration of the experiment.

The upper 3 cm of both the clean sediments and the sediments containing drilling mud solids remained aerobic throughout the experiment, as indicated by their light color. Occasionally, small patches of darker sediment, possibly indicating anaerobic conditions, were observed at depths greater than 3 cm in the sediments.

Mortality of juvenile winter flounder during the 98-day experiment was only 4% and was equally distributed between control and exposure tanks. Fish appeared healthy and remained active throughout the experiment in all tanks.

The concentrations of barium and chromium in surficial sediments were measured in composite samples taken from the tanks containing sediments contaminated with drilling mud solids at the start, mid-point and termination of the experiment (Table 1). The concentration of barium in the sediment dropped by about 50% during the first 52 days of the experiment, and its concentration was halved again between days 52 and 98. Thus, the half-time for wash-out of barium from the sediment was about 50 days. There was a three-fold increment in chromium concentration above background level in the sediment containing the drilling mud settleable fraction. During the 98-day experiment, approximately two-thirds of the chromium originally present was lost from the contaminated sediments and sediment chromium concentration dropped to background values.

TABLE 1

Mean concentrations of barium and chromium in control sediments and drilling mud/sediment mixtures used in bioaccumulation experiments with winter flounder and lobster ($n = 4$)

Material	Exposure time (days)	Mean metal concentration ($\mu g/g$) $\pm$ SD	
		Barium	Chromium
Control sediment	—	314 ± 48	7.1 ± 0.9
Drilling mud/sediment	1	$5\,347 \pm 120$	21.2 ± 1.1
	52	$2\,749 \pm 66$	14.5 ± 4.2
	98	$1\,124 \pm 34$	6.8 ± 0.3

Sand worms, *Neanthes virens*, used as food for both the flounder and lobster were obtained from the commercial supplier approximately twice monthly. Concentrations of barium and chromium were measured in composite samples from each batch of worms collected at the time of feeding (Table 2). Concentrations of barium and chromium in the uncontaminated worms were quite variable from batch to batch, as indicated by the large standard deviations, possibly reflecting different collection sites for the worms. There was an approximately three-fold mean increase in barium and a 1·6-fold mean increase in chromium concentration in worms exposed to the settleable fraction of drilling mud. Virtually all of the accumulated barium and chromium were lost from the worms during the depuration period. The difference in mean barium concentration in uncontaminated and contaminated worms was highly

TABLE 2

Mean concentrations of barium and chromium in sand worms, *Neanthes virens*, fed to winter flounder and lobsters throughout the bioaccumulation/biomagnification experiments. Each value is the mean of composite samples taken at weekly intervals (total of 15 samples) during the experiments

Treatment	Concentration in tissues ($\mu g/g$ dry weight $\pm$ SD)	
	Barium	Chromium
Uncontaminated	7.1 ± 5.4	0.62 ± 0.36
Contaminated	22 ± 11^a	1.02 ± 1.08
Contaminated/ depurated	7.6 ± 4.6	0.63 ± 0.39

[a] Significantly higher concentration than in tissues of uncontaminated and contaminated/depurated worms ($p < 0.005$, Mann–Whitney U-test).

significant ($p < 0.005$) based on the Mann–Whitney U-test. This reflects the fact that, in every batch of worms tested, the barium concentration in the contaminated worms was always higher than that in the uncontaminated and depurated worms. The mean concentrations of chromium in the types of worms were not significantly different. These results indicate that during exposure to the drilling mud settleable fraction, the worms accumulated some, but not all drilling mud components.

Mean concentrations of barium and chromium in somatic muscle tissue of 15 replicate flounder from each treatment group are tabulated in Table 3. There was no statistically significant difference in mean chromium concentrations among treatment groups. The concentration of chromium in different somatic muscle samples from flounder ranged from 0·10 to 0·24 µg/g dry weight, with a mean of about 0·17 µg/g in all treatment groups. The mean concentrations of barium in muscle tissue of the two groups of flounder exposed to sediments containing the settleable fraction of drilling mud were significantly higher than those in muscle of flounder exposed to clean sediments (Table 3). The group of flounder exposed to contaminated sediment and fed contaminated food had a significantly higher mean muscle barium concentration than the group exposed to contaminated sediment and fed uncontaminated worms. However, among flounder exposed to uncontaminated sediment, the type of food did not have a significant effect on muscle barium concentration. Among all the fish analyzed, muscle barium concentration ranged from 0·1 to 6·2 µg/g dry weight, with means for different treatment groups ranging from 0·35 to 2·5 µg/g (Table 3). The bioenrichment in barium concentration between muscle of the 3 groups of flounder fed clean food and muscle of flounder fed contaminated food and exposed to contaminated sediment was 5·0- to 7·1-fold. The mean estimated bioaccumulation factor of barium from sediment was 0·001. This value was estimated by dividing the mean concentration of barium in muscle of flounder exposed to contaminated sediment and fed uncontaminated food by the mean concentration of barium in sediments at the end of the exposure period.

There was little difference in the responses of winter flounder to exposure to sediments containing the settleable fraction of drilling mud and to feeding on sand worms contaminated with drilling mud (Table 3). Total length increase and size-specific length increment were unaffected by any treatment. Weight gain also was not affected significantly by treatment. Although there were no significant differences in length and weight increase, an interesting trend emerged. The flounder with the lowest mean length increase and weight gain were those exposed to uncontaminated

TABLE 3

Comparisons of mean values of variables measured in juvenile flounder (*Pseudopleuronectes americanus*) subjected to five treatments for 98 days. Treatments connected by a vertical line are not significantly different ($n = 15$)

Length increase (cm)	Final length / initial length (ratio)	Weight increase[b] (g)	Final weight[b] / initial weight (ratio)	Liver glycogen[a] (mg/g)	Hepato-somatic index[a] (%)	Muscle barium concentration[a] (µg/g)	Muscle chromium concentration[a] (µg/g)
US,DF[c] 8·0 ± 0·4	US,DF 1·60 ± 0·03	US,DF 117 ± 7	US,DF 5·00 ± 0·31	US,DF 104 ± 11	US,DF 1·93 ± 0·08	US,UF 0·35 ± 0·03	CS,CF 0·16 ± 0·01
CS,UF 8·7 ± 0·4	CS,UF 1·64 ± 0·03	CS,CF 134 ± 10	CS,UF 5·38 ± 0·33	CS,UF 106 ± 9	US,CF 2·21 ± 0·08	US,DF 0·50 ± 0·06	US,CF 0·16 ± 0·01
CS,CF 8·9 ± 0·4	CS,CF 1·68 ± 0·04	US,CF 135 ± 12	CS,CF 5·92 ± 0·33	CS,CF 107 ± 5	US,UF 2·33 ± 0·13	US,CF 0·50 ± 0·06	US,DF 0·17 ± 0·01
US,CF 8·9 ± 0·4	US,CF 1·69 ± 0·03	CS,UF 139 ± 11	US,CF 6·09 ± 0·47	US,CF 125 ± 9	CS,CF 2·35 ± 0·10	CS,UF 1·6 ± 0·4	CS,UF 0·17 ± 0·01
US,UF 9·2 ± 0·5	US,UF 1·71 ± 0·05	US,UF 147 ± 11	US,UF 6·89 ± 0·66	US,UF 131 ± 5	CS,UF 2·42 ± 0·11	CS,CF 2·5 ± 0·6	US,UF 0·18 ± 0·01

[a] Statistically significant differences among treatments ($P < 0.05$; ANOVA or Kruskal–Wallis test).

[b] Cochrans test on homogeneity of variances is significant ($P = 0.02$), making ANOVA invalid. After log-transformation of data, ANOVA yielded $P = 0.08$; Kruskal–Wallis test also yielded $P = 0.08$; the result shown using the Student–Newman–Keuls procedure is therefore significant at the 92% level of confidence.

[c] CS, contaminated sediment; CF, contaminated food; US, uncontaminated sediment; UF, uncontaminated food; DF, depurated food.

sediment and depurated food. The group with the greatest length and weight gain were those exposed to uncontaminated sediment and uncontaminated food (controls). The ratio of final weight to initial weight (relative weight gain) was significantly higher in flounder exposed to uncontaminated sediment and food (controls) than in the treatment group exposed to uncontaminated food and fed depurated worms.

Mean liver glycogen concentrations were significantly higher in flounder exposed to uncontaminated sediment and fed either contaminated or uncontaminated food than in flounder in the other three treatment groups. Lowest mean liver glycogen concentration was observed in flounder exposed to uncontaminated sediment and fed depurated food.

The hepatosomatic index of winter flounder exposed to uncontaminated sediment and fed depurated food was significantly lower than that of all other treatment groups except the group exposed to uncontaminated sediment and fed contaminated food (Table 3). The two groups of fish exposed to contaminated sediments had the highest hepatosomatic indices.

3.2. Lobster

During Experiment 2, the temperature of the seawater flowing through the lobster chambers ranged between 10 and 20°C. During the 99-day experiment, extending from July 28 to November 4, the mean seawater temperature was 16·5°C (SD = 3·2; n = 36).

Seawater salinity ranged between 30 and 32 ppt. Oxygen concentration in the lobster chambers was maintained at saturation by continuous gentle bubble aeration.

There were some mortalities among juvenile lobsters during the 99-day experiment in all but one treatment group. The percentage mortality was higher among lobsters exposed to contaminated sediments than among those exposed to clean sediments. The mortalities in the different treatment groups at the end of the experiment were as follows:

Uncontaminated sediments:	
Uncontaminated food	8·7%
Contaminated food	13·0%
Depurated food	0.0%
Contaminated sediments:	
Uncontaminated food	26.1%
Contaminated food	21.7%

These results strongly suggest that drilling mud-contaminated sediments and possibly contaminated food were stressful to the lobsters.

Lobsters were sampled for chemical analysis after 56 and 99 days. Mean concentrations of barium and chromium in edible tail muscle of 12–23 replicate lobsters from each treatment group are tabulated in Table 4.

At 56 days, there were no statistically significant differences in mean chromium concentrations in tail muscle of lobster from different treatment groups. Mean chromium concentrations in abdominal muscle of animals from the different treatment groups ranged from 0·33 to 1·4 μg/g dry weight. Because of the small size of individual samples, there was much variability in chromium concentrations among replicates, and chromium concentrations were below detection limits in several samples at day 99. Thus chromium values at day 99 were not evaluated statistically.

After both 56 and 99 days, abdominal muscle of lobsters exposed to contaminated sediments contained significantly higher concentrations of barium than did muscle of lobsters exposed to clean sediment. In addition, at 99 days, lobsters from contaminated sediments and receiving uncontaminated food had a significantly higher concentration of muscle barium than did lobsters from contaminated sediments and receiving contaminated food. Food type had no effect on mean barium concentrations in tail muscle of lobsters exposed to uncontaminated sediments. In four out of the five treatment groups, there was no significant change in the concentration of barium in tail muscle between day 56 and day 99. In lobsters exposed to contaminated sediments and fed contaminated food, the mean concentration of barium in tail muscle dropped from 11·1 μg/g at day 56 to 5·76 μg/g at day 99. Mean barium concentrations in lobster tail muscle ranged from 0·49 to 12·9 μg/g dry weight in different treatment groups.

The bioenrichment in muscle barium concentration of the 3 groups of lobsters exposed to uncontaminated sediment compared to lobsters fed contaminated food and exposed to contaminated sediment for 56 and 99 days ranged from 5·05 to 22·7 and from 5·14 to 8·72, respectively. The mean estimated bioaccumulation factor of barium from sediment at both 56 and 99 days was 0·004 and 0·007, respectively. At 99 days, the mean concentration of barium in abdominal muscle of control lobsters (US, UF) was about 3·2 times that in somatic muscle of control flounder. In addition, lobsters accumulated about 2–8 times as much barium as flounder when both were exposed to contaminated sediment and fed contaminated or uncontaminated food.

Weight increase of juvenile lobsters, *Homarus americanus*, was not affected significantly by any treatment after 56 and 99 days (Table 4). At both time intervals, the greatest weight gain was observed in lobsters that

TABLE 4

Comparisons of mean values of variables measured in juvenile lobsters (*Homarus americanus*) subjected to five treatments for 56 and 99 days. Treatments connected by a vertical line are not significantly different (n = 12 at 56 days and 18–23 at 99 days)

56 days exposure (n = 12)					99 days exposure (n = 18–23)[b]			
Weight increase (g)	Final weight/ initial weight (ratio)	Liver glucose and glycogen[a] (mg/g)	Muscle barium concentration[a] (µg/g)	Muscle chromium concentration[a] concentration (µg/g)	Weight increase (g)	Final weight[a]/ initial weight (ratio)	Liver glucose and glycogen[a] (mg/g)	Muscle barium concentration[a] (µg/g)
CS,UF[c] 2·85 ± 0·35	US,DF 6·86 ± 0·39	US,UF 27·0 ± 2·8	US,DF 0·49 ± 0·15	US,CF 0·33 ± 0·07	US,DF 6·03 ± 0·35	US,DF 11·3 ± 0·6	US,CF 6·63 ± 1·02	US,DF 0·66 ± 0·06
CS,CF 3·15 ± 0·39	US,CF 7·03 ± 0·44	CS,UF 27·5 ± 2·8	US,UF 1·3 ± 0·5	US,UF 0·50 ± 0·07	CS,CF 6·03 ± 0·42	CS,UF 12·7 ± 0·9	US,DF 8·38 ± 1·58	US,UF 1·11 ± 0·11
CS,CF 3·17 ± 0·27	US,UF 7·04 ± 0·48	US,CF 29·2 ± 2·8	US,CF 2·2 ± 0·6	CS,CF 0·73 ± 0·28	CS,CF 6·16 ± 0·41	US,UF 13·1 ± 0·8	CS,UF 10·5 ± 1·58	US,CF 1·12 ± 0·16
US,DF 3·78 ± 0·34	CS,CF 7·28 ± 0·82	US,DF 31·5 ± 2·9	CS,CF 11·1 ± 3·1	US,DF 0·92 ± 0·44	US,CF 6·69 ± 0·24	US,CF 13·6 ± 0·6	US,UF 12·1 ± 1·9	CS,CF 5·76 ± 1·07
US,UF 4·00 ± 0·43	CS,UF 7·58 ± 0·70	CS,CF 39·6 ± 4·2	CS,UF 12·9 ± 3·9	CS,UF 1·4 ± 0·6	US,UF 6·87 ± 0·48	CS,CF 14·4 ± 1·0	CS,CF 14·8 ± 2·1	CS,UF 10·3 ± 2·9

[a] Statistically significant differences among treatments (P < 0·05; ANOVA or Kruskal–Wallis test).
[b] Chromium concentrations after 99 days of exposure are not reported because most are less than detection limit.
[c] CS, contaminated sediment; CF, contaminated food; US, uncontaminated sediment; UF, uncontaminated food; DF, depurated food.

were exposed to uncontaminated sediments and food. The highest relative weight increase (final weight/initial weight) was for lobsters exposed to contaminated sediment and fed contaminated food. At 99 days, the relative weight increase of the lobsters exposed to uncontaminated sediment and fed depurated food was significantly lower than that of the lobsters exposed to contaminated sediment and fed contaminated food. This latter group of lobsters also had the highest mean concentration of hepatopancreatic glucose plus glycogen (Table 4) at 56 and 99 days.

4. DISCUSSION

We have shown in this investigation that two species of demersal/benthic marine animals, lobster and winter flounder, are able to accumulate small amounts of barium but not chromium in somatic muscle tissue during exposure for up to 99 days to sandy sediments heavily contaminated with the settleable fraction of a used lignosulfonate drilling mud. The extent of bioaccumulation of metals during exposure can be expressed as a bioenrichment factor or a bioaccumulation factor from sediment. In this context, the bioenrichment factor is defined as the ratio of the concentration of the metal in muscle tissues of exposed animals to the concentration of the metal in muscle tissues of control animals. The bioaccumulation factor from sediments can be defined as the ratio of the concentration of the metal in the animal tissues to the total concentration of the metal in the sediments to which the animals are exposed. No effort was made in this investigation to define a 'bioavailable' metal fraction in sediment by selective extraction techniques. In the present investigation, bioenrichment and bioaccumulation factors for barium in muscle of winter flounder were 5·0–7·1 and 0·001, respectively. The corresponding values for barium in muscle tissue of lobster after 56 and 99 days were 5·05–22·7 and 0·004 and 5·14–8·72 and 0·007, respectively. Chromium was not bioaccumulated from any of the sediments by either species, reflecting the lack of substantial contamination of the sediment with chromium by addition of the settleable fraction of drilling mud.

These results indicate a very low order of bioavailability of these metals from the settleable fraction of drilling mud. Lobsters had higher background concentrations of barium in muscle tissue and accumulated more barium from drilling mud than did winter flounder.

The sand worm, *Neanthes virens*, used in this and similar investigations (Neff *et al.*, 1985) as food for the flounder and lobsters accumulated substantially more barium and slightly more chromium from drilling mud settleable fraction than did the flounder and lobster. However, virtually all the accumulated barium and chromium were lost when the worms were placed in clean sand and allowed to depurate for 24–48 h. Thus, it is likely that much of the accumulated metal was associated with ingested sediments in the gut of the worms and was evacuated with fecal material during the depuration period.

The results of the present investigation agree quite well in most cases with results of earlier studies of the bioaccumulation of barium and chromium from drilling fluids (see reviews of Neff, 1982; National Academy of Sciences, 1983; Petrazzuolo, 1983). Biological enrichment factors of 8·8 and 10 were reported by Liss *et al.* (1980) for barium in kidney and adductor muscle, respectively, of scallops, *Placopecten magellanicus*, exposed to a 1000-ppm suspension of a whole unused lignosulfonate drilling mud. Biological enrichments of up to 350 for barium were reported by Brannon & Rao (1979) for grass shrimp exposed to solid barite powder. The shrimp depurated 90% of accumulated barium in 14 days when removed from the barite substrate. Biological enrichment factors for chromium in several species of marine animals following exposure to suspensions of whole drilling mud or soluble or suspended particulate fractions of drilling mud usually are much lower than those for barium. All but two of the 40 biological enrichment factors for chromium, summarized by Petrazzuolo (1983), are below 10. The highest values were for juvenile oysters, *Crassostrea gigas*, exposed to 60 000 and 80 000 ppm suspended particulate phase preparations of a 12·7 lb/gal lignosulfonate drilling mud (McCulloch *et al.*, 1980).

The biological significance of such a relatively minor bioaccumulation of barium from sediments contaminated with high concentrations of the settleable fraction of drilling mud in this investigation is difficult to assess. There is little evidence of acute or chronic toxicity of barium. Conklin *et al.* (1980) reported histological damage to the posterior gut epithelium in grass shrimp consuming pure particulate barite. It is uncertain whether the damage was caused by abrasion or chemical toxicity. A variety of other chronic and sublethal effects have been reported in marine animals following exposure to solid barite at concentrations orders of magnitude in excess of its seawater solubility (see review of Neff, 1982). These studies have limited environmental relevance. Although some soluble barium salts are acutely toxic when ingested by man, there is no evidence that barium salts in either deficient or excessive quantities cause any chronic disease in man (Schroeder *et al.*, 1972).

This is the first investigation in which the transfer of metals from drilling fluids to a prey organism and from the prey organism to predators was examined. We were able to demonstrate a statistically significant apparent transfer of barium from drilling-fluid-contaminated worms to winter flounder muscle. Flounder exposed to contaminated sediments and fed contaminated worms for 98 days contained significantly higher concentrations of barium in muscle than flounder exposed for the same length of time to contaminated sediments and fed uncontaminated food. However, flounder did not accumulate any barium from contaminated food when the fish were exposed to uncontaminated sediment. Thus, it is uncertain whether the increased accumulation of barium in the former case was directly from the contaminated food or from ingestion of contaminated sediment with the food.

All experiments were continued for sufficiently long that the predators ate several times their body weight of contaminated food. Therefore, if food chain transfer or biomagnification of these metals was to occur, we should have detected it. There is substantial evidence that concentrations of barium (Ng & Patterson, 1982) and chromium (Phelps *et al.*, 1975), and probably all inorganic metal salts and ions, actually decrease with higher trophic level (build down) rather than biomagnify (Metayer *et al.*, 1980; Brown *et al.*, 1982). Our results indicate that, although limited food chain transfer of barium may occur, the transfer is so small that biomagnification of barium in marine food webs is extremely unlikely. Because the worms used as food became minimally contaminated with chromium during exposure to drilling mud, we were unable to demonstrate convincingly whether food chain transfer of chromium from drilling mud might occur.

The flounder were not severely stressed by exposure for 98 days to sediments containing high but environmentally realistic concentrations of the settleable fraction of drilling mud. Mortality among flounder was low. However, juvenile lobsters showed higher mortality in the experimental than in the control treatments. Lobsters were moderately stressed by exposure for 99 days to sediments contaminated with drilling mud.

Several biochemical and physiological indices were used to detect pollutant stress in flounder and lobsters exposed to the settleable fraction of drilling mud. These measurements showed that the experimental animals may have been mildly stressed by exposure to the settleable fraction of drilling mud. However, the stress was not sufficient to affect the rate of growth of juvenile flounder and lobsters. In fact, the most severely stressed flounders were those consuming depurated worms.

Biochemical analysis of the depurated worms revealed that they depleted their glycogen reserves during the depuration period and therefore were of lesser nutritional quality than undepurated worms to the flounders (Neff *et al.*, 1985). The lower nutritional value of the depurated worms probably was responsible for the significantly lower specific weight increments in flounder and lobster and the significantly lower hepatosomatic index in flounder subjected to this treatment.

Growth or relative weight increment (final weight/initial weight) in juvenile lobsters were not significantly affected by exposure to sediments containing the settleable fraction of drilling mud, by feeding the lobsters food contaminated with drilling mud settleable fraction or by a combination of both, with two exceptions. Relative weight increment and concentration of glycogen plus glucose in the liver of lobsters exposed to contaminated sediments and fed contaminated food were significantly higher than those of lobsters exposed to uncontaminated sediment and fed depurated food. These differences probably are indicative of sublethal (probably nutritional) stress in lobsters fed depurated food (low in glycogen).

ACKNOWLEDGEMENTS

The research reported here was supported by a contract from the American Petroleum Institute to Battelle Memorial Institute. The final report for this project, including the investigations reported here, is available as API Publication No. 4397 from the American Petroleum Institute, 1220 L Street, N.W., Washington, D.C. 20005.

REFERENCES

BRANNON, A.C. & RAO, K.R. (1979). Barium, strontium and calcium levels in the exoskeleton, hepatopancreas and abdominal muscle of the grass shrimp, *Palaemonetes pugio*: Relation to molting and exposure to barite. *Comp. Biochem. Physiol.*, **63A**, 261–74.

BROWN, D.A., GOSSETT, R.W., HERSHELMAN, G.P., SCHAFER, K.A., JENKINS, K.D. & PERKINS, E.M. (1982). Bioaccumulation and detoxification of contaminants in marine organisms from Southern California coastal waters. In *Waste Disposal in the Oceans*, ed. D.F. Soule & D. Walsh. Westview Press, Boulder, CO., pp. 171–93.

BURTON, J.D., MARSHALL, N.J. & PHILLIPS, A.J. (1968). Solubility of barium sulfate in sea water. *Nature*, **217**, 834–5.

CARR, R.S. & NEFF, J.M. (1984). Quantitative semi-automated enzymatic assay for tissue glycogen. *Comp. Biochem. Physiol.*, **77B**, 447–9.

CONKLIN, P.J., DOUGHTIE, D.G. & RAO, K.R. (1980). Effects of barite and used drilling muds on crustaceans, with particular reference to the grass shrimp *Palaemonetes pugio*. In *Symposium on Research on Environmental Fate and Effects of Drilling Fluids and Cuttings*. Courtesy Associates, Washington, D.C., pp. 912–43.

LISS, R.G., KNOX, F., WAYNE, D. & GILBERT, T.R. (1980). Availability of trace elements in drilling fluids to the marine environment. In *Symposium on Research on Environmental Fate and Effects of Drilling Fluids and Cuttings*. Courtesy Associates, Washington, D.C., pp. 691–722.

DIXON, W.J. (1981). *BMDP Statistical Software 1981*. University of California Press, Berkeley, 725 pp.

MCCULLOCH, W.L., NEFF, J.M. & CARR, R.S. (1980). Bioavailability of heavy metals from used offshore drilling fluids to the clam *Rangia cuneata* and the oyster *Crassostrea gigas*. In *Symposium on Research on Environmental Fate and Effects of Drilling Fluids and Cuttings*. Courtesy Associates, Washington, D.C., pp. 964–83.

METAYER, C., AMIARD, J.C., AMIARD-TRIQUET, C. & MARCHAND, J. (1980). On the transfer of several trace elements in neritic and estuarine food chains: bioaccumulation in omnivorous and carnivorous fishes. *Helg. Meeresunters*, **34**, 179–91.

NATIONAL ACADEMY OF SCIENCES (1983). *Drilling Discharges in the Marine Environment*. Panel on Assessment of Fates and Effects of Drilling Fluids and Cuttings in the Marine Environment. Marine Board, Commission on Engineering and Technical Systems, National Research Council, Washington, D.C., 180 pp.

NEFF, J.M. (1982). Fate and biological effects of oil well drilling fluids in the marine environment: a literature review. US EPA, Environmental Research Laboratory, Gulf Breeze, FL. EPA-600/53-82-064, 151 pp.

NEFF, J.M., BRETELER, R.J., SAKSA, F.I. & CARR, R.S. (1985). Chronic effects of drilling fluids discharged to the marine environment, with emphasis on bioaccumulation/biomagnification potential of drilling fluid metals. API Publication No. 4397, American Petroleum Institute, Washington, D.C., 151 pp.

NG, A. & PATTERSON, C.C. (1982). Changes of lead and barium with time in California off-shore basin sediments. *Geochim. Cosmochim. Acta*, **46**, 2307–21.

NIE, N.H., HULL, C.H., JENKINS, J.G., STEINBRENNER, K. & BENT, D.H. (1975). *SPSS: Statistical Package for the Social Sciences*. McGraw-Hill, New York, 675 pp.

PETRAZZUOLO, G. (1983). Draft final technical report document: Environmental assessment: drilling fluids and cuttings released onto the OCS. Submitted to Office of Water Enforcement and Permits, US EPA, Washington, D.C. by Technical Resources, Inc., Bethesda, MD, 184 pp. plus tables and figures.

PHELPS, D.K., TELEK, G. & LAPON, R.L. (1975). Assessment of heavy metal distribution within the food web. In *Marine Pollution and Waste Disposal*, ed. E.A. Pearsona & E.D. Frangipane. Pergamon Press, New York, pp. 341–8.

PLUMB, R.H. Jr (1981). Procedure for handling and chemical analysis of sediment and water samples. Technical Report EPA/CE-81-1, Prepared by Great Lakes Laboratory, State University College at Buffalo, Buffalo, NY, for the US Environmental Protection Agency/Corps of Engineers Technical Committee on Criteria for Dredged and Fill Material. Published by the US Army Engineer Waterways Experiment Station, CE, Vicksburg, Mississippi.

SCHROEDER, H.A., TIPTON, I.H. & NASON, A.P. (1972). Trace metals in man: strontium and barium. *J. Chronic Dis.*, **25**, 491–517.

ZAR, J.H. (1974). *Biostatistical Analysis.* Prentice Hall, Englewood Cliffs, N.J., 620 pp.

23

Bioaccumulation of Trace Metals from Drilling Mud Barite by Benthic Marine Animals

JERRY M. NEFF, ROBERT E. HILLMAN and J. J. WAUGH

Battelle Ocean Sciences, 397 Washington Street, Duxbury, Massachusetts 02332, USA

ABSTRACT

Discharge to the ocean of water-based drilling muds containing impure barite is suspected of contaminating benthic marine animals with metals. To evaluate this concern, we exposed four species of demersal or benthic marine animals in flow-through aquaria to natural marine sediments containing approximately 100 000 mg/kg of either a relatively pure grade of barite or an impure barite. The pure barite contained less than 1 mg/kg of cadmium and mercury; the impure barite contained 11 and 15 mg/kg of cadmium and mercury, respectively. After 1, 3 and 13 weeks of exposure, tissues were analyzed for arsenic, cadmium, copper, lead and mercury. Because of the natural variability over time in the concentrations of metals in the tissues of control animals, comparisons between metals concentrations in tissues of control and barite-exposed animals at different exposure times were used most frequently as an indication of bioaccumulation of metals from the different barite-contaminated sediments. Based on this analysis, winter flounder, Pseudopleuronectes americanus, *failed to accumulate any metals during exposure. There was some indication that soft shell clams,* Mya arenaria, *accumulated small amounts of mercury, cadmium and lead from the impure barite. Concentrations of mercury, cadmium, copper and lead increased slightly in tissues of sand worms,* Nereis virens, *exposed to both forms of barite. Concentrations of lead were significantly higher in the tissues of grass shrimp,* Palaemonetes pugio, *at 1 and 4 weeks compared to the corresponding controls. Our results support the hypothesis that much of the metals apparently accumulated from barite-contaminated sediment in tissues of marine animals was actually in the gut or gills as unassimilated barite particles. We conclude that metals associated with drilling mud barite are virtually nonbiovailable to marine organisms that might come in contact with discharged drilling fluid solids.*

1. INTRODUCTION

Water-based drilling fluids are mixtures of natural clays and/or polymers, weighting agents, and other materials suspended in fresh or salt water. Barite (barium sulfate) is the major weighting agent in most water-based drilling fluids used to drill wells in US and Canadian coastal and outer continental shelf waters (National Academy of Sciences, 1983). Barite is a soft, fine-grained mineral with a high density ($4 \cdot 50 \, g/cm^3$) and a very low aqueous solubility. Burton *et al.* (1968) estimated the solubility of barite in fresh water and seawater at $1360 \, \mu g$ Ba/litre and $50 \, \mu g$ Ba/litre respectively. Its low solubility and high density make barite an ideal weighting agent in drilling fluids. Barite is added to drilling fluids as a fine powder in amounts ranging from about 1 wt % up to 60–80 wt %, the amount added usually increasing with the depth of the drilled hole.

Natural deposits of barite ore are rarely pure. Several metals may be associated with the impurities in barite. These include iron, lead, zinc, mercury, arsenic, chromium, cadmium, nickel, and copper (Kramer *et al.*, 1980; Trefry *et al.*, 1986). Many of these metals are associated with sulfide mineral inclusions and have very low aqueous solubilities.

Proposed new source performance standards (NSPS) for the Offshore Subcategory of the Oil and Gas Extraction Point Source Category (US EPA, 1985) set an upper limit of 1 mg/kg dry weight for mercury and cadmium in whole drilling fluid destined for offshore disposal. EPA feels that, by controlling these metals, they are controlling all the metals of concern in drilling mud. Because the major source of cadmium and mercury in drilling fluids is barite, the proposed NSPS, in effect, limit concentrations of these metals in drilling mud barite to these levels or lower. Although many commercial barite deposits seem to meet these criteria, few data are available on mean and range of variation of metal concentrations in barite from different sources or even from a single source over time. The objective of this investigation was to determine the availability to marine animals of several metals associated with barite. These metals are arsenic, cadmium, copper, lead and mercury. Use of impure barite would represent a potential hazard to the marine environment only if these barite-associated metals are mobile and bioavailable following discharge of drilling mud to the marine environment.

2. MATERIALS AND METHODS

2.1. Test Materials
The barite samples used in the exposures were applied by the Offshore Operators Committee. They consisted of two samples of low trace metal barite (LTMB) from Battle Mountain, Nevada, a sample of high trace metal barite (HTMB) from Peru, and two HTMB samples from Rocky Mountain, Utah. The two LTMB samples were blended in a 50:50 proportion in a Teflon-coated cement mixer for 30 min such that the final mixture contained approximately 0·12 mg/kg mercury and 0·03 mg/kg cadmium. Equal portions of each of the two Rocky Mountain barite samples and the Peruvian sample were blended so that the final HTMB mixture contained approximately 10 mg/kg of both mercury and cadmium. Concentrations of arsenic, copper and lead were several-fold higher in the HTMB than in the LTMB mixtures. Sufficient LTMB or HTMB was added to and mixed thoroughly with clean natural marine sediment from Duxbury Bay, Massachusetts, to produce a nominal increase in the concentration of barium in the sediment of about 100 000 mg/kg, more than twice the greatest increment observed in the field (Petrazzuolo, 1983). A sample of the clean sediment and each sediment/barite mixture was removed for chemical analysis prior to the exposures. The sediment/barite mixtures were prepared by adding 59·4 kg barite to 246 kg clean natural sediment and mixing them in a Teflon-lined cement mixer for several hours.

2.2 Test Animals
Four species of benthic/demersal marine animals, representing different trophic groups, were used in the study. Juvenile winter flounder, *Pseudopleuronectes americanus*, were collected by otter trawl from Duxbury Bay, Massachusetts. Grass shrimp, *Palaemonetes pugio*, were collected in Centerville, Massachusetts. Sand worms, *Nereis (Neanthes) virens*, were purchased from Maine Bait Company. Soft-shell clams, *Mya arenaria*, were purchased from a local fish market.

All animals were held in large Fiberglas-lined tanks supplied with running natural seawater (flow rate > 10 litres/min) and fed in excess from procurement until the start of the exposure. The flounder were fed sand worms, and the worms and shrimp were fed Tetramin fish food daily during the week and once on weekends during both the pre-exposure acclimation period and the exposures. The clams were supplied continuously with phytoplankton in the unfiltered seawater. Samples of the food were taken periodically for analysis for the metals of interest in the study.

2.3. Exposure System

The test animals were exposed to a reference sediment of clean, fine sand collected from Duxbury Bay, Massachusetts, or to reference sediment containing added clean (LTMB) or impure (HTMB) barite, for up to 13 weeks (91 days). The three large (122 × 244 × 30·4 cm) Fiberglass exposure tanks were supplied continuously with Duxbury Bay seawater flowing at a rate of approximately 5–10 litres/min. Periodic analysis of the local seawater has revealed that metals concentrations are in the range expected for uncontaminated estuarine waters.

Each tank had a layer of approximately 4–6 cm of one of the three experimental sediments. Each tank was subdivided into four compartments, the first compartment being approximately 122 × 91·4 cm. The remaining space was subdivided into three equal-sized compartments. The dividers were perforated (3 mm diameter holes) polypropylene sheets, about 3–4 mm thick. The sediments in the tanks were equilibrated in flowing seawater for 3 days before any animals were added.

After 3 days, 35 juvenile winter flounder were added to the larger compartment in each tank. In addition, 45 sand worms, 45 clams approximately 2–3 in in length, and 120 grass shrimp were added to each of the three remaining compartments in each tank. Exposures were begun on August 6, 1985. At that time, 10 winter flounder, 15 clams, 15 worms and 30 shrimp were removed from the stock tanks and frozen for later chemical analysis as zero-time controls.

After 1, 4 and 13 weeks, 10 winter flounder were removed from each of the flounder compartments. At the same times, the sediment in one compartment of each tank were sieved to remove all clams, worms and shrimp. The animals were placed in a tank with running seawater but no sediment for 24 h to allow them to purge their digestive tracts of sediments. Gastric evacuation time of flounder is substantially less than 24 h at the temperatures used in these experiments (Huebner & Langton, 1982). Food turnover rate is nereid polychaetes is of the order of 1·7–8·9 times body weight per day, depending on temperature (Cammen, 1980). Gut evacuation rates of clams and shrimp were expected to fall in the same range as those of the other two species used here. Thus, a 24 h depuration period was considered adequate to allow most of the ingested sediment to be purged from the gut. All specimens from each group and an HTMB-sediment and LTMB-sediment sample, collected at each sampling time, were labeled and frozen for later chemical analysis. No shrimp remained in any tank at the time of the 13-week sample. Grass shrimp tend to be cannibalistic, especially upon individuals that have just molted. Therefore, disappearance of the shrimp was probably due to cannibalism and not mortality from some other cause.

Throughout the exposure period, water quality parameters (salinity, temperature and dissolved oxygen) were recorded at least weekly using a Hydrolab water quality monitoring system. Water pH was measured weekly using a Ag/Cl electrode and pH meter. Observations of mortality or abnormal behaviour were made daily. Dead or moribund animals were immediately removed from the tanks.

2.4. Chemical Analyses

2.4.1. Preparation of Barite and Barite/Sediment Samples

Three replicates of each barite or sediment/barite sample were analyzed. All samples were prepared using a concentrated nitric acid/hydrochloric acid digestion (Trefry *et al.*, 1986). An initial sample weight of approximately 0·05 g was used for barite samples because of the elevated metal concentrations present. Approximately 0·2 g of barite/sediment mixtures was used. The weighed samples were placed in Teflon bombs with 3–4 ml of concentrated nitric acid in 1–2 ml of concentrated hydrochloric acid and autoclaved for 12 h. The digested samples were then brought to a final volume of 15 ml by weight in acid. This digestion technique was intended to extract the 'total' metals from the barite and barite/sediment mixtures. However, barite has a low solubility in concentrated acid and a small amount of residue, probably undissolved barite, remained after digestion of the barite samples.

2.4.2. Tissue Preparation

For each exposure group and sampling time, three replicate samples of each species, each representing a composite of several animals, were analyzed. Tissue samples were thawed and digested with ultrapure concentrated nitric acid. The whole soft parts of clams and worms were digested initially by mixing with concentrated nitric acid equivalent to 0·25 times the tissue sample wet weight. The flounder muscle tissue was digested initially by mixing with concentrated nitric acid equivalent to 0·5 times the wet weight of the sample. A 4·0-ml aliquot of each tissue/acid slurry was then added to a Teflon bomb and autoclaved for 12 h. Each shrimp sample was small enough to be digested *in toto*. Each shrimp sample was placed in a Teflon digestion bomb with concentrated nitric acid equivalent to 0·25 times the tissue sample wet weight and autoclaved for 12 h. After digestion, the worm, clam and flounder samples were brought to a final volume of 10 ml and the shrimp samples were brought to a final volume of 20 ml with concentrated nitric acid.

Week-13 worm and flounder samples were freeze-dried. The freeze-dried tissues were powdered and approximately 1 g of powdered sample was added to a Teflon bomb with 10 ml of concentrated nitric acid. The mixture was autoclaved for 12 h and then brought to a volume of 100 ml with concentrated nitric acid. This method of tissue preparation was easier than the previous method and was not expected to significantly affect the recovery of analytes.

2.4.3. Analysis

All samples were analyzed for mercury using an LDC mercury analyzer; and for cadmium, copper, lead and arsenic using a Perkin-Elmer Model 703 Graphite Furnace Atomic Absorption Spectrophotometer. All analyses by Zeeman AA were done according to specifications recommended by the instrument manufacturer, with modifications as necessary to optimize quantification of metals in the different sample matrices (Neff *et al.*, 1986).

Samples containing low concentrations of arsenic as revealed by atomic absorption were reanalyzed by a hydride generation technique (Trefry *et al.*, 1986).

Sediment samples were analyzed for barium by instrumental neutron activation analysis at North Carolina State University, Nuclear Services Laboratory.

2.5. Statistical Analysis

A multivariate analysis approach was followed to determine if the concentration of the metals covaried in the animal tissues. Initially, pairwise correlations between metals concentrations in each species were examined. A principal components analysis was then performed to investigate these correlations.

Next, a two-way multivariate analysis of variance (MANOVA) was carried out. In this study, the model used was designed to test for significant effects of individual variables (i.e. WEEK and LEVEL) and for a significant interaction effect (i.e. WEEK*LEVEL). A significant WEEK effect would suggest that trace metal concentrations in the animals were changing over time. A significant LEVEL effect would suggest that trace metal concentrations in the animals were affected by the barite treatment

TABLE 1

Mean trace metals concentrations in samples of low trace metal barite (LTMB), and high trace metal barite (HTMB), and control sediments used in experiments reported here. All values are given in $\mu g/g$ dry weight (ppm). Selected values from analyses by Trefry *et al.* (1986) are included

Metal	Sample		
	LTMB	*HTMB*	*Control*
Barium	536 000[a]	567 000	174·3
Mercury	<2·72 (0·12)[a]	15·23	0·02
Cadmium	<0·081 (0·03)[a]	11·17	0·02
Copper	9·110	55·95	1·25
Lead	6·840	664·0	3·96
Arsenic	<3·075 (2·2)[a]	97·772	0·77

[a] Trefry *et al.* (1986).

to which the animals were subjected. A significant WEEK*LEVEL interaction effect would suggest that concentrations were influenced by the combination of week and barite treatment exposure. Thirdly, an analysis of variance (ANOVA) was carried out on all data sets.

3. RESULTS AND DISCUSSION

Concentrations of mercury, cadmium, copper, lead and arsenic were much lower in LTMB than in HTMB (Table 1). Concentrations of all metals analyzed were lower in control sediments than in the LTMB.

Mean concentrations of metals measured in the three treatment sediments at different times during the 13-week exposure period are summarized in Table 2. At the start of the exposure, the control sediment contained 174 $\mu g/g$ (ppm) barium, compared to mean barium concentrations of 137 667 and 139 667 ppm in the LTMB- and HTMB-sediment mixtures, respectively. There was little drop in the concentration of barium in the HTMB-sediment mixture during the 13-week exposure period. However, the mean barium concentration in the LTMB-sediment mixture dropped to just over one-third of the zero time value in the same time period. The difference in the behavior of barite in the two systems is unexplained. The decrease in the concentration of barium in the LTMB-sediment

TABLE 2

Mean (plus standard deviations) concentrations of metals in the two drilling mud/sediment mixtures at different times during the exposure period. All values are $\mu g/g$ dry weight

Metal	Week	LTMB/sediment	HTMB/sediment
Barium	0	$137\,667 \pm 13\,500$	$139\,667 \pm\ 3\,510$
	13	$49\,667 \pm\ 7\,371$	$109\,333 \pm 18\,930$
Mercury	0	0.09 ± 0.03	$4.48 \pm\ 0.25$
	1	0.04 ± 0.01	$0.90 \pm\ 0.52$
	4	0.05 ± 0.00	$1.07 \pm\ 0.21$
	13	0.03 ± 0.00	$2.48 \pm\ 1.30$
Cadmium	0	0.02 ± 0.00	$2.82 \pm\ 0.14$
	1	0.04 ± 0.01	$0.86 \pm\ 0.56$
	4	0.05 ± 0.01	$0.94 \pm\ 0.07$
	13	0.03 ± 0.00	$1.79 \pm\ 0.57$
Copper	0	3.84 ± 1.00	$19.0 \pm\ 2.75$
	1	3.98 ± 0.43	$4.87 \pm\ 2.75$
	4	5.89 ± 1.06	$4.29 \pm\ 0.35$
	13	2.27 ± 0.48	$8.82 \pm\ 4.29$
Lead	0	2.98 ± 0.16	105.7 ± 12.1
	1	3.78 ± 0.25	76.0 ± 68.7
	4	4.78 ± 0.41	$5.79 \pm\ 0.35$
	13	3.42 ± 0.19	140.6 ± 25.5
Arsenic	0	1.01 ± 0.16	$24.5 \pm\ 6.56$
	1	0.83 ± 0.08	$8.63 \pm\ 3.93$
	4	—	$7.27 \pm\ 1.86$
	13	0.61 ± 0.08	$12.61 \pm\ 6.04$

mixture probably was due to washout or redistribution of the barite in the mixture.

Mercury, copper and arsenic concentrations dropped slightly and cadmium and lead concentrations increased slightly in the LTMB-sediment mixture during the 13-week exposure period. However, in the HTMB-sediment mixture, all five metals analyzed showed some evidence of a lower concentration at some times, usually at weeks 1 and/or 4, during the 13-week exposure period. None of the decreases between week 0 and week 13 were greater than about 50% . These temporal differences in metals concentrations were statistically significant. The changes may reflect inhomogeneity in the distribution of barite in the sediment. At all sampling times, the mean concentration of any metal was higher in the HTMB-sediment than in the LTMB-sediment with one exception. The concentration of copper was slightly higher in the LTMB-sediment mixture than in the HTMB-sediment mixture at 4 weeks. Therefore, experimental animals were being exposed to higher concentrations of the five metals in the HTMB-sediment mixture than in the LTMB-sediment mixture.

Tables 3 and 4 summarize mean concentrations of metals in the tissues of test animals from the different treatments at different times during exposure. Because of the natural variation over time in the concentrations of metals in tissues of control animals, most assessments of metals uptake were based on comparisons at any sampling time between concentrations in control animals and animals exposed to the two barite-sediment mixtures.

There was not a statistically significant increase in the concentration of mercury, cadmium, copper, lead or arsenic in the muscle tissue of winter flounder during 13 weeks of exposure to LTMB- and HTMB-sediment mixtures (Table 3). In addition, there was not a statistically significant difference in the concentration of any metal in muscle tissue of flounder from the three treatments at any sampling time. The flounder did not accumulate metals from the barite-sediment mixtures.

There was a high degree of variability among replicates at different sampling times and in the three treatment groups in the concentrations of the five metals in the soft tissues of soft shell clams (Table 3). This variability made it difficult to detect trends in tissue metal concentrations over time in any treatment group. Clams exposed to the HTMB-sediment mixture accumulated a small amount of cadmium and a larger amount of lead in comparison to clams in other treatment groups at 1 and 13 weeks. Cadmium and lead concentrations in soft tissues of clams from the LTMB-sediment treatment were not significantly higher than those in control clams at any time. Mercury concentrations showed similar, though less marked, trends to those described for cadmium and lead. Mercury concentrations in clams exposed to HTMB-sediment were significantly higher than those in control clams at all sampling times. Mercury concentrations in clams exposed to LTMB-sediment were not significantly different from those in controls at any sampling time. Mean concentrations of copper and arsenic did not vary greatly within or between treatment groups over time. Although there was a trend for some metals concentrations to increase with time in clams from the HTMB treatment, this trend was inconsistent and not statistically significant.

TABLE 3

Mean concentrations (µg/g) of five metals in tissues of winter flounder (*Pseudopleuronectes americanus*) and soft shell clams (*Mya arenaria*) at different times during exposure to three sediment mixtures. $n = 3$ unless indicated otherwise

Week	Mercury			Cadmium			Copper			Lead			Arsenic		
	Control	LTMB	HTMB	Control	LTMB	HTMB	Control	LTMB	HTMB	Control	LTMB	HTMB	Control	LTMB	HTMB
Flounder															
0	0·28	—	—	0·09	—	—	0·70	—	—	0·53	—	—	0·68	—	—
1	0·31	0·24	0·26	0·09	0·09	0·09	0·53	0·53	0·92	0·53	0·53	0·70	0·77	0·61	0·44
4	0·18	0·22	0·20	0·09	0·09	0·09	0·53	0·53	1·49	0·53	0·53	0·52	—	0·42[a]	—
13	0·24	0·23	0·21	0·05	0·05	0·05	1·03	0·73	0·65	0·29	0·27	0·29	—	0·21[a]	0·27
Clam															
0	0·46	—	—	0·18	—	—	17·08	—	—	4·24	—	—	17·44	—	—
1	0·39	0·62	0·98	0·27	0·31	0·64	16·88	22·36	17·37	5·35	6·74	43·08	12·35	17·30	19·77
4	0·68	0·46	0·92	0·62	0·60	0·50	26·73	29·77	28·11	37·29	7·93[a]	9·74[a]	21·18	23·25	20·68
13	0·17	0·30	0·69	0·34	0·31	0·75	17·96	17·32	24·10	4·57	5·57	27·06	12·12	13·40	24·31

LTMB, low trace metal barite. HTMB, high trace metal barite.
[a] Two replicates.

Mercury and lead concentrations in clams exposed to the HTMB-sediment actually decreased during the 13-week exposure.

Variability in metal concentrations among replicates was much less marked in sand worms than in soft shell clams (Table 4). Concentrations of arsenic in worm tissues remained relatively constant among treatments and over time. The same was true for mercury in control worms and worms exposed to the LTMB-sediment mixture. At week 1, but not at later sampling times, worms exposed to the HTMB-sediment mixture contained significantly higher mercury concentrations than the zero-time controls or other treatment groups at week 1. Cadmium, copper and lead were present in tissues of worms exposed to LTMB- and HTMB-sediment mixtures at concentrations significantly higher than those in zero-time controls or in worms exposed to control sediments at some sampling times. The differences in values for exposed and control or zero-time control animals were not large. The largest difference (about 20-fold compared to controls) was for lead in worms exposed for one week to HTMB-sediment.

There was a high degree of variability in metal concentrations among replicates in grass shrimp, particularly for copper and lead at some sampling times (Table 4). Arsenic concentrations are not available for control shrimp and shrimp exposed to the LTMB-sediment mixture at weeks 1 and 4. Arsenic concentrations in shrimp exposed to the LTMB-sediment were similar to those in zero-time controls. Mean mercury concentrations in shrimp from both LTMB- and HTMB-sediment treatments were not statistically significantly higher than in zero-time control shrimp or in shrimp exposed to control sediment for 1 or 4 weeks. Lead concentrations were significantly higher than controls in shrimp exposed to the HTMB-sediment mixture at both 1 and 4 weeks.

As a general rule, one or more metals, not necessarily the same metals for all species, was present at significantly higher concentrations in the tissues of animals, except flounder, exposed to the HTMB-sediment at one or more sampling times than in the corresponding control animals. However, the large variability observed among replicate samples (Table 5) within a sampling period may have obscured some real trends. Results for the quality control samples (Table 6) indicate that this variability was not introduced by the analytical methods.

Wide variability in body burdens of metals in natural populations of benthic infaunal animals particularly bivalve molluscs, has been reported both among replicates collected at one time from a site (Wright *et al.*, 1985) and within a population over time (Luoma *et al.*, 1985). This

TABLE 4

Mean concentrations (µg/g) of five metals in tissues of sand worms (*Neanthes virens*) and grass shrimp (*Palaemonetes pugio*) at different times during exposure to three sediment mixtures. $n = 3$ unless indicated otherwise

Week	Mercury			Cadmium			Copper			Lead			Arsenic		
	Control	LTMB	HTMB	Control	LTMB	HTMB	Control	LTMB	HTMB	Control	LTMB	HTMB	Control	LTMB	HTMB
Worms															
0	0·17	—	—	0·15	—	—	6·45	—	—	1·34	—	—	11·31	—	—
1	0·14	0·16	0·52	0·30	0·30	0·70[a]	5·74	10·54	7·31	1·45	2·24	22·31	10·82	11·15	11·73
4	0·22	0·20	0·24[a]	0·29	0·37	0·49[a]	8·45	12·54	13·43[a]	1·38	3·40	4·95[a]	12·02	12·21	13·50[a]
13	0·10	0·10	0·15	0·22	0·31	0·36	19·65	19·63	18·13	1·42	2·39	6·68	8·97	8·10	9·52
Shrimp															
0	0·15	—	—	0·47	—	—	178·8	—	—	2·86	—	—	0·82	—	—
1	0·16	0·20[a]	0·23[a]	0·07	0·24[a]	0·15[a]	77·33	88·06[a]	52·7[a]	0·81	1·07[a]	8·55[a]	—	1·42[a]	—
4	0·15	0·35	0·27	0·13	0·25	0·20	184·9	173·8	118·6	2·31	4·37	5·89	—	1·41	—

LTMB, low trace metal barite/sediment. HTMB, high trace metal barite/sediment.
[a] Two replicates.

TABLE 5

Average background concentrations (μg/g) and associated precision statistics of trace metals in tissues of experimental animals prior to exposure

Species		Hg	Cd	Cu	Pb	As
Shrimp	$\bar{x}$	0·15	0·47	178·78	2·86	0·82
	s	0·057	0·193	81·078	2·075	0·131
	cv	38·0	41·1	45·4	72·6	16·0
Worm	$\bar{x}$	0·17	0·15	6·45	1·34	11·31
	s	0·068	0·06	0·921	0·202	2·009
	cv	40·0	40·0	11·0	15·1	17·8
Clam	$\bar{x}$	0·46	0·18	17·08	4·24	17·44
	s	0·243	0·0	7·102	0·371	4·544
	cv	52·8	0·0	34·3	8·8	26·1
Flounder	$\bar{x}$	0·28	0·09	0·70	0·53	0·68
	s	0·036	0·0	0·297	0·006	0·409
	cv	12·9	0·0	42·4	1·1	60·1

Note: $\bar{x}$ is the average concentration (μg/g) with $n = 3$.
  s is the standard deviation (μg/g).
  cv is the coefficient of variation (percent) where $cv = (s/x) \times 100$.

variability could be due to a wide variety of environmental and biological factors. Outlier values seem to occur most often on the high side of the mean, suggesting that variable retention of sediment materials (usually higher than tissues in metals concentrations) in the gut and gills may account for some of this variability (Wright *et al.*, 1985). As a consequence of this variability, very large numbers of replicates may be needed to detect all but massive changes in body burdens of metals.

For the most part, the mean zero-time control concentrations of metals in the tissues of the experimental animals (Table 4) were similar (within an order of magnitude) to mean values for marine animals from uncontaminated environments (Spaargaren, 1985).

For each species, concentrations of five metals were measured in each replicate sample composed of either 10 flounder, 15 clams, 15 worms or 30 shrimp. Because these metals may behave similarly, either within the barite or within the animals, a multivariate statistical analysis approach was followed. Initially, the pairwise correlations between the metals for each species were examined (Table 7). For soft shell clams, 8 out of a possible 10 correlations were significant ($p < 0.05$). A significant correlation means that the correlation coefficient is significantly different from zero. This suggests that these metals covary in the animals under these exposure

TABLE 6
Precision statistics from analysis of the quality control samples

Species	Barite level	Week	Replicate		Hg	Cd	Cu	Pb	As
Flounder	CNTL	13	2,3	s	0·017	0·006	0·148	0·022	0·322
				cv	7·3	12·7	18·4	7·9	12·8
				n	4	4	4	4	4
Clam	CNTL	13	1	s	0·014	0·021	1·65	0·028	1·032
				cv	7·8	5·1	7·6	0·5	7·2
				n	2	2	2	2·	2
Clam	LTMB	4	2	s	0·023	0·021	1·255	1·42	0·574
				cv	8·3	5·8	4·0	19·4	2·9
				n	3	3	3	3	3
Clam	HTMB	4	1	s	0·028	0·007	2·95	1·005	0·749
				cv	1·9	2·7	9·8	12·2	5·1
				n	2	2	2	2	2
Worm	CNTL	13	2,3	s	0·017	0·028	2·10	0·44	2·09
				cv	21·9	11·8	12·0	28·3	24·0
				n	4	4	3	4	4

CNTL, control; clean sediment.
LTMB, low trace metal barite/sediment mixture.
HTMB, High trace metal barite/sediment mixture.

s is the standard deviation ($\mu g/g$).
cv is the coefficient of variation (percent).
n is the number of samples.

TABLE 7
Pairwise correlations between metals for each species

Species	Hg	Cd	Cu	Pb	As
Clam					
Hg	1·00				
Cd	0·44[a]	1·00			
Cu	0·16	0·39[a]	1·00		
Pb	0·68[a]	0·88[a]	0·14	1·00	
As	0·43[a]	0·88[a]	0·64[a]	0·73[a]	1·00
Flounder					
Hg	1·00				
Cd	0·08	1·00			
Cu	−0·07	−0·08	1·00		
Pb	0·13	0·85[a]	0·13	1·00	
As	−0·20	−0·29	0·33	−0·28	1·00
Shrimp					
Hg	1·00				
Cd	0·58[a]	1·00			
Cu	0·21	0·58[a]	1·00		
Pb	0·38	0·53[a]	−0·08	1·00	
As	−0·17	0·02	0·36	0·11	1·00
Worm					
Hg	1·00				
Cd	0·79[a]	1·00			
Cu	−0·26	0·01	1·00		
Pb	0·91[a]	0·82[a]	−0·10	1·00	
As	0·47[a]	0·51[a]	−0·22	0·33	1·00

[a] Denotes that the coefficient is statistically different from zero. Thus the metals react similarly in the species.

conditions. For sand worms, 5 out of a possible 10 correlations were significant ($p < 0.05$). For sand shrimp, 3 out of a possible 10 correlations were significant ($p < 0.05$), and for winter flounder, 1 out of a possible 10 correlations was significant ($p < 0.05$). For clams and worms, but not for flounder and shrimp, the metals that covaried were also the metals that were bioaccumulated at some sampling times. These results suggest that the five trace metals behave similarly to each other in clams and, to some extent, in worms, but do not covary in shrimp and flounder. However, copper (the only one of the metals studied that is an essential trace nutrient) is dissimilar to the other metals in its behavior in the animals.

These results suggest that the clams and, to a lesser extent, the worms were accumulating some of the metals in the form of metal-containing barite particles from the barite-sediment mixtures. The barite particles probably were present in the gut or gills (clams) in an unassimilated form. Unfortunately, barium was not analyzed in the animal tissues, so this hypothesis cannot be verified.

A principal components analysis was performed to investigate these correlations. Principal components is a multivariate technique for identifying associations or patterns between variables that tend to react similarly in their responses to the design factors (i.e. barite treatment or week effect). In effect, all response variables are grouped into combinations of variables (the total number of which is less than the original number of responses), which are then used to better interpret the results. For instance, the behavior of the five trace metals can potentially be explained by modeling two interrelated groups of metals. In this experiment, for each species, the first two principal components accounted for over 80% of the variability in the data. In three of the four species (i.e. clams, flounder and shrimp), mercury, cadmium and lead acted as a group, and copper and arsenic acted as a separate group. In worms, one of the first two principal components showed mercury, cadmium, lead and arsenic acting as a group, with copper responding differently.

Values for water quality parameters, temperature, dissolved oxygen concentration, salinity and pH, were quite uniform in all treatments during the 13-week exposure period. Temperature dropped from a high of 22–23°C in mid-August to 10·4°C in November. Dissolved oxygen concentration ranged between 3·6 and 8·6 mg/litre and tended to increase with decreasing ambient temperature, as expected. Salinity remained relatively constant between 30·5 and 35·5°/oo during the experiment. The pH varied only slightly between 7·72 and 8·12 during the 13 weeks of the exposure. All these values are within the optimal tolerance range of the test species, and so the test animals probably were not stressed by any adverse water quality parameter during the experiment.

In a parallel study to this one, Trefry *et al.* (1986) investigated the forms and reactivity (leachability) of trace metals associated with the LTMB and HTMB samples. Copper and, to a lesser extent, cadmium were the metals most readily leached from the barite samples by mild acid solutions. Mercury was the metal most tightly bound to the barite samples. Following equilibration of barite samples with seawater for 48 h, only cadmium and arsenic concentrations in seawater equilibrated with HTMB increased to above normal ambient levels. Lead behaved differently from the other metals. Larger amounts of lead were released at higher rather than at lower pH, and under reducing not under oxidizing conditions. Most of the metals leached from barite were released into the water very quickly (within 24–48 h) after mixing barite with seawater.

There was no clear relationship between the amount of metal leached from different barite samples in the experiments performed by Trefry *et al.* (1986) and the apparent bioavailability of the metals, as indicated by the bioaccumulation studies described here. This is probably because test animals were not exposed to the test barite/sediment mixtures until the mixtures had equilibrated with flowing seawater for three days. This equilibration time was sufficient, according to the results of Trefry *et al.* (1986), to allow all or most of the readily leached metals to be removed. In the natural environment, a similar phenomenon will occur. As the drilling mud solids settle through the water column, the easily leached metals will leave the particles and remain in the water column. The barite settling on the bottom will have equilibrated with seawater and will no longer contain the rapidly leached portion of metal contaminants. The metals leached into the water column will be diluted rapidly to background concentrations.

Although none of the metals showed consistent and statistically significant accumulation by all four species of test animals in the present study, cadmium and lead were the two metals that showed some indication of bioaccumulation at some times in some species. Arsenic and mercury appeared to be the least bioavailable of the metals. There was a very slight indication of some accumulation of copper at some times by one or more species. These trends in bioaccumulation do not seem to relate to the physical behavior of the metals in barite as reported by Trefry *et al.* (1986). It is probable that much of the apparent accumulation of metals was due to retention of particulate materials (sediment or barite) in the gut or gills of the test animals. The winter flounder, which is the least likely of the test species to ingest sediments, showed no evidence of bioaccumulation of any metals.

We can conclude that, even during chronic exposure to very high concentrations of heavily contaminated barite, bottom-living marine animals show only a slight tendency to accumulate heavy metals from barite. Under more realistic exposure conditions likely to occur in sediments near exploration or development platforms (in which sediment barite concentrations would be one-tenth or less of the concentrations used here), it is extremely unlikely that demersal or benthic animals will accumulate any of the trace metal impurities associated with barite. This will be true irrespective of whether the barite contains greater or lesser amounts of trace metals.

ACKNOWLEDGEMENTS

This research was supported by a contract from the Offshore Operators Committee. We would like to thank Dr Bruce Cox, Marathon Oil Company, and Dr Robert Ayers, Exxon Production Research Company, for their contributions to the design of this project and for their critical reviews of the final report.

REFERENCES

BURTON, J.D., MARSHALL, N.J. & PHILLIPS, A.J. (1968). Solubility of barium sulfate in sea water. *Nature*, **217**, 834–5.

CAMMEN, L.M. (1980). A method for measuring ingestion rate of deposit feeders and its use with the polychaete *Nereis succinea*. *Estuaries*, **3**, 55–60.

HUEBNER, J.D. & LANGTON, R.W. 1982. Rate of gastric evacuation for winter flounder, *Pseudopleuronectes americanus. Can. J. Fish. Aquat. Sci.*, **39**, 356–60.

KRAMER, J.R., GRUNDY, H.D. & HAMMER, L.G. (1980). Occurrence and solubility of trace metals in barite for ocean drilling operations. In *Symposium on Research on Environmental Fate and Effects of Drilling Fluids and Cuttings*. Courtesy Associates, Washington, D.C., pp. 789–98.

LUOMA, S.M., CAIN, D. & JOHANSSON, C. (1985). Temporal fluctuations of silver, copper, and zinc in the bivalve *Macoma balthica* at five stations in south San Francisco Bay. *Hydrobiology*, **129**, 109–20.

NATIONAL ACADEMY OF SCIENCES (1983). Drilling discharges in the marine environment. Panel on Assessment of Fates and Effects of Drilling Fluids and Cuttings in the Marine Environment. Marine Board, National Research Council. National Academy Press, Washington, D.C., 180 pp.

NEFF, J.M., HILLMAN, R.E., LECZYNSKI, B. & BERNER, T. (1986). Bioavailability of trace metals from barite to benthic marine animals. Final report to the Offshore Operators Committee, New Orleans, LA. Battelle Ocean Sciences, Duxbury, MA, 41 pp.

PETRAZZUOLO, G. (1983). Environmental assessment: Drilling fluids and cuttings released onto the OCS. Draft final technical support document submitted to Office of Water Enforcement and Permits (US EPA, Washington, D.C., 184 pp.

SPAARGAREN, D.H. (1985). Elemental composition and interrelations between element concentrations in marine finfish, molluscs and crustaceans. *Neth. J. Sea. Res.*, **19**. 30–7.

TREFRY, J.H., TROUNE, R.P., METZ, S & SISLER, M.A. (1986). Forms, reactivity and availability of trace metals in barite. Draft final report prepared for the Offshore Operators Committee, Taskforce on Environmental Science. March 1986, 50 pp.

US ENVIRONMENTAL PROTECTION AGENCY (1985). Development document for effluent limitations guidelines and standards for the offshore segment of the oil and gas extraction Point Source Category. US EPA, Industrial Technology Division, Washington, D.C. EPA 400/1-85/055, 408 pp.

WRIGHT, D.A., MIHURSKY, J.A. & PHELPS, H.L. (1985). Trace metals in Chesapeake Bay oysters: intra-sample variability and its implications for biomonitoring. *Mar. Environ. Res.*, **16**, 181–97.

24

Acute and Chronic Toxicity of Base Oil and Cuttings from Three Wells Drilled in the Beaufort Sea

J. OSBORNE and C. LEEDER

Environmental Protection, Environment Canada, PO Box 5037, St John's, Newfoundland, Canada A1C 5V3

ABSTRACT

This study was conducted to evaluate the acute and chronic toxicity of drilling fluids, base oils and oiled cuttings used during the drilling of exploratory wells in the southern Beaufort Sea in 1983. Testing was conducted at the Environment Canada laboratory in St John's Newfoundland, during 1985/86.

The testing included a long-term (32-day) exposure of an invertebrate species to pre-determined concentrations of oiled cuttings and sand in seawater and 96 h and 192 h exposures of fish to the 100% water-soluble fraction of the base oils in freshwater. The cuttings/sand mixtures were prepared on a volume basis and simulated exposures at increasing distance from the rig discharge. The proportion of cuttings to sand ranged from 10% cuttings/90% sand to 90% cuttings/10% sand. The mixtures were spread evenly on the bottom of the exposure tanks to a depth of 3·5 cm and were overlayed with 20 cm of continuously replaced saltwater at a flow rate of 0·5–3 litre/min. The test species (Echinarachnius parma) was distributed evenly over the sediment surface.

The cuttings/sand mixtures were toxic to the invertebrate test organisms at concentrations of 10% cuttings by volume. Some sub-lethal effects were recorded (e.g. reduced burrowing and movement).

The 100% water-soluble fraction of base oils and oil released from cuttings were prepared according to the EPS (1985) protocol (Laboratory Procedure for Determining the Acute Lethality of a Water Soluble Fraction of Mineral Oil to Rainbow Trout. Environment Canada, EPS, Atlantic Region, Dartmouth, N.S.). The validity of the protocol was also examined by varying the mixing, settling and exposure times. Screening of base oils alone using the water-soluble fraction technique was not indicative of any toxic effects of oily cuttings on invertebrates.

1. INTRODUCTION

The acute toxicity of diesel oil associated with discharged cuttings became a major concern in the North Sea in the late 1970s. A different base oil, with a lower aromatic hydrocarbon content and lower acute lethality, began to be used to replace diesel oil in drilling muds in the early 1980s. Hinds *et al.* (1983) demonstrated that when fish and marine invertebrates were exposed to water-soluble fractions (WSF) of low toxicity oils the 96 h LC_{50}s were typically greater than 10 000 ppm as compared to LC_{50}s of less than 2000 ppm for diesel oil. Studies conducted by Barchard & Doe (1984), Doe *et al.* (1984) and Hutcheson *et al.* (1984) on cuttings, base oils and drilling muds, respectively, collected from the Alma F-67 well drilled near Sable Island, indicated that discharged cuttings with their retained oil were toxic to selected marine invertebrates.

The use of oil-based drilling mud in Canadian offshore areas has been approved on a well-by-well basis, on the basis of guidelines for this drilling method issued by the Canada Oil and Gas Lands Administration (1986). Permits have occasionally included requirements for monitoring as well as specific conditions on the disposal of oiled cuttings. Some of the monitoring results are presented in other papers in these conference proceedings. The drilling conditions afforded an opportunity to collect large samples of oiled cuttings, base oils and mud for testing purposes. During the period of 1982–1984 Environment Canada subjected a number of oil-based mud formulations and components to toxicity testing (Doe *et al.*, 1984; Barchard *et al.*, 1985). The work presented in this paper continued that program and deals with cuttings from three wells drilled in the southern Beaufort Sea—Nipterk L-19A, Adgo G-24 and Minuk I-53.

The studies were designed to evaluate the acute and chronic effects of selected concentrations of cuttings in sediments on invertebrates, and the acute lethality of the WSF of base oils and oil released from cuttings on rainbow trout (*Salmo gairdneri*). The applicability of the protocol for preparation of the WSF and the acute toxicity evaluations were examined in relation to the effects recorded on invertebrates.

2. METHODS AND MATERIALS

2.1. Sample Collection
The oily cuttings were collected from three wells, that is, from storage of undischarged cuttings at the Nipterk L-19A and Adgo G-24 sites and directly from the shale shakers at the Minuk I-53 site. The cuttings were sealed in plastic containers and shipped from the Arctic to St John's, Newfoundland within 10 days, where they were stored at $4 \pm 0.2°C$ until used in testing. Conoco Vista ODC base oil containing 2% ESSO DMO-75 base oil was also collected at the Minuk I-53 site. Shell Canada provided a base oil from stock, Shell Sol DMS Mineral Oil.

2.2. Water-soluble Fraction—Acute Tests
Water-soluble fractions of the base oils and oil released from cuttings were prepared according to the EPS (1985) protocol. A diesel WSF was prepared for comparative purposes. Additional WSFs were prepared based on a modified procedure of the Environmental Protection Service (EPS, 1985) protocol. The WSFs were prepared from each sample by using different combinations of mixing time (20, 25 and 29 h), settling time (2, 4 and 6 h), and sample-to-water ratio (0.5:9.5, 1:9, 1.5:8.5 and 2:8) (Table 1). The acute toxicity of the WSF to rainbow trout was determined and the parameters resulting in the lowest LT_{50} (time to 50% mortality in fixed concentration) was used to prepare subsequent WSFs.

TABLE 1
Summary of the water soluble fraction (WSF) preparation regime

Mixing time (h)	Settling time (h)	Ratio sample:water
20	4	0.5:9.5
25	4	0.5:9.5
29	4	0.5:9.5
Time at which greatest toxic effect or oil/grease content was observed	2	0.5:9.5
	6	0.5:9.5
Time at which greatest toxic effect or oil/grease content was observed	Time at which greatest toxic effect or oil/grease content was observed	1:9
		1.5:9.5
		2:8

Toxicity tests using WSFs of base oils were conducted following a standard protocol for testing the toxicity of liquid effluents (EPS, 1982); the exposure period was 192 h instead of the prescribed 96 h. The test organism was laboratory-acclimated rainbow trout with a mean weight of 1·47 g. The test solutions were kept at $15 \pm 1°C$, and not aerated during the test. The test organisms were exposed to each of the WSFs and controls were maintained concurrently; dissolved oxygen and pH were determined at the start and finish of each test. The LT_{50} values were calculated according to the method of Litchfield (1949).

2.3. Chronic Tests

Chronic effects of cuttings were monitored by observing behaviour and mortality among laboratory-acclimated sand dollars (*Echinarachnius parma*) exposed to cuttings for 32 days. The methods used by Hutcheson *et al.* (1984) were followed.

The cuttings and sand were thoroughly mixed in the test tanks to simulate sea-floor conditions after a number of seasons when ice scour and wave action would result in mixing and redistribution of cuttings with local sediment. Each test tank had approximately 3·5 cm of the cuttings and No.1 silica sand covering the bottom (Table 2). The control tanks contained 3·5 cm of washed No.1 silica sand. Ambient seawater (average salinity 27 ppt) flowed continuously into the tanks at a replacement rate of 0·5–3 litre/min, to maintain a water depth of 20 cm over the cuttings. No aeration was supplied since replacement seawater was at the O_2 saturation level. Dissolved oxygen and pH was determined at the start and finish of each test (Table 3).

Five to ten sand dollars were evenly distributed over each of the tanks. The number of dead sand dollars was recorded every 2 days; death was defined as a lack of movement of tube feet when examined using a dissecting microscope. Dead individuals were removed. Observations of number of burrowed individuals and distance travelled were made daily. Test organisms were not fed during the test period.

2.4. Oil Analyses

The total oil and grease content of the WSF and sediment/cuttings mixtures was determined at the beginning and end of each test (Table 4). Analysis was performed by the partition gravimetric method (APHA/AWWA/WPCF, 1980) with freon extraction.

TABLE 2
Chronic assay tank and test species information

Sample	Depth	Proportion of cuttings/ sand (V/V)	Size of tank (cm)	Numbers of E. parma
Phase 1				
Nipterk L-19A				
	800 m	100% cuttings on sand	44 × 100	5
	1 900 m	100% cuttings on sand	44 × 100	5
	3 100 m	100% cuttings	44 × 100	5
Control	NA	0/100	44 × 100	5
Phase 2				
Nipterk L-19A				
	NA	10/90	44 × 100	10
	NA	25/75	44 × 100	10
	NA	50/50	44 × 100	10
	NA	75/25	36 × 56	5
Control	NA	0/100	44 × 100	10
ADGO G-24				
	500 m	25/75	75 × 77	10
		50/50	75 × 77	10
	1 100 m	25/75	75 × 77	10
		50/50	75 × 77	10
	3 087 m	25/75	75 × 77	10
		50/50	75 × 77	10
Control	NA	0/100	75 × 77	10
Minuk I-53				
	2 350 m	10/90	44 × 100	10
		25/75	44 × 100	10
		50/50	44 × 100	9
		75/25	44 × 100	10
		100% cuttings on sand	44 × 100	10
Control	NA	0/100	44 × 100	10

NA = not applicable.

J. Osborne and C. Leeder

TABLE 3
Summary of experimental conditions of chronic toxicity tests

Sample		pH	DO[a]	Initial Temp (°C)	Salinity (ppt)	pH	DO	Final Temp (°C)	Salinity (ppt)
Phase 1									
Nipterk L-19A									
	800 m	8·0	8·5	12	NA	8·1	9·0	11	27
	1900 m	8·0	8·9	12	NA	8·1	10·2	11	27
	3100 m	8·0	8·5	13	28	8·0	10·3	9	28
Control		8·0	8·7	12	28	8·0	10·0	10	27
Phase 2									
Nipterk L-19A									
	10/90	7·9	12·1	2	29	7·9	11·8	2	24
	25/75	7·9	11·9	2	29	7·9	11·9	2	24
	50/50	7·9	11·9	2	29	7·9	11·0	2	24
	75/25	7·9	12·3	0	29	7·9	12·3	2	24
Control	0/100	7·9	12·2	2	29	7·9	12·2	2	24
ADGO G-24									
500 m	25/75	7·9	12·3	2	29	7·7	11·4	3	27
500 m	50/50	7·9	12·1	2	29	7·8	11·9	3	27
1100 m	25/75	7·9	12·7	2	30	7·8	11·7	3	28
1100 m	50/50	7·9	12·0	1	30	7·8	11·9	3	27
3087 m	25/75	7·9	12·1	2	29	7·7	11·7	3	27
3087 m	50/50	7·9	12·5	2	29	7·8	11·3	3	28
Control	0/100	7·9	12·2	2	29	7·9	12·1	3	28
Minuk I-53									
	10/90	7·8	12·4	2	30	7·9	12·2	2	24
	25/75	7·8	12·5	2	30	7·9	11·5	2	24
	50/50	7·6	12·5	2	30	7·6	11·6	1	27
	75/25	7·8	12·4	2	30	7·9	12·3	2	24
100% cuttings on sand		7·8	12·5	2	30	7·9	12·5	2	24
Control		7·9	12·2	2	29	7·9	11·9	2	27

[a] Dissolved oxygen content (ppm).
NA = not available.

TABLE 4

Initial and final oil/grease contents of sediments of the chronic toxicity test tanks

Test tank	Initial oil/ grease content (g/100 g)	Final oil/ grease content (g/100 g)	Aromatics content (g/100 g)
Minuk I-53			
10/90	0·52	0·6	0·03
25/75	1·3	0·9	0·05
50/50	2·6	NA	NA
75/25	3·9	2·3	0·12
100% cuttings			
on sand	5·2	3·1	0·16
Sand beneath			
100% cuttings	0	0	0
Nipterk L-19A			
10/90	2·4	2·4	0·64
25/75	6·0	5·2	1·40
50/50	12·0	9·8	2·63
75/25	18·0	11·9	3·19
ADGO G-24 500 m			
25/75	3·1	2·6	0·005
50/50	6·1	4·5	0·008
ADGO G-24 1 100 m			
25/75	3·2	7·5	0·015
50/50	6·4	6·7	0·013
ADGO G-24 3 087 m			
25/75	2·1	2·4	0·005
50/50	4·1	4·6	0·009

NA = not available.

3. RESULTS

3.1. Water-soluble Fraction—Acute Tests

The oil and grease content of the 96 h water-soluble fraction tests are presented in Table 5. None of the WSFs was acutely lethal to rainbow trout in 96 h tests. In 192 h tests, however, the WSF of oil released from Adgo G-24 cuttings was acutely lethal to all test organisms in 32–142 h. All other combinations of mixing and settling times were not acutely lethal (Table 5). It should be noted that the pH in the Adgo G-24 series was initially 8·6–9·8. This could account for the lethal response, although it had decreased to 6·1–6·5 by the end of 96 h.

TABLE 5

Summary of results of mixing regimes, LT_{50}s, oil/grease content, and total suspended solids content (fresh water tests)

Mixing time (h)	Settling time (h)	Ratio (oil:water)	% Mortality at 192 h	LT_{50} (h)	Initial O/G (mg/litre)	TSS (JTU)
Shell Sol DMS Mineral Oil						
20	4	0·5:9·5	0	>192	0·59	0
25	4	0·5:9·5	0	>192	0·55	0·88
29	4	0·5:9·5	0	>192	2·17	0·60
29	2	0·5:9·5	20	>192	1·06	0
29	6	0·5:9·5	0	>192	0·79	0
29	4	1:9	0	>192	0·50	0·32
29	4	1·5:8·5	0	>192	0·73	0
29	4	2:8	20	>192	0·83	0·55
Conoco Vista ODC-Rich Drilling Fluid (ADGO G-24)						
20	4	0·5:9·5	100	33·9	1·42	NA
25	4	0·5:9·5	100	33·9	1·59	0·42
29	4	0·5:9·5	100	53·0	0·46	0
25	2	0·5:9·5	100	49·1	0·97	1·56
25	6	0·5:9·5	80	63·0	0·33	0·51
25	4	1:9	80	141·9	2·38	0·07
Esso DMO75-Rich Drilling Fluid (Nipterk L-19A)						
20	4	0·5:9·5	0	>192	0·77	2·19
25	4	0·5:9·5	0	>192	0·83	0
29	4	0·5:9·5	0	>192	1·11	0
29	2	0·5:9·5	0	>192	2·70	1·08
29	6	0·5:9·5	0	>192	0·52	2·22

3.2. Chronic Tests

Exposure to the various cuttings/sand mixtures of Niptek L19-A and Adgo G-24 samples caused 100% mortality of sand dollars within 27 and 21 days, respectively (Figs 1 and 3). The Minuk mixtures, however, were not acutely lethal; average mortality was 16% after 32 days, with a range of 10–30% (Fig. 2).

There was no movement or burrowing activity of sand dollars in those tests where eventual 100% mortality occurred. In the Minuk I-53 series

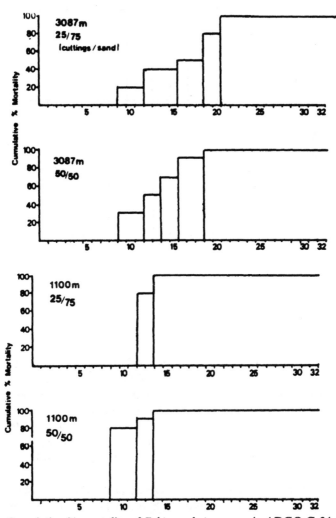

FIG. 1. Cumulative % mortality of *Echinarachnius parma* in ADGO G-24 assays.

of tests, none of the sand dollars burrowed in the 100% cuttings layered on sand and distance travelled was negligible, although only 30% mortality was recorded. The number of burrowed individuals in the Minuk I-53 cuttings/sand mixtures 10/90, 25/75, 50/50 and 75/25 was significantly less than that recorded in controls (Mann–Whitney U-test), and movement along the surface was reduced for all but the 10/90 cuttings/sand mixture. Distance travelled in the 10/90 mixture was comparable to controls.

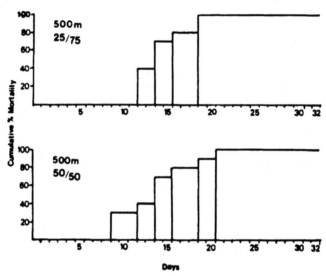

FIG. 1. —*contd.*

4. DISCUSSION

In 1986, the Canada Oil and Gas Lands Administration issued guidelines for the use of oil-base drilling mud (COGLA, 1986). These guidelines require that the base oil used be not acutely lethal to rainbow trout in 96 h LT_{50} tests of the 100% water-soluble fraction. The preparation of the water-soluble fraction for these tests was to follow the EPS (1985) protocol. This protocol was based on the work of Anderson *et al.* (1974) and required the stirring of a 1:9 oil:water ratio for 20 h, followed by 4 h of settling.

The work of Anderson *et al.* (1974) was based on crude oils and fuel oils, composed of different classes of hydrocarbons from the mineral oil used for today's 'low toxicity' oil-based muds. Diesel oil can contain approximately 20% mono- and di-aromatic hydrocarbons whereas the permitted mineral oils contain less than 5% of these compounds. Thoresen & Hinds (1983) attributed the toxicity of base oils to their mono- and di-aromatic hydrocarbon content. The aromatics are more soluble in water than the aliphatics which comprise the majority of the hydrocarbon content in mineral oils. The observed low toxicity in acute lethal tests of mineral oils may be in part due to low solubility and thus bio-availability of the aliphatics (Rice *et al.*, 1976).

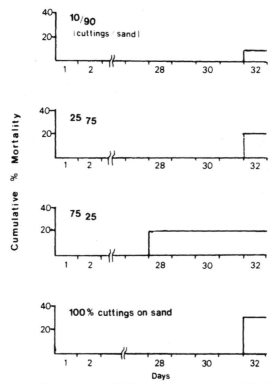

FIG. 2. Cumulative % mortality of *Echinarachnius parma* in Minuk I-53 assays.
(NB: 0% mortality in 50:50)

The oil and grease content of the WSF of mineral oil and oil released from cuttings was less than 3 ppm, and no evidence exists to correlate toxicity with the concentration of oil and grease. It would appear that the present protocol for preparation of the WSF will result in a concentration of oil in water that is sufficient to measure the acute toxicity of the base oil, but not necessarily its effects after discharge as part of the cuttings.

The oiled cuttings from Nipterk L-19A and Adgo G-24 were toxic to *E. parma* within 26 days and 50% mortality occurred within 20 days. Hutcheson *et al.* (1984) recommended a 20-day exposure regime for the species tested; this appears to be a reasonable time period to determine chronic effects.

E. parma normally feeds as it moves over and through sediment.

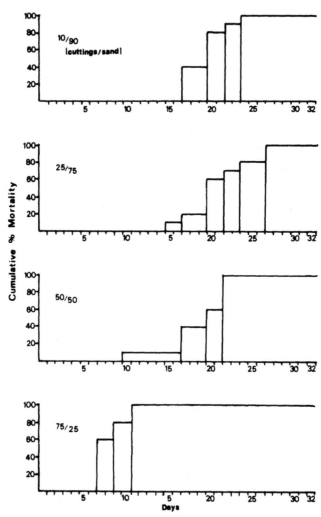

FIG. 3. Cumulative % mortality of *Echinarachnius parma* in Nipterk L-19A assays.

Reduced movement would affect food consumption and the ability of the animal to survive. It was not possible to determine if starvation was a contributing factor in those tests where 100% mortality occurred, but survival of individuals in the Minuk I-53 tests where burrowing was inhibited would seem to discount starvation. However, anoxic conditions due to biodegradation of the cuttings may have decreased oxygen concentration at the sediment:water interface even though incoming sea water was at O_2 saturation. The presence of anoxic conditions in the top few millimetres of cuttings was noted by Gillam *et al.* (1986) in work conducted in the North Sea.

The results of our tests indicated that oily cuttings/sediment mixtures with as little as 10% oily cuttings although not necessarily acutely lethal could affect survival of benthic organisms. Since the chronic effects were not observed in all test regimes, the quality of oil on the cuttings, the chemical additives, and the rate of biodegradation of oil may be an influence on the potential for chronic toxicity.

ACKNOWLEDGEMENTS

This study was funded through the Northern Oil and Gas Action Program, Indian and Northern Affairs Canada. Test materials were provided by ESSO Canada Resources Limited and Shell Oil Canada. The authors would like to thank EPS, St John's for the use of their facilities, Suzanne Roussel for her technical assistance, and Camille Mageau, David Milburn and Richard Martin for their comments on the draft paper.

REFERENCES

ANDERSON, J.W., NEFF, J.M., COX, B.A., TATEM, H.E. & HIGTOWER, G.M. (1974). Characteristics of dispersions and water-soluble extracts of crude and refined oils and their toxicity to estuarine crustaceans and fish. *Marine Biology*, **27**, 75–88.

APHA/AWWA/WPCF (1980). *Standard Methods for the Examination of Water and Waste Water*. Amerian Public Health Association, American Water Works Association, Water Pollution Control Federation, Washington, D.C.

BARCHARD, W.W. & DOE, K.G. (1984). Preliminary results of bioassays of oily cuttings from the Alma F-67 exploration drilling programme. In *Report of Workshop on Environmental Considerations in the Offshore Use of Oil-based Drilling Muds*, ed. G.D. Greene & F.R. Engelhardt. Canada Oil and Gas Lands Administation, Environmental Protection Branch, Ottawa, Ontario, pp. 58–9.

BARCHARD, W.W., DOE, K.G., MAHON, S.E., MOORES, R.B., OSBORNE, J.M. & PARKER, W.R. (1986). Environmental implications of release of oil based drilling fluids and oily cuttings into waters of Canadian North West Atlantic. EPS Surveillance Report EPS-5-AR-7-1, 73 pp.

CANADA OIL AND GAS LANDS ADMINISTRATION (COGLA) (1986). *Drilling for Oil and Gas on Frontier Lands, Guidelines and Procedures*. Energy, Mines and Resources Canada, Indian and Northern Affairs Canada.

DOE, K.G., OSBORNE, J.M., MOORS, R.B., PARKER, W.R. & BARCHARD, W.W. (1984). Acute lethality of oil-based drilling muds and components. Internal Environment Canada Report.

EPS (1982). *Standard procedures for testing the acute lethality of liquid effluents.* Environment Canada, Environmental Protection Service, Report EPS 1-WP-80-1.

EPS (1985). *Laboratory Procedure for Determining the Acute Lethality of a Water-Soluble Fraction of Mineral Oil to Rainbow Trout.* Environment Canada, Environmental Protection Service, Atlantic Region, Dartmouth, N.S.

GILLAM, A.H., O'CARROLL, K. & WARDELL, J.N. (1986). Biodegradation of oil adhering to drill cuttings. In *Proceedings of Conference, Oil Based Drilling Fluids, Cleaning and Environmental Effects of Oil Contaminated Drill Cuttings,* Trondheim, Norway, Norwegian State Pollution Control Authority (SFT) and Statfjord Unit and Mobil Exploration Inc., Oslo, Norway, pp. 123–36.

HINDS, A.A., SMITH, S.P.T. & MORTON, E.K. (1983). A comparison of the performance, cost, environmental effects of diesel-based and low-toxicity oil mud systems. Society of Petroleum Engineers of AIME SJPE11891/1.

HUTCHESON, M.S., STEWARD, P.L., ODENSE, R., FOWLER, B.F. & GREEN, D. (1984). *Development of Toxicity Testing Guidelines for Oiled Cuttings – Final Report.* Atlantic Oceanic Company Ltd, for: Environment Canada, Dartmouth, N.S.

LITCHFIELD, J.T. JR (1949). A method for rapid graphic solution of time percent effect curves. *Journal of Pharmacology and Experimental Therapeutics,* **96,** 399–408.

RICE, S.D., SHORT, J.W. & KARINEN, J.F. (1976). Comparative oil toxicity and comparative animal sensitivity. In *Fate and Effects of Petroleum Hydrocarbons in Marine Ecosystems, Proceedings of a Symposium.* November 10–12, 1976, ed. D.G. Wolfe. Pergamon Press, Toronto, pp. 78–94.

THORESEN, K.M. & HINDS, A.A. (1983). A Review of the Environmental Acceptability and the Toxicity of Diesel Oil Substitutes in Drilling Fluid Systems. IADC/SPE 11401.

25

Effects of Low-aromatic Drill Cuttings on an Enclosed Deepwater Sediment Community

TORGEIR BAKKE, JOHN A. BERGE

Norwegian Institute for Water Research, PO Box 33 Blindern, N-0313
Oslo 3, Norway

MORTEN SCHAANNING

Institute for Georesources and Pollution Research,
PO Box 9, N-1432 Oslo, Norway

GEIR M. SKEIE

Cooperating Marine Scientists a.s, Billingstadsletta 19A,
N-1361 Billingstadsletta, Norway

and

FRØYDIS ORELD

Center for Industrial Research, PO Box 124 Blindern,
N-0314 Oslo 3, Norway

ABSTRACT

Low-aromatic drill cuttings were added to undisturbed $0.25m^2$ box core sections from a mud bottom community at 200 m depth in the Oslofjord, Norway. The boxes were kept in indoor flow-through mesocosms, and the effects of the addition on community functions followed for 6 months. The total hydrocarbon concentration at the sediment surface, representative for the 'faunal transition zone' around platforms, was 80 100, 41 200, 8400, and 1300 ppm in the four treated boxes. A control box had 99 ppm, suggesting some urban influence on the parent community. Significant loss of hydrocarbons with time, from 40 to 95%, was found in all boxes except the most heavily loaded. Biodegradation of the oil was demonstrated by elevated sediment respiration and bacterial production, and reduced ratio between straight-chain and branched alkanes. Only respiration was clearly dose dependent. Mass balance considerations suggested incomplete mineralization of the hydrocarbons and production of metabolites in all but the least contaminated box, where mineralization appeared to be essentially complete after 6 months. Changes in pore water were marginal, but reduction in sediment phosphate accounted for most of the phosphate demand of the oil degraders. A slight reduction in redox potential was detected at the end, but there was no production of H_2S. Negative effects on macrofauna structure and bioturbation were small, and the species negatively affected were surface deposit feeders. The additions did not trigger any increase in opportunistic species. The reasons for the slight effects found in comparison to a field situation are thought to be that the cuttings did not reduce exchange of pore water, and that recruitment by settling larvae were essentially excluded by the design. Direct toxic effects on adults were negligible for most species except in the highest dosed box.

1. INTRODUCTION

About 90% of the discharge of oil into the North Sea from offshore operations results from the use of oil-based drilling mud (Anon., 1987). Much of the oil is closely bound to the cuttings, which settle to the bottom. The area of observable biological effects is mainly limited to a distance of less than 1–2 km from the platform (Davies et al., 1984), but subtle effects have been found as far out as 3 km (Hannam et al., 1987). The effects observed can be characterized as sediment smothering resulting in reduced bioturbation and exchange of pore water, suboxic or anoxic sediments, general reduction in faunal diversity, and selection for opportunistic species of animals. The rate of removal or degradation of oil from the sediment and the recovery of the fauna after drilling has terminated is only sparsely known because only a few of the relevant oil fields have completed their drilling programs, but resuspension and outward transport of already settled material have been recorded (cf. Mair et al., 1987).

The investigations around platforms are very much a state-of-the-art description of benthic community structure, whereas information on effects mechanisms is essential to improve the environmental quality (toxic effects, recovery rates) of drilling mud formulations, and to optimize treatment and deposition procedures. Experimental work is required to strengthen the knowledge of functional responses to the contamination such as microbial stimulation, changes in metabolic rates and chemical processes in the sediment, as well as animal activity (e.g. larval settling, bioturbation).

Several experiments on the effects of oil-based drill cuttings on whole benthic communities have been performed recently (e.g. Dow, 1984; Bakke *et al.*, 1986*a*, *b*; present volume; Dixon, 1987; and others). In general these experiments have been performed on shallow water communities, accessible to divers. In the present experiment we wanted to address the effects of cuttings on a community from 200 m depth. The purpose was to test the functional responses to cuttings in sections of a sediment community which came from the same depth and was influenced by similar water masses as bottom communities in the North Sea.

2. MATERIAL AND METHODS

2.1. Experimental Design
The experiment was performed during the period 7 November 1985 to 21 May 1986 at the mesocosm facility of the Marine Research Station Solbergstrand (MRSS), Norway. Five large box core samples (50 × 50 cm sediment surface, 40 cm depth) were taken with an USNEL spade corer (Rosfelder & Marshall, 1967) from a mud and clay community at 200 m depth outside MRSS. By fitting internal removable PVC boxes into the stainless steel housing of the box-corer, and lowering these gently into a container upon retrieval of the sample, sediment sections could be obtained and transplanted to the mesocosm with very little disturbance. At MRSS the PVC boxes were placed in a large flow-through basin with seawater from 45 m depth adjusted to a stable current of about 1 cm/s across the sediment surfaces, which were at a water depth of 15 cm. This caused no disturbance or resuspension of the sediments. The boxes were kept in darkness and allowed to stabilize for 18 days. Berge *et al.* (1986) have given a thorough description and evaluation of the mesocosm and its performance, as well as the procedure to obtain sediment sections.

The cuttings applied were produced from drilling with low-aromatic, oil-based mud at the Statfjord field in 1983 (courtesy of Mobil Exploration Norway Inc.). The cuttings had been recovered from a shale shaker (screen size 60/60). Reported retort value was 31·1% of dry cuttings, which is higher than the normal range. The cuttings were added to four of the boxes. The fifth box was used as control. At addition, the water level was raied to 60 cm above the sediment surfaces and the water column above each surface was enclosed by a PVC box. The cuttings were mixed vigorously with 1 litre of seawater in a high speed mixer, and the slurry was quickly decanted off and gently stirred into the surface of the enclosed water column. This process was repeated 4–6 times until only the coarse mineral grains of the cuttings remained (less than 10% of original volume), which were discarded. This procedure intended to simulate the exposure on the sea floor where the fine or suspendable fraction of the drill cuttings are settling to the bottom. The procedure gave a reasonably even distribution of a flocculent layer of cuttings in the boxes. The volumes of original cuttings applied per box were 500 ml, 250 ml, 75 ml and 12·5 ml respectively, corresponding to a nominal layer of cuttings on top of the sediment of 2, 1, 0·3 and 0·05 mm.

After addition of the cuttings, 100 ml solid glass beads (60 to 90 μm diameter Ballotini grade 15, Jencons (Sci) Ltd) were stirred into the surface water of all boxes as a tracer for monitoring bioturbation. The cuttings and beads were allowed to settle to the sediment for 2 h before the upper PVC box was removed and the water level again reduced to 15 cm above the sediment surface.

The boxes were identified as Box-0·05, Box-0·3, Box-1, and Box-2 after nominal thickness of cuttings, and the control as Box-C.

2.2. Sampling Program

Core samples for the various purposes were taken from a random number grid. Each time a core was taken a liner was pushed into the sediment outside the corer, and left in position to prevent disturbance of the surrounding sediment.

2.2.1. Environmental Conditions

Water temperature was logged automatically every hour throughout the experiment. Salinity was measured once every day by use of a salinoterm.

2.2.2. Hydrocarbon Concentrations

After 2 and 175 days, 3 replicate cores (diameter 45 mm) were withdrawn from each box, and the upper 10 mm section removed for analysis. The three replicate sections were pooled prior to analysis. After addition of deuterized internal standards the samples were extracted 3 times with dichloromethane at sonication (Sonicator Cell Disruptor Model W-375). After centrifugation the supernatant extract was decanted off, pooled, dried with anhydrous sodium sulphate, and the extract volume adjusted. The total amounts of hydrocarbons (THC) were determined by capillary column gas chromatography. The sum of naphthalenes, phenanthrenes and dibenzothiophenes (NPD) and ratios between n-C_{17}/pristane and n-

C_{18}/phytane were determined by coupled gas chromatography–mass spectrometry. The amounts were related to dry weight of the samples by drying an aliquot at 110°C for 48 h.

2.2.3. Sediment Chemistry

An electrode assembly consisting of a Radiometer platinum Eh electrode, a Radiometer sulphide ion selective electrode and an Orion Ross pH electrode with an internal common reference electrode ($E_{ref} = 432$ mV at 10°C) was inserted to selected depths in the submerged sediments to obtain profiles of pH, Eh and S^{2-} ion activity 4 days before and 1, 10, 21, 42, 62, 102 and 175 days after the addition of drill cuttings.

For pore water analysis, 59 mm diameter cores were drawn at day 1 and day 175. The cores were sectioned immediately in 1 and 2 cm slices, and the pore water separated by centrifugation. Alkalinity was determined by acidification of a subsample and back titration with dilute NaOH solution. The remaining pore water was stored with a few drops of chloroform at −20°C until simultaneous nutrient determination of all samples could be made by the end of the experiment. Silicate, phosphate, nitrate, nitrite and ammonia were determined using standard automated procedures.

Oxygen consumption was measured 4 days before and 1, 10, 21, 42, 62, 102 and 175 days after the addition of drill cuttings. The measurements were performed on the whole of the sediment surface by closing the boxes with lids and measuring the decrease in concentration of dissolved O_2 in the enclosed water every 4 h during a total period of isolation of 16 h. The concentration was never allowed to decline below 50% saturation. The water was circulated through a system which gave a constant laminar type current across the sediment, and adjusted to prevent concentration gradients in the water or resuspension of sediment particles. Samples for oxygen determination were drawn from the PVC tube, constituting the external part of the circulation system. Measurement was by the standard Winkler procedure.

2.2.4. Microbiology

All plastic and glassware was acid washed with 20% HNO_3, rinsed twice in distilled water, twice in particle-free (0·2 μm filtered) distilled water, autoclaved (120°C, 20 min) and stored under sterile conditions until use.

After 84 days, five replicate 12·5 mm ∅ cores were taken from each of the five boxes as described by Warwick *et al.* (1986). The cores from each box were pooled and made into a slurry immediately. Subsamples of 0·2–

0.5 g wet weight were added to 18 ml polypropylene centrifuge tubes, containing 30 μCi ^{3}H-methyl thymidine (42 Ci/mmol, Amersham, UK) in 0.500 ml sterile filtered seawater, and incubated at in-situ temperature for 20 min. The incubation was terminated by addition of 10 ml 80% ethanol. Blank samples were obtained by mixing isotope and ethanol before the addition of sediment. Seven replicate samples and 3 blanks were processed from each of the slurries. Samples for estimation of bacterial numbers were preserved in sterile filtered, 4% buffered formaldehyde solution.

DNA was extracted following a modification of the method by Moriarty & Pollard (1982), using an alkaline extraction followed by dialysis (for detailed description, see Skeie & Gray, 1988). After dialysis the volume of the dialysate was adjusted to 5 ml, mixed, and 1 ml of the solution transferred to a scintillation vial containing 9 ml scintillation cocktail (Opti-Fluor, Packard). Radioactivity was measured on a Packard 2450 Liquid Scintillation Spectrometer at 70% gain, and quench correction calculated using the SCR (Sample Channels Ratio) method.

Previous experiments with similar sediments in the mesocosms had shown that the period of linear incorporation of thymidine exceeded 30 min, and that no isotope dilution occurred under the conditions specified. These values were adopted for this study. A conversion factor of 2×10^9 cells produced per nmol thymidine incorporated (Moriarty, 1986) was used.

Bacterial abundance was measured by staining with DAPI following the procedure given by Porter & Feig (1980). Bacteria on the filters were counted at 1500× magnification (Leitz Orthoplan, UV-excitation BP 340-380, RKF 400 mirror, LP 430 filter). All measurements were standardized against sediment surface (m^2) using values obtained from measurements of porosity and water content. Statistical tests were done on $\log_{10}$ transformed values.

2.2.5. Bioturbation

The glass beads represented an easily quantifiable tracer for bioturbative and other mixing of surface sediment particles. Three replicate cores were taken from each box with a suction corer (inner diam. 20 mm) at 2, 30, 90 and 167 days for analysis of vertical distribution of glass beads. The cores were frozen undisturbed, and then sectioned at 5, 20, 40 and 100 mm depth. At sectioning, the outer 1 mm of the core sample was carved off to prevent contamination from beads pushed down during sampling.

The sections from days 2, 30 and 90 were thawed, washed 3–4 times with 4 times the sediment volume of distilled water. freeze-dried for 24 h.

and gently homogenized in a mortar. A weighed subsample (1–3% of the material) was taken for inverted microscope count of beads by a modified procedure to that used for phytoplankton (Utermöhl, 1958), and the counts converted to number of beads per g dry sediment. Each section from day 167 was suspended in 1 litre tap water in a meiofauna sample divider with 8 compartments (Elmgren, 1973), repeatedly divided until the count of beads in a subsample was below 1000 (usually 1–3 times), then 2 aliquots were analyzed for bead content by counting under a dissecting microscope. The bead content was expressed as number/ml sediment.

2.2.6. *Macrofauna*
At termination of the experiment the sediment of each box down to a depth of 20 cm was washed through a sieve with 1 mm circular holes. The material on the sieve was preserved in 10% formalin and stained with Rose Bengal (Thiel, 1966). After sorting, the animals were stored in 70% ethanol until identified under a dissecting microscope. Numbers of species, species abundance and total individual densities per box were obtained, and community diversity expressed as rarefaction curves (Hurlbert, 1971).

3. RESULTS

3.1. Environmental Conditions
Temperature decreased rapidly from about 10 to 5°C during the first 50 days, and then remained at 5–6°C for the rest of time (Fig. 5(B)). Salinity was stable at 34–35 ppt during most of the experiment (Fig. 5(A)).

3.2. Hydrocarbon Concentrations
Total hydrocarbon content (THC) at day 2 ranged from 80 100 ppm (dry wt basis) in Box-2 to 1290 ppm in Box-0·05, against a background of 99 ppm in Box-C (Table 1). The net concentrations of THC and NPD were strongly correlated with the volume of cuttings used to make up the slurries added to the boxes ($r^2 = 0.99$), showing that the amount of oil settling to the bottom of the boxes was proportional to the amount of cuttings applied.

After 175 days the surface content of both THC and NPD was reduced in all except Box-2 (Table 1), where only a loss in NPD (50%) was detected. The percentage loss of hydrocarbons from Box-1 and Box-0·3 corresponded well at 40% for THC and 75% for NPD.

Box-0·05 lost nearly all oil hydrocarbons during the 6 months: 95% of

TABLE 1

Content of total hydrocarbons (THC), sum of naphthalenes, phenanthrenes/
anthracenes, and dibenzothiophenes (NPD), and percentage loss with time in the
upper 10 mm of the sediments at day 2 and 175 (mg/kg dry sediment).

Treatment	Volume (ml) of cuttings added	THC			NPD		
		Day 2	Day 175	% loss	Day 2	Day 175	% loss
Box-2	500·0	80 100	89 100	−11	145·0	73·4	49
Box-1	250·0	41 200	25 400	38	76·0	18·0	76
Box-0·3	75·0	8 360	4 850	42	13·9	3·4	76
Box-0·05	12·5	1 290	130	95	2·9	0·2	100
Box-C	0·0	99	73	—	0·3	0·2	—

the net concentration of THC had disappeared, and the NPD concentration was at the level of Box-C. The chromatogram of Box-0·05
(Fig. 1(E)) showed that the typical pattern of oily cuttings had almost
disappeared after 6 months, but were still present in the other boxes (Fig.
1(B–D)).

The ratios C_{17}/pristane and C_{18}/phytane at the end of the experiment
was only slightly reduced in Box-2 (Table 2), indicating that no significant
biodegradation of the medium size n-alkanes on the cuttings had occurred.
The ratios were lower in Box-1 and more so in Box-0·3, indicating
significant degradation of the n-alkanes. In Box-0·05 the ratios were close
to zero at the end suggesting that nearly all the petrogenous n-alkanes
had been lost.

3.3. Sediment Chemistry

The sulphide ion electrode did not indicate any significant S^{2-} activity.
Thus H_2S production appeared to have been insignificant. The variations
in Eh and pH at the 0·5 cm depth level were small (Fig. 2). Elevated pH
in the high dose boxes for the first 4 days probably reflected a direct
chemical influence of the cuttings on pH. In deeper sediments the increase
in pH was more gradual, but also more persistent (Fig. 3). The redox
potential was always high (250–400 mV) in the upper (0–4 cm) layer (Fig.
3). In general the potential decreased to a level of about 100 mV at greater
sediment depths. A small decrease in Eh of Box-2 was observed during
the last survey, but on the whole the Eh of the contaminated boxes did
not deviate significantly from the control.

The water content was in general higher at the end of the experiment
(Fig. 3), most pronounced in Box-C, least in Box-2. Such increase has

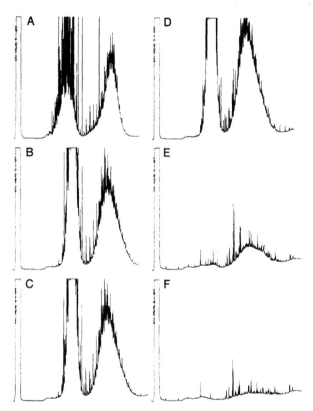

FIG. 1. Capillary gas chromatograms of the hydrocarbons in the upper 10 mm of the boxes at day 175. (A) original low-toxicity cuttings; (B) Box-2; (C) Box-1; (D) Box-0·3; (E) Box-0·05; (F) Box-C.

TABLE 2

Ratio between n-C_{17}: pristane, n-C_{18}: phytane and pristane: phytane in the upper 10 mm of the sediments at days 2 and 175.

Treatment	n-C_{17}:pristane		n-C_{18}:phytane		Pristane:phytane	
	Day 2	Day 175	Day 2	Day 175	Day 2	Day 175
Box-2	2·91	2·72	2·13	2·00	0·75	0·81
Box-1	3·02	2·51	2·06	1·82	0·75	0·76
Box-0·3	3·11	1·56	2·21	1·13	0·75	0·81
Box-0·05	3·01	<0·1	2·21	<0·1	0·81	1·54

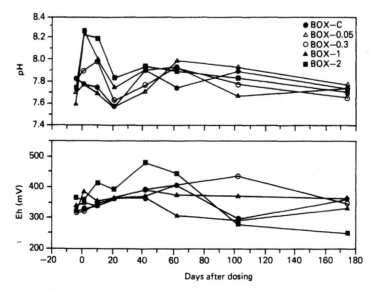

FIG. 2. pH and *Eh* variations at 0·5 cm sediment depth.

been observed in similar mesocosm boxes before (Berge *et al.*, 1987) and was not an effect of the treatment.

The profiles of alkalinity (Fig. 3), ammonia and nitrate (Fig. 4) did not reveal any persistent trend with regards to time or dose. Most of the carbonate released by respiration of an active bacterial population in a surface oily layer would be lost to the overlying water, so that any increase of the alkalinity of the pore water becomes small compared to background variations. The profiles of silicate and phosphate (Fig. 4) were less scattered. They indicated no net uptake or release of silicate, but a net loss of phosphate from the pore water of the upper 3 cm of the contaminated boxes. If the loss of PO_4^- was due to bacterial uptake, silicate should remain unaffected as was in fact observed. A concomitant uptake of nitrogen may have occurred from sources other than the relatively depleted initial pore water concentrations of NH_4^+, NO_3^- or NO_2^-. Such sources might be various inorganic or organic components of the interstitial or overlying watermass.

The development of the sediment oxygen consumption (SOC) in each of the five boxes is shown in Fig. 5(C). The two high dose boxes showed a rapid increase of SOC to maximum rates 40–60 days after addition, and still high values at the end of the experiment. Any lag time between

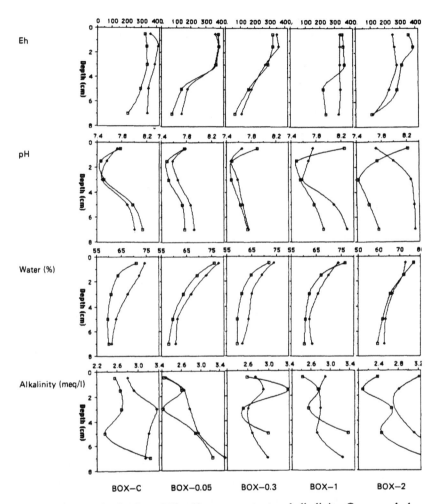

FIG. 3. Vertical distribution of *Eh*, pH, water content and alkalinity. Open symbols represent day 1, closed symbols day 175.

addition and stimulated SOC must have been less than 24 h. The SOC of the two low dose boxes decreased during the first 10 days. Thereafter Box-0·3 behaved similar to Box-1 and Box-2, whereas Box-0·05 showed a pronounced maximum at 20–40 days and a level not significantly different from Box-C by the end. Also the SOC of the control box decreased between day 1 and 10. After that the control SOC was fairly

Torgeir Bakke et al.

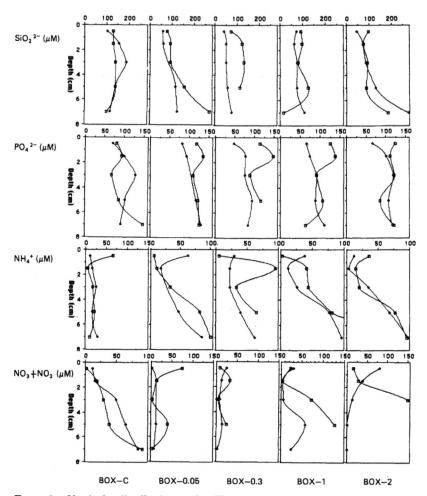

FIG. 4. Vertical distribution of silicate, phosphate, ammonium and nitrate + nitrite. Open symbols represent day 1, closed symbols day 175.

stable at about $200\,\mu mol/m^2/h^1$. The simultaneous decrease of SOC in the control and the two low dose boxes is partially explained by the sudden decline in temperature, but a Q_{10} as high as 5 suggests other causes too. A slight decrease in SOC in all boxes from day 60 to 100, as well as in temperature, suggested a mean Q_{10} of 1.7 which is similar to values found before for whole sediment respiration at this temperature range (Smith, 1973; Davies, 1975; Hartwig, 1978).

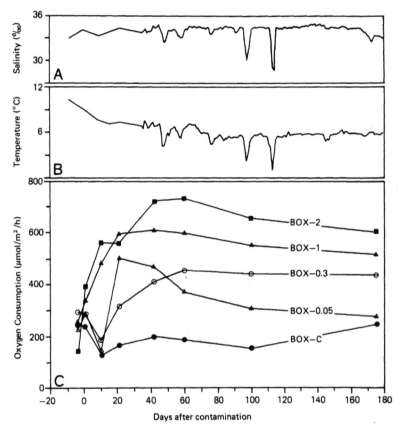

FIG. 5. Variation of salinity ($S^0/_{00}$), temperature (°C), and sediment oxygen consumption during the experiment.

The dose-dependent response in SOC (Fig. 6), indicates a second-order relationship with the amount of cuttings added. The mean profile shown in Fig. 6 fitted the polynomial function: $SOC = 276 + 1.78x - 0.002x^2$ (x is ml of cuttings added) at a significance of $r = 0.96$. From this equation a maximum SOC of approximately $680 \, \mu mol/m^2/h^1$ was predicted at an addition of 450 ml. Hence, addition of more cuttings than were given to Box-2 (500 ml) would have produced no further increase in SOC, presumably because the available surface area for bacterial attack would not be increased. This implies that thicker layers of cuttings will remain relatively undegraded unless physically disturbed (e.g. resuspension), at least for a

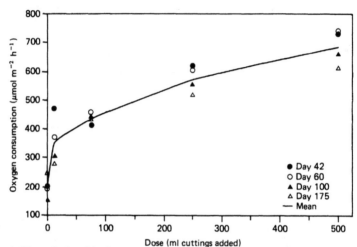

FIG. 6. The relationship between amounts of drill cuttings added and sediment oxygen consumption during the period 42–175 days after dosing. The line drawn is the best fit polynomial function of SOC on ml cuttings.

time span of 6 months. Other experiments (Bakke *et al.*, this volume) indicated significant biodegradation of layers as thick as 10 mm after 1–5 years.

3.4. Bacterial Abundance and Production
There were no significant differences ($p > 0.05$) in bacterial abundances among the different boxes (Fig. 7(a)). Although the bacterial production did not differ among the contaminated sediments, and hence showed no dose-related response, they all showed significantly higher production ($p < 0.001$, ANOVA) than in the control sediment (Fig. 7(b)). When standardizing the bacterial abundance against the bacterial production (Fig. 7(c)) it is evident that the addition of cuttings resulted in an increased specific growth rate (cell division rate) of the bacterial community.

3.5. Bioturbation
Between 59 and 98% (mean 85%) of the glass beads remained in the upper 5 mm of the sediments until the end of the experiment, and only about 2·5% were found below 4 cm depth (Table 3). This shows that sediment reworking was low and confined to the sediment surface in all boxes. A slight downwards transport of beads was indicated during the

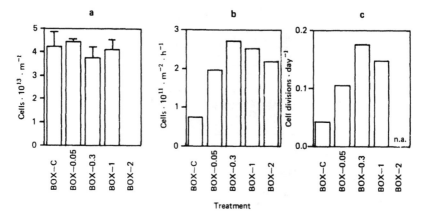

FIG. 7. Mean bacterial abundance (a), bacterial production (b), and specific growth rate (cell division rate) (c) of the sediment surfaces at day 84. n.a.: not analyzed.

TABLE 3

Mean percentage distribution of glass beads with sediment depth at days 2, 30, 90 and 167, and distribution in Box-C alone at day 167.

Sediment depth (mm)	Occurrence of beads (%)				
	Mean of all boxes				Box-C
	Day 2	Day 30	Day 90	Day 167	Day 167
0–5	98·5	92·5	93·5	84·7	72·8
6–20	0·9	6·5	4·9	8·9	16·2
21–40	0·5	0·7	0·9	3·7	6·4
41–60	0·1	0·3	0·8	2·7	4·6

first month, but the change in mean vertical distribution between day 30 and 167 was insignificant when the data from all boxes were pooled.

At days 30 and 90 there was no systematic difference in bead distribution among the boxes. At day 167 the data indicated that Box-C had less beads at the surface than the others and higher densities in the deeper layers (Fig. 8), but the difference was only statistically significant with respect to Box-1 ($p < 0.05$, regression analysis). This suggested that the bioturbation intensity had been slightly reduced in the contaminated boxes, but not systematically related to the treatment.

3.6. Macrofauna Composition

Generally the effect of the treatment on gross macrofauna community structure was marginal. At day 175 the community in the control box

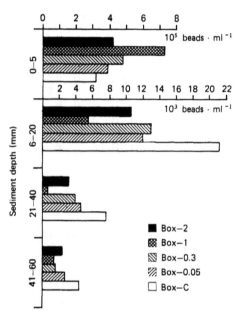

FIG. 8. Mean concentration of glass bead tracer at 4 depths in the sediment after 167 days based on 3 replicate cores from each box.

was most diverse and the community in the highest dosed box showed the lowest diversity (Fig. 9). No boxes gave a clear dose related response. Generally all major phyla were reduced in density in Box-2 (Fig. 10(a)). The macrofauna community was dominated by bivalves with *Thyasira equalis* and *Kelliella miliaris* as the two most abundant. These species are typical for deep water soft sediment and are predominant in other deep water (300 m) areas (Josefson, 1981).

Linear regression between abundance and dose for the 14 most dominant species (represented with at least 20 individuals totally in all boxes), showed a negative relationship for 7 species and positive for 7 and, thus, does not indicate any overall negative effect. Omitting Box-2 resulted in a negative correlation for 4 species and a positive for 9, indicating this even more. For all the regressions the correlation coefficient was low.

The two species which showed the most convincing negative correlation with dose, all boxes included, were (Fig. 10(b)): *Abra nitida* ($r = 0.65$, 0.68 when omitting Box-2) and *Spiophanes krøyeri* ($r = 0.94$, 0.96). For the species showing a positive correlation between density and dose, only the polychaete *Euchone papillosa* was convincing ($r = 0.44$, 0.81). Only *S.*

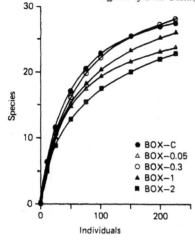

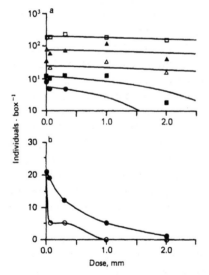

FIG. 9. Macrofauna diversity (Hurlbert's rarefaction curves) at day 175.

FIG. 10. Abundance (density per box) of macrofauna as function of thickness of cuttings. (a) abundance of major faunal groups: bivalves (□), polychaetes (▲), crustaceans (△), echinoderms (●), others (■). The curves are linear regression lines for each group (note exponential ordinate axis). (b) Abundance of *Abra nitida* (○) and *Spiophanes kroeyeri* (●). The curves are drawn as interpolations of the data.

krøyeri, however showed a significant ($p < 0.05$) correlation between dose and density.

The only feeding group which seemed to respond negatively to the treatment was the interface feeders, i.e. animals feeding on particles in the sediment water interface. Subsurface deposit feeders, predators and filter feeders showed no systematic response apart from indications of a maximum abundance at 1 mm dosing.

4. DISCUSSION

The mode of application of cuttings was designed to simulate the settling of fine or resuspended cuttings in a wide area around a platform. This area is important both because it is large compared to the area containing the coarse fraction in the immediate vicinity of the discharge point and because the finer particle fraction may be resuspended and transported from a physically unstable pile of cuttings to more distant locations by bottom currents. Recent estimates have suggested that up to 30 km^2 of the bottom around single platforms in the Norwegian sector of the North Sea may be contaminated (Reiersen et al., this volume).

The THC level in the control box was 99 mg/kg which is high compared to background offshore North Sea sediments (10–30 mg/kg). This is probably caused by a larger influence from urban activity than in pristine areas of the North Sea. The faunal community or porewater chemistry did not suggest any effect of this increase, but the rapid SOC response to the addition of cuttings could indicate the existence of a microbial community adapted to slight hydrocarbon input. The addition of cuttings gave initial levels of THC ranging from 80 000 to 1300 mg/kg in the upper 10 mm. Hydrocarbon data around platforms will in general represent the upper 4–6 cm of the sediment, which is the average penetration depth of the grabs used. If the experimental concentrations found in the upper 10 mm were diluted to a thickness of 5 cm, the experimental levels applied would represent 'grab' concentrations between 16 000 and 260 mg/kg. This is equivalent to distances from about 1 km and inwards towards a platform with reasonably long drilling history (cf. Sjögren et al., 1986), covering most of the 'community transition zone' (Davies et al., 1984; Hannam et al., 1987).

A substantial loss of hydrocarbons from the sediment surface was detected during the 6 months. The processes involved would be biodegradation, burial, and leakage to the water column. Biodegradation was indicated by the reduced normal-to-branched alkane ratios, enhanced SOC, and increase in bacterial DNA production relative to the control. SOC expresses total community respiration, and an open question is whether the dose–response relationship found for SOC most of the time was due to bacteria, fauna or both. The bacterial component of respiration in shallow water sediments has been shown to vary from 25 to 95% of the total (references in Hartwig, 1978), and an estimate of the bacterial contribution to SOC can only be tentative. Animal respiration may be enhanced by oil, but the response is not systematic, and could therefore either add to or counteract an increase in bacterial respiration. Considering the low temperature, the significant bacterial contribution to benthic respiration in general, and the fact that animal respiration may shift either way due to oil stress, we find it most likely that the dose-dependent increase of SOC above the control was primarily a response to the oil by bacteria and possibly the microbial predators on the bacteria. The increase in bacterial production shown by synthesis of new DNA at day 82 was not clearly dose related, but at this time the SOC had reached a steady state, and could reflect elevated bacterial respiration, without a concomitant increase in cell division rates.

The persistently elevated SOC suggested that biodegradation proceeded steadily after about 1·5 months, only influenced by a further slight reduction in temperature. The highest SOC was found in the high dose Box-2, where the THC and normal-to-branched alkane ratio suggested that no biodegradation had occurred. The lack of response in the alkane ratio must be because only a small fraction of the total *n*-alkane pool of the cuttings had been degraded at the end of the experiment. This is supported by the persistent high level of THC in Box-2. The fact that the SOC leveled off in this box too, suggests that much of the oil was physically inaccessible to biodegradation, or the process was limited by supply of oxygen and nutrients.

Assuming mineralization to follow the equation:

$$C_nH_{2n} + 1 \cdot 5nO_2 = nCO_2 + nH_2O$$

1·5 mol O_2 are required to turn 1 mol C into CO_2. In Table 4 the loss of carbon estimated from the observed SOC elevation above control is compared to the observed loss in THC, converted from ppm to moles of carbon by assuming a water content of 75% (cf. Fig. 3) and a density of 2·6 g/cm^3 of the dry sediment. In Box-0·05 the SOC alone would account for 50% higher oxidation than was recorded as loss in THC. SOC has therefore involved oxidation processes other than mineralization of THC. The drill cuttings may have contained reducing agents other than those measured by THC, and the oil degraders may have been able to utilize other inherent sources of organic carbon as the pool of THC was depleted. The decline in SOC to control level towards day 175 suggests that these sources were not able to sustain a large population of oil degraders for any length of time.

In Box-0·3 and Box-1 the SOC would explain 75 and 28% of the THC loss respectively. The other likely losses would be leakage and burial. Leakage to the overlying water was not measured, but earlier experiments

TABLE 4

Various mass balance considerations of the experimental boxes. $\triangle$SOC: total oxygen consumption, HC_{OX}: hydrocarbons mineralized.

Dose (nominal mm)	0·05	0·3	1	2
$\triangle$SOC (mmol O_2 m²) in 175 days	620	966	1 638	1 974
HC_{OX} (mmol C/m²) in 175 days	434	644	1 092	1 316
THC_2–THC_{175} (mg/kg or ppm)	1 160	3 510	15 800	−9 000
THC_2–THC_{175} (mmol C/m²)	286	864	3 889	−2 217
Fraction of loss of THC accounted for by oxidation (%)	152	75	28	—
Observed loss of PO_4^{3-} from pore water (μmol/m²)	530	980	840	560
Estimated PO_4^{3-} fixation (μmol/m²)	409	621	1 016	1 228

(Poley & Wilkinson, 1983; Bakke *et al.*, 1985) suggest that this would account for less than 1 ppm THC per day or maximum 5% loss in Box-0·3 and 1% in Box-1. The bioturbation results suggested a further loss of about 10% of the THC from the 10 mm sampling zone by burial in deeper sediment strata. Thus the processes described can account for all the hydrocarbon loss in Box-0·05, 90% of the loss from Box-0·3, but only 40% of the observed loss in Box-1. This suggests that the hydrocarbons in the latter two boxes were not completely mineralized, but degraded to metabolites intermediate between the original hydrocarbon and CO_2, requiring less oxygen to remove 1 mol C from the matter analyzed as THC, than according to the equation above. The fraction of loss accounted for by oxidation (Table 4) also suggested that the hydrocarbons became more completely oxidized as the THC concentration decreased. In Box-2 the THC did not decrease, and although the SOC was higher than in the other boxes, it would account for a loss only half as large as the variability in THC between day 2 and day 175.

The observed loss of PO_4^{3-} from the pore water (Table 4) was estimated from the profiles shown in Fig. 4. The fixation in organic tissue was estimated from the general ecological assumptions that tissue production corresponded to 10% of the carbon mineralized, and that the C:P ratio in the organisms corresponded to the Redfield ratio of 106:1. The results (Table 4) show that estimated P-fixation was of the same order of magnitude as the observed loss of PO_4^{3-} from the pore water. This strongly suggests that the pore water reservoir was the major source of PO_4^{3-} for the oil consumers during this experiment.

Most earlier experiments with similar levels of cuttings (Bakke *et al.*, this volume; Gillam *et al.*, 1986; Leaver *et al.*, 1986; and others) and with oil (Berge *et al.*, 1987), as well as investigations around platforms (Hannam *et al.*, 1987), have shown that the contaminated sediments rapidly become more reducing, and even anoxic. Several factors may play a role: reduced flux of oxygen through the hydrophobic oily layer, stabilization of the sediments through reduced bioturbation, and increase in oxygen demand due to oil degradation. Significant reduction in *Eh* was not detected in the present experiment, although the hydrocarbon level, mode of application, and batch of cuttings were similar to those which caused a significant drop in *Eh* in an earlier field experiment (Bakke *et al.*, 1986*b*). Other features of the pore water were only affected close to the surface. The results suggest that the degradation of the oil occurred at the very top of the sediment, and that the supply of oxygen and nitrogen to the process occurred by eddy and molecular diffusion between bacterial sites and overlying water, with little or no contribution from the pore water below. In other experiments, as well as around platforms, the cuttings have generally been mixed down to some depth in the sediment, either deliberately or as a function of physical or biological activity. Degradation of buried oil would utilize oxygen and nutrients in the pore water, with decreased *Eh* and nutrient level as the result. The stability in pore water characteristics therefore shows that the amounts of oil mixed into the sediment was too small to deprive the pore water of oxygen. It seems further that the cuttings did not prevent the normal exchange processes across the water–sediment boundary (no lid effect).

Macrofauna analysis in general demands large samples of sediments, and total box analysis could only be performed at the end of the experiment. Core samples for macrofauna were taken at the start and at the end, but due to time constraint they could not be treated for this presentation. Similarity in faunal composition between boxes at transplantation have therefore not been tested, but were assumed to be small based on sediment appearance and previous experience with similar sections. In order to reflect significant dose-related responses, the effect of the treatment would have to override the initial variability in the faunal composition in the boxes. If the treatment effect was marginal, it would be masked by the inherent variability between boxes. The faunal composition in the boxes did not differ much at the end of the experiment.

The high dose Box-2 was the only one demonstrating a subnormal community structure (Fig. 10). It thus shows that any negative effects of doses less than 80 000 mg/kg were less than the variability in the faunal composition when the boxes were brought in. The apparent lack of macrofauna response contradicts findings in the transition zone around platforms, but it is important to note that the experiment was performed on an intact mature community dominated by adults, and in an experimental system and during a period of the year where larval settlement was not facilitated. This implies that potential effects on recruitment (e.g. habitat selection, larval survival) were limited to species with non-pelagic larval development. Other experiments have shown that oil-based cuttings may seriously influence the habitat selection process of settling larvae (Bakke *et al.*, 1986*a*). The most probable fauna response to be seen in this experiment was therefore mortality and not increased densities.

Several opportunistic species increase in abundance immediately outside the heavily polluted zone around platforms (e.g. Hannam *et al.*, 1987) in hydrocarbon concentrations similar to ours. One of these species, *Chaetozone setosa*, was present also in our boxes, but did not respond with increased density. It has been claimed that reduced competition or increased availability of food may trigger the bloom of opportunists (cf. Branch, 1984). If one accepts these two mechanisms, it follows that the carbon enrichment caused by the increased bacterial production in all contaminated boxes and a proposed reduced competition caused by the slight faunal reduction in the Box-2, was not sufficient to elicit an increase in density of opportunistic species. Other explanations are that opportunists did not have the time to respond or were not available as larvae. The first explanation is unlikely since one of the properties characterizing opportunistic species is that they can respond on a short time scale. We therefore claim lack of larvae as the main cause for opportunists not to prosper in the contaminated boxes.

Since the cuttings were confined to a layer at the sediment/water interface, animals dependent on this interface for food were most likely to be affected. Effects may be negative if the added material is toxic, or stimulative if the hydrocarbons stimulate bacterial production as a food source for macrofauna. The only two species showing a systematic negative response to the treatment were *Abra nitida* and *Spiophanes kroeyeri* (Fig. 10(*b*)), both dependent on particles in or near the sediment surface for food, and both characterized as non-tolerant to pollution (Rygg, 1985). For these species it seems that toxicity, or negative effect on feeding activity, have been more pronounced than a positive effect of high bacterial production. *A. nitida* feeds solely at the surface (Wikander, 1980) and its larvae have previously been shown not to settle in sediments contaminated with oil based cuttings (Bakke *et al.*, 1985), which may indicate an efficient ability to reject unfavorable surface conditions.

The only species which indicated a significant positive correlation with dose was *Euchone papillosa*. This species is a filter feeder and the individuals found in the boxes were adults which feed 2–6 cm above the sediment and are thus not likely to be affected by the surface cuttings. The genus *Euchone* is considered tolerant to pollution in general (Rygg, 1985), and we assume that the density distribution of *E. papillosa* in the boxes reflects initial density differences rather than a positive response to the cuttings.

We believe that the reasons for the slight effect of the treatment found in the experiment compared to a field situation are that (1) the discharged cuttings have not affected important sediment parameters at similar concentrations which affect the bottom around oilfields, possibly because they were not mixed into the sediment to any extent, (2) the effects of reduced recruitment caused by habitat selection in settling larvae were excluded, and (3) toxic effects on adults have been negligible for most species except in the highest dosed box.

ACKNOWLEDGEMENTS

We would like to thank BP Petroleum Development (Norway) Ltd for support of this project through Grant No. '1039—Shoreline Pollution'. We also acknowledge the comments from two anonymous reviewers. This paper is contribution No. 26 from the Marine Research Station Solbergstrand.

REFERENCES

ANON. (1987). Annual report on discharges from platforms in 1985. The Paris Commission, GOP 11/3/1-E.

BAKKE, T., GREEN, N.W., NAES, K., PEDERSEN, A., SPORSTØL, S. & ORELD, F. (1985). Drill cuttings on the sea bed. Field experiments on recolonization and chemical changes. Phase 1. Thick (10 mm) layers of cuttings 1982–1983. Norwegian Institute for Water Research report No. 1752, Oslo, 202 pp.

BAKKE, T., GREEN, N.W., NAES, K. & PEDERSEN, A. (1986a). Drill cuttings on the sea bed. Phase 1 and 2. Field experiment on benthic recolonization and chemical changes in response to various types and amounts of cuttings. In

Oil based drilling fluids. Cleaning and environmental effects of oil contaminated drill cuttings. Proceedings. Norsk Petroleumsforbund. Trondheim, pp. 17–31.

BAKKE, T., GREEN, N.W., NAES, K. & PEDERSEN, A. (1986*b*). Drill cuttings on the sea bed. Phase 3. Field experiment on benthic community response and chemical changes to thin (0·5 mm) layers of cuttings. In *Oil based drilling fluids. Cleaning and environmental effects of oil contaminated drill cuttings. Proceedings.* Norsk Petroleumsforbund. Trondheim, pp. 33–41.

BAKKE, T., BERGE, J.A., NAES, K., ORELD, F., REIERSEN, L.O. & BRYNE, K. (1988). Long term recolonization and chemical change in sediments contaminated with oil-based drill cuttings. This volume, pp. 521–44.

BERGE, J.A., SCHAANNING, M., BAKKE, T., SANDØY, K.A., SKEIE, G.M. & AMBROSE, W.G. JR (1986). A soft bottom sublittoral mesocosm by the Oslofjord: description, performance and examples of application. *Ophelia,* **26**, 37–54.

BERGE, J.A., LICHTENTHALER, R.G. & ORELD, F. (1987). Hydrocarbon depuration and abiotic changes in artificially oil contaminated sediments in the subtidal. *Estuarine, Coastal and Shelf Science,* **24**, 567–83.

BRANCH, G.M. (1984). Competition between marine organisms: Ecological and evolutionary implications. *Oceanogr. Mar. Biol. Ann. Rev..,* **22**, 429–593.

DAVIES, J.M. (1975). Energy flow through the benthos in a Scottish sea loch. *Mar. Biol.,* **31**, 353–62.

DAVIES, J.M., ADDY, J.M., BLACKMAN, R.A.A., BLANCHARD, J.R., FERBRACHE, J.E., MOORE, D.C., SOMERVILLE, H.J., WHITEHEAD, A. & WILKINSON, T. (1984). Environmental effects of the use of oil-based drilling muds in the North Sea. *Mar. Pollut. Bull.,* **15**(10), 363–70.

DIXON, I.M.T. (1987). Experimental application of oil-based muds and cuttings to seabed sediments. In *Fate and Effects of Oil in Marine Ecosystems,* ed. J. Kuiper & W.J. Van den Brink. Nijhoff, Amsterdam, pp. 133–50.

DOW, F.K. (1984). Studies on the environmental effects of production water and drilling cuttings from North Sea offshore oil installations. Ph.D. Thesis, University of Aberdeen, 243 pp.

ELMGREN, R. (1973). Methods of sampling sublittoral soft bottom meiofauna. *Oikos,* **15**, 112–20.

GILLAM, A.H., O'CARROLL, K. & WARDELL, J.N. (1986). Biodegradation of oil adhering to drill cuttings. In *Oil based drilling fluids. Cleaning and environmental effects of oil contaminated drill cuttings. Proceedings.* Norsk Petroleumsforbund. Trondheim, pp. 123–36.

HANNAM, M.D., ADDY, J.M. & DICKS, B. (1987). Seabed biological monitoring of drill cuttings discharges in the Thistle oil field. In *Fate and Effects of Oil in Marine Ecosystems,* ed. J Kuiper & W.J. Van den Brink. Nijhoff, Amsterdam.

HARTWIG, E.O. (1978). Factors affecting respiration and photosynthesis by the benthic community of a subtidal siliceous sediment. *Mar. Biol.,* **46**, 283–93.

HURLBERT, S.H. (1971). The non-concept of species diversity: A critique and alternative parameters. *Ecology,* **52**, 577–86.

JOSEFSON, A.B. (1981). Persistence and structure of two deep macrobenthic communities in the Skagerak (West coast of Sweden). *J. exp. mar. Biol. Ecol.,* **50**. 63–97.

LEAVER, M.J., MURISON, D., DAVIES, J.M. & RAFAELLI, D. (1986). Experimental studies of the effects of drilling discharges. *Phil. Trans. R. Soc. Lond.* **B316.**

MAIR, J. McD., MATHESON, I. & APPELBEE, J.F. (1987). Offshore macrobenthic recovery in the Murchison field following the termination of drill-cuttings discharges. *Mar. Pollut. Bull.,* **18,** 628–34.

MORIARTY, D.J.W. (1986). Measurement of bacterial growth rates in aquatic systems from rates of nucleic acid synthesis. *Adv. Microb. Ecol.,* **9,** 1–41.

MORIARTY, D.J.W. & POLLARD, P.C. (1982). Diel variation of bacterial productivity in seagrass (*Zostera capricorni*) beds measured by rate of thymidine incorporation into DNA. *Mar. Biol.,* **72,** 165–73.

POLEY, J.P. & WILKINSON, T.G. (1983). Environmental impact of oil-base mud cuttings discharges—a North Sea perspective. IADC/SPE Drilling Conference, New Orleans: 335–42.

PORTER, K.G. & FEIG, Y.S. (1980). The use of DAPI for identifying and counting aquatic microflora. *Limnol. Oceanogr.,* **25,** 943–8.

REIERSEN, L.O., GRAY, J.S. PALMORK, K.H. & LANGE, R. (1988). Monitoring in the vicinity of oil and gas platforms; results from the Norwegian sector of the North Sea and recommended methods for forthcoming surveillance. This volume, pp. 91–117.

ROSFELDER, A.M. & MARSHALL, N.F. (1967). Obtaining large undisturbed, and orientated samples in deep waters. In *Marine Geotechnique.* ed. A.F. Richards. University of Illinois Press, pp. 243–63.

RYGG, B. (1985). Distribution of species along pollution-induced diversity gradients in benthic communities in Norwegian fjords. *Mar. Pollut. Bull.,* **16,** 469–74.

SJÖGREN, C.E., ORELD, F., DRANGSHOLT, H., ØFSTI, T. & NORDENSEN, S. (1986). *Statfjord environmental survey 1986.* SI Report no 860704, Center for Industrial Research, Oslo, Norway.

SKEIE, G.M. & GRAY, J.S. (1988). Evaluating the incorporation rate of ^{3}H-thymidine into bacterial DNA as a measure of bacterial production in a subtidal silt-clay sediment of the Oslofjord (in preparation).

SMITH, K.L. JR (1973). Respiration of a sublittoral community. *Ecology,* **54,** 1065–75.

THIEL, H. (1966). Quantitative Untersuchungen über die Meiofauna des Tiefseebodens. *Veröffentlichungen des Institut für Meeresforschung in Bremerhaven,* **2,** 131–48.

UTERMOHL, H. (1958). Zur Vervollkommnung der quantitativen Phytoplankton-Methodik. *Mitt. int. Ver. Theor. Agnew. Limnol.,* **9,** 1–30.

WARWICK, R.M., GEE, J.M. & AMBROSE, W. JR (1986). Effects of the feeding activity of the polychaete *Streblosoma bairdi* (Malmgren) on meiofaunal abundance and community structure. *Sarsia,* **71,** 11–16.

WIKANDER, P.B. (1980). Biometry and behaviour in *Abra nitida* (Müller) and *A. longicallus* (Scacchi) (Bivalvia, Tellinacea). *Sarsia,* **65,** 255–68.

26

Long Term Recolonization and Chemical Change in Sediments Contaminated with Oil-based Drill Cuttings

T. BAKKE, J. A. BERGE, K. NÆS

Norwegian Institute for Water Research, P.O. Box 33 Blindern, N-0313 Oslo 3, Norway

F. ORELD

Center for Industrial Research, P.O. Box 124 Blindern, N-0314 Oslo 3, Norway

L.-O. REIERSEN

Norwegian State Pollution Control Authority, P.O. Box 8100 Dep, N-0032, Oslo 1, Norway

and

K. BRYNE

STATOIL, Div. SVK, P.O. Box 300, N-4001 Stavanger, Norway

ABSTRACT

Field experiments were conducted to test the suitability of different types and amounts of cuttings as substrate for a benthic community. Trays with defaunated sea-bed sediment and overlaid with test layers of water-based, low aromatic and diesel-based cuttings, were exposed on the sea floor at 11 m depth for 4 and 5·3 years. During this time a significant loss of oil hydrocarbons, 65–99%, was measured from the surface of the oil-based trays. Significant biodegradation of oil was observed in all trays within 2 years, with lag time positively correlated to dose. The oil contaminated sediments rapidly developed reduced conditions, but the redox values

521

*gradually increased, and the redox potential discontinuity layer shifted
downwards as animal recolonization and bioturbation increased. Macrofauna
recolonization during the first year indicated no recovery from the oil
contamination, but the fauna structure after 4 years showed that slight
contamination stimulated colonization by opportunistic species, presumably
as an effect of organic enrichment overrode any toxic effect. The heavily
loaded sediments (10 mm layer of cuttings) prevented macrofauna recolonization for 5·3 years, and since the layer of cuttings still formed a physical
barrier to burrowing, an extremely slow recovery can be expected. A
10 mm layer of water-based cuttings gave no apparent negative effect on
recolonization, but redox conditions were subnormal. Low aromatic cuttings
showed only slight improvement above diesel cuttings with respect to redox
conditions and recolonization.*

1. INTRODUCTION

Oil discharged together with drill cuttings constitutes nearly 90% of the
input of oil to the North Sea from offshore operations (Anon., 1987). The
drilling of an average well produces 800–1000 t of cuttings, which when
discharged contains about 7–25% oil from the oil-based mud. Since the
number of wells drilled from a platform may be in the range of 20–40,
the amount of oily material deposited on the sea floor is substantial. Close
to the platform the bottom may be completely covered by several meters
of anoxic cuttings, and devoid of any higher life. In general platform
monitoring has shown that the mode of disposal gives an abrupt reduction
in the cover of cuttings with distance from the platform (Davies *et al.*,
1984), but for platforms with a long drilling history, the zone where
subnormal community structure may be detected can extend beyond 2–
3 km (Matheson *et al.*, 1986; Hannam *et al.*, 1987; Reiersen *et al.*, this
volume).

In 1982 when the present project started, none of the North Sea
platforms using oil-based muds had terminated their drilling program,
and little was known on the recovery rates of the contaminated bottoms.
The aim of the study was therefore to experimentally assess the potential
recovery rate of medium polluted bottom areas after drilling has stopped,
and to test if recovery was related to the amount of cuttings and whether
the base oil was diesel or low aromatic.

In the period June 1982 to October 1987, two experiments were
conducted on the sea floor at a depth of 11 m to test changes in sediment

characteristics and rate of recolonization by benthic animals in trays with sediments contaminated with drill cuttings. The main sampling program of the experiments lasted to May 1984, and the results have been summarized by Bakke *et al.* (1986*a*). The main conclusion was that oil-based cuttings at a nominal thickness from 0·5 mm and upwards on the sediment surface ($\geq 12\ 000$ ppm total hydrocarbons in the upper 5 mm of the sediment) resulted in adverse redox conditions and seriously reduced invasion by animals. Differences in effects of diesel and low-aromatic cuttings were small. Water-based cuttings had no deleterious effect on the recolonization. The effects correlated positively with contamination level. No signs of recovery of the communities were apparent during this 2-year period, but biodegradation of the oil was indicated in all trays by the *n*-alkane: iso-alkane ratio, suggesting that long-term recovery might occur.

Since then the sediment trays, still in position on the sea floor, have been subjected to further natural environmental change and invasion by animals. On the basis of the indications of recovery represented by the decrease in hydrocarbons and signs of biodegradation of the oil, it was decided to reinvestigate the sediment trays in October 1987, 48 and 64 months after they had been deployed. The present paper compares the state of the tray communities at the end of the experiment.

2. MATERIAL AND METHODS

2.1. Experimental Design

Clean seabed sediment was collected at 5–15 m depth from an area close to the experimental site, and with the same sediment type. The sediment was frozen and thawed to kill the existing fauna, passed through 10 mm sieve to remove large objects, and mixed with clean granite sand and seawater in the proportion 3(sediment): 1(sand): 1(water) in a cement mixer to produce the control sediment. The median grain size of this sediment was 0·34 mm. PVC trays (1 × 1 m bottom area, 20 cm deep) were filled with a 12-cm layer of the control sediment. On top of this a layer of the appropriate cuttings was distributed with a nominal thickness ranging from 0·5 to 10 mm. The 5 and 10 mm trays were prepared by spreading a slurry of seawater and cuttings (2:3 by volume) in an even layer on top of the control sediment. When preparing the 0·5, 1 and 3 mm layers, the cuttings were mixed with control sediment in proportions which, when laid out in a 5 mm thick slurry, would represent the appropriate nominal thicknesses. The cuttings used, all from the Statfjord field, were: (1) water-

TABLE 1

Overview of experimental trays and analyses. WBM: water based cuttings, LAC: low aromatic cuttings, DOC: diesel cuttings

Tray no.	Treatment	Start date	Analyzed from last survey		
			Redox	Hydrocarbons	Macrofauna
1–4	Control	22 May 1982	—	—	—
5–8	10 mm WBM	22 May 1982	—		
9–12	10 mm LAC	22 May 1982	—	—	—
13–16	10 mm DOC	22 May 1982	—	—	—
17–18	Control	30 September 1983	—	—	—
19–20	0·5 mm LAC	30 September 1983	—	—	—
21–22	1 mm LAC	30 September 1983	—	—	—
23–24	3 mm LAC	30 September 1983	—	—	—
25–26	5 mm LAC	30 September 1983	—		
27–28	10 mm LAC	30 September 1983	—		
29–30	1 mm DOC	30 September 1983	—	—	—
31–32	3 mm DOC	30 September 1983	—	—	—
33–34	5 mm DOC	30 September 1983	—		
35–36	10 mm DOC	30 September 1983	—		

based cuttings from screen 80/100 (WBM-cuttings, 10 mm layers only), (2) low-aromatic oil-based cuttings from screen 60/60, reported retort value 31·1% of dry cuttings (LAC-cuttings, 0·5, 1, 3, 5 and 10 mm layers), and (3) diesel oil-based cuttings from screen 100/100, reported oil content 10·1% of dry cuttings (DOC-cuttings, 1, 3, 5 and 10 mm layers). Altogether 36 experimental trays were prepared for the two experiments (including 6 trays containing only control sediment (Table 1).

The trays were transported undisturbed to the experimental site, a semi-sheltered level bottom at 11 m depth in Raunefjorden, SW of Bergen, western Norway, and positioned 4 meters apart on the sea floor. In the center of the tray matrix, a rig with an Aanderaa RCM 4 current meter and two replaceable sediment traps was positioned 1·5 m above the bottom.

2.2. Sampling Program and Analysis

The main sampling and analysis program during 1982–4 has been described by Bakke *et al.* (1986*a*). The final survey was a modification of this. The trays sampled in the final survey (Table 1) were the control trays, all trays with layers of cuttings less than 10 mm, and 3 parallel trays from 10 mm WBM, 10 mm LAC and 10 mm DOC. All samples were taken by divers using hand held corers.

2.2.1. Redox Profiles

Two cores (inner diameter 45 mm) were taken from each tray and sealed off by rubber bungs for measurement of redox (*Eh*) profiles. Immediately upon sampling the cores were mounted vertically on a stand and redox values obtained by gently lowering a platinum electrode (calomel reference) at 1 cm intervals down to 6 cm in the sediment. The electrode was allowed to stabilize for 1 min before each reading was taken. Profiles from 3 cores were obtained for each substrate type.

2.2.2. Hydrocarbons

Two cores (inner diameter 45 mm) were taken from each tray and sectioned at 10, 20 and 30 mm sediment depth. The sections 0–10 and 20–30 mm were wrapped separately in cyclohexane (p.a. grade) washed aluminium foil and kept frozen until analyzed. Analysis was performed on a selected set of samples by capillary gas chromatography for total hydrocarbon concentration (THC), and coupled gas chromatography–mass spectrometry for the sum of naphthalene, phenanthrene/anthracene, dibenzothiophene, pyrene/fluoranthene, and their C_1–C_3 alkyl derivatives (NPD), and ratios between n-C_{17}:pristane and n-C_{18}:phytane. Some samples were analyzed for THC only. Prior to analysis the two parallel core sections from each tray were homogenized and a subsample from each pooled for a combined analysis. After addition of deuterized internal standards, the samples were extracted 3 times with dichloromethane at sonication (Sonicator Cell Disruptor Model W-375). After centrifugation the supernatant extract was decanted off, pooled, dried with anhydrous sodium sulphate, and the extract volume adjusted. The amounts of THC and NPD were related to dry weight of the samples by drying an aliquot at 110 °C for 48 h.

2.2.3. Benthic Macrofauna

Four cores (88 mm inner diameter) were taken to the bottom of each tray and pooled two by two in separate containers, giving 2 samples each covering a sediment surface area of 121 cm^2. The samples were washed on 1 mm sieve (circular holes) and the material on the sieve preserved in 10% neutralized formalin. Prior to sorting and identification of the fauna the samples were transferred to 70% ethanol and stained with Rose Bengal solution (Thiel, 1966).

3. RESULTS

3.1. Sediment Appearance

The appearance of the tray surfaces differed little from observations made after 1 and 2 years. The control and WBM tray surfaces were sandy and very uneven due to bioturbation, primarily by *Arenicola marina* and *Pectinaria koreni*, and much like the surrounding bottom. None of the LAC and DOC trays had been colonized by these species. Their surfaces were even and with patches of algal debris, which were not found in the control. The oiled surfaces were partially covered with grey and white mats of heterotrophic growth, and the degree of cover increased with contamination. In some of the trays, sea urchins (primarily *Echinus acutus*) and starfish *Asterias rubens* crawled on the surface, but their occurrence appeared not to be related to contamination. The core sampling showed that the 10 mm layers of LAC and DOC still formed a well-defined layer on top of the seabed sediment, but it appeared less oily than before. In the cores from the 0·5–3 mm trays the surface sediment was generally not discernible from the sediment beneath. The most contaminated trays (5 and 10 mm) had a distinct smell of oil and H_2S.

3.2. Hydrocarbons

The hydrocarbon content of control and WBM substrates was quite low, 3–8 ppm (mg/kg) for THC and 0·02–0·19 ppm for NPD. There was no increase in the concentrations with time, showing that no secondary contamination from the oiled trays had occurred.

All the contaminated sediments showed lower mean THC values than initially in the upper 1 cm (Table 2). In the 1 mm LAC trays the THC concentration was reduced significantly ($p < 0·05$, t-test) from 10 300 ppm to 117 ppm, i.e. 99% of the THC had disappeared from the surface. Most of the loss (72%) occurred during the first year. NPD concentrations were not analyzed at the end, but the NPD loss the first year was 85%. The 1 mm DOC trays had less THC initially: 3300 ppm. After 48 months the mean level was 680 ppm, but the decrease was not significant due to a very large difference in concentration between the last two replicate samples (cf Table 3). The mean value alone would represent a loss of 79%, out of which 56% disappeared the first year. The corresponding loss in NPD the first year was 48%.

The 3 mm trays showed the same trends (Table 2): a somewhat higher fraction of THC lost from the LAC than the DOC trays, the loss was only significant in the LAC ($p < 0·05$), and most of the loss occurred during the first year.

TABLE 2

Mean total hydrocarbon concentrations (THC, ppm of dry sediment) and percentage loss of initial THC in the upper 1 cm of the tray sediments. The mean concentrations of selected aromatic compounds: naphthalene, phenanthrene/anthracene, dibenzothiophene, pyrene/fluoranthene and their C_1–C_3 alkyl derivatives (NPD, ppm of dry sediment) for the 10 mm trays are also given. LAC: low aromatic oil cuttings, DOC: diesel oil cuttings, n.a.: not analyzed.

Tray	Initial ppm	12 months ppm	12 months % loss	48 months ppm	48 months % loss	64 months ppm	64 months % loss
THC							
1 mm LAC	10 300	2 880	72	117	99	n.a.	—
1 mm DOC	3 300	1 450	56	682	79	n.a.	—
3 mm LAC	48 800	15 000	69	2 490	95	n.a.	—
3 mm DOC	14 500	6 300	57	2 260	84	n.a.	—
10 mm LAC	86 400	82 100	5	n.a.	—	30 600	65
10 mm DOC	126 000	92 500	27	n.a.	—	27 400	78
NPD-concentrations (ppm)							
10 mm LAC	1 330	668	50	n.a.	—	158	88
10 mm DOC	3 430	2 120	38	n.a.	—	899	74

TABLE 3

Total hydrocarbon concentrations after 48 months (THC, ppm of dry sediment) measured in the 0–1 and 2–3 cm level of individual sediment cores, and mean levels of THC calculated for the 0–3 cm sediment depth. Percentage loss of THC not accounted for by burial is also given.

Treatment	Control		1 mm LAC		1 mm DOC		3 mm LAC		3 mm DOC	
Corer No.	a	b	a	b	a	b	a	b	a	b
Section:										
0–1 cm	<20	<20	144	90	1 340	23·8	3 290	1 690	4 260	262
2–3 cm	<20	<20	94·1	150	25·3	40·0	1 820	1 380	926	1 290
Mean THC 0–3 cm	<20		120		357		2 050		1 690	
Theoretical surface THC[a]			360		1 070		6 140		5 060	
% loss from day 1			97%		68%		87%		65%	

[a]If all the THC were concentrated in the surface 1 cm.

The THC concentrations in the 2–3 cm depth of the 1 and 3 mm trays showed clearly that oily material had been transported down into the sediment (Table 3). All the sections from this depth had higher THC than at corresponding depths in the control. The cores from 1 and 3 mm LAC indicated reasonably well mixed sediments down to 3 cm. The 1 and 3 mm DOC trays had a very patchy distribution of cuttings at the surface, but less patchy at 2–3 cm depth.

Final hydrocarbon concentrations at the surface of the 10 mm LAC and DOC trays were only analyzed from the sediments which had been out for 64 months (Table 2). During the first 2 years these trays showed no significant reduction in THC (Fig. 1). Between 12 and 64 months the mean THC level was reduced significantly from 82 100 to 30 600 ppm in the 10 mm LAC ($p < 0.05$, t-test) and from 92 400 to 27 400 ppm in the 10 mm DOC ($p < 0.01$). Hence, over the total experimental period the chemical analyses showed that the highest dosed sediments lost significant amounts of THC from the upper 1 cm, even though a distinct layer of cuttings was retained at the sediment surface until the end.

A significant reduction in NPD ($p < 0.01$) was observed during the first year in both LAC and DOC 10 mm (Table 2), mainly occurring during the first 3 months of the experiment. For the period 12–64 months the LAC trays had a further significant decrease ($p < 0.01$), but not the DOC due to patchiness at the end of the experiment. Table 2 also shows that although the NPDs were most rapidly lost from the cuttings, the total loss after 64 months was about the same for THC and NPD.

The ratios of n-C_{17}: pristane and n-C_{18}: phytane in the 10 mm LAC and DOC trays remained high and unchanged for the first 10 months (Fig. 2). Between 10 and 24 months a substantial biodegradation of alkanes was indicated in both substrates by the ratios, gradually decreasing to 0·3–0·6. The ratios after 64 months were similar to this and to the control sediment (0·7–0·8), indicating an almost complete biodegradation of the normal alkanes within 2 years. In the 0·5–5 mm trays of both types of cuttings, biodegradation of alkanes was indicated after 3–6 months and the ratios were clearly below 1 within 1 year (Bakke *et al.*, 1986a).

3.3. Redox Profiles

At the start of the experiment all redox profiles were similar with positive *Eh* values down to a depth of at least 6 cm. During the first year all contaminated trays showed a shift towards negative *Eh* values with no apparent concomitant change in the control, and at 12 months a redox

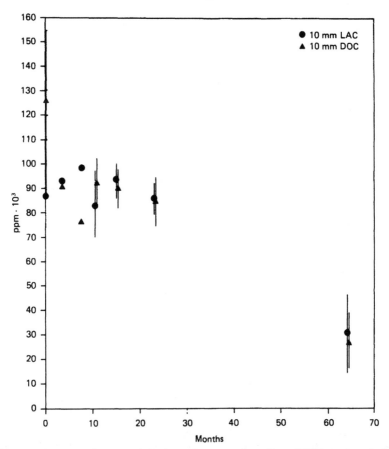

FIG. 1. Mean surface total hydrocarbon concentration (THC, ppm of dry sediment) in the 10 mm low aromatic (LAC) and diesel (DOC) cuttings trays from June 1982 to October 1987. Each point is based on 8 replicate cores. Vertical bars give ±1 SD. Where no bars are given the core samples were pooled. Control sediment values were below 10 ppm.

potential discontinuity (RPD) layer was generally found in the upper 0–1 cm in all LAC and DOC trays (Fig. 3). From 12 to 48 months this trend was reversed, and the RPD layer shifted down to between 1 and 4 cm depth. The shift was least pronounced in the 10 mm trays.

Representative *Eh* readings in the upper 2–3 cm of the sediment were not always easy to obtain due to the RPD layer creating large differences

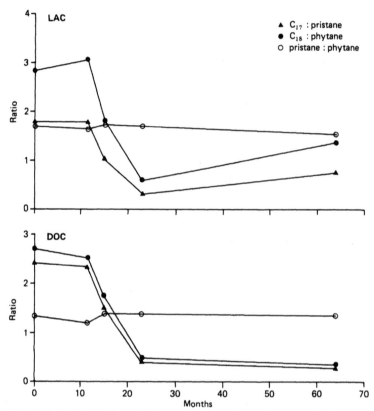

FIG. 2. Reduction in ratios of n-C_{17}: pristane, n-C_{18}: phytane, and pristane: phytane in the 10 mm low aromatic (LAC) and diesel (DOC) trays from June 1982 to October 1987.

in *Eh* over small vertical distances. Hence, we also focused on values below 3 cm sediment depth which were more stable and discriminating.

After 12 months all the oil contaminated substrates had significantly lower redox potential at 5·5 cm depth than the control ($p < 0·05$, Mann–Whitney U-test, Fig. 4). There were only small differences among trays loaded with the same cuttings, e.g. 0·5 mm LAC was not significantly different from 10 mm LAC. In general the redox conditions of the LAC trays were significantly better than the corresponding DOC trays at 5·5 cm, but the 10 mm LAC was not different from the 10 mm DOC.

Improved redox conditions were observed in the deeper strata of all

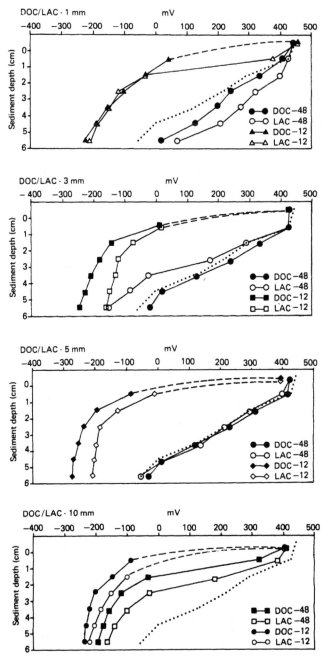

FIG. 3. Vertical profiles of *Eh* (mV) at 12 and 48 months in corresponding trays with diesel (DOC) and low aromatic (LAC) cuttings. The control sediment profile at 48 months is included in the graphs (dotted line).

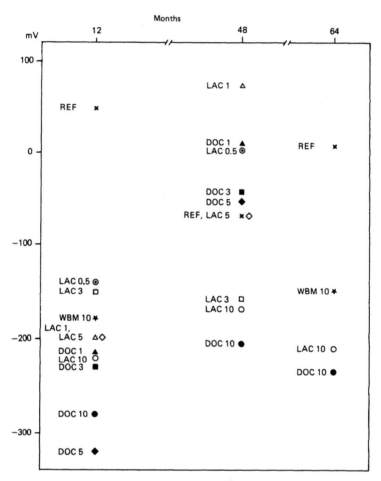

Variation in redox potentials with exposure time
and substrate type at 5.5 cm sediment depth.

FIG. 4. Redox (*Eh*) values at 5·5 cm sediment depth in the trays after 12, 48 and
64 months. The values at 12 and 48 months are from Exp. 2, those from 64
months from Exp. 1 (cf. Table 1).

oiled substrates after 48 months, and this was most pronounced in the low and medium contaminated trays (except 3 mm LAC). These substrates now grouped around the control with redox values around 0 to −50 mV at 5·5 cm sediment depth. There was a slight, but significant, shift towards better *Eh* values in the LAC trays compared with the DOC tray group. The *Eh* value of the WBM trays at 5·5 cm was also clearly below the control after 12 months and did not improve after 64 months.

3.4. Benthic Macrofauna

After 48 months the pattern in faunal response to the two types of cuttings was essentially the same when related to the initial amount of material added (0–3 mm trays, Figs 5–6). For both LAC and DOC, a significant stimulation of the fauna was seen for most parameters at the lowest concentrations (0·5 and 1 mm). Even at 3 mm, of the frequently occurring species, only *Mediomastus fragilis* (Fig. 6(B)) and *Scoloplos armiger* (Fig. 6(C)) were less abundant than in the control. This stimulation was seen for total density of individuals (Fig. 5(A)), number of species (Fig. 5(B)), polychaete biomass (Fig. 5(C)), and for densities of most species and faunal groups which were abundant in the samples (Fig. 6(A–D)). Only *Scoloplos armiger* showed a systematic reduction in density with dose, especially pronounced in the DOC trays (Fig. 6(C)). The results further showed that the colonization intensity in the 1 mm trays was stronger in the LAC substrates than in the DOC substrates, both with respect to number of species and individual densities.

The numbers of species and individual densities were further used to calculate Hurlbert's rarefaction curves (Hurlbert, 1971) as a measure of community diversity (Fig. 7). In the DOC substrates diversity was negatively related to dose (Fig. 7(B)), in the LAC substrates the 0·5 mm trays had the same diversity as the control, the 1 mm and 3 mm trays had less and about equal diversity (Fig. 7(A)). Superimposing the endpoints of the curves in Fig. 7 upon a classification diagram for the general distribution of Hurlbert curves at perturbed or unperturbed bottoms in Norwegian fjords (Fig. 8), suggested that the control and 0·5 mm LAC trays had reached normal diversity after 48 months, the 1 and 3 mm LAC and 1 mm DOC substrates were at the border between normal and moderate diversity, and the 3 mm DOC substrates clearly moderate.

The trays with 10 mm LAC and DOC layers had very little recolonization during the first 12 months. The animals were only found in the detritus at the surface, and LAC contained a higher number of individuals and species than DOC (Bakke *et al.*, 1986a). After 24 months the only

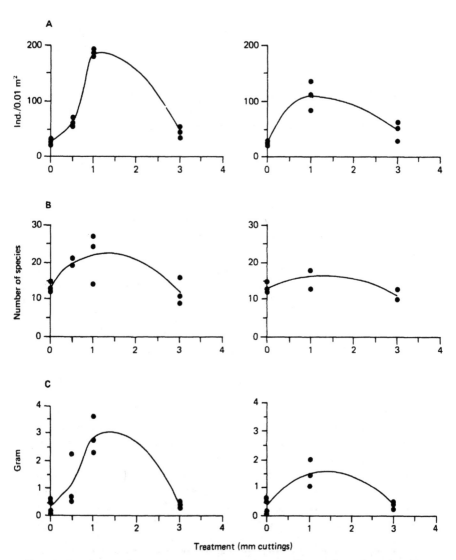

FIG. 5. Total macrofauna abundance (A) (ind. $0.01\,\text{m}^2$), number of species (B), and polychaete biomass (C), (g wwt $0.01\,\text{m}^2$) after 48 months as function of initial thickness of cuttings. Points show individual sample values. Left column: low aromatic (LAC) trays, right column: diesel (DOC) trays. Control trays represent zero thickness.

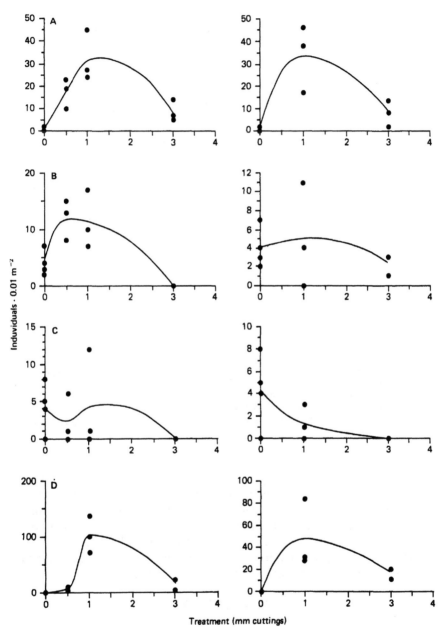

FIG. 6. Densities (ind. $0.01\,\mathrm{m}^{-2}$) of cirratulidae (A), *Mediomastus fragilis* (B), *Scoloplos armiger* (C), and *Polydora socialis* (D) after 48 months as a function of initial thickness of cuttings. Points show individual sample values. Left column: low aromatic (LAC) trays, right column: diesel (DOC) trays. Control trays represent zero thickness.

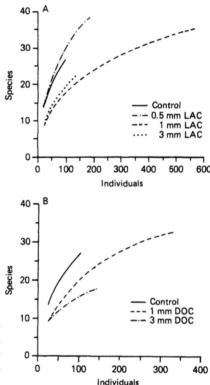

FIG. 7. Hurlbert's rarefaction curves for macrofauna in the 0·5–3 mm low aromatic (LAC) and diesel (DOC) cuttings trays after 48 months in comparison with the control trays.

real macrofauna species found in these trays was *Capitella capitata*, which did not differ in density among the trays. After 64 months the 10 mm LAC and DOC substrates still contained an extremely reduced macrofauna (Table 4) where *C. capitata* and a few individuals of cirratulids were the only infauna species recorded. In the control trays which had been out for the same time, the number of species and individual densities were in the same range as the control trays exposed for 48 months.

4. DISCUSSION

The concentrations of THC in the upper 10 mm after 12 months ranged from 2500 ppm and upwards in the LAC trays and from 1500 ppm and upwards in the DOC trays. Analyses showed no downwards migration of oil from the cuttings. If one assumes a complete mixing of this oil into

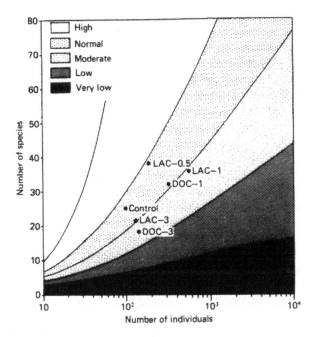

FIG. 8. Number of tray species as function of total individual density superimposed on a diversity classification graph (Rygg, 1984) based on 68 Norwegian perturbed and unperturbed fjord communities.

TABLE 4

Fauna composition in the trays dosed with 10 mm of diesel (DOC), and 10 mm of low aromatic (LAC) cuttings after 64 months of exposure on the sea floor, compared to control trays (two trays analyzed for each treatment).

Total number of species	10 mm LAC	10 mm DOC	Control
	3	6	16
Individual densities of infauna			
Capitella capitata	7	13	0
Scoloplos armiger	0	0	8
Cirratulids	1	1	0
Other polychaetes	0	1	16
Molluscs	1	1	12
Echinoderms	0	0	3

the upper 5 cm of the sediment, which is the average depth of an offshore grab sample, the concentration would correspond to what has been found from about 1 km and inwards towards a platform in the main current direction for both types of cuttings (e.g. Sjögren *et al.* (1986) for Statfjord A). In general, the tray conditions after 1 year, in which no apparent fauna recovery had occurred, represented a transect from the outer part of the transition zone and well into the zone of 'highly modified benthic community' according to Davies *et al.* (1984).

After 48 months the THC of the 1 mm trays (Table 3) would correspond to the THC at stations between 1000 and 2500 m downstream from Statfjord A, and the 3 mm trays to 500 m downstream. A comparison of the expected number of macrofauna species per 100 individuals ($E(S_{n=100})$) derived from the curves in Fig. 7 with $E(S_{n=100})$ values derived from Hurlbert curves of sediment fauna found around Statfjord A in 1984 (IOE, 1985), further showed that the 1 mm and 3 mm substrates had diversities similar to stations about 750 m from the platform, and the 0.5 mm LAC similar to the stations at 1500 m. Hence the relation between THC and diversity of the experimental trays showed a reasonably good correspondence with similar relationship around a typical North Sea platform.

About 80–90% of the initial THC were lost over the course of the experiment. The mechanisms involved would be leakage, resuspension, burial, biodegradation, and loss through the sampling. The total number of samples withdrawn from one tray during the experiment removed about 10% of the surface THC, and although resampling of an area was avoided, this loss has to be included in the total THC budget since redistribution of the surface sediments with time was indicated. Burial was not directly measured, but strongly indicated in the 1 and 3 mm trays. Oil hydrocarbons were found at 2–3 cm sediment depth, the RPD layer had shifted down from 1 to about 3 cm depth suggesting bioturbation, and some of the larger macrofauna recorded are strong burrowers (e.g. young *Mya arenaria*). On the basis of the THC levels found in the 0–1 and 2–3 cm sediment strata (Table 3) and the simple model of even distribution of the THC down to the RPD layer at approximately 3 cm depth, the theoretical THC residues above the RPD layer were calculated (Table 3). Comparison with Table 2 shows that burial could explain from 2 to 20% of the loss of surface THC over the years. Based on these considerations, the loss due to the other factors: biodegradation, resuspension, and leakage to the overlying water would constitute at least 70–90% of the reduction in THC. Based on measurements during the first year, Bakke *et al.* (1985) suggested that leakage had removed less than 2% of the oil lost during that time from the 10 mm LAC and DOC trays. Hence, the main factors were biodegradation and possibly resuspension.

Bioturbation facilitates exchange of pore water, gases, etc., between the bottom water and the upper layers of the sediment, and thus prevents development of stagnant anoxic conditions. Successful sediment recoloniz-ation is therefore dependent on the invasion by bioturbators. This process occurred during the first year in the control sediments, and the main bioturbators were the polychaetes *Pectinaria* (primarily *P. koreni*) and *Arenicola marina*. At the same time, recolonization of the contaminated trays was only superficial and did not penetrate the cuttings. After 4 years animal invasion of deeper sediments had also occurred in the medium contaminated substrates, as shown by the downwards movement of the RPD layer and the burial of hydrocarbons. The lag phase could be due to toxicity of the cuttings, negative stimuli from the cuttings in the settling process of macrofauna larvae, or their influence on sediment physical structure. The latter could obstruct the burrowing of animals which had settled. Direct oil toxicity is less likely since both LAC and DOC had similar influence on the recolonization, in spite of different acute toxicity, but toxicity of other components of the low aromatic drill mud was suggested (Bakke *et al.*, 1986*a*).

The shift in RPD down to about 3 cm in the low and medium contaminated trays after 48 months corresponded reasonably well with the depth distribution of the macrofauna species encountered. This depth range in bioturbation would also explain the lack of correspondence between fauna development and *Eh* values at 5·5 cm depth, especially expressed in the WBM substrates. Still the redox conditions, even at 5·5 cm sediment depth, was to some extent influenced by the bioturbation. Redox separated the substrates into two groups, with the low and medium contaminated trays in one group and the 10 mm trays in the other, but with the 3 mm LAC as an exception to the trend. There were no significant differences within these groups. The significant lowering of the control redox at 5·5 cm between 12 and 48 months is difficult to explain, especially since the tray appearance and fauna content suggested strong bioturbative activity in the control sediments. In the 10 mm trays the surface layer of oily material was intact after 64 months, and not different in appearance from the 12 months samples. The redox results showed that this layer still represented an efficient barrier to animal penetration and exchange of pore water.

The supply of organic carbon (sedimentation, photo- and chemo-

T. Bakke et al.

autotrophic production) to marine sediments regulates the secondary production in the benthos. Hydrocarbons associated with drill cutting may function as substrate for bacterial production (Gillam *et al.*, 1986), and are thus potentially available for benthic consumers. In areas with natural oil seeps and high concentrations of hydrocarbons in the sediment, it has been documented that both macrofauna and meiofauna increase in abundance (Spies & Davis, 1979; Montagna *et al.*, 1987). This increase is hypothesised to be a consequence of increased bacterial activity (Spies & Davis, 1979; Montagna *et al.*, 1987).

In an area where the supply of organic carbon is a limiting factor for benthic production, it is expected that moderate amounts of hydrocarbons (up to the level where toxic effects become predominant) may enrich the sediment and thus increase secondary production. In our experiment a significant stimulation of the fauna was seen at the low (0·5–1 mm) concentrations after 48 months compared to both the control sediment and the sediment dosed with 3 mm of cuttings. This stimulation was not apparent after 12 months. Sampling along transects from offshore drilling platforms show distinct macrofaunal zones (Hannam *et al.*, 1987) and several opportunistic macrofaunal species have maximum density at a certain distance from the platform. In the Thistle oilfield, peak densities of cirratulid polychaetes have been found in the range 0·5–1 km from the platform (Hannam *et al.*, 1987), and similar distribution has been recorded for the Murchison field (Mair *et al.*, 1987). The increase both in our experiment and in the field may be explained by trophic enrichment via microbial hydrocarbon degradation in the sediment, but it can also be a result of biological interaction, e.g. reduced predation pressure. In our experiment the accumulation of organic detrital matter on the sediment during the phase when recolonization was suppressed would also have a stimulatory effect on colonization when the negative influence of the cuttings ceased.

A reduction in negative effects of cuttings with age, either due to loss of compounds or change in toxicity is consistent with recent observations that the zones of macrobenthic impact surrounding offshore installations may shrink when the drilling is terminated (Hannam *et al.*, 1987; Mair *et al.*, 1987). It is in this respect interesting to note that the levels and percentage reduction of THC from 12 to 48 months in the 1 mm trays were similar to the levels and percentage reduction detected at the inner stations around the Murchison oilfield during the first 16 months after drilling was terminated (Mair *et al.*, 1987), concomitant with a significant improvement in macrofauna. Most of the species found to increase in abundance in the transition zone around a platform are species which live at or near the sediment surface. They feed on the surface and should be well adapted to utilize bacteria associated with particulate material at or near the sediment surface. Surface restricted behaviour is optimal in areas where the sediment subsurface conditions are unfavorable because of reducing conditions or toxic compounds (e.g. H_2S), but not where toxic compounds are deposited. On the other hand the sediment surface is the part of the habitat where toxic compounds are most likely to be depurated (Berge *et al.*, 1987) and thus most rapidly regain an acceptable habitat for benthic animals. Hence, we believe that subsurface deposit feeders are less likely to be stimulated in an area contaminated with cuttings because (1) bacterial degradation takes place mainly at the sediment surface and (2) subsurface conditions are not within their range of tolerance.

The total fauna densities of the trays were less than desired to obtain fully reliable rarefaction curves. Still the curves showed that the stimulation of abundance and numbers of species in the 1 mm LAC and DOC trays, did not give higher diversity than in the control, i.e. the stimulation in the 1 mm trays was more an increase in abundance than in number of species. The results also showed that the increase in individual density was due primarily to the 'new' species, mainly polychaetes such as *Cirratulus cirrata*, *Chaetozone setosa*, other cirratulids, and *Polydora socialis*. These species, all surface deposit feeding opportunists, were therefore not only attracted to the 1 mm trays more than to the control, but they were also very successful colonizers in terms of abundance. Hence, the stimulation of the 1 mm trays had all indications of opportunist invasion, which in this context is only the first step towards a normal community. Furthermore, harmful effects after 48 months were only recognized for *Scoloplos armiger* and *Mediomastus fragilis* both of which are recognized as subsurface deposit feeders (Fauchald & Jumars, 1979).

An increasing effect of toxic compounds or smothering was indicated at contamination above 1 mm, but even at 3 mm these effects were not sufficient to reduce the positive effect of the organic enrichment to a level below that of the control sediment for parameters other than the density of *S. armiger* and *M. fragilis*. It is however important to note that the density and number of species in the control samples were sparse which might indicate that the net flux of organic carbon to these sediments was small and limited benthic production/biomass. The sediment trap measurements showed an annual gross sedimentation rate of 102 g carbon/m^2/yr during 1982/3 (Bakke *et al.*, 1985), which is normal in comparison with trap data from other shallow fjords (Wassmann, 1983), but the rates of resuspension from the traps are considered to be less than from the sediment trays, and estimates of real or net flux of sedimented material available for the benthos would be speculative. Still the range below our experimental control fauna values was narrow, which to some extent would limit the possibility of detecting subnormal conditions.

The very low density and number of species in the 10 mm LAC and DOC trays after 64 months represented no improvement over the findings at 24 months (Bakke *et al.*, 1986*a*) and show that the conditions in these trays were very stable, in spite of the fact that a significant 70–80% reduction in hydrocarbons had occurred. Recolonization was extremely slow and supports our earlier view that areas which receive sufficient amounts of cuttings to produce a smothering surface layer will not be recolonized until this layer is broken down or buried. Even though the integrity of the 10 mm layers had been broken for at least 3 years by the samples taken within the first 2 years, this was not sufficient to elicit any recolonization.

One of the prime purposes of the experiment was to compare the environmental acceptability of low-aromatic versus diesel cuttings. The results from 12 and 24 months revealed no significant improvement of the LAC over the DOC (Bakke *et al.*, 1986*a*). After 48 and 64 months LAC was slightly superior to DOC with respect to redox conditions and percentage loss of THC with time, and the fauna abundance stimulation in the 1 mm trays was slightly stronger in the LAC. Other than that, the recovery from LAC contamination was not found to be better than from DOC. Hence, although the initial acute toxicity of the low-aromatic cuttings was 1–2 orders of magnitude less than for the diesel cuttings (Bakke *et al.*, 1986*b*), this had only a marginal influence on the long-term effects of the cutting on the sea floor.

ACKNOWLEDGEMENTS

Thanks are due to Mobil Exploration Norway Inc. on behalf of the Statfjord Unit Owners for providing the cuttings and funding the main phase of the project in 1982–4, to the State Pollution Control Authority, Norway and Den norske stats oljeselskap a.s STATOIL for funding the last sampling survey and analysis, and to two anonymous reviewers for constructive criticism of the manuscript.

REFERENCES

ANON. (1987). Annual report on discharges from platforms in 1985. The Paris Commission. GOP 11/3/1-E.

BAKKE, T., GREEN, N.W., NAES, K., PEDERSEN, A., SPORSTØL, S. & ORELD, F. (1985). Drill cuttings on the sea bed. Field experiments on recolonization and chemical changes. Phase 1. Thick (10 mm) layers of cuttings 1982–1983. Norwegian Institute for Water Research report no. 1752, Oslo, 202 pp.

BAKKE, T., GREEN, N.W., NAES, K. & PEDERSEN, A. (1986a). Drill cuttings on the sea bed. Phase 1 and 2. Field experiment on benthic recolonization and chemical changes in response to various types and amounts of cuttings. In *Oil based drilling fluids. Cleaning and environmental effects of oil contaminated drill cuttings. Proceedings.* Norsk Petroleumsforbund, Trondheim, pp. 17–31.

BAKKE, T., BLACKMAN, R.A.A., HOVDE, H., KJØRSVIK, E., NORLAND, S., ORMEROD, K. & ØSTGAARD, K. (1986b). Drill cuttings on the sea bed. Toxicity testing of cuttings before and after exposure on the sea floor for 9 months. In *Oil based drilling fluids. Cleaning and environmental effects of oil contaminated drill cuttings. Proceedings.* Norsk Petroleumsforbund, Trondheim, pp. 79–84.

BERGE, J. A., LICHTENTHALER, R. G. & ORELD, F. (1987). Hydrocarbon depuration and abiotic changes in artificially oil contaminated sediments in the subtidal. *Estuarine, Coastal and Shelf Science*, **24**, 567–83.

DAVIES, J.M., ADDY, J.M., BLACKMAN, R.A.A., BLANCHARD, J.R., FERBRACHE, J.E., MOORE, D.C., SOMERVILLE, H.J., WHITEHEAD, A. & WILKINSON, T. (1984). Environmental effects of the use of oil-based drilling muds in the North Sea. *Mar. Pollut. Bull.*, **15** (10), 363–70.

FAUCHALD, K. & JUMARS, P. A. (1979). The diet of worms: a study of polychaete feeding guilds. *Oceanogr. Mar. Biol. Ann. Rev.*, **17**, 193–284.

GILLAM, A. H., O'CARROLL, K. & WARDELL, J. N. (1986). Biodegradation of oil adhering to drill cuttings. In *Oil based drilling fluids. Cleaning and environmental effects of oil contaminated drill cuttings. Proceedings.* Norsk Petroleumsforbund, Trondheim, pp. 123–36.

HANNAM, M.D., ADDY, J.M. & DICKS, B. (1987). Seabed biological monitoring of drill cuttings discharges in the Thistle oil field. In *Fate and Effects of Oil in Marine Ecosystems*, ed. J. Kuiper & W.J. Van den Brink. Nijhoff, Amsterdam.

HURLBERT, S.N. (1971). The non-concept of species diversity. *Ecology*, **53**, 577–

IOE (1985). Statfjord environmental survey, June 1984. Final Report. Institute of Offshore Engineering, Edinburgh. Report No. IOE/84/251, 220 pp.

MAIR, J.McD., MATHESON, I. & APPELBEE, J.F. (1987). Offshore macrobenthic recovery in the Murchison field following the termination of drill-cuttings discharges. *Mar. Pollut. Bull.*, **18**, 628–34.

MATHESON, I., KINGSTON, P.F., JOHNSTON, C.S. & GIBSON, M.J. (1986). Statfjord Field environmental study. In *Oil based drilling fluids. Cleaning and environmental effects of oil contaminated drill cuttings. Proceedings.* Norsk Petroleumsforbund, Trondheim, pp. 3–15.

MONTAGNA, P.A., BAUER, J.E., YOAL, J., HARDIN, D. & SPIES, R.B. (1987). Temporal variability and the relationship between benthic meiofaunal and microbial populations of a natural coastal petroleum seep. *J. Mar. Res.*, **45**, 761–89.

REIERSON, L.-O., GRAY, J.S., PALMORK, K.H. & LANGE, R. (1988). Monitoring in the vicinity of oil and gas platforms; results from the Norwegian sector of the North Sea and recommended methods for forthcoming surveillance. This volume, pp. 91–117.

RYGG, B. (1984). Bløtbunnsfauna som indikatorsystem på miljøkvalitet i fjorder. Bruk av diversitetskurver til å beskrive faunasamfunn og anslå forurensningspåvirkning (Soft bottom fauna as indicator system for environmental quality of fjords. The use of diversity curves to describe animal communities and judge impact of pollution.) Norwegian Institute for Water Research Report No. F. 480, 39 pp. (in Norwegian).

SJÖGREN, C.E., ORELD, F., DRANGSHOLT, H., ØFSTI, T. & NORDENSEN, S. (1986). Statfjord environmental survey 1986. SI Report no 860704, Center for Industrial Research, Oslo, Norway.

SPIES, R.B. & DAVIS, P.H. (1979). The infauna benthos of a natural oil seep in the Santa Barbara Channel. *Mar. Biol.*, **50**, 227–37.

THIEL, H. (1966). Quantitative Untersuchungen über die Meiofauna des Tiefseebodens. Veröffentlichungen des Institut für Meeresforschung in Bremerhaven, **2**, 131–48.

WASSMANN, P. (1983). Sedimentation of organic and inorganic particulate material in Lindåspollene, a stratified, land locked fjord in western Norway. *Mar. Ecol. Prog. Ser.*, **13**, 237–48.

27

The Fate and Partitioning of Hydrocarbon Additives to Drilling Muds as Determined in Laboratory Studies

PAUL D. BOEHM*, JOHN BROWN and ADOLFO G. REQUEJO‡

Battelle Ocean Sciences, 397 Washington Street, Duxbury, Massachusetts, 02332, USA

ABSTRACT

A series of laboratory partitioning experiments was conducted to determine the fate and partitioning of individual and total hydrocarbon components in diesel oil–drilling mud mixtures. A generic mud (No. 8) was mixed in several concentrations with a high sulfur (5%) diesel oil, and this mixture was in turn introduced to seawater in several concentrations. After agitation, settleable solids, aqueous ('dissolved'), and suspended particulate phases were isolated and detailed analysis were conducted by high resolution gas chromatography and gas chromatography/mass spectrometry.

Results indicate that under extremely high particle loadings (7000 mg/ liter) and quiescent conditions about 20% of the diesel oil was associated with the settleable particles. With time and more realistic mixing conditions, the diesel oil was partitioned into the aqueous and suspended particulate phases. At greater dilutions (700 mg/liter) and lower initial diesel oil concentrations, much of the hydrocarbon component was accommodated and/or solubilized in the aqueous phase. Very small amounts of the total diesel oil added (less than 2%) were associated with the settleable solid phase. Excellent agreement between two sets of experiments support the validity of the results. The results indicate that in the field a significant amount of diesel oil (and presumably other similar petroleum material) can be expected to be volatilized initially (about 50%) upon discharge and that transport of these petroleum additives to the bottom sediment can be expected

* Present address: Marine Sciences Unit, Arthur D. Little Inc., Acorn Park, Cambridge, Massachusetts 02140, USA.
‡ Present address: ARCO Oil and Gas Company, 2300 West Plano Parkway, Plano, Texas 75075, USA.

to be a very minor process. Detailed results addressing individual saturated and aromatic hydrocarbon compound partitioning are discussed. Application of these results to other potential hydrocarbon additives are discussed as well.

1. INTRODUCTION

The US Environmental Protection Agency (EPA) will be publishing Best Conventional Technology (BCT), Best Available Technology (BAT), and New Source Performance Standard (NSPS) guidelines pertaining to offshore effluents of the oil and gas industry. A comparative study on the environmental chemistry and toxicity to marine organisms of typical lignosulfonate drilling fluids containing mineral oil and diesel fuel additives was conducted to address issues related to the development of new discharge regulations (Breteler *et al.*, 1985).

The primary objective of the part of the study reported here was to assess the potential environmental fate of a typical drilling fluid heavily treated with barite and lignosulfonate (generic mud No. 8) and containing diesel oil, upon discharge into the marine environment. The objective was addressed by measuring the partitioning of hydrocarbons among liquid, suspended particulate, and solid phase components of 1% and 0·1% mixtures of drilling mud–diesel oil mixtures in seawater in the laboratory experiments.

Diesel fuel sometimes is added to water-based drilling fluids for fluids control or to improve the lubricating properties of the mud when drilling a slanted hole. As much as 2–10% diesel fuel may be added under some circumstances. Because much of the added oil may quickly become associated with drilling mud particulates and with surfactant material, ocean discharge of drilling muds containing this amount of oil rarely results in an oil sheen on the water surface. Sometimes, a 'pill' of oil or oil-based drilling mud is used in an emergency to free stuck pipe. Usually this pill is kept separate from the bulk mud system and is recovered and disposed of onshore. Nevertheless, a small amount of oil from the pill may get into and become mixed with the bulk drilling mud, and may be ultimately discharged to the marine environment.

Because of the concern about the toxicity of diesel fuel additives in water-based drilling fluids destined for ocean disposal, the oil industry has adopted the use of alternative lubricants in some cases. Vegetable oils and various mineral oils have been recommended as being more environmentally acceptable than diesel fuel.

A critical factor in assessing the environmental impact of discharge to the oceans of drilling fluids containing hydrocarbon additives is the fate of the hydrocarbons in the ocean. Following discharge to the ocean, a drilling fluid rapidly separates into several discrete fractions which behave differently from one another (Ayers *et al.*, 1980). A dense particulate fraction (approximately 90% of the mud solids) settles rapidly to the bottom. Finer and lighter clay-size particles remain suspended in a surface plume and may eventually settle to the bottom over a wide area at some distance from the discharge. Water-soluble components of the mud remain with the surface plume initially and are diluted over a wide area in the water column. Dilution of the surface plume is considered to be sufficiently rapid that no adverse effects are expected on water column organisms. The rapidly-settling fraction may impact benthic communities near the discharge site. If hydrocarbon additives remain associated with this rapidly-settling fraction of drilling mud, they could contribute to its impact on the benthos. If they remain with the upper plume, their impact is likely to be negligible. Previous studies on the partitioning of No. 2 fuel oil between solid and aqueous phases in seawater have been conducted by Zurcher & Thuer (1978) and Gearing *et al.* (1980) while preliminary studies of the partitioning of diesel oil in drilling mud dispersed in seawater were described in an unpublished masters thesis by Anne (1981).

The research program described herein was designed to determine the partitioning of hydrocarbon additives among dissolved and different particulate phases following discharge of drilling fluids to the ocean.

2. METHODS AND MATERIALS

2.1. Experimental Design

A series of laboratory partitioning experiments was conducted to determine the fate of total hydrocarbons, of selected *n*-alkanes and the aromatic hydrocarbon series in a drilling mud–diesel additive formulation upon mixing with seawater. The mud-additive stock material formulation used for these experiments was the generic mud No. 8 containing 5% high sulfur diesel fuel. The composition of generic mud No. 8 is presented in Table 1. The composition of the high sulfur diesel fuel is presented in the Results section. This formulation was chosen because it contains the highest concentration of a diesel additive with a high concentration of potentially toxic aromatic hydrocarbons and thus represents a 'worst case' situation. The mud formulation was transferred from its sealed 4-gallon

TABLE 1

Composition of generic mud no. 8 (lignosulfonate freshwater mud) and formulation of the base mud used in this program

Components	Generic mud no. 8 (lbs/bbl)
Bentonite	10–50
Barite	0–450
Caustic	2–5
Lignosulfonate	4–15
Lignite	2–10
Drilled solids	20–100
Cellulose polymer[a]	0–2
Soda ash/sodium bicarbonate	0–2
Lime	0–2
Fresh water	As needed[b]

[a]Sodium carboxymethyl cellulose.
[b]Deionized water.

plastic container to a covered 10-gallon plastic carboy and mixed vigorously with a Lightnin Mixer (Model ND-1) at 1725 rpm for 30 min. Approximately 2·5 liters of homogenized mud was then transferred to a 4 liter glass beaker, which was covered with aluminum foil and stored at 4 °C. Aliquots of this 2·5-liter subsample were subsequently used in all laboratory partitioning experiments.

Partitioning experiments were conducted in triplicate in 20-liter glass carboys which were modified by cutting and removing the necks and tops. All carboys were previously washed and rinsed with methanol and methylene chloride to avoid hydrocarbon contamination. Prior to initiating individual partitioning experiments, the 2·5-liter drilling mud subsample was rehomogenized by mechanical stirring for 30 min. Fifteen liters of sand-filtered seawater was added to each of six carboys and a weighed amount of drilling mud introduced beneath the seawater surface using a large glass funnel. Care was taken to minimize transfer losses to the walls of the funnel during addition of the mud. Two concentrations of mud-diesel seawater mixtures were prepared: triplicate 1% solutions (by adding ~150 g of stock mud-diesel formulation to the 15 liters of seawater) and triplicate 0·1% solutions (~15 g of stock material in 15 liters). After addition, the drilling mud was suspended in the water column by mechanical stirring for 10 min. Agitation produced by the stirring was sufficient to suspend barite particles. Immediately after mixing ceased, a solvent-rinsed glass jar was placed on the bottom of each carboy in order to collect a sample of settleable solids. The solutions were then allowed to settle for 1 h. At the end of this time aqueous and particulate phase samples were collected by siphoning 150 ml aliquots from three depths (surficial, mid-depth and immediately above the settleable solid phase) of each carboy (~450 ml total) directly into solvent-rinsed centrifuge tubes. The glass jars containing the settleable solids were then removed.

The remaining solutions were mixed again for 4 h, then allowed to settle for 1 h. Another glass jar was placed at the bottom of each carboy immediately after mixing to collect settleable solids and removed after the 1-h settling time. Identical aqueous and particulate phase samples were collected from each carboy by siphoning 150 ml of solution from the three depths. Throughout the entire experiment, and through the sample collection procedures, seawater temperatures remained between 25 and 27 °C. These experiments were conducted with the experimental glass carboy containers open to the atmosphere.

Immediately after collection, suspended particulate matter was isolated from the 150-ml samples obtained at each depth by centrifugation at 3000 rpm for 10 min. The resulting aqueous phases were decanted, pooled and the exact volume measured in preparation for chemical determination (i.e. solvent extraction). The suspended particulate phases isolated were also pooled in preparation for solvent extraction. The settleable solid samples were centrifuged, the aqueous phases decanted and discarded and samples extracted individually.

2.2. Analytical Methods

Immediately after isolation, pooled aqueous phase samples were extracted and concentrated as follows: The 450 ml pooled aqueous phases were transferred to 2-liter separatory funnels and were extracted with methylene chloride (three times, 65 ml each). The combined methylene chloride extracts were dried over Na_2SO_4, a 4 ml aliquot of the combined extracts was transferred to a 5 ml glass vial and the *o*-terphenyl internal quantification standard added directly to the vial.

The diesel fuel additives used in preparing the drilling mud formulations tested in this study contain significant amounts of volatile hydrocarbons. To ensure quantitative recovery, a method was selected to minimize evaporative losses of these components during concentration of methylene chloride extracts for gas chromatographic analysis. A two step procedure was selected for use in which the extracts were concentrated from approximately 4000 μl to 125 μl by evaporation under a stream of nitrogen. In a separate experiment, recoveries of two volatile hydrocarbons, n-decane (n-C_{10}) and toluene, were determined. The results obtained indicate that losses of either compound at each evaporative step were minor ($<10\%$).

Suspended particulate samples were initially extracted by shaking the isolated particulate material in a Teflon jar with 25 ml of methanol (three times, 10 min each) followed by 25 ml of 1:9 methanol:methylene chloride (three times, 10 min each). Since a very small amount of material was available for extraction (generally 2 g or less), the solvent was isolated following each extraction by allowing particles to settle for 10 min, then transferring the solution to another container using a Pasteur pipette. The combined methanol:methylene chloride extracts were added to an equal volume of organic-free deionized water and partitioned versus methylene chloride (three times, 25 ml each). The methylene chloride extracts were combined and dried over Na_2SO_4 and concentrated as described above.

Settleable solids were extracted using the same extraction procedure described above for suspended particulate material. The greater amount of settleable solids collected (1–4 g) enabled centrifugation to be used to isolate the solvent following each extraction. Some samples required addition of Na_2SO_4 to remove water prior to extraction. The remainder of the analytical procedure was identical to that described above.

Analysis of all sample extracts focused on: (1) selected aromatic hydrocarbon series (benzene, naphthalene, phenanthrene and their alkyl substituted homologues), (2) selected n-alkanes (n-C_{10}, n-C_{16} and n-C_{21}, representing a range of aqueous solubility and volatility) and (3) total hydrocarbons. The extract of each sample, containing o-terphenyl as an internal standard, was injected directly into a Finnigan MAT Model 4530 quadrupole GC/MS. GC conditions were: splitless injection on a WCOT SE-54 fused silica column (30 m × 0·25 mm) temperature programmed from 40 to 270 °C at 8 °C/min. MS conditions were: electron impact mode, scan range 50–450 atomic mass units, scan rate 1 scan/s, ionizing voltage 70 electron volt (eV).

Concentrations of individual aromatic hydrocarbons were quantified by computerized measurement of peak areas (i.e. ion currents) in chromatograms of selected masses corresponding to the specific compounds of interest, and relating these areas to that of the o-terphenyl internal standard. Response factors relative to the internal standard were calculated for each aromatic compound from analysis of authentic standard mixtures and applied to obtain the ultimate quantitative results. When standard compounds were not available, response factors were assigned by extrapolating from calculated values of available standards.

Concentrations of individual *n*-alkanes and total hydrocarbons were calculated by a similar measurement of specific peak areas or the total area, respectively, in total ion chromatograms (rather than mass chromatograms) and relating these areas to that of the internal standard. *n*-Alkanes were identified both by relative retention times and by mass chromatograms of m/e 57.

Aqueous phase concentrations are reported in units of μg/liter (ppb); suspended phase and settleable solid concentrations are reported in units of μg/g dry weight. Dry weights were determined directly by weighing the extracted sample after drying in an oven at 40 °C for 24–48 h. Concentration values were used in conjunction with the GC/MS characterization of the drilling mud-additive formulation to calculate recoveries (mass balances) and phase distributions of the selected hydrocarbons.

Measures taken to establish and document the quality of analytical results in this study included: (1) blank extractions conducted with each set of solid phase samples and drilling mud-additive formulations analyzed, (2) blind spikes of authentic aromatic hydrocarbons into drilling mud extracts to determine recovery efficiencies and quantitative accuracy, and (3) triplicate determinations of recoveries of high sulfur diesel additive after spiking a known amount into 1 liter of deionized water.

3. RESULTS AND DISCUSSION

3.1. Hydrocarbon Additive and Mud Formulations

3.1.1. Composition of Diesel Oil Additives
Figure 1 shows the reconstructed ion chromatogram (RIC) obtained from the GC/MS analysis of the high-sulfur diesel fuel additive. The high sulfur diesel oil was used in combination with generic mud No. 8 in these experiments. For comparison, concentrations of total hydrocarbons and selected *n*-alkane and aromatic hydrocarbons in several hydrocarbon additives are shown in Table 2. Total hydrocarbon concentrations reported in Table 2 represent an estimate based on the amount of internal standard added and do not incorporate factors such as differential GC/MS response among the different constituents of each additive. Concentrations of the target aromatic hydrocarbons in the high sulfur diesel oil were found to be approximately 29% of the total hydrocarbons in the high-sulfur diesel

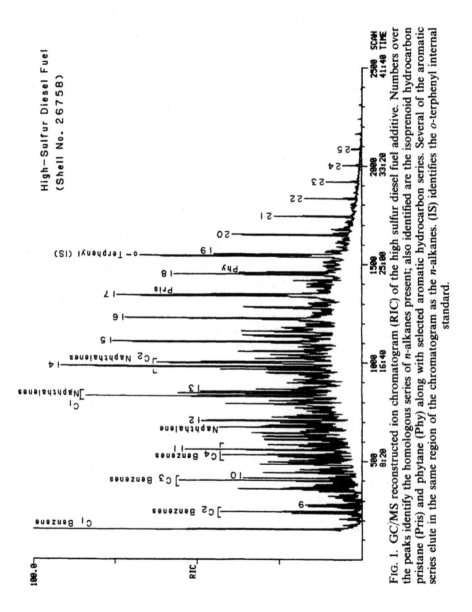

FIG. 1. GC/MS reconstructed ion chromatogram (RIC) of the high sulfur diesel fuel additive. Numbers over the peaks identify the homologous series of *n*-alkanes present; also identified are the isoprenoid hydrocarbon pristane (Pris) and phytane (Phy) along with selected aromatic hydrocarbon series. Several of the aromatic series elute in the same region of the chromatogram as the *n*-alkanes. (IS) identifies the *o*-terphenyl internal standard.

TABLE 2
Concentrations of total hydrocarbons and selected saturated and aromatic hydrocarbons in three drilling fluid additives

	High S diesel oil	Low S diesel oil	Mineral oil	Drilling mud 5% high S diesel formulation (μg/g dry wt)
	(mg/ml additive)			
	Total hydrocarbons[a]			
	850	1 100	960	24 200
	n-Alkanes			
n-C_{10}	3·5	12	ND	100
n-C_{16}	11	13	11	230
n-C_{21}	2·7	4·2	ND	68
	Aromatic hydrocarbons			
1. C_1-benzene	8·3	ND	ND	32
2. C_2	0·2	2·7	ND	73
3. C_3	27	17	ND	280
4. C_4	31	23	ND	370
5. C_5	13	11	ND	180
6. C_6	1·9	4·6	ND	120
7. C_0-naphthalene	25	2·2	0·1	83
8. C_1	30	7·1	0·4	370
9. C_2	43	10	0·8	610
10. C_3	34	7·2	1·4	550
11. C_4	12	0·9	0·7	230
12. C_0-phenanthrene	2·1	0·4	0·4	41
13. C_1	5·3	0·4	0·3	100
14. C_2	4·3	0·2	ND	92
15. C_3	1·5	ND	ND	43
Total aromatic hydrocarbons (No. 1–15)	250	87	4·1	3 174

ND = not detected.
[a]Total area in reconstructed ion chromatogram (RIC) quantified versus *o*-terphenyl internal standard.

(Table 2). In contrast, these aromatic compounds make up only 8% of the total hydrocarbons in the low-sulfur diesel oil. The composition of the three aromatic series in each diesel oil was also found to be different. Figure 2 shows that C_3–C_5 alkylated benzenes dominate the aromatic distribution in the low-sulfur diesel oil, whereas both alkylated benzenes and naphthalenes are nearly equally abundant in the high-sulfur diesel

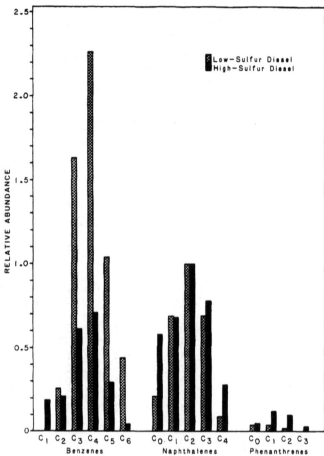

FIG. 2. Distributions of benzene, naphthalene and phenanthrene aromatic hydrocarbon series in low-sulfur and high-sulfur diesel fuels. For comparison, the series
have been normalized relative to C_2 naphthalene.

oil. Both additives contain a relatively low phenanthrene content.

   n-Alkane distributions in these two diesel additives also differ. Figure
3 shows the normalized *n*-alkane distributions of both high- and low-
sulfur diesel oils. The distribution in the low-sulfur diesel oil maximizes
at n-C_{11} while the maximum in the high-sulfur diesel oil appears to be
somewhat biomodal. In both additives the relative abundance of *n*-alkanes
drops off rapidly after n-C_{15} and *n*-alkanes beyond n-C_{25} are scarce or

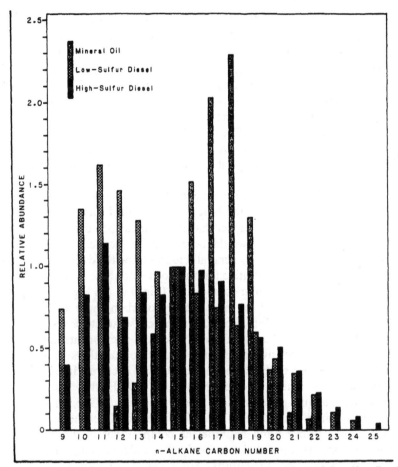

FIG. 3. Distributions of *n*-alkanes in the low-sulfur diesel fuel, high sulfur diesel fuel, and mineral oil. Distributions have been normalized relative to *n*-pentadecane (n-C_{15}).

absent. In conjunction with the aromatic distributions, the hydrocarbon composition of the low-sulfur diesel fuel appears to be dominated by a relatively greater abundance of lower molecular weight, more volatile saturated and aromatic compounds than the high-sulfur diesel.

The hydrocarbon composition of a selected mineral oil additive is shown in Table 2 and in Fig. 3 and is characteristically different from those of the diesel additives. The concentrations of total and individual component hydrocarbon parameters in the drilling mud–diesel formulation used in this study are presented in Table 2 as well. Table 2 shows that total concentrations of fifteen compounds comprising the benzene, naphthalene and phenanthrene aromatic series are 1–2 orders of magnitude lower in the mineral oil than in the two diesel fuels. Within the three aromatic series quantified, alkylated benzenes were not detected and only very low concentrations of the naphthalene series and C_0- and C_1-phenanthrenes are present. The mineral oil n-alkane distribution (Fig. 3) is restricted to a narrow range within the diesel oil range and is dominated by n-C_{17} and n-C_{18}. Significant quantities of pristane and phytane are also evident.

RICs of the mud-formulation and the base mud analyzed by GC/MS are shown in Fig. 4. Since the base mud contained such a low level of hydrocarbons, one would expect that the hydrocarbon composition of a mud-additive formulation would be essentially identical to that of its respective additive. This was generally found to be the case. Comparison of the RICs of the mud-additive formulations with those of the respective additive reveal a very similar composition (Fig. 4 versus Fig. 1). However, a detailed comparison of the aromatic hydrocarbon and n-alkane distribution of formulations with those of the respective additives reveals some subtle differences. In general, the more volatile components are present in slightly lower relative abundance in the formulations than in the additive, suggesting that some evaporative losses occurred during the hot-rolling process, which involved heating at 150 °F, that was used to formulate the mud–diesel mixture (see Breteler *et al.*, 1985, for details of mud preparation procedures). This effect is especially pronounced in the C_1–C_4 alkyl benzene and naphthalene compositions of the high-sulfur diesel oil formulation and also in its C_9–C_{16} n-alkane composition of the diesel oil formulation. Evaporative losses therefore appear to contribute minimally to compositional differences among individual additives and additive-mud formulations, with the effect most pronounced for the most volatile aromatic hydrocarbons and n-alkanes.

3.2. 1% Drilling Mud–Seawater Experiment

3.2.1. Hydrocarbon Mass Balances
A summary of the amount of total hydrocarbons and the amounts of individual alkane and aromatic hydrocarbons recovered during the first sampling (10 min mix, 1 h settling) and second sampling (4 h mix, 1 h settling) is presented in Table 3.

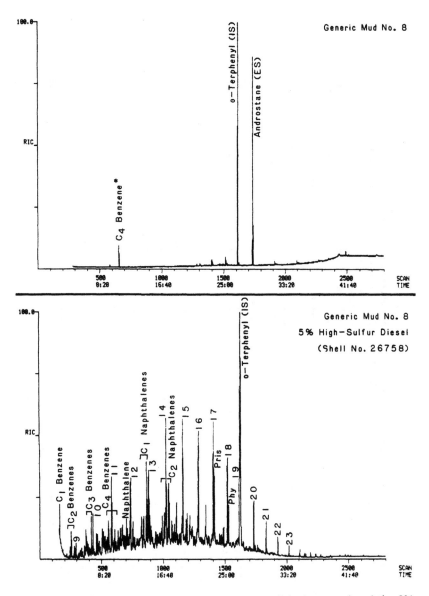

FIG. 4. GC/MS reconstructed ion chromatograms of the base mud and the 5% high-sulfur diesel formulations. C_4-benzene has been added to the base mud extract as a QA/QC analytical check (ES = external standard; IS = internal standard).

TABLE 3

Mass balance of total hydrocarbons and selected *n*-alkane and aromatic hydrocarbons in both first and second samplings of the 1% drilling mud-seawater partitioning experiment

	Amount added (mg)	Total amount recovered during first sampling (all phases) (mg)	% of total added	Total amount recovered during second sampling (all phases) (mg)	% of total added
Total hydrocarbons	2 513	1 130	45	1 050	42
***n*-Alkanes**					
n-C_{10}	10·6	3·4	32	4·7	44
n-C_{16}	23·7	20·6	87	20·0	44
n-C_{21}	7·0	3·9	56	9·7	139
Aromatic hydrocarbons					
C_1-Benzene	3·4	0·6	18	0	0
C_2	7·5	3·4	45	0	0
C_3	29·2	12·4	42	1·2	4
C_4	38·8	16·2	42	6·5	17
C_5	18·9	8·0	42	5·0	26
C_6	12·2	5·5	45	3·6	30
C_0-Naphthalene	8·6	5·8	67	2·0	23
C_1	37·8	26·0	69	15·5	41
C_2	63·0	45·6	72	36·2	57
C_3	57·5	38·2	66	33·6	58
C_4	24·0	12·9	54	11·0	46
C_0-Phenanthrene	4·3	3·6	84	4·5	105
C_1	10·4	8·0	77	8·6	83
C_2	9·5	5·4	57	5·2	55
C_3	4·5	1·2	27	1·2	27

Overall, at the time of the first sampling, 45% of the total hydrocarbons added were accounted for in the various liquid and solid phases of the system. This finding most likely implies that about half of the diesel oil added in this 1% experiment volatilized prior to this initial sampling sequence. Thereafter, only an additional 3% was 'lost' or unaccounted for.

Of the alkane and aromatic hydrocarbons quantified, n-C_{10} and C_1-benzene (toluene) were apparently lost rapidly (60% and 77% losses) within the first hour. Thereafter, the C_1–C_3 benzenes were nearly totally lost due to volatilization as were significant amounts of the C_4–C_6 benzenes and the C_0–C_3 naphthalenes. The total concentrations of higher molecular weight alkylated naphthalenes and phenanthrenes apparently remained stable between the first and second samplings, with no further losses due to evaporation.

3.2.2. Hydrocarbon Phase Distributions—First Sampling

The phase distributions of total hydrocarbons and individual components obtained in this initial sampling are shown in Table 4. During this initial sampling, which involved only a 10 min mixing time followed by a 1 h settling period, roughly three quarters (70%) of the total hydrocarbons isolated were found associated with the settleable material, while the remainder was found primarily in the aqueous phase (25%). Only a minor amount (5%) was found in the fine grain particulate phase. The distribution of n-alkanes generally followed those of the total hydrocarbons, having a major fraction (69–74%) associated with the settleable solids.

The effect of decreasing aqueous solubility with increasing alkylation of aromatic hydrocarbons was especially pronounced in the phase distributions of individual alkyl homologues within each aromatic series. The constituents of the benzene aromatic series exhibited a steady decrease in the proportion found associated with the aqueous phase with increasing alkylation, and a concomitant increase in the fraction associated with the settleable solids. This trend is best depicted in Fig. 5 (upper figure), which shows diagrammatically the phase distribution of total hydrocarbons and individual hydrocarbon constituents for the first samplings. The naphthalene aromatic series exhibits a similar, but less pronounced trend, with constituents decreasing in abundance in the aqueous phase and increasing in the settleable solids with increasing degree of alkylation (Fig. 5 and Table 4). Figure 6 shows representative RICs of the three phases from individual replicate experiments obtained during the first sampling. It is evident from these chromatograms that the aqueous phase is enriched

Paul D. Boehm, John Brown and Adolfo G. Requejo

TABLE 4
Phase distribution of total hydrocarbons and selected n-alkanes and aromatic hydrocarbons recovered in the first sampling of the 1% drilling mud-seawater partitioning experiment

	Total amounts (μg)				Relative amounts (%)		
	Suspended particulate phase	Aqueous phase	Settleable solids	Total	Suspended particulate phase	Aqueous phase	Settleable solids
Total hydrocarbons							
	61 373	277 000	793 000	1 131 000	5	25	70
n-Alkanes							
n-C$_{10}$	0	1 050	2 310	3 360	0	31	69
n-C$_{16}$	1 700	3 720	15 100	20 520	8	18	74
n-C$_{21}$	300	900	2 740	3 940	8	23	69
Aromatic hydrocarbons							
C$_1$-Benzene	0	555	0	555	0	100	0
C$_2$	0	2 820	590	3 410	0	83	17
C$_3$	0	7 040	5 370	12 410	0	57	43
C$_4$	0	5 800	10 400	16 200	0	36	64
C$_5$	0	2 180	5 850	8 030	0	27	73
C$_6$	0	1 320	4 190	5 510	0	24	76
C$_0$-Naphthalene	752	3 380	1 660	5 790	13	58	29
C$_1$	5 610	8 370	12 000	25 980	22	32	46
C$_2$	12 300	9 360	24 000	45 600	27	20	53
C$_3$	10 800	6 510	21 000	38 310	28	17	55
C$_4$	2 100	2 380	8 430	12 910	16	18	66
C$_0$-Phenanthrene	1 350	560	1 660	3 570	38	16	46
C$_1$	3 160	1 110	3 700	7 970	40	14	46
C$_2$	1 600	820	3 010	5 430	29	15	56
C$_3$	50	260	860	1 170	4	22	74

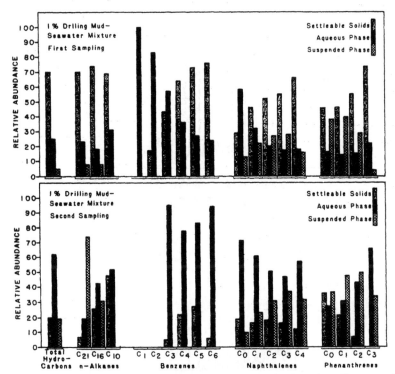

FIG. 5. Phase distribution of total hydrocarbons and selected *n*-alkanes and aromatic hydrocarbons in the 1% drilling mud-seawater partitioning experiment.

in aromatic hydrocarbons and, in particular, constituents of the benzene series (C_1–C_3) and C_0 and C_1 naphthalenes. These same constituents are absent from or are lesser constituents in both the particulate phase and settleable solids. These figures also illustrate that the bulk of hydrocarbons are initially associated with the settleable solids.

The results of the first sampling in the 1% experiments indicate that under relatively high particle loadings and a very limited mixing time (10 min), the major fraction of hydrocarbons not lost by volatilization become associated with particles (i.e. adsorption) which tend to settle.

3.2.3. Hydrocarbon Phase Distributions—Second Sampling
Hydrocarbon phase distributions obtained during the second sampling in the 1% drilling mud experiment are shown in Table 5. Results are strikingly different when compared to the initial short-term mixing. After

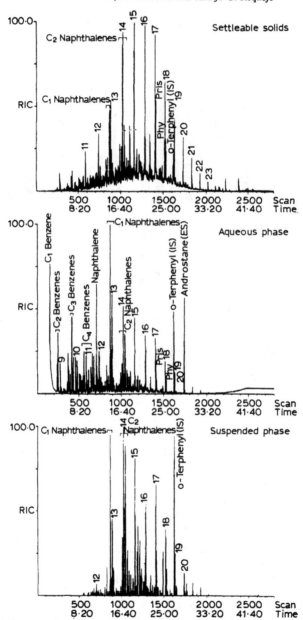

FIG. 6. Representative GC/MS reconstructed ion chromatograms (RIC) of the three phases sampled during the first sampling of the 1% drilling mud – seawater partitioning experiment.

a 4 h mixing time, the bulk of the remaining hydrocarbons (those not lost due to volatilization) were found primarily in the aqueous phase (62%), with the remainder distributed nearly equally between the particulate phase and settleable solids (Table 5). The distribution of *n*-alkanes exhibited a very interesting trend. With increasing molecular weight (and decreasing solubility and volatility), compounds appear to adsorb to the fine-grained suspended particulate phase. Thus n-C_{10} is found exclusively in the aqueous phase and coarser settleable solids while the bulk of n-C_{21} is found in the suspended particulate phase. The trend is highlighted diagrammatically in Fig. 5 (lower figure).

The general trend in the aromatic hydrocarbon series parallels that of the total hydrocarbons. Almost all of the benzenes are found in the aqueous phase (C_1 and C_2 benzene were not detected in the second sampling) as are the naphthalene series (Table 5 and Fig. 5). The phenanthrene series are divided almost equally between the particulate and aqueous phases. The effect of increased partitioning into the solid phases with increasing degree of aromatic alkylation is not as pronounced in the phase distributions from the second sampling. However, with decreasing solubility within a homologous aromatic series and hence increasing affinity for particles, the compounds tend to become increasingly associated with the fine-grained clay-like suspended particles which have a greater surface area than the coarser settled material (e.g. barite).

The results from the second sampling of the 1% experiment indicate that upon more extensive but realistic turbulent mixing, the hydrocarbons are transferred from settleable solids (as evidenced by the first sampling) to the particulate and aqueous phases.

3.3. 0·1% Drilling Mud–Seawater Experiment

3.3.1. Hydrocarbon Mass Balances
The quantities of hydrocarbons recovered during the first and second samplings of the 0·1% addition experiment are shown in Table 6. In contrast to the 1% experiment, it appears that the diesel oil hydrocarbons are not as rapidly volatilized when concentrations of added diesel oil are much lower. The results of the first sampling indicate that a large percentage ($\sim 90\%$) of the total hydrocarbons can be accounted for. This amount decreases to 53% of the total added after the second sampling. Thus, the extent of the apparent loss due to volatilization is the same in both experiments, with the rates of loss apparently lower when less diesel oil is added.

TABLE 5

Phase distribution of total hydrocarbons and selected n-alkanes and aromatic hydrocarbons recovered in the second sampling of the 1% drilling mud-seawater partitioning experiment

	Total amounts (μg)				Relative amounts (%)		
	Suspended particulate phase	Aqueous phase	Settleable solids	Total	Suspended particulate phase	Aqueous phase	Settleable solids
Total hydrocarbons							
	188 000	652 000	206 000	1 046 000	18	62	20
n-Alkanes							
$n\text{-}C_{10}$	0	2 430	2 230	4 660	0	52	48
$n\text{-}C_{16}$	6 170	8 600	5 320	20 090	31	43	26
$n\text{-}C_{21}$	7 190	1 840	720	9 750	74	19	7
Aromatic hydrocarbons							
C_1-Benzene	0	0	0	0	—	—	—
C_2	0	0	0	0	—	—	—
C_3	0	1 120	50	1 170	0	95	0
C_4	0	5 080	1 400	6 480	0	78	22
C_5	0	4 140	860	5 000	0	83	17
C_6	0	3 350	220	3 570	0	94	6
C_0-Naphthalene	190	1 430	380	2 000	10	71	19
C_1	3 520	9 450	2 580	15 550	23	61	16
C_2	11 270	18 640	6 340	36 250	31	51	18
C_3	12 560	15 740	5 320	33 620	37	47	16
C_4	3 490	6 300	1 240	11 030	32	57	11
C_0-Phenanthrene	1 650	1 220	1 610	4 480	37	27	36
C_1	4 130	2 650	1 770	8 550	48	31	21
C_2	2 580	2 240	380	5 200	50	43	7
C_3	420	820	0	1 240	34	66	0

The detailed chemical nature of these losses is similar to that previously seen in the 1% experiment. At the end of the second sampling none of the benzene series and only low levels of the naphthalene series apparently remain in the system (Table 6). Loss of the n-C_{10} alkane also appears to be substantial as well.

3.3.2. Hydrocarbon Phase Distribution—First Sampling

Table 7 shows the phase distribution of total hydrocarbons and individual components obtained during the first sampling of the 0·1% experiment. In contrast to the results of the first sampling from the 1% experiment, total hydrocarbons recovered are found almost exclusively in the aqueous phase. Similarly, individual n-alkanes and all constituents of both the benzene (99–100%) and the naphthalene (83–99%) aromatic series were recovered in the aqueous phase. The major portion of the phenanthrene series also was recovered in the aqueous phase, but a significant component (~ 20–30%) was present in the particulate phase. The aromatic and n-alkane content of the settleable solid phase was found to be a negligible fraction of the total. These distributions are depicted diagrammatically in Fig. 7 (upper figure). The gradual decrease in the aqueous phase abundance of the individual aromatic hydrocarbons proceeding through the benzene, naphthalene and phenanthrene series and the concomitant increase in the portion associated with the suspended phase is pronounced.

Figure 8 shows representative reconstructed ion chromatograms (RICs) of the settleable solids, aqueous phase and suspended phase from the first sampling of the 0·1% experiment. The prevalence of both the bulk hydrocarbons and individual aromatic components in the aqueous phase is clearly evident.

The results of the first sampling in the 0·1% experiment indicate that at greater drilling mud-seawater dilutions, and thus lower total hydrocarbon concentrations, the bulk of the hydrocarbon can be accommodated initially (i.e. colloidal material) or solubilized in the aqueous phase and a relatively small fraction of the total is either volatile or becomes associated with the suspended particulate or settleable solid phases.

3.3.3. Hydrocarbon Phase Distribution—Second Sampling

The phase distribution obtained during the second sampling is shown in Table 8. Although the bulk of total hydrocarbons still remains in the aqueous phase (65%), a significant portion is now associated with the particulate phase. This trend holds for the n-alkanes, which are present

TABLE 6

Mass balance of total hydrocarbons and selected n-alkane and aromatic hydrocarbons in both first and second samplings of the 0·1% drilling mud-seawater partitioning experiment

	Amount added (mg)	Total amount recovered during first sampling (all phases) (mg)	% of total added	Total amount recovered during second sampling (all phases) (mg)	% of total added
Total hydrocarbons	252	228	91	133	53
***n*-Alkanes**					
n-C_{10}	1·1	0·8	73	0·3	27
n-C_{16}	2·4	3·4	142	2·5	104
n-C_{21}	0·7	0·9	129	0·5	71
Aromatic hydrocarbons					
C_1-Benzene	0·3	0·3	100	0	0
C_2	0·8	0·6	75	0	0
C_3	2·9	2·4	83	0	0
C_4	3·9	3·0	77	0	0
C_5	1·9	1·1	58	0	0
C_6	1·2	0·5	42	0·1	4
C_0-Naphthalene	0·9	1·2	133	<0·1	2
C_1	3·8	6·0	158	0·2	5
C_2	6·3	7·9	125	0·6	10
C_3	5·8	5·7	98	1·4	24
C_4	2·4	1·8	75	1·0	42
C_0-Phenanthrene	0·4	0·7	175	0·3	75
C_1	1·0	1·3	130	0·6	60
C_2	1·0	0·6	80	0·6	60
C_3	0·4	0·1	25	0·2	50

TABLE 7

Phase distribution of total hydrocarbons and selected n-Alkanes and aromatic hydrocarbons recovered in the first sampling of the 0·1% drilling mud-seawater partitioning experiment

	Total amounts (μg)				Relative amounts (%)		
	Suspended particulate phase	Aqueous phase	Settleable solids	Total	Suspended particulate phase	Aqueous phase	Settleable solids
Total hydrocarbons							
	10 100	214 000	4 720	228 820	4	94	2
n-Alkanes							
$n\text{-}C_{10}$	0	760	0	760	0	100	0
$n\text{-}C_{16}$	500	2 760	220	3 480	14	80	6
$n\text{-}C_{21}$	110	720	30	860	13	84	3
Aromatic hydrocarbons							
C_1-Benzene	0	270	0	270	0	100	0
C_2	0	570	0	570	0	100	0
C_3	0	2 400	0	2 400	0	100	0
C_4	4	2 960	6	2 970	0	100	0
C_5	4	1 080	0	1 084	0	100	0
C_6	4	450	0	454	1	99	0
C_0-Naphthalene	20	1 220	6	1 246	2	98	0
C_1	180	5 790	40	6 010	3	96	1
C_2	590	7 230	120	7 940	7	91	2
C_3	830	4 770	140	5 740	14	83	3
C_4	370	1 350	50	1 770	21	76	3
C_0-Phenanthrene	170	460	40	670	25	69	6
C_1	380	820	60	1 260	30	65	5
C_2	250	500	40	790	32	63	5
C_3	40	80	0	120	35	65	0

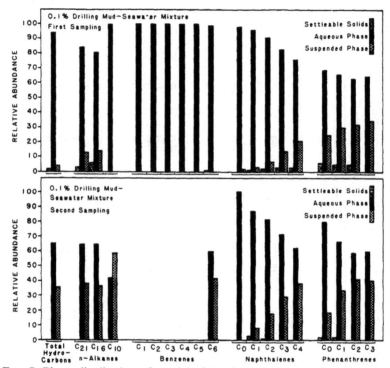

FIG. 7. Phase distribution of total hydrocarbons and selected *n*-alkanes and aromatic hydrocarbons in the 0·1% drilling mud-seawater partitioning experiment.

in significant relative proportions in the suspended phase. Only the C_6 constituent of the benzene aromatic series was recovered in the second sampling; the remaining homologues probably were lost to evaporation during the 4-h mixing time. Evaporative losses of the benzene aromatic series would be consistent with the overall decreased recoveries of hydrocarbons obtained during the second sampling (Table 6). Distribution of the naphthalene and phenanthrene aromatic series remained skewed toward the aqueous phase, but significant quantities of the higher alkylated homologues are found in the particulate phase. Again, only negligible amounts of hydrocarbons were found associated with the settleable solids. This apparent increase in the relative abundance of hydrocarbons in the particulate phase reflects a loss, probably due to evaporation of those compounds from the aqueous phase, rather than an increase in the amounts associated with the particulate phase, as confirmed by the similar

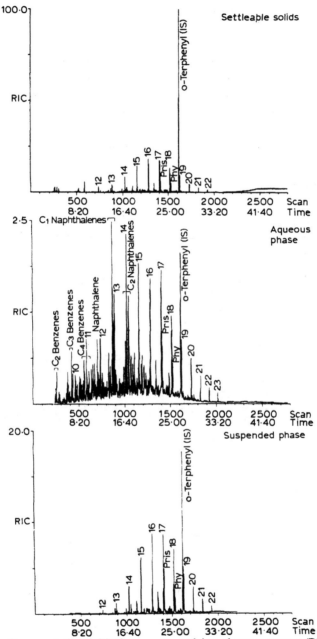

FIG. 8. Representative GC/MS reconstructed ion chromatograms (RIC) of the three phases sampled during the first sampling of the 0·1% drilling mud-seawater partitioning experiment.

TABLE 8

Phase distribution of total hydrocarbons and selected n-alkanes and aromatic hydrocarbons recovered in the second sampling of the 0·1% drilling mud-seawater partitioning experiment

	Total amounts (µg)				Relative amounts (%)		
	Suspended particulate phase	Aqueous phase	Settleable solids	Total	Suspended particulate phase	Aqueous phase	Settleable solids
Total hydrocarbons							
	46 200	86 500	440	133 140	35	65	0
n-Alkanes							
n-C$_{10}$	200	140	0	340	58	42	0
n-C$_{16}$	910	1 580	20	2 510	36	63	1
n-C$_{21}$	200	350	0	550	37	63	0
Aromatic hydrocarbons							
C$_1$-Benzene	0	0	0	0	—	—	—
C$_2$	0	0	0	0	—	—	—
C$_3$	0	0	0	0	—	—	—
C$_4$	0	0	0	0	—	—	—
C$_5$	0	0	0	0	—	—	—
C$_6$	20	30	0	50	41	59	0
C$_0$-Naphthalene	0	20	0	20	0	100	0
C$_1$	10	140	10	160	8	87	4
C$_2$	110	490	10	610	18	81	1
C$_3$	400	970	10	1 380	29	71	0
C$_4$	390	650	0	1 040	38	62	0
C$_0$-Phenanthrene	50	220	10	280	18	80	2
C$_1$	210	410	10	630	33	66	1
C$_2$	260	360	0	620	41	59	0
C$_3$	90	130	0	220	40	60	0

quantities of hydrocarbons associated with the particulate phase after both samplings (Tables 7 and 8). This result is particularly evident in the case of the *n*-alkanes for which, for example, between the first and second samplings, twice the quantified concentrations of *n*-alkanes are lost (due to evaporation) from the aqueous phase than are partitioned into the particulate phase.

The results from the second sampling of the 0·1% experiment indicate that after extensive mixing at dilute hydrocarbon concentrations, the majority of hydrocarbons remain associated with the aqueous phase, but due to apparent evaporative losses from the aqueous phase, a significant portion of the remainder are now associated with the fine-grained clayey suspended phase. At these hydrocarbon concentrations, little or no amount is associated with settleable solids.

3.4. Analytical Quality Assurance/Quality Control

Gas chromatograms of procedural blanks carried out in conjunction with analyses of mud-additive formulations and solid phase bioassay samples revealed no interference from contaminants within the analytical range of interest (n-C_9–n-C_{25}). Very low level phthalate ester contamination was observed in the blanks, but the interfering compounds were present well outside the analytical range ($> n$-C_{25}).

Tetramethylbenzene (C_4-benzene) was added to the base mud methanol extract (prior to final extraction with methanol:methylene chloride) as a blind spike to determine whether hydrocarbon recoveries were quantitative. GC/MS analysis of the base mud extract yielded a 92% recovery of this aromatic hydrocarbon, confirming that total hydrocarbon recoveries using this extraction procedure were indeed quantitative and complete.

The results of the spiked diesel recovery experiments, for which recoveries of a known amount of high-sulfur diesel oil added to 1 liter of deionized water were calculated, indicate that the mean recovery obtained was greater than 96%. Although this experiment does not establish the efficiency of the methylene chloride extraction for diesel fuel in actual aqueous samples, it does document that little or no loss occurs during analytical work-up of sample extracts and that the internal standard technique employed (addition to extracts prior to gas chromatographic analysis) is accurate.

3.5. Extrapolation of Partitioning Results to Field Conditions

How do the results from this partitioning study apply to possible discharges of water-based drilling fluids containing diesel fuel in the marine environment? Clearly, the conditions under which these experiments were conducted differ from those which might be encountered in actual field discharge situations, most noticeably in the absence of a decrease in the suspended solid concentration with time. However, the results obtained do lend themselves to some interesting interpretations.

Given an average moisture content of 30% for these laboratory mud formulations, the suspended solids concentrations in the 1% and 0·1% addition experiments are 7000 and 700 mg/liter, respectively. Several field studies have shown that drilling muds discharged to the ocean are diluted rapidly within 1000–7000 m downcurrent of the discharge pipe and 0·1–4 h of discharge (Ecomar, 1978, 1983; Ayers *et al.*, 1980; Northern Technical Services, 1983). Figure 9 depicts the results of five mud dispersion studies conducted in different outer continental shelf waters of the United States. The data show that dilutions of 1000-fold or more are generally encountered almost immediately after discharge. A comparison of the suspended solids concentrations used in the partitioning experiments to these results indicates that the experimental values are representative of those found in the field within 0·5–5 min after discharge (Fig. 9). All partitioning experiments incorporated a 1 h particle settling time prior to sampling, and thus no samples were collected earlier than 70 min following introduction of drilling muds. However, the experimental conditions up until the time of the first samplings, which involved 10 min mixing followed by the 1-h settling period, would come closest to simulating field conditions. Therefore, extrapolating the first sampling results to field discharge, it would appear that:

1. Under high particle loadings and moderate to vigorous mixing conditions, a significant amount of the diesel hydrocarbons could be lost due to evaporative processes, depending on the degree of contact of the discharged mud with the atmosphere. At lower particle loadings or minimal atmospheric exchange, solubilization or water accommodation will be the prevalent processes. These latter processes appear to have the greatest influence on the fate of hydrocarbons discharged in muds.

2. The extent to which diesel hydrocarbons become associated with settleable particles will vary with particle density. However, taking into account solubilization and evaporative losses, our results indicate that a maximum of about 2% of the total would be associated with particles which would settle close to the point of discharge. Transport of hydrocarbons on fine-grain suspended particles which would tend to sediment at greater distance from the discharge point, appear to

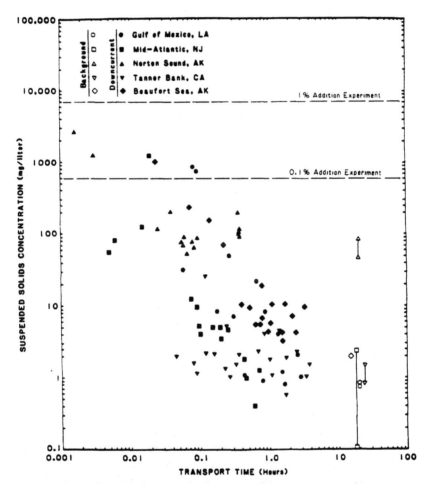

FIG. 9. Graphical presentation of results of five drilling fluid dispersion studies performed in US Outer Continental Shelf waters. Concentration of suspended solids is used as an indicator of drilling fluid solids concentration and is plotted against transport time (distance from rig divided by current speed). Undiluted drilling fluids contained from 200 000 to 1 400 000 ppm suspended solids before discharge. Dashed horizontal lines indicate drilling mud concentrations used in the present partitioning studies (from Neff, 1987).

be a minor process. The degree to which observations (1) and (2) apply in the field probably will be highly dependent on the additive content of the mud formulation and on the nature of the additive itself, together with other factors such as discharge rate and the magnitude of prevailing currents.

3. Aromatic hydrocarbons (represented in this case by the benzene, naphthalene and phenanthrene alkylated series) will exhibit a greater tendency to become solubilized (or volatilized) than the bulk hydrocarbons. As a result, sedimenting particles containing discharged hydrocarbons will be depleted in the more toxic components relative to their original composition.

The second part of each experiment incorporated a total mixing time of 4 h which represents a maximum of those in the field, was conducted to examine longer term partitioning of the remaining hydrocarbons (those not lost to evaporative phases) between the respective phases. The results of the 1% addition experiment showed a mass shift of hydrocarbons from the settleable and particulate phases to the aqueous phase relative to the first sampling. In the 0·1% experiment, the mass balance results (see Tables 7 and 8) would seem to indicate a shift in hydrocarbons from the aqueous to the particulate phase. However, note that the absolute concentrations of hydrocarbons in the suspended particulate phase are similar in the two samples, thus indicating no quantitative shift of material to the fine-grain particles. Instead, we believe that the apparent mass shift is due more to an evaporative loss of aqueous hydrocarbons from the system (approximately a 40% loss occurred between the first and second sampling) rather than to an actual quantitative phase shift. We interpret the results to indicate that given additional mixing time, diesel hydrocarbons will become even further solubilized in seawater (or, conversely, particles will become more depleted of hydrocarbons with time). This solubilization process is probably a combination of true aqueous solubility and solubilization by surface active agents in the drilling muds (Boehm & Quinn, 1973).

Under conditions that simulate artificially high particle loadings (> 7000 mg/liter in the 1% drilling mud: seawater experiment) and quiescent conditions (10 min mixing time), a significant fraction (> 20%) of additive hydrocarbons not lost to volatilization may be found associated with settleable material (however, volatilization losses were found to be as high as 50% of the total). With time and more extensive mixing, these hydrocarbons become solubilized and the majority of hydrocarbons are found in the aqueous phase. Under conditions simulating more dilute and realistic, but still high (700 mg/liter), particle loadings, a larger fraction of the hydrocarbons can be solubilized and thereby associated with aqueous phases. Only a small fraction is found in either suspended or settleable material (~0–2%).

In summary, one must conclude from the results of the experiment that given realistic conditions of dilution and settling, diesel oil discharged as part of drilling mud formulations will largely partition into the aqueous phase with lesser amounts in the settled phase and lesser still in the particulate phase.

ACKNOWLEDGEMENTS

The authors wish to acknowledge the important technical input of Dr Jerry Neff of Battelle and Dr Ronald Breteler of Springborn Bionomics to the study. This work was funded by the American Petroleum Institute, Washington, DC.

REFERENCES

ANNE, D.C. (1981). Characterization and partitioning of petroleum in drilling fluid dispersed in seawater. M.S. Thesis, Florida Institute of Technology.

AYERS, R.C., JR, SAYER, T.C., STEUBNER, D.O. & MEEK, R.P. (1980). An environmental study to assess the effect of drilling fluids on water quality parameters during high rate, high volume discharges to the ocean. In *Symposium on Research on Environmental Fate and Effects of Drilling Fluids and Cuttings*. Courtesy Associates, Washington, D.C., pp. 351–81.

BOEHM, P.D. & QUINN, J.G. (1973). Solubilization of hydrocarbons by dissolved organic matter in seawater. *Geochim. Cosmochim. Acta*, **37**, 2459–77.

BRETELER, R.J., BOEHM, P.D., NEFF, J.M. & REQUEJO, A.G. (1985). Acute toxicity of drilling muds containing hydrocarbon additives and their fate and partitioning among liquid, suspended and solid phases. Final Report, American Petroleum Institute, Washington, DC.

ECOMAR, INC. (1978). Tanner Bank fluids and cuttings study. Conducted for Shell Oil Company, January through March 1977. Ecomar Inc., Goleta, CA, 495 pp.

ECOMAR, INC. (1983). Mud dispersion study Norton Sound Coast Well No. 2, Report for ARCO Alaska, Inc., 91 pp.

GEARING, P.J., GEARING, J.N., PRUELL, R.J., WADE, T.L. & QUINN, J.G. (1980). Partitioning of No. 2 fuel oil in controlled estuarine ecosystems, sediments, and suspended particulate matter. *Environ. Sci. Technol.*, **14**, 1129–36.

NEFF, J.M. (1987). Biological effects of drilling fluids, drill cuttings and produced waters. In *Long-term Environmental Effects of Offshore Oil and Gas Development*, ed. D. F. Boesch & N. N. Rabalais. Elsevier Applied Science, New York, pp. 469–538.

NORTHERN TECHNICAL SERVICES (1983). Beaufort Sea drilling effluent disposal study. Performed for Reindeer Island Stratigraphic Test Well participants under direction of SOHIO Petroleum Company, Anchorage, AK, 329 pp.

ZURCHER, F. & THUER, M. (1978). Rapid weathering process of fuel oil in natural waters: Analyses and interpretations. *Environ. Sci. Technol.*, **12**, 838–43.

28

Evidence of Oil Contamination in North Sea Cod

C. E. SJØGREN, H. DRANGSHOLT, F. ORELD, T. ØFSTI and
S. P. SPORSTØL

Senter for industriforskning (SI), P.O. Box 124, N-0314 Blindern, Oslo 3, Norway

ABSTRACT

Environmental surveys have been carried out annually in the North Sea to assess the extent of petroleum contamination around oil platforms using oil-based drilling mud. In order to investigate the effect of the oil discharge on fish, samples of liver from cod caught 0–250 m from Norwegian North Sea drilling and production platforms were analysed for hydrocarbons by means of GC and GC/MS.

The following compounds were determined: naphthalene, phenanthrene/anthracene, and dibenzothiophene, and their C_1-, C_2-, C_3-alkylated homologues (NPD). In addition, the total hydrocarbon content (THC) as well as the sum content of C_5–C_8 alkyl decalins, C_8-alkylcyclohexanes and C_9–C_{12} alkylbenzenes were determined in some of the samples.

The analysis of several hydrocarbon compound groups in cod liver samples from the North Sea shows that all parameters studied may be used to detect contamination due to the use of oil-based muds at drilling platforms. In extreme cases, fingerprint oil chromatograms are clearly observed in GC traces of cod liver extracts.

The higher specificity and sensitivity of the NPD analysis, and the fact that naphthalenes are known to be readily absorbed by fish, still makes the NPD's our prime choice as a monitoring parameter for the detection of subtle hydrocarbon cantamination effects.

1. INTRODUCTION

Environmental surveys have been carried out annually in the North Sea to assess the extent of petroleum contamination around drilling and

577

production platforms. In order to investigate the effect of the discharged oil on fish, liver from cod caught <250 m from Norwegian North Sea oil platforms is regularly analysed for total hydrocarbon content and selected hydrocarbons.

The hydrocarbon parameters which have been regularly monitored in the Norwegian sector of the North Sea are naphthalenes, phenanthrenes/anthracenes and dibenzothiophenes and their C_1–C_3 alkylated homologues (NPD) (Reiersen *et al.*, 1988; Grahl-Nielsen *et al.*, 1988). Additional compounds have also been determined, i.e. cyclohexanes, decalins and benzenes, as well as the total hydrocarbon content. Mono- and bicyclic naphthenes are known to be major components of the 'low toxicity' base oils used in drilling muds which have low aromatic content. It has been shown that bicyclic hydrocarbons, such as decalins, are present in heavily biodegraded crudes (Philp *et al.*, 1981). This resistance towards biodegradation suggested that these bicyclic naphthenes might be useful as monitoring parameters in the North Sea environmental surveys.

Accumulation of polycyclic aromatic hydrocarbons (PAH) in fish has been reported by several authors (Lee, in Wolfe, 1977; Neff, 1979) and it has been shown that naphthalenes in particular are absorbed (McCain *et al.*, 1978). A recent investigation of the accumulation and distribution of hydrocarbons in fish from the Mediterranean (Albaiges *et al.*, 1987) reports accumulation of hydrocarbons and PCB in fish liver and muscle, and suggests that liver and muscle may act as indicator organs of acute and chronic exposure to hydrocarbon pollutants, respectively.

Results obtained in these laboratories (Drangsholt *et al.*, 1987) have shown instances of cod liver from the Norwegian sector of the North Sea being very clearly contaminated by the input of oil from platforms using oil-based drilling muds. Similar results have recently been reported in the British sector (McGill *et al.*, 1987). In the following an account will be given of the results obtained from the analyses of fish liver carried out to date in these laboratories.

2. MATERIALS AND METHODS

2.1. Sampling and Work up Procedures

Liver samples were obtained from cod (*Gadus morhua*) caught by angling from a supply vessel close to (0–250 m) North Sea drilling and production platforms. Reference samples were obtained by angling in an unpolluted area, more than 10 000 m from the nearest oil installation. For practical reasons, reference samples were sometimes obtained from commercial sources, i.e. cod caught in the coastal waters near Stavanger or Oslo (Drangsholt *et al.*, 1987). The samples were wrapped in clean aluminium foil, frozen in the field and stored at $-20\,°C$ until needed for analysis.

Individual cod liver samples (10 g) were homogenized and known amounts of deuterated internal standards were added (naphthalene-d_8, phenanthrene-d_{10} and pyrene-d_{10}). The samples were then saponified in 80 ml methanol, 10 ml water and 12 g KOH under reflux for 22 h. The mixture was filtered under suction and extracted twice with pentane or dichloromethane. The combined extracts were then washed with water/methanol and dried over Na_2SO_4. Polar components were removed by chromatographing on Florisil and the eluate was concentrated and analysed for hydrocarbons.

2.2. GC/MS Analysis of Selected Hydrocarbons

The selected compounds were determined using computerized gas chromatography/mass spectrometry (Finnigan 4023 GC/MS) operated in the selected ion monitoring mode. The following analytical conditions were employed: Ionization: 70 eV (EI); 0.6 s/scan; GC column: $30\,m \times 0.25\,mm$ i.d., $0.25\,\mu m$ DB-5; GC temperature programming: 60^0 (1 min)–6 °C/min–300 °C (10 min). The following compounds were determined: naphthalene, phenanthrene/anthracene, dibenzothiophene, and their C_1–C_3 alkyl homologues (NPD). In addition, the sum content of C_5–C_8 alkylated decalins, C_8-alkylcyclohexanes and C_9–C_{12} alkylbenzenes was determined in some of the samples. The selected compounds were quantified by comparing their integrated molecular ion fragmentograms with those of the internal standards. The values were corrected using multi-level response factor curves compiled from analyses of standards containing known amounts of the target compounds.

2.3. GC Analysis of Total Hydrocarbon Content

The total hydrocarbon content (THC) was determined by capillary gas chromatography (HP 5880 GC). Quantification was carried out by computer measurement of the flame ionization detector response within the boiling point range n-C_{10}–n-C_{40} alkane and comparing the result with the corresponding response of known amounts of a marine diesel oil. The following GC chromatographic conditions were used: Column: $12.5\,m \times 0.20\,mm$ i.d. fused silica crosslinked with dimethylsilicon; Temperature program: 50 °C (3 min)–25 °C/min–350 °C (10 min).

3. RESULTS AND DISCUSSION

3.1. General

Table 1 contains a review of the hydrocarbon analysis of individual cod liver samples in the 1985 and 1986 North Sea environmental surveys. The corresponding results from the 1987 surveys are shown in Table 2. Samples from fish caught close (<250 m distance) to a platform are termed P, and the reference samples (>10 000 m distance) are given the code R. An additional letter identifies the samples belonging to the same area or platform. In the early environmental surveys only NPDs were measured, and additional parameters have been added afterwards. The total hydrocarbon content, as determined by GC, was included when it was found that a fingerprint chromatogram of the base oil could in fact be clearly seen in the chromatogram of a North Sea cod liver sample (Drangsholt et al., 1987). This is also shown in Fig. 1, depicting the GC traces of extracts of cod liver caught close to a North Sea oil platform, and the base oil used in the drilling mud.

3.2. Total Hydrocarbons (THC)

The results shown in Table 1 indicate a poor correlation between the THC and NPD results which may be caused by the higher specificity of the NPD analysis. The GC chromatogram of a cod liver extract (Fig. 1) may contain interfering peaks of non-petrogenic origin which will influence the quantitative results.

The GC chromatograms shown in Fig. 1 illustrate, on the other hand, that while the quantitative results may be uncertain, the qualitative results contained in a GC fingerprint may be used to detect high levels of contamination.

3.3. Aliphatic Hydrocarbons

3.3.1. Selected Cyclohexanes and Decalins

The results given in Tables 1 and 2 show reference levels of 0·02–15 ppm for C_5–C_8 alkyldecalins and 0·01–3·1 ppm for C_8-alkylcyclohexanes, respectively. The VP and SP samples, from platforms using oil-based muds, have corresponding concentration levels in the 1·0–250 ppm and 0·1–30 ppm ranges, respectively. The difference between samples and background is less pronounced in samples which were taken near a platform where only water-based muds have been used (Table 2, UP and UR).

TABLE 1

Total hydrocarbon content and selected hydrocarbons in individual cod liver samples from the Norwegian North Sea, 1985 and 1986 (mg/kg dry matter).

Sample[a]	THC	NPD	C_5–C_8 decalins	C_8-alkyl cyclohexanes	C_9–C_{12} benzenes
OP1		0·637			
OP2		0·467			
OP3		0·831			
OP4		10·2			
OP5		1·71			
OR1		0·405			
OR2		0·090			
OR3		0·132			
OR4		0·235			
OR5		0·311			
EP1		0·53			
VP1		1·54			
VP2		5·64			
VP3		0·57			
VP4		2·45			
VP5		0·23			
VP6	1 880	0·905	78	30	13
VP7	402	1·98	8·0	2·3	3·7
VP8	870	1·94	29	16	11
VP9	1 420	1·50	43	30	24
VP10	446	1·04	14	4·5	6·7
VR1		0·176			
VR2		0·137			
VR3	315	0·520	2·17	0·60	0·74
VR4	575	0·305	0·095	0·022	0·10
VR5	377	0·296	1·67	0·48	0·42
VR7	434	0·274	2·00	0·61	0·65
VR8	190	0·218	0·016	0·005	0·024
SP1	850	0·768	23		
SP2	2 300	0·323	44		
SP3	1 600	0·241	19		
SP4	600	0·122	39		
SP5	8 000	0·423	250		
SR1	100	0·223	1·9		
SR2	50	0·300	1·4		
SR3	450	0·284	1·9		
SR4	500	0·303	1·2		
SR5	550	0·294	1·3		

[a]Sample designations: O, water-based drilling mud used; E, mainly water-based drilling mud used, some oil-based drilling mud; V, S, mainly oil-based drilling mud used; P, platform sampling site; R, reference sampling site.

TABLE 2
Total hydrocarbon content and selected hydrocarbons in individual cod liver samples from the Norwegian North Sea, 1987 (mg/kg dry matter)

Sample[a]	THC	NPD	C_5–C_8 decalins	C_8-alkyl cyclohexanes	C_9–C_{12} benzenes
UP1	1 050	0·69	18	3·6	
UP2	1 110	0·51	12	2·6	
UP3	1 080	0·49	12	3·2	
UP4	820	0·21	8·3	1·7	
UP5	830	0·084	9·9	2·2	
UR1	1 170	0·019	7·9	2·5	
UR2	1 020	0·054	7·2	1·7	
UR3	880	0·004	7·6	1·6	
UR4	770	0·048	7·3	2·7	
UR5	560	0·088	15	3·1	
VP11	1 890	2·35	65	26	27
VP12	950	3·23	12	5·6	11
VP13	580	1·47	1·0	0·11	0·26
VP14	510	0·436	1·8	1·1	0·5
VP15	710	1·54	2·0	1·3	1·6
VR9	660	0·097	2·1	0·9	0·5
VR10	430	0·058	1·6	0·4	0·2
VR11	670	0·060	1·4	0·5	0·14
VR12	480	0·067	2·0	1·1	0·4
VR13	770	0·11	2·7	1·1	0·24
EP2	740	0·36			
EP3	450	0·29			
EP4	350	0·40			
EP5	640	0·43			
EP6	350	0·61			
ER1	490	0·086			
ER2	270	0·038			
ER3	c. 400	0·10			
ER4	300	0·089			
ER5	410	0·10			

[a]Sample designations: U, water-based drilling mud used; E, mainly water-based drilling mud used, some oil-based drilling mud; V, oil-based drilling mud used; P, platform sampling site; R, reference sampling site.

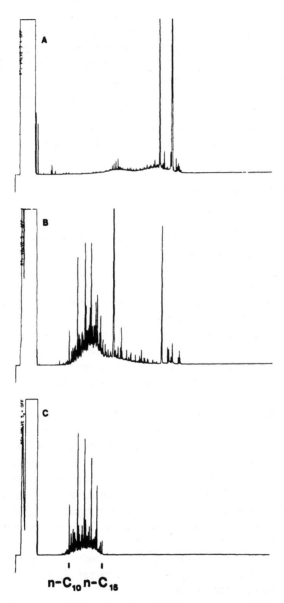

FIG. 1. Gas chromatographic traces of cod liver extracts and base oil used in drilling mud. (A) Reference cod (Oslo Fjord). (B) Cod caught 100 m from a North Sea drilling platform.

No statistical analysis of the cyclohexane and decalin data has been carried out and it is therefore not known how well these parameters may be used to detect very low contamination levels. The GC/MS chromatograms of both decalins and cyclohexanes are very broad, unresolved contours, and detailed information is therefore lost in contrast to the NPD chromatograms where single isomers may be detected. The results indicate that although useful information is obtained, the same, and more detailed data, may be extracted from the NPD analysis.

3.4. Aromatic Hydrocarbons

3.4.1. Selected Alkylbenzenes
The analysis of C_9–C_{12} alkylbenzenes in cod liver yields reference levels and sample levels in the range of 0·02–0·74 ppm and 0·3–27 ppm, respectively. The statistical material is rather limited since all the samples studied come from two surveys only. However, the quite unambiguous contamination effects encountered in these two surveys is clearly reflected also in the alkylbenzene results.

The situation is quite analogous to that of the cyclohexanes and decalins in that the benzenes also show broad, unresolved GC/MS chromatographic contours. The benzenes are therefore not expected to yield additional information not already contained in the NPD data.

3.4.2. Selected Di- and Triaromatics (NPD)
The NPD hydrocarbons are among the main aromatic constituents of petroleum products and they have been regularly monitored in the North Sea environmental surveys. The studies of oil uptake in fish cited above, also point towards NPDs and naphthalenes in particular as a suitable diagnostic tool for determination of oil contamination in fish. The results given in Tables 1 and 2 show typical reference levels of 0·04–0·3 ppm NPD although the range is 0·004–0·52 ppm. The NPD concentration range for the platform samples is 0·1–10·2 ppm where total NPD concentrations above 1 ppm have been measured close to platforms using oil-based muds. This is often accompanied by gas chromatographic traces showing typical oil profiles as depicted in Fig. 1.

In order to detect more subtle contamination effects statistical methods have been applied. The details of the statistical analysis will be given elsewhere (Vogt *et al.*, 1988), but a brief presentation of the most important results will be given here. Principal Component Analysis (PCA) of the results given in Table 1 and of the sum content of the different NPD alkyl isomers, revealed a systematic difference in the pattern of naphthalenes in the platform and reference sample groups (Vogt *et al.*, 1988). The results showed that the cod liver samples from the platforms contain increased amounts of naphthalenes whereas the phenanthrenes/anthracenes and dibenzothiophenes are not significantly accumulated.

Thus, the available analytical results may be taken to indicate that naphthalene and its alkyl homologues are well suited as indicators of petroleum contamination in North Sea cod. On the other hand it may be argued that the current use of low aromatic content base oils might advocate the use of aliphatic instead of aromatic contamination marker compounds. Results obtained in these laboratories indicate that a low aromatic content base oil may contain as much as 5–10% decalins, while the naphthalene content may range from *ca.* 0·1 to 1%. However, the large number of individual C_5–C_8 alkyldecalin isomers results in very broad, unresolved GC/MS molecular ion fragmentograms which may be compared with the relatively simple fragmentograms of the C_0–C_3 alkyl NPDs, where single isomers are clearly resolved. Thus, NPDs may be detected both in considerably lower amounts and with greater certainty than the decalins. We therefore feel that the higher sensitivity and selectivity of the NPD analysis more than compensates for the higher amounts of decalins available in the present day base oils.

In the statistical analysis cited above (Vogt *et al.*, 1988) the sum content within each NPD alkyl isomer group did not prove sufficient to discriminate between samples from platforms using low aromatic oil-, diesel oil- or water-based muds. The fact that individual NPD isomers may be detected and quantified in the GC/MS chromatograms indicates that the NPD analyses may be carried one step further and that the isomer patterns may be used to identify different sources of oil input.

4. CONCLUSIONS

The analysis of several hydrocarbon compound groups in cod liver samples from the North Sea shows that all parameters studied may be used to detect contamination due to the use of oil-based muds at drilling platforms. In extreme cases fingerprint oil chromatograms are clearly observed in GC traces of cod liver extracts.

The higher specificity and sensitivity of the NPD analysis, and the fact that naphthalenes are known to be readily absorbed by fish, still makes the NPDs our prime choice as a monitoring parameter for the detection of more subtle hydrocarbon contamination effects.

REFERENCES

ALBAIGES, J., FARRAN, A., SOLER, M., GALLIFA, A. & MARTIN, P. (1987). Accumulation and distribution of biogenic and pollutant hydrocarbons, PCB's and DDT in tissues of Western Mediterranean fishes. *Mar. Environ. Res.*, **22**, 1–18.

DRANGSHOLT, H., ORELD, F., SPORSTØL, S.P. & SJØGREN, C.E. (1987). Evidence of oil contamination in North Sea cod. Poster paper in the *Proceedings of the International Conference on Environmental Protection of the North Sea*, London 24–27 March, ed. P.J. Newmann & A.R. Agg, Heinemann.

GRAHL-NIELSEN, O. (1988). The five-year fate of sea-floor petroleum hydrocarbons from discharged drill cuttings. This volume, pp. 667–81.

MCCAIN, B.B., HODGINS, H.O., GRONLUND, W.D., HAWKES, J.W., BROWN, D.W., MYERS, M.S. & VANDERMEULEN, J.H. (1978). Bioavailability of crude oil from experimentally oiled sediments to English sole (*Parophrys vetulus*) and pathological consequences. *J. Fish. Res. Bd Can.*, **35**, 657–64.

MCGILL, A.S., MACKIE, P.R., HOWGATE, P. & MCHENERY, J.G. (1987). The flavour and chemical assessment of dabs (*Limanda limanda*) caught in the vicinity of the Beatrice oil platform. *Mar. Pollut. Bull.*, **18**(4), 186–9.

NEFF, J.M. (1979). *Polycyclic Aromatic Hydrocarbons in the Aquatic Environment*. Applied Science Publishers, London.

PHILP, R.P., GILBERT, T.D. & FRIEDRICH, J. (1981). Bicyclic sesquiterpenoids and diterpenoids in Australian crude oils. *Geochim. Cosmochim. Acta*, **45**, 1173–80.

REIERSEN, L.-O., GRAY, J.S., PALMORK, K.H. & LANGE, R. (1988). Monitoring in the vicinity of oil and gas platforms: results from the Norwegian sector of the North Sea and recommended methods for forthcoming surveillance. This volume, pp. 91–117.

VOGT, N.B., DAVIDSEN, N.B. & SJØGREN, C.E. (1988). Multivariate and statistical analysis of di- and triaromatic hydrocarbons in fish from the North Sea. *Oil Chem. Pollut.*, **4**(3) 217–42.

WOLFE, D.A. (Ed.) (1977). Fate and effects of petroleum hydrocarbons in marine organisms and ecosystems. Proceedings of a symposium November 10–12. 1976, Seattle, Washington. Pergamon Press, Oxford.

29

Sediment Deposition, Biological Accumulation and Subcellular Distribution of Barium Following the Drilling of an Exploratory Well

Kenneth D. Jenkins, Stavros Howe, Brenda M. Sanders and
Charles Norwood

*Molecular Ecology Institute, California State University, Long Beach,
California 90840, USA*

ABSTRACT

This paper reports the results of a field study which took place around an exploratory well located in the Santa Barbara Channel. The distribution of Ba in the sediments, and its bioaccumulation and subcellular distribution in three benthic species were examined before and after drilling. Statistically significant increases in the accumulation of Ba were found in the sediments down current from the well site after drilling. Statistically significant increases in the bioaccumulation of Ba were also observed in two of the three species examined, Cyclocardia ventricosa and Pectinaria californiensis. Within these organisms the majority of the Ba was localized in the granular pellets (>97%) and less than 0·1% accumulated in the cytosol. These data indicate that although significant bioaccumulation of Ba occurs in some species immediately down current from the well, most of it remains in an insoluble form, presumably as $BaSO_4$. The labile fraction of the accumulated Ba, which is associated with the other subcellular fractions, represents less than 3% of the total and does not appear to be present in sufficient quantities to cause toxicity.

1. INTRODUCTION

During offshore drilling operations drilling fluids and cuttings are released into the marine environment and may accumulate in the surrounding sediments (NRC, 1983). These fluids often contain measurable concentrations of a number of potentially toxic heavy metals including Ba, Cr, Cd, Hg, Pb and Zn (NRC, 1983; Trefry *et al.*, 1986). These observations have led to concerns regarding the potential biological availability and toxicity of these metals to marine organisms.

Based on data from laboratory studies (Neff *et al.*, 1985–1986), it appears that the bioavailability of metals from drilling fluids is generally limited. Also, with the possible exception of Cr in fluids containing chrome lignosulfonate, there is little correlation between the metal content of drilling fluids and their toxicity (Chaffee & Spies, 1982; Conklin *et al.*, 1983; NRC, 1983; Carls & Rice, 1984). Although field studies have generally supported laboratory results (NRC, 1983), they are often difficult to interpret because of: (1) a lack of appropriate 'control' sites and populations in areas of high anthropogenic activity; (2) the high variability in metal content of both sediments and organisms collected at field sites (Tillery & Thomas, 1980; EG&G, 1982); and (3) the difficulty in establishing a relationship between the total concentrations of metals in sediments and the fraction of those metals that are biologically available (Thompson *et al.*, 1984).

In addition, it is often difficult to predict the toxicity of metals accumulated by organisms *in situ* because they actively regulate the distribution of metals at both the tissue and subcellular levels (Mason & Nott, 1981; George, 1982; Jenkins & Brown, 1984). This regulation is mediated through metal-binding ligands including: soluble proteins such as metallothionein (Sanders & Jenkins, 1984), insoluble vesicles (George, 1982), and granules (Mason & Nott, 1981). These ligands can be used by the organism to control the tissue and subcellular distributions of metals and limit their nonspecific binding to macromolecules, a major mechanism of toxicity. Because of this capacity to compartmentalize metals and the impact of this compartmentalization on metal toxicity, measurements of total metal accumulation often provide little insight into the toxicity of the accumulated metal. Rather, the biological effects of metal accumulation are more accurately assessed by characterizing the subcellular distribution of those metals (Sanders *et al.*, 1983).

The determination of subcellular distributions is particularly important for Ba, the major metal constituent of drilling fluids. Barium is added to drilling fluids as particulate barite ore where the major Ba species is the highly insoluble salt, $BaSO_4$ (Liss *et al.*, 1980). Although insoluble metal salts are usually not considered to be biologically available, George *et al.* (1976) demonstrated that mussels can accumulate insoluble iron salts by pinocytosis and retain these salts in intracellular membrane bound secondary lysosomes. Interestingly, even though Fe is an essential element, data from this study indicated that as much as 30% of this absorbed Fe was eliminated from the organism without ever being released from the vesicle into the soluble cytoplasm of the cell (George *et al.*, 1976). It is important to determine if a similar mechanism is involved in the accumulation of Ba salts; specifically whether Ba is retained in an insoluble form or released into the cell in a soluble form which might then interact with macromolecules and result in metal toxicity.

In this study our basic hypotheses were: (1) the release of drilling fluids into the marine environment would result in a temporal increase in the concentrations of drilling fluid-associated metals in the post-drilling sediments relative to pre-drilling concentrations; (2) the metal concentrations in the sediments would follow a declining spatial gradient which correlates with distance from the well site, and is dependent upon the strength and direction of prevailing water currents; and (3) to the degree that these metals are bioavailable, the benthic fauna will reflect both the temporal and spatial trends seen in sediment metal concentrations.

In order to test these hypotheses, we designed this field experiment in such a way as to minimize the limitations inherent in in-situ studies. For both sediment and faunal samples, ten replicates were analyzed from each station to characterize the natural variability in the metal concentrations. Sediment metal concentrations were characterized using a weak acid leachable (WAL) extraction procedure which has been found to yield data that closely correlate with metal bioaccumulation by benthic organisms (Luoma & Bryan, 1981). Finally, the potential biological effects of any increased accumulation of metals in the invertebrate fauna were evaluated by examining the subcellular distribution of those metals.

The overall study was designed to assess the biological availability and potential toxicity of a number of trace metals associated with drilling fluids (Ba, Cd, Cr, Cu, Hg, Ni, Pb and Zn). This paper summarizes the Ba data exclusively. Data for the other metals will be summarized in subsequent papers that are currently in preparation (Howe *et al.*, 1988; Jenkins *et al.*, 1988; Sanders *et al.*, 1988).

2. EXPERIMENTAL DESIGN

2.1. Station Locations

This field study focused on an exploratory well that was drilled by a jack-up rig during the period of August–December of 1984. The well site was

located 3 miles offshore at a depth of 73 m in the Molino lease site of the
Santa Barbara Channel (Fig. 1). The station locations were selected along
a transect down-current from the well to allow us to examine sediments
and fauna for spatial changes in the concentrations of drilling fluid
associated metals after drilling. In order to determine the optimal locations
of these stations, preliminary studies were implemented to characterize
both the hydrography and sediments.

An initial current meter study was carried out prior to drilling to obtain
a preliminary estimate of current direction and speed. In this study, an
array of three current meters was maintained at the drilling site from
April 17th to June 20th, 1984. Meters were placed at 10 m (near-surface),
39 m (mid-depth) and 64 m (near-bottom) depths. Data indicated a net
westerly flow at all depths with a mean direction ranging from 283° near
surface to 255° near bottom (Howe *et al.*, 1988). Mean current speeds
ranged from 28 ± 15 cm/s near surface to 14 ± 7 cm/s near bottom.
Based upon these hydrographic data, two reference stations were chosen
for the temporal component of the study: a near-field station immediately
down-current (180 m) from the drilling site and a far-field station located
1500 m up-current and relatively isolated from other recent drilling
activities (Fig. 2(a)). Both stations were sampled for sediments and benthic
fauna prior to the placement of the exploratory rig at the site (July 16–
21, 1984).

Drilling was initiated on August 19, 1984 and the final drilling fluids
were released from the rig on December 12 of that year. Drilling fluids
were introduced into the water column at a depth of 18 m via a shunt
pipe. A second current meter study was carried out while drilling was in
progress (September 25 to December 11) to define the hydrography as
fluids were being released. Because of the depth of release of drilling
fluids, only two meters were deployed in this study, one at mid-depth
(35 m) and a second located near-bottom (67 m). Data from this study
indicated a net westerly flow consistent with results from the first study,
with mean directions of 301° at mid-depth and 247° near-bottom (Howe
et al., 1988). Current speeds ranged from 20 ± 11 cm/s at mid-depth to
12 ± 6 cm/s near-bottom.

Based on the results of the current meter studies we hypothesized that
the major component of drilling fluid deposition would follow the westerly
axis defined by the prevailing currents. In order to test this hypothesis,
an intensive sediment coring program was carried out in late November,
after final well depth was achieved but prior to the final release of drilling
fluids. Over 95% of the total barite used for this well had been added to

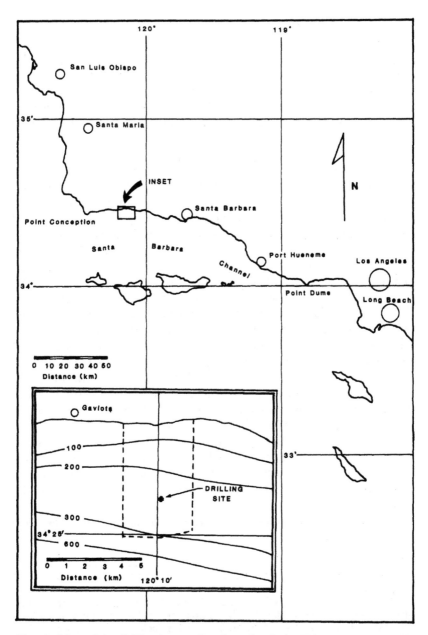

FIG. 1. Map of the California coastline from San Luis Obispo to Long Beach. The inset depicts the Molino Lease site in the Santa Barbara Channel and the location of the drilling site.

the drilling fluids by this time. Sixty sediment cores were sampled (box core samples $10 \times 24 \times 30$ cm) in a radial pattern around the well site with more dense sampling to the west (Fig. 2(a)). Aliquots of surface sediments (upper 2 cm) were removed from each core, sieved (100 μm pore size), and the salt-free fine fraction was analyzed for Ba by neutron activation analysis (NAA).

A contour summary of barium data (Fig. 2(b)) (SYMAP, nearest neighbor interpolation = 3), indicates that the major gradient of Ba deposition was located in a westerly direction thus confirming our hypothesis. Total concentrations of Ba in the sediments ranged from a maximum of 34 000 μg/g in the vicinity of the drilling and to a background concentration of 700–900 μg/g some 500–2500 m from the well (Howe *et al.*, 1988). Based upon these data, five stations were selected along a NNW transect (285°; Fig. 2(b)). This transect was chosen because the stations which lie along it reflect a clearly defined declining gradient in Ba deposition which could be related to drilling activities and provide a basis for examining spatial changes in Ba bioaccumulation. Stations 1 through 5 were located along this axis at distances of 120 m, 360 m, 610 m, 880 m and 1500 m down-current from the well, respectively. Concentrations of total Ba in the sediments from the gradient stations for the November sampling period are shown in Table 1. The most westerly transect (270°) had more extensive accumulation of Ba at the stations beyond 1500 m than did the axis chosen for this study but was not used because the gradient was more poorly defined.

The final release of drilling fluids occurred on December 13, 1984. In total, some 10 720 barrels of drilling fluids were released during the drilling of this well. This fluid included 19 092 bags of barite, resulting in a total mass loading of $8 \cdot 66 \times 10^5$ kg of barite or approximately $7 \cdot 36 \times 10^5$ kg of Ba.

The post-drilling sampling took place from January 14 to 19, 1985

FIG. 2. (a) Station locations (solid circles) for samples collected during the November sediment coring program. Cores were also taken at the near-field and far-field reference stations (solid triangles). The well site is represented by the solid diamond at the center of the grid. (b) Contour plot of barium data derived from neutron activation analysis of the sediment cores depicted in (a). Concentrations of barium for the various contours are represented by shades of gray as reflected in the accompanying scale. The figure depicts the axis of the gradient stations (solid line), and the gradient stations (solid circles). The well location is represented by the solid diamond.

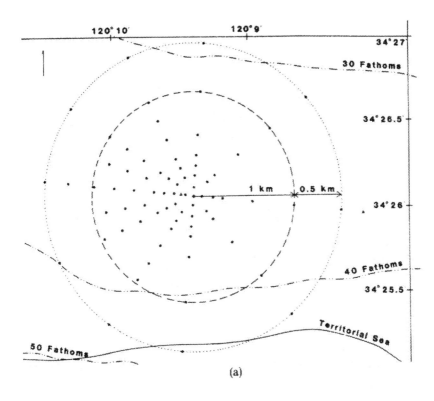

120°10′   120°9′

34°27

30 Fathoms

34°26.5

1 km   0.5 km

34°26

40 Fathoms

34°25.5

Territorial Sea

50 Fathoms

(a)

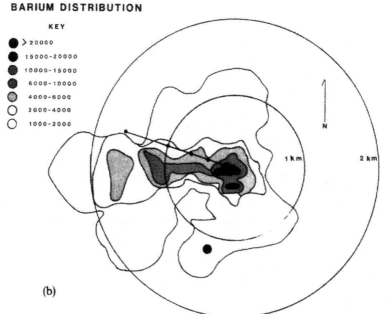

BARIUM DISTRIBUTION

KEY

- > 20000
- 15000-20000
- 10000-15000
- 8000-10000
- 4000-6000
- 2000-4000
- 1000-2000

N

1 km   2 km

(b)

TABLE 1

Concentrations of barium in sediments of gradient stations as determined by neutron activation (NAA) and weak acid leachable/atomic absorption method (μg/g dry weight)

	Core samples (11/84)	Post-drilling samples (1/85) (Mean and SD)		WAL as a percentage of NAA
	NAA (n = 1)	NAA (n = 3)	WAL (n = 10)	
Station 1	18 607	16 985 + 1 515·3	230·39 + 22·97	1·36
Station 2	6 689	9 487 + 2 462·5	136·10 + 17·04	1·43
Station 3	5 093	3 094 + 358·5	133·79 + 21·59	4·32
Station 4	3 510	1 910 + 14·8	71·59 + 11·73	3·75
Station 5	1 692	1 183 + 181·0	68·87 + 7·14	5·82

approximately 1 month after the final release of drilling fluids. Sediment and fauna were collected from the five gradient stations and the near-field and far-field reference stations.

2.2. Choice of Species

Species which exhibit different feeding strategies were chosen for this study because they reflect different exposures to drilling fluid components. The infaunal species selected were the cardit clam, *Cyclocardia ventricosa*, the ice cream cone polychaete, *Pectinaria californiensis*, and two nepthyid polychaetes, *Nephtys ferruginea* and *Nephtys caecoides*. *Cyclocardia* is a filter feeder that uses a siphon to access surface particulates and would thus be particularly exposed to recently deposited drilling fluids. *Pectinaria* is a deposit feeding polychaete with limited mobility that uses a 'conveyor belt' feeding strategy to ingest sediments and detritus and rework the sediments. The *Nephtys* species are both deposit feeders and are active burrowers responsible for much sediment reworking. Both *Nephtys* species co-occur in this area and visual taxonomic distinctions between species could not be made on board ship. These species are morphologically similar, exhibit similar burrowing, feeding and reproductive behaviors and, as such, are considered functional analogs. Therefore, organisms from both species were pooled in this study to provide a total *Nephtyid* composite.

TABLE 2
Barium concentrations (μg/gm) for drilling fluid samples collected at three well depths

		Well depth (m)	
	2 365	*3 506*	*3 838*
NAA[a] Mean	60 727	351 610	443 931
SD	4 130	10 120	13 097
WAL[b] Mean	501	413	395
SD	18	27	4
WAL as a % of NAA	0·83	0·12	0·09

[a]Analysis by Neutron Activation (NAA).
[b]Analysis by Weak Acid Leachable/Atomic Absorption Method
(WAA).

2.3. Sampling

Samples of drilling fluids and major drilling fluid components were
collected from the rig mud system at well depths of 2365 (9/19/84), 3506
(10/17/84) and 3838 (11/18/84) m for metal analysis. These three samples
represent depths at which 11·6%, 47·3% and 95·9% of the total barite
had been utilized, respectively (Table 2). All samples were placed in acid
washed plastic containers and stored frozen ($-20\,°C$) until digestion and
analysis.

Ten replicates of the sediment and all three of the benthic species were
collected at each of the following sites: near-field and far-field stations
prior to drilling, near-field, far-field and the five gradient stations following
drilling. A total of 871 benthic grab samples were collected using teflon-
coated Van Veen grabs ($0·1\,m^2$). Due to the low density of organisms, 8–
12 grabs were normally required to obtain the minimum biomass of 1
gram wet weight per replicate for each species. Sediment samples were
collected from each grab sample and cross referenced to the organisms
from that same grab. In the laboratory, the percentage contribution of
each grab sample to the faunal replicate was determined. The sediments
from each grab were subsampled and composited following these ratios
to provide a final replicate sample. In this way the relative contribution
of each grab sample to a specific composite were identical for both
sediments and fauna. With this compositing, the concentrations of Ba
accumulated by a given composite of organisms could be correlated with
the metal concentrations in the specific sediments from which those
organisms were collected. Aliquots of composited sediments were also

taken from three replicates at each station (both pre- and post-drilling) for analysis of total Ba by NAA.

Clean techniques were used in all phases of sample collection, preparation and analysis. Sediment samples from the top 2 cm were collected in acid-washed petri dishes and frozen at $-20\,°C$ until digestion and analysis. Fauna were collected on acid washed Nitex sieves (1 mm), and intact whole organisms were rinsed with filtered ambient seawater and allowed to depurate for 12 h in a flow-through system on board ship. Organisms were then placed in acid-washed plastic containers, frozen on dry ice and stored at $-80\,°C$ until fractionation and analysis.

2.4. Sample Preparation and Analysis

In order to obtain a more accurate estimate of the biologically available fraction of metals, sediment metal concentrations were characterized using a weak-acid leachable (WAL) extract on the fine sieved fraction (Flegal et al., 1981). Sediment metal concentrations measured by this method had previously been shown to correlate closely with metal bioaccumulation by benthic organisms (Luoma & Bryan, 1981). Flegal's modifications to this procedure (1981), which were followed in our present study, were developed for the State Water Quality Control Board and are endorsed by that group. In this method, sediment samples are filtered through a $100\,\mu m$ sieve and the fine fraction weighed and subjected to a weak acid leach (WAL) in 1N HCl at ambient temperature for 24 h. Barium was analyzed by flame AAS using a Instrument Laboratories 551 AAS. Sediment metal concentrations are expressed as μg metal normalized to gram wet weight. Based on WAL analysis of fine ($<100\,\mu m$) and course ($>100\,\mu m$) fractions, $85\cdot02 \pm 0\cdot56\%$ of the leachable Ba was associated with the fine fraction. Relative to a near-total digestion (hot aqua regia taken to dryness), Ba recovery of the WAL extract was $10\cdot51 \pm 1\cdot54\%$ in unsieved sediments and $15\cdot91 \pm 2\cdot55\%$ in fine sieved sediments ($<100\,\mu m$).

Drilling fluid and sediment samples ($<100\,\mu m$) to be analyzed by NAA were washed twice in milli-Q water to remove interfering Na and $0\cdot1\,g$ aliquots were placed in acid washed plastic vials and brought to dryness. Samples were then placed in the reactor for $1\cdot0\,h$ at $1\cdot0 \times 10^{12}\,n\,cm^2\,s$. Following activation, Ba concentrations were determined at the 166 keV photopeak of ^{139}Ba at 83·3 min (Guinn et al., 1977).

Faunal samples were thawed, weighed, homogenized and fractionated by differential centrifugation to provide granular, nuclear, mitochondrial, microsomal and cytosolic fractions (Nash et al., 1981). The cytosol was further fractionated for high molecular weight (HMW) and metallo-thionein/low molecular weight fractions (MT/LMW) by gel permeation high performance liquid chromatography (GP-HPLC) using a 0·25M NaCl, 0·05M Tris-HCl running buffer at pH 7·4. All other parameters for GP-HPLC were as described by Sanders *et al.* (1983). The HMW and MT/LMW fractions were defined by molecular weight markers and preliminary metal analysis on individual fractions. In subsequent chroma-tograms these fractions were pooled for final metal analysis. The HMW and MT/LMW fractions were analyzed directly for metals by Induction Coupled Plasma Mass Spectroscopy (ICPMS; VG Plasmaquad). The remaining subcellular fractions underwent vigorous acid digestion (concen-trated HNO_3, refluxed to near dryness) and were analyzed for metals by either AAS or ICPMS. For each subcellular fraction, Ba concentrations are expressed as μg metal normalized to gram wet weight of the tissues of the intact replicate sample.

Quality control was maintained with the use of standard reference materials (MESS sediments) to confirm metal recovery and reproducibility for each digestion (WAL and near-total) and analytical (ICPMS and AAS) procedure. Blank samples and controls were routinely processed to aid in identifying and minimizing potential sources of metal contamination. Lastly, metal sensitivity standards (aqueous metal spikes in sample matrix) were routinely analyzed to address matrix interferences and document analytical precision and reproducibility.

2.5. Statistical Analysis

Changes in metal concentrations in sediment and biota following drilling were evaluated for statistical significance using the nonparametric Kruskal–Wallis chi-square (KW) test to minimize concern over assumptions for use of parametric statistics. Pre-drilling data (composites of near- and far-field data) for both sediments and fauna were compared with post-drilling data based on: (1) composite of data for all down-current stations; and (2) individual data for each of the five gradient stations. Spatial relationships for Ba accumulation in the sediments and the various subcellular fractions of the biota were evaluated using the Kendall's-Tau rank correlation (KT) and regression analyses with both linear and logarithmic models. Only Ba data which demonstrated both significant temporal increases following drilling and decreases in concentrations along the gradient stations are presented in this manuscript. Significance is assessed based on comparisons with pre-drilling data ($p \leq 0\cdot05$).

3. RESULTS AND DISCUSSION

3.1. Metal Concentrations in Drilling Fluids

Total Ba in drilling fluids, as measured by NAA, increased with well depth from a concentration of approximately $60\,000\,\mu g/g$ at 2365 m to $>440\,000\,\mu g/g$ at 3838 m (Table 2). In contrast, the concentrations of Ba determined by WAL/AAS were two to three orders of magnitude lower ($398-501\,\mu g/g$ or $0.09-0.83\%$ of NAA data). These dramatic differences in the concentrations of Ba by the two methods reflect the relative insolubility of $BaSO_4$ in $1N$ HCl. Since the NAA method measures activated isotopes of Ba directly in the solid phase, solubility is not an issue. Measurements by AAS, however, require prior solubilization and are thus limited by the solubility of Ba in $1N$ HCl at ambient temperature which was used in the WAL extraction. The lack of apparent increases in Ba concentration in the drilling fluids with depth in the WAL/AAS data probably reflects the variability in recovery associated with using the WAL/AAS procedure to evaluate the high concentrations of $BaSO_4$ associated with the drilling fluids.

3.2. Distribution of Barium in Sediments

The mean concentrations of total (NAA) and WAL/AAS Ba in sediments for the five gradient stations (post-drilling) are presented in Table 1 as are the NAA data from the cores taken at those five stations in November. Although 4.1% (approximately 3.0×10^4 kg) of the total Ba used for this well was released between the November core sampling and the final sampling in January the total concentrations of Ba (NAA) from the January sampling were lower at all stations except Station 2 when compared to the November core samples. These results suggest that redistribution and dilution of Ba had occurred during the period between samplings. These changes could be due to resuspension and reworking of the sediments (NRC, 1983) but data should be interpreted with caution since there were no replicates for the November coring samples.

Distributions of total Ba in the sediments, as determined by NAA, showed two distinct trends. After drilling, there was a significant increase in the concentrations of Ba in sediments of all down-current stations relative to pre-drilling stations (Fig. 3) (KW chi-square ≥ 5.400; $p < 0.020$ for each comparison). Also, Ba concentrations in the sediments of the gradient stations declined sharply with distance from the well (Table 1 and Fig. 3). The shape of this gradient for total Ba accumulation is best described by a power function of the form $y = -15\,516 \log X + 48\,674$

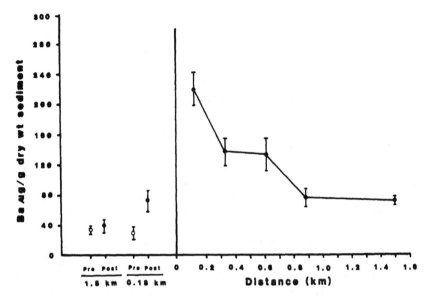

FIG. 3. Barium concentrations (mean ± SD) in the sediments of the reference stations and gradient stations as determined by the weak acid leachable/AAS technique. The data for the gradient stations (right of axis) are presented in relation to the distance of the stations down-current from the well. The concentrations of barium for near-field and far-field reference stations (left of axis) are presented for pre- and post-drilling. The near-field station is located 0·18 km down-current from the well while the far-field station was located 1·5 km upcurrent.

($F = 116\cdot2$ [1,13 d.f.]; $p < 0\cdot001$). This equation is similar to that used by Boothe & Presley (1983) to describe the distributions of Ba around a rig in the Gulf of Mexico.

The concentrations of 'bioavailable' Ba in the sediments, as determined by the WAL/AAS technique, correlated linearly with the total (NAA) data ($r^2 = 85\cdot8$; $F = 158\cdot7$ [d.f. $= 1, 25$]). As in the case of the drilling fluids, Ba concentrations determined by the WAL/AAS procedure were substantially lower than those determined by NAA (1·4–5·8%; Table 1). Again, this appears to reflect the relative insolubility of the Ba salts in the weak acid.

As the above correlation data suggest, the distributions of 'bioavailable' Ba in the January sediment samples follow a pattern similar to that of the NAA data (Table 1 and Fig. 3). Concentrations of Ba at the down-current post-drilling stations were significantly higher than those at the pre-drilling stations (Fig. 3; KW chi-square $\geq 19\cdot35$; $p < 0\cdot001$ for each comparison) and the concentrations of Ba at the gradient stations declined sharply with distance from the well. These data can be described by a power function similar to that presented above where $y = -151 \cdot 813 \log X + 539 \cdot 229$ ($F = 298 \cdot 5$, [1,48 d.f.]; $p < 0 \cdot 0001$).

This pattern of distribution for both total and 'bioavailable' Ba in the sediments fits the temporal and spatial hypotheses set forward in this study and is clearly a consequence of drilling activities. Although this study was not designed to define the spatial extent of Ba deposition in the sediments, some insights can be gained from these data. For the gradient stations, the concentrations of Ba in the sediments following drilling decline with the log of the distance, indicating that the most significant Ba deposition occurred near-field over the first 2–3 stations (e.g. 300–600 m down current; Fig. 3). However, relative to pre-drilling concentrations, both total and 'bioavailable' Ba were significantly elevated after drilling at Station 5 (1500 m down current from the well). These observations indicate a more extensive low level deposition of Ba and are consistent with the November coring data where samples exceeding background concentrations were observed at the farthest sampling points 2500 m to the west.

Data on the extent of Ba distribution from other field studies are highly variable and depend on a number of factors including the water column depth, depth of release of drilling fluids, and water current speed and direction (NRC, 1983). In studies carried out in the Gulf of Mexico it was estimated that for a well at a water depth similar to the present study (e.g. 75–80 m) elevated concentrations of Ba in the sediments may extend as far as 3000 m down current (Boothe & Presley, 1983). In the same study, however, Ba concentrations approached background levels 500 m down-current from a well in shallower water depths (13–34 m; Boothe & Presley, 1983).

3.3. Bioaccumulation and Subcellular Distribution of Barium

Statistically significant spatial and temporal changes in the concentrations of Ba were observed in *Cyclocardia* and *Pectinaria* but not in the *Nephtys* species. In *Cyclocardia* the highest concentrations of Ba were observed in the granular pellets (Table 3). In clams from the gradient stations, the concentrations of Ba in the granular pellets declined with the log of the distance from the well ($y = -10 \cdot 27 \log X + 39 \cdot 1$; $F = 86 \cdot 51$ [d.f. = 1,48]) and were significantly greater than in pre-drilling samples at Stations 1, 2 and 3 (KW chi-square = $16 \cdot 02$, $10 \cdot 89$, $6 \cdot 51$, respectively; $p < 0 \cdot 001$, $p = 0 \cdot 001$, $p = 0 \cdot 01$; respectively; Fig. 4). Moreover, accumulation of Ba

TABLE 3

Relative concentrations of barium in the subcellular fractions of two invertebrate species from Station 1

Subcellular fraction	N	Barium (µg/g) Mean	SD	Distribution (%)
Cyclocardia ventricosa				
Granular	10	23·083	7·088 3	97·99
Nuclear	10	0·124	0·055 7	0·53
Mitochondrial	10	0·083	0·045 4	0·35
Microsomal	10	0·249	0·184 6	1·06
HMW	8	0·007	0·013 2	0·03
MT/LMW	8	0·012	0·016 9	0·05
Total		23·558		100·00
Pectinaria californiensis				
Granular	10	118·247	45·825 0	97·02
Nuclear	10	3·017	1·993 9	2·48
Mitochondrial	9	0·049	0·086 7	0·04
Microsomal	10	0·515	0·575 0	0·42
HMW	8	0·015	0·024 8	0·01
MT/LMW	8	0·039	0·025 8	0·03
Total		121·884		100·00

in the granular pellets of *Cyclocardia* from the various stations correlated linearly with concentrations of 'bioavailable' Ba in the sediments from the same stations ($F = 68·6$; [1,48 d.f.]; $p < 0·0001$). These data strongly suggest that the significant increases in the concentrations of Ba in the granular pellets of clams from Stations 1, 2 and 3 are a consequence of drilling activities.

Spatial and temporal changes in the accumulation of Ba were also observed in the cytosolic fractions of *Cyclocardia* (Fig. 4). Detectable concentrations of Ba were limited to the HMW and MT/LMW fractions in organisms from Stations 1 and 2. Barium accumulation in both cytosolic fractions correlated directly with Ba in the sediments ($KT = 0·294$ for HMW ($p = 0·002$) and $0·323$ for MT/LMW ($p = 0·001$); $N = 88$) and appear to be attributable to drilling activities. The accumulation of Ba in the remaining subcellular fractions of *Cyclocardia* showed no significant spatial or temporal changes which could be related to drilling activities.

Significant temporal and spatial changes in Ba accumulation were also observed in *Pectinaria*. As with *Cyclocardia*, most of the Ba was found in

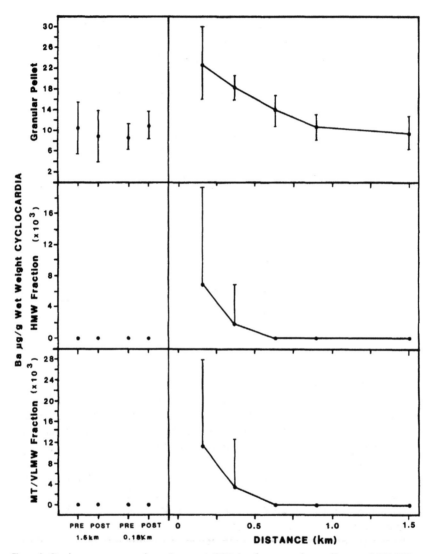

FIG. 4. Barium concentrations (mean ± SD) in the granular pellets and HMW and MT/LMW fractions of *Cyclocardia* from the reference stations and gradient stations. See Fig. 3 for interpretation of *X*-axis.

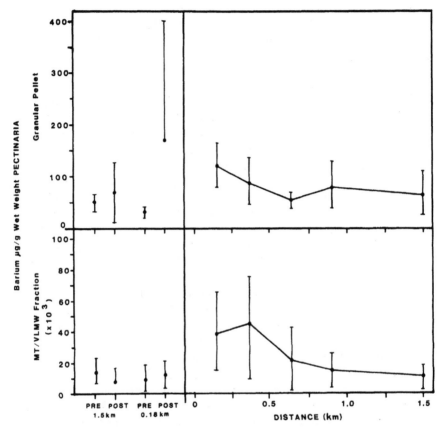

FIG. 5. Barium concentrations (mean ± SD) in the granular pellets and MT/LMW fractions of *Pectinaria* from the reference stations and gradient stations. See Fig. 3 for interpretation of *X*-axis.

the granular pellet (Table 3) and declined with the log of the distance from the well ($F = 16.50$; 1,48 d.f.; $p < 0.001$). Significant increases in Ba concentrations for the granular pellets were observed in organisms from Stations 1 and 2 compared to pre-drilling organisms (Fig. 5). These data correlated with the distributions of 'bioavailable' Ba in the sediment (KT = 0.20, $p = 0.006$, $N = 87$). Spatial and temporal changes in the accumulation of Ba were also observed in the MT/LMW fractions of *Pectinaria* with high concentrations of Ba occurring in the MT/LMW fraction of organisms from Stations 1 and 2. The concentrations of Ba in this fraction declined with the log of the distance from the well (Fig. 5) and showed significant linear correlation with WAL Ba in the sediment (KT = 0·172; $p = 0·022$). The accumulation of Ba in the remaining subcellular fractions of *Pectinaria* showed no significant changes which could be related to drilling activities.

There were no significant differences in the concentrations of Ba in any of the subcellular fractions of *Nephtys*. Two factors may contribute to a lack of an increase in bioaccumulation of Ba by *Nephtys* following drilling. First, unlike the other species, *Nephtys* are active deposit feeders that burrow and feed in the subsurface layers of the sediments (5–10 cm depths) and would have less access to recently deposited drilling fluids localized in superficial layers of the sediments. Second, *Nephtys* were most effective in clearing their gut content during the 12-h depuration period used in this study. This increased rate of depuration could result in a reduction in the residual Ba in the gut which would reduce the apparent accumulation of Ba in these organisms due to this mechanism.

3.4. Biological Significance of Barium Accumulation and Compartmentalization in *Cyclocardia* and *Pectinaria*

As is clear from the previous discussion, drilling activities have resulted in significant increases in the concentrations of Ba in *Cyclocardia* and *Pectinaria*. The vast majority of the Ba accumulated in both species was found in the granular pellets. This subcellular fraction contains all of the materials in the homogenate that have sufficient density and mass to precipitate in a 10-min centrifugation at 200 × G, including subcellular components such as phosphate rich granules and barite particles taken up by pinocytosis. In addition it would contain extracellular materials such as barite particles adhering to epithelial surfaces or retained in the gut. Depuration of the gut contents by *Pectinaria* and *Cyclocardia* were not as complete as in *Nephtys*. *Pectinaria*, in particular, passes prodigious quantities of sediments through its gut. Microscopic evaluation of these organisms prior to homogenization revealed that significant quantities of sediments remained in the gut following the depuration period. A similar but less extreme situation was seen for *Cyclocardia*. From these observations it appears that for *Pectinaria* and *Cyclocardia*, much of the Ba accumulated in the granular pellet could be derived from sediment particles retained in the gut cavity. Because the Ba in the granular pellet represents over 97% of the Ba accumulated in these organisms, Ba particles entrained in the gut could substantially inflate the estimates of total Ba accumulation in the tissues of these organisms.

These observations do not preclude intracellular accumulation of Ba particles in pinocytotic vesicles, but this issue can only be resolved by quantitative electron microscopy (George *et al.*, 1976). Regardless of whether this granular Ba is compartmentalized within the tissues of the organism or in the gut cavity, the high concentrations of Ba in the granular pellet compared with the other subcellular fractions (>97%) suggest that most of the Ba accumulated by these organisms remains in an insoluble form, presumably $BaSO_4$ (Table 3).

From these data we estimate that no more than 2–3% of the total Ba accumulated by *Cyclocardia* and *Pectinaria* could be in a soluble form which is available for binding to intracellular macromolecules and thus has the potential for impairing cellular metabolism (Reeves, 1986). Although this labile Ba is distributed among all of the subcellular fractions (Table 3), only Ba associated with the cytosol showed both the temporal and spatial changes which could be related to drilling. The important question here is whether these increased concentrations of Ba in the cytoplasmic pools are sufficient to cause perturbations in cellular metabolism of these species.

Because of its similar chemistry, Ba^{2+} is capable of competing with Ca^{2+} in cellular processes. Calcium is an important secondary messenger in cells and thus plays a major role in regulating cellular metabolism. The competitive binding of Ba^{2+} to Ca^{2+}-binding sites can modify cellular metabolism (Berggren *et al.*, 1982) and appears to represent the major mechanism for chronic Ba toxicity. As this is a competitive interaction, the degree to which Ba^{2+} interferes with Ca^{2+} metabolism is dependent upon the relative concentrations of these ions in the cytosol and their relative affinities for Ca^{2+}-binding proteins. For most Ca^{2+}-binding proteins, the affinity for Ca^{2+} is several orders of magnitude higher than that for Ba^{2+} (Berggren *et al.*, 1982). Therefore, for Ba^{2+} to compete significantly for these ligands, the concentration of Ba^{2+} in the cytosol would have to be in substantial excess of that for Ca^{2+}.

The concentration Ca^{2+} in animal cells is in the range of 10^{-7} M and is highly regulated. In the present study the subcellular Ba^{2+} concentration was not measured. Free Ba^{2+} however, would fractionate with the MT/LMW fraction of the cytosol on the TSK column. As the MT/LMW fraction also contains Ba bound to smaller ligands, the total Ba measurements for this fraction should overestimate the concentration of the Ba^{2+} ion and thus represent a worst case situation for these organisms. The concentrations of Ba in the MT/LMW fractions of organisms from Station 1 were 2×10^{-8} M and 9×19^{-9} M for *Pectinaria* and *Cyclocardia*, respectively. These concentrations are one to two orders of magnitude lower than the ambient concentration of Ca^{2+}. Thus, even under this worst case situation, the concentrations of Ba in the cytosols of organisms from Station 1 should not be sufficient to perturb Ca metabolism. Assuming that perturbations in Ca metabolism represent the rate limiting step in chronic Ba toxicity, we conclude that the statistically significant increases in cytosolic Ba which have occurred as a result of drilling activities, will not have toxic impact on these species.

ACKNOWLEDGEMENTS

This research was conducted under the auspices of the California Regional Water Quality Control Board (Central Coast) through its Oceanographic Technical Advisory Panel and funded by Shell California Production Inc. We thank W. Nash, A. Mason, P. Sullivan, L. Fausak and G. Shaw for their contributions to various phases of this project. We also wish to thank the numerous undergraduate and graduate students that participated in sample collection, analysis and data processing.

REFERENCES

BERGGREN, P., ANDERSSON, B. & HELLMAN, B. (1982). Stimulation of insulin secretory mechanisms following barium accumulation in pancreatic beta-cells. *Biochim. Biophys. Acta*, **720**, 320–8.

BOOTHE, P.N. & PRESLEY, B.J. (1983). Distribution and behavior of drilling fluids and cuttings around Gulf of Mexico drilling sites. Draft final report, API projects 243, 253, 65 pp.

CARLS, M.G. & RICE, S.D. (1984). Toxic contributions of specific drilling mud components to larval shrimp and crabs. *Mar. Environ. Res.*, **12**, 45–62.

CHAFFEE, C. & SPIES, R.B. (1982). The effects of used ferrochrome lignosulphonate drilling muds from a Santa Barbara channel oil well on the development of starfish embryos. *Mar. Environ. Res.*, **7**, 265–77.

CONKLIN, P.J., DRYSDALE, D., DOUGHTIE, D.G., RAO, K.R., KAKAREKA, J.P., GILBERT, T.R. & SHOKES, R.F. (1983). Comparative toxicity of drilling muds: role of chromium and petroleum hydrocarbons. *Mar. Environ. Res.*, **10**, 105–25.

EG&G ENVIRONMENTAL CONSULTANTS. (1982). A study of environmental effects of exploratory drilling on the mid-Atlantic outer continental shelf—final report of the block 684 monitoring program (for Exxon Production Research Company).

FLEGAL, A.R., CUTTER, L.S. & MARTIN, J.H. (1981). A study of the chemistry of marine sediments and waste water sludge. Final Report to The California State Water Resources Control Board, May, 1981.

GEORGE, S.G. (1982). Subcellular accumulation and detoxication of metals in aquatic animals. In *Physiological Mechanisms of Marine Pollutant Toxicity*, ed. W.B. Vernberg, A. Calabrese, F.P. Thurburg & F.J. Vernberg. Academic Press, New York, pp. 3–52.

GEORGE, S.G., PIRIE, B.J.S. & COOMBS, T.L. (1976). The kinetics of accumulation and excretion of ferric hydroxide in *Mytilus edulis* (L.) and its distribution in the tissues. *J. Exp. Mar. Biol. Ecol.*, **23**, 71–84.

GUINN, V.P., CHRISTENSEN, E.R., de LANCEY, K. & WADMAN, W. W. (1977). Neutron activation analysis trace metal studies in connection with offshore drilling for oil. In *Nuclear Methods in Environmental and Energy Research*. University of Missouri, Columbia, MO, pp. 303–11.

HOWE, S.R., FAUSAK, L. & JENKINS, K.D. (1988). Hydrography and deposition of drilling fluids in the Santa Barbara Channel relating to exploratory drilling activities (in progress).

JENKINS, K.D. & BROWN, D.A. (1984). Determining the biological significance of contaminant bioaccumulation. In *Concepts in Marine Pollution Measurements*, ed. Harris H. White. Maryland Sea Grant College.

JENKINS, K.D., SANDERS, B.M., HOWE, S.R. & NASH, W. W. (1988). Biological accumulation and subcellular distribution of trace metals following the drilling of an exploratory well (in preparation).

LISS, R.G., KNOX, F., WAYNE, D. & GILBERT, T. R. (1980). Availability of trace elements in drilling fluids to the marine environment. In *Research on Environmental Fate and Effects of Drilling Fluids and Cuttings*. Lake Buena Vista, Florida, Symposium, Vol. II, Courtesy Associates, Washington, DC, 691–719.

LUOMA, S. & BRYAN, G. (1981). A statistical assessment of the form of trace metals in oxidized estuarine sediments employing chemical extractants. *Sci. Total Environ.*, **17**, 165–96.

MASON, A.Z. & NOTT, J.A. (1981). The role of intracellular biomineralized granules in the regulation and detoxification of metals in gastropods with special reference to the marine prosobranch *Littorina littorea*. *Aquat. Toxicol.*, **1**, 239–56.

NASH, W.W., POOR, B.W. & JENKINS, K.D. (1981). The uptake and subcellular distribution of lead in developing sea urchin embryos. *Comp. Biochem. Physiol.*, **69C**, 205–11.

NEFF, J.M., BRETELER, R.J., SAKSA, F.I. & CARR, R. S. (1985). Chronic effects of drilling fluids discharged to the marine environment, with emphasis on bioaccumulation/biomagnification potential of drilling fluid metals. Final report to American Petroleum Institute (API Publication No. 4397).

NEFF, J.M., HILLMAN, R.E., LECZYNSKI, B. & BERNER, T. (1986). Bioavailability of trace metals from barite to benthic marine animals. Final Report to the Offshore Operators Committee, October 1986.

NRC (1983). *Drilling Discharges in the Marine Environment*. National Academy Press, Washington, D.C., 180 pp.

REEVES, A.L. (1986) Barium. In *Handbook on the Toxicology of Metals*, 2nd edn, ed. L. Friberg, G.F. Nordberg & V. Vouk. Elsevier Scientific Publishers, B.V., pp. 84–94.

SANDERS, B.M. & JENKINS, K.D. (1984). Relationships between free cupric ion concentrations in sea water and copper metabolism and growth in crab larvae. *Biol. Bull.*, **167**, 704–12.

SANDERS, B.M., JENKINS, K.D., SUNDA, W.G. & COSTLOW, J. D. (1983). Free cupric ion activity in seawater: Effects on metallothionein and growth in crab larvae. *Science*, **222**, 53–5.

SANDERS, B.M., HOWE, S.R., JENKINS, K.D. & NASH, W. W. (1988). Sediment deposition of trace metals following the drilling of an exploratory well (in preparation).

THOMPSON, E.A., LUOMA, S.N., JOHANSSON, C.E. & CAIN, D.J. (1984). Comparison of sediments and organisms in identifying sources of biologically available trace metal contamination. *Water Res.*, **18**, 755–65.

TILLERY, J.B. & THOMAS, R.E. (1980). Heavy metal contamination from petroleum production platforms in the Central Gulf of Mexico. In *Research on Environmental Fate and Effects of Drilling Fluids and Cuttings*, Lake Buena Vista, Florida, Symposium, Vol. I, Courtesy Associates, Washington, DC, pp. 562–83.

TREFRY, J.H., TROCINE, R.P., METZ, S. & SISLER, M. A. (1986). Forms, reactivity and availability of trace metals in barite. Report to the Offshore Operators Committee Task Force on Environmental Science. March. 1986.

30

Microcosm Studies on the Effects of Drilling Fluids on Seagrass Communities*

R. Dana Morton‡

Bro-Environmental Services Division, Department of Health and Welfare, 421 W. Church Street, Jacksonville, Florida 32202-4111, USA

and

Stacy Owsley Montgomery

Technical Resources Inc., c/o US Environmental Protection Agency, Environmental Research Laboratory, Gulf Breeze, Florida 32561, USA

ABSTRACT

*Four experiments were conducted between May 1983 and November 1984 to study effects of drilling fluids on seagrass communities (*Thalassia testudinum *Konig et Sims) maintained in the laboratory. Effects on the macrobenthos community and potential for bioaccumulation of barium (Ba) and chromium (Cr) were examined. Plant and animal components of the laboratory microcosms were compared with the seagrass community of their origin.*

Microcosms were 15·3 cm diameter seagrass cores that were maintained in the laboratory in flowing seawater. Treated microcosms were exposed to either 190 parts per million (v:v) of the suspended particulate phase of a used drilling fluid, or an equivalent amount of montmorillonite clay. Untreated microcosms received only flowing seawater. Each treated and untreated set consisted of sixteen replicates during each 6-week experiment. Analysis by Neutron Activation Analysis quantified concentrations of Ba and Cr in macrofauna, seagrass and sediments.

*Contribution No. X566, Environmental Research Laboratory, Gulf Breeze, Florida; ERC-169, Ecosystems Research Center, Cornell University, Ithaca, New York.
‡ To whom correspondence should be addressed.

There were statistically significant reductions in the number of individuals for numerically dominant macroinvertebrate species among microcosms receiving the clay and drilling fluid in relation to untreated microcosms. Concentrations of Ba and Cr in the seagrass macrofauna were a clear indication of bioaccumulation; bioaccumulation factors (concentration in tissues/concentration in sediment) for Ba in the benthos ranged from two to sixteen, factors for Cr were zero to four. The experiments quantified some effects of chronic exposure to drilling fluid on an experimental macrobenthic seagrass community.

1. INTRODUCTION

Seagrass communities are highly dynamic and productive ecosystems inhabiting the shallow coastal waters of all but the most polar seas. Productivity of seagrasses may range from 5 to 15 gC/m^2 per day and, when other primary producers (benthic algae, epiphytes and phytoplankton) are included, daily production may exceed over 20 gC/m^2 per day (McRoy & Helfferich, 1977). Seagrass communities have important functions in coastal waters: they stabilize sediment; accumulate material from the water; and serve as a food resource, a nursery and shelter (Hartog, 1977).

Since they inhabit only the shallow marginal waters, seagrass beds are especially vulnerable to degradation by human-induced stresses. Due to their mode of growth, seagrasses are subject to damage from stress or pollutants both in the water column and in the sediments. Although there have been few pollution studies in tropical seagrass beds, Zieman (1975) addressed the more important problem areas. Dredging activities are the leading cause of seagrass bed destruction. Physical disruption of the bed is only part of the problem. Increased turbidity from suspended particulates, and smothering of seagrasses by these particulates reduces light available for photosynthesis leading to the decline of the grasses. Oil spills have been responsible for the destruction of seagrass beds in at least two documented incidents (Diaz-Piferer, 1962; Nadeau & Berquist, 1977). Fauna associated with the seagrasses are likely to be even more sensitive to oil pollution than the vegetative parts of the seagrass (Hartog & Jacobs, 1980; Chan, 1977).

In recent years, concern has arisen over the possible adverse effects of discharged cuttings and drilling fluid from oil and gas well drilling platforms on seagrass beds. The functions of drilling fluids have been described by Simpson (1975). Water-based drilling fluids are used in US marine drilling operations and usually consist of bentonite and other clays, barium sulfate (barite) as a weighting agent, lignosulfonates as thinners, and various 'specialty' items to control pH or bacteria. Diesel and other oils have been added to some drilling fluids to enhance lubrication (National Research Council, 1983).

If drilling fluids are discharged onto or near seagrass beds the effects are likely to be more pronounced than discharges in deeper offshore waters. In the shallower coastal waters inhabited by seagrasses there is less chance for dilution, and tidal exchanges may be slow, allowing for longer exposure of the seagrass community to drilling fluid. Drilling fluids may pose two distinct problems to seagrass communities. Physical alteration of the environment by increased turbidity would reduce light available for photosynthesis, and in the immediate vicinity of a discharge drilling fluids and cuttings covering the grass or adhering to it would certainly have a negative effect on the health of the seagrass bed. If drilling fluids are toxic to the seagrass or some associated fauna a discharge could pose an added threat, possibly leading to more significant effects on the structure of the seagrass community.

To study the effects of a drilling fluid discharge on a seagrass community an experimental seagrass microcosm was developed using turtle grass, *Thalassia testudinum* Konig et Sims (Morton *et al.*, 1986; Morton, 1987). During the Spring and Fall of 1983 and 1984 four experiments were conducted to determine the effects of used drilling fluids on seagrass communities. Each test was 6 weeks in duration. Two different drilling fluids taken from operating rigs in the Gulf of Mexico were used in the experiments. Several different endpoints were measured: health of the *Thalassia* and its epiphytes; decomposition rates of *Thalassia* leaves; structure of the macroinvertebrate community; and fate of barium (Ba) and chromium (Cr). This paper will summarize effects on the macroinvertebrate community and fate of Ba and Cr in test microcosms. Effects on other endpoints have previously been addressed (Morton *et al.*, 1986; Price *et al.*, 1986; Kelly *et al.*, 1987).

2. METHODS

Four experiments were conducted at the US EPA Research Laboratory on Sabine Island, Gulf Breeze, Florida (Long. 87° 9′ 28″, Lat. 29° 20′ 25″). The 42 day experiments were conducted over a two year period: Test 1, May 31–July 11, 1983; Test 2, October 5–November 15, 1983; Test 3, June 11–July 23, 1984; Test 4, October 2–November 12, 1984. A brief

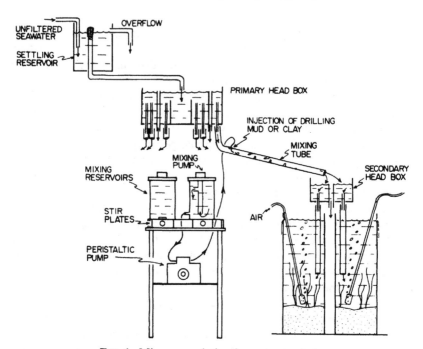

FIG. 1. Microcosm design for exposure tests.

summary of methods follows. A more detailed treatment of procedures and analytical methods already exists (Morton *et al.*, 1986; Morton, 1987).

2.1. Sample Collection

Thalassia seagrass cores (15·3 cm in diameter and 10·0 cm deep) were collected by divers from Santa Rosa Sound adjacent to Sabine Island in 2 m of water. The cores were placed into acrylic cylinders (15·9 cm inside diameter, 50·0 cm height) and transferred to the laboratory. Thus a microcosm consisted of a seagrass core with associated fauna contained within an acrylic cylinder. In the laboratory all microcosms received flowing seawater. A rarefaction analysis on the macroinvertebrate population indicated that 90% of the species present were contained in ten cores. In the experiments described, sixteen replicate microcosms were provided for each treatment.

2.2. Test Procedure

In the laboratory the microcosms were randomly assigned to one of twelve clusters, each consisting of four microcosms (Fig. 1). The twelve

TABLE 1
Characteristics of 'whole' and suspended particulate phase (SPP) drilling fluids

	P-1	P-2
Drilling fluid type	Lightly treated lignosulfonate	Freshwater lignosulfonate
Date received	04/15/82	04/16/82
Whole drilling fluid		
'Diesel'[a] (μg/g)	9·43	2·14
Ba[b] (percentage dry weight)	36·9	37·2
Cr[b] (μg/g)	814	483
Suspended particulate phase		
Ba[c] (mg/liter)	60 800	78 300
Cr[c] (mg/liter)	299	308
Total suspended matter[d] (g/liter)	111	161

[a]New England Aquarium, 1984.
[b]Science Applications Inc., 1984.
[c]Present study, neutron activation analysis by University of Florida Training Reactor.
[d]Residue retained on a 0·4 μm membrane filter.

clusters were then randomly assigned one of three treatments: control (received seawater only), clay, and drilling fluid. The microcosms were acclimated to laboratory conditions for approximately 7 days prior to the start of each experiment. Dosing was continuous during the 42 day exposure period. At the conclusion of the test, microcosms were harvested and samples preserved for analysis. At this time ten field cores were taken for comparison to laboratory controls. All animals retained on a 0·5 mm sieve were preserved and identified to the lowest taxonomic level possible.

2.3. Drilling Fluid and Clay Preparation and Delivery
Two drilling fluids designated P-1 and P-2 were used in the experiments. A characterization of both can be found in Duke & Parrish (1984). The drilling fluids were collected from operating rigs in the Gulf of Mexico in April of 1982, and stored at 4 °C until used. The two drilling fluids were similar in composition, the major difference being the concentration of diesel oil, P-1 9·43 μg/g and P-2 2·19 μg/g. P-1 was used in Tests 1 and 2, P-2 was used in Tests 3 and 4. The suspended particulate phase (SPP) was used in all experiments and was prepared according to Petrazzuolo (1983), except that a 4:1 ratio of seawater to drilling fluid was used rather than a 9:1 ratio. Table 1 outlines some pertinent qualities of both drilling fluids. The toxicity of the SPP of both drilling fluids as determined by

Mysidopsis bahia 96-h LC_{50} was 1936 ppm for P-1 and 18 830 ppm for P-2 (Duke & Parrish, 1984). The clay treatment consisted of suspension of Migel®, a common drilling fluid additive consisting of sodium montmorillonite clay. The clay was included to serve as a sediment control, thus simulating the physical properties of drilling fluid while possessing no toxic properties. The clay was made into a suspension the same density as the SPP, both were delivered at a rate yielding a concentration of 190 parts per million (ppm) (v: v). This resulted in 15 mg of solids per liter of water.

The concentration of drilling fluid used was calculated partly from information presented at the Adaptive Environmental Assessment workshop (Auble *et al.*, 1982), and from ecotoxicology experience from colleagues at our laboratory. The concentration of 190 ppm (15 mg/liter) SPP used may only occur near the discharge point, and discharges would not be continuous. However, discharges from multiple platforms within a field under development in shallow water (2–4 m), may result in an exposure similar to the one used in our experiments. Nevertheless, exposure of the laboratory seagrass communities may have been longer in duration and higher in concentration than what a field community may actually be exposed to.

2.4. Metals Analysis

Six microcosms per treatment in addition to field cores were harvested for analysis of Ba and Cr. Due to the small quantity of animal tissue in each microcosm, tissues from three microcosms of the same treatment were pooled to form one replicate. Therefore, the six microcosms provided two replicates per treatment. To further increase animal tissue sample size organisms were grouped by phylum for analysis. Three of the six microcosms provided adequate sample size for sediment and *Thalassia* samples. The net Ba and Cr values presented were calculated by subtracting average Ba and Cr values from control samples not treated with drilling fluids, these values averaged less than 5 μg/g for Ba and Cr. Neutron Activation Analysis (NAA) was employed for analysis of Ba and Cr. Analysis was performed at the University of Florida Training Reactor (UFTR) in Gainesville, Florida.

2.5. Statistical Analysis

The ten dominant species chosen for statistical analysis in each test were the ten most abundant species when all laboratory microcosms were combined. In each test these 'dominant' species represented at least 60% of the total number of organisms identified. Faunal data were analyzed by analysis of variance (ANOVA) on transformed data log $(x + 1)$, $\alpha = 0.05$. When significant differences among the three treatment groups were detected, Duncan's multiple comparison procedure was used to determine which pairs of means differed (Winer, 1971).

3. RESULTS AND DISCUSSION

3.1. Effects on Seagrass Macrofauna

3.1.1. Test 1: May 31–July 11, 1983; P-1 Drilling Fluid
Five of the ten dominant species were significantly reduced in numbers by drilling fluid exposure (Table 2). Three of these species were annelids, *Aricidea philbinae*, *Lietoscoloplos robustus*, and *Pectinaria gouldii*; one arthropod, *Elasmopus laevis*; and one mollusc, *Crepidula maculosa*. Three of these species were affected by drilling fluid alone, *A. philbinae*, *L. robustus* and *C. maculosa*. This type of response would appear to indicate a toxic effect of the drilling fluid. The two other species, *P. gouldii* and *E. laevis* were significantly reduced in numbers by clay and drilling fluid. However, there were significantly fewer numbers in drilling fluid microcosms than in clay. This type of result indicates sensitivity to the physical and possible toxic properties of drilling fluid. When all organisms beyond the ten dominant species are combined into the overall total it is clear that the drilling fluid significantly impacted the macrofauna, and there was a toxic effect.

3.1.2. Test 2: October 5–November 15, 1983; P-1 Drilling Fluid
In Test 2, six of the ten dominant species were significantly reduced in numbers by drilling fluids (Table 3). Two species were annelids, *Armandia maculata* and *Aricidea philbinae*; three arthropods, *Cymadusa compta*, *Grandidierella bonneroides*, and *Elasmopus laevis*; and one mollusc, *Crepidula maculosa*. For the first five species listed it was apparently the physical alteration of their environments that led to reduced numbers. However, *C. maculosa* was apparently affected by some toxic properties of the drilling fluid, since there were significantly fewer numbers in the drilling fluid treatment than the clay. As represented by the overall total, the seagrass macrofauna in Test 2 were reduced equally by clay and drilling fluid, both significantly less than control microcosms. The response of the

TABLE 2

Test 1: Mean number of macroinvertebrates per microcosm for ten numerically abundant species after exposure to drilling fluid or clay (standard deviation in parentheses)

Organism	Control	Clay	Drilling fluid
Annelida			
Aricidea philbinae	23·9 (11·0) A[a]	25·6 (20·5) A	7·2 (6·2) B
Lietoscoloplos robustus	0·9 (1·2) A	0·8 (1·5) A	0·0 (0·0) B
Nereis pelagica	23·3 (10·6) A	25·6 (17·8) A	26·0 (13·8) A
Pectinaria gouldii	1·9 (2·5) A	0·7 (1·3) B	0·1 (0·3) C
Prionospio heterobranchia	17·3 (9·7) A	11·9 (1·4) A	16·0 (8·1) A
Total annelids	67·3 (16·0) A	64·6 (28·9) A	49·3 (20·9) A
Arthropoda			
Cymadusa compta	3·3 (6·2) AB	4·6 (7·6) A	0·0 (0·0) B
Elasmopus laevis	92·8 (46·2) A	21·2 (19·4) B	3·4 (4·0) C
Grandidierella bonnieroides	10·7 (31·8) AB	26·4 (48·7) A	0·0 (0·0) B
Total arthropods	106·7 (67·6) A	52·1 (57·3) B	3·4 (4·0) C
Mollusca			
Crepidula maculosa	15·8 (27·6) A	4·0 (6·3) A	3·3 (12·5) B
Brachiodontes exustus	1·3 (2·0) A	1·8 (2·0) A	1·6 (2·3) A
Total molluscs	17·1 (28·6) A	5·8 (6·9) A	4·9 (12·3) A
Overall total[b]	281·8 (135·5) A	193·8 (99·5) A	114·6 (47·0) B

[a]Letters A, B and C denote significantly different means detected by Duncan's multiple comparison procedure on transformed data $\log(x + 1)$, ($\alpha = 0·05, n = 16$).
[b]Includes all organisms identified at least to phyla, beyond the ten listed.

macrofauna in Test 2 indicates that the physical properties of the drilling fluid were responsible for the reduced numbers of organisms.

3.1.3. Test 3: June 11–July 23, 1984; P-2 Drilling Fluid

This is the first of two experiments using the second drilling fluid, designated P-2. In Test 3, four of the dominant ten species were significantly reduced in numbers by drilling fluid (Table 4). Three of the affected species were annelids, *Armandia maculata*, *Fabriciola* spp. and *Pista sp.* A; the fourth was an arthropod, *Elasmopus laevis*. *A. maculata* was the only species significantly reduced in numbers only by drilling fluid, expressing an apparent toxic response. The other three species were similarly affected by clay and drilling fluid, indicating a sensitivity to the physical properties of drilling fluid. Effects on the macrofauna as measured by response of

TABLE 3
Test 2: Mean number of macroinvertebrates per microcosm for ten numerically abundant species after exposure to drilling fluid or clay (standard deviation in parentheses)

Organism	Control	Clay	Drilling fluid
Annelida			
Armandia maculata	14·4 (13·2) A[b]	4·4 (10·9) B	0·3 (1·0) B
Aricidea philbinae	20·4 (10·2) A	13·0 (13·5) B	6·0 (5·5) B
Nereis pelagica	10·6 (3·6) A	6·3 (4·5) B	12·2 (4·4) A
Prionospio heterobranchia	51·9 (14·4) A	24·2 (15·4) B	40·1 (22·5) AB
Total annelids	97·3 (26·8) A	47·9 (33·8) B	58·6 (24·7) B
Arthropoda			
Cymadusa compta	2·9 (5·4) A	0·6 (1·8) AB	0·0 (0·0) B
Elasmopus laevis	35·9 (39·4) A	0·6 (1·4) B	0·0 (0·0) B
Grandidierella bonnieroides	2·4 (3·1) A	1·3 (3·8) AB	0·0 (0·0) B
Total arthropods	41·3 (44·7) A	2·5 (6·8) B	0·0 (0·0) B
Mollusca			
Crepidula maculosa	4·8 (5·6) A	1·8 (2·7) B	0·1 (0·3) C
Marginella apicina	1·3 (0·9) A	1·5 (1·4) A	1·3 (1·1) A
Mitrella lunata	1·4 (2·5) A	0·6 (1·1) A	1·9 (3·6) A
Total molluscs	7·4 (6·2) A	3·9 (3·5) B	3·3 (3·5) B
Overall total[b]	199·9 (77·9) A	82·7 (51·9) B	85·4 (31·7) B

[a]Letters A, B and C denote significantly different means detected by Duncan's multiple comparison procedure on transformed data $\log(x+1)$, ($\alpha = 0.05$, $n = 16$).
[b]Includes all organisms identified at least to phyla, beyond the ten listed.

the overall total showed no significant reduction in numbers among treatments.

3.1.4. Test 4: October 2–November 12, 1984; P-2 Drilling Fluid
In Test 4 only two of the ten dominant species were significantly affected by drilling fluid (Table 5). Two arthropods were the sensitive species, *Cymadusa compta* and *Elasmopus laevis*. *C. compta* was affected by drilling fluids alone, a toxic response, while *E. laevis* was affected by clay and drilling fluid, a physical response. The response of the community as a whole, measured by overall total indicated no significant differences among treatments.

TABLE 4

Test 3: Mean number of macroinvertebrates per microcosm for ten numerically abundant species after exposure to drilling fluid or clay (standard deviation in parentheses)

Organism	Control	Clay	Drilling fluid
Annelida			
Aricidae philbinae	17·3 (13·8) A[a]	16·6 (10·2) A	15·3 (9·6) A
Armandia maculata	2·4 (1·6) A	3·3 (3·8) A	1·1 (3·3) B
Fabriciola spp.	13·7 (12·3) A	4·2 (9·8) B	2·3 (5·3) B
Mediomastus californiensis	3·7 (3·0) A	2·8 (2·3) A	2·9 (3·3) A
Nereis pelagica	4·6 (3·0) A	7·0 (10·3) A	4·9 (3·0) A
Pista sp. A	9·1 (5·0) A	3·1 (4·4) B	3·1 (3·4) B
Prionospio heterobranchia	19·9 (12·7) A	16·4 (9·7) A	15·9 (13·7) A
Tharyx marioni	4·6 (6·7) A	1·8 (1·5) A	1·9 (2·5) A
Total annelids	75·3 (27·1) A	55·2 (27·1) A	47·3 (25·1) A
Arthropoda			
Elasmopus laevis	46·5 (38·6) A	11·3 (18·1) B	4·6 (7·3) B
Mollusca			
Bittium varium	0·7 (0·7) A	5·0 (10·9) A	2·2 (3·7) A
Overall total[b]	211·4 (76·7) A	152·1 (56·8) A	129·6 (63·4) A

[a]Letters A, B and C denote significantly different means detected by Duncan's multiple comparison procedure on transformed data $\log(x + 1)$, ($\alpha = 0.05, n = 16$).
[b]Includes all organisms identified at least to phyla, beyond the ten listed.

3.2. A Comparison of Laboratory and Field Communities

At the conclusion of each experiment ten field cores were taken for comparison with laboratory control microcosms. The ten most abundant species for laboratory controls and field cores for each test are compared in Table 6. This comparison was undertaken to assess the effect of holding *Thalassia* cores in the laboratory would have on the macrofauna.

When the ten most abundant species for laboratory control microcosms and field cores were combined for all four tests, a total of 27 species are represented. Nine of these species ranked among the ten most abundant in both laboratory controls and field cores in a given test at least once. When the four tests were compared individually, a substantial degree of similarity between control microcosms and the field was demonstrated.

There were seven common species among the ten most abundant for laboratory controls and field cores: three annelids, three arthropods and one mollusc in Test 1. Seven of the ten species dominant in the field were

TABLE 5

Test 4: Mean number of macroinvertebrates per microcosm for ten numerically abundant species after exposure to drilling fluid or clay (standard deviation in parentheses)

Organism	Control	Clay	Drilling fluid
Annelida			
Aricidea philbinae	34·3 (17·4) A[a]	27·1 (18·5) A	24·1 (13·6) A
Capitella capitata	11·5 (13·5) A	15·4 (15·9) A	18·2 (17·2) A
Nereis pelagica	7·4 (5·4) A	6·7 (3·1) A	7·0 (4·3) A
Prionospio heterobranchia	47·1 (19·8) A	37·3 (17·2) A	38·8 (26·4) A
Total annelids	100·2 (31·1) A	86·4 (39·3) A	88·1 (50·1) A
Arthropoda			
Cymadusa compta	22·6 (20·3) A	18·6 (19·3) A	2·7 (3·7) B
Elasmopus laevis	82·3 (70·9) A	27·7 (27·8) B	24·3 (27·3) B
Gitanopsis spp.	4·4 (3·4) A	4·8 (3·3) A	3·9 (2·0) A
Grandidierella bonnieroides	8·1 (13·3) AB	12·3 (13·5) A	1·6 (4·7) B
Total arthropods	117·3 (83·7) A	63·4 (35·7) A	32·6 (32·1) B
Mollusca			
Bittium varium	3·3 (4·7) A	5·9 (9·0) A	4·1 (7·0) A
Crepidula maculosa	6·4 (8·0) A	3·7 (5·4) A	1·7 (2·2) A
Total molluscs	9·6 (8·9) A	9·6 (12·9) A	5·8 (8·2) A
Overall total[b]	287·8 (111·5) A	225·1 (82·9) A	194·6 (89·6) A

[a]Letters A, B and C denote significantly different means detected by Duncan's multiple comparison procedure on transformed data $\log(x + 1)$, $(\alpha = 0·05, n = 16)$.
[b]Includes all organisms identified at least to phyla, beyond the ten listed.

also found to be among the ten most abundant overall for the laboratory. These seven species were included in the statistical analysis for effects of drilling fluids.

Test 2 was similar to Test 1, with regard to laboratory and field comparisons. There were seven common species (two annelids, four arthropods, and one mollusc) between the ten most abundant species for control microcosms and field cores. Again, seven of the dominant field species were included in the statistical analysis for sensitivity to drilling fluids.

In Test 3 six common species (four annelids, one arthropod, and one mollusc) were among the ten most abundant for laboratory controls and field cores. Six of the ten most abundant species in the field were included in the statistical analysis for effects of drilling fluids.

TABLE 6

Combined list of all species which ranked in the top ten for abundance in laboratory control microcosms (L) or field cores (F), for Experiments 1 through 4.

Organism	Test 1	Test 2	Test 3	Test 4
Annelida				
Nereis pelagica	L, F	L, F	L, F	L, F
Prionospio heterobranchia	L, F	L, F	L, F	L, F
Aricidea philbinae	L, F	L	L, F	L, F
Pectinaria gouldii	L			
Armandia maculata		L		
Fabriciola spp.			L	
Pista sp. A	F		L, F	
Tharyx marioni			L	
Mediomastus californiensis			L	
Capitella capitata				L
Lietoscoloplos spp.				L
Axiothella mucosa				F
Arthropoda				
Elasmopus laevis	L, F	L, F	L, F	L
Grandidierella bonnieroides	L, F	L, F		L
Cymadusa compta	L, F	L, F	L	L, F
Gitanopsis spp.				L
Hippolyte pleuracantha		L, F	F	F
Anaplodactylus spp.		F		
Pagurus spp.		F		
Processa bermudensis				F
Mollusca				
Crepidula maculosa	L, F	L, F	L, F	L, F
Brachiodontes exustus	L			
Marginella apicina	L		F	F
Mitrella lunata	F	F	F	F
Bittium varium			F	
Laevicardium mortoni		L		
Phyllaplysia spp.	F			

In Test 4 there were five common species (three annelids, one arthropod, and one mollusc) among the ten most abundant for control microcosms and field cores. Five of the dominant field core species were included in the statistical analysis for drilling fluid effects.

3.3. Ba and Cr Levels in the Macrofauna

Uptake of Ba and Cr by the seagrass-associated macroinvertebrates was apparent in all four tests (Table 7). Ba levels were the highest in annelid

TABLE 7
Net Ba and Cr levels in animal tissue samples from
drilling fluid exposed microcosms[a]

Drilling fluid	P-1		P-2	
Test number	1	2	3	4
Annelida				
Ba	1 060	10 200	53 600	35 616
	4 856	12 825	11 720	39 432
Cr	19	70	142	113
	33	79	167	128
Arthropoda				
Ba	1 356	7 830	1 705	8 650
	1 053	9 800	5 282	2 926
Cr	27	62	16	39
	5	77	30	19
Mollusc shell				
Ba	ND[b]	c	1 842	c
	ND	c	1 303	c
Cr	ND	c	3	c
	ND	c	3	c
Mollusc meat				
Ba	12	c	ND	c
	ND	c	6 684	c
Cr	6	c	ND	c
	ND	c	58	c

[a]Values are $\mu g/g$ dry weight (ppm); tissues from three
microcosms per treatment comprise each replicate. For
each test mean Ba and Cr levels from animal tissue
in all microcosms not exposed to drilling fluids were
subtracted from levels in microcosms exposed to drilling
fluids to yield net Ba and Cr levels. Tests 1 and 2 were
dosed with drilling fluid P-1, 3 and 4 were dosed with
drilling fluid P-2.
[b]Below limits of detection, Ba = $1.0\,\mu g/g$, Cr = $1.0\,\mu g/g$
or unacceptable analytical error.
[c]Insufficient tissue for analysis.

tissue for Tests 3 and 4, averaging 35 092 μg/g. Ba levels in annelid tissue from Tests 1 and 2 averaged 7235 μg/g. The Cr levels in annelid tissue followed the trend set by Ba, although there was less of a difference between the two sets of tests, Tests 1 and 2 averaging 50 μg/g and Tests 3 and 4, 138 μg/g.

Levels of Ba and Cr in arthropods did not follow the trend set by annelids. The Ba and Cr levels were relatively consistent over both sets of tests. Ba levels averaged 5010 μg/g for Tests 1 and 2, and 4641 μg/g for Tests 3 and 4. Cr levels for Tests 1 and 2 averaged 43 μg/g, and Tests 3 and 4 averaged 26 μg/g.

Determining Ba and Cr levels in mollusc tissue was more difficult than that of annelids or arthropods. There was only sufficient tissue for analysis in Tests 1 and 3, and results were somewhat inconsistent. In Test 1, Ba and Cr were detected only once, a mollusc meat sample with levels of 12 and 6 μg/g respectively. Test 3 indicated relatively consistent levels of Ba and Cr in mollusc shell samples, averaging 1573 and 3 μg/g, respectively. Mollusc meat levels were less consistent, Ba and Cr were detected in only one replicate, levels were 6684 and 58 μg/g respectively.

The nature of this study and the fauna included for analysis of metals did not allow for voiding of any material from the digestive tract. Thus, Ba and Cr concentrations could be from material in the gut, not sorbed to specific tissues. However, if consumed by a predator, the entire animal, gut contents and all would be passed on to the predator. If the Ba and Cr measured was merely due to material in the digestive tract, levels should be similar to those in sediment or *Thalassia* samples. Average Ba levels in sediment samples for Tests 1 and 2 were 667 μg/g, Cr levels were 6 μg/g. Sediment Ba levels for Tests 3 and 4 averaged 2534 μg/g, Cr averaged 6 μg/g. In *Thalassia* shoot samples Ba averaged 1241 μg/g for Tests 1 and 2, Cr was detected only in Test 1 at 8 μg/g. Barium and chromium in Tests 3 and 4 samples averaged 2324 and 19 μg/g, respectively. These levels of Ba and Cr in material likely to be present in animal digestive tracts are considerably below animal tissue levels reported. For example, in Tests 3 and 4, Ba levels in annelids averaged 35 092 μg/g while sediment levels were 2534 μg/g. If the Ba was from homogenized sediment, this represents a fourteen-fold increase over sediment levels. However, if the animals ingested a more concentrated amount of settled SPP material than represented by homogenized sediment samples, Ba levels may represent SPP solids (much higher in Ba) in the gut rather than sediment or tissue levels.

3.4. Ecological Relevance

Simulating in the laboratory an event in which a seagrass community is exposed to a drilling fluid discharge is at best a difficult undertaking. The actual exposure endured by a seagrass bed will always be determined by the specific hydrologic parameters in operation at the time of drilling. Seagrasses occur under a wide range of hydrologic conditions, however, generally in water less than 10 m in depth. The area from which the *Thalassia* cores were taken for the study described here is very shallow (1–2 m), of moderate energy, and with less than 1 m tidal variation. Models created to predict fate of discharged drilling fluids in the marine environments have not addressed the conditions described. Additionally, to the authors knowledge there have been no field studies conducted determining the fate of discharged drilling fluid under similar conditions. Given this, it is difficult to estimate how well our laboratory microcosm exposure simulated a drilling fluid discharge on or near a seagrass bed. The exposure regime created in the microcosms may have been closer to a 'worst case' situation, than to a 'typical case'. This was primarily due to the continuous dosing (platforms do not discharge continuously) and absence in the laboratory of currents present in the field. In the environment, the physical structure of seagrasses reduces current velocity at the sediment–water interface, and traps particulates from the water column, but drilling fluid particulates would gradually be transported out of the grass bed. No doubt, this transport was much less in the microcosm than could occur in the environment. The accumulation of drilling solids at the sediment–water interface in the microcosm naturally created a different type and amount of exposure for benthic infauna and epifauna, compared to fauna in the water column which were exposed to the 190 ppm. It is worth noting however, that in the field, benthic macrofauna would experience a similar type of exposure to drilling fluid solids accumulated in the seagrass bed. The ecological relevance of drilling fluid testing with seagrass microcosms has also been addressed by Morton *et al.* (1986).

4. CONCLUSIONS

The results of the four experiments indicate some possible effects to seagrass-associated macrofauna, if exposed in nature to a drilling fluid similar to those tested.

The seagrass microcosm fauna compared well with field core macrofauna collected at the end of each test. At least half of the numerically dominant species in the field were also dominant in the laboratory control microcosms. These results indicate a fairly high degree of similarity between field and laboratory macrofauna.

Macrofauna in all four experiments were significantly affected by exposure to drilling fluid. From the nineteen species included for statistical analysis in the four tests, ten were significantly affected by drilling fluid in at least one test. The results present evidence that arthropod species are the most sensitive to drilling fluid exposure. When analysed by phylum, arthropods were significantly affected in all four experiments and appeared to be the most sensitive species tested. Examination of the data also seems to indicate that drilling fluid P-1 (used in Tests 1 and 2) was more toxic than P-2 (in Tests 3 and 4). A total of eleven species were significantly reduced in number after drilling fluid exposure in Tests 1 and 2. In Tests 3 and 4, six species were affected by the drilling fluid treatment. This trend also follows to the 'overall total' category: in Tests 1 and 2, the overall total was significantly reduced by drilling fluid while there was no significant reduction in Tests 3 and 4. These results may be explained by the higher levels of diesel present in P-1, 9·43 μg/g versus 2·14 μg/g in P-2. In other toxicity tests with marine organisms and drilling fluids, there was a direct correlation between toxicity and concentration of diesel oil (Duke & Parrish, 1984).

In all four experiments animal tissue analysis provided evidence for the potential bioaccumulation of Ba and Cr. Whether the levels reported represent actual tissue levels or gut content levels remains unanswered. The significance of the levels reported may not lie in any direct threat to the organisms health, but as an entry point for Ba and Cr to move into the food web.

In summary, the experiments conducted portray an approximation of what may occur should a *Thalassia* bed experience this type of drilling fluid exposure. The results should be useful in predicting effects on the macrofauna community, but are particularly helpful in identifying certain responsive species to toxic or physical stress from drilling fluid exposure.

ACKNOWLEDGEMENTS

We thank T. W. Duke for his guidance and support throughout this research project, C. Deans for suggestions on experimental design and statistical analysis, and several laboratory colleagues for their assistance in processing benthic invertebrate samples. Neutron activation analysis was performed by the University of Florida with the support of the US Department of Energy (DOE), Grant No. DE-FG05-83ER751; however, opinions, findings, conclusions, or recommendations expressed herein are those of the authors and do not necessarily reflect the views of DOE or EPA. This research was sponsored in part through Cooperative Agreement CR-811649 between the US Environmental Protection Agency and the University of West Florida, Dr C. N. D'Asaro, Principal Investigator. This publication is ERC-169 of the Ecosystems Research Center (ERC), Cornell University. Support during writing and analysis was provided by the US Environmental Protection Agency Cooperative Agreement Number CR812685, and CR812685-02. Additional funding was provided by Cornell University.

REFERENCES

AUBLE, G.T., ANDREWS, A.K., ELLISON, R.A., HAMILTON, D.B., JOHNSON, R. D., ROELLE, J.E. & MARMUREK, D.R. (1982). Results of an adaptive environmental assessment workshop concerning potential impacts of drilling muds and cuttings on the marine environment, Report No. EPA-600/9-82-019, US Environmental Protection Agency, Gulf Breeze, FL, 64 pp.

CHAN, E.I. (1977). Oil pollution and tropical littoral communities: biological effects of the Florida Keys oil spill. In *Proceedings, 1977 Oil Spill Conference, American Petroleum Institute*, Washington, DC, pp. 539–42.

DIAZ-PIFERER, M. (1962). The effects of oil on the shore of Guanica, Puerto Rico. *Deep-Sea Res.*, 11, 855–6.

DUKE, T.W. & PARRISH, P.R. (1984). Results of the drilling fluid research program sponsored by the Gulf Breeze Environmental Research Laboratory, 1976–84, and their application to hazard assessment. Report No. EPA-600/4-84-055, US Environmental Protection Agency, Gulf Breeze, FL, pp. 27–31.

HARTOG, C. DEN (1977). Structure, function and classification in seagrass communities. In *Seagrass Ecosystems a Scientific Perspective*, ed. C. P. McRoy & C. Helfferich. Marcel Dekker, New York, pp. 89–121.

HARTOG, C. DEN & JACOBS, R.P.W.M. (1980). Effects of the Amoco Cadiz oil spill on an eelgrass community at Roscoff (France) with special reference to the mobile benthic fauna. *Helgolaender Wissenschaftliche Meeresuntersuchungen*, 33, 182–91.

KELLY, J.R., DUKE, T.W., HARWELL, M.A. & HARWELL, C. C. (1987). An ecosystem perspective on potential impacts of drilling fluid discharges on seagrasses. *Environ. Managmt*, 11 (4), 537–62.

MCROY, C.P. & HELFFERICH, C. (1977). *Seagrass Ecosystems a Scientific Perspective*. Marcel Dekker, New York.

MORTON, R.D. (1987). Effects of drilling fluids on an experimental seagrass (*Thalassia testudinum*) community and potential for bioaccumulation of barium and chromium. University of West Florida, Pensacola, Florida, Thesis, pp. 9–22.

MORTON, R.D., DUKE, T.W., MACAULEY, J.M., CLARK, J. R., PRICE, A. W., HENDRICKS, S.J., OWSLEY MONTGOMERY, S.L. & PLAIA, G. R. (1986). Impact of drilling fluids on seagrasses: An experimental approach. In *Community Toxicity Testing, ASTM STP 920*, ed. J. Cairns, American Society for Testing and Materials, Philadelphia, pp. 199–212.

NADEAU, R.J. & BERQUIST, E.T. (1977). Effects of the March 18, 1973 oil spill near Cabo Rojo, Puerto Rico on tropical marine communities. In *Proceedings, 1977 Oil Spill Conference, American Petroleum Institute*, Washington, D.C., pp. 535–8.

NATIONAL RESEARCH COUNCIL (US) (1983). Drilling discharges in the marine environment. *Panel on Assessment of Fate and Effects of Drilling Fluids and Cuttings in the Marine Environment*. National Academy Press, Washington, D.C., pp. 13–21.

PETRAZZUOLO, G. (1983). *Proposed Methodology Drilling Fluids Toxicity Tests for Offshore Subcategory; Oil and Gas Extraction Industry*. Technical Resources Inc., Bethesda, MD, 145 pp.

PRICE, A.W., MACAULEY, J.M. & CLARK, J.R. (1986). Effects of drilling fluid on *Thalassia testudinum* and its epiphytic algae. *Exp. Environ. Bot.*, **26**, 321–30.

SIMPSON, J.P. (1975). Drilling fluid principles and operations. In *Environmental Aspects of Chemical Use in Well Drilling Operations*. US EPA, Office of Toxic Substances, Washington, D.C., pp. 61–100.

WINER, B.J. (1971). *Statistical Principles in Experimental Design*. McGraw-Hill, New York, 907 pp.

ZIEMAN, J.C. (1975). Tropical seagrass ecosystems and pollution. In *Tropical Marine Pollution*, ed. E.J. Ferguson-Wood & R.E. Johannes. Elsevier Scientific Publishing Co., New York, pp. 63–74.

31

Modelling Drill Cuttings Discharges

Sylvain de Margerie

ASA Consulting Ltd, Argo Building, Bedford Institute of Oceanography, P.O. Box 2025, Dartmouth East, Nova Scotia, Canada B2W 3X8

ABSTRACT

A computer model has been developed to predict the accumulation of drill cuttings discharged from solids control equipment during offshore drilling operations. This model has been applied to the evaluation of cuttings accumulation and persistence at both the Hibernia and Venture sites, with qualitative agreement with observations.

Recently, under funding from the Environmental Studies Research Funds (ESRF), a more quantitative assessment of cuttings fate at two exploratory sites on the Scotian Shelf (South Des Barres and West Venture) has been performed. This study provided more detailed information on discharge budgets, particle size distribution, concurrent environmental conditions and benthic distributions than is normally available. These sites, which are in different water depths and significantly different wave and current regimes, have been used to test the computer model. The simulations result in reasonable estimates of cuttings mound dimensions (within a factor of two). Model sensitivity tests indicate that the most sensitive parameters are the effective particle size distribution, i.e. the particle size as it actually occurs in the water column, and the total volumetric discharge. At the shallower site (West Venture), extensive sediment reworking during the drilling period is also indicated as a controlling process.

1. INTRODUCTION

Concern over hydrocarbon pollution of the benthos by oil-based drilling fluids has prompted studies on the environmental impact of drill cuttings disposal. In an effort to provide a quantitative basis for environmental impact analysis, ASA Consulting Ltd has developed a simple numerical model for simulating cuttings dispersion and fate, under funding by the Environmental Studies Research Funds (ESRF). Two sites on Sable Island Bank, South Des Barres and West Venture, for which bottom survey data were available, were considered for model application. Each of these two sites represents a very different environment, providing an ideal opportunity to test the model methodology. The modelling strategy is presented first, followed by discussions of its application to the South Des Barres and West Venture sites.

2. MODELLING PRINCIPLES

Two processes are considered by the model:

(1) the initial deposition of the cuttings on the bottom during the discharge; and
(2) the subsequent weathering and ultimate obliteration of the material accumulated on the bottom.

These two processes are considered independently. Given the paucity of information on cuttings behaviour, a very simple modelling procedure is justified. As will be seen, a dominant factor affecting prediction uncertainty is not a detailed representation of environmental processes, but rather the assessment of the quantity and properties of released materials.

2.1. Initial Settling and Distribution

Cuttings are typically released near the water surface. The initial cuttings distribution on the bottom will be determined principally by the movement of the particles in the water column as they settle. This movement is affected by two factors: a particle's settling velocity, and the motion imparted to the particle by the water currents.

The size of released cuttings particles typically varies between 0·000 1 m and 0·001 m (Addy et al., 1983), with corresponding fall velocities in the range of 0·01 to 0·1 m/s. For particles finer than this, it can be argued that the turbulent velocities induced by a moderate bottom current (~ 0.15 m/s) would overcome the settling velocity and inhibit deposition, obviating the need for a cuttings dispersion model. In 50 m of water, the particles of concern will settle to the bottom in an hour or less. The discharged cuttings will reach their terminal velocities within a few seconds of their release, so that for all practical purposes they can be considered to settle at a constant rate, and are carried horizontally at the same velocity as ocean currents. From this we can infer that the initial distribution of the cuttings on the bottom is governed by:

—the variability of ocean currents, which at different times will transport the particles in different directions, dispersing the cuttings on the bottom;
—the range of particle size, with finer portions taking a longer time to sink through the water column and therefore being transported over a greater distance by the current.

On time scales of 1 h and less, ocean currents (tides, wind driven circulation, mean flows, oceanic eddies, etc.) can be considered constant, while environmental turbulence will induce dispersion on the order of metres (Okubo, 1962) and can be neglected. Furthermore, since the particles descend at constant velocity, the effect of the ocean currents on the particle will be felt equally at all depths, and the net horizontal displacement corresponds to the depth-averaged current velocity. Given that the discharge of cuttings takes place more or less continuously over a long period, a necessary and sufficient description of the ocean climate to compute the dispersion of cuttings while they settle is the joint distribution of depth-averaged currents by direction and speed, $P'(\theta, V)$. It can be noted that the joint distribution of depth-averaged currents is easily obtained from measurements where ocean currents are relatively uniform with depth; in a stratified ocean the evaluation of depth-averaged currents is generally more difficult and requires more detailed measurements.

For a particle of known size, ϕ, and settling velocity, $W(\phi)$, the horizontal displacement is given by:

$$x = UH/W(\phi) \qquad (1)$$

where x is the displacement,
    U is the depth-averaged velocity, and
    H is the water depth.

Characterizing the cuttings by their grain size distribution, Π_ϕ, the corresponding distribution of fall velocity, $\Pi_W = \Pi_\phi(\phi(W))d\phi/dW$, can be estimated from the formula of Gibbs *et al.* (1971) for $\phi(W)$, which relates the grain size to fall velocity. The thickness of the material deposited on the bottom, $T(\theta, x)$, at a distance x and in the direction θ from the release point, is given by:

$$T(\theta, x) = V \frac{1}{2\pi x} \int_0^\alpha P'(\phi, v) \Pi_w \left(\frac{Hv}{x} \right) \frac{Hv}{x^2} \, dv \qquad (2)$$

where V is the volume of discharged material. This thickness neglects the voids between settled cutting particles, i.e. it represents only the volume of pure cuttings, and the true thickness of the deposit is likely to be about twice this quantity, i.e. void ratio of 1. This expression is evaluated numerically by the model, and contours of thickness are plotted as a final result. This thickness is typically very small, and often the released material will be partially integrated with the natural sediments, so that hydrocarbon concentration is the only practical field measurement. Given the fraction of hydrocarbons in the released cuttings, one can transform the predicted thickness to an equivalent hydrocarbon concentration per m^3, if one assumes the cuttings get integrated into the sediments over a certain depth, i.e. the mobile layer. Given an initial hydrocarbon concentration in the released material, C_0, and a mixing depth of D, the concentration distribution in the sediment is:

$$C = C_0 \frac{T(\phi, x)}{D} \qquad (3)$$

As will be seen later, the depth D can be a function of sedimentary processes in the area and/or of the sampling strategy for bottom samples; in either case it is estimated outside of the model proper.

2.2. Subsequent Obliteration
Once the cuttings material has settled on the bottom, we can expect that natural processes which disturb the sediment will gradually disperse the cuttings until, after sufficient time has passed, no trace of the discharge can be detected. We would expect the persistence of the cuttings deposit in any instance to be proportional to the quantity of material dumped. Thus, in a highly dynamic sediment transport environment, and if the dimension of the initial deposit was small, we would expect that the persistence time would be short. Based on simple dimensional analysis (ASA, 1983), the following relationship can be obtained for the persistence time:

$$\Pi = \frac{V}{BQ} \qquad (4)$$

where V is the total volume of discharged material (m^3),
  B is the horizontal dimension of the deposit (m), and
  Q is a characteristic sediment transport rate for the environment (m^2/s).

The total volume, V, is known *a priori*, and B can be estimated from the deposition model, discussed previously, as the distance over which the deposit thickness decreases by a factor of 2. The characteristic transport rate, Q, depends on wave and current regime as well as natural sediment properties; it is calculated using current/wave boundary layer theory and sediment transport formulae implemented into a sediment transport model (de Margerie, 1982; Davidson, 1986). This model is used to compute a mean Q, based on the available environmental data. It should be stressed that eqn (4) is only intended to provide an order of magnitude estimate of the persistence time. Given the uncertainties in cuttings behaviour, and sediment transport in general, a more detailed analysis cannot be justified, and calculations are not carried into the deposition model.

3. MODEL APPLICATION

Two sites, South Des Barres and West Venture, shown in Fig. 1, were selected for model application. The South Des Barres site is in 72 m of water, offering the possibility for substantial dispersion of cuttings, by ocean currents, before they reach the bottom. The other site, West Venture, is in only 15 m of water, so that cuttings rapidly reach the bottom, but are then subject to a very dynamic sediment transport environment.

The input parameters required by the initial deposition model are:

—the total volume of discharged materials,
—the particle size distribution of the released materials,
—the composition of the discharged material (hydrocarbon fraction), and
—the joint distribution of depth-averaged currents by speed and direction.

The persistence model requires the additional specification of:

—the wave climate, and
—the near bottom current climate.

The determination of these data will be discussed in turn, followed by modelling results.

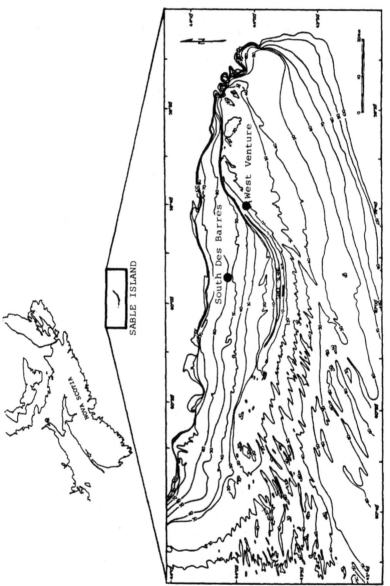

FIG. 1. Site location.

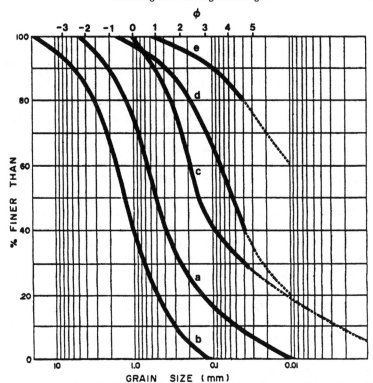

FIG. 2. Grain size distribution for cuttings: (a) used as working hypothesis; (b) coarser distribution from Addy *et al.* (1983); (c) finer distribution from Addy *et al.* (1983); (d) grain size analysis for West Venture; (e) grain size analysis for South Des Barres.

3.1. Discharge Properties

Cuttings are the rock fragments cut by the drill bit. These fragments are carried back to the surface by the circulating drill mud, and are removed from the mud by screens and passage through a centrifuge, to be finally discharged into the ocean.

Grain size properties of cuttings, obtained from laboratory analysis, are shown in Fig. 2 as curves (d) and (e) for West Venture and South Des Barres, respectively. To obtain these curves, the cuttings were first subjected to vigorous washings in various solvents to rid them of their coats of drilling mud. This treatment of the material prior to the measurement immediately raises the question of relevance of the laboratory determinations for our purposes. Two curves obtained for unwashed cuttings in the North Sea show considerably coarser sizes. Furthermore, the solids control logs for South Des Barres and West Venture indicate that approximately 90% of the released cuttings are intercepted by a 62 μm screen ($\phi = 4$). We must therefore conclude that the laboratory determinations are not representative, and the precise grain size distributions are unknown. As a working hypothesis, we will use a curve intermediate to those of Addy *et al.* (1983), adjusted so that 90% of the material is coarser than 62 μm (see curve (a), Fig. 2). This is an educated guess and its implication will be considered later by running the model for coarser and finer sizes (curves (b) and (c), Fig. 2).

One could assume that each cuttings grain is coated with a layer of mud and settles as an individual particle. However, Fig. 2 shows that the coat of mud required would be very much larger than the particle itself. The volume of mud so discharged would exceed by many times the total amount used to drill a well, and the hypothesis must be discarded. A more probable hypothesis is that the released particles are a conglomerate of cuttings bound in a mud matrix. A reasonable assumption is that the mud fills the voids which would naturally exist in a clean sample of unsorted cuttings grains. We have used a void ratio of 33% (1/3 mud to 2/3 cuttings by volume) for all calculations. This corresponds to the void ratio of very tightly packed soil, and may in fact be an underestimate (typical loose soil void ratios are 60%).

Typical mud composition is 40% oil, 5% water and 55% barite by weight, giving a total density of 2·1. This is close enough to that of typical rock (2·2), so that no adjustment is made for the particle density. In calculating hydrocarbon concentrations in the sediments, the slight density difference between the cuttings and the natural sediments is also neglected.

We were unable to accurately assess the volume of discharged cuttings material from the solids management logs, which in some cases implied total volumes either smaller or many times larger than the size of the hole. We therefore assessed the hole volume from the diameter and depth of the various well sections, and augmented this volume by 50% to account for the mud, as explained above. The total discharged volumes thus computed are 760 m³ and 700 m³ for South Des Barres and West Venture, respectively, corresponding to 1600 t and 1500 t of material. The volume of the cuttings is by necessity very close to that of the hole itself, and the uncertainty in the well geometry is no more than 10%; the estimate of accompanying mud volume, however, could be out by a factor of 2, resulting in a total uncertainty of about 50% in the discharged volume.

3.2. Oceanographic Conditions

Two current meter records are available for the drilling period at South Des Barres, one at mid depth (34 m) and the other near the bottom (60 m). They both show similar velocity fluctuations (tidal and others) but the mid depth record exhibits a mean flow of 0·11 m/s, which is absent in the bottom data. For the present purpose, we have assumed that the mid depth current meter is representative of depth-averaged conditions, and the distribution of currents used for the model is given in Fig. 3(a). Any error in this assumption would likely be in the mean flow, causing a net displacement of the centre of mass of the bottom deposit, without otherwise affecting its distribution.

At West Venture, only one current meter was moored during the drilling period. This instrument was 1 m off the bottom in 14 m of water, and is likely strongly affected by the bottom boundary layer. We estimate that depth mean flows could be 25–100% larger than the measured velocities, due to the frictional effect of the bottom. An intermediate value of 66% was selected for the model runs, acknowledging a wide margin of uncertainty. The joint distribution of unadjusted currents at West Venture is shown in Fig. 3(b).

The wave climates for South Des Barres and West Venture are represented as joint distributions of wave height and direction, in Figs 4 and 5, respectively. The bottom orbital velocities, computed from linear wave theory, are also shown in the figures, together with a shaded area representing the waves capable of mobilizing the natural sediments in the area ($D_{50} \sim 0·5$ mm).

It is clear that the South Des Barres site is relatively quiet, and no significant sediment transport is predicted, even when considering currents in addition to the waves. The deposit persistence model was not applied to the site, as it would simply predict no erosion. For West Venture, however, waves alone are sufficient to mobilize sediments over 50% of the time, and a mean transport rate of 200 m²/yr was computed from the sediment transport model.

3.3. Simulation Results

Simulation results, in terms of accumulated material thicknesses, are presented in Fig. 6 for South Des Barres and West Venture (note that Figs 6–9 show the results of the deposition model only, and that the persistence of the cuttings deposit on the bottom will be considered separately). For South Des Barres, the maximum thickness is about 0·1 m, near the release site, decreasing to less than 0·01 m within 100 m from the

Sylvain De Margerie

(a) Current percent frequency by speed (m/s) and direction

	0–0·10	0·10–0·20	0·20–0·30	0·30–0·40	0·40–0·50	Sum
N	1·2	0·7	0·1	0·1	0·0	2·0
NNE	1·8	2·5	0·6	0·0	0·1	4·9
NE	2·7	5·8	2·5	0·5	0·1	11·5
ENE	3·5	11·7	5·5	1·7	0·6	23·0
E	3·7	12·1	4·2	2·1	0·9	23·0
ESE	3·1	5·9	2·1	0·5	0·0	11·7
SE	2·6	2·7	0·5	0·1	0·0	5·9
SSE	2·0	1·7	0·2	0·1	0·0	4·0
S	1·3	1·1	0·1	0·1	0·0	2·7
SSW	1·5	0·8	0·1	0·1	0·0	2·5
SW	0·8	0·9	0·0	0·1	0·0	1·8
WSW	0·5	0·9	0·0	0·1	0·0	1·5
W	0·4	0·7	0·1	0·1	0·0	1·2
WNW	0·5	0·3	0·2	0·1	0·0	1·0
NW	0·7	0·9	0·1	0·0	0·0	1·7
NNW	0·9	0·4	0·1	0·1	0·0	1·5
Sum	27·2	49·1	16·4	5·7	1·6	100·0

(b) Current percent frequency by speed (m/s) and direction

	0–0·6	0·6–0·12	0·12–0·18	0·18–0·24	0·24–0·30	0·30–0·36	Sum
N	1·77	2·49	0·55	0·01	0·00	0·00	4·82
NNE	3·01	4·61	3·62	1·66	0·39	0·06	13·35
NE	3·72	3·02	3·99	1·98	0·71	0·24	13·66
ENE	2·79	2·43	0·97	0·04	0·00	0·00	6·23
E	1·36	0·89	0·00	0·00	0·00	0·00	2·25
ESE	1·42	0·69	0·03	0·00	0·00	0·00	2·14
SE	2·02	0·35	0·00	0·00	0·00	0·00	2·37
SSE	1·51	0·77	0·01	0·00	0·00	0·00	2·29
S	1·39	1·00	0·07	0·00	0·00	0·00	2·46
SSW	2·81	2·40	0·95	0·22	0·00	0·00	6·38
SW	4·11	4·02	3·51	1·93	0·20	0·00	13·77
WSW	3·36	5·52	3·39	0·97	0·12	0·00	13·36
W	1·97	3·31	0·62	0·00	0·00	0·00	5·90
WNW	2·01	2·07	0·24	0·00	0·00	0·00	4·32
NW	2·57	0·71	0·06	0·00	0·00	0·00	3·34
NNW	1·90	1·34	0·12	0·00	0·00	0·00	3·36
Sum	37·72	35·62	18·13	6·81	1·42	0·30	100·00

Fig. 3. Currents used in the model: (a) South Des Barres; (b) West Venture

% OCCURRENCE

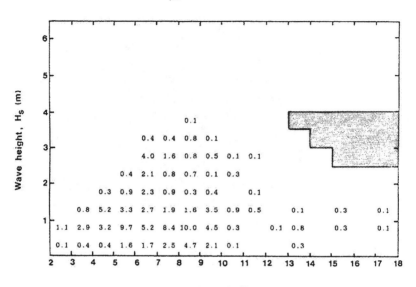

Wave period, T (s)

BOTTOM (72 m) ORBITAL VELOCITIES

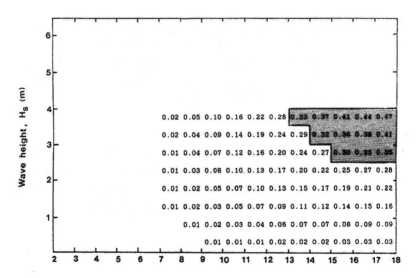

Wave period, T (s)

FIG. 4. Wave climate at South Des Barres. Shaded areas represent waves with sufficient energy to mobilize bottom sediments.

% OCCURRENCE

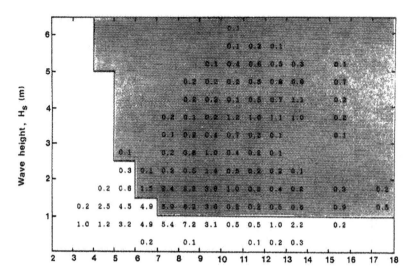

Wave period, T (s)

BOTTOM (16 m) ORBITAL VELOCITIES

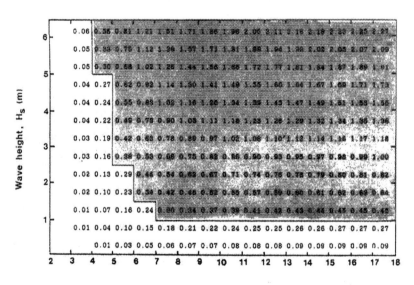

Wave period, T (s)

FIG. 5. Wave climate at West Venture. Shaded areas represent waves with sufficient energy to mobilize bottom sediments.

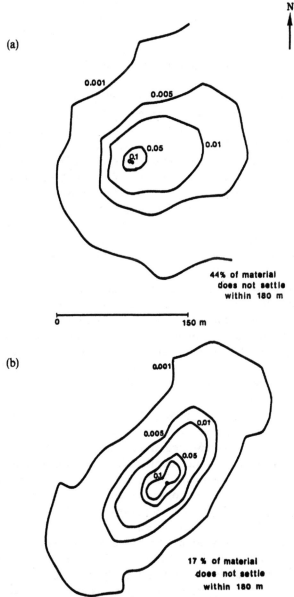

MOUND THICKNESS DISTRIBUTION

– contours in meters

N

(a)

0.001

0.005

0.05

0.01

0.01

0.1

44% of material
does not settle
within 180 m

0   150 m

(b)

0.001

0.005   0.01

0.05

0.1

17 % of material
does not settle
within 180 m

FIG. 6. Predicted deposit thickness distribution at: (a) South Des Barres; (b) West
Venture.

discharge. The effect of the (northeast) mean flow is evident, as the centre of distribution is displaced to the northeast. At West Venture the maximum thickness is about 0·3 m, decreasing to less than 0·01 m within 75 m of the discharge point. The dominance of the oscillatory tidal motion is evidenced by the symmetric spreading of the materials along a SW to NE axis.

Of particular interest is the model's prediction of hydrocarbon concentrations, which can be compared to observations. At the South Des Barres site, the bottom materials are largely undisturbed by waves and currents, and we would expect little mixing of the released material with the underlying natural sediments. However, the field sampling was undertaken with a small grab, estimated to collect the first 0·05 m of materials, and the analysis results are given in terms of hydrocarbon concentration for the whole sample. Therefore, the sampling strategy effectively mixes the first 0·05 m of sediments; this mixing depth was used to transform modelled deposit thickness to hydrocarbon concentration. As seen in Fig. 7, the predicted shape and dimension of the hydrocarbon distribution reproduces observations well.

The model does underestimate material concentrations near the release site. The difference is largely explained by the uncertainty in the initial hydrocarbon concentration of the released materials. Integrating the observed hydrocarbon distribution, we find that a total of 300 t of oil are present in the sediment, while the model input implies a total release of about 200 t; using a higher mud to cuttings ratio would have corrected this discrepancy. Another possibility is that the 0·05 m mixing depth is an overestimate.

The sediments at the West Venture site are highly mobile, and we can expect that the released material will get mixed with natural sediments. Hodgins *et al.* (1986) suggest a mixed layer of 0·1–0·2 m for a nearby site in 30 m of total water depth. We have used a value of 0·1 m for our calculations, which may be an underestimate, especially for a shallower site. The model predictions are shown in Fig. 8, and it is evident that the deposition model has overestimated the hydrocarbon concentration. The large difference cannot be explained by an error in the mobile layer thickness or the amount of released hydrocarbons.

To explain the West Venture results, we must consider the persistence of the bottom deposit in this very dynamic sediment transport environment. With the mean sediment transport rate $Q = 200 \, \text{m}^2/\text{s}$ (derived earlier), a horizontal mound dimension $B = 30 \, \text{m}$ (Fig. 6), and a total discharged volume $V = 700 \, \text{m}^3$, we obtain from eqn (4) a time scale of 1·5

-contours in µg/g

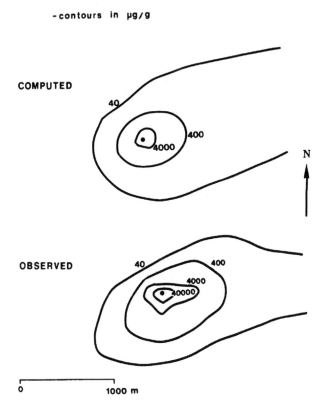

FIG. 7. Hydrocarbon concentration for South Des Barres.

months for the persistence of the deposit. Since this is shorter than the release period, we would expect extensive mixing and dilution of the cuttings into the sediments even as drilling progresses. Further dispersion of the released material would have taken place before the field survey, undertaken approximately 1 month after drilling was completed. It should therefore come as no surprise that our estimates of mound thickness and hydrocarbon concentrations, neglecting sediment dynamics, are much higher than observations. For West Venture our persistence estimates can explain the observations and we conclude that sediment transport mechanisms play a major role even in the short term fate of oiled cuttings.

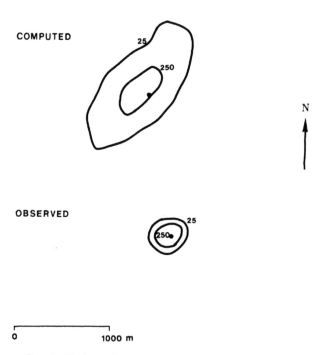

FIG. 8. Hydrocarbon concentration for West Venture.

4. CONCLUSIONS

We have seen that a simple deposition model has adequately predicted the distribution of cuttings in a relatively quiet sediment environment, at South Des Barres. It is interesting to consider the sensitivity of the model predictions to input parameters. In Fig. 9, we have changed the input parameters for ocean conditions and grain size distribution, over a range commensurate with the uncertainties of these inputs. Changing currents by $\pm 20\%$ changes the results by less than a factor of 2. Varying the grain size over the range suggested by Addy *et al.* (1983), however, causes changes by more than an order of magnitude. Improved predictions are therefore dependent primarily on a better quantification of discharge

MOUND THICKNESS DISTRIBUTION

– contours in meters

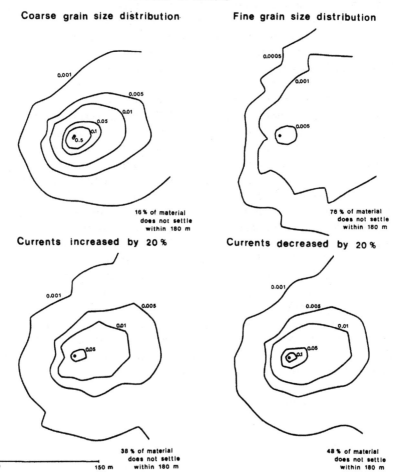

FIG. 9. Sensitivity of model results to specified mean currents and grain size distribution for South Des Barres.

properties with respect to grain size distribution and, as noted before, fraction of mud accompanying the cuttings. This could be accomplished by directly measuring the settling velocity distribution of the materials (unadulterated from their condition at the time of discharge), and analysing their composition in terms of hydrocarbon content and mud fraction.

In a shallower environment, such as West Venture, where sediments are highly mobile, the deposition model is not directly applicable. However, this was expected from a simple persistence calculation. A model considering the simultaneous deposition and dispersion into the sedimentary environment would be necessary to fully consider this problem. The feasibility of such an analysis is checked by our limited ability to predict sediment transport and dispersion. This is compounded by the fact that the introduction of significant amounts of oiled cuttings in the sediment could modify its properties in an unknown manner. For the present, therefore, estimates of deposit stability, characterized by a persistence time, may be as far as one can practically go. In this respect it would be interesting to obtain a time series of hydrocarbon concentrations after drilling, to verify the decay time of the cuttings deposit. The tracer experiments of Hodgins *et al.* (1986) are highly relevant here, although they consider natural sediments, unaffected by cuttings.

The models and calculations presented here are very simple and one is easily tempted to evolve more complex and seemingly more realistic simulations. However, the practical usefulness of a more complex treatment is limited by a lack of characterization of the physical properties of the cuttings themselves, i.e. their composition, total quantity, in situ particle size, cohesive properties once on the bottom, etc., properties which are themselves very variable, i.e. dependent on rock formation, type of mud and mud treatment equipment. Without some control and monitoring of cuttings discharge properties, more precise predictions of impacts of released cuttings are not possible.

ACKNOWLEDGEMENTS

This work was funded by the Environmental Studies Research Funds, under a project administered by Dobrocky Seatech Ltd. We are grateful to Dobrocky Seatech, and particularly to Warren Drinnan, for allowing us the opportunity to work on this project and providing the field observations for model verification.

REFERENCES

ADDY, J., BLACKMAN, R., BLANCHARD, J., DAVIES, J., FERBRACHE, J., MOORE, D., SOMERVILLE, H., WHITEHEAD, A. & WILKINSON, T. (1983). Environmental effects of oil-based mud cuttings. *Offshore Europe Conference*, Aberdeen SPE 1182.

APPLIED SCIENCE ASSOCIATES, INC. (ASA) (1983). A preliminary ocean waste disposal site designation manual, prepared for the US EPA Criteria and Standards Division.

DAVIDSON, S. (1986). SED1D: A microcomputer version, unpublished report by ASA Consulting Ltd, submitted to The Geological Survey of Canada under contract no. 11SC.23420-6-M736.

DE MARGERIE, S. (1982). Sediment transport on a continental shelf, unpublished report by Martec Limited, submitted to The Geological Survey of Canada under contract no 10SC.23420-1-M571, 91 pp.

GIBBS, R.J., MATTHEWS, M.D. & LINK, D.A. (1971). The relationship between sphere size and settling velocity. *J. Sedimentary Petrology*, **41**(1), 7–18.

HODGINS, D.O., DRAPEAU, G. & KING, L.H. (1986). Field measurements of sediment transport on the Scotian Shelf, ESRF Report No. 41.

OKUBO, A. (1962). A review of theoretical models of turbulent diffusion in the sea. Technical Report 30, Chesapeake Bay Institute, Johns Hopkins University.

32

Field Verification of the OOC Mud Discharge Model

J. E. O'REILLY, T. C. SAUER*, R. C. AYERS, JR

Exxon Production Research Company, PO Box 2189, Houston, Texas 77252-2189, USA

M. G. BRANDSMA

Consulting Engineer, PO Box 378, Durango, Colorado 81302, USA

and

R. MEEK

Ecomar, Inc., 158 Santa Felicia Drive, Goleta, California 93117, USA

ABSTRACT

The original version of the Offshore Operators' Committee (OOC) mud discharge model was released to both operators and regulatory agencies in 1983. Since then, it has been widely used to predict the initial fate of drilling mud and cuttings discharged to the marine environment. The OOC model has undergone continuous modification to improve its capabilities and ease of use. A revised version of the OOC model is now available which not only predicts the initial dynamics and passive diffusion for drilling mud, but does the same for produced waters.

A mud discharge study was conducted offshore California to verify the model with field data. The location was particularly desirable because the water was deep enough (~18 m) for plume development to occur and shallow enough for divers to work comfortably on the seafloor. Near-field water column and bottom sediment were monitored during a 30 min constant rate mud discharge on January 28, 1984.

*Present address: Battelle Ocean Sciences, 397 Washington Street, Duxbury, Massachusetts 02332, USA.

647

The field data collected in this study represents one of the most intensive sampling efforts ever reported for verification of a discharge model. A unique water column sampling apparatus was developed to collect more than 180 samples from four depths over a 2000 m² area within 50 min. Settled solids were collected in 96 specially designed sediment traps positioned over a 12 500 mm² area extending between 50 m upcurrent and 100 m downcurrent of the discharge pipe. The traps were opened before and closed after discharge by divers. All water column and trap samples were analysed for total dry weight, barium, chromium, iron and cobalt.

Model calculations of total suspended solids concentrations in the water column were representative of solids concentrations measured in the field. The field data were also used to estimate the total amount of solids present in the water column at one specific time following the discharge. The model estimate was in excellent agreement with the field data. The sediment trap data suggest that the model accurately predicted the amount of settled mud solids in most of the deposition area.

1. INTRODUCTION

A drilling mud discharge model was developed by the Offshore Operators Committee (OOC) and Exxon Production Research Company (EPR) from the US Army Corps of Engineers/Environmental Protection Agency (EPA) Dredge Spoil Model (Brandsma *et al.*, 1980). The model predicts concentrations of solids and soluble components in the water column and the initial deposition of solids on the seafloor. The OOC Mud Discharge Model was distributed to OOC member companies and Federal and state agencies concerned with offshore drilling discharge regulation in September 1983. The model has had extensive use in NPDES permitting and impact assessment.

The dynamic plume portion of the model has been verified with laboratory data (Brandsma & Sauer, 1983c). However, until now, no field test had been conducted to specifically address all aspects of model verification. To meet this need, a mud discharge study was conducted from Unocal's Platform Eva off Huntington Beach, California on January 28, 1984.

The field verification study was designed to address the following objectives:

● To determine drilling mud concentrations in the water column at

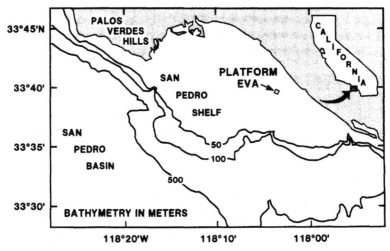

FIG. 1. Platform Eva is located about 3 km off Huntington Beach, California in 18 m (60 ft) of water.

various times, distances, depths and directions from the discharge pipe.
● To determine the total amount of drilling mud solids deposited on the seafloor after discharge.

2. METHODS

2.1. Study Site

The test was conducted from a production platform about 3·2 km (2 miles) off Huntington Beach, California (Fig. 1). This test site was chosen because of excellent visibility for divers and photographic documentation. Unocal's Platform Eva was a particularly desirable location because the water is deep enough for plume development to occur and shallow enough for divers to work comfortably at the sea bottom. Water depths around the platform averaged approximately 18 m (60 ft). Bottom type was generally sandy with areas of shell debris near the platform.

Platform Eva is a production platform and all development drilling was completed approximately 20 years ago. A mud pump and tanks were rented, and a shunt pipe was installed for this test because they were unavailable on site.

Drilling mud was barged to Platform Eva for the test from an Exxon USA site in federal waters. The California Water Quality Control Board required a toxicity test and chemical analyses on the mud and pre- and post discharge monitoring of the macrobenthos near the platform, before it would allow discharge in state waters.

2.2. Mud Discharge
Mud was pumped at a constant rate from the mud tanks through a 15 cm (6 in OD) pipe into a 25·4 cm (10 in OD) vertical pipe which was open to the air at the upper end. On January 28, 1984, 268 barrels of mud were discharged in 30 min at a constant rate of 536 bbls/h.

2.3. Study Design
The object of the study was to verify the mud portion of the OOC Mud and Produced Water Discharge Model by comparing measurements of water column concentrations and initial deposition of solids on the bottom with model predictions. The concentrations of solids measured in the water column during a test needed to be averaged both spatially and temporally, for comparison to the model which calculates average solids concentrations in the water column.

Water column samples were to be collected at least three times during the discharge from each location, without significantly perturbing the dispersing plume. Samples needed to be collected along the centerline of the discharge plume, primarily at depths were most of the discharge plume is predicted to travel, and at the edges of the plume so that the lateral extent could be determined. In addition, the sampling apparatus needed to align with the changing ambient currents, so that samples would be taken at a constant position with respect to the discharge plume, even when the plume direction varied.

Sediment traps were used to collect solids reaching the seafloor during the discharge test. The traps had to be flush with the sediment surface because traps placed above the seafloor would miss solids in the discharge plume's turbid layer moving along the bottom. In addition, the number of deployed traps had to adequately cover the primary area of solids deposition. The traps needed to have covers so that solids accumulation would occur only during and immediately following the discharge.

2.4. Sampling

2.4.1. Water Column Sampling
A sample collection system was constructed using bundles of plastic tubes which extended into the water column to different depths and distances

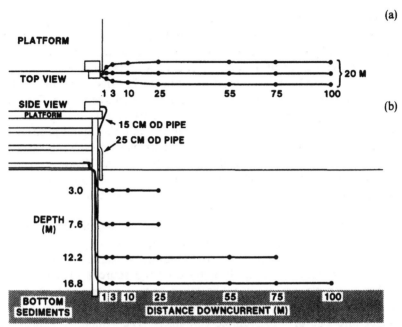

FIG. 2. Water column sampling apparatus used in field verification test. (a) Top view; (b) side view.

downcurrent, from a common vacuum pumping system located near the discharge source. Ports at the terminus of each tube string allowed water to flow easily into the tubes. Water was aspirated through the tube strings into sample containers. Care was taken to ensure that the flow rate of seawater within the tubes (4·6 litres/min) was high enough to prevent solids from settling within the tubes.

Top views and side views of the configuration of sample tube strings in the water column are shown in Fig. 2. Four sampling tube arrays were used, one at each of four depths (3, 7·6, 12·2 and 16·8 m). Each sampling array was attached near the source and allowed to move independently of the other. Each array had 3 bundles of tubes (left, middle and right) that were separated horizontally by a 10 m spreader bar at distances of 1, 3 and 10 m, and a 20 m spreader bar at greater distances.

Separate tubes were used to collect samples from each point in the water column. In the two surface arrays (at depths of 3 and 7·6 m) the tube strings extended to and had open ports at distances 1, 3, 10 and 25 m downcurrent. The tube ports of the mid-depth (12·2 m) array were located at the same distances downcurrent as the surface array ports with additional ports at distances of 55 and 75 m downcurrent. The deep (16·8 m) array had tube strings with open ports at the same distances as at 12·2 m with an additional port located 100 m downcurrent.

The sampling tubes were fixed to support wires that were attached to, but free to rotate about, an anchor wire at the discharge source. The tubes arranged laterally in the water column were attached to stiff spreaders at the source and at their port terminus to maintain the lateral positions at each depth. An orientation drag vane was attached to each tube array to keep the tubes extended in the direction of the current. However, currents during the test were too light for the drag vane to work properly, and the arrays for each depth were anchored in position based on measured current directions for each depth.

Three discrete 2-litre samples were collected from each tube during the test. Water was continuously pumped through each sampling tube. When samples were not being collected, the water was discarded through a hose to a point 200 m crosscurrent from the discharge source ($\sim$ 121 litres/min). When a sample was ready to be taken from a particular tube, valves connected to the sample container were switched to divert the pumped water into one of three sample containers. The sample container was flushed for 3–10 min depending on the length of the tube string and filled in approximately 30 s. This procedure was repeated three times for each tube at preselected times.

The time at which samples were collected depended on the position of the discharge plume in the water column and the location of the terminus of the tube string (distance and depth). The transit times required to pump water from the open ports to the sample container were calculated for each tube. Since water was being pumped continuously through each tube string, many samples were collected simultaneously during the discharge. Replicate samples for each tube were usually collected once every 5–10 min.

The location of the sampling ports downstream of the discharge source were fixed by the length of the tube string. Based on pre-test current measurements, the centerlines of each sample array were aligned to 325° (magnetic).

2.4.2. Sea Bottom Sampling

Settled solids were collected with sediment traps constructed so they could be placed in the sediment flush with the sea bottom (Fig. 3). Each trap had a grid mesh over the opening (10·5 cm diameter). Two traps were

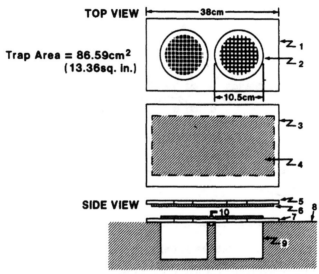

FIG. 3. Schematic diagram of sediment traps. Each sediment trap is capable of collecting replicate samples with a trap area of $86 \cdot 59 \, cm^2$ ($13 \cdot 36 \, in^2$). Divers remove and replace the lids immediately prior to and after the discharge. Key: 1, canister sediment trap array (top view); 2, canister type sediment trap with $1 \, cm \times 1 \, cm$ grid; 3, lid for sediment trap array (top view); 4, neoprene seal; 5, lid for sediment trap (side view); 6, neoprene seal; 7, sediment trap array (side view); 8, sea bottom; 9, canister inserted flush with sea bottom; 10, threaded bolt for lid retention.

attached together. A lid covered each pair of trap openings. The traps were designed to have the lids removed after trap deployment just before the start of the discharge test and replaced after the test. However, in this case the divers were able to close and recover only two traps before dark. The remaining traps were not closed until the following morning. Divers performed all trap deployment, lid covering and trap retrieving activities.

Traps were placed at pre-determined locations upcurrent and downcurrent of the discharge source (Fig. 4). Results of model simulations of anticipated water column conditions were used to position the traps. Six distances downcurrent (0, 3, 10, 25, 50 and 100 m) and 2 distances upcurrent (-10 and -50 m) from the discharge pipe were selected. At each distance, 6 pairs of traps were positioned along a line perpendicular to the predicted direction of the current. Locations of the traps were determined from measurements by divers and range finder/theodolite measurements of surface buoys placed above the trap lines. Based on pre-

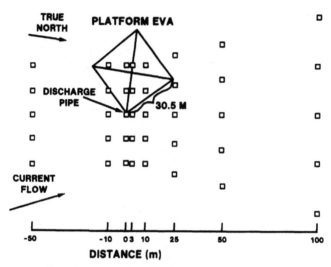

FIG. 4. Location of sediment traps used in field verification test.

test current measurements, the centerline of the sediment sampling system was aligned at 337° magnetic.

2.5. Measurements and Analyses

2.5.1. *Current and Hydrographic Measurements*

Current data were collected for 2 weeks before the discharge to provide information on current flow patterns that could be anticipated during the field test. This information was used in model simulations to position the sediment traps and water column arrays. ENDECO Model 105 recording current meters were placed at four depths (3, 7, 12 and 17 m) on a taut wire mooring near the discharge.

During the discharge, the same recording current meter array remained in place. To determine the optimal time of day to conduct the discharge tests, 'realtime' current measurements were collected every daylight hour with an ENDECO Model 110 deck-readout profiling meter. This meter also supplied information to verify pre-test current predictions, confirm data generated from the recording current meters during the discharge tests, and help finalize the layout of the water column sampling arrays and sea bottom traps.

Temperature and salinity profile measurements were made with a Hydrolab 6D Surveyor. These data were used to determine water column density and stratification required for model simulations of the tests. Densities were calculated at 1 m intervals from the surface to the bottom.

2.5.2. Sample Analyses

Most water column samples were filtered through pre-weighed 0·45 μm poresize polycarbonate Nuclepore® filters. Trap and water column samples with large amounts of solids were not filtered. The solids were allowed to settle for 1 week in the sample containers and the liquid portion was decanted and centrifuged at 3000 rpm for 10 min. The supernatant was siphoned off to recover the remaining solids. After drying at 105 °C, the weights of the settled and centrifuged solids were combined to determine total solids concentrations. Solids from water and sediment trap samples were analysed for barium by instrumental neutron activation analysis (INAA) as described in Kalil (1980).

Drilling fluids discharged during the test were also analysed for barium by INAA. Mud density, rheological properties and volume percent solids were measured using standard API procedures (API, 1984). Fall velocities and particle size distribution were determined in artificial seawater (Brandsma & Sauer, 1983a). Approximate grain sizes for each solids class were calculated from Stokes' law according to the fall velocities and density of the particles.

3. RESULTS AND DISCUSSION

3.1. Mud Properties/Rheology

The mud discharged in this study was a low-density ($1·131\ 6\ \text{g/cm}^3$ or 9·44 lbs/gal) lignosulfonate mud (Table 1). Fall velocities and the weight

TABLE 1
Properties of mud discharged during verification test no. 2

Mud density	
(g/ml)	1·131 6
(lb/gal)	9·44
Insoluble solids in mud	
(wt %)	15·75
(vol %)	5·28
Average particle density (g/cm³)	3·377 0
Density of suspending fluid (g/cm³)	1·023 4
Solids concentration (mg/litre)	179 000
Mud discharge rate (bbl/h)	536
Duration of discharge (min)	30

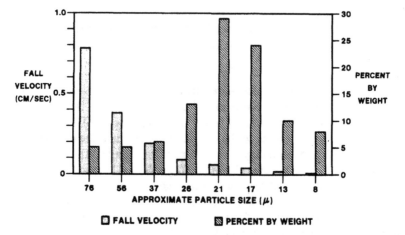

FIG. 5. Particle settling rates (cm/s) and percentage weight distribution for each solids class. Fall velocities were calculated using the modified Andreason pipette method (Brandsma & Sauer, 1983a). The approximate particle size for each solids class was calculated from Stokes' law using the measured fall velocities and an average particle density of $3\cdot377\,\mathrm{g/cm^3}$.

percent distribution of the solids classes are shown in Fig. 5. Fall velocities in seawater ranged from $0\cdot01$ to $0\cdot76$ cm/s ($0\cdot000\,3$–$0\cdot025$ ft/s). Seventy percent of the mud solids had fall velocities less than $0\cdot06$ cm/s ($0\cdot002$ ft/s).

3.2. Currents/Hydrology
Currents at the site were generally low and appeared to be primarily tidally forced. Pre-test results indicated that a northerly flow (about 300° magnetic) was slightly more dominant at the site.

Currents during this test tended more toward 320–330° magnetic (Table 2). Surface flows ranged from $2\cdot5$ to $7\cdot2$ cm/s while bottom flow ranged from $1\cdot5$ to $4\cdot1$ cm/s. The data indicate a very slight counterclockwise rotation of current with depth.

Hydrographic profiles of the water column showed a slight thermal stratification (temperature ranged from $15\cdot0$ to $15\cdot9$ °C), but essentially no density stratification (σ_t ranged from $24\cdot7$ to $24\cdot8$). Salinity was uniform with depth ($33\cdot4\,^{0}/_{00}$), as shown in Table 3.

3.3. Comparison of Model Predictions with Field Measurements

3.3.1. Water Column Solids Concentration
The field data collected here represent one of the most intensive sampling efforts reported for verification of a discharge model. The water column

TABLE 2
Current velocities measured during verification test no. 2. Mud discharge started
at 1520 hours and ended at 1550 hours

Time	Depth (m)	Current	
		Speed (cm/s)	Direction (° magnetic)
1430	3·0	3·6	338
	7·6	2·6	321
	12·2	4·1	320
	16·8	1·5	313
1500	3·0	6·7	333
	7·6	2·1	325
	12·2	5·2	316
	16·8	2·6	319
1530	3·0	4·1	328
	7·6	4·6	329
	12·2	5·7	315
	16·8	3·6	321
1600	3·0	7·2	331
	7·6	3·6	318
	12·2	5·2	313
	16·8	3·6	322
1630	3·0	7·2	322
	7·6	4·6	308
	12·2	4·1	313
	16·8	4·1	323
1700	3·0	5·2	328
	7·6	5·7	313
	12·2	7·2	308
	16·8	2·6	320
1730	3·0	2·6	328
	7·6	4·6	315
	12·2	7·2	316
	16·8	2·6	322

TABLE 3
Hydrographic conditions during verification test no. 2 (Date: 1/28/84; Time:
1500)

Depth (m)	Temperature (°C)	Salinity (⁰/₀₀)	Density (σₜ)
2	15·1	33·4	24·7
4	15·1	33·4	24·7
6	15·0	33·4	24·8
8	15·0	33·4	24·8
10	15·0	33·4	24·8
12	15·0	33·4	24·8
14	15·0	33·4	24·8
16	15·9	33·4	24·8
18	15·9	33·4	24·8
19	15·8	33·4	24·8

sampling apparatus collected over 180 samples from four depths over an area 100 m long and up to 20 m wide, within a 50 min time period.

A comparison of model predicted values (contours) and field measured values (solid circles) for vertical cross-sections of the plume at three distances downcurrent (1, 10 and 75 m) from the discharge pipe are shown in Fig. 6. Keep in mind that the model presents an averaged view of water column concentrations while the field measurements reflect instantaneous values (i.e. measurements of a greatly variable plume at specific points and times). Even so, Fig. 6 demonstrates how well the model predicts the development of an upper and lower plume with roughly the same concentrations as measured in the field.

The agreement between model and field data is better illustrated by comparing model predictions and field measured values for each array (left, middle and right) at specific distances downcurrent (Fig. 7) and at specific depths (Fig. 8). As a result of the heterogeneity of the plume, field data are best represented by the median, minimum and maximum concentrations measured at any one point during the test. In each of these figures, the minimum and maximum values for the model were calculated by outputting model data over the times the field data was collected and at points 0·61 m (2 ft) above and below the actual sample location in the field.

The model predictions and field results agree in both behavior (the presence of an upper and lower plume) and in solids concentrations

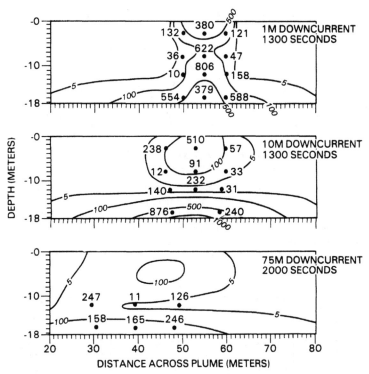

FIG. 6. Vertical cross-sections of the plume comparing water column solids concentrations (mg/litre) predicted by the model (contours) and measured at field sample points after discharge for (top) 1300 seconds at 1 m downcurrent, (middle) 1300 s at 10 m downcurrent, and (bottom) 2000 s at 75 m downcurrent, respectively. Model predictions are average concentrations.

(Fig. 7). Model predictions for 75% of all water column sample locations fell within the range of concentrations measured in the field. The model successfully predicted the suspended solids concentrations at 86% of the sample points in the lower water layer (16·8 m or 55 ft), where most of the mud solids are found (Fig. 8).

Figure 9 compares the median concentrations predicted by the model and measured in the field at specific locations downcurrent. Nearly one-third of the median model predicted concentrations were within ±50% of the median concentrations measured in the field for 62 locations sampled in the water column. Nearly 40% of the predictions were within a factor of 2 and 81% were within a factor of 5.

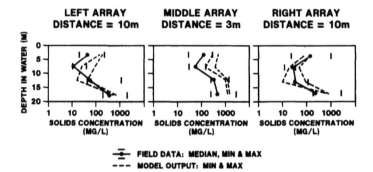

FIG. 7. Comparison of water column solids concentration (mg/litre) vs depth (m) measured in field verification test and predicted by the model for (right) right side of plume at a distance of 10 m; (middle) mid-line of plume at a distance of 3 m, (left) left side of plume at a distance of 10 m. Minimum and maximum model values were calculated by outputting data above and below actual sample locations over the time period that field samples were taken.

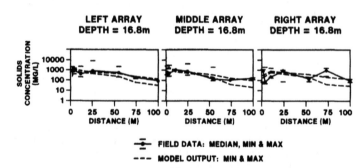

FIG. 8. Comparison of water column solids concentration (mg/litre) versus distance downcurrent (m) measured in field verification test and predicted by the model at a depth of 16·8 m (55 ft); (left) left side of plume; (middle) mid-line of plume; (right) right side of plume. Minimum and maximum model values were calculated by outputting data above and below actual sample locations over the time period that field samples were taken.

The deviations shown at the higher solids concentrations in Fig. 9 reflect the large variation observed in the field within 10 m of the discharge pipe. Better agreement between model and field data occurred at greater distances from the discharge pipe (> 10 m) where the overall concentrations were lower and discontinuities less severe.

Two points plot offscale in Fig. 9, each with predicted concentrations

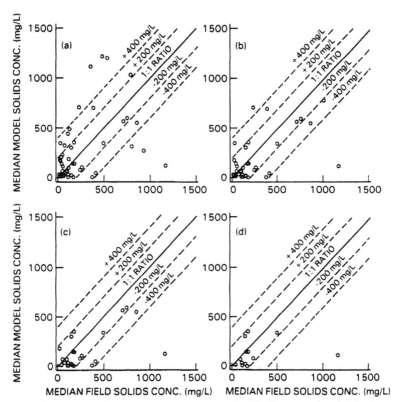

FIG. 9. Comparison of the median solids concentrations (mg/litre) measured in the field with median concentrations predicted by the model at the same time and locations. The median solids concentration for the model was calculated with model data output above and below our field sample locations over similar time periods. Two points located at a distance of 1 m from discharge with predicted concentrations of 2000+ and 4000+ mg/litre are not shown. (a) Data for all distances downcurrent. (b) Distances ≥ 10 m. (c) Distances ≥ 25 m. (d) Distances ≥ 55 m.

of 2000+ and 4000+ mg/litre. Both of these were located 1 m from the discharge and the field sample could easily have missed the maximum concentration at that point, as a result of the 'puffy' nature of plumes. A functional regression (Ricker, 1973) was used to fit the data shown in Fig. 9, since it is more suitable in cases where both variables are subject to measurement error. The slope of the fitted regression line is not statistically different from one, indicating that overall, the model does a good job

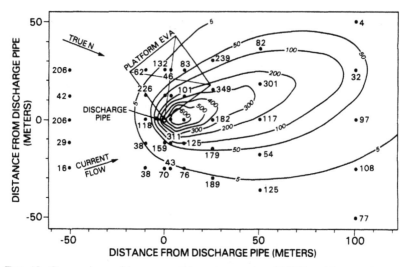

FIG. 10. Comparison of bottom solids concentration (lb/20 ft grid square) measured in the field (points) and predicted by the model (contours).

predicting concentrations of water column solids measured in the field, with no particular bias for under- or overpredicting.

The total amount of mud solids in the water column after discharging mud for approximately 8 min was calculated from the field measured concentrations using a contouring routine developed at EPR. These calculations show that 97% of the mud solids discharged up to that point were still in the water column. This agrees well with the model which predicts 97% of the discharged mud solids to be present in the water column. A comparison of total mass in the water column at longer times was not possible because the mud solids were too dispersed, in relation to our sample locations, to allow reliable estimates of total mass in the field.

3.3.2. Sediment Trap Solids Concentration

Ninety-six sediment trap samples were collected between 50 m upcurrent of the discharge and 100 m downcurrent (12 500 m²).

Agreement between the settled mud solids predicted by the model and measured by the sediment traps (Fig. 10) is not as good as the agreement observed in the water column. The error appears to lie with the field data and not with the model predictions. Most of the traps were not closed until the morning following the discharge and therefore collected resuspended bottom sediments and settled mud solids during the night.

The sediment trap data were corrected for resuspended bottom sediments by using a barium mass balance equation to determine the amount of mud solids in the sediment traps:

Mud solids in trap =

$$\text{Total solids in trap} \times \frac{(\text{Conc. Total Ba in traps} - \text{Conc. Sediment Ba})}{(\text{Conc. Mud Ba} - \text{Conc. Sediment Ba})}$$

Background sediment barium concentrations were estimated to range from 0·8 to 1·0 mg/g dry weight. The average drilling fluid barium concentration was 260 mg/g dry weight. We cannot correct for resuspended mud solids which entered the traps during the night. The sediment trap data are therefore an overestimate of the actual amount of mud solids which settled to the bottom immediately following the discharge.

For the most part, the model predictions for traps along the centerline and west of the centerline (to the left looking downcurrent) and downcurrent from the discharge show a good agreement with the field data. Differences occur downcurrent and east of the centerline (to the right looking downcurrent), and also upcurrent of the discharge pipe (Fig. 10).

These differences may be explained in part by noting that divers, 2 h following the discharge, observed a 1 ft deep 'cloud of mud' on the bottom east of the sediment trap centerline at distances of 1, 3 and 10 m. Sediment traps west of the centerline were reported to be clear of the cloud. No observations were made beyond 25 m downcurrent. If bottom currents had continued in the same direction after the test, as they were during the test, unsettled mud solids would have been carried out of the sampling area, and would not have been seen by the divers.

A significant amount of mud solids may have been present in the cloud observed by the divers. The model predicts that 36% of the discharged mud remains suspended 1 h after the start of the discharge. Thus, a better understanding of the movement of this 'cloud of mud' may help to explain some of the discrepancies observed in our sediment trap data.

During a tidal cycle, a suspended particle moves along an elliptical path in a clockwise direction. If these tidal currents had already begun to reverse (as suggested for the bottom current meter data shown in Table 2), then the 'cloud of mud' observed by the divers may have been unsettled mud solids which were possibly carried into the eastern part of the sampling area. An elliptical path would have carried the suspended mud across the eastern side of our sampling area. Since the sediment traps were still open, they may have collected additional mud solids contained in the reversing currents.

Unfortunately this hypothesis cannot be verified with the field data, since the current meters were retrieved 2 h following the discharge. We can, however, test the hypothesis by making a longer model run which include such tidal currents. These results and the results from another verification test will be the subject of a future paper.

5. CONCLUSIONS

Overall, the model predictions agree well with solids concentrations measured in the field verification study.

Model predictions of total suspended solids concentrations in the water column are representative of those measured in the field. Seventy-four percent of the concentrations predicted by the OOC model were within the range of concentrations measured in the field. Nearly 33% of the median concentrations predicted by the model fell within $\pm 50\%$ of the median mud solids concentrations measured at 62 locations in the water column. Nearly 40% were within a factor of 2 and 81% were within a factor of 5.

The amount of mud solids predicted by the model to be in the water column after discharging mud for about 8 min, accounts for 100% of the solids found in the water column at that time. A comparison of total mass in the water column at longer times was not possible because the mud solids were too dispersed, in relation to our sample locations, to allow reliable estimates of total mass in the field.

The initial deposition of mud solids predicted by the model correlated with those measured in the traps, except for some traps that may have collected unsettled mud solids from water that returned to the sample site as a result of changing tidal currents.

ACKNOWLEDGEMENTS

The work presented here was supported by the Offshore Operators Committee and by Exxon Production Research Company. We wish to thank Unocal for allowing us to use their platform to conduct the study. We especially wish to thank M. Kinworthy, of Unocal, for his assistance in conducting the field work. The authors are also indebted to L. C. Davis and R. C. Y. Koh for reviewing the field data. We also wish to acknowledge P. D. Cook for making numerous model runs and providing the computer graphics.

REFERENCES

API (1984). API Recommended Practice for Field Testing Drilling Fluids, 10th edn, API RP 13B. American Petroleum Institute.

AYERS, R.C. JR, MEEK, R.P., SAUER, T.C. JR & STUEBNER, D.O. (1980). An environmental study to assess the effect of drilling fluids on water quality parameters during high-rate, high volume discharges to the ocean. *Journal of Petroleum Technology* (January 1982), 165–73.

BRANDSMA, M.G. & SAUER, T.C. (1983a). Mud discharge model. A model which predicts the fate of drilling fluid discharges in the marine environment: Report and User's Guide (Version 1), available from Exxon Production Research Co., P.O. Box 2189, Houston, Texas, 77252-2189.

BRANDSMA, M.G. & SAUER, T.C. (1983b). The OOC Model: Prediction of short term fate of drilling mud in the ocean. Part 1: Model description. *Proceedings of a Minerals Management Service Workshop entitled, An Evaluation of Effluent Dispersion and Fate Models for OCS Platforms*, February 7–10, 1983, Santa Barbara, California, Minerals Management Service, Los Angeles, LA, pp. 58–84.

BRANDSMA, M.G. & SAUER, T.C. (1983c). The OOC Model: Prediction of short term fate of drilling mud in the ocean. Part 2: Model results. *Proceedings of a Minerals Management Service Workshop entitled An Evaluation of Effluent Dispersion and Fate Models for OCS Platforms*, February 7–10, 1983, Santa Barbara, California, Minerals Management Service, Los Angeles, CA, pp. 86–106.

BRANDSMA, M.G., DAVIS, L.R., AYERS, R.C. JR & SAUER, T.C. JR (1980). A computer model to predict the short term fate of drilling discharges in the marine environment. In *Proceedings of the Symposium: Research on Environmental Fate and Effects of Drilling Fluids and Cuttings*, Lake Buena Vista, Florida, pp. 351–81.

KALIL, E.K. (1980). Chemical analysis of drill muds and discharge plumes. In *Proceedings of the Symposium: Research on Environmental Fate and Effects of Drilling Fluids and Cuttings*, Lake Buena Vista, Florida, pp. 799–811.

RICKER, W.E. (1973). Linear regressions in fishery research. *Journal of the Fisheries Research Board Canada*, 30(3), 409–34.

33

The Five-year Fate of Sea-floor Petroleum Hydrocarbons from Discharged Drill Cuttings

O. GRAHL-NIELSEN

Department of Chemistry, University of Bergen, N-5007 Bergen, Norway

S. SPORSTØL, C. E. SJØGREN and F. ORELD

Center for Industrial Research, P.O. Box 350, Blindern, N-0314 Oslo, Norway

ABSTRACT

An exploratory well was drilled in the North Sea, at a water depth of 60 m, between September 1981 and February 1982, partly with the use of diesel oil-based mud. A total of 200 m³ contaminated cuttings, containing approximately 25 t of oil, were released to the sea. After completion of the drilling, the well was capped, and no further drilling activity has taken place in the area since then. To study the fate of the hydrocarbons associated with the cuttings deposited on the sea-floor, samples of the sediment have been collected yearly, the first time 1 month after the discharging was finished, along 4 radial transects at distances between 100 and 3000 m from the wellsite. Total petroleum hydrocarbons, as well as selected aliphatics and aromatics were determined. The discharges resulted in an inner heavily polluted area with concentrations of petroleum hydrocarbons between 1000 and > 50 000 μg/g (wet weight). This area had an ellipsoidal extension of 500 m in the direction of the wind-driven current during the discharge period. An outer area with concentrations from 1000 μg/g down to the background level of approximately 10 μg/g initially had a longitudinal extension of 2500 m. The changes during 5 years were small: no significant decrease in concentrations in the inner heavily polluted area was detected. The extension of the polluted area decreased somewhat during the first year, but remained thereafter at approximately 1500 m. The ratios between normal alkanes and unresolved complex mixtures, between n-C_{17} and pristane, between n-C_{18} and phytane and between naphthalenes and dibenzothiophenes, all indicate an *increasing rate of degradation with decreasing concentrations of petroleum hydrocarbons. The degradation had started during the first year in areas with concentrations of hydrocarbons below 1000 µg/g.*

1. INTRODUCTION

The offshore exploration and exploitation of oil and gas resources leads to inevitable discharges, both controlled and accidental, of petroleum hydrocarbons, and other components, extraneous to the environment. One of the major discharges is drilling waste which leaves large quantities of cuttings and used mud (NAS, 1983). In contrast with US offshore oil and gas activities, oil-based drilling mud has been used extensively in the North Sea. Petroleum hydrocarbons from drilling have been estimated to account for 75–85% of the total oil input to the North Sea from offshore operations (Grogan, 1985). Offshore oil and gas operations have been estimated to contribute between 15 and 20% of the total input of hydrocarbons to the North Sea (Side *et al.*, 1985). Before 1984, diesel oil was used as a base for drilling mud in the Norwegian sector, of the North Sea. Since 1984, specially formulated 'low toxicity' oils have been obligatory.

Under Norwegian legislation, drill cuttings should not contain more than 100 g/kg of hydrocarbons when discharged to the sea. Furthermore, the marine environment surrounding the platforms has to be regularly monitored. Consequently, a number of surveys have been carried out, although only one of them has been reported in the open literature. (Grahl-Nielsen *et al.*, 1980). In addition, several investigations in the British sector of the North Sea have been published (Davies *et al.*, 1984).

The first objective with surveys for discharges around platforms is to map the distribution of the extraneous material. When the discharges are drill cuttings the surveys are focussed on the sea-floor. Knowledge of amounts and distribution of the discharges in such cases is substantial, as reviewed by Davies *et al.* (1984). Very high concentrations of hydrocarbons have been found within 250 m of platforms where oil-based drilling mud is used. The distributions are usually ellipsoidal along the axis of the most persistent current and the oil concentration falls rapidly with increasing distance, usually reaching background levels within 3000 m.

In addition to the distribution of the hydrocarbons, the other important aspect is their fate on the sea-floor. Near-shore sediment-bound petroleum hydrocarbons from point-sources have half-lives ranging from months, in exposed locations, to decades in sheltered areas (NAS, 1985). In the surveys of the sediments around North Sea platforms, where oil-based drilling mud has been used, the fate of the hydrocarbons has been more difficult to investigate, because continuous discharging has been going on. Nevertheless, Saltzmann (1982) has demonstrated increased weathering of aliphatic and aromatic hydrocarbons with increasing distance from 3 North Sea platforms.

The present case, where one well was drilled with mud based on diesel oil over a short period of time, with the nearest installation 12 000 m away, and with no further activity in the area for the following 5 years, presented a unique possibility to study the fate of the petroleum on the sea-floor. Two hundred cubic metres of drill cuttings were discharged. The cuttings had been cleaned on shakers, leaving an average amount of 16% diesel oil, by volume, so that the total amount of discharged diesel was 25 t. The cuttings were released from a shute just above the sea surface. The purpose of the present investigation was to carry out yearly surveys of the sea-floor around the wellsite to investigate the long-term fate of the hydrocarbons from the diesel oil used in the mud.

2. MATERIALS AND METHODS

2.1. Sampling Strategy

The well is located in the central part of the North Sea (56° 10′ 35·2″ N, 03° 27′ 36·62″ E). After the drilling was completed in February 1982, the well was capped and the only installation left on the site was a template on the sea-floor. Its position was marked with a surface buoy, anchored at the template. Sampling was carried out along 4 perpendicular transects away from the template. With no residual in the central part of the North Sea, the current is wind driven. Figure 1 shows the direction and velocity of the wind during the discharge period. The prevailing wind has been towards the eastern side of the north-south direction with a slight predominance in the north-eastern direction. The main transect for sampling was therefore chosen at 74°.

For the first 2 years, 1982 and 1983, the surface marker buoy was expected to be at the correct location of the well and was used as the centre of the radial transects. The results of the analyses showed that the buoy had not been in the right position. In the 1984 survey the position

670   *O. Grahl-Nielsen, S. Sporstøl, C. E. Sjøgren and F. Oreld*

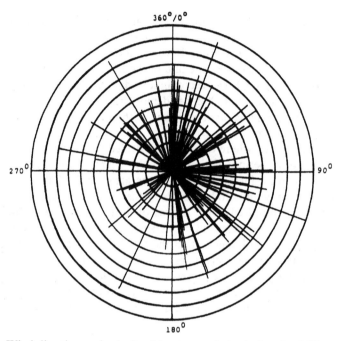

FIG. 1. Wind direction and velocity, 5 knots per circle, during the drilling period
from October 1981 through February 1982.

of the template was found by a standby vessel using sophisticated
navigation equipment. In the 1985 survey the template was located by a
remotely operated vehicle connected to a standby vessel. In the 1986 and
1987 surveys a Sonar 8 navigation system was used for positioning. Even
with this high degree of sophistication of the navigation equipment it has
been estimated that the uncertainty in the reported positions is at least
50 m.

The sampling positions, located relative to the template, for all surveys
are given in Fig. 2. The number of sampling positions and the distance
between them varied somewhat from year to year, as shown in the figure.

Reference samples were collected from locations expected to be beyond
the influence of the discharges. In the 1982 to 1985 surveys the reference
station was 6000 m to the north, along the 344° transect. In 1986 and
1987 the reference station was 16 000 m to the north-west, along the 300°
bearing.

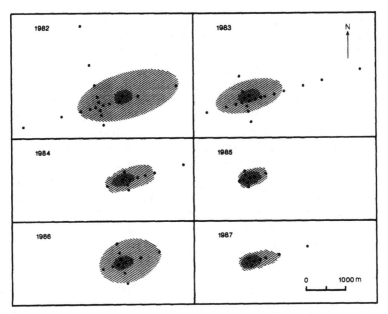

FIG. 2. Sampling positions and polluted area. The sampling positions are shown by closed circles, the size of the circles represents the estimated uncertainty in positioning. The wellsite is shown by a cross. The cross-hatched areas contain total hydrocarbons in excess of 1000 μg/g, wet sediment, the hatched areas contain total hydrocarbons from 1000 μg/g down to background level. Note that the outer limit of this area could not be estimated in 1984, 1985 and 1987, see text.

2.2. Sampling

Samples were collected with a 0·2 m^2 van Veen grab, operated from a supply vessel. The van Veen grab gives fairly undisturbed samples of the top layer of the hard-packed sandy bottom in the area. The grab was usually 1/4 full. While the vessel was kept at location, three grab replicates were taken. From each of them about 0·1 m^2 of the upper 3 cm was transferred with a stainless steel spoon to a stainless steel bucket. After thoroughly mixing the material from the 3 replicate grabs, 3 subsamples, of about 100 g each, were wrapped in aluminium foil and stored in the deep freezer until they were subjected to analysis.

2.3. Chemical Analysis

The analytical procedure follows the previously reported method (Grahl-Nielsen *et al.*, 1980), which has been recommended by the Intergovernmental Oceanographic Commission (IOC, 1982). After thawing the samples, subsamples of 5–50 g of wet sediment were homogenized together with 1–500 μg of the internal standards biphenyl-d_{10}, phenanthrene-d_{10} and pyrene-d_{10}. The samples were then saponified in 80 ml of 0·5 M methanolic KOH under reflux for 2 h. This extraction method has been shown to have the same efficiency as soxhlet and ultrasonic extraction (Lichtenthaler et al., 1986). The mixture was filtered under suction. The sediment material remaining on the filter was washed with 50 ml 1 M HCl in methanol (1:3), methanol and finally with n-pentane. The combined filtrates were extracted twice with 50 ml n-pentane. The combined extracts were washed with water, dried over Na_2SO_4 and carefully evaporated to 0·5 ml using a modified Kuderna-Danish concentrator. Polar components were removed by applying the concentrated extract to a 7 × 0·5 cm column of florisil, dried and deactivated with 18% water. Non-polar components were eluted with 10 ml n-pentane. The eluate was concentrated, using the modified Kuderna-Danish concentrator, and analysed by gas chromatography.

For determination of total hydrocarbons (THC) the chromatography was performed with hydrogen as carrier gas on a 12·5 m × 0·2 mm fused silica column with cross-linked methyl silicone as stationary phase, and flame ionization detection. By comparing the areas of the chromatograms between the n-C_{10} and n-C_{32} alkanes, with the corresponding area of chromatograms of an external standard, marine gas oil, a diesel oil ranging in normal alkanes from n-C_8 to n-C_{35}, the amounts of THC in the samples were obtained. The areas were electronically integrated, except for the 1982 survey, when the 'cut-and-weight' method was used for integration of the chromatograms between n-C_{14} and n-C_{32}. In the 1982 survey, a different external standard was used: an extract of oiled drill cuttings. By spiking sediment samples from reference locations with diesel oil, the method has been shown to give a recovery in excess of 90% (Lichtenthaler et al., 1986).

For determination of selected aromatic hydrocarbons, naphthalene, phenanthrene, dibenzothiophene and their C_1-, C_2- and C_3-alkylated derivatives (NPD) as well as the 3 deuterated internal standards, chromatography was performed with helium as carrier gas on a 30 m × 0·25 mm fused silica column with cross-linked 95% methyl 5% phenyl silicone as stationary phase and multiple ion detection of the molecular ions in a mass spectrometer. The various components were quantified by comparing with the closest eluting internal standard and taking the total response factors of the components and standards into consideration.

TABLE 1

Background levels, mean ± standard deviation, in μg/g wet sediment, of total hydrocarbons, THC and selected aromatics, NPD, in the upper 3 cm. For 1982–1985 the reference location was 6000 m from the well at 344°, for 1986 and 1987 the location was 16 000 m from the well at 300°

Year	THC			NPD		
	Amount	n	Limit[a]	Amount	n	Limit[a]
1982	11·0 ± 1·1	9	16	0·097 ± 0·009	9	0·14
1983	3·5 ± 1·1	4	14·7	0·018 ± 0·004	4	0·06
1984						
1985	5·4 ± 0·9	10	9·3	0·024 ± 0·006	5	0·07
1986	6·5 ± 1·1	5	14·4	0·028 ± 0·004	5	0·05
1987	4·7 ± 1·0	5	12·0	0·031 ± 0·003	5	0·05

[a] 99·9% confidence limit of average value. Samples with concentrations above these limits are considered polluted by petroleum.

3. RESULTS AND DISCUSSION

3.1. Background Levels

In order to find the outer limits of the polluted area, the general background level of petroleum hydrocarbons in this region of the North Sea must be known. The amounts of THC and NPD detected at the reference locations are given in Table 1. The values for 1982 are higher than the others due to a somewhat different method of quantification. For the 1983 survey the quantification method was changed and the new method was standardized for the succeeding surveys. The background values from these surveys can therefore be compared.

The NPDs are regarded as good markers for pollution by petroleum, being typical constituents of petroleum, and not naturally present in sediments (Teal *et al.*, 1978). The THCs are not as specific, since this fraction contains all components with a polarity allowing elution from the deactivated florisil column with pentane. This means that not only petroleum hydrocarbons, but also components resulting from their biodegradation and naturally occurring components, will be included in the unresolved complex mixture typical of the THC-chromatograms (Jones *et al.*, 1983).

Table 1 shows a year-to-year change in the background values. The standard deviations of the measurements are rather high compared to the difference between them, so no firm conclusions can be drawn. Still, the data may imply that a rise in the background levels of petroleum hydrocarbons in the area has occurred. The NPDs have increased continuously since 1983, while the THCs increased until 1986 but decreased the following year.

Other investigations in the North Sea show that the background level of THC, or 'total oil', is between 1 and 20 μg/g dry weight. The lowest levels were found in the Dogger Bank area in the central North Sea (Law & Fileman, 1985). In two locations close to the area covered in the present investigation, Law & Fileman (1985) found a THC concentration of 8 μg/g dry weight. Further north the level was somewhat higher (Davies *et al.*, 1981; Davies *et al.*, 1984). The composition and particle size of the sediment apparently influences the background levels of hydrocarbons; finer sediments are more rich in organic material (Davies *et al.*, 1984). Comparison between different investigations should be exercised with care, due to differences in work-up and analytical procedures. The need for standardization of the procedures is obvious.

In the present work, a sample is defined to be significantly polluted by the discharged drill cuttings if it contains THC and NPD in amounts exceeding the 99·9% confidence limit of the average background levels. The limits are given in Table 1.

3.2. Concentrations and Distribution

The results of the determinations of THC and NPD in the samples are given in Table 2. In the 1982 survey the NPD values were obtained from flame ionization detection of the aromatic fraction of the extracts. The values can therefore not be directly compared with the GC/MS results from later surveys, and they are not included in the table.

The true positions of the sampling locations, i.e. with reference to the well, have been estimated on the basis of the results each year, and they are given in Fig. 2. As seen, the centre of the transects has deviated substantially from the desired position, i.e. up to 500 m.

In addition to the large uncertainty in the determination of the correct position, comes the fact that the discharged cuttings do not have an even concentration gradient away from the discharge point (Grahl-Nielsen *et al.*, 1980). To obtain exact information of the distribution of the pollution in the area, and the change from year to year, a much larger number of samples, and a better positioning system, would be needed. One solution would be to collect the samples by means of a remotely operated vehicle on the sea floor.

TABLE 2

Concentrations, in μg/g wet sediment, of total hydrocarbons, THC, and selected aromatics, NPD, in the upper 3 cm. The values for 1982, 1983 and 1984 are averages of 2 parallel analyses, the others are results of a single analysis.

Dir[a]	Dist[b]	1982	1983		1984		1985		1986		1987	
		THC	THC	NPD	THC	NPD	THC	NPD	THC	NPD	THC	NPD
East	100	46	12 200	600	5 770	139	90 000	1 920	52 300	1 240	10 200	174
	200						52 000	1 770				
	250	550	2 800	105	550	13			4 650	125	1 950	23
	300											
	400		970	37								
	500	10 200	130	2·7	75	0·5	102	1	318	3	750	8
	750		110	2·6	9	0·05			27	0·1	19	0·1
	1 000	560	14	0·08								
	1 200		7	0·03								
	1 500		3	0·04	6	0·02						
	2 000	56	3	0·04							19	0·04
	3 000		2	0·05								
South	100	34	300	12	910	13	41 000	840	6 200	270	6 470	143
	200				176	2	1 500	28				
	250	36	64	7					270	8		
	500	31	6	0·13					30	0·1		
West	100	78	1 000	61	33 800	990	61 000	1 930	17 900	522	24 700	610
	200						2 300	46				
	250	54	540	31					860	14		
	500	56	35	0·7	22	0·1			46	0·2		
	1 000	9	6	0·01								
	2 000	7										
North	100	128	10 300	350	4 950	163	4 600	106	8 040	220	4 240	110
	200				222	5	700	9				
	250	49	1 700	68					76	0·2		
	500	18	4	0·06					18	0·06		
	1 000	14										
	2 000	16										

[a] The exact bearings of the transects are: east—74°, south—164°, west—254°, north—344°.
[b] Distances in metres from the centre of the transects. The position of the centre relative to the well is given in Fig. 2.

As seen in Table 2, increasingly higher concentrations of petroleum hydrocarbons were found from 1982 to 1985. This must be due to the deviation of the centre of the sampling transects from the true position of the well, as seen in Fig. 2. The maximum concentrations in 1986 and 1987 were lower than in 1985.

The area containing the highest concentrations, i.e. in excess of 10 000 μg/g, is obviously very small, probably with a diameter of the order of 100 m, or less. The uncertainty in the determination of the position where the grab touches the bottom is estimated to be 50 m, as stated above. The lower values in 1986 and 1987 are, therefore, not sufficient evidence of a real decrease of the amounts in the inner, most polluted, zone.

With 10 000 μg/g as a limit for *heavy* pollution, the limit for *moderate* pollution is set at 1000 μg/g. The area with moderate pollution is shown crosshatched in Fig. 2. Since there were sampling points both inside and outside this area in all surveys, it is fairly well characterized. The area is approximately 500 m wide, and the extension has apparently not decreased during the 5 years of study.

To find the outer limit for the distribution of the discharges, the values in Table 2 were compared with the background levels given in Table 1. In the 1982, 1983 and 1986 surveys there were enough sampling locations to estimate the *total* extension of the polluted area, i.e. the outer locations along all 4 transects contained background levels or concentrations just above this level. The latter was the case for the outermost location towards the east in 1982 and for all 4 outermost locations in 1986.

In Fig. 2 the moderately to zero polluted area, containing from 1000 μg/g to background levels of petroleum hydrocarbons, is shown by hatching. Degradation of the hydrocarbons in the outer parts of the area with the lowest levels of pollution has occurred from 1982 to 1983. A further decrease of the polluted area has apparently not taken place; the area had about the same extension in the east-west direction, and even a somewhat wider extension in the north-south direction, in 1986.

For the 1984, 1985 and 1987 surveys, the total polluted area cannot be outlined, since too few locations were sampled. In 1984 the outer limit could only be detected towards the east; the pollution did not reach as far as the 750 m location, which means that the extension in this direction was approximately the same as in 1983. In 1985 only locations close to the well were sampled, and the outer limit could not be established. Finally, in 1987 the location 750 m towards east contained THC and NPD slightly above the pollution limit, while the outermost location contained THC above, and NPD below, the limit. With the NPD result being judged the most trustworthy, the 1500 m location is considered free of pollution.

The vertical distribution of hydrocarbons was examined in all 1985 stations and in some of the 1986 stations by collecting subsamples from the 0–3 cm layer, the 3–6 cm layer and from the 6–9 cm layer of the contents in the grab. The results are given in Table 3. They show that the pollution reached at least 9 cm depth. At some of the locations an indication of a vertical gradient may be seen, but since the gradient was absent at most of the locations, it is difficult to draw any conclusions in this respect. In the 1984 survey it was observed that a thin layer of redistributed clean sediment had covered the polluted sediment. It was, however, hard to evaluate if the clean layer had increased in thickness in the subsequent surveys.

3.3. Degradation

Even if the gross changes have been small during the 5 years covered by this investigation, a more thorough study of selected components indicates that degradation processes have indeed taken place, at least in the area which initially contained less than 1000 μg/g hydrocarbons. This is particularly noticeable in the decrease of the polluted area from 1982 to 1983, as discussed above, and seen in Fig. 2.

The THC chromatograms of the samples containing the highest loading of petroleum each year (Fig. 3) show that the normal alkanes in the newly discharged cuttings were present in all samples until 1986. The most polluted 1987 sample had lost most of the normal alkanes. This sample was, however, probably not obtained from the spot with the very highest load. In the areas outside the significantly polluted zone, i.e. concentrations below 1000 μg/g, a relative decrease of the *n*-alkanes was observed in the 1983 and later surveys.

The most used measures of microbial degradation of petroleum, the *n*-C_{17}/pristane and *n*-C_{18}/phytane ratios, also indicated a degradation in the moderately polluted regions after 1 year.

However, a better measure of degradation was the ratio of dibenzothiophenes versus naphthalenes. The dibenzothiophenes and naphthalenes have been shown to differ considerably in resistance towards degradation (Grahl-Nielsen, 1978). The ratio of the sum of the naphthalenes versus the sum of the dibenzothiophenes, N/D ratio, is plotted against the logarithm of the concentration of THC in Fig. 4.

The correlation between the N/D ratio and log THC is fairly good in

TABLE 3
Concentrations, in μg/g wet sediment, of total hydrocarbons, THC, and selected aromatics, NPD, at various depths

Dir[a]	Dist[a]	Depth[b]	1985		1986	
			THC	NPD	THC	NPD
East	100	0–3	90 000	1 920	52 300	1 240
		3–6	65 000	1 770	54 800	1 200
		6–9	72 000	1 930	23 200	630
	200	0–3	52 000	1 230		
		3–6	23 000	580		
		6–9	23 500	585		
	250	0–3			4 650	125
		3–6			8 690	294
		6–9			7 330	258
	500	0–3	102	1·2	318	2·9
		3–6	88	1·2	111	1·0
		6–9	100	1·1	35	0·2
	1 000	0–3			27	0·1
		3–6			27	0·1
		6–9			28	0·1
South	100	0–3	41 000	840		
		3–6	58 000	1 330		
		6–9	59 000	1 140		
	200	0–3	1 500	28		
		3–6	590	12		
		6–9	500	11		
West	100	0–3	61 000	1 930		
		3–6	51 000	1 930		
		6–9	24 000	560		
	200	0–3	2 300	46		
		3–6	4 500	108		
		6–9	4 500	164		
North	100	0–3	4 600	106		
		3–6	2 200	47		
		6–9	1 200	26		
	200	0–3	700	9		
		3–6	270	4		
		6–9	290	3		

[a]Directions and distances as in Table 2.
[b]Depths in centimetres.

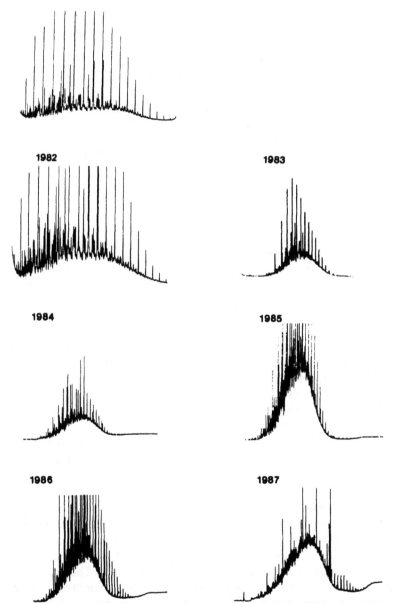

FIG. 3. Chromatograms of the total hydrocarbons in the drill cuttings before discharging and in the sediment samples containing the heaviest load of hydrocarbons each year.

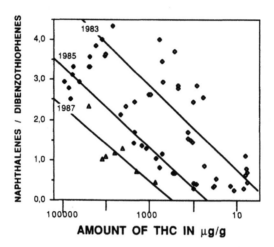

FIG. 4. Plot of the ratio of naphthalenes to dibenzothiophenes against the logarithm of the amount of total hydrocarbons. The plots for 1984 and 1986 had the same slope and correlation coefficients of about 0·9, as the others, and fell close to the 1985 plot. They have not been included in the figure for the sake of clarity.

all surveys, with correlation coefficients around 0·9. This shows that the microbial degradation increases in extent when the concentration of hydrocarbons becomes lower. The concentration-dependent rate of bio-degradation has been experimentally documented by Fusey & Oudot (1984). The plots also show that the degradation has increased from 1983 to 1987, regardless of the concentration of hydrocarbons. The relatively few in situ studies of biodegradation indicate that naphthalenes are degraded more quickly than the larger three, four and five ring aromatics (Davies *et al.*, 1984).

3.4. Conclusions

The persistence of the petroleum hydrocarbons from the discharges in the top layer of the bottom sediments was very high, as no large changes had taken place during the first 5 years. Still, single components, such as the normal alkanes and naphthalenes, had been degraded. The extent of the degradation increased with decreasing load of petroleum hydrocarbons.

With large differences in the concentration of hydrocarbons within small distances, and with uneven distribution, two conditions must be fulfilled to obtain a good description of the state around a discharge site.

First, a proper control of the position where the sampling equipment touches the bottom, i.e. with a maximum uncertainty of 10 m and second, a relatively large number of sampling locations, both inside and outside the polluted area.

ACKNOWLEDGEMENT

Amoco Norway Oil Company is acknowledged for their help in obtaining the samples and for financial support.

REFERENCES

DAVIES, J.M., HARDY, R. & MCINTYRE, A.D. (1981). Environmental effects of North Sea oil operations. *Mar. Pollut. Bull.*, **12**, 412–16.

DAVIES, J.M., ADDY, J.M., BLACKMAN, R.A., BLANCHARD, J.R., FERBRACHE, J.E., MOORE, D.C., SOMMERVILLE, H.J., WHITEHEAD, A. & WILKINSON, T. (1984). Environmental effects of the use of oil-based drilling muds in the North Sea. *Mar. Pollut. Bull.*, **15**, 363–70.

FUSEY, P. & OUDOT, J. (1984). Relative influence of physical removal and biodegradation in the depuration of petroleum-contaminated seashore sediments. *Mar. Pollut. Bull.*, **15**, 136–41.

GRAHL-NIELSEN, O. (1978). The Ekofisk Bravo blowout: petroleum hydrocarbons in the sea. In *Assessment of Ecological Impacts of Oil Spills. Proceedings of the Conference.* 14–17 June 1978, Colorado. American Institute of Biological Sciences, Washington, DC, pp. 476–87.

GRAHL-NIELSEN, O., SUNDBY, S., WESTRHEIM, K. & WILHELMSEN, S. (1980). Petroleum hydrocarbons in sediment resulting from drilling discharges from a production platform in the North Sea. In *Research on Environmental Fate and Effects of Drilling Fluids and Cuttings. Proceedings of the Symposium.* 21–24 January 1980, Florida, American Petroleum Institute, Washington, DC, pp. 541–61.

GROGAN, W.C. (1985). Input of contaminants to the North Sea from the United Kingdom. Final report prepared for the Department of the Environment. Institute of Offshore Engineering, Heriot-Watt University, Edinburgh, pp. 1–203.

IOC (1982). The determination of petroleum hydrocarbons in sediment. *Manuals and Guides no. 11 1982.* Intergovernmental Oceanographic Commission, UNESCO.

JONES, D.M., DOUGLAS, A.G., PARKES, R.J., TAYLOR, J., GIGER, W. & SCHAFFNER, C. (1983). The recognition of biodegraded petroleum-derived aromatic hydrocarbons in recent marine sediments. *Mar. Pollut. Bull.*, **14**, 103–8.

LAW, R.J. & FILEMAN, T.W. (1985). The distribution of hydrocarbons in surficial sediments from the central North Sea. *Mar. Pollut. Bull.*, **16**, 335–7.

LICHTENTHALER, R.G., ORELD, F., SPORSTØL, S. & VOGT, N.B. (1986). Evaluation of chemical methods for monitoring of hydrocarbon discharges from offshore installations. *Oil Based Drilling Fluids. Proceedings.* SFT/Statfjord Unit Joint Research Project, Trondheim, Norway.

NAS (1983). *Drilling Discharges in the Marine Environment.* National Academic Press, Washington DC, pp. 1–180.

NAS (1985). *Oil in the Sea. Input, Fates, and Effects.* National Academic Press, Washington DC, pp. 326–30.

SALTZMANN, H.A. (1982). Biodegradation of aromatic hydrocarbon in marine sediments of three North Sea oil fields. *Mar. Biol.,* **72**, 17–26.

SIDE, J., HERD, C. & GROGAN, W.C. (1985). Oil spill pollution: The North sea experience of cooperative measures. *1985 Oil Spill Conference. Proceedings.* API publ. no. 4385, Washington DC, pp. 615–20.

TEAL, J.M., BURNS, K. & FARRINGTON, J. (1978). Analysis of aromatic hydrocarbons in intertidal sediments resulting from two spills of no. 2 fuel oil in Buzzards Bay, Massachusetts. *J. Fish. Res. Bd Can.,* **35**, 510–20.

PART V
Testing Results and Protocols

34

Review of Biodegradability Test Protocols for Drilling Fluid Base Oils

A. H. GILLAM, S. R. T. SEVERN

CBR International, PO Box 2010, 101-9865 West Saanich Road, Sidney, British Columbia, Canada V8L 3S3

and

F. R. ENGELHARDT

Canada Oil and Gas Lands Administration, Environmental Protection Branch, 355 River Road, Ottawa, Ontario, Canada K1A 0E4

ABSTRACT

The offshore discharge in Canadian waters of waste formation rock contaminated with oil-based drilling mud is regulated currently by guidelines which, among other requirements, describe methods for short-term acute toxicity testing to screen those chemicals of high toxicity. This screening test may not correlate, however, with recent evidence from the North Sea, which suggests that the major environmental effect associated with drilling discharges is one of organic enrichment. A measure of the relative biodegradabilities of drilling fluid base oils may be a more relevant indication of ecological effects in that it can assess the potential for recovery of sediments surrounding offshore drilling operations.

This paper reviews the development of biodegradability test protocols in North America (Canada and the United States) and Western Europe (European Economic Community). The majority of aquatic biodegradation protocols described previously have been concerned with fresh water testing of soluble substrates. This paper provides information on biodegradation test protocols for poorly soluble substrates in marine waters. Based on this review, a novel test protocol using a continuous culture technique is being developed to determine kinetic biodegradation data for drilling fluid base oils.

1. INTRODUCTION—USAGE AND DISCHARGE OF DRILLING FLUIDS

Drilling muds are of two main types—water-based muds (WBMs) and oil-based muds (OBMs). One of the major components of OBM is the drilling fluid base oil. Diesel oil and the new, or lower toxicity, base oils are produced initially from the kerosene and light gas oil fractions of crude oil. The aromatic content of marine diesel oil is usually in the 20–30% range (Brandes IR analysis) whilst the lower toxicity base oils generally have lower aromatic levels (1–10%), and are often referred to as mineral oils. Mineral oils used in drilling muds are mixtures of alkanes, alkenes, cycloalkanes and aromatic hydrocarbons produced by petroleum refining.

Diesel oils were used in the initial OBM formulations but are known to be toxic to a wide range of marine organisms, due to the higher concentrations of the potentially toxic materials (2-, 3- and 4-ring aromatics) in diesel. More recently, a large number of highly refined, white mineral oils, with a much smaller percentage of aromatics than diesel, have been developed. These lower toxicity, or alternative base, oils are mixtures of alkanes, alkenes and cycloalkanes, and have been found to be less toxic to marine organisms.

Base oils intended for use in Canadian offshore drilling operations are subject to the Canada Oil and Gas Lands Administration (COGLA) 'Guidelines for the Use of Oil-Based Drilling Fluids' (COGLA, 1987). The general objective of these guidelines is to ensure that OBM is used in a safe and environmentally-acceptable manner. The base oil is a critical and large volume component in the formulation of an environmentally-acceptable OBM. The COGLA guidelines recommend that the base oil should have a total aromatic hydrocarbon content of <5% (method of analysis not specified) and the content of polycyclic aromatic hydrocarbons (PAHs) should be as low as possible. Higher molecular weight (4–7 ring) PAHs should not be present in the base oil. In addition, base oils should be non-acutely toxic in a standard 96 h LC_{50} test with *Salmo gairdneri* and the 100% water soluble fraction (WSF) of the base oil. By comparison, similar short-term (96 h LC_{50}) bioassay with brown shrimp (*Crangon crangon*) is used to regulate base oils for OBM in the UK Sector of the North Sea (DEn, 1984). In the Norwegian Sector of the North Sea a range of organisms (algae, barnacle larvae and fish larvae) is used to assess the short-term toxicity of base oils (SFT, 1984).

A United Kingdom government assessment (Blackman *et al.*, 1983) of the UK toxicity testing procedure correlated 96 h LC_{50} toxicity data with possible toxic components in base oils. Based on the results from the UK study, it was concluded that acute toxicity could not be correlated with the total aromatic content, nor with levels of specific aromatic compounds. Some correlation between high total naphthalene levels (> 2 g/litre) in the base oils and high acute toxicities (< 1000 ppm) was found. In the UK, substantial emphasis is attached to short-term acute toxicity at the expense of the other attributes of the discharges, such as long-term biodegradability. The treatise of this review is that such an emphasis may need to be reconsidered as more evidence becomes available from field studies. Some of this evidence is reported elsewhere in these proceedings.

It is increasingly apparent that the major environmental effect of the discharge of drill cuttings contaminated with OBM is organic enrichment of the sediments surrounding an offshore drilling operation, rather than a direct toxic effect. Further field studies have not been able to differentiate between the ecological effects of cuttings contaminated with either diesel OBM or lower toxicity OBM (Addy *et al.*, 1984; Davies *et al.*, 1984; Hannam *et al.*, 1987). Recent evidence from the Norwegian sector of the North Sea goes so far as to suggest that the discharge of cuttings contaminated with diesel OBM may have less detrimental effect on the benthic fauna immediately surrounding the platform than cuttings contaminated with the lower toxicity mineral oil-based muds (Matheson *et al.*, 1986).

Recent information from the UK and Norway also indicates that the relative biodegradabilities of base oils used in drilling fluids may be more relevant in determining the long-term recovery of the sediments surrounding an offshore drilling operation than a short-term acute toxicity test. This leads to our consideration that a biodegradability test would be a useful addition to the regulatory permitting system, and may actually have a greater environmental relevance than the acute toxicity tests used currently. The problem at hand is to develop a test protocol which is sensitive, accurate, usable by a wide range of laboratories, and of course, economical.

2. EXISTING BIODEGRADABILITY TEST PROTOCOLS

2.1. Introduction
The fate and effects of synthetic organic chemicals, including drilling fluid base oils discharged to the aquatic environment, is an area of concern as

these materials are considered to have negative effects on aquatic ecosystems. Knowledge of the environmental fate of these materials is required to assist regulatory decisions on the use and disposal of these compounds.

The inability of microorganisms to break down certain chemicals can lead to the persistence or bioaccumulation of these materials in the aquatic environment. Knowledge of the potential biodegradability of the individual chemicals or groups of chemicals would enable regulatory agencies to assess more fully the impact of these materials. Thus, biodegradability determinations in conjunction with acute and chronic toxicity testing could be used to rank materials based on their environmental impact and persistence (Liu, 1986).

From the viewpoint of modern ecotoxicology, toxicity and biodegradability assessments are closely related. Both should be carried out to determine the environmental behaviour of existing and future chemicals that may be discharged to the aquatic environment (Liu, 1986). It is noteworthy that biodegradability testing is already required as an integral part of the process of environmental hazard evaluation of new and existing chemicals in many of the world's industrialized nations, e.g. United States, European Economic Community (EEC) and Japan.

Biodegradability testing originated in response to problems with the discharge of detergents into river systems in Europe. The first test protocols were simple shake-flask tests, also known as 'river die-away', or RDA, tests. In subsequent years they became known as 'repetitive die-away' tests. Consequently, many different test protocols have been developed to provide information on the biodegradabilities of chemicals in the aquatic environment.

More emphasis has been placed in the EEC than in North America on determining biodegradability of new chemicals. Several methods have been assessed by the EEC for the determination of biodegradability in aquatic environments, but the majority of this work has concentrated on tests with water-soluble materials in fresh water. Work has been carried out on biodegradation of petroleum hydrocarbons but there have been no detailed reports of the development of standard biodegradability test methods for these materials. The following sections outline recent developments in biodegradability test protocols in Canada, the US and the EEC.

2.2. Canada

There is no 'standard' method for the assessment of aquatic biodegradability of organic chemicals in Canada. However, work in this area has been carried out since the mid-1970s at the National Water Research Institute (NWRI), Burlington, Ontario.

In 1980, a standardized method was proposed by NWRI to the International Standards Organization (ISO). The objective was to describe a method for the determination of 'primary' biodegradability of organic compounds at a known concentration. The disappearance of the test contaminant from the culture broth under controlled laboratory environmental conditions is followed by gas chromatographic (GC) or other appropriate analytical techniques. This proposed standard method involves incubation of a chemically-defined liquid medium (containing test compound at 1 or 10 mg/litre) with a mixture of microorganisms from activated sludge, soil and sediments. The reaction mixture is maintained in an all glass/teflon cyclone fermentor at room temperature ($22 \pm 2\,°C$) under aerobic and anaerobic conditions, with and without co-metabolites. The data are then corrected for abiotic degradation processes, plotted and used to determine the biodegradability ($t_{\frac{1}{2}}$) of the test compound. It is of interest that Canada is planning to introduce this type of testing requirement in 1988 as part of the Canadian Environmental Protection Act (CEPA).

2.3. United States
The creation of the Environmental Protection Agency (EPA) in 1970 enabled the protection of aquatic resources in the US to be focussed on a single government regulatory agency. In 1972, the Marine Protection Research and Sanctuaries Act extended the authority of the EPA to regulate the dumping of materials into the oceans under the statutory limits of US authority. Regulatory control over the use and disposal of commercial chemicals was brought within the jurisdiction of the EPA in 1976 by enactment of the Toxic Substances Control Act (TSCA).

In 1977 the Organization for Economic Cooperation and Development (OECD), of which the US is a member, recognized the need for a concerted effort to provide internationally-consistent data for new chemicals. Agreement on test methodologies and general good laboratory practices was reached in 1981 and resulted in the publication of a recommended premarket base set of data called the Minimum Premarket Data (MPD) set to be used in the assessment of chemicals. Included in this base set were tests to acquire data on the chemical degradation and accumulation of test materials (Foster, 1985).

Under Section 5 (manufacturing and processing notices—premanufacturing notification) of the TSCA, the US EPA has recommended

that manufacturers subject to premanufacture notification (PMN) requirements utilize this OECD base set as a starting point for designing a testing
program (EPA, 1981). The EPA has no regulatory authority to request
the submission of test data on environmental effects of new chemicals but
it participates in the OECD international standardization effort and has
issued a test guideline for 'aerobic aquatic biodegradation' (EPA, 1982).
This test method is based on the Gledhill shake-flask test (US Federal
Register, 1985).

2.4. European Economic Community (EEC)

Following the development of the OECD testing strategy in 1978/79 and
the publication of OECD guidelines in 1981, the EEC Directive on
the Classification, Labelling and Packaging of Dangerous Substances
(79/831/EEC) was amended to include a requirement for data on the
biodegradability of new chemicals. Following the OECD format, the EEC
Directive suggests three tiers of OECD biodegradability testing (Base Set,
Level 1 and Level 2), equivalent to 'ready biodegradability tests', 'inherent
biodegradability tests' and 'simulation tests'.

In 1983, an Aquatic Biodegradation Working Group was established
by the European Chemical Industry Ecology and Toxicology Centre
(ECETOC) to review the present status of biodegradability testing and
outline any future needs in the EEC (ECETOC, 1983).

Four methods were selected by ECETOC as being acceptable for Base
Set (ready biodegradability) determinations, carried out within the EEC.
These are the Modified Association Francaise de Normalisation (AFNOR)
Test, the Modified Sturm Test, the Closed Bottle Test and the Modified
OECD Screening Test. The Japanese Ministry of International Trade and
Industry (MITI) test method, originally rejected because of difficulties
with both the source and preparation of the inoculum, was subsequently
modified and reevaluated in a ring-test (Painter & King, 1985). The results
of the ring test confirmed the validity of the modified MITI method,
subsequently renamed MITI(II). The MITI(II) method is expected to be
accepted by the EEC Sub-Group on Degradability and Accumulation. All
of the methods assessed were for fresh water aquatic biodegradability testing
of soluble substrates, and thus may be limited for marine applications.

The 1983 review of biodegradation test protocols by ECETOC was
directed towards modifying existing tests to increase their predictive
abilities, i.e. to yield fewer false negative results. Recommendations were
made concerning preacclimatization procedures, harmonization of test
procedures and ammonia nitrification during biodegradability tests. Possi-

bilities for harmonization of the biodegradability tests have been discussed by Blok *et al.* (1985). From this study, two sets of harmonized test conditions for seven ready biodegradability test protocols were proposed (Blok *et al.*, 1985).

2.5. Test Protocols for Poorly Soluble Materials

The aquatic biodegradation test methods in the OECD guidelines (OECD, 1981) and in Annex V of the 6th Amendment to the EEC Directive 79/831/EEC (EEC, 1984) were assessed by ECETOC again in 1985 from the perspectives of the applicability, limitations in use, reproducibility and significant technical weaknesses. It was noted that modifications to these methods would be required for successful testing of poorly soluble organic materials.

The methods discussed by the OECD (1981), EPA (1982) and the EEC (1984) require incubation of a test chemical at concentrations of 2–50 mg/litre in a synthetic mineral solution. Such concentrations often exceed the solubility of the test substance in water (ECETOC, 1985; Painter & King, 1985). To date, none of the biodegradability test methods based on CO_2 respirometry, DOC analysis or BOD determinations have been adapted for the testing of insoluble or poorly soluble materials.

It has also been determined that the test results of a poorly soluble material will be influenced by the method used to accommodate the chemical in the test medium (Fogel *et al.*, 1985). For example, Fogel *et al.* (1985) found that the ultimate degradation of hexadecane was inhibited by the presence of dimethyl sulphoxide or zeolites used as a carrier for the addition of the hexadecane to the test flask.

Gutnick & Rosenberg (1977) reviewed the available literature on the biodegradability of poorly soluble hydrocarbons. These authors concluded that bacteria can utilize solid and liquid hydrocarbons in the dissolved phase and liquid hydrocarbons directly at the water/oil interface. Therefore, a simple dispersion of oil-in-water would be suitable for bacterial utilization in a biodegradability test. This area has also been the subject of several other reviews (Ruffo *et al.*, 1984; Blok & Booy, 1984; ECETOC 1985).

Gerike (1984) reviewed the biodegradation testing literature and scrutinized the test protocols for 'ready biodegradability' for their compatibility with insoluble substrates. The test protocols examined were the Muller–Tittizer test (MT Test); the Sturm test; the Closed Bottle test and the Blok test. The modified Blok test was found to be the most suitable method for testing of poorly soluble materials.

Attempts to modify existing methods for testing of soluble materials have not resulted in a widely-accepted test protocol for poorly soluble substrates. However, ECETOC carried out an experimental evaluation of three test methods for their applicability to poorly soluble materials. The methods tested were the modified MITI(II) test, the Sturm test and the RDA test. Experiments with calcium stearate, anthraquinone and beeswax demonstrated that the methods tested could be adapted for use with poorly water soluble substrates. ECETOC recommended that the three test procedures could form the basis of a general chapter in the existing Guidelines (OECD/EEC) setting out the special requirements for the biodegradation testing of poorly soluble materials (ECETOC, 1985).

2.6. Test Protocols for Use in Marine Waters
The recommended guidelines of the OECD, EEC and EPA all refer to biodegradability testing of organic substrates released to fresh water systems. Until very recently, there had been no work carried out on biodegradation testing of industrial chemicals in seawater (VKI, 1987). The applicability of results from biodegradability studies in fresh water or activated sludge model systems for the prediction of the fate of chemicals in seawater is unknown. This is due primarily to the different populations of biodegrading organisms in the respective systems, and to a lesser degree to differences in the physico-chemical properties of fresh and marine waters.

The test medium for a marine test is basically the same as a medium prepared for a fresh water test; it is a collection of inorganic and organic chemicals designed to simulate the aquatic medium to which the test substrate is discharged. However, the exotoxicological impact of a contaminant may be very different in fresh and marine waters. Thus, it is pertinent when discussing the ecotoxicological behaviour of oil-based drilling fluids discharged in sea water to obtain information on biodegradability of these materials with a marine test protocol.

The choice of the bacterial inoculum for marine tests is a significant problem. There appears to be no readily available marine equivalent to the 'sewage sludge' inoculum used in fresh water tests. In the interlaboratory comparison described by the Water Quality Institute (VKI), Denmark, for example, natural microbial populations were used without further inoculation.

The interlaboratory comparison (VKI, 1987) demonstrated the technical feasibility of both a marine RDA and a closed bottle test. A test period for the marine RDA test of 60 days, compared with the standard 28 days, was suggested by some participants.

There has been no previous description of biodegradation testing of petroleum hydrocarbon mixtures, such as oils. However, there are some reports of testing of individual hydrocarbons. Shiaris *et al.* (1980) examined the biodegradation of phenanthrene in reservoir water samples in Tennessee and found that biodegradation ranged from 66 to 91% over a 60-day period. Fogel *et al.* (1985) discussed the potential of a modified shake-flask technique and a modified soil biometer technique to measure the 'aerobic aquatic biodegradation' of hexadecane, an alkane common to hydrocarbon oils. These authors recommended the modified soil biometer test as a convenient, inexpensive and reproducible technique.

3. RATIONALE FOR DEVELOPMENT OF A NOVEL TEST PROTOCOL FOR USE WITH DRILLING FLUID BASE OILS

3.1. Current Test Protocols

The RDA or shake-flask test protocol (Gledhill, 1975; EPA, 1982; US Federal Register, 1985) has been recommended as part of the EPA guidelines for compliance with Section 4 of the TSCA in the United States. The EPA has also assessed the applicability of a modified RDA test protocol for the assessment of the biodegradability of poorly soluble chemicals, e.g. hexadecane in fresh water (Fogel *et al.*, 1985). A modified soil biometer flask and a $Ba(OH)_2$ absorption method to measure CO_2 evolution was used to assess the biodegradation of the C_{16} alkane. Presumably future recommendations for the biodegradability testing of insoluble or poorly soluble organics by the EPA will include this method.

A modified RDA test (test chemical added at the beginning of the experiment, which runs for 28 days) was assessed also by ECETOC (1985) for use in the EEC with poorly soluble substrates. However, the EEC method requires the measurement of DOC and oxygen uptake to determine the extent of biodegradation. Neither of these analytical techniques is entirely suitable for use with poorly soluble substrates (Gerike, 1984). In a study aimed at harmonization of 'ready' biodegradability tests, Blok *et al.* (1985) recommended that a modified RDA test be accepted for use with poorly soluble chemicals. Thus, the RDA method appears to have potential as a standard in the biodegradation testing of drilling fluid base oils, providing a suitable means for measuring the rate of removal of petroleum hydrocarbons is used.

3.2. Development of a Novel Protocol for Biodegradability Testing

From an ecotoxicological viewpoint, biodegradation and short-term and long-term toxicity assessments of chemical substances should be closely interrelated. A more accurate understanding of the environmental behaviour of existing and new chemicals can only be achieved after all three assessments have been made.

It must be noted with some caution that the majority of work on microbial degradation of organic pollutants has been carried out under controlled laboratory conditions. To completely mimic a certain aquatic environment into which a material will be discharged, adequate knowledge is required of all aspects of the interfaces in the aquatic environment, the test substrate and the biota in that system (Liu, 1986).

It is not possible to reflect in a laboratory study all of the complex interrelationships found in the natural environment. Therefore, in the same manner that the acute toxicity of drilling fluid base oils is assessed using a 'standard' test organism, the approach taken here is of the same 'yardstick' nature.

Test protocols have been designed to provide reproducible estimates of the biodegradability of certain soluble and, more recently, insoluble substrates (ECETOC, 1985; US Federal Register, 1985). While European test protocols follow closely those of the OECD in classifying tests as either for 'ready' or 'inherent' biodegradability or as 'simulations', all three series of tests produce values for 'percentage biodegradability' rather than rates of biodegradability. The EPA test guidelines for compliance with the TSCA (US Federal Register, 1985) also suggest that 'percentage removal' of the test substrate be reported. However, it is important not only to identify how much and which fraction of the oil will be removed, but also to determine the length of time the oil will be present in a particular environment.

Test protocols which produce information on the rates of biodegradation of test substrates are becoming more widely accepted by European and North American groups. Painter & King (1985) recommend that 'studies are required to discover better quantitative ways of expressing the metabolic breakdown of organic compounds other than by the simple recording of percentage removal'. These authors recommend the use of Michaelis–Menten/Monod growth kinetics to produce data on the specific growth rates and saturation/affinity constants for certain substrates.

A draft document prepared for the American Society for Testing and Materials (ASTM) describes a biodegradation test method which allows for the development of first- and second-order rate constants. The method is based on an RDA-type test, but it provides for the effects of natural sediments on the transformation of the test compound and uses shaking to assure a dissolved oxygen supply.

The biodegradation kinetics of priority pollutants have been described by Philbrook & Grady (1985). Their approach was from an environmental/ process engineering perspective in that the methods of evaluating bio-degradation kinetics were divided by the authors into both batch and continuous procedures. Continuous culture techniques were considered superior to batch techniques. Philbrook & Grady (1985) reviewed several continuous culture techniques, including the continuous stirred tank reactor (CSTR) technique; the infinite dilution technique and the modified infinite dilution technique (MIDT). These authors concluded that, even under carefully controlled conditions, biodegradation kinetics were subject to considerable variations, notwithstanding use of the most favourable technique, MIDT. The additional information gained from these kinetic test methods allows a more complete determination of the potential hazard of the test substrate to the environment than does a simple 'percentage degradability' measurement.

A novel testing protocol is being developed in our laboratory which can be readily adapted for use in a wide range of laboratories and which requires limited special expertise. The protocol is based on a continuous dilution-styled technique using a mixed bacterial culture obtained from marine sediments in the Canadian Beaufort Sea. The protocol will allow the determination of biodegradation kinetics of drilling fluid base oils in marine aquatic environments.

The continuous dilution test procedure proposed incorporates a fermen-tor/reactor vessel which can be continuously diluted. The oil concentration is fixed at the beginning of the experiment. The microbial population is then allowed to grow as a batch culture until it has reached a predetermined population size. A water sample is then taken from the fermentor vessel and the oil concentration and cell density are determined. The system is then diluted at a predetermined rate, which will not wash out the microbial population. The dilution medium in these experiments is seawater modified to contain 0.1 mM NO_3^- and 1.0 mM PO_4^{3-}. Samples are taken periodically and the oil concentration and cell numbers are determined. The change in oil concentration during the experiment reflects the rate of oil degra-dation at a specified dilution rate. By varying the dilution rate, the residence time required for degradation can be determined. Easily degraded oils will have a high dilution rate (low residence time), whereas recalcitrant oils will have a lower dilution rate (high residence time).

4. SUMMARY

As the major environmental effect associated with the discharge of OBM drill cuttings is one of organic enrichment, a measure of the relative biodegradabilities of oil-based drilling fluids is of relevance when assessing their approval for use in offshore drilling operations.

The majority of biodegradability test protocols have been developed for use with soluble substrates in fresh water. A review of test protocols indicated that a repetitive die-away (RDA) or shake-flask test would be suitable for development of a marine biodegradability test protocol for drilling fluid base oils. In addition, continuous-culture, or chemostat, techniques appear to be suitable for this purpose. A novel test protocol using a continuous culture with a mixed bacterial population isolated from the Canadian Beaufort Sea is being developed to determine kinetic biodegradation data for drilling fluid base oils.

REFERENCES

ADDY, J.M., HARTLEY, J.P. & TIBBETTS, P.J.C. (1984). Ecological effects of low toxicity oil-based mud drilling in the Beatrice oilfield. *Mar. Pollut. Bull.*, **15**, 429–36.

BLACKMAN, R.A.A., FILEMAN, T.W. & LAW, R.J. (1983). The toxicity of alternative base-oils and drill muds for use in the North Sea. ICES CM 1983/E: 11.

BLOK, J. & BOOY, M. (1984). Biodegradability test results related to quality and quantity of the inoculum. *Ecotoxicol. Environ. Safety*, **8**, 410.

BLOK, J., DE MORSIER, A., GERIKE, P., REYNOLDS, L. & WELLENS, H. (1985). Harmonization of ready biodegradability tests. *Chemosphere*, **14**, 1805–20.

COGLA (1987). Guidelines for the use of oil-based drilling muds. Canada Oil and Gas Lands Administration, Ottawa. 8 pp.

DAVIES, J.M., ADDY, J.M., BLACKMAN, R.A., BLANCHARD, J. R., FERBRACHE, J.E., MOORE, D.C., SOMERVILLE, H.J., WHITEHEAD, A. & WILKINSON, T. (1984). Environmental effects of the use of oil-based drilling muds in the North Sea. *Mar. Pollut. Bull.*, **15**, 363–70.

DEn (1984). Regulation of oil containing discharges resulting from offshore drilling operations. UK Department of Energy, Petroleum Engineering Division. 13 pp.

ECETOC (1983). Technical Report No. 8. Biodegradation Testing: An Assessment of the Present Status. 19 pp. European Chemical Industry and Ecotoxicology Centre, Brussels.

ECETOC (1985). Technical Report No. 20. Biodegradation tests for poorly-soluble compounds. 21 pp. European Chemical Industry and Ecotoxicology Centre, Brussels.

EEC (1984). European Economic Communities Directive 79-831. Annex V-Part

C-Methods for the determination of ecotoxicity. Test methods C 3, 4, 5, 6, 7, 8, 9. *Off. J.*, **1251**, 1.

EPA (1981). New chemical substances: premanufacture testing policy. *Fed. Regist.*, **46**, 8986–93.

EPA (1982). Chemical Test Fate Guidelines, EPA 560/6-82-003. NTIS PB 82-233008. Washington, D.C.

FOGEL, S., LANCIONE, R., SEWALL, A. & BOETHLING, R. S. (1985). Application of biodegradability screening tests to insoluble chemicals: hexadecane. *Chemosphere*, **14**, 375–82.

FOSTER, R.B. (1985). Environmental Legislation. In *Aquatic Toxicology*, Chap. 20, ed. G.M. Rand & S.R. Petrocelli. McGraw-Hill International, New York, pp. 587–600.

GERIKE, P. (1984). The biodegradability testing of poorly water soluble compounds. *Chemosphere*, **13**, 169–90.

GLEDHILL, W.E. (1975). Screening test for assessment of ultimate biodegradability: linear alkylbenzene sulphonates. *Appl. Microbiol.*, **30**, 922–9.

GUTNICK, D.I. & ROSENBERG, E. (1977). Oil tankers and pollution: a microbiological approach. *Ann. Rev. Microbiol.*, **31**, 79.

HANNAM, M.D., ADDY, J.M. & DICKS, B. (1987). Ecological monitoring of drill cuttings discharges to the seabed in the Thistle oilfield. In *Fate and Effects of Oil in Marine Ecosystems*, ed. J. Kuiper & W.J. Van den Brink, Martinus Nijhoff, Dordrecht.

LIU, D. (1986). Biodegradation tests. In *Toxicity Testing Using Microorganisms*, Vol. 2, Chap. 7, ed. B.J. Dutka & G. Bitton. CRC Press, Boca Raton, Florida, pp. 157–73.

MATHESON, I., KINGSTON, P.F., JOHNSTON, C.S. & GIBSON, M. J. (1986). Statfjord field environmental study. In *Proc. of Symposium on Oil-based Drilling Fluids-Cleaning and Environmental Effects of Oil Contaminated Drill Cuttings*, Trondheim, Norway, Statens forurensningstilsyn, Statfjord Unit Joint Research Project Committee, Statfjord, Norway, pp. 1–16.

OECD (1981). OECD Guidelines for testing of chemicals. Section 3, Degradation and Accumulation. TGs 301. Organization for Economic Cooperation and Development, Paris, France.

PAINTER, H.A. & KING, E.F. (1985). A respirometric method for the assessment of ready biodegradability: results of a ring test. *Ecotoxicol. Environ. Safety*, **9**, 6–16.

PHILBROOK, D.M. & GRADY, C.P.L. JR (1985). Evaluation of biodegradation kinetics for priority pollutants. *Proc. 40th Industrial Waste Conf.*, pp. 795–804.

RUFFO, C., GALLI, E. & ARPINO, A. (1984). Comparison of methods for the biodegradability evaluation of soluble and insoluble organo-chemicals. *Ecotoxicol Environ. Safety*, **8**, 273.

SFT (1984). SFT's environmental test procedure for oil based drilling muds. Guidelines and requirements. In *Proceedings of Conference on Drilling Fluids*, Fagernes, Norway, March 5–7, Norwegian Society of Chartered Engineers, Oslo, Norway, 8 pp.

SHIARIS, M.P., SHERRILL, T.W. & SAYLER, G.S. (1980). Tenax-GC extraction technique for residual polychlorinated biphenyl and polyaromatic hydrocarbon analysis in biodegradation assays. *Appl. Environ. Microbiol.*, **39**, 165–71.

US Federal Register (1985). Toxic Substances Control Act Test Guidelines: Final Rule, **50**, 39252.

VKI (Vand Kvalitets Institutett) (1987). Screening test methods for assessment of biodegradability of chemical substances in sea water. Report prepared for Commission of the European Communities—Degradation/Accumulation Sub Group.

35

Toxicity Evaluations of Drilling Sump Fluids: Microtox versus Fish Toxicity Tests

KELLY J. MOYNIHAN, JOHN B. MACLEOD, YVONNE V. HARDY,
ANDREW R. TEAL

*Esso Resources Canada Limited, Research Department, 339–50th Avenue
S.E., Calgary, Alberta T2G 2B3 Canada*

MERL L. KORCHINSKI, DONALD C. ROBERTS, IRENE B. ZABORSKI,
and J. ROGER CREASEY

*Energy Resources Conservation Board, Chemical Research Laboratory,
3512-33rd Street N.W., Calgary, Alberta T2L 2A6 Canada*

ABSTRACT

*A two-phase Esso Resources Canada Limited (ERCL)/Energy Resources
Conservation Board (ERCB) research project that examined nine fresh
water-based drilling sump fluids failed to produce a meaningful correlation
between Microtox and trout toxicity data. For example, the coefficient of
correlation between multiple dilution trout and Microtox toxicity data
for the sump fluid samples from Phase II of this study was in the range
0·14–0·44. Notwithstanding, a Western Canada Microtox Users Committee
(WCMUC) Microtox round robin study involving these same samples and
twelve participating laboratories demonstrated that the Microtox toxicity
test possesses very good precision for the toxicity evaluation of drilling
sump fluids (RSD for EC(50) = 22–25%; RSD for EC(20) = 37–40%
(data quoted are after rejection of outliers)). Sample turbidity and colour
were found to be more detrimental to the Microtox test than to fish toxicity
tests, and the Microtox test was observed to be more sensitive than the fish
tests for the toxicity evaluation of liquid drilling wastes. Sample age and
handling were found to be important factors with respect to the toxicity
data produced by the Microtox test, and the response of the Microtox
reagent bacteria was not significantly affected by sample pH as long as it
was in the range 6·5–9·0. Finally, routine sample analyses are not deemed
to be reliable for the identification of potential toxicants in a sump fluid
sample.*

1. INTRODUCTION

The generation of liquid and solid wastes is an unavoidable consequence
of the drilling of oil and gas wells. Drilling wastes are typically contained
on or near the drilling site in an excavated and lined open pit referred to
as a sump. In the sump, the solid and liquid components of the produced
drilling wastes are allowed to separate prior to their treatment/disposal.

In addition to water, wellbore cuttings and spent drilling mud, drilling
wastes can contain a variety of chemical additives that are used to improve
the performance of the drilling fluid. A partial list of typical drilling fluid
additives is given below: (Canadian Petroleum Association, 1976)

1. pH modifiers: $NaOH$, KOH, $Ca(OH)_2$, soda ash
2. Biocides: paraformaldehyde
3. Corrosion inhibitors: NH_4HSO_3, organic phosphates, filming amines
4. Defoamers: alcohols, sulphonated oils
5. Lubricants: diesel fuel
6. Thinners: inorganic phosphates, tannins, lignite
7. Emulsifiers: Ca salts of organic acids, surfactants
8. Weight adjusters: $BaSO_4$ (barite)

Other non-drilling-related wastes such as surplus formation fracturing
fluids, well workover fluids, industrial wastes, herbicides, pesticides and
sewage have also been known to be disposed of in drilling sumps
(Korchinski & Roberts, 1987). Some of these additives and non-drilling-
related wastes are toxic in nature, and therefore it is obvious that the
proper disposal of drilling sump fluids is of great importance.

The Energy Resources Conservation Board (ERCB) regulates the
disposal of drilling wastes in the Canadian province of Alberta. Since
1975, the ERCB drilling sump fluid disposal guidelines have requested
that single dilution (3) fish (rainbow trout *Salmo gairdneri*) toxicity test
data be provided in order that the toxicities of these wastes can be assessed
(Energy Resources Conservation Board, 1975). In recent years, however,
the above-cited increase in the complexity of drilling waste fluids together
with a heightened public awareness concerning the potential impact that
the improper disposal of liquid drilling wastes can have on the environment
have resulted in the need for more precise, sensitive, and rapid toxicity
assessment procedures.

A relatively new toxicity testing procedure, namely the Microtox
method, is attracting considerable interest as an alternative to the fish
toxicity test procedure for the toxicity assessment of drilling sump fluids

TABLE 1

Microtox test EC(20)	EC(50)	Toxicity	Trout test[a]
0–30%	0–50%	Toxic	All fish dead
30–50%	50–90%	Slightly toxic	Some fish dead at x h; disorientation
>50%	>90%	Non-toxic	All fish survive for 96 h

Note: EC(50) = concentration of sample at which the light output of the bacterial population is reduced by 50%.
EC(20) = concentration of sample at which the light output of the bacterial population is reduced by 80%.
[a]Single dilution test (neat sample used), 3 fish, 96 h in duration.

(see for example Strosher, 1984; Hagen, 1985; Hagen & Halmo, 1985). The Microtox procedure (Microbics Corporation, 1982) uses a suspension of photoluminescent marine bacteria (*Photobacterium phosphoreum*) to measure the toxicity of aqueous-type samples. The bacterial suspension is exposed to different concentrations of the potentially toxic solution, and the Microtox unit (in reality a photomultiplier-based instrument) is used to monitor the change in the light output of the test organisms after each 'challenge' episode. A reduction of the light output of the bacterial population indicates that the test solution possesses some toxic properties. The identified benefits of the Microtox test as it relates to the evaluation of drilling sump fluids are given below:

1. Increased statistical reliability (c. 10^6 organisms utilized per test)
2. Rapid (test takes 15 min)
3. Cost (c. $100 per test)
4. Small sample size (<25 ml)
5. Instrument portability (potential for on-site testing)

The ERCB has been conducting comparative drilling sump fluid toxicity studies involving single dilution (3) fish and Microtox toxicity tests for several years, and some correlation between the two test methods has been observed (Korchinski *et al.*). This finding is in agreement with the work of Strosher (Strosher, 1984). In view of these findings, the ERCB recently proposed some *preliminary* drilling sump fluid toxicity categories based on Microtox EC(50) and EC(20) values (McDonald, 1986). These categories are given in Table 1.

It has also been observed by the ERCB (Korchinski *et al.*) and others (Strosher, 1984) that the Microtox toxicity procedure is more sensitive than fish toxicity tests for the evaluation of drilling sump fluids. Consequently, it is important that this fact be taken into consideration if the ERCB intends to establish new drilling sump fluid disposal guidelines that would include a Microtox testing component.

In the summer of 1986, the Research Department of Esso Resources Canada Limited (ERCL) purchased a Microtox toxicity analyser and began to examine the relative toxicities of some drilling sump fluids. It was found that although there is some relationship between the Microtox and single dilution (3) fish toxicity tests for the determination of drilling sump fluid toxicities, the ERCB Microtox toxicity guidelines presented above are probably inappropriate (MacLeod *et al.*, 1986). After lengthy discussions with the ERCB research laboratory personnel concerning this matter, it was decided to organize a joint ERCL/ERCB research project whose objectives are given below:

- Attempt to identify and then quantify a correlation between toxicity data for drilling sump fluids from Microtox and fish (rainbow trout) toxicity tests.
- Determine the precision of the Microtox toxicity test as it applies to drilling sump fluids.
- Identify parameters that complicate the Microtox toxicity assessment of drilling sump fluids and make appropriate testing protocol recommendations.
- Generate additional data that will assist the ERCB in establishing appropriate Microtox-based disposal guidelines for drilling sump fluids.

The joint ERCL/ERCB study was designed to proceed in two phases. Phase I was intended to be a preliminary study involving a number of analytical issues pertaining to drilling sump fluid toxicities. Comparative Microtox and single dilution (3) fish toxicity tests together with appropriate chemical analyses of a small number of sump liquids formed the basis of this study phase. An outline of Phase I is given below:

1. Sump fluid sampling (6 freshwater-based sumps in total)
2. *Routine* analyses of sump fluids
3. Microtox toxicity tests (non-pH-adjusted and pH-adjusted aliquots)
4. Single dilution (3) fish toxicity tests

Note: *Routine* analyses = pH, conductivity, potassium, sodium, calcium, magnesium, chloride, sulphate, bicarbonate, carbonate, phosphate, total organic carbon (TOC).

Phase II of the joint study was designed to gather additional (wider scope) comparative fish and Microtox toxicity data on three (new) fresh water-based sump fluids. This study phase made use of the Western Canada Microtox Users Committee (WCMUC) round robin organization as a vehicle for the evaluation of the precision of the Microtox test method as it applies to the determination of drilling sump fluid toxicities. It should be noted here that this WCMUC round robin study was the first collaborative study involving the Microtox toxicity test to examine 'real world' (as opposed to synthetic) samples (Qureshi *et al.*, 1987). Multi-dilution trout bioassays were also conducted in order to determine the LC(50) value, i.e. the concentration of a sample that produces a 50% mortality in a population of test organisms, for each waste fluid under study. In addition, *routine* chemical analyses of the sump fluids were performed. The Drilling Waste Management Sub-Committee of the Canadian Petroleum Association (CPA) agreed to provide funds for the chemical analyses and fish toxicity tests associated with this phase of the study.

This paper will summarize the results obtained from this two-phase joint study, and some conclusions and recommendations pertaining to the toxicity assessment of drilling sump fluids will be presented and discussed.

2. MATERIALS AND METHODS

2.1. General

All the glassware that was used in this study was thoroughly cleaned and then rinsed with copious amounts of deionized, distilled water. In addition, small quantities of the appropriate drilling sump fluids were used to rinse all glassware prior to their use.

Standard chemical analytical procedures were used to characterize the drilling wastes that were the focus of this study. All chemical analyses were performed on unaltered sample aliquots of the sump fluids.

With regard to sampling, it was not the objective of this study to necessarily obtain a representative sample at each sump location visited. Whenever possible, however, every attempt was made to obtain samples that were reasonably representative. All the samples that were examined in this study were obtained from *untreated* drilling sumps in the winter months, and therefore no special sample preservation practices were employed between the time of sampling and the time of analysis. Once

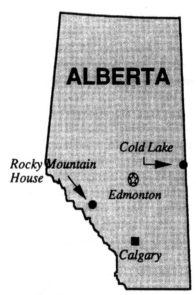

FIG. 1. Sampling locations for drilling sump fluids.

the samples were returned to the laboratory, they were kept at room temperature ($20 \pm 3\,°C$) and typically analysed within 1 week.

2.2. Phase I—Sampling

Sumps A, B and C were located in a conventional oil-type oilfield near Rocky Mountain House, Alberta, while sumps D, E and F were situated in a heavy oil-type oilfield near Cold Lake, Alberta (see Fig. 1). All the sump fluids that were sampled were fresh water-based. Approximately 18 litres of fluid were collected at each sump location, and all sump fluids were accumulated, transported, and stored in 20 litre glass containers that were capped with Parafilm-wrapped stoppers.

Sumps A, B and C were sampled on 12 February, 1987. At the time of sampling, these single well sumps (liquid depths = 50–200 cm) were covered by 30–60 cm of ice. The required samples were therefore obtained through holes that were drilled in the ice. The pattern of the holes drilled in the ice sheet was chosen so as to enable the sampling of as much of the liquid portion of the sump as was practical. Approximately equal volumes of liquid were obtained from each hole at depths of 10–100 cm below the ice/liquid interface.

Sumps D, E and F were sampled on 22 April, 1987. These multi-well (remote) sumps were not frozen over at the time of sampling, and therefore the required liquid samples were obtained from the edge of each sump using weighted plastic containers. Fluid samples were obtained at a number of sites around each sump, and approximately equal volumes (*c.* 1–3 litres) were gathered at each sampling point.

2.3. Phase I—Sample Preparation
In the laboratory, each sump fluid sample was homogenized for *c.* 1 h. Aliquots of each sample were then removed from the bulk sample. The fluids from sumps A, B, C and D were extremely thick and turbid due to the presence of large amounts of suspended solids, and therefore it was decided to filter aliquots from these sump fluid samples prior to their analysis. The sample aliquots from sumps A, B, C and D were first pressure filtered in a mud press at *c.* 800 kPa through a *rinsed* No. 50 hardened Whatman filter disk and then gravity filtered through a *rinsed* No. 42 Whatman filter paper. This filtering process produced pale yellow, clear liquid samples. The fluids from sumps E and F did not contain appreciable quantities of suspended solids, and consequently aliquots of these samples were sequentially gravity filtered through *rinsed* No. 2 and No. 42 Whatman filter papers. Clear, pale yellow liquids were produced via this filtration process. Most sample aliquots that were destined for toxicity testing were pH-adjusted to *c.* 8·0 using reagent grade nitric acid.

2.4. Phase II—Sampling
Sump A was located in a heavy oil-type oilfield near Cold Lake, Alberta while sumps B and C were located in conventional oil-type oilfields near Rocky Mountain House, Alberta and Calgary, Alberta respectively (see Fig. 1). All the sump fluids that were sampled were fresh water-based. Approximately 80 litres of fluid were obtained at each sump location, and all sump liquids were accumulated, transported, and stored in 20 litre glass containers that were capped with Parafilm-wrapped stoppers.

Sump A was sampled on 2 December, 1987. At the time of sampling, this multi-well (remote) sump (liquid depth = 180–290 cm) was covered by 19–22 cm of ice. The required liquid samples were therefore obtained through holes that were drilled in the ice. The pattern of the holes drilled in the ice sheet was chosen so as to enable the sampling of as much of the liquid portion of the sump as was practical. Approximately equal volumes of sample were obtained from each hole at depths of 10–125 cm below the ice/liquid interface.

Sump B was sampled on 1 December, 1987. This single well sump (liquid depth = 50–150 cm) was also frozen over at the time of sampling, and therefore a similar approach to that described above was utilized to obtain the required liquid samples. The ice covering this sump was 9–12 cm in thickness, and samples were obtained from each hole at depths of 10–125 cm below the ice/liquid interface.

The fluid sample from sump C was obtained via a waste disposal contractor, and therefore little is known about the sampling procedure used. This sump was not frozen over at the time of sampling (23 November, 1987), and it is likely that the sample was obtained from the edge of the excavation at several points.

2.5. Phase II—Sample Preparation

In the laboratory, the four 20 litre sample portions from each sump location were blended together and then homogenized for *c.* 1 h. Aliquots of each sample were then removed from the bulk sample. Sample A was brown-black in colour and somewhat turbid. It is interesting to note that if an aliquot of this sample was filtered through a 0·45 μm filter, the filtrate retained its dark colour and slight turbidity. Conversely, a clear, colourless liquid could be obtained from this sample if an aliquot was pressure filtered (*c.* 800 kPa) through several (i.e. stacked) 0·45 μm filters in a mud press. This indicates that the observed colour in this sample is due to the presence of very finely divided solid particles suspended in the fluid. Sample B was clear and very pale yellow in colour. Sample C was extremely turbid. The grey solids associated with this sample were noticed to settle out when an aliquot was allowed to sit undisturbed for several minutes, giving a clear and colourless supernatant.

Non-filtered samples were used for most chemical analyses and toxicity tests. Sample aliquots that were destined for the Microtox and fish toxicity tests were adjusted to *c.* pH 8·0 using reagent grade sulphuric acid and/or calcium hydroxide. The toxicities of some non-pH-adjusted sample aliquots were also evaluated.

2.6. Non-WCMUC Microtox Toxicity Testing (Phases I and II Samples)

Microtox analyses of the Phases I and II sump fluid samples were performed using modified versions of the standard test procedure outlined in the Microbics Microtox system operating manual (Microbics Corporation, 1982). Sample aliquots were osmotically adjusted as per the conventional test method protocol, and in most cases, both pH adjusted and unadjusted samples were examined. Toxicity values were obtained via linear regression analyses of the experimental data, i.e. log gamma versus log concentration plots, using BASIC or LOTUS 1-2-3-based computer routines (Moynihan, 1988). Unless otherwise noted, all Microtox toxicity values are expressed in a percentage format based on undiluted sample concentration.

2.7. WCMUC Microtox Round Robin Study (Phase II Samples only)

Fifteen laboratories agreed to participate in this round robin study, and twelve laboratories actually submitted test data to the study coordinator. Participating laboratories were provided with 50 ml aliquots of each sample that had been pH adjusted (*vide supra*). In addition, one reference solution was supplied. It is important to note here that round robin participants were supplied with *diluted* versions of samples A and C for evaluation. A consensus decision of the WCMUC members dictated that this type of sample preparation procedure be used to deal with the problems associated with highly coloured and/or turbid samples. With hindsight, this may not have been a good decision—sample supernatants/centrifugates/filtrates may have been better candidates for evaluation in this collaborative study. A 1:1 dilution, i.e. 1 part deionized, distilled water added to 1 part of a sample aliquot, of sample A was prepared for the Microtox round robin study while a 1:4 dilution, i.e. 4 parts deionized, distilled water added to 1 part of a sample aliquot, of sample C was prepared. All samples were shipped via courier on 7 December, 1987 to each laboratory, and no provision was made to keep the samples cool during transport. Once received, round robin participants were told to store their samples at 4 °C until the time of analysis. All laboratories were instructed to perform the toxicity tests on Samples A, B and C *in duplicate* on 10 December, 1987. Each laboratory was required to supply their own reagent bacteria and other test supplies, and each laboratory was allowed to use their own version of the standard Microtox test procedure (Microbics Corporation, 1982). The sump fluid samples were also to have been vigorously agitated by hand for 1 min immediately prior to their analysis. The reference sample (a 131·6 mg/l aqueous phenol solution) was permitted to be analysed on any date in December, 1987.

The data that were obtained from this collaborative study were processed in two ways. In the first case, the toxicity values 'as reported' were tabulated and analysed. Subsequently, the *raw* test data that were submitted by the round robin participants were used to recalculate sample toxicity values using a LOTUS 1-2-3-based routine (Moynihan, 1988). This routine is attractive in that in addition to calculating toxicity values, i.e. EC(50) and EC(20) values at 5 and 15 min, it also calculates the degree of linearity, i.e. Goodness of Fit (GOF) for the linear least squares fits of the test data at 5 and 15 min. Individual data points that produced negative log gamma values were excluded from the recalculations (see the Appendix). By recalculating the sample toxicity values along with their corresponding GOF values, a greater insight into the overall performance of the laboratories participating in this study was obtained. Statistical data for both the 'as reported' and 'recalculated' toxicity values for each sample were calculated using the computer program Stat-Packets (Walonick Associates, 1987).

For the 'as reported' toxicity values, the skewedness and kurtosis values associated with each toxicity parameter were used as the basis for the identification/rejection of outliers (see ASTM E178-80). This method is known to be superior to the well known Grubbs (Grubbs, 1950) and Dixon (Dixon, 1953) tests for the identification/rejection of multiple outliers in data from collaborative studies (ASTM E178-80). In addition, two Alberta-based analytical quality control/quality assurance organizations (Alberta Water Analysts Committee (AWAC), Alberta Hydrocarbon Analysts Committee (AHAC)) have adopted this method for the statistical analysis of their round robin data. The guidelines developed by Ferguson (ASTM E178-80) were used to identify skewedness and kurtosis-based non-normality. When these guidelines were exceeded, data furthest from the mean were sequentially deleted until new statistical calculations revealed acceptable skewedness and kurtosis values. All data rejections were based on a 95% confidence level for the toxicity value data, and all statistical calculations excluded those data indicated by 'NA', 'N/A' or 'ND'.

As far as the recalculated toxicity values are concerned, the first criterion used to identify outliers was based on GOF values. Data that were associated with least squares regression lines possessing GOF values less than 0·90 were rejected *regardless of their value*. This approach to outlier identification is appropriate since the Microtox toxicity test is based on the assumption that there is a linear relationship between the logarithm of gamma (a parameter based on bacterial light output) and the logarithm of sample dilution concentration (Microbics Corporation, 1982). After using this approach for the identification of outliers, the skewedness and kurtosis-based method discussed above was applied to the toxicity value data in order to identify additional outliers. The skewedness and kurtosis approach was not applied to the GOF values for each sample, since there is no need to optimize the statistical parameters associated with these data quality estimates. It is important to note that if *both* data points at one particular time were determined to be outliers by the skewedness/kurtosis method, then the corresponding GOF value was also classified as an outlier. The Microtox round robin experimental and statistical data are given in the Appendix.

2.8. Multiple Dilution Fish Toxicity Tests (Phase II Samples only)
The multiple dilution fish toxicity tests were performed by three commercial laboratories located in Calgary, Alberta. These tests were conducted as per the Alberta Environmental Wastewater Effluent Guidelines (Alberta Environment, 1985) using 10 rainbow trout (*Salmo gairdneri*) fingerlings (average weight = 0·7–1·0 g; average length = 3–5 cm) per sample dilution. A minimum of 4 dilutions was used to determine the LC(50) value for each sample, i.e. concentration of sample at which 50% mortality is achieved. It is important to note here that an undiluted aliquot of each sample was (unfortunately) not always examined as part of these fish toxicity tests. The fish toxicity tests were performed at 10·0 ± 1·5 °C, and the fish were not fed during the 96-h test period.

2.9. Single Dilution (3) Fish Toxicity Tests (Phases I and II Samples)
The single dilution (3) fish toxicity tests were performed by the ERCB laboratory and Chemex Laboratories, Calgary, Alberta, using a modified version of a standard procedure (Environment Canada, 1980). This static 96-h test utilizes a *neat* aliquot (500–700 ml) of a sump fluid sample. Three rainbow trout (*Salmo gairdneri*) fingerlings (0·7–1·5 g in weight; 3–5 cm in length) were used for each test, and the fish were not fed during the test period. The test is carried out at a constant temperature (10 °C), and the sample aliquot is aerated during the entire test period.

3. RESULTS

A summary of the toxicity test data for the drilling sump fluid samples examined in Phases I and II can be found in Tables 2 and 3 respectively while the results of the sump fluid chemical analyses are summarized in Table 4. Additional Microtox toxicity data for the Phase II sump fluid samples can be found in Table 5. The WCMUC Microtox round robin data are gathered in the Appendix.

TABLE 2
Comparative toxicity data for Phase I drilling sump fluid samples
(a) Single dilution trout toxicity test data (3 fish per sample aliquot, neat sample)

Sample	Observations
Sump A, pH adj. to 8·0	No deaths after 96 h
Sump B, no pH adj.	2 dead at 102 h
Sump B, pH adj. to 8·0	No deaths after 96 h
Sump C, pH adj. to 8·0	No deaths after 96 h
Sump D, no pH adj.	3 dead at 71 h
Sump D, pH adj. to 8·0	1 dead, others stressed at 113 h
Sump E, no pH adj.	No deaths after 113 h
Sump E, pH adj. to 8·0	No deaths after 113 h
Sump F, no pH adj.	No deaths after 113 h
Sump F, pH adj. to 8·0	No deaths after 113 h

(b) Microtox toxicity test data (averaged data from ERCB and ERCL laboratories)

Parameter	Sample A (%)		Sample B (%)		Sample C (%)		Sample D (%)		Sample E (%)		Sample F (%)	
	Non.	Adj.	Non.	Adj.	Non.	Adj.	Non.	Adj.	Non.	Adj.	Non.	Adj.
EC(50) 5 min	41	31	>100	>100	>100	>100	31	>100	>100	>100	>100	>100
EC(20) 5 min	15	10	47	50	50	40	14	80	>100	55	>100	>100
EC(50) 15 min	36	34	75	>100	>100	>100	25	>100	>100	>100	>100	>100
EC(20) 15 min	13	10	40	41	30	40	12	55	55	65	>100	>100

Non. = Non-pH-adjusted sample aliquot.
Adj. = Sample aliquot adjusted to pH 8·0.

TABLE 3
Comparative toxicity data for Phase II drilling sump fluid samples

(a) Multiple dilution trout toxicity test data (10 fish per dilution, multiple dilutions; pH adjusted samples)

Laboratory No.	Sample A LC(50) (%)	Sample B LC(50) (%)	Sample C LC(50) (%)	Sample R LC(50) (%)
1	No toxicity	55	25	5
2	No toxicity	53	Sample lost	6·3
3	85·2	37	40	3·2
'Average'	95·1[a]	48·3	32·5	4·8

[a]Calculation based on assumption that 'no toxicity' = 100%.
Sample R = 131·6 mg/litre aqueous phenol solution.

(b) Single dilution trout toxicity test data (3 fish per sample aliquot, neat sample; performed by ERCB)

Sample	Observations
Sump A, no pH adj.	3 dead at 19 h
Sump A, pH adj. to 8·6	1 dead at 42 h, 1 dead at 91 h
Sump B, no pH adj.	3 dead at 19 h
Sump B, pH adj. to 7·1	2 dead at 42 h, 1 dead at 67 h
Sump C, no pH adj.	3 dead at 19 h
Sump C, pH adj. to 8·9	3 dead at 19 h
Sump A ice, no pH adj.	No deaths after 114 h
Sump C ice, no pH adj.	1 dead at 42 h

(c) Microtox toxicity test data (WCMUC Round Robin Data: recalculated data, after outlier rejection)

Parameter	Sample A (WCMUC data) (1:1 dilution) (%)	Sample A 'True' toxicity (%)	Sample B (%)	Sample C (WCMUC data) (1:4 dilution) (%)	Sample C 'True' toxicity (%)	Sample R (%)
EC(50) 5 min	34·6 ± 5·0	17·3	31·7 ± 10·5	20·8 ± 3·1	4·2	18·8 ± 2·3
EC(20) 5 min	12·7 ± 3·8	6·4	8·7 ± 4·4	6·5 ± 2·5	1·3	4·6 ± 0·8
EC(50) 15 min	31·7 ± 4·2	15·9	34·2 ± 10·4	16·5 ± 5·2	3·3	20·5 ± 2·2
EC(20) 15 min	11·2 ± 2·9	5·6	8·7 ± 4·1	5·2 ± 2·6	1·0	5·4 ± 0·9

Notes: 'True' toxicities for samples A and C were obtained by dividing the WCMUC-obtained toxicity value by the appropriate sample dilution-based factor.
Sample R = 131·6 mg/litre aqueous phenol solution.

TABLE 4
Comparative chemical analyses data summary

(a) Phase I drilling sump fluid samples

Sample	pH	Conduct. (µS/cm)	K (mg/litre)	Na (mg/litre)	Ca (mg/litre)	Mg (mg/litre)	Cl (mg/litre)	SO_4 (mg/litre)	HCO_3 (mg/litre)	CO_3 (mg/litre)	PO_4 (mg/litre)	TOC (mg/litre)
Sump A	9·3	2 070	6	425	8	0·4	397	178	152	63	ND	46
Sump B	8·5	3 400	10	700	20	3	790	195	380	17	ND	78
Sump C	9·0	1 430	4	310	3	0·3	140	140	340	38	ND	90
Sump D	9·7	1 145	4	235	2	0·3	130	80	161	105	ND	16
Sump E	7·5	470	6	85	9	0·3	72	45	110	—	ND	8·6
Sump F	7·2	362	2	61	11	4	22	72	95	—	ND	9·8

(b) Phase II drilling sump fluid samples

Sample	pH	Conduct. (µS/cm)	K (mg/litre)	Na (mg/litre)	Ca (mg/litre)	Mg (mg/litre)	Cl (mg/litre)	SO_4 (mg/litre)	HCO_3 (mg/litre)	CO_3 (mg/litre)	PO_4 (mg/litre)	TOC (mg/litre)
Sump A	8·6	2 350	124	580	28	116	418	71	643	27	114	250
Sump B	9·3	11 135	2 607	548	101	5	3 359	111	65	48	—	430
Sump C	9·7	5 365	70	1 034	430	2	100	2 917	1	55	—	16

Notes: Data quoted for Phase I samples are mean values from 2 sets of independent analyses.
Data quoted for Phase II samples are mean values from 3 sets of independent analyses.
ND = Not determined.
TOC = Total organic carbon.

TABLE 5

Additional Microtox data for Phase II drilling sump fluid samples

(a) ERCL/ERCB sample dilutions and pH alterations data

Sample	EC(50) 5 min (%)	EC(20) 5 min (%)	EC(50) 15 min (%)	EC(20) 15 min (%)
Sump A, no dil., no pH adj.; ERCL	16·2	5·4	16·8	5·6
Sump A, no dil., no pH adj.; ERCB	17·6	5·7	14·5	4·3
Sump A, no dil., pH adj. to 7·8; ERCL	17·3	5·4	17·8	5·5
Sump A, no dil., pH adj. to 8·6; ERCB	17·0	5·6	16·1	5·2
Sump A, 1:1 dil., pH adj. to 7·8; ERCL	31·1	10·6	30·6	10·1
Sump A, 1:1 dil., no pH adj.; ERCB	33·5	11·7	34·2	12·8
Sump B, no dil., no pH adj.; ERCL	19·4	4·6	23·7	5·9
Sump B, no dil., no pH adj.; ERCB	42·3	8·3	34·7	6·1
Sump B, no dil., pH adj. to 7·9; ERCL	23·2	5·0	27·1	5·9
Sump B, no dil., pH adj. to 7·1; ERCB	>100	>100	>100	>100
Sump C, no dil., no pH adj.; ERCL	5·5	2·9	4·4	2·3
Sump C, no dil., no pH adj.; ERCB	1·8	0·5	<0·5	<0·5
Sump C, no dil., pH adj. to 8·1; ERCL	3·8	1·0	3·2	0·7
Sump C, no dil., pH adj. to 8·9; ERCB	2·3	0·3	2·5	0·4
Sump C, 1:4 dil., pH adj. to 8·1; ERCL	18·7	4·3	14·8	3·4
Sump C, 1:4 dil., no pH adj.; ERCL	12·7	4·4	4·6	1·5

(b) WCMUC colour corrected and 'filtrate' data

Sample	WCMUC Lab No.	EC(50) 5 min (%)	EC(20) 5 min (%)	EC(50) 15 min (%)	EC(20) 15 min (%)
Sample A (1:1 dil.) (colour corrected)	2	>100	94; >100	>100	>100
Sample A 'filtrate' (1:1 dil.)	1	200·5	70·0	122·5	56·7
Sample C (1:4 dil.) (colour corrected)	2	82; 68	13; 15	37; 30	7·5; 8·1
Sample C 'filtrate' (1:4 dil.)	1	84·0	59·2	83·2	51·1
Sample C 'filtrate' (1:4 dil.)	4	>100	>100	>100	>100

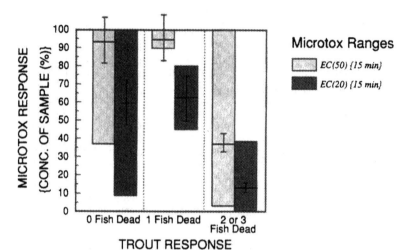

FIG. 2. Comparison of Microtox and single dilution (3) fish toxicity test data: Phases I and II. Mean values and estimated errors shown for each Microtox range bar. Microtox values >100% assigned a value of 100% in calculations. 'True' Microtox toxicity values used for Phase II sumps A & C. EC(50) error taken to be ± 25%; EC(20) error taken to be ± 40%.

4. DISCUSSION

An inspection of Figs 2 and 3 together with the data in Tables 2 and 3 reveals that overall, there is no intuitive or meaningful numeric correlation between the Microtox and fish toxicity data for the drilling sump fluids examined in this joint study. With regard to the Phase II samples only, the comparison of the toxicity data obtained from the multiple dilution fish and Microtox toxicity tests (see Fig. 3) yielded the following correlation coefficients: EC(50) (5 min), 0·22; EC(20) (5 min), 0·44; EC(50) (15 min), 0·14; EC(20) (15 min), 0·35. This poor correlation may possibly be due to the fact that only a small number of sump fluids were examined in this phase of the study and/or that undiluted sample aliquots of each sump fluid sample were not examined in all the multiple dilution trout toxicity tests (*vide supra*). This lack of a correlation between the data from the two types of toxicity tests is in contrast to the observations of Strosher (Strosher, 1984). In his study (that examined 48 drilling sump fluid samples), Strosher found that the coefficient of correlation between single dilution (2) fish and Microtox (EC(50)) toxicity data was 0·78–0·87, while that between multiple dilution fish and Microtox (EC(50)) toxicity data was 0·77–0·89 (Strosher, 1984).

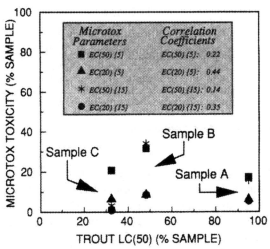

FIG. 3. Comparison of Microtox and multiple dilution fish toxicity test data: Phase II.

TABLE 6

Level of toxicity	Microtox	Multidilution fish	Single dilution (3) fish
Most toxic	Sump C	Sump C	Sump C
Less toxic	Sump A	Sump B	Sump B
Least toxic	Sump B	Sump A	Sump A

Notes: 'True' Microtox toxicity values used for sumps A and C.
Values for pH adjusted samples used for both fish tests.

With regard to Phase II of this study, the order of sump fluid toxicity for each toxicity test is given in Table 6.

While all the toxicity tests found sump C to be the most toxic, the Microtox ranking of the toxicity of this sample must be viewed as potentially suspect due to the large quantity of solids associated with it (*vide supra*). The same argument holds true for sample A. Support for this suspicion can be found in Table 5. As one can see, the Microtox toxicity values for sumps A and C increase substantially if a colour correction is applied to the experimental data or if a clear filtrate of these samples is analysed. It is therefore obvious that large amounts of suspended solids associated with a fluid sample are detrimental to the Microtox assessment

TABLE 7

	EC(50) (5) (%)	EC(20) (5) (%)	EC(50) (15) (%)	EC(20) (15) (%)
RSD range[a]	14·5–34·1	29·2–50·6	12·1–31·5	25·9–50·0
Average RSD[a]	22·2	36·6	24·7	40·4

[a]RSD = relative standard deviation = coefficient of variation = standard deviation/mean.
Data quoted are after the rejection of outliers.

of its toxicity. In contrast, the fish toxicity tests appear to be less adversely affected by suspended solids in a liquid sample. Despite these findings, it is worth noting that the data for Phases I and II of this joint ERCL/ERCB study showed that for the sump fluids studied, a toxic Microtox response is *always* associated with a toxic fish response. The converse of this statement, however, is not true.

The data in Table 5 indicate that the dilution of a sump fluid sample does not necessarily produce a linear change in the toxicity response of the Microtox reagent bacteria. This is an important discovery since it was thought by the members of WCMUC that the detrimental effects associated with dark sample colour or the presence of suspended solids may be able to be overcome by sample dilution (*vide supra*). It now appears as though filtration may be the most appropriate means of dealing with a turbid sump fluid sample. Centrifugation and long-term sample settling are also viable alternatives. As far as dark-coloured samples are concerned, a colour correction must be applied to the experimental Microtox data. With regard to the filtering of turbid drilling sump fluid samples prior to Microtox analyses, some preliminary tests have recently been conducted in the ERCL laboratories (Moynihan et al., 1988). These tests revealed that *unrinsed* cellulose acetate filters can render a sample more toxic than it actually is and that this effect is accentuated if multiple filters are used to clarify a single sample aliquot. Conversely, Whatman No. 1, No. 42, and No. 50 filters along with nylon and glass fibre filters introduce little or no additional toxicants into a filtrate. It was also discovered that suction filtration is more prone to introduce additional toxicants into a filtrate than is gravity filtration.

Perhaps the most significant discovery made in this study is that the Microtox test possesses very good precision for the toxicity evaluation of drilling sump fluids. An examination of the WCMUC round robin data

TABLE 8

	EC(50) (5) (%)	EC(50) (15) (%)
RSD range	4–66	5–85
Average RSD	18	21

Note: 48 sump fluid samples used to generate these data.

(see the Appendix) together with the data summary provided in Table 7 support this claim.

This level of performance is remarkable in view of the nature of samples (*vide supra*), the number of laboratories involved (12) and their geographic distribution (Saskatoon, Saskatchewan to Vancouver, British Columbia) and the fact that the laboratories involved in the WCMUC collaborative study used their own analytical supplies (reagent bacteria, diluent, osmotic adjustment solution) and Microtox testing protocols. It is our opinion that this WCMUC study should be viewed as a 'worst case' or 'real world' scenario related to the Microtox evaluation of drilling sump fluids.

In addition to demonstrating that the Microtox test possesses very good precision, the WCMUC data for the reference sample (a 131·6 mg/l aqueous phenol solution) (see Appendix) show that the round robin participants performed accurately (Qureshi *et al.*, 1987). It is worthwhile to note that the RSD values quoted above compare favourably with those for other standard instrumental analytical procedures for the analysis of aqueous samples (AWAC, 1988). These RSD values are also very similar to those obtained by Strosher in his comparison of Mictotox sump fluid toxicity data generated by three laboratories (Strosher, 1984). The Strosher data are summarized in Table 8.

Further examination of the WCMUC round robin data indicates (not surprisingly) that Microtox toxicity data for drilling sump fluids are often a function of sample age and handling. In addition, one can see from Tables 2, 3 and 5 that the pH of a drilling sump fluid sample has little affect on Microtox toxicity data provided the pH of the sample is in the range 6·5–9·0. This observation is in agreement with those reported by Hagen (Hagen, 1985) and Hagen and Halmo (Hagen & Halmo, 1985). The data in Tables 2 and 3, however, show that rainbow trout are much more sensitive to sample pH.

The analytical data for the sump fluid samples examined in this study (see Table 4) are basically unspectacular and in most cases do not really offer any insight into the identity of the toxic agents in these samples. The data do show that sumps B and C (Phase II) warrant attention from a toxicity/disposal point of view due to the high levels of potassium chloride and sodium sulphate respectively in these fluids. In addition, sumps A and B (Phase II) also possess elevated concentrations of organic carbon. The high magnesium content in the Phase II sump A sample was shown to be due to the suspended solids in the fluid.

5. CONCLUSIONS

The Microtox test holds a great deal of promise as a tool for the screening of drilling sump fluid toxicities. In addition to being rapid and cost effective, the Microtox test possesses very good analytical precision for the toxicity assessment of drilling sump fluids. The establishment of this fact is crucial for the more widespread use of this test for the toxicity evaluation of drilling sump fluids and other types of wastes. A comparison of the Microtox and fish toxicity data for the sump fluids examined in this study revealed that the Microtox test is more sensitive than fish toxicity tests. This observation is in agreement with the findings of other researchers (Strosher, 1984; Korchinski et al.). Further examination of the experimental data reveals that no meaningful correlation exists between the Microtox and fish toxicity values for the drilling sump fluids studied. A generalization of this statement, however, must await further experimental data. Notwithstanding, it is perhaps more prudent to acknowledge the fact that each toxicity test should be judged on its own merit. It was found that the Microtox test is adversely affected by sample turbidity and colour and that appropriate sample clarification procedures or data corrections must be used when necessary. Sample dilution should be avoided as a means of dealing with these problems. Sample pH appears to have less of an impact on the Microtox test than on fish toxicity tests, and sample age and handling appear to be important parameters with respect to the toxicity assessment of drilling sump fluids. Finally, it is concluded that *routine* chemical analyses should not be relied upon for the identification of potentially toxic materials in drilling sump fluids. More sophisticated (and expensive) analyses, e.g. gas chromatography/mass spectrometry, would be required for the identification of organic toxicants in sump fluids while sample digestion followed by inductively coupled plasma or atomic absorption analyses would be necessary for the identification of inorganic, i.e. metallic, species.

6. RECOMMENDATIONS

The following recommendations pertain to the toxicity evaluation of drilling sump fluids:

1. Each toxicity test should be judged according to its own merit—correlations between toxicity data from different tests are often inappropriate or misleading.
2. Turbid samples must be filtered prior to Microtox evaluation—gravity filter the sample if possible, use the minimum number of filters, rinse filters prior to use with copious amounts of deionized, distilled water, and avoid using cellulose acetate-type filters. Other sample clarification techniques, e.g. centrifugation, settling plus decantation, may also be appropriate.
3. A colour correction should be applied to the Microtox experimental data corresponding to a clear, dark-coloured sample.
4. Samples should be analysed as soon as possible after the time of sampling—cool samples if possible until the time of analysis. Further studies aimed at examining the effects that sample age/preservation have on toxicity would be valuable.
5. *Routine* analytical data should not be relied upon for the identification of toxicants in a sump fluid sample.

ACKNOWLEDGEMENTS

The authors would like to thank the managements of the Research Department of Esso Resources Canada Limited and the Chemical Research Laboratory of the Energy Resources Conservation Board for allowing this work to be published. We would also like to thank Mr Mel Strosher for the use of sample containers and for useful discussions. The analytical assistance of Ms Kelly Pisiak (ERCB), Ms Michelle Sieben (ERCL), and Mr Doug Rancier (ERCL) throughout this study is very much appreciated. Finally, the financial support of this project by the Canadian Petroleum Association is gratefully acknowledged.

The following individuals are thanked for their participation in the WCMUC collaborative study that was discussed in this paper:
S. Cryer and R. Wittenberg, Beta Research Laboratories Ltd, Calgary, Alberta.
B. Cudby, Chevron Canada Ltd, Burnaby, British Columbia.

G. Elliott and B. Bednar, Environment Canada, Edmonton, Alberta.
J. Fujikawa and S. McIntosh, Alberta Environment, Lethbridge, Alberta.
M. Korchinski, D. Roberts, K. Pisiak and I. Zaborski, ERCB, Calgary, Alberta.
D. LaBerge and R. Venzi, Chemex Labs Alberta Ltd, Calgary, Alberta.
K. Moynihan, J. MacLeod and Y. Hardy, ERCL, Calgary, Alberta.
A. Qureshi and I. Gaudet, Alberta Environmental Centre, Vegreville, Alberta.
J. Ribo, University of Saskatchewan, Saskatoon, Saskatchewan.
R. Salahub, Chemical & Geological Laboratories Ltd, Edmonton, Alberta.
A. Thakore, B.C. Forest Products Ltd, Crofton, British Columbia.
G. Van Aggelen, B.C. Ministry of the Environment, Vancouver, British Columbia.

REFERENCES

ALBERTA ENVIRONMENT (1985). Waste effluent guidelines for Alberta petroleum refineries.
ASTM Method E178-80. Standard practice for dealing with outlying observations.
AWAC (Alberta Water Analysts Committee) (1988). M. L. Korchinski, Round Robin Coordinator. Round Robin Reports 1–12.
CANADIAN PETROLEUM ASSOCIATION (1976). Drilling Fluid Additives Datafile, pp. 17–39.
DIXON, W.J. (1953). Processing data for outliers. *Biometrics*, BIOMA, **9**, 74–89.
ENERGY RESOURCES CONSERVATION BOARD (1975). Interim Directive No. ID-OG 75-2.
ENVIRONMENT CANADA, ENVIRONMENTAL PROTECTION SERVICE, WATER POLLUTION CONTROL DIRECTORATE (1980). Standard procedure for testing the acute lethality of liquid effluents. Report No. EPS 1-WP-80-1.
GRUBBS, F.E. (1950). Sample criteria for testing outlying observations. *Annals of Mathematical Statistics*, AASTA, **21**, 74–89.
HAGEN, I. (1985). Microtox Testing of Drilling Fluids. The influence of sample preparation, pH, and salt content ($CaCl_2$) on the toxicity. Senter for Industriforskning. Report No. 84 03 22-3.
HAGEN, I. & HALMO, G. (1985). Toxicity testing of ten drilling fluids measured by the Beckman Microtox method. Foundation for Scientific and Industrial Research, Norwegian Institute of Technology. Report No. STF21 A85089.
KORCHINSKI, M.L. & ROBERTS, D.C. (1987). Personal communication.
KORCHINSKI, M.L., McDONALD, I., ROBERTS, D.C., SHAW, D.R. & ZABORSKI, I. B. Unpublished results.
MACLEOD, J.B., MOYNIHAN, K.J. & TEAL, A.R. (1986). Unpublished results.
McDONALD, I. (1986). Internal ERCB communication.

MICROBICS CORPORATION (1982). *Microtox System Operating Manual.* Carlsbad, California.

MOYNIHAN, K.J. (1988). A LOTUS 1-2-3-based routine for the analysis of Microtox toxicity data. Manuscript in preparation.

MOYNIHAN, K.J., HARDY, Y.V. & HOOEY, M.C. (1988). Toxicity evaluations of various filter media using the Microtox method. Manuscript in preparation.

QURESHI, A.A., SHARMA, A.K. & PARAN, J.H. (1987). Microtox quality control collaborative study: a unique and enlightening experience. 3rd International Symposium on Toxicity Testing Using Microbial Systems. Valencia, Spain.

STROSHER, M.T. (1984). A comparison of biological testing methods in association with chemical analyses to evaluate toxicity of waste drilling fluids in Alberta. Canadian Petroleum Association.

WALONICK ASSOCIATES (1987). Stat-Packets statistical analysis package for LOTUS 1-2-3 worksheets. Minneapolis, Minnesota.

APPENDIX: WCMUC MICROTOX ROUND-ROBIN DATA FOR PHASE II SUMP FLUID SAMPLES

Note: The data for samples A and C in the following tables are for 1:1 and 1:4 dilutions of these samples respectively. 'True' toxicity values for these samples can be obtained by dividing by the appropriate dilution-based factor.

(appendix overleaf)

Kelly J. Moynihan et al.

SAMPLE: SAMPLE A

COMMENTS: ALL ANALYSES (ie. AM + PM); DATA "AS REPORTED"
          (ie. NO RECALCULATIONS)

Lab #	Date of Test	Reagent Lot #	EC(50) 5 min. (%)	EC(20) 5 min. (%)	EC(50) 15 min. (%)	EC(20) 15 min. (%)
1	10/12/87	M708	28.7	8.8	26.2	8.4
1	10/12/87	M708	29.7	10.7	26.4	9.3
2	10/12/87	M610	34.0	15.0	33.1	13.9
2	10/12/87	M610	34.2	13.3	31.4	11.7
3	10/12/87	M705	28.8	18.3	27.7	17.7
3	10/12/87	M705	33.5	12.5	29.9	10.9
4	10/12/87	M610	29.0	9.1	28.5	8.8
4	10/12/87	M610	31.7	9.1	30.8	8.5
5	15/12/87	-	42.7	20.2	36.1	R 22.1
5	15/12/87	-	35.9	12.3	37.8	12.4
6	10/12/87	M708	15.3	6.5	ND	ND
6	10/12/87	M708	18.8	7.5	ND	ND
7	10/12/87	M708	29.8	11.0	27.6	9.7
7	10/12/87	M708	32.1	11.8	31.7	13.2
8	10/12/87	M708	31.2	10.1	28.8	9.2
8	10/12/87	M708	28.0	7.7	28.0	7.2
9	10/12/87	M606	32.0	10.7	31.7	10.9
9	10/12/87	M606	32.5	10.8	31.5	9.8
10	10/12/87	-	35.9	13.6	33.8	10.8
10	10/12/87	-	41.3	17.6	40.5	18.4
11	10/12/87	M708	36.9	14.8	36.4	16.5
11	10/12/87	M708	32.5	12.4	32.0	12.2
12	10/12/87	M610	42.3	ND	R 49.3	ND
12	10/12/87	M606	43.7	ND	R 49.6	ND

LEGEND

ND: Not Determined
R: Outlier

STATISTICAL DATA SUMMARY

* Before Outlier Rejection

NO. DATA PTS.	24	22	22	20
MEAN	32.5	12.0	33.1	12.1
MEDIAN	32.3	11.4	31.6	10.9
STANDARD DEVIATION	6.5	3.5	6.3	3.8
VARIANCE	42.2	12.1	39.8	14.4
COEFF. OF VARIATION (%)	20.0	29.2	19.0	31.4
MAX VALUE	43.7	20.2	49.6	22.1
MIN VALUE	15.3	6.5	26.2	7.2
RANGE	28.4	13.7	23.4	14.9
SKEWEDNESS	-0.57	0.66	1.43	1.10
KURTOSIS	3.89	2.87	4.40	3.43

* After Outlier Rejection

NO. DATA PTS.	24	22	20	19
NO. OUTLIERS	0	0	2	1
MEAN	32.5	12.0	31.5	11.6
MEDIAN	32.3	11.4	31.5	10.9
STANDARD DEVIATION	6.5	3.5	3.8	3.1
VARIANCE	42.2	12.1	14.5	9.6
COEFF. OF VARIATION (%)	20.0	29.2	12.1	26.7
MAX VALUE	43.7	20.2	40.5	18.4
MIN VALUE	15.3	6.5	26.2	7.2
RANGE	28.4	13.7	14.3	11.2
SKEWEDNESS	-0.57	0.66	0.66	0.87
KURTOSIS	3.89	2.87	2.72	2.81

SAMPLE: SAMPLE A

COMMENTS: ALL ANALYSES (ie. AM + PM); RECALCULATED DATA
ONLY DATA PRODUCING NEGATIVE GAMMA VALUES REJECTED IN TOXICITY CALCULATIONS

Lab #	Date of Test	Reagent Lot #	EC(50) 5 min. (%)	EC(20) 5 min. (%)	G-O-F 5 minute line	EC(50) 15 min. (%)	EC(20) 15 min. (%)	G-O-F 15 minute line	Comments
1	10/12/87	M708	28.7	8.7	0.9809	26.1	8.2	0.9759	6 pt. reg.; 2 bl.
1	10/12/87	M708	29.7	10.7	0.9967	26.4	9.3	0.9945	6 pt. reg.; 2 bl.
2	10/12/87	M610	R 39.0	R 13.0	R 0.8802	37.3	12.0	0.9051	5 pt. reg.; 1 bl.; data pt. @ 1.42% rejected
2	10/12/87	M610	36.8	18.1	0.9304	30.0	12.8	0.9805	5 pt. reg.; 1 bl.; data pt. @ 1.42% rejected
3	10/12/87	M705	36.7	14.8	0.9915	33.9	13.7	0.9940	5 pt. reg.; 1 bl.; data pt. @ 2.8% rejected
3	10/12/87	M705	33.6	12.5	0.9938	30.0	10.9	0.9908	5 pt. reg.; 1 bl.
4	10/12/87	M610	29.0	9.1	0.9976	28.5	8.8	0.9925	x 8 pt. reg.; 2 bl.
4	10/12/87	M610	31.7	9.1	0.9831	30.9	8.5	0.9874	x 8 pt. reg.; 2 bl.
5	15/12/87	---	43.5	20.3	0.9924	R 39.1	R 22.9	R 0.9567	7 pt. reg.; 1 bl.
5	15/12/87	---	36.1	12.3	0.9811	ND	12.2	N/A	7 pt. reg.; 1 bl.
6	10/12/87	M708	R 8.3	R 5.2	R 0.8123	ND	ND	0.9778	7 pt. reg.; 1 bl.
6	10/12/87	M708	R 8.9	R 5.6	R 0.8727	ND	ND	R N/A	7 pt. reg.; 1 bl.
7	10/12/87	M708	30.1	11.0	0.9854	27.9	9.7	0.9839	x 8 pt. reg.; 2 bl.
7	10/12/87	M708	33.0	11.7	0.9614	33.1	13.1	0.9419	x 8 pt. reg.; 2 bl.
8	10/12/87	M708	31.4	10.1	0.9954	28.8	9.2	0.9969	5 pt. reg.; 1 bl.
8	10/12/87	M708	26.9	5.7	0.9865	27.6	6.5	0.9927	5 pt. reg.; 1 bl.
9	10/12/87	M606	31.5	11.4	0.9975	29.8	13.3	0.9536	5 pt. reg.; 1 bl.
9	10/12/87	M606	34.1	10.1	0.9775	33.1	9.0	0.9735	5 pt. reg.; 1 bl.
10	10/12/87	---	36.2	13.3	0.9746	34.0	10.6	0.9727	7 pt. reg.; 1 bl.
10	10/12/87	---	42.7	17.2	0.9752	R 18.3	R 20.5	R 0.9673	7 pt. reg.; 1 bl.
11	10/12/87	M708	35.0	16.5	0.9755	33.0	R 20.0	0.9065	7 pt. reg.; 1 bl.
11	10/12/87	M708	33.9	9.3	0.9241	R 34.0	R 8.0	R 0.9490	7 pt. reg.; 1 bl.
12	10/12/87	M610	R 42.9	19.6	0.9866	R 40.2	R 22.2	R 0.9842	7 pt. reg.; 1 bl.
12	10/12/87	M606	43.9	15.0	0.9955	R 51.4	16.2	0.9768	7 pt. reg.; 1 bl.

STATISTICAL DATA SUMMARY

* Before Outlier Rejection

	EC(50) 5 min.	EC(20) 5 min.	G-O-F 5 minute	EC(50) 15 min.	EC(20) 15 min.	G-O-F 15 minute
NO. DATA PTS.	24	24	24	22	22	22
MEAN	33.8	12.1	0.9641	33.8	12.5	0.9616
MEDIAN	34.8	11.6	0.9833	33.1	11.5	0.9761
STANDARD DEVIATION	6.7	4.1	0.0422	6.6	4.5	0.0401
VARIANCE	74.9	16.9	0.0022	43.0	21.3	0.0016
COEFF. OF VARIATION (%)	26.6	33.9	4.8	19.5	36.8	4.2
MAX VALUE	43.9	20.3	0.9969	51.4	22.9	0.9969
MIN VALUE	8.3	5.2	0.8123	26.1	6.5	0.8490
RANGE	35.6	15.1	0.1853	25.3	16.4	0.1479
SKEWEDNESS	-1.47	0.28	-1.90	1.19	0.97	-1.49
KURTOSIS	5.39	2.39	5.83	3.80	2.87	4.18

* After Outlier Rejection

	EC(50) 5 min.	EC(20) 5 min.	G-O-F 5 minute	EC(50) 15 min.	EC(20) 15 min.	G-O-F 15 minute
NO. DATA PTS.	21	21	21	18	18	19
NO. OUTLIERS	3	3	3	3	3	3
MEAN	34.6	12.7	0.9797	31.7	11.2	0.9707
MEDIAN	34.0	11.7	0.9855	30.1	10.8	0.9768
STANDARD DEVIATION	5.0	3.8	0.0204	4.2	2.9	0.0267
VARIANCE	24.6	14.6	0.0004	17.1	8.5	0.0007
COEFF. OF VARIATION (%)	14.5	29.9	2.1	13.2	25.9	2.8
MAX VALUE	43.9	20.3	0.9976	41.3	18.3	0.9969
MIN VALUE	26.9	5.7	0.9241	26.1	6.5	0.9051
RANGE	17.0	14.6	0.0735	15.2	11.8	0.0918
SKEWEDNESS	0.56	0.41	-1.56	0.74	0.69	-1.38
KURTOSIS	2.33	2.32	4.51	2.74	3.00	3.96

LEGEND

ND: Not Determined
N/A: Not Applicable
R: Outlier
G-O-F: Goodness of Fit
x: Duplicate analysis of 1 dilutions

SAMPLE: SAMPLE B

COMMENTS: ALL ANALYSES (ie. AM + PM); DATA "AS REPORTED"
          (ie. NO RECALCULATIONS)

Lab #	Date of Test	Reagent Lot #	EC(50) 5 min. (%)	EC(20) 5 min. (%)	EC(50) 15 min. (%)	EC(20) 15 min. (%)
1	10/12/87	M708	27.7	4.6	31.0	5.0
1	10/12/87	M708	26.8	5.6	28.5	6.5
2	10/12/87	M610	20.2	4.5	21.6	4.7
2	10/12/87	M610	26.0	7.4	26.7	6.8
3	10/12/87	M705	34.2	10.2	33.5	12.4
3	10/12/87	M705	31.6	8.8	30.9	8.4
4	10/12/87	M610	21.3	4.0	25.1	4.7
4	10/12/87	M610	22.2	4.1	24.1	4.5
5	15/12/87	—	36.3	12.6	39.3	12.9
5	15/12/87	—	44.2	10.9	40.1	9.0
6	10/12/87	M708	14.6	4.3	ND	ND
6	10/12/87	M708	NA	NA	NA	NA
7	10/12/87	M708	23.0	4.3	25.9	4.9
7	10/12/87	M708	26.3	5.2	31.7	6.3
8	10/12/87	M708	22.4	5.7	26.5	7.3
8	10/12/87	M708	23.6	5.6	20.9	3.0
9	10/12/87	M606	38.5	8.5	45.5	10.0
9	10/12/87	M606	53.5	13.2	R 65.0	16.2
10	10/12/87	—	28.8	9.0	33.5	11.7
10	10/12/87	—	38.1	11.3	41.3	12.4
11	10/12/87	M708	40.8	8.1	48.3	12.2
11	10/12/87	M708	58.8	14.9	60.1	R 24.1
12	10/12/87	M610	38.2	ND	50.7	ND
12	10/12/87	M606	R 79.2	ND	R 74.6	ND

LEGEND

STATISTIAL DATA SUMMARY

ND: Not Determined
NA: Not Available
R: Outlier

* Before Outlier Rejection

NO. DATA PTS.	23	21	22	20
MEAN	33.8	7.8	37.5	9.2
MEDIAN	28.8	7.4	32.6	7.9
STANDARD DEVIATION	14.3	3.3	14.3	4.9
VARIANCE	205.5	10.9	204.0	24.2
COEFF. OF VARIATION (%)	42.3	42.3	38.1	53.3
MAX VALUE	79.2	14.9	74.6	24.1
MIN VALUE	14.6	4.0	20.9	3.0
RANGE	64.6	10.9	53.7	21.1
SKEWEDNESS	1.49	0.59	1.08	1.32
KURTOSIS	5.24	2.15	3.31	4.82

* After Outlier Rejection

NO. DATA PTS.	22	21	20	19
NO. OUTLIERS	1	0	2	1
MEAN	31.7	7.8	34.3	8.4
MEDIAN	28.3	7.4	31.4	7.3
STANDARD DEVIATION	10.8	3.3	10.4	3.6
VARIANCE	116.7	10.9	107.3	13.1
COEFF. OF VARIATION (%)	34.1	42.3	30.3	42.9
MAX VALUE	58.8	14.9	60.1	16.2
MIN VALUE	14.6	4.0	20.9	3.0
RANGE	44.2	10.9	39.2	13.2
SKEWEDNESS	0.83	0.59	0.83	0.44
KURTOSIS	3.16	2.15	2.89	2.09

SAMPLE: SAMPLE B
COMMENTS: ALL ANALYSES (i.e. AM + PM): RECALCULATED DATA
ONLY DATA PRODUCING NEGATIVE GAMMA VALUES REJECTED IN TOXICITY CALCULATIONS

Lab #	Date of Test	Reagent Lot #	EC(50) 5 min. (2)	EC(20) 5 min. (2)	G-O-F 5 minute line	EC(50) 15 min. (2)	EC(20) 15 min. (2)	G-O-F 15 minute line	Comments
1	10/12/87	M708	27.7	4.5	0.9929	31.0	5.0	0.9950	6 pt. reg.; 2 bl.
1	10/12/87	M708	26.7	5.1	0.9781	28.5	6.3	0.9747	6 pt. reg.; 2 bl.
2	10/12/87	M610	20.8	4.6	0.9940	24.6	8.1	0.9934	6 pt. reg.; 2 bl.
2	10/12/87	M610	R 22.8	R 10.2	R 0.9908	24.5	8.1	0.9482	5 pt. reg.; 1 bl.
3	10/12/87	M705	34.2	10.2	0.9972	33.5	12.1	0.9961	5 pt. reg.; 1 bl.
3	10/12/87	M705	31.7	8.8	0.9884	31.0	8.4	0.9848	8* pt. reg.; 2 bl.
4	10/12/87	M610	21.3	4.0	0.9953	25.1	4.7	0.9869	8* pt. reg.; 2 bl.
4	10/12/87	M610	22.2	4.1	0.9642	24.1	4.5	0.9840	7 pt. reg.; 1 bl.
5	15/12/87	–	37.3	12.5	0.9623	39.8	12.9	0.9796	7 pt. reg.; 1 bl.
5	15/12/87	–	45.8	10.7	0.9974	40.8	8.9	N/A	7 pt. reg.; 1 bl.
6	15/12/87	–	14.7	4.3	N/A	ND	ND	N/A	Test aborted; 3 of 4 data pts. -> neg. gamma values
6	10/12/87	M708	23.1	4.3	0.9940	26.0	4.8	0.9926	8* pt. reg.; 2 bl.
7	10/12/87	M708	26.1	5.2	0.9959	31.9	6.3	0.9923	8* pt. reg.; 2 bl.
7	10/12/87	M708	22.6	5.8	0.9922	26.5	7.2	0.9889	5 pt. reg.; 1 bl.
8	10/12/87	M708	37.0	4.7	0.9985	41.0	3.1	0.9982	5 pt. reg.; 1 bl.; data pt. @ 15% rejected
8	10/12/87	M606	37.4	9.1	0.9755	50.4	11.7	0.9343	5 pt. reg.; 1 bl.
9	10/12/87	M606	29.0	15.4	0.9740	33.6	11.6	0.9642	5 pt. reg.; 1 bl.
9	10/12/87	–	38.3	8.9	0.9797	41.9	12.3	0.9817	5 pt. reg.; 1 bl.; data pt. @ 2.8% rejected
10	10/12/87	–	36.3	11.3	0.9996	48.2	13.6	0.9909	7 pt. reg.; 1 bl.
10	10/12/87	M708	40.7	8.1	0.9996	60.1	13.6	1.0000	7 pt. reg.; 1 bl.
11	10/12/87	M708	58.8	14.9	0.9453	N/A	N/A	N/A	3 pt. reg.; 1 bl.; data pt. @ 5.63% rejected
11	10/12/87	M610	37.9	17.4	0.9968	N/A	N/A	N/A	
12	10/12/87	M606	R 77.5	17.5	0.9850	R 109.2	R 43.9	R 0.8704	1 pt. reg.; 1 bl.; 2 of 4 data pts. @ 15 min. rej.

STATISTICAL DATA SUMMARY

* Before Outlier Rejection

	EC(50) 5 min.	EC(20) 5 min.	G-O-F 5 minute	EC(50) 15 min.	EC(20) 15 min.	G-O-F 15 minute
NO. DATA PTS.	23	23	23	21	21	21
MEAN	33.4	8.8	0.9763	37.8	11.1	0.9761
MEDIAN	29.0	8.8	0.9908	31.9	8.4	0.9869
STANDARD DEVIATION	13.6	4.3	0.0341	18.9	8.9	0.0299
VARIANCE	191.6	18.7	0.0012	358.3	79.9	0.0009
COEFF. OF VARIATION (%)	41.3	49.2	3.5	50.0	80.2	3.1
MAX VALUE	77.5	17.5	0.9996	109.2	43.9	1.0000
MIN VALUE	14.7	4.0	0.8500	21.0	3.1	0.8704
RANGE	62.8	13.5	0.1496	88.2	40.8	0.1296
SKEWNESS	1.48	0.63	-2.51	2.51	2.39	-2.22
KURTOSIS	5.34	2.23	9.02	9.85	8.95	7.86

* After Outlier Rejection

	EC(50) 5 min.	EC(20) 5 min.	G-O-F 5 minute	EC(50) 15 min.	EC(20) 15 min.	G-O-F 15 minute
NO. DATA PTS.	21	22	22	20	19	20
NO. OUTLIERS	2	1	1	1	2	1
MEAN	31.7	8.7	0.9821	31.2	8.7	0.9814
MEDIAN	29.0	8.5	0.9915	31.5	8.1	0.9879
STANDARD DEVIATION	10.5	4.1	0.0214	10.1	4.1	0.0187
VARIANCE	109.2	19.1	0.0005	108.6	17.2	0.0003
COEFF. OF VARIATION (%)	33.1	50.6	2.2	30.1	47.1	1.9
MAX VALUE	58.8	17.5	0.9996	60.1	19.7	1.0000
MIN VALUE	14.7	4.0	0.9075	21.0	3.1	0.9343
RANGE	44.1	13.5	0.0921	39.1	16.6	0.0657
SKEWNESS	0.69	0.67	-2.08	0.80	0.82	-1.32
KURTOSIS	3.08	2.20	7.32	2.82	3.23	3.55

LEGEND

ND: Not Determined
N/A: Not Applicable
R: Outlier
G-O-F: Goodness of Fit
*: Duplicate analysis of 4 dilutions

726   *Kelly J. Moynihan* et al.

SAMPLE: SAMPLE C

COMMENTS: ALL ANALYSES (ie. AM + PM); DATA "AS REPORTED"
(ie. NO RECALCULATIONS)

Lab #	Date of Test	Reagent Lot #	EC(50) 5 min. (%)	EC(20) 5 min. (%)	EC(50) 15 min. (%)	EC(20) 15 min. (%)
1	10/12/87	M708	22.7	5.9	15.9	3.2
1	10/12/87	M708	20.1	4.5	13.1	2.1
2	10/12/87	M610	22.3	6.6	15.6	5.2
2	10/12/87	M610	20.3	5.3	15.0	4.3
3	10/12/87	M705	25.6	8.5	20.4	8.6
3	10/12/87	M705	21.1	5.8	13.2	4.1
4	10/12/87	M610	22.5	4.3	16.3	4.1
4	10/12/87	M610	24.4	5.2	18.4	5.1
5	15/12/87	-	R 65.1	R 24.5	R 35.9	R 24.0
5	15/12/87	-	R 67.6	R 17.3	R 42.4	11.0
6	10/12/87	M708	13.6	5.7	ND	ND
6	10/12/87	M708	16.4	6.3	ND	ND
7	10/12/87	M708	18.6	5.2	10.7	3.1
7	10/12/87	M708	16.6	3.3	9.6	2.2
8	10/12/87	M708	18.9	5.0	13.6	3.0
8	10/12/87	M708	16.1	2.6	8.7	1.3
9	10/12/87	M606	22.3	6.4	17.3	5.3
9	10/12/87	M606	22.0	5.9	18.9	5.6
10	10/12/87	-	20.3	7.0	15.2	5.9
10	10/12/87	-	R 34.8	R 11.9	27.4	6.5
11	10/12/87	M708	R 34.8	R 12.1	25.2	8.4
11	10/12/87	M708	27.0	10.0	24.6	9.4
12	10/12/87	M610	R 43.2	ND	R 50.0	ND
12	10/12/87	M606	R 57.1	ND	R 70.6	ND

LEGEND

STATISTICAL DATA SUMMARY

ND: Not Determined
R: Outlier

* Before Outlier Rejection

NO. DATA PTS.	24	22	22	20
MEAN	28.1	7.7	22.6	6.1
MEDIAN	22.3	5.9	16.8	5.2
STANDARD DEVIATION	14.9	4.9	14.7	4.8
VARIANCE	221.1	24.1	215.1	23.1
COEFF. OF VARIATION (%)	53.0	63.6	65.0	78.7
MAX VALUE	67.6	24.5	70.6	24.0
MIN VALUE	13.6	2.6	8.7	1.3
RANGE	54.0	21.9	61.9	22.7
SKEWEDNESS	1.63	2.09	1.87	2.50
KURTOSIS	4.46	7.10	6.05	9.76

* After Outlier Rejection

NO. DATA PTS.	18	18	18	19
NO. OUTLIERS	6	4	4	1
MEAN	20.6	5.8	16.6	5.2
MEDIAN	20.7	5.8	15.8	5.1
STANDARD DEVIATION	3.4	1.7	5.1	2.6
VARIANCE	11.5	2.8	25.7	6.6
COEFF. OF VARIATION (%)	16.5	29.3	30.7	50.0
MAX VALUE	27.0	10.0	27.4	11.0
MIN VALUE	13.6	2.6	8.7	1.3
RANGE	13.4	7.4	18.7	9.7
SKEWEDNESS	-0.16	0.59	0.55	0.63
KURTOSIS	2.52	3.86	2.62	2.61

SAMPLE: SAMPLE C

COMMENTS: ALL ANALYSES (ie. AM + PMD); RECALCULATED DATA
ONLY DATA PRODUCING NEGATIVE GAMMA VALUES REJECTED IN TOXICITY CALCULATIONS

Lab #	Date of Test	Reagent Lot #	EC(50) 5 min. (%)	EC(20) 5 min. (%)	G-O-F 5 minute line	EC(50) 15 min. (%)	EC(20) 15 min. (%)	G-O-F 15 minute line	Comments
1	10/12/87	H708	20.0	5.9	0.9904	15.8	3.0	0.9876	6 pt. reg.; 2 bl.
1	10/12/87	H708	20.5	8.1	0.9908	12.9	2.0	0.9858	6 pt. reg.; 2 bl.
2	10/12/87	M610	22.0	8.1	0.9869	16.3	6.6	0.9853	6 pt. reg.; 1 bl.
2	10/12/87	M610	20.5	4.8	0.9883	14.9	4.1	0.9921	5 pt. reg.; 1 bl.
3	10/12/87	H705	25.6	8.5	0.9277	20.5	8.6	0.9394	6 pt. reg.; 1 bl.
3	10/12/87	H705	21.1	5.8	0.9993	13.2	4.1	0.9969	5 pt. reg.; 1 bl.
4	10/12/87	M610	R 30.0	R 5.4	R 0.6360	R 21.2	R 5.1	R 0.6652	x 8' pt. reg.; 2 bl.
4	10/12/87	M610	R 24.4	R 5.2	0.9098	R 18.1	R 5.1	R 0.8992	x 8' pt. reg.; 2 bl.
5	15/12/87	-	R 68.6	R 24.9	R 0.9605	R 48.1	R 21.5	R 0.9891	7 pt. reg.; 1 bl.
6	15/12/87	-	R 70.6	R 17.1	R 0.9691	R 44.5	10.7	R 0.9492	7 pt. reg.; 1 bl.; data pt. @ 3.1% rejected
6	10/12/87	H708	R 16.6	R 6.8	R 0.8102	ND	ND	N/A	8' pt. reg.; 2 bl.
7	10/12/87	H708	16.3	6.3	0.9985	ND	ND	N/A	8' pt. reg.; 2 bl.
7	10/12/87	H708	18.6	5.2	0.9986	10.6	2.9	0.9638	7 pt. reg.; 1 bl.
8	10/12/87	H708	16.6	3.3	0.9957	9.6	2.2	0.9976	5 pt. reg.; 1 bl.
8	10/12/87	H708	16.1	2.1	0.9866	8.1	2.3	0.9591	5 pt. reg.; 1 bl.
9	10/12/87	M606	23.4	2.6	0.9908	18.3	1.1	0.9876	5 pt. reg.; 1 bl.
9	10/12/87	M606	22.9	7.0	0.9984	20.1	5.8	0.9904	5 pt. reg.; 1 bl.
10	10/12/87	M606	20.1	5.9	0.9549	14.8	5.4	0.9531	7 pt. reg.; 1 bl.; data pt. @ 9% rejected
10	10/12/87	-	R 35.6	6.6	0.9838	27.4	6.5	0.9908	7 pt. reg.; 1 bl.
11	10/12/87	H708	R 34.8	11.6	0.9765	25.2	8.4	0.9914	5 pt. reg.; 1 bl.
11	10/12/87	H708	25.7	12.1	R 0.9764	22.8	8.2	0.9518	5 pt. reg.; 1 bl.
12	10/12/87	M610	R 44.5	R 28.3	R 0.9795	N/A	N/A	N/A	3 pt. reg.; 1 bl.; data pts. @ 12.5% & 25% suspect
12	10/12/87	M606	R 64.1	R 26.8	R 1.0000	N/A	N/A	N/A	3 pt. reg.; 1 bl.; data pts. @ 12.5% & 25% suspect

STATISTICAL DATA SUMMARY

* Before Outlier Rejection

	EC(50) 5 min.	EC(20) 5 min.	G-O-F 5 min.	EC(50) 15 min.	EC(20) 15 min.	G-O-F 15 min.
NO. DATA PTS.	24	24	24	20	20	20
MEAN	29.1	9.4	0.9560	19.8	6.0	0.9545
MEDIAN	22.8	6.5	0.9853	17.3	5.3	0.9855
STANDARD DEVIATION	16.1	7.2	0.0786	10.2	4.3	0.0716
VARIANCE	259.6	52.3	0.0062	103.9	18.8	0.0051
COEFF. OF VARIATION (%)	55.3	76.6	8.2	51.5	71.7	7.5
MAX VALUE	70.6	28.3	1.0000	48.1	21.5	0.9969
MIN VALUE	16.1	2.6	0.6360	8.1	1.1	0.6652
RANGE	54.5	25.7	0.3640	40.0	20.1	0.3317
SKEWNESS	1.65	1.64	-3.03	1.63	2.17	-3.25
KURTOSIS	1.42	4.37	12.01	5.07	8.52	13.36

* After Outlier Rejection

	EC(50) 5 min.	EC(20) 5 min.	G-O-F 5 min.	EC(50) 15 min.	EC(20) 15 min.	G-O-F 15 min.
NO. DATA PTS.	16	18	18	17	18	18
NO. OUTLIERS	8	6	6	3	2	2
MEAN	20.8	6.5	0.9772	16.5	5.2	0.9686
MEDIAN	20.8	5.9	0.9694	15.8	5.3	0.9855
STANDARD DEVIATION	3.1	2.5	0.0276	5.2	2.6	0.0280
VARIANCE	9.8	6.4	0.0008	27.5	6.8	0.0008
COEFF. OF VARIATION (%)	14.9	38.5	2.8	31.5	50.0	2.9
MAX VALUE	25.7	12.1	0.9993	27.4	10.7	0.9969
MIN VALUE	16.1	2.6	0.9088	8.1	1.1	0.8992
RANGE	9.6	9.5	0.0905	19.0	9.6	0.0977
SKEWNESS	-0.06	0.77	-1.37	0.42	0.33	-1.02
KURTOSIS	1.83	2.96	3.47	2.34	2.23	3.05

LEGEND

ND: Not Determined
N/A: Not Applicable
R: Outlier
G-O-F: Goodness of Fit
x: Duplicate analysis of 1 dilutions

SAMPLE: SAMPLE R   (Phenol Solution: 131.6 mg/L)

COMMENTS: ALL ANALYSES; DATA "AS REPORTED"
(ie. NO RECALCULATIONS)

Lab #	Date of Test	Reagent Lot #	EC(50) 5 min. (%)	EC(50) 5 min. (mg/L)	EC(20) 5 min. (%)	EC(20) 5 min. (mg/L)	EC(50) 15 min. (%)	EC(50) 15 min. (mg/L)	EC(20) 15 min. (%)	EC(20) 15 min. (mg/L)
1	15/12/87	M708	16.4	21.6	3.9	5.1	17.9	23.6	4.6	6.1
2	11/12/87	M610	17.0	22.1	4.1	5.4	20.4	26.8	5.5	7.2
3	11/12/87	M705	r 24.0	31.6	5.4	7.1	25.6	33.7	6.5	8.6
4	11/12/87	M610	17.2	22.6	4.5	5.9	18.6	24.5	5.3	7.0
5	11/12/87	M610	20.2	26.6	5.1	6.7	22.7	29.9	6.5	8.6
6	14/12/87	-	18.7	24.6	6.2	8.2	21.7	28.6	5.4	7.1
7	11/12/87	M708	21.8	28.7	4.4	5.8	ND	N/A	ND	N/A
8	11/12/87	M708	17.2	22.7	3.3	4.3	18.8	24.7	4.7	6.2
9	14/12/87	M606	16.0	21.1	4.1	5.4	18.0	23.7	3.8	5.0
10	14/12/87	M606	17.8	23.1	4.6	5.8	20.5	27.0	5.2	6.8
11	11/12/87	-	19.2	25.3	4.6	6.1	20.8	27.4	5.1	6.7
12	17/12/87	M708	18.7	24.6	4.6	6.1	20.7	27.2	5.7	7.5
			R 30.7	R 40.1	R 11.5	R 15.1	R 31.2	R 41.1	R 13.6	R 17.9
			NA	NA	N/A	N/A	NA	N/A	NA	N/A

LEGEND

ND: Not Determined
N/A: Not Applicable

R: Outlier — Outlier according to S & K criteria;
r: judged to be a valid data point, however, after visual inspection of all data

STATISTICAL DATA SUMMARY

* Before Outlier Rejection

	EC(50) 5 min. (%)	EC(50) 5 min. (mg/L)	EC(20) 5 min. (%)	EC(20) 5 min. (mg/L)	EC(50) 15 min. (%)	EC(50) 15 min. (mg/L)	EC(20) 15 min. (%)	EC(20) 15 min. (mg/L)
NO. DATA PTS.	13	13	13	13	12	12	12	12
MEAN	19.6	25.8	5.1	6.7	21.4	28.2	6.0	7.9
MEDIAN	18.7	24.6	4.4	5.8	20.6	27.1	5.4	7.0
STANDARD DEVIATION	3.9	5.1	2.0	2.6	3.6	4.7	2.4	3.2
VARIANCE	14.9	25.9	3.9	6.8	13.0	22.6	5.8	10.0
COEFF. OF VARIATION (%)	19.9	19.9	39.2	39.2	16.8	16.8	40.0	40.0
MAX VALUE	30.7	40.1	11.5	15.1	31.2	41.1	13.6	17.9
MIN VALUE	16.0	21.1	3.3	4.3	17.9	23.6	3.8	5.0
RANGE	14.7	19.3	8.2	10.8	13.3	17.5	9.8	12.9
SKEWNESS	1.77	1.77	2.54	2.54	1.57	1.57	2.54	2.54
KURTOSIS	5.49	5.49	8.60	8.60	4.85	4.85	8.43	8.43

* After Outlier Rejection

	EC(50) 5 min. (%)	EC(50) 5 min. (mg/L)	EC(20) 5 min. (%)	EC(20) 5 min. (mg/L)	EC(50) 15 min. (%)	EC(50) 15 min. (mg/L)	EC(20) 15 min. (%)	EC(20) 15 min. (mg/L)
NO. DATA PTS.	12	12	12	12	11	11	11	11
NO. OUTLIERS	1	1	1	1	1	1	1	1
MEAN	18.7	24.6	4.5	5.9	20.5	27.0	5.3	7.0
MEDIAN	18.3	24.0	4.4	5.8	20.5	27.0	5.3	7.0
STANDARD DEVIATION	2.3	3.0	0.7	0.9	2.2	2.9	0.8	1.1
VARIANCE	5.1	8.8	0.5	0.7	4.7	8.2	0.6	1.0
COEFF. OF VARIATION (%)	12.3	12.3	15.6	15.6	10.7	10.7	15.1	15.1
MAX VALUE	24.0	31.6	6.2	8.2	25.6	33.7	6.5	8.6
MIN VALUE	16.0	21.1	3.3	4.3	17.9	23.6	3.8	5.0
RANGE	8.0	10.5	2.9	3.8	7.7	10.1	2.7	3.6
SKEWNESS	1.03	1.03	0.75	0.75	0.86	0.86	-0.09	-0.09
KURTOSIS	3.22	3.22	3.38	3.38	3.27	3.27	2.71	2.71

SAMPLE: SAMPLE R (Phenol Solution: 131.6 mg/L)

COMMENTS: ALL ANALYSES; RECALCULATED DATA
NO DATA POINTS REJECTED IN TOXICITY CALCULATIONS

Lab #	Date of Test	Reagent Lot #	EC(50) 5 min. (%)	EC(50) 5 min. (mg/L)	EC(20) 5 min. (%)	EC(20) 5 min. (mg/L)	G-O-F 5 minute line	EC(50) 15 min. (%)	EC(50) 15 min. (mg/L)	EC(20) 15 min. (%)	EC(20) 15 min. (mg/L)	G-O-F 15 minute line	Comments
1	15/12/87	H708	16.4	21.6	3.9	5.1	0.9968	17.8	23.4	4.6	6.1	0.9979	6 pt. reg.; 2 bl.
2	11/12/87	M610	17.2	22.6	4.5	5.7	0.9983	20.3	26.7	6.7	8.8	0.9866	6 pt. reg.; 1 bl.
3	11/12/87	H705	24.0	31.6	5.1	7.1	0.9897	25.6	33.7	6.6	8.7	0.9848	6 pt. reg.; 1 bl.
4	11/12/87	M610	17.2	22.6	4.1	5.9	0.9981	18.6	24.5	5.3	7.0	0.9958	x 8* pt. reg.; 2 bl.
5	11/12/87	M610	20.2	26.6	5.1	6.7	0.9876	22.7	29.9	6.5	8.6	0.9827	x 8* pt. reg.; 2 bl.
6	11/12/87	-	18.7	24.6	4.2	5.8	0.9931	21.7	28.6	5.4	7.1	0.9945	1 pt. reg.; 1 bl.
7	11/12/87	H708	22.3	29.3	6.3	8.3	0.9877	ND	N/A	ND	N/A	N/A	1 pt. reg.; 1 bl.
8	11/12/87	H708	17.3	22.8	4.4	5.8	0.9926	18.6	24.7	4.7	6.2	0.9958	x 8* pt. reg.; 2 bl.
9	11/12/87	H708	15.6	20.5	3.1	4.1	0.9935	18.0	23.7	3.9	5.1	0.9998	5 pt. reg.; 1 bl.
10	14/12/87	M606	18.4	24.2	3.9	5.1	0.9988	21.0	27.6	5.6	7.4	0.9997	5 pt. reg.; 1 bl.
11	14/12/87	M606	19.5	25.7	4.6	6.1	0.9921	21.2	27.8	5.6	7.1	0.9933	1 pt. reg.; 1 bl.
12	17/12/87	H708	R 30.8	R 40.4	R 11.6	R 15.3	R 0.9734	R 31.2	R 41.1	R 13.7	R 18.0	R 0.9934	1 pt. reg.; 1 bl.
-	-	-	NA	NA	NA	NA	N/A	NA	NA	NA	NA	N/A	No data submitted

LEGEND

ND: Not Determined   R: Outlier
NA: Not Available   G-O-F: Goodness of Fit
N/A: Not Applicable   x: Duplicate analysis of 4 dilutions

STATISTICAL DATA SUMMARY

* Before Outlier Rejection

	EC(50) 5 min. (%)	EC(50) 5 min. (mg/L)	EC(20) 5 min. (%)	EC(20) 5 min. (mg/L)	G-O-F 5 min	EC(50) 15 min. (%)	EC(50) 15 min. (mg/L)	EC(20) 15 min. (%)	EC(20) 15 min. (mg/L)	G-O-F 15 min
NO. DATA PTS.	13	13	13	13	13	12	12	12	12	12
MEAN	19.7	25.9	5.1	6.7	0.9925	21.4	28.2	6.1	8.0	0.9852
MEDIAN	18.7	24.6	4.5	5.9	0.9931	20.6	27.1	5.5	7.2	0.9939
STANDARD DEVIATION	3.9	5.1	2.0	2.6	0.0058	3.6	4.7	2.4	3.2	0.0131
VARIANCE	15.1	26.1	4.1	7.1	0.0000	13.1	22.7	5.9	10.2	0.0002
COEFF. OF VARIATION (%)	19.8	19.8	39.2	39.2	0.7	16.8	16.8	39.3	39.3	1.3
MAX VALUE	30.7	40.4	11.6	15.3	0.9995	31.2	41.1	13.7	18.0	0.9998
MIN VALUE	15.6	20.5	3.1	4.1	0.9734	17.8	23.4	3.9	5.1	0.9494
RANGE	15.1	19.9	8.5	11.2	0.0261	13.4	17.7	9.8	12.9	0.0504
SKEWEDNESS	1.66	1.66	2.47	2.47	-1.51	1.56	1.56	2.42	2.42	-2.24
KURTOSIS	5.23	5.23	8.38	8.38	5.12	1.81	1.81	8.03	8.03	7.23

* After Outlier Rejection

	EC(50) 5 min. (%)	EC(50) 5 min. (mg/L)	EC(20) 5 min. (%)	EC(20) 5 min. (mg/L)	G-O-F 5 min	EC(50) 15 min. (%)	EC(50) 15 min. (mg/L)	EC(20) 15 min. (%)	EC(20) 15 min. (mg/L)	G-O-F 15 min
NO. DATA PTS.	12	12	12	12	12	11	11	11	11	11
NO. OUTLIERS	1	1	1	1	1	1	1	1	1	1
MEAN	18.8	24.7	4.6	6.1	0.9941	20.5	27.0	5.4	7.1	0.9928
MEDIAN	18.6	24.4	4.5	5.9	0.9937	20.5	27.0	5.4	7.1	0.9945
STANDARD DEVIATION	2.3	3.0	0.8	1.1	0.0041	2.2	2.9	0.9	1.2	0.0054
VARIANCE	5.1	9.0	0.6	1.0	0.0000	4.8	8.3	0.7	1.3	0.0000
COEFF. OF VARIATION (%)	12.2	12.2	17.4	17.4	0.4	10.7	10.7	16.7	16.7	0.5
MAX VALUE	24.0	31.6	6.3	8.3	0.9995	25.6	33.7	6.7	8.8	0.9998
MIN VALUE	15.6	20.5	3.1	4.1	0.9876	17.8	23.4	3.9	5.1	0.9827
RANGE	8.4	11.1	3.2	4.2	0.0119	7.8	10.3	2.8	3.7	0.0171
SKEWEDNESS	0.84	0.84	0.45	0.45	-0.22	0.83	0.83	-0.02	-0.02	-0.65
KURTOSIS	2.93	2.93	3.41	3.41	1.70	3.21	3.21	2.10	2.10	2.07

36

Drilling Fluid Toxicity Test: Variability in US Commercial Laboratories

J. P. RAY

Shell Oil Company, PO Box 4320, Houston, Texas 77210, USA

K. W. FUCIK

T.H.E. Consultants, 4850 Fairlawn Court, Boulder, Colorado 80301, USA

J. E. O'REILLY

Exxon Production Research, PO Box 2189, Houston, Texas 77252-2189, USA

E. Y. CHAI

Shell Development Company, Westhollow Research Center, PO Box 1380, Houston, Texas 77251, USA

and

L. R. LaMOTTE

2739 Barbara Lane, Houston, Texas 77005, USA

ABSTRACT

Since January 1, 1987, the US offshore petroleum industry in the Gulf of Mexico has been faced with compliance toxicity testing of discharged drilling fluids using a test method specified by the Environmental Protection Agency (EPA). The toxicity of the suspended particulate phase (SPP) is evaluated using juvenile mysid shrimp (Mysidopsis bahia) *in a static 96-h LC_{50} test. For the discharge to be in compliance, the measured LC_{50} value must be greater than 30 000 SPP.*

This round robin study examined the intra- and inter-laboratory variability of the EPA drilling fluid toxicity test for two drilling fluids, a KCL mud

(Generic Mud No. 1) and a lignosulfonate mud (Generic Mud No. 8) containing 2% mineral oil. These were carefully prepared, hot rolled and shipped to 13 laboratories which were randomly selected from the population of laboratories that indicated they could provide the service.

A broad variety of operational problems were noted amongst the test laboratories. This study demonstrates the need for careful review of toxicity test reports, whether from an experienced or an inexperienced laboratory, to ensure that the tests are properly conducted.

The results of this study fully support industry's concern that toxicity test variability alone may cause an operator to violate the 30 000 ppm toxicity limitation. Variability for a potassium chloride drilling fluid (generic mud No. 1—based on the same formulation that EPA used in setting the toxicity limitation), was some 3–6 times larger than EPA regulations accounted for in setting the toxicity limitation at 30 000 (95% Confidence multiplier = 3·5, Coefficient of variation = 61·9% versus approximately 1·1 and 9% respectively). These results demonstrate an 83% failure rate, even though EPA apparently intended to allow the use of all formulations of generic mud No. 1. Variability for the more commonly used lignosulfonate mud containing mineral oil was even higher (95% Confidence multiplier = 9·03, Coefficient of variation = 131%). The experience level of the testing laboratories did not appear to be related to either intra- or inter-laboratory variability.

Measurements of biological oxygen demand (BOD), total organic carbon (TOC) and total suspended solids (TSS) in whole mud were also highly variable with coefficients of variation ranging from 63 to 275%.

1. INTRODUCTION

1.1. Background Information
On July 2, 1986, the US petroleum industries offshore operations entered a new era of regulation. For the first time, all operators in the Gulf of Mexico would be required to meet a toxicity limit on their drilling fluid discharges under the provisions of the Environmental Protection Agencies (EPA) General Permit (51 FR 24897, July 9, 1986). The permit conditions required that the drilling fluid toxicity be tested by EPA defined protocols (50 FR 34631, August 26, 1985). The supply of test organisms for conducting drilling fluid toxicity tests was inadequate to meet the estimated demand. As a result, EPA granted industry's request to delay the requirement for compliance with the toxicity limitation until January 1, 1987.

On a case-by-case basis, operators are allowed to request Alternate Toxicity Limits (ATL) to allow for the use of additives such as mineral oil. These special requests require the submission of substantial additional data explaining why the additives are required and their toxicity when formulated into a drilling mud.

The offshore industry has not previously been required to conduct toxicity tests under compliance conditions. Under the current regulations, if the drilling mud is discharged and later found to exceed the allowable toxicity limit, the operator can be fined up to $25 000 per day of violation.

In drafting its regulations, EPA exercised Best Professional Judgement in deciding that the best control for both type and concentration of toxic pollutants, and non-conventional pollutants, was through the control of final mud toxicity. The General Permit requires that discharges meet a daily and monthly average toxicity limitation. In addition, an end of well test is required. The limit is an LC_{50} value for juvenile mysid shrimp (*Mysidopsis bahia*) based on a 96-h static test of the suspended particulate phase of the drilling fluid.

Industry expressed its concern over using the toxicity test as a compliance tool in written comments and in discussions with the agency prior to the implementation of the General Permit (API, 1985; OOC, 1985; O'Reilly, 1985). These comments addressed the lack of validation of the test protocol, availability of the test species, purity of test species cultures, variability inherent in the test, the lack of statistical basis for the limit implemented, and the lack of a laboratory certification program.

Prior to the commencement date of the permit, only a few laboratories were available in the US that had mysid cultures, and had experience conducting drilling mud toxicity tests. In anticipation of the demand for drilling mud toxicity tests, numerous small bioassay labs opened around the country to supply the service mandated by the permit. Industry became increasingly concerned about the variability of drilling fluid toxicity test results as estimates of test variability became available (O'Reilly, 1985, 1986a, b, Bailey & Eynon, 1986: Parrish & Duke, 1986; LaMotte, 1987; O'Reilly & LaMotte, 1987), and particularly with the establishment of many small, less experienced laboratories.

To better define the magnitude of the variability question caused by methodology problems and laboratory experience, the Offshore Operators Committee (OOC), an association of 65 operators in the Gulf of Mexico, funded a project to test randomly a cross-section of laboratories. This paper describes the design of this 'round robin' study, the problems encountered, and the results of the tests. This project was not designed as a laboratory screening or certification program. For this reason, the identity of all companies participating in the round robin will remain confidential; laboratories are designated only as A, B, C, etc.

2. MATERIALS AND METHODS

2.1. Selection of Contractors
Based on a survey of oil and gas operators and drilling mud companies, approximately 30 laboratories providing drilling mud toxicity testing services were identified. A telephone survey was conducted, and 23 of the laboratories on the list verified that they were capable of conducting the drilling fluid toxicity test using mysids and reporting the results within 1 month after receipt of mud samples. Of these 23 laboratories, thirteen were randomly selected (SAS, 1986).

The 13 laboratories which agreed to participate in this study reflect a broad range in experience levels among the commercial toxicity testing laboratories available to an operator. Over half of the laboratories in this study (7/13) have conducted commercial toxicity tests for more than 6 years, and nearly half (5/13) have conducted commercial mysid toxicity tests for more than 5 years. Their experience with the drilling mud toxicity tests in particular, however, is more limited (as would be expected with any newly required test method), with only half of the labs having conducted the drilling mud toxicity test for more than 1 year.

2.2. Selection of Muds and Preparation
Two muds were chosen for testing. A potassium chloride mud (generic mud No. 1), since it had previously been identified as one of the most toxic muds and was used by EPA to set the toxicity limitation used in the permits (Tables 1a, 1b). The second was a freshwater lignosulfonate mud (generic mud No. 8) containing 2% mineral oil (Tables 2a, 2b). This was the same formulation used by EPA in addressing variability of the drilling fluid toxicity tests (Bailey & Eynon, 1986; Parrish & Duke, 1986).

The ingredients were added in the order listed, stirred, and adequate time allowed between additions for mixing and hydration. Following the initial mixing phase, mud was placed in cans and hot rolled for 16 h at 150 °F. Following an 8 h cooling period, samples were stirred, and placed in 1 gallon acid-rinsed plastic pails. As required by the EPA toxicity test method, head space was purged with nitrogen, and the pails sealed for shipping. Pre-shipping storage was at a temperature of 2–4 °C.

TABLE 1a
Formulation of generic mud No. 1

Seawater	117·18 litres
Max Pac Superlo	0·5 lb
Yellow starch	2 lb
XC Polymer	1 lb
KCl	50 lb
Barite	285 lb
KOH	0·5 lb

TABLE 1b
Mud properties of generic mud No. 1

Weight (lb/gal)	14
PV @ 120 °F (cP)	21
YP @ 120 °F (lb/100 ft^2)	23
10 s gel (lb/100 ft^2)	7
10 min gel (lb/100 ft^2)	9
pH	8·4
API filtrate (ml)	13·2
P_f/M_f	0·5
P_m	0·4
Ca$^+$ (mg/litre)	1 680
Cl$^-$ (mg/litre)	98 000
Solids (% vol. retort)	24
Oil (% vol. retort)	0
Water (% vol. retort)	76
K$^+$ (mg/litre)	90 000 mg/litre

2.3. Shipping of the Mud Samples

The thirteen participating laboratories were sent duplicate samples of the two mud types (total of four samples). Samples with coded labels were sent in two batches at approximately 1 week intervals. Laboratories were not told what kind or how many muds they were receiving. Samples were sent in 1 gallon plastic pails that were packed two to a five gallon bucket with blue ice to keep the samples chilled. Samples were sent via overnight express delivery, with laboratories notified of arrival time. Samples included a return receipt postcard and a Chain of Custody form. Laboratories were contacted by phone to confirm receipt.

Of the thirteen laboratories selected initially two declined to participate after receipt of the samples. Two additional laboratories were chosen randomly from the original list.

TABLE 2a
Formulation of generic mud No. 8 with 2%
mineral oil

Deionized water	99·88 litres
Bentonite	15 lb
Chrome lignosulfonate	15 lb
Lignite	10 lb
CMC regular	0·25 lb
Sodium bicarbonate	1 lb
Barite	550 lb
Caustic soda (NaOH)	2·5 lb
Mineral oil	3 179·4 ml

TABLE 2b
Mud properties of generic mud No. 8

Weight (lb/gal)	17·6
PV @ 120 °F (cP)	83
YP @ 120 °F (lb/100 ft^2)	21
10 s gel (lb/100 ft^2)	10
10 min gel (lb/100 ft^2)	14
pH	9·5
API filtrate (ml)	1·8
P_f/M_f	0·6/4·8
P_m	2·2
Ca$^+$ (mg/litre)	200
Cl$^-$ (mg/litre)	2 500
Solids (% vol. retort)	37
Oil (% vol. retort)	2
Water (% vol. retort)	61
Oil (% vol by wt)	2·14

2.4. Testing Protocols

All laboratories were instructed to conduct the toxicity tests as prescribed
by the EPA method (51 FR 24897, July 9, 1986). The laboratories were
also requested to conduct the standard toxicant test using the reference
toxicant, dodecyl sodium sulfate (Lot No. 864239, Fisher Scientific Co.,
Houston, TX, USA) provided with their samples. The laboratories were
instructed to contact EPA if clarification of the published method was
needed.

The laboratories were requested to submit all data from their four tests
on data sheets supplied by the OOC. Selected laboratories were requested
to provide additional information on the muds. These included dissolved

solids, suspended solids and total solids. Biological oxygen demand (BOD) was measured using Method 415.1 of the EPA procedures manual (EPA, 1979). Total organic carbon (TOC) was measured by following Method 415.2 of the same manual. Total suspended solids (TSS) were to be measured by Method 209-C, which requires drying at 103–5 °C (APHA, 1985).

2.5. Statistical Analysis

The objectives of this study were to determine the intra- and inter-laboratory variability inherent in the EPA drilling fluid toxicity test for these two muds. As controlled in this experiment, there were two sources of variation of the LC_{50}. The first was intra-lab variation, the difference between the LC_{50}s reported by one laboratory for two samples of the same mud. Any differences would result from a combination of many factors, including sample-to-sample differences in the mud, variation in the preparation of the SPP phase, variation in tolerances of the mysids, and even the experience of the testing laboratory.

Second, LC_{50} determinations could differ from laboratory to laboratory. Such differences could be due to slight differences in interpretation of procedures such as SPP preparation or the numerical method used to compute the LC_{50}. Lab-to-lab differences among mysid cultures may also be a source of variation.

Analysis of variance (ANOVA) was used to estimate the intra- and inter-laboratory variance components based on the natural logarithm of the LC_{50} reported by each laboratory. The LC_{50}s are expected to follow a lognormal relative frequency distribution. The total variance of $\ln(LC_{50})$ for a given mud is calculated as the sum of the estimates of the two variance components (intra plus inter).

Five estimates of variability were calculated for these data:

—H/L ratio—the ratio of highest to lowest LC_{50};
—coefficient of variation (CV) based on the $\ln(LC_{50})$.
   $(CV = \sqrt{\exp(\hat{\sigma})^2 - 1} \times 100)$
—'95% confidence multiplier' as calculated by Bailey & Eynon (1986) (included here only for comparison with EPA data, since it is not considered appropriate) ('CM1' = $\exp(1\cdot96 \times \hat{\sigma})$);
—'95% confidence multiplier' as calculated by O'Reilly & LaMotte (1987) for calculating 95% confidence limits ('CM2' = $\exp(t_{0.025,df} \times \hat{\sigma})$);
—'95% confidence multiplier' as calculated by O'Reilly & LaMotte (1987) for calculating Alternate Toxicity Limit

('CM3' $= \exp(t_{0.025,\text{df}} \times \hat{\sigma} \times \sqrt{2})$)
 where: df $=$ degrees of freedom
          $\hat{\sigma} = \sqrt{\text{variance}}$.

Three 95% confidence multipliers are shown in Table 4, labeled CM1, CM2 and CM3. CM1 is the 95% confidence multiplier used by Bailey & Eynon (1986). CM2 and CM3 follow discussion in O'Reilly & LaMotte (1987). Where CM1 employs a standard normal critical value, CM2 uses a 'Student's t' critical value in order to properly adjust for the fact that the population variance of $\ln(LC_{50})$ was not known and was estimated from the available data. This is the correct 95% confidence multiplier when applied to one LC_{50} determination. However, in the way that EPA uses Bailey & Eynon's 95% confidence multiplier, 2·33, the LC_{50} for the test mud is to be compared to the LC_{50} in the Alternate Toxicity Request (EPA variance procedure to allow operators with special mud additive needs to apply for a lower permissible toxicity limit). In this case, the 95% confidence multiplier is to be applied to a ratio of LC_{50}s (equivalently, a difference in the logarithms of the two LC_{50}s). Consequently, the logarithm of the 95% confidence multiplier needs to be multiplied by $\sqrt{2}$. This results in CM3. In summary, CM1 is not really the right answer to any question, CM2 is the appropriate 95% confidence multiplier to use on one LC_{50}, and CM3 is the appropriate multiplier to use when comparing two LC_{50}s.

Variability for the solids, BOD and TOC measurements were also estimated using ANOVA as described above.

3. RESULTS AND DISCUSSION

3.1. Operational Problems

The laboratories included in this study reflect a broad range of experience levels. The extent of variety and number of problems encountered by the laboratories in this study was surprisingly high (Table 3), but do not appear to be related to a laboratory's level of experience. These problems demonstrate that toxicity test results, whether from an experienced or inexperienced laboratory, need to be thoroughly reviewed before they are accepted.

Many, if not all of these problems, were corrected by each of the laboratories before submittal of their final data report, and hence do not affect the levels of variability calculated from these data. Although the

TABLE 3

Tabulation of bioassay problems encountered with commercial laboratories during the round-robin study

Problem	A	B	C	D	E	F	H	J	K	L	M	P	Q
1. Mysids													
(a) Availability													
(b) Culturing			×	×									
(c) Control survival						×		×			×		
2. Shipping													
(a) Federal Express Cards not returned		×	×			×					×		
3. Procedures													
(a) Didn't use OOC Reference Toxicant	×									×	×		×
(b) Didn't complete all tests			×	×									
(c) Lack of communication on test failures				×		×		×					
(d) Missing description of methodology		×			×	×		×	×	×	×	×	
4. Data reporting													
(a) Delays in returning/missing chain of custody forms						×	×	×	×	×	×		
(b) Data sheets inconsistent with reports							×	×			×	×	
(c) Data missing			×			×					×		
(d) Inconsistency between Lab/OOC Sample No.	×									×			
(e) Data supplied in different form than requested		×				×			×		×	×	

original intent was to measure the total variability inherent in the data as submitted by the toxicity testing laboratories, a quality check was required to ensure that the tests were conducted as required by the EPA method. Checking the data report in detail as done here was a necessary step. Doing so, however, may have reduced the total amount of variation from that present in the originally submitted data report.

The problems detailed below represent problems that operators should be aware of when evaluating toxicity test data.

3.1.1. Shipping
No major problems were encountered in shipping the samples. All laboratories received their samples the day after shipping.

3.1.2. Mysids
The availability of mysids, although not as severe a problem as it was during late 1986 (when EPA granted industry's request to delay mandatory toxicity testing), was still an apparent problem. Four of the 15 laboratories that participated in this study (13 laboratories received mud samples plus two laboratories that were originally selected and subsequently replaced because they were unable to obtain mysids), were unable to complete the required drilling fluid toxicity tests as a result of mysid related problems.

The unavailability of mysids prevented two laboratories from participating in this study (replaced at the onset of the study) and another from completing the required tests (Lab C). In addition, results from Lab Q were delayed because of a mysid supply problem. Survival of mysids in controls (no drilling fluid added) was such a severe problem for one of the 13 laboratories (D) that they failed to complete any of the required tests.

Two other laboratories (E and H) experienced control mortalities in excess of the allowable limit of 10%. One of these had to repeat the test several times to get an acceptable mortality rate in the controls.

3.1.3. Procedures
Problems with general laboratory practices (e.g. not performing or reporting additional tests as requested, missing description of methods, etc.) were evident with all of the laboratories in this round-robin study and may represent a universal problem among all toxicity testing laboratories.

One of the main procedural problems was that of developing data for dissolved solids, suspended solids and total solids. The laboratories supplied the results, but for the most part did not provide any reference to the detailed procedure followed. In addition, it appears that some of the laboratories followed the requested test method, while others simply added the dissolved and suspended solids data together to get total solids.

Each laboratory was requested to conduct the standard toxicant tests with the reference toxicant supplied by OOC. Despite this, some laboratories decided to use in-house reference toxicant without prior approval.

Some laboratories reported that they corrected for control mortality in calculating the 96-h LC_{50}, others did not. The written reports did not provide any discussion of the corrections used, therefore making data difficult to compare.

3.1.4. Data Reporting

Chain of Custody forms were supplied with all mud samples and all laboratories were instructed at the beginning of the program to return them. Despite the importance of such documentation, most of the Chain of Custody forms were returned unsigned, incomplete, late, or were lost entirely.

Of the twelve laboratories that submitted final data reports for this project, half of them (A, E, F, H, L and M) submitted final reports which either did not reference the mud sample tested or reported incorrect confusing information. Laboratories F, H, L and M submitted final reports which were inconsistent with the data sheets. Laboratories E and H reported test results which initially did not meet the EPA requirement of less than 10% mortality in control mysids. Laboratories E and H were each requested to repeat the toxicity test. One of them had to repeat the test several times in order to meet this requirement. Laboratories A, F and L reported test results using their own internal laboratory identification number, with no cross reference to the sample number provided with the mud sample.

These problems necessitated numerous phone calls to the laboratories to cross-check and verify that the data were correct and that they were associated with the proper sample.

3.2. Toxicity

The data collected encompassed over half (12/23) of the commercial laboratories that confirmed they could provide the mysid toxicity test procedure as specified by EPA. The round-robin study was designed to measure the intra- and inter-laboratory variability of the toxicity test for

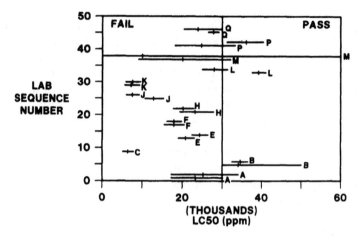

FIG. 1. LC_{50}s with 95% confidence limits for generic mud No. 1 (as represented by horizontal lines). Plus (+) indicates the estimated LC_{50}. Letters indicate the different test laboratories.

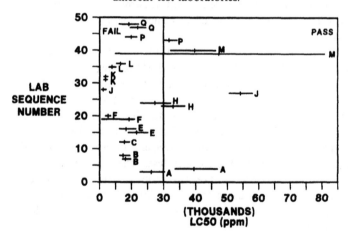

FIG. 2. LC_{50}s with 95% confidence intervals for each laboratory in reference to generic mud No. 8.

two mud types. Variation in toxicity between samples within each mud type was expected to be minimal since all samples of each mud type were taken from one 'batch' of well mixed mud. The study was not designed to study the additional variation expected when multiple 'batches' are prepared either by the same or several mud companies.

Figures 1 and 2 show the 95% confidence intervals and LC_{50}s obtained from each laboratory for each mud. Each horizontal line shows the range of the 95% confidence interval on the LC_{50} for each laboratory. For example, the long line labeled M in Fig. 1 shows the confidence interval extending from 0 to 60 000 ppm SPP. Each line is labeled with the identifying letter of the laboratory. A plus (+) marks the estimated LC_{50} within each confidence interval. In Fig. 1, note that the two intervals for each laboratory are fairly close together, while intervals for different laboratories are farther apart. These data show that for mud No. 1, inter-laboratory variation accounts for an estimated 84·5% of total variation of LC_{50} (Table 4).

Figure 1 also shows industry's ability to comply with the Gulf of Mexico permit limitations of 30 000 ppm SPP. EPA set this toxicity limitation based on the results of one sample of KCl mud tested once at a single laboratory. This test showed a mean LC_{50} of 33 000 ppm SPP with a lower 95% confidence limit of 30 000 ppm SPP. EPA chose 30 000 as the toxicity limit because they wanted to allow for variability in this mud type. The amount of variation accounted for by this confidence interval approximately represents a coefficient of variation of only about 9%, or a 95% confidence multiplier of about 1·1. This confidence limit does not include variability either within or between laboratories, between replicate samples of the same KCl mud, or even for KCl muds prepared by different companies. As shown in Fig. 1, only four out of 23 of these test results would pass the 30 000 limitation (or 17%), even though the mud sample was prepared using the same formulation as used for EPA's test. These results show that EPA regulators underestimated the variability inherent in this test.

The pattern is somewhat different for mud No. 8 (Fig. 2 and Table 4). Here, while lab-to-lab differences are similar to mud No. 1, there is considerably more intra-laboratory variation. Note the results in Fig. 2 for laboratories A, J and P. Because of these three large within-laboratory variations, for mud No. 8 inter-laboratory variation accounts for only an estimated 23·5% of total variation of $\ln(LC_{50})$. These data show that for the most part, intra-laboratory variation, as represented by the ratio of highest to lowest LC_{50} (H/L), ranged from 1 to 2·2, even though a broad range in experience levels is represented by the laboratories. These findings agree remarkably well with the intra-laboratory H/L ratio of 2 reported for multiple tests conducted on the same sample of mud by EPA Gulf Breeze, Florida laboratory (Parrish & Duke, 1985, 1986). The agreement between these two estimates suggests that variation between the samples of each mud type used in this round-robin was minimal. Only two of the

TABLE 4

Statistics variability of reference toxicant and drilling fluids variability estimates for lower 95% confidence interval of the LC_{50} estimate

| | Reference toxicant | | | Drilling fluid | | |
| | OOC prep. | In-house preps. | | KCl (No. 1) | CLS (No. 8) w/2% M Oil | |
	OOC-RR	EPA(B&E)	OOC-RR	OOC-RR	EPA(B&E)	OOC-RR
Variance Estimates						
Intra-laboratory	0·001 6	0·015 1	0·002 6	0·050 2	0·120	0·764 3
Inter-laboratory	0·242 9	0·349 6	0·190 3	0·274 1	0·066	0·235 3
Total	0·244 5	0·364 7	0·192 9	0·324 3	0·186	0·999 6
% Due to inter-Lab variability	99·3	95·9	98·6	84·5	35·5	23·5
High/low	3·4	3·8	222	6·3	42·8	40·1
CV (%)	52·6	66·3	46·1	61·9	45·2	131·0
95% confidence multipliers						
95% CM1	2·64	3·27	2·36	3·05	2·33	7·10
95% CM2	8·39	13·44	2·93	3·50	2·87	9·03
95% CM3	20·26	39·44	4·57	5·89	5·62	22·47
Whole mud solids (% vol. retort)	NA	NA	NA	24	NA	37
Oil (% vol. retort)	0	0	0	0	2	2

TABLE 5
Comparison of inter-laboratory H/L ratios by level of experience

Experience with	No. years	No. labs	No. tests	H/L ratio for generic muds		Laboratory groupings
				No. 1	No. 8 w/oil	
Drilling fluid	<1	6	11	5·8	20·3	B, C, K, M, P, Q
Toxicity test	≥1	6	12	5·3	40·1	A, E, F, H, J, L
Mysid toxicity	<1	5	10	5·1	20·3	B, K, M, P, Q
tests	≥1	7	13	6·3	40·1	A, C, E, F, H, J, L
Other toxicity	<5	6	12	5·1	20·3	B, H, K, M, P, Q
tests	≥5	6	11	6·3	40·1	A, C, E, F, J, L

thirteen laboratories reporting test results in this study had intra-laboratory H/L ratios which fell outside this range (3·6 and 40). Both of these were for the lignosulfonate mud containing 2% oil. These two laboratories had in excess of seven and nine years experience, respectively.

The effect of three types of experience on the intra-laboratory H/L ratios were examined by splitting the laboratories into two groups of nearly equal size, based on the number of years a laboratory had conducted a toxicity test on a commercial basis:

(1) experience with the drilling fluid toxicity test (<1 and ≥1 year);
(2) experience with mysid toxicity tests (<5 and ≥5 years);
(3) experience with toxicity testing in general (<5 and ≥5 years).

The laboratories comprising each group were similar, whether the groups were based on experience with the drilling fluid test, mysid toxicity tests, or other toxicity tests.

Experience level had little or no effect on inter-laboratory H/L ratios for either the KCl mud or the lignosulfonate mud containing oil (Table 5). Note the large differences in the inter-laboratory H/L ratio for the two types of mud. The larger inter-laboratory H/L ratios calculated for the lignosulfonate mud containing oil (20–40) probably resulted from the presence of oil and increased solids content of this mud (the effect of oil and solids cannot be separated in this study). The presence of oil content may also be responsible for the apparent increase in the H/L ratio for the lignosulfonate mud tested at the more experienced laboratories, which is not thought to be meaningful.

In Bailey & Eynon's (1986) study, the small variances which were measured reflected the way the laboratories were chosen, and the fact that they were given the results of the EPA range finding test. The selection of test concentrations for the definitive toxicity test can significantly affect the calculated LC_{50} value (Rue *et al.*, in press). The raw data were supplied to EPA so that they could calculate all LC_{50}s using the same method. Our results indicate that their variance estimates are not representative of laboratory performance likely to be encountered in practice (for example, when selecting from commercially available laboratories).

Results for their study and our study can be compared in Table 3 for generic mud No. 8 and the reference toxicants. Variance estimates for $\ln(LC_{50})$ for mud No. 8 in this study are 0·764 3 and 0·235 3, compared to 0·120 and 0·066 (Bailey & Eynon, 1986), for intra- and inter-laboratory variances, respectively. In other words, we found over six times as much intra-laboratory variability and over three times as much inter-laboratory variability as did Bailey & Eynon.

With these more realistic variance estimates, we found 95% confidence multipliers much greater than we found with Bailey & Eynon's data. The 95% confidence multiplier calculated by Bailey & Eynon (CM1 in Table 4), even if it were the correct form, was too small by a factor of three in comparison to the results for this study (2·33 versus 7·1 respectively). If it is not the right form as we suggest, then it is too small by a factor of ten (2·33 versus 22·47, respectively). This is not a difference to be taken lightly when it may lead to toxicity limitations that are unrealistically too stringent.

The variability shown in LC_{50}s is considerably less for generic mud No. 1 (total variance 0·324 3) than for generic mud No. 8 (total variance 0·999 6). However, this difference is due mostly to the much smaller intra-laboratory variance for mud No. 1 (0·050 2 compared to 0·764 3); inter-laboratory variances are about the same for both muds (0·274 1 compared to 0·235 3). Inter-laboratory variability found in this study is in the same range for both muds and the reference toxicants. This suggests the possibility that inter-laboratory variability in LC_{50} may stay at about the same level for all muds, while intra-laboratory variability differs with different muds.

Analyses of results for LC_{50}s of reference toxicants are shown in Fig. 3 and Table 4. Reference toxicant LC_{50} was reported with each mud LC_{50}, but the reference toxicant was not actually assayed each time a mud was assayed. Apparently the reference toxicant was assayed every few days or once a week, and the LC_{50} of the most recent determination

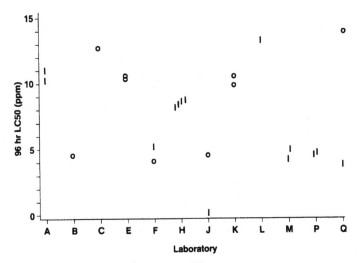

FIG. 3. Estimated LC_{50}s of reference toxicants are plotted with letters corresponding to laboratories. I = in-house reference toxicant; O = OOC supplied reference toxicant.

was recorded whenever a mud was assayed. In order to avoid multiple appearances of the same reference toxicant assay, distinct reference toxicant LC_{50}s for each laboratory were used. Further, there were different reference toxicants used. Some laboratories used the OOC-supplied reference toxicant and some prepared their own. Even so, note in Table 4 and Fig. 3 that variability of reference toxicant LC_{50} is similar for in-house and OOC preparations in this study, and also similar to, though somewhat less than variability found in reference toxicant assays by Bailey & Eynon (1986). In each case, inter-laboratory variability is very large (95·9–99·3% of total variance) in comparison to intra-laboratory variability.

The results from this study suggest that toxicity test variability increases with either an increase in oil content (Parrish & Duke, 1986) or solids content (see Table 4). This increase is almost entirely due to the large increases measured for the intra-laboratory variance component. Note that for the reference toxicant (no solids), the intra-laboratory variance was 0·001 6 and 0·002 6. As the solids content increased to 24% (volume) in the KCl mud (No. 1), the variance component increased to 0·050 2. With further increases of solids to 37% (volume) and the addition of 2% mineral oil (volume) to mud No. 8, the intra-laboratory variance component increased to 0·764 3. The effects of oil and solids cannot be separated from one another in the lignosulfonate mud.

Interestingly, although we see increases in the intra-laboratory variance, there is no corresponding increase in the inter-laboratory variance. It remains fairly constant for both the reference toxicant and each of the muds. A similar trend is seen in the Bailey & Eynon (1986) data. The inter-laboratory variance component, as a percentage of the total variance, was similar for generic mud No. 8 containing 2% mineral oil in both the Bailey & Eynon study and this study (35·5 and 23·5%).

Together, the data for both the standard toxicant and the muds suggest that extraction of the SPP for toxicity testing may be a factor in increasing intra-laboratory variability as the solids and oil concentration of the whole mud sample increase. However, they have no apparent effect on inter-laboratory variability. This is important because it suggests that even if a better test is developed to get around the problems with SPP preparation, the total variance will still be large due to the inter-laboratory variance component.

3.3. Conventional Pollutants

This study also addressed variability in measuring possible conventional pollutants in drilling fluids. These include: suspended particulate characteristics—dissolved solids, suspended particulate phase (DSOL-SPP): suspended solids, suspended particulate phase (SSOL-SPP); biological oxygen demand (BOD-mud); total organic carbon (TOC-mud); and total suspended solids (TSS-mud). (Although total solids in SPP were reported, we have not analysed them because only two laboratories measured total solids independently. All the other laboratories simply added dissolved solids and suspended solids to obtain total solids, contrary to instructions.) Results for these analyses are shown in Table 5.

In Table 6, the large values of the coefficient of variation (CV) and proportion of total variation due to inter-laboratory variation indicate that these mud characteristics are not measured precisely and that most of the imprecision is due to lab-to-lab differences. However, most of the large inter-laboratory variance is, in many cases, due to values from one or two laboratories that are wildly different from the others. For example, both of laboratory B's values for SSOL-SPP are about ten times the values obtained from the other laboratories (Fig. 4). Similar anomalies were encountered with most of the other measurements. The complete data for the bioassays and mud properties are presented in Tables 7 and 8.

TABLE 6
Statistics on generic muds No. 1 and No. 8 for DSOL, SSOL, BOD, TOC, TSS
for the original data

Measure of variability	Results for mud No. 1				
	DSOL	SSOL	BOD	TOC	TSS
CV	70·6%	183·5%	129·9%	274·5%	62·6%
Variance:					
Intra-laboratory	0·008 2	0·035 4	0·006 4	0·356 9	0·023 9
Inter-laboratory	0·395 8	1·438 4	0·982 3	1·787 5	0·306 8
Total	0·404 0	1·473 9	0·988 8	2·144 3	0·330 7
% Due to laboratories	98·0%	97·6%	99·3%	83·4%	92·8%

Measure of variability	Results for mud No. 8				
	DSOL	SSOL	BOD	TOC	TSS
CV	86·7%	106·9%	206·4%	136·0%	66·6%
Variance:					
Intra-laboratory	0·014 4	0·043 2	0·033 3	0·277 5	0·022 6
Inter-laboratory	0·546 0	0·719 1	1·626 8	0·769 8	0·344 3
Total	0·560 4	0·762 3	1·660 0	1·047 3	0·366 8
% Due to laboratories	97·4%	94·3%	98·0%	73·5%	93·9%

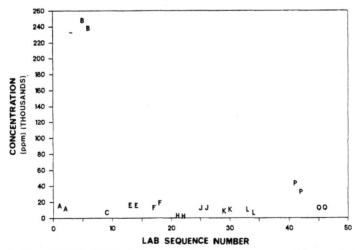

FIG. 4. Plot of SSOL-SPP by laboratory, showing the extreme difference between measurements by laboratory **B** and all others.

TABLE 7
Mortality data from the round-robin toxicity test

Lab	Mud type	Calculation method	Control mortality correction	No. of mysids	96 h LC_{50} (ppm SPP)	Lower 95% CI	Upper 95% CI	Std. Tox LC_{50} (ppm)
A	1	MOVAVG	Y	60	23 000	17 000	30 000	10·80
A	1	PROBIT	Y	60	25 000	17 000	34 000	10·10
A	8	MOVAVG	Y	60	19 000	15 000	24 000	10·10
A	8	MOVAVG	Y	60	35 000	28 000	44 000	10·80
B	1	BINOM	Y	60	34 500	30 000	50 000	4·60
B	1	MOVAVG	Y	60	34 500	32 400	36 400	4·60
B	8	BINOM	Y	60	9 500	8 300	11 500	4·60
B	8	PROBIT	Y	60	8 600	7 300	11 200	4·60
C	1	—		0	ND	ND	ND	ND
C	1	MOVAVG	N	52	6 254	5 072	7 684	12·80
C	8	—		0	ND	ND	ND	ND
C	8	MOVAVG	N	60	9 150	7 405	11 065	12·80
E	1	MOVAVG	Y	60	20 683	18 884	22 868	10·60
E	1	PROBIT	Y	60	24 203	22 260	26 301	10·60
E	8	MOVAVG	Y	60	13 786	10 701	17 720	10·40
E	8	MOVAVG	Y	60	9 779	7 344	13 518	10·40
F	1	MOVAVG	N	60	17 800	16 000	19 700	5·24
F	1	MOVAVG	N	60	17 700	15 300	20 300	5·24
F	8	MOVAVG	N	60	3 100	2 200	4 100	4·10
F	8	MOVAVG	N	60	11 000	800	13 000	4·10
H	1	MOVAVG	N	60	20 050	17 595	22 738	8·40
H	1	MOVAVG	N	60	23 030	19 250	27 687	8·70
H	8	PROBIT	N	60	20 637	15 321	26 481	8·30
H	8	PROBIT	N	60	27 484	23 058	31 985	8·50
J	1	MOVAVG	Y	60	7 387	5 913	9 100	4·65
J	1	MOVAVG	Y	60	12 637	10 569	15 149	4·65
J	8	MOVAVG	Y	60	1 323	714	2 460	0·06
J	8	MOVAVG	Y	60	53 049	48 806	57 701	4·65
K	1	PROBIT	N	60	7 340	9 538	5 771	10·70
K	1	PROBIT	N	60	7 079	5 525	9 254	10·70
K	8	PROBIT	N	60	2 211	1 510	2 958	9·90
K	8	PROBIT	N	60	2 227	1 510	2 995	9·90
L	1	MOVAVG	N	60	27 948	24 918	31 375	13·31
L	1	MOVAVG	N	60	39 328	37 733	41 146	13·31
L	8	MOVAVG	N	60	7 877	6 355	9 799	13·31
L	8	MOVAVG	N	60	4 646	3 566	6 196	13·31
M	1	BINOM	N	60	10 000	0	60 000	4·90
M	1	MOVAVG	N	60	20 000	9 070	32 200	4·20
M	8	PROBIT	N	60	44 900	6 000	85 000	4·90

TABLE 7—*contd.*

Lab	Mud type	Calculation method	Control mortality correction	No. of mysids	96 h LC₅₀ (ppm SPP)	Lower 95% CI	Upper 95% CI	Std. Tox LC₅₀ (ppm)
M	8	PROBIT	N	60	35 600	26 600	43 600	4·20
P	1	MOVAVG	Y	60	36 200	31 300	40 600	4·60
P	1	MOVAVG	Y	60	24 800	18 100	33 400	4·60
P	8	MOVAVG	Y	60	26 000	24 000	29 000	4·70
P	8	MOVAVG	Y	60	12 000	9 800	14 000	4·70
Q	1	PROBIT	Y	60	27 917	26 595	29 242	13·99
Q	1	PROBIT	Y	60	23 949	20 511	30 050	13·99
Q	8	MOVAVG	Y	60	14 267	11 596	17 275	3·80
Q	8	MOVAVG	Y	60	10 954	7 420	14 655	3·80

ND = not determined.

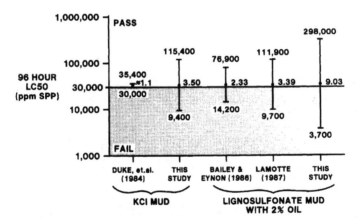

FIG. 5. Comparison of variability results between studies and in relationship to the US compliance limit of 30 000 ppm suspended particulate phase.

4. CONCLUSIONS

Over one-half of the laboratories that conduct the drilling fluid toxicity tests on a commercial basis were included in this study. These laboratories represented a broad range of experience levels since they were randomly selected from the entire population of available laboratories. All of the laboratories, however, encountered problems with conducting and reporting toxicity test results, regardless of their level of experience. Some

TABLE 8
Mud characteristics analyses from the OOC round-robin toxicity test

Lab	Mud type	Suspended particulate phase				Whole mud			
		Diss. solids (mg/litre)	Suspd. solids (mg/litre)	Tot solids (mg/litre)	BOD (mg/litre)	TOC No. 1 (mg/litre)	TOC[a] No. 2 (mg/litre)	TSS No. 1 (mg/litre)	TSS[a] No. 2 (mg/litre)
A	1	35 800	15 050	50 850	2 550	11 200	ND	565 000	ND
A	1	36 075	11 400	47 475	2 000	9 900	ND	564 000	ND
A	8	31 300	16 800	48 100	2 200	39 100	ND	700 000	ND
A	8	23 975	18 375	42 350	2 390	49 700	ND	701 000	ND
B	1	21 400	ND	ND	2 400	6 350	ND	820 000	ND
B	1	15 200	ND	ND	2 310	6 250	ND	800 000	ND
B	8	24 000	220 800	244 800	4 770	50 800	ND	1 260 000	ND
B	8	17 000	250 000	267 000	7 280	47 000	ND	1 350 000	ND
C	1	ND	6 550	ND	10 800	11 000	ND	610 167	ND
C	1	30 067	ND	ND	8 100	11 000	ND	651 667	ND
C	8	ND	ND	ND	ND	ND	ND	847 000	ND
C	8	22 228	11 843	ND	12 900	11 500	ND	1 128 500	ND
E	1	32 900	15 400	48 300	ND	ND	ND	ND	ND
E	1	34 300	15 000	49 300	ND	ND	ND	ND	ND
E	8	20 400	21 500	41 900	ND	ND	ND	ND	ND
E	8	21 000	14 900	35 900	ND	ND	ND	ND	ND
F	1	29 760	18 092	47 852	ND	ND	ND	ND	ND
F	1	34 745	12 047	46 792	ND	ND	ND	ND	ND
F	8	20 368	17 456	37 824	ND	ND	ND	ND	ND
F	8	18 667	26 589	45 256	ND	ND	ND	ND	ND
H	1	41 580	1 670	43 250	401	325	ND	301 800	ND
H	1	43 600	2 260	45 860	390	287	ND	342 000	ND
H	8	32 240	55 700	87 940	396	2 880	ND	551 000	ND
H	8	28 880	60 300	89 180	402	2 760	ND	455 000	ND

Lab	N								
J	1	32 640	11 920	44 560	335	35 564	ND	1 641 000	ND
J	1	30 140	11 660	41 800	399	16 128	ND	2 852 000	ND
J	8	14 740	22 470	37 210	327	8 272	ND	2 018 000	ND
J	8	19 150	20 640	39 790	335	48 950	ND	3 444 000	ND
K	1	43 100	9 790	43 531	ND	ND	ND	ND	ND
K	1	45 800	7 917	40 183	ND	ND	ND	ND	ND
K	8	25 400	24 200	33 356	ND	ND	ND	ND	ND
K	8	26 670	23 000	48 180	ND	ND	ND	ND	ND
L	1	39 467	6 180	51 230	1 983	3 600	ND	570 000	ND
L	1	39 167	10 150	49 740	1 870	3 580	ND	65 000	ND
L	8	34 360	13 240	46 080	5 388	21 820	ND	495 000	ND
L	8	32 240	15 520	46 840	7 294	20 610	ND	592 000	ND
M	1	ND	ND	ND	1 883	7 500	ND	570 000	ND
M	1	ND	ND	ND	1 800	7 500	ND	560 000	ND
M	8	ND	ND	ND	2 881	41 600	ND	690 000	ND
M	8	ND	ND	ND	2 530	52 500	ND	666 000	ND
P	1	228 870	32 240	261 110	4 440	33 519	33 285	1 283 400	1 280 700
P	1	216 900	42 770	259 670	4 440	42 619	41 686	1 017 100	1 018 300
P	8	236 420	9 180	245 600	1 620	8 974	7 691	1 836 500	ND
P	8	242 250	18 270	260 520	1 995	10 613	10 114	1 864 600	ND
Q	1	35 740	12 140	47 880	3 818	2 210	ND	472 000	ND
Q	1	33 275	12 312	45 587	3 552	20 660	ND	510 000	ND
Q	8	22 840	19 340	42 180	7 937	19 500	ND	650 000	ND
Q	8	23 775	21 750	45 525	12 248	56 400	ND	640 000	ND

ND = not determined.

[a] Replicate analysis.

even failed to complete the requested tests altogether. As demonstrated by this study, toxicity test results, whether from an experienced or an inexperienced laboratory, need to be carefully reviewed to ensure that the tests were conducted properly. If compliance testing is to be required via regulation, then either the regulatory agencies or industry are going to have to establish laboratory minimal standards and certification programs. This will improve the reliability of data, both for agency use, and for the protection of the permittee.

These results reinforce industry's concern that toxicity test variability alone may cause an operator to violate the 30 000 ppm SPP toxicity limitation specified in the Gulf of Mexico NPDES permit and proposed in the BAT/NSPS Offshore Guidelines (see Fig. 5). Variability for the KCl mud (No. 1) used in this study was some 3–6 times larger than EPA accounted for in setting the 30 000 toxicity limit (CM2 = 3·5, CV = 61·9% in this study versus 1·1 and 9% respectively). Variability for the more commonly used lignosulfonate mud containing mineral oil was even higher (CM2 = 9·03, CV = 131%). Neither intra- nor inter-laboratory variability appeared to be related to the experience level of the laboratories.

This study also examined the variability in measurement of BOD, TOC and TSS in whole mud; and dissolved and suspended solids in the suspended particulate phase. Inter-laboratory variability was high for all of these analyses. Occasional major discrepancies between laboratories occurred rendering some measurements in apparent error by factors ranging from 10 to 100. The utility of analyses such as these for possible use in monitoring compliance in the future is questionable.

ACKNOWLEDGEMENTS

This project was made possible through funding from the Offshore Operators Committee. We would like to acknowledge the assistance provided by M-I Drilling Fluids in preparation of the drilling fluids used for testing.

REFERENCES

APHA (1985). *Standard Methods for the Examination of Water and Wastewater*, 16th edn, ed. A.E. Greenberg, R.R. Trussell, L.S. Clesceri & M.H. Franson. American Public Health Association, Washington, D.C.

API (1985). American Petroleum Institute comments on the Environmental Protection Agencies Gulf of Mexico offshore BAT General NPDES Permit. Administrative Record for General Permit File No. GMG280000, pp. 1983–2082; 2520–2531, EPA, Region VI, Dallas, Texas, USA.

BAILEY, R.C. & EYNON, B.P. (1986). Toxicity testing of drilling fluids: Assessing laboratory performance and variability. Submitted to: Papers on Symposium on Chemical and Biological Characterization of Sludges, Sediments, Dredge Spoils, and Drilling Muds, ASTM.

EPA (1979). *Methods for Chemical Analysis of Water and Wastes*. US EPA Environmental Monitoring and Support Laboratory, Cincinnati, Ohio. EPA-600-4-79-020.

LA MOTTE, L.R. (1987). Comments on variability among toxicity assays. Comments prepared for American Petroleum Institute's 'Permit Modification Request', submitted to Region VI EPA, Dallas, TX, re: Gulf of Mexico General Permit, January 1987.

OOC (1985). Offshore operators Comments on the Environmental Protection Agencies Gulf of Mexico offshore BAT General NPDES Permit. Administrative Record for General Permit File No. GMG280000, pp. 562–1973. EPA, Region VI, Dallas, Texas, USA.

O'REILLY, J.E. (1985). American Petroleum Institute Comments on the Environmental Protection Agencies Gulf of Mexico offshore BAT General NPDES Permit. Administrative Record for General Permit File No. GMG280000, pp. 606–28. EPA, Region VI, Dallas, Texas, USA.

O'REILLY, J.E. (1986a). Variability in drilling fluid toxicity tests. Comments submitted for the BAT-NSPS Guidelines. EPA Headquarters, Washington, D.C. (Also submitted as addendum to O'Reilly, 1986b.)

O'REILLY, J.E. (1986b). Variability in drilling fluid toxicity tests. A reply to EPA's response to industry's comments. In Request of the American Petroleum Institute *et al.* to Modify Permit Terms, Conditions, and Effluent Limitations, before U.S. Environmental Protection Agency, Regions IV and VI, April, 1987.

O'REILLY, J.E. & LAMOTTE, L.R. (1987). Variability in drilling fluid toxicity test results. *Proceedings of the 10th Annual Analytical Symposium, US Environmental Protection Agency, Office of Water*. May 13–14, 1987, Norfolk, VA, pp. 713–51.

PARRISH, P.R. & DUKE, T.W. (1985). Acute toxicity of a laboratory-prepared generic drilling fluid to mysids (*Mysidopsis bahia*), and evaluation of test results from ten commercial laboratories. US Environmental Protection Agency, Environmental Research Laboratory, Sabine Island, Gulf Breeze, Fl. EPA/600/X-85/388.

PARRISH, P.R. & DUKE, T.W. (1986). Variability of the acute toxicity of drilling fluids to mysids (*Mysidopsis bahia*). To be published in *Proceedings of the Symposium on Chemical and Biological Characterization of Municipal Sludges, Sediment, Dredge Spoils, and Drilling Muds*. ASTM, Philadelphia, PA, 19103.

RUE, W.J., FAVA, J.A. & GROTHE, D.R. (in press). A review of inter- and intra-laboratory effluent toxicity test method variability. To be published in ASTM's *Aquatic Toxicology and Hazard Assessment: 10th Symposium*.

SAS (1986). *SAS user's guide: Statistics*. 1986 edn, SAS Institute Inc., Cary, North Carolina.

37

The EPA/API Diesel Pill Monitoring Program

R. C. AYERS, JR, J. E. O'REILLY

Exxon Production Research Company, PO Box 2189, Houston, Texas 77252-2189, USA

W. A. TELLIARD, D. C. RUDDY

US Environmental Protection Agency, Industrial Technology Division, 401 M Street, S.W., Washington, DC 20460, USA

D. J. WEINTRITT

Weintritt Testing Laboratories, Inc., PO Box 30162, Lafayette, Louisiana 70503, USA

and

T. H. FIELDSEND

Data Technology Research, Inc., 301 Route 17N, Rutherford, New Jersey 07070, USA

ABSTRACT

In the Gulf of Mexico General NPDES Permit issued in July 1986, the discharge of mud to which diesel had been added was prohibited unless, (1) the oil was added as a pill to free stuck drill pipe, (2) the pill plus at least 50 bbl of drilling fluid on either side of the pill was removed from the active mud system and not discharged, and (3) the operator participated in the Diesel Pill Monitoring Program (DPMP). The objective of the DPMP was to evaluate the effectiveness of diesel recovery practices after a diesel pill had been spotted to free stuck pipe. This involved measuring the diesel content and toxicity of mud samples taken before and after pill recovery.

In this study mean and median recovery levels for diesel were found to be 76·5% and 83·0%, respectively. Mud toxicity was observed to be a strong *function of diesel content, especially at low diesel concentrations. Diesel content appeared to have a much greater influence on LC_{50} than difference in basic mud type. The overall success rate for freeing stuck pipe, based on data from the first pill per incident, was 40%. Success rates in muds with densities less than 12 ppg were considerably better than those in muds with densities greater than 12 ppg.*

1. INTRODUCTION

Differential pressure sticking is a major drilling problem. To maintain well control, the drilling mud density is adjusted such that the pressure in the wellbore exceeds the formation pressure. Under certain conditions, the differential pressure seals the drillstring (usually the drill collars) against a permeable zone in the wellbore (see Fig. 1). If the pipe cannot be mechanically jarred free, the most common remedy is to spot a pill. This entails pumping a certain volume or 'pill' of oil-based mud down the drillstring and up the annulus to the point where the pipe is stuck so that the oil mud remains in contact with the filter cake. The oil weakens the bond between the pipe and filter cake and after a few hours the pipe may be pulled free. After the pipe is free, the oil mud pill and some of the oil contaminated mud surrounding it can be removed from the mud system and set aside for disposal.

If the oil mud pill is unsuccessful, a second or third pill may be tried. If the pipe is still stuck, the operator's only recourse is to leave the stuck drillstring in the hole and sidetrack (drill) around it.

Most operators prefer diesel for the oil mud pill. A recent industry

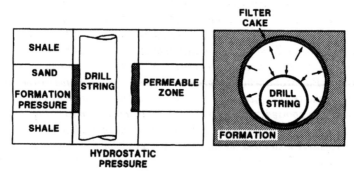

FIG. 1. The differential pressure sticking phenomenon.

survey covering 506 stuck pipe incidents in the Gulf of Mexico during the period 1983–6, reported that diesel pills were used more often and were more effective in freeing the pipe than mineral oil pills (Ayers *et al.*, 1987).

The preference for diesel has generated some controversy between the EPA and Industry. With current pill recovery techniques, some residual oil will remain in the mud system. The EPA has encouraged the use of mineral oil pills over diesel pills because muds containing mineral oils are normally less toxic than muds containing diesel oils at the same oil content (Bretler *et al.*, 1984; Duke *et al.*, 1984). Furthermore, there is still some disagreement on whether or not diesel is really superior to mineral oil for freeing stuck pipe.

1.1. Program Background

The diesel Pill Monitoring Program (DPMP) was a jointly funded effort by the EPA's Industrial Technology Division, the American Petroleum Institute (API), and Gulf of Mexico permittees to investigate the practice of recovering diesel pills. The program involved the collection and analysis of samples from active mud systems prior to use and after removal of diesel pills. A primary purpose of the DPMP was to provide a mechanism to collect data for consideration in developing waste discharge regulations for US offshore oil and gas operations.

The DPMP was implemented as part of EPA Region IV and VI's General NPDES Permit for Oil and Gas Operations in the Gulf of Mexico (US EPA Permit No. GMG 280000, 1986*a*) that became effective on July 2, 1986. The Regions recognized that there was minimal information on pill recovery and determined that an opportunity should be made to demonstrate whether a diesel pill can be effectively removed from a mud system.

The permit established conditions for the discharge of drilling muds, in addition to other waste streams, that included a prohibition on the discharge of oil-based muds, free oil as evidenced by a sheen on the receiving water, and limitations on diesel oil. A toxicity limitation on muds was also established at a minimum 96-h LC_{50} value of 30 000 ppm (for *Mysidopsis bahia*) on the suspended particulate phase, based on monthly and end-of-well compliance monitoring.

The permit prohibited the discharge of mud to which diesel was added unless:

1. The diesel was added as a pill in an attempt to free stuck pipe,
2. The diesel pill and at least 50 bbl of drilling fluid on either side were

removed from the active drilling fluid system and not discharged to the waters of the United States, and

3. Samples of the drilling fluid after pill removal and other additional data were provided to the EPA in accordance with the Diesel Pill Monitoring Program (DPMP).

For participants in the DPMP, compliance with the permit's end-of-well toxicity limitation was demonstrated by analysing the mud samples taken just prior to the introduction of the pill.

If a permittee used a diesel pill and did not participate in the DPMP, then all waste mud generated after introduction of the pill had to be hauled ashore for disposal. Also, if a permittee used a mineral oil pill, he could continue to discharge provided the discharged mud met the other permit limitations (e.g. toxicity limitation and no free oil).

1.2. Elements of the Program

The program participants and their functions are illustrated in Fig. 2. The Oversight Committee, composed of representatives from the EPA, Industry, and environmental organizations, developed and managed the program and had the responsibility for preparing the final documentation.

Permittees stored special sampling kits on the platforms. When a sticking event occurred, samples of the pill, the diesel used in the pill, and the mud before pill addition and after pill recovery were taken and shipped under ice to the Central Control Laboratory.

The Central Control Laboratory received the samples and field data, distributed subsamples to subcontractors for bioassay and chemical analyses, compiled the test results, and issued reports to the participating permittees, the EPA Sample Control Center, and the Oversight Committee. The Central Control Laboratory also archived subsamples and sold sampling kits to the permittees.

The EPA's Sample Control Center tracked the program status, maintained a computerized data management system and compiled data reports for the Oversight Committee. The EPA's Environmental Research Laboratory in Gulf Breeze, Florida conducted toxicity tests for quality review purposes.

2. METHODS

The mud and pill samples were tested by standard API RP 13B procedures for rheology, pH, and oil and water content by 10 ml retort. The retort

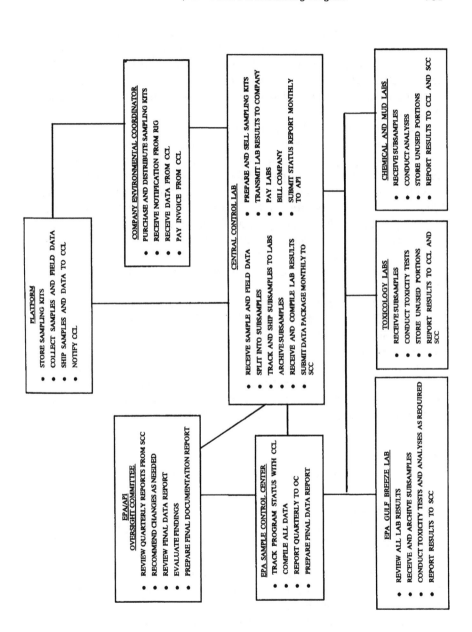

PLATFORM
- STORE SAMPLING KITS
- COLLECT SAMPLES AND FIELD DATA
- SHIP SAMPLES AND DATA TO CCL
- NOTIFY CCL

COMPANY ENVIRONMENTAL COORDINATOR
- PURCHASE AND DISTRIBUTE SAMPLING KITS
- RECEIVE NOTIFICATION FROM RIG
- RECEIVE DATA FROM CCL
- PAY INVOICE FROM CCL

CENTRAL CONTROL LAB
- RECEIVE SAMPLE AND FIELD DATA
- SPLIT INTO SUBSAMPLES
- TRACK AND SHIP SUBSAMPLES TO LABS
- ARCHIVE SUBSAMPLES
- RECEIVE AND COMPILE LAB RESULTS
- SUBMIT DATA PACKAGE MONTHLY TO SCC
- PREPARE AND SELL SAMPLING KITS
- TRANSMIT LAB RESULTS TO COMPANY
- PAY LABS
- BILL COMPANY
- SUBMIT STATUS REPORT MONTHLY TO API

CHEMICAL AND MUD LABS
- RECEIVE SUBSAMPLES
- CONDUCT ANALYSES
- STORE UNUSED PORTIONS
- REPORT RESULTS TO CCL AND SCC

TOXICOLOGY LABS
- RECEIVE SUBSAMPLES
- CONDUCT TOXICITY TESTS
- STORE UNUSED PORTIONS
- REPORT RESULTS TO CCL AND SCC

EPA/API OVERSIGHT COMMITTEE
- REVIEW QUARTERLY REPORTS FROM SCC
- RECOMMEND CHANGES AS NEEDED
- REVIEW FINAL DATA REPORT
- EVALUATE FINDINGS
- PREPARE FINAL DOCUMENTATION REPORT

EPA SAMPLE CONTROL CENTER
- TRACK PROGRAM STATUS WITH CCL
- COMPILE ALL DATA
- REPORT QUARTERLY TO OC
- PREPARE FINAL DATA REPORT

EPA GULF BREEZE LAB
- REVIEW ALL LAB RESULTS
- RECEIVE AND ARCHIVE SUBSAMPLES
- CONDUCT TOXICITY TESTS AND ANALYSES AS REQUIRED
- REPORT RESULTS TO SCC

procedure was modified slightly, in that solvent extracted glass wool was used instead of the oil treated steel wool supplied by the manufacturer. Total oil contents in the mud samples were also determined gravimetrically using a 20 ml retort. The retorted oil was extracted from the liquid phase with methylene chloride, and weighed after evaporation of the methylene chloride.

Diesel was determined by gas chromatography (GC) using a method described in the DPMP Program Manual (USEPA, 1986a). (The diesel method is currently under review by the EPA and Industry. It has been proposed by the Agency as part of the effluent limitations guidelines for US offshore oil and gas operations.) In brief, oil from the 20 ml retort was diluted 1:1 with a 1% solution of 1,3,5-trichlorobenzene (internal standard) in methylene chloride and analysed by GC using a 30 m capillary column, a nitrogen carrier and flame ionization detection. To estimate the diesel content, the chromatogram would be compared with one obtained on the diesel oil used in the pill, using a technique described in the DPMP Manual.

The drilling fluid bioassay tests were conducted according to the EPA's proposed Drilling Fluids Toxicity Test (50 FR 34592, August 26, 1985). In this test, acute toxicity is determined on the suspended particulate phase by exposure of *Mysidopsis bahia* to the phase for 96 h.

The EPA collected additional data on the levels of priority pollutant organics, metals, and conventional pollutants in some sampled muds. Analyses and conclusions derived from these data are beyond the scope of this paper, but are provided in the final DPMP documentation (USEPA, 1988).

3. RESULTS AND DISCUSSION

3.1. Data Bases
During the period the DPMP was in effect, 105 sampling kits were submitted to the program, representing 105 pills spotted in 56 wells. However, not all the data collected were used for evaluating the results of the program. Table 1 identifies the datasets that were considered valid for each dataset's specified use. Generally, we tried to use as much valid data as possible for each type of correlation.

As seen in Table 1, Dataset No. 1 was used for examining relationships between diesel concentration and toxicity and between methods used to measure total oil content and diesel content. This dataset included all the

TABLE 1
Description of DPMP datasets

	Dataset 1[a]	Dataset 2[b]	Dataset 3[c]
Used for	Correlations of toxicity and oil content with diesel content	Calculation of frequency of success for freeing stuck pipe	Diesel recovery calculations
Includes:	All valid data collected while program was in effect	All valid first pill per stuck pipe incident data	All valid diesel recovery data in Dataset 2
No. of kits	100	70	58
No. of wells	55	55	48

The overall DPMP dataset (105 kits used on 56 wells) consists of all data collected while the DPMP was in effect under the Gulf of Mexico Permit (7/2/86–9/30/87). Each of the three datasets above were derived from the overall DPMP dataset by deleting kits which were not considered 'valid' for the dataset's specified use. The term 'valid' is defined as data relevant to the correlation for which we had no known reason for discarding it. Deletions from the overall DPMP for each of the three datasets are given below:
[a]Deleted: 3 kits where mineral oil was spotted, 1 kit where mud sampled improperly, and 1 kit where samples were mislabeled.
[b]Deleted: 3 kits where mineral oil spotted, 29 kits as multiple pills per incident, 1 kit where pipe was never stuck (all mud and cuttings hauled), 1 kit which spotted an unacceptably low pill volume (10 bbls) and 1 kit which reused a pill a second time (only 10% oil as a result of mixing with mud).
[c]Deleted: same kits as for Dataset 2, plus 6 kits where part or all of pill was either lost to formation or cemented in the hole, 3 kits where samples were mislabeled, 2 kits for which recovery could not be calculated.

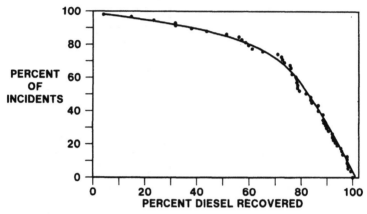

FIG. 3. Percentage of incidents with diesel recovery exceeding a given value: Dataset 3.

information collected except for three sampling kits later found to have involved mineral oil pills and two kits for which samples were found to be mislabeled.

Dataset No. 2 was used in calculating success rates for freeing stuck pipe. In this dataset, only the first pill per stuck pipe incident was considered; that is, if multiple pills were used on a single incident, only the first pill was considered in success rate calculations. This approach is the same as that used in the recent industry survey mentioned earlier (Ayers *et al.*, 1987).

Dataset No. 3 was used in correlation with diesel recovery levels. This dataset consisted of Dataset No. 2 with an additional 11 pills (5 wells) deleted for which meaningful recovery calculations could not be performed. For example, cases where all or part of the pill was lost to the formation or was cemented in the hole were dropped from Dataset No. 3.

3.2. Diesel Oil Recovery
Diesel oil recovery was determined from the difference between the amount of diesel oil added to the mud system and the amount of diesel oil remaining in the active system after two complete circulations of the mud system following pill recovery. Figure 3 shows the percentage of incidents where diesel recovery levels exceeded a given value (based on Dataset No. 3). For example, in 75% of the incidents, diesel recovery exceeded 68%. Figure 4 shows the percentage of cases in Dataset No. 3 which had

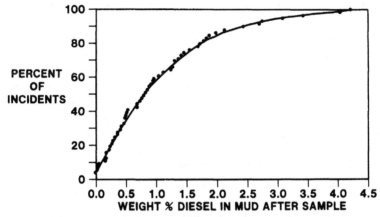

FIG. 4. Percentage of incidents with residual diesel in mud less than a given value: Dataset 3.

TABLE 2

Percentage diesel recovered versus quantity of extra buffer[a] hauled ashore for disposal: Dataset 3

Extra buffer (bbls)	*Number of incidents*	*% Percentage diesel recovered*			
		Mean	*Median*	*Minimum*	*Maximum*
0	11	73·4	77·1	32·1	96·0
0 < bbls < 100	18	75·0	87·8	4·2	100·0
100 < bbls < 200	10	78·0	83·9	44·1	96·2
200 < bbls < 300	13	77·3	82·3	24·0[b]	97·9
bbls ≥ 300	6	82·8	79·5	72·9	98·0
	58	76·5	83·0	4·2	100·0

[a]Volume of extra buffer hauled ashore is equal to: Volume hauled − Volume spotted − 100.
[b]Next lowest value is 61·4.

residual diesel levels in the mud after pill recovery, less than a specified value. For example, in 72% of the cases, residual diesel in the mud after pill recovery was less than 1·5% by weight.

Table 2 shows how diesel recovery varies with extra buffer [Volume Mud Hauled (bbl) − Pill Volume (bbl) − 100]. For the overall program (based on Dataset No. 3), diesel recovery ranged from 4·2% to 100%.

The mean recovery level was 76·5% while the median recovery level was 83%. Increasing buffer size had little or no effect on the mean, median or maximum recoveries. On the other hand, increasing the buffer volume appeared to increase the minimum recovery level (from 32·1 to 72·9% over the entire extra buffer interval).

These results may be explained, in part, by considering that recovery should depend on how much pipe movement occurred prior to circulating out the pill and also on the accuracy of the estimate of when the pill and buffer arrive at the surface during the circulation procedure. If little movement occurs and the recovery is started at the right time, one might expect high recovery levels even if the extra buffer is low. On the other hand, if substantial pipe movement has occurred (which could result in considerable mixing of the pill and surrounding mud) or if the calculation for pill arrival at the surface is wrong (due to an inaccurate estimate of hole diameter), recovery might not be high even with substantial extra buffer. In any case, collecting extra buffer, while not guaranteeing a high recovery, should lessen the chance of getting very poor recovery.

Mud test data (e.g. density, rheological properties) are not shown here, but will be presented in the final program documentation. Generally, for any given sticking incident, the pill density closely matched the mud density. We found no correlation between mud rheological properties and diesel recovery.

3.3. Toxicity versus Diesel Content

Figure 5 shows how mud toxicity depends on diesel content. At low diesel concentrations, mud LC_{50} values decrease rapidly with increasing diesel content. At higher diesel concentrations, mud LC_{50} values decrease less rapidly. Similar trends can be seen in data presented in other studies (Bretler et al., 1984; Duke & Parrish, 1984; ERCO, 1984, 1987) using individual laboratory muds containing varying concentrations of oil. Note that while most of the LC_{50}s of the muds sampled before spotting are higher than those sampled after spotting (the median LC_{50} values of the mud before and mud after samples were 52 000 ppm and 6 000 ppm respectively), some are not. The mud before samples with low LC_{50} values represent muds which already contained diesel (or mineral oil). In most cases, these mud samples were obtained before spotting a second or third pill, after one or two pills had already been spotted.

Figure 5 also shows that there can be a wide spread between LC_{50} values at approximately the same diesel content. We believe these differences are due primarily to (1) toxicity test variability (see Bailey &

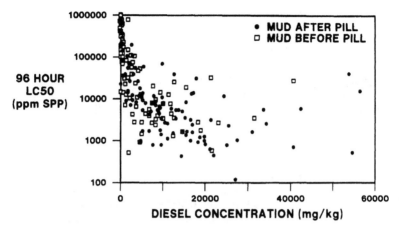

FIG. 5. 96 h LC$_{50}$ versus diesel concentration: Dataset 1.

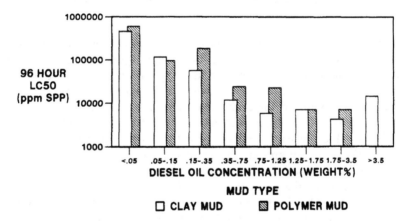

FIG. 6. Mean mud toxicity versus diesel concentration: Dataset 1.

Eynon, 1986; O'Reilly & LaMotte, 1987), (2) the presence or absence of other materials that can significantly affect toxicity (e.g. mineral oils), and (3) differences in diesel composition (either differences in basic composition or differences due to weathering).

Water-based muds may be broadly classified as either clay muds (those that depend on clay for viscosity) or polymer muds (those that depend on a polymer for viscosity). To examine the effect of diesel on the toxicity of different mud types, the DPMP muds were classified as such. Figure 6 shows the mean LC$_{50}$s of the two basic mud types as a function of

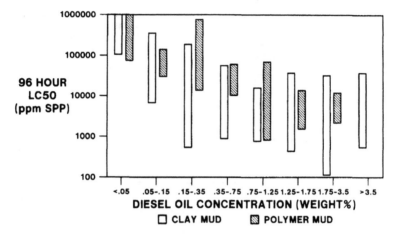

FIG. 7. Mud toxicity range versus diesel concentration: Dataset 1.

approximate diesel content. Note that at very low diesel concentrations the mean LC_{50} values for both basic mud types are greater than 400 000 ppm. Mean LC_{50}s for both mud types decrease in essentially the same way with increasing diesel content. Figure 7 presents the range of LC_{50} values obtained for each basic mud type as a function of diesel concentration. Note that at each diesel concentration interval, the range of LC_{50} values for each mud type greatly exceeds the difference in the mean values. Thus, it appears that diesel content is much more important than mud type in determining mud toxicity.

Figure 8 shows the percentage of mud samples, taken after recovering the first pill, with LC_{50}s greater than a given value (Dataset No. 3 was used in this analysis). For example, based on this dataset, about 39% of the samples had LC_{50}s greater than 10 000 ppm, and 87% of the samples had LC_{50}s greater than 1000 ppm.

3.4. Diesel and Total Oil Content Measurements
Figure 9 shows the correlation between diesel content as measured by GC and total or gravimetric oil (wt % oil determined by methylene chloride extract of 20 ml retort fluids). The diesel concentration ranges from essentially nil to about 55 000 mg/kg. Note that the diesel content averaged about 72% of the total oil content. While the correlation is relatively tight ($R^2 = 0.89$), there was some variation, especially at low diesel concentrations. In some cases the total oil content exceeded the

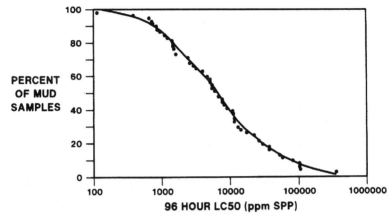

FIG. 8. Percentage of mud samples less toxic than a given value: Dataset 3 (mud after pill recovery).

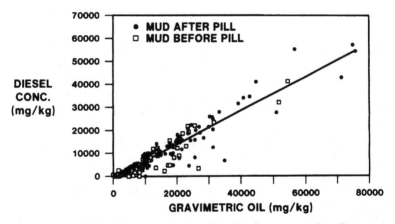

FIG. 9. Diesel concentration versus gravimetric oil concentration: Dataset 1.

diesel content by almost an order of magnitude. This may be reflecting the presence of oils other than diesel in the mud.

Figure 10 shows the correlation between total oil measured gravimetrically and measured as vol. % oil (by retort without extraction). This comparison is made because the vol. % oil by retort is commonly measured by field personnel on the platform. The fitted regression accounts for 90% of the observed variation in these data. A likely source of this variation is the difference in detection limits for these two methods.

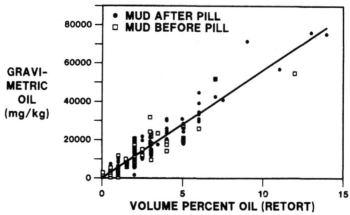

FIG. 10. Gravimetric oil concentration versus volume percentage oil (retort): Dataset 1.

3.5. Pipe Freeing Success Rates

The overall rate of success for freeing stuck drill pipe was 40·0% using Dataset No. 2 (first pill per sticking incident). This determination is based on 28 successes in 70 incidents. Six of the incidents involved stuck casing rather than stuck drill pipe. The pipe was not successfully freed in any of these cases. The success rate for freeing stuck drill pipe was 43·8% (28 successes in 64 incidents).

Good operating practice involves spotting a pill equal in density to the mud density for well control and to prevent gravity migration of the pill away from the interval where the drill pipe is stuck. Generally, the pill density was closely matched to the mud density for each of the sticking incidents in this program.

To examine the effect of density on success rate, we divided the incidents in Dataset No. 2 into two groups based on pill density. Approximately half of the incidents were in the group with pill densities less than 12·0 ppg and the other half were in the greater than 12·0 ppg group. Figure 11 shows that success rate is related to pill (and therefore mud) density. The success rate for those cases where the pill density was less than 12 ppg was 62·5%, while the success rate for those cases where the pill density exceed 12 ppg was only 21%.

There are several reasons why success might decrease as mud density increases. The higher mud density increases the chance that excessive differential pressure might occur. More dense muds usually have a higher

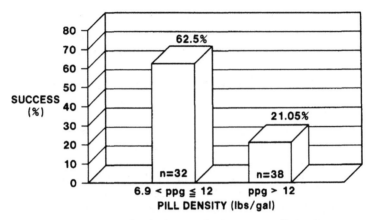

FIG. 11. Percentage success for freeing stuck pipe versus pill density: Dataset 2.

solids content and tend to build thicker filter cakes having more contact area with the drillpipe. Also, the net force required to pull free has been shown to increase with barite content of the mud (Annis & Monaghan, 1962). In addition, higher density pills contain more weighting agent and material to suspend the weighting agent and therefore less diesel. These pills are harder to prepare and in practice may be less stable than lower density pills and less likely to remain intact down hole.

Some other measured pill properties appeared to correlate with success. Diesel content appeared to correlate directly with success while yield point, 10 s and 10 min gel strengths and plastic viscosity exhibited inverse correlations. However, all of these properties had an even stronger direct correlation with mud density. We were unable to ascertain if these additional correlations were anything more than a reflection of the strong inverse relationship between mud density and success discussed above.

4. CONCLUSIONS

● The mean and median recovery levels for diesel in this study were 76·5% and 83·0% respectively, based on Dataset No. 3. Over the range of extra buffer volume considered, increasing buffer volume had little or no effect on mean, median or maximum diesel recovery levels; however, minimum diesel recovery levels appeared to increase with extra buffer.

- Mud toxicity was observed to be a strong function of diesel content, especially at low diesel concentrations. Diesel content appeared to have a much greater influence on LC_{50} than difference in basic mud type.
- The difference in LC_{50} values for muds with approximately the same diesel content ranged from nil to more than an order of magnitude.
- The correlation of diesel content with total oil content (retort-gravimetric) was fairly good, though there was some variation, especially at low diesel contents. Measurements in total oil as determined by retort agreed well with those obtained gravimetrically.
- The overall success rate for freeing stuck pipe based on data from the first pill per incident was 40%. Success rates in muds with densities less than 12 ppg were considerably better than those in muds with densities greater than 12 ppg.

ACKNOWLEDGEMENTS

The DPMP was funded by the US Environmental Protection Agency, the American Petroleum Institute and participating permittees. The authors would like to acknowledge the contributions of D. D. Caudle of Conoco, and J. P. Ray and T. M. Randolph of Shell, all of whom served on the Oversight Committee. In addition, we would also like to acknowledge the contribution of Peter Cook of Exxon Production Research Company, who carried out many of the calculations used in this paper.

REFERENCES

ANNIS, M.R. & MONAGHAN, P.H. (1962). Differential pressure sticking—laboratory studies, friction between steel and mud filter cake. *J. Petrol. Technol.*, May 1962.

API (1985). Standard Procedure for Testing Drilling Fluids. API RP 13B, May 1, 1985.

AYERS, R.C., JR, O'REILLY, J.E. & HENRY, L.R. (1987). Offshore Operators Committee Gulf of Mexico Spotting Fluid Survey, April 4, 1987 (see Request of the American Petroleum Institute *et al.* to Modify Permit Terms, Conditions and Effluent Limitations, Before US Environmental Protection Agency Regions IV and VI, April 1987).

BAILEY, R.C. & EYNON, B.P. (1986). Toxicity testing of drilling fluids: assessing laboratory performance and variability. Symposium on Chemical and Biological Characterization of Sludges, Sediments, Dredge Spoils, and Drilling Muds, ASTM.

BRETLER, R.J., BOEHM, P.D., NEFF, J.M. & REQUEJO, A.G. (1984). Acute toxicity of drilling muds containing hydrocarbon additives and their fate and partitioning between liquid, suspended, and solid phases. Report prepared for the American Petroleum Institute by Battelle New England Marine Research Laboratory, Duxbury, Mass.

DUKE, T.W. & PARRISH, P.R. (1984). Results of drilling fluids program sponsored by the Gulf Breeze Research Laboratory, 1976–1984, and their application to hazard assessment. US Environmental Protection Agency, Environmental Research Laboratory, Gulf Breeze, FL, EPA-600/4-84-005.

DUKE, T.W., PARRISH, P.R., MONTGOMERY, R.M., MACAULEY, S.D., MACAULEY, J.M. & CRIPE, G.M. (1984). Acute toxicity of eight laboratory-prepared generic drilling fluids to mysids (*Mysidopsis bahia*). US Environmental Protection Agency, Environmental Research Laboratory, Gulf Breeze, Fla., EPA-600/3-84-067.

ERCO (1984). Acute toxicity of the suspended particulate phase of drilling fluids containing diesel fuels. Prepared for the US EPA by ERCO, a Division of Enseco, Cambridge, Mass.

ERCO (1987). Acute toxicity of water based drilling fluids containing mineral oils. Final data report prepared for the Offshore Operators Committee by ERCO, a Division of Enseco, Cambridge, Mass.

O'REILLY, J.E. & LAMOTTE, L.R. (1987). Variability in drilling fluid toxicity test results. 10th Annual EPA Conference on Analysis of Pollutants in the Environment, Norfolk, VA, May 13–14.

US EPA (1985). Drilling fluids toxicity test method, Appendix 3 of Proposed Effluent Limitations Guidelines and New Source Performance Standards for the Offshore Subcategory of the Oil and Gas Extraction Point Source Category. 50 Federal Register 34592, August 26.

US EPA (1986a). *Program Manual*. EPA/API Diesel Pill Monitoring Program, Offshore Oil and Gas Extraction Industry, January, 1986 (available from EPA Region IV and VI).

US EPA (1986b). Permit EPA No. GMG 280000. Final NPDES General Permit for the Outer Continental Shelf (OCS) of the Gulf of Mexico. 51 Federal Register 24897–24927, July 9.

US EPA (1988). EPA/API Diesel Pill Monitoring Program, Final Documentation. 1988 (in preparation).

38

Evaluation of an Organic Chemical Method for Drilling Fluid Determination in Outer Continental Shelf Sediments

T. C. SAUER JR,* J. S. BROWN, A. G. REQUEJO‡ and P. D. BOEHM

Battelle Ocean Sciences, Duxbury, Massachusetts 02332, USA

ABSTRACT

Lignosulfonates, which are the major organic additives in most water-based drilling fluids discharged in the marine environment, are candidate environmental tracers of drilling fluids in marine sediments. A cupric oxide oxidation method, which is specific for the analysis of lignin-derived organic matter (e.g. lignosulfonate), was tested as part of a methodology evaluation of the use of lignosulfonate additives as drilling fluid tracers. Three lignin parameters—vanillyl phenol content, lignin composition and vanillyl acid/aldehyde ratios—were identified as the most suitable tracers of drilling mud discharges. In this study, sediments from various Outer Continental Shelf (OCS) regions were analysed to determine their lignin composition and content. The range of baseline values for the lignin parameters in OCS sediments was evaluated in relation to the potential application of the cupric oxide method in detecting inputs of lignosulfonate-containing drilling fluids in sediments of each region. The results indicate that all three lignin parameters are suitable as drilling fluid tracers, but to different extents, depending on the nature of the existing background. The three parameters exhibit different degrees of sensitivity within the sediments of any given region.

1. INTRODUCTION

Presently, barium sulfate is the predominate fate tracer of drilling fluids discharged into the marine environment. This compound is popular

*Present address: Marine Sciences Unit, Arthur D. Little, Inc., Acorn Park, Cambridge, Massachusetts 02140, USA.
‡Present Address: Exxon Production Research Company, 3120 Buffalo Speedway, Houston, Texas 77252-2189, USA.

because the ore, barite, is the major ingredient in most drilling fluids, the compound is stable in the marine environment, and the element barium (Ba) is relatively easy to measure. The density of barium sulfate ($4.5\,\mathrm{g/cm^3}$) is, however, considerably higher than the other constituents in drilling fluids, such as the low gravity solids (clays, $2.6\,\mathrm{g/cm^3}$) and organic constituents (approximately $1.0\ \mathrm{g/cm^3}$). Because of its different density, the fate of barium sulfate may be different from other compounds and, therefore, may not represent the fate of the other drilling fluid constituents.

Lignosulfonates, which are principal organic additives in water-based drilling fluids and represent up to 5% by weight of the major drilling fluid formulations, are candidates for use as environmental tracers of drilling fluids. Their association with low-gravity solid particles (e.g. bentonite) and other organic constituents in drilling fluids may make them more representative of the fate of these less dense particles in the marine environment.

Until recently, the determination of lignosulfonate in environmental samples, especially marine sediment, was not possible and, therefore, the use of lignosulfonate as a tracer was not previously pursued. In Sauer *et al.* (1988), a method for the quantitative determination of lignosulfonate in drilling muds was presented. Based on the technique of Hedges & Ertel (1982) for determination of lignin residues in geochemical samples, the method involves mild oxidation of lignosulfonate to its component phenolic monomers, which are quantified by capillary gas chromatography. Preliminary results of Sauer *et al.* (1988) indicate that this technique may be suitable for detecting inputs to marine sediments of lignosulfonates derived from drilling fluids. The sensitivity of this application, however, is highly dependent on the background contributed by natural lignin and lignin-derived geopolymers present in the sediments, in terms of both their concentrations and their compositions.

This paper presents the results of a study in which the cupric oxide oxidation method was applied to sediments from various Outer Continental Shelf (OCS) regions to determine their lignin composition and content, as part of a methodology evaluation of the use of lignosulfonate additives as environmental tracers of drilling fluid in marine sediments. The objective of this study was to establish the sedimentary lignin background in areas of ongoing or future offshore oil and gas exploration, in order to assess which geographic regions are most suitable for determination of lignosulfonate by this analytical method. In this process, new geochemical data have been generated on the natural lignin distributions found in OCS sediments. In several instances, these results represent the first attempts to characterize the lignin geochemistry of the OCS.

2. ANALYTICAL METHODS

2.1. Samples

Sediments from six regions of the OCS were obtained from either archival collections or ongoing programs sponsored by the Minerals Management Service of the US Department of the Interior. The six regions were the Georges Bank-North Atlantic OCS, Mid-Atlantic OCS, Gulf of Mexico OCS, Southern California OCS, Norton Sound and Beaufort Sea in Alaska. The exact location and water depth of each sediment sample station are listed in Table 1.

2.2. Analysis

The cupric oxide oxidation technique for characterization of lignin residues in sediments described by Hedges & Ertel (1982) was modified (Sauer *et al.*, 1988) and used to analyse sediments of the six regions of the OCS. The Hedges & Ertel method was modified for specific application to determine lignosulfonates in drilling fluid formulations and lignosulfonate-containing sediment.

2.2.1. Sample Preparation

Frozen sediment samples were thawed, acidified to pH 1 with 6N HCl, filtered, rinsed with deionized water, air-dried, and ground to pass through a 1-mm sieve.

2.2.2. Cupric Oxide Oxidation

Oxidations were carried out in custom-made 10-ml Monel 400 'minibombs'. Into each minibomb was added 20 g sediment, along with 1·0 g methylene chloride-extracted cupric oxide (CuO) powder, 0·1 g hydrous ferric ammonium sulfate ($Fe(NH_4)_2(SO_4)_2.6H_2O$), 7·0 ml 2N sodium hydroxide (NaOH) and a stainless steel ball to agitate the sample and reagents. Four oxidations were conducted simultaneously by placing the minibombs inside a 200-ml Parr bomb (Model 4753). The sediment samples and reagents were initially weighed into 4-dram borosilicate vials and placed with the minibombs and Parr bomb inside a glove box, which was flushed with N_2 for 3 h to remove all traces of O_2. Samples and reagents were then transferred to the minibombs, sealed, and inserted inside the Parr bomb along with 1·0 g CuO, 0·1 g $Fe(NH_4)_2(SO_4)_2.6H_2O$ and enough 2N NaOH to cover the topmost minibomb. At this point the Parr bomb was sealed, removed from the glove box, and heated with agitation for 3 h at 170 °C.

TABLE 1

Station locations and water depths of sediment samples analysed by cupric oxide oxidation method

Station No.	Location		Depth (m)
	Latitude (N)	Longitude (W)	
George's Bank-N. Atlantic OCS			
13A	40°30.08′	71°00.79′	80
11	40°30.73′	68°33.58′	86
5	40°39.46′	67°46.12′	84
20	40°37.30′	68°00.67′	89
3	41°01.39′	66°20.16′	1 350
7	40°27.54′	67°40.34′	560
8	40°10.37′	67°37.43′	2 175
12	39°54.32′	70°55.09′	558
13	39°48.45′	70°54.86′	1 245
14	39°40.90′	70°54.20′	2 105
Mid-Atlantic OCS			
3	38°36.84′	72°51.35′	2 055
6	39°05.54′	72°02.97′	2 090
11	38°40.17′	72°56.37′	1 515
Gulf of Mexico OCS			
7	26°16.98′	82°43.66′	32
23	25°16.92′	83°37.78′	74
55	24°36.17′	82°41.96′	27
23RP	28°29.00′	89°42.60′	620
21RP	28°46.70′	89°36.50′	230
CF7	28°44.45′	89°25.12′	105
CF17	28°06.24′	91°09.48′	102
CF12	28°11.90′	89°59.10′	245
CF21	27°36.20′	93°35.57′	395
S. California OCS			
58	34°34.35′	120°45.18′	101
60	34°33.25′	120°48.34′	279
62	34°30.46′	120°52.13′	591
73	34°28.21′	120°36.80′	100
75	34°26.08′	120°39.65′	297
78	34°18.78′	120°49.30′	774
89	34°13.79′	120°16.55′	478

TABLE 1—*contd.*

Station No.	Location		Depth (m)
	Latitude (N)	Longitude (W)	
Beaufort Sea			
3B	70°17.9′	147°02.0′	3·7
4A	70°18.4′	147°40.0′	4·3
5A	70°29.9′	148°45.8′	11·6
5D	70°24.3′	148°32.9′	2·0
6A	70°32.2′	149°56.7′	3·0
6B	70°33.3′	150°24.9′	5·2
7C	70°54.8′	152°00.7′	14·3
Norton Sound			
131	64°23.60′	161°49.27′	17
121	63°52.99′	163°01.34′	20
154	63°45.08′	164°37.43′	18
156	63°28.39′	165°19.28′	17
170S	63°41.72′	165°45.81′	25
166S	63°14.62′	167°02.21′	26
39	64°07.09′	171°18.00′	34
43	63°57.85′	167°48.03′	35

2.2.3. Isolation of Oxidation Products

Following heating, the individual minibombs were opened and 100 μg of ethyl vanillin added as an internal standard. The contents of each minibomb were washed into a 30-ml trifluoroethylene resin centrifuge tube with 1N NaOH and centrifuged at 2500 rpm for 10 min. The supernatant was decanted and saved. This isolation procedure was repeated twice. The combined aqueous phases were acidified to pH 1 with 6N HCl and extracted three times with 20 ml peroxide-free ethyl ether to isolate the phenolic compounds of interest. The ether extracts were dried over sodium sulfate (Na_2SO_4), concentrated by rotary evaporation under vacuum, transferred to a 1-dram vial, and concentrated by dryness under a stream of nitrogen.

2.2.4. Derivatization and Gas Chromatography

The dried sample extracts were dissolved in 50 μl of pyridine and reacted with 50 μl of 1% Regisil (Regis Chemical Co.) at 70 °C (10 min) to form the trimethylsilyl derivatives of the oxidation products and internal standard. The resultant solution was allowed to cool, then analysed immediately by fused silica capillary gas chromatography/flame ionization detection. Analytical conditions were as follows: Split injection (10:1 split ratio) on a 30 m × 25 mm DB-5 bonded phase column that was temperature programmed from 100 °C to 290 °C at 4°/min after an initial 0·1 min isothermal period. Identifications of oxidation products were made by retention time comparison versus identically silated standard mixtures that were prepared and analysed prior to the actual samples. Concentrations were calculated relative to the internal standard and are uncorrected for differences in response.

2.3. Quality Control

2.3.1. Method Validation
Recovery experiments were conducted on standard mixtures of lignin oxidation products to validate results obtained with ethyl vanillin used as an internal standard. The analyses of the standard mixtures showed that the percentage recovery of most of the oxidation products was approximately 80–90. Only the recoveries of the syringyl phenols were low, 33–70%, compared to the standard (Sauer et al., 1988).

The amount of sample required to produce the best precision in the analysis is at least 500 mg for sediments. Samples low in organic content will encounter the problem of superoxidation, which is the chemical conversion of aldehydes to corresponding carboxylic acids in each of the oxidation product classes.

2.3.2. Reference Sediments
A Lake Washington sediment provided by Dr John Hedges of the University of Washington was analysed to verify the analytical precision and accuracy of the method. There was excellent agreement between our results and those generated previously. Because the Lake Washington sediment is in short supply, another sediment from the upper Narragansett Bay was 'certified'. Results for both the Lake Washington 'intercalibration' and the Narragansett Bay sediment analyses are discussed in Sauer et al. (1988). The Narragansett Bay reference sample was analysed periodically throughout this study to verify continuity among the data generated internally. In general, the reference sample results obtained consistently fell within the concentration ranges of ±20% for the various oxidation products.

2.3.3. Blanks

One analytical blank was analysed for every twelve samples analysed. Contamination interferences of oxidation products of interest were not evident in these blanks.

2.4. Total Organic Carbon (TOC)

TOC analyses on OCS sediments were carried out by high-temperature combustion on a Carlo Erba Model 1106 elemental analyser. All samples were dried and ground prior to analysis. Concentrations are reported in units of mg/g dry sample weight.

3. DATA INTERPRETATION FRAMEWORK

Before presenting the results of the sediment analyses, the framework in which the data are interpreted is discussed.

3.1. Lignin Parameters

The cupric oxide oxidation method yields as its primary products eleven phenolic oxidation products that can be grouped in the four classes. The compound classes include (1) vanillyl phenols, consisting of vanillic acid, vanillin, and acetovanillone; (2) syringyl phenols, consisting of syringic acid, syringaldehyde, and acetosyringone; (3) cinnamyl phenols, consisting of the compounds p-coumaric acid and ferulic acid; and (4) p-hydroxyl phenols, consisting of p-hydroxybenzoic acid, p-hydroxybenzaldehyde, and p-hydroxyacetophenone. The amounts of these lignin oxidation product classes are used with the sediment organic carbon content to calculate the following series of lignin parameters:

P = mg of p-hydroxyl phenols produced from 100 mg of organic carbon
V = mg of vanillyl phenols produced from 100 mg of organic carbon
S = mg of syringyl phenols produced from 100 mg of organic carbon
C = mg of cinnamyl phenols produced from 100 mg of organic carbon
Λ = sum of V, S and C.

The lignin parameter Λ (lambda) relates the total yield of lignin-derived phenols to total organic carbon concentrations and is useful as an indicator of the relative lignin content of sedimentary organic matter (Hedges & Mann, 1979a). The parameter P can also be indicative of lignin-derived organic matter, but is excluded from this calculation because non-lignin compounds also yield p-hydroxyl phenols. The ratios P/V and, in particular, S/V and C/V can be used as compositional parameters to characterize different lignin sources in sediments. Plots of S/V versus C/V can resolve oxidation product data for vascular plant tissues into four distinct compositional regions: gymnosperm woods (G), non-woody gymnosperm tissues (g), angiosperm woods (A), non-woody angiosperm tissues (a). Figure 1 illustrates the typical 'Hedges plot' of vascular plant lignin compositions and the range of values for the different compositional types (Hedges & Mann, 1979b).

The final lignin parameter that can be calculated from the oxidation product data is the acid/aldehyde ratio (V_a/V_h, P_a/P_h and S_a/S_h, corresponding to the acid/aldehyde ratios of vanillyl, p-hydroxyl, and syringyl phenols, respectively). Recent studies have shown that this parameter indicates the degree of oxidative diagenesis undergone by natural lignin polymers, with elevated ratios (relative to those found for vascular plant tissues) indicating increased aerobic degradation (Hedges et al., 1982; Ertel & Hedges, 1984; Ertel et al., 1984).

3.2. Lignin Parameters in Lignosulfonate-containing Drilling Fluids

The lignin parameters described above can be applied not only to the detection of lignin sources in sediments, but also to the environmental detection of drilling fluids in sediments. In Sauer et al. (1988), five lignosulfonates that are commonly used as additives in drilling fluids were analysed by the cupric oxide oxidation method.

The study produced three principal results. (1) The major oxidation products of lignosulfonates are vanillyl phenols. The high concentration of vanillyl phenols relative to the other oxidation products (approximately two orders of magnitude higher) indicate that lignosulfonates are derived primarily from the sulfonation of soft (gymnosperm) woods and would plot in the Hedges plot in the region designated for these woods (Fig. 1). (2) The vanillyl phenol products from oxidation of drilling fluid formulations vary linearly with lignosulfonate content (correlation coefficient 0·99–0·97). (3) One percent addition of typical lignosulfonate-containing drilling fluid to a marine sediment could be detected by using the cupric oxide oxidation method. The actual detection limits of the vanillyl phenol products will vary according to the natural background sediments.

Given these results, the vanillyl phenols are seemingly the best tracers

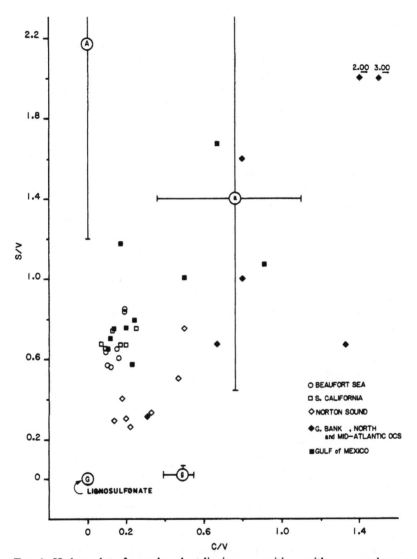

FIG. 1. Hedges plot of vascular plant lignin compositions with mean and range values of gymnosperm woods (G), non-woody gymnosperm tissues (g), angiosperm woods (A), non-woody angiosperm tissues (a). Also shown are lignin compositions of OCS sediments and drilling fluid lignosulfonates.

of drilling fluid inputs to sediment. These tracers can be applied by first determining an absolute or relative increase in the sedimentary concentration of vanillyl phenol products (as either an increase in the concentration of individual vanillyl phenols or an increase in the lignin parameter V); and second, documenting a shift in the region where sediment data points fall in a compositional Hedges plot towards typical lignosulfonate values relative to background sediment values.

The two ways of using the vanillyl phenols as tracers are illustrated with the data generated in Sauer *et al.* (1988), in which incremental additions of mud formulations were mixed with sediments from Georges Bank. The results of these experiments have been arranged into a graph (vanillyl phenol concentration versus drilling fluid content of the sediment) and a Hedges plot, and are presented in Fig. 2.

The upper diagram in Fig. 2 shows a linear increase in the concentration of the sum of the vanillyl phenols with increasing drilling fluid content of the sediment. The same results could be illustrated with the lignin parameter V. The lower diagram shows the same results plotted on a Hedges plot. It is evident that with increasing drilling fluid content, the lignin composition of the mixtures shifts toward the value calculated for the lignosulfonate in the drilling fluid formulation. The greater the difference between the composition of natural lignin in a sediment and that of lignosulfonates, the more sensitive this approach is in detecting drilling fluid inputs. Changes in lignin composition of sediments would also be the preferred tracer technique in cases in which the natural lignin background is relatively high, i.e. where the value for the lignin parameter Λ is large.

Although not shown in Fig. 2, acid/aldehyde ratios were also calculated for the various drilling fluid–sediment mixtures and results revealed that, even at the lowest addition, a significant change toward a ratio characteristic of lignosulfonates could be detected. These values are characteristic of relatively unaltered gymnosperm woods (range 0·11–0·15, J. Hedges, personal communication). This parameter could represent a third means by which to detect drilling fluid inputs to sediments, although significant variability was found between the acid/aldehyde ratios of individual lignosulfonates and lignosulfonate-mud formulations (Sauer *et al.*, 1988).

In light of these results, the natural lignin distributions of sediments from six regions of the OCS (Table 1) were evaluated in relation to (1) their relative vanillyl phenol content, i.e. the lignin parameter V; (2) their lignin compositions, i.e. Hedges plots; and (3) their vanillyl acid/aldehyde ratios.

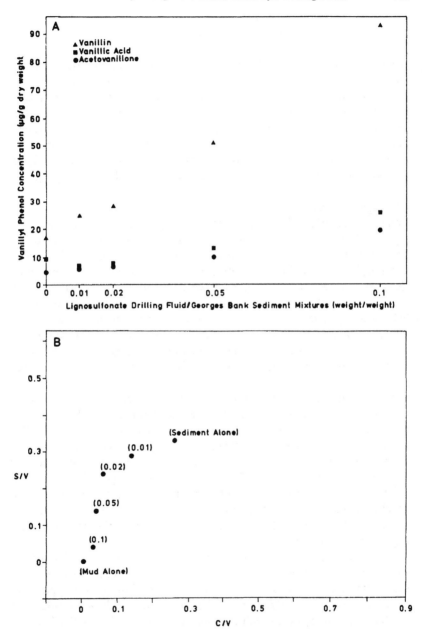

FIG. 2. Cupric oxide oxidation results of incremental addition of a lignosulfonate drilling fluid with Georges Bank sediment (Sauer *et al.*, 1988).

4. RESULTS AND DISCUSSION

Results of the analysis of OCS sediment by the cupric oxide oxidation method are presented in Table 2. Although all the lignin composition parameters are shown in the table, only the parameters and ratios (i.e. V, C/V and S/V, and V_a/V_h) that relate to the use of lignosulfonate as a tracer in sediments are discussed. The results in Table 2, however, represent new data on natural lignin distributions in OCS sediments. This section of the paper is divided into three major headings: lignin parameter, V; lignin composition, C/V and P/V; and vanillyl acid/aldehyde ratios, V_a/V_h.

4.1. Lignin Parameter

Table 3 summarizes the values of the lignin parameter V in the six OCS regions. Regions exhibiting low V values would contain the lowest background interferences and would thus be highly suited for the application of the cupric oxide oxidation technique; those sediments containing elevated V values would be considered less suitable. This evaluation of regions is admittedly simplistic, in that it neglects spatial gradients and heterogeneity in sediment lignin composition and content. It also ignores factors such as net erosional and depositional patterns. However, the results are useful as a first-order evaluation of the natural background contributions of the lignin parameters in sediments from the various regions.

The Atlantic OCS is the region where the lignin parameter V would be most applicable. The V values for the Georges Bank and the Mid-Atlantic range from <0.01 to 0.05 at all stations except Station 13A. At this station, which is located in the 'Mud-Patch' region of Georges Bank, the V value is 0.16 and corresponds to the highest value for total yield of lignin-derived phenols, Λ, of 0.26. Sediments of this depositional area also exhibit elevated concentrations of hydrocarbons and total organic carbon (Boehm, 1984; Boehm & Requejo, 1986). No lignin-derived phenols were detected at Stations 11, 20 and 5; these stations also exhibited the lowest organic carbon contents.

The Gulf of Mexico OCS was found to be the next most suitable region, with V values at several sites comparable to those found in the Atlantic sediments. Only in the sediments located west of the Mississippi River delta (Stations 23RP, 21RP and CF7) and on the shallow Florida shelf (Stations 23 and 55) were V values significantly greater than in the Atlantic. The same trend was observed in Λ values of the region.

Following the Gulf of Mexico, the Southern California OCS and Norton

TABLE 2

Lignin compositional parameters for sediments in six OCS regions

Station No.	TOC (mg/g dry wt)	P	V	S	C	Λ	P/V	C/V	S/V	V_a/V_h	P_a/P_h	S_a/S_h
Georges Bank-N. Atlantic OCS												
13A	19·2[a]	0·35	0·16	0·05	0·05	0·26	2·18	0·31	0·31	0·60	1·06	—
11	1·7[a]	1·63	0	0	0	0	—	—	—	—	2·61	—
5	0·8[a]	0	0	0	0	0	—	—	—	—	—	—
20	ND[b]	0	0	0	0	0	—	—	—	—	—	—
3	14·1	0·12	0·05	0·05	0·04	0·14	2·40	0·80	1·00	0·63	1·18	2·72
7	5·3	0·10	0·03	0·02	0·02	0·07	1·00	0·67	0·67	0·54	1·06	3·23
8	11·0	0·32	0	0	0·02	0·02	—	—	—	—	2·11	—
12	8·5	0·13	0·01	0·02	0·03	0·06	13·0	3·00	2·00	—	2·11	—
13	9·6	0·29	0	0	0	0	—	—	—	—	2·61	—
14	14·1	0·04	0	0·01	0·01	0·02	—	—	—	—	1·77	—
Mid-Atlantic OCS												
3	9·7	0·11	0·01	0·02	0·02	0·05	11·0	2·00	2·00	1·45	1·00	—
6	13·3	0·12	0·03	0·04	0·02	0·09	4·00	1·33	0·67	0·42	1·05	1·42
11	18·4	0·12	0·05	0·08	0·02	0·17	2·40	0·80	1·60	0·51	1·11	2·00
Gulf of Mexico OCS												
7	5·8	0·17	0·04	0·04	0·02	0·10	4·25	0·50	1·00	—	0·89	—
23	94·3	0·56	0·14	0·11	0·03	0·28	4·00	0·21	0·79	0·65	0·92	4·22
55	137	1·13	0·24	0·13	0·05	0·42	4·73	0·21	0·54	0·34	0·86	1·18
23RP	11·2	0·07	0·12	0·09	0·02	0·23	0·58	0·17	0·75	3·55	1·32	7·01
21RP	13·7	0·25	0·15	0·17	0·03	0·45	0·60	0·12	0·68	0·60	1·03	0·74
CF7	15·6	0·18	0·30	0·19	0·03	0·52	0·60	0·10	0·63	0·58	1·03	0·72
CF17	14·1	0·10	0·07	0·05	0·01	0·13	1·43	0·14	0·71	0·64	0·83	2·95
CF12	14·6	0·05	0·03	0·05	0·02	0·10	1·67	0·67	1·67	0·09	1·44	—
CF21	16·2	0·06	0·03	0·03	0·03	0·09	2·00	1·00	1·00	0·45	1·23	3·83
27RP	12·1	0·02	0·06	0·07	0·01	0·14	0·33	0·17	1·17	1·07	1·00	—

(continued)

TABLE 2—contd.

Station No.	TOC (mg/g dry wt)	P	V	S	C	A	P/V	C/V	S/V	V_a/V_h	P_a/P_h	S_a/S_h
Southern California OCS												
58	9·8	0·36	0·46	0·34	0·06	0·86	0·78	0·13	0·74	0·41	2·06	0·39
60	13·1	0·19	0·23	0·16	0·02	0·41	0·83	0·09	0·70	0·51	1·72	0·44
62	20·9	0·21	0·15	0·10	0·03	0·28	1·40	0·20	0·67	0·41	2·06	0·39
73	12·5	0·14	0·15	0·10	0·01	0·26	0·93	0·07	0·67	0·50	0·96	0·43
75	17·0	0·17	0·12	0·09	0·03	0·24	1·42	0·25	0·75	0·49	0·85	0·60
78	43·9	0·17	0·06	0·04	0·01	0·11	2·83	0·17	0·67	0·57	0·79	1·10
89	39·5	0·18	0·08	0·06	0·02	0·16	2·25	0·25	0·75	0·76	0·80	1·14
Beaufort Sea												
3B	9·2[c]	0·89	0·62	0·40	0·09	1·11	1·43	0·15	0·65	0·47	2·22	0·48
4A	7·4	0·94	0·79	0·45	0·08	1·32	1·19	0·10	0·57	0·48	1·64	0·41
5A	10·3	0·71	0·59	0·33	0·07	0·99	1·20	0·12	0·56	0·44	1·61	0·42
5D	30·2	1·54	0·77	0·64	0·15	1·56	2·00	0·19	0·83	0·65	3·14	0·55
6A	11·5	0·74	0·63	0·38	0·10	1·11	1·17	0·16	0·60	0·53	1·84	0·54
6B	16·6	0·62	0·31	0·26	0·06	0·63	2·00	0·19	0·84	0·70	2·65	0·60
7C	14·1	0·66	0·56	0·35	0·05	0·96	1·18	0·09	0·63	0·70	1·93	0·69
Norton Sound												
131	4·4	0·20	0·04	0·03	0·02	0·09	5·00	0·50	0·75	0·49	0·97	—
121	11·8	0·13	0·10	0·03	0·02	0·15	1·30	0·20	0·30	0·83	3·65	3·28
154	9·9	0·70	1·18	0·59	0·56	2·33	0·59	0·47	0·50	0·59	2·45	0·52
156	13·0	0·27	0·50	0·20	0·09	0·79	0·54	0·18	0·40	0·57	2·99	0·52
170S	5·2	0·26	0·23	0·06	0·05	0·34	1·13	0·22	0·26	0·64	2·87	2·34
166S	11·6	0·06	0·03	0·01	0·01	0·05	2·00	0·33	0·33	0·52	1·72	—
39	3·8	0·05	0	0	0	0	—	—	—	—	—	—
43	6·0	0·16	0·07	0·02	0·01	0·10	2·28	0·14	0·29	0·35	1·50	—

[a]Average of six replicate box core samples.

[b]ND = not determined; in spite of the fact that no TOC data were available for this sample, no phenolic oxidation products were detected, thus the values shown.

[c]Average of six replicate samples.

TABLE 3
Summary of lignin parameters V (means and ranges) in OCS sediments
and lignosulfonate drilling mud additives

OCS region (No. of samples)	Range of values for lignin parameter V	$\bar{x} \pm SD$
Georges Bank North Atlantic OCS (10)	0–0·16	0·03 ± 0·05
Mid-Atlantic OCS (3)	0·01–0·05	0·03 ± 0·02
Gulf of Mexico OCS		
Eastern (3)	0·04–0·24	0·14 ± 0·10
Western (7)	0·03–0·30	0·11 ± 0·10
Southern California OCS (7)	0·06–0·46	0·18 ± 0·14
Norton Sound (8)	0–1·18	0·27 ± 0·40
Beaufort Sea (7)	0·31–0·79	0·61 ± 0·16
Lignosulfonates (5)[a]	12–35	—

SD = standard deviation.
[a]Calculated based on a mean organic carbon content of 37% for neat chrome lignosulfonates.

Sound regions contained the next highest vanillyl phenol background (mean V values are 0·18 and 0·27, respectively). In the sediments of the Santa Maria Basin off the Southern California coast (Stations 58 and 60), background vanillyl phenol concentrations were generally higher than in either Atlantic or Gulf sediments and were reflected in the highest Λ values (0·41–0·85) for the region. In Norton Sound, however, with exception of the two sites near the opening of the sound (Stations 154 and 156), most sediment V values compared favorably with those of the Gulf sediments. The two exceptional sites had V values of 1·18 and 0·50, respectively, which corresponded to Λ values of 2·33 and 0·79.

The highest V values throughout a region were found in the nearshore Beaufort Sea sediments (range 0·31–0·79). The sediment samples constitute a longshore transect in <15 m of water from the area west of the Sagavanirktok River to Cape Halkett. These elevated values probably reflect the significant continental drainage by various rivers and the persistence of ice cover and frigid temperature factors that result in enhanced preservation of organic matter. The lignin parameter V would be least sensitive as a tracer of drilling mud discharges in this region.

It should be noted that the present evaluation focuses strictly on the lignin background in the various OCS regions and not on other factors that could also notably affect the 'tracer potential' of any given region. For example, although the vanillyl phenol background found in the Georges Bank-North Atlantic OCS is low, this area is known for its dynamic physical oceanography (especially the shallow regions of the Bank proper), which could result in a net erosional environment over any particular time period being considered. Thus, little accumulation of discharged drilling mud lignosulfonates would be expected for this region. All such factors should be considered in evaluating the ultimate sensitivity of this technique.

4.2. Lignin Compositions (S/V versus C/V Plots)

The approach adopted in evaluating OCS lignin compositions is analogous to that employed above for their vanillyl phenol contents, i.e. compositions determined for the various regions are contrasted to those determined for the lignosulfonates in drilling muds. The criterion selected to determine the extent of differences between lignosulfonate and sediment lignin compositions is the difference in the locations of data points in a Hedges compositional plot (S/V versus C/V) for the various regions. Those regions where sediment lignin compositions exhibit the greatest differences from the drilling muds are judged the most suitable for the application of this specific technique. Our approach is highly empirical; we have refrained from using more quantitative mathematical techniques, e.g. clustering in two-dimensional space to quantify differences in lignin composition, because of the limitations imposed by the small sample sets for each region.

The lignin compositions of the six OCS regions are shown in Fig. 1. Note that lignosulfonates exhibit compositional values identical to those of gymnosperm woods and that all of the OCS sediments analysed contain lignin compositions different from the lignosulfonates. Thus, it appears that this lignin parameter may serve as a sensitive drilling mud tracer in all OCS regions, but to different degrees.

The OCS sediments are seen to loosely fall into the following three compositional groupings: (1) some of the Norton Sound samples, (2) the samples from the Beaufort Sea and Southern California OCS and the majority of samples from the Gulf of Mexico, and (3) the remaining samples, including most of the Atlantic OCS samples. The latter grouping consists primarily of sediments exhibiting S/V and C/V values >0.9 and relatively low Λ values (generally <0.10). These data points contain a large uncertainty because of the inherent variability in the values for the compositional ratios at these low lignin contents. However, it seems likely that incorporation of even small amounts of lignosulfonates associated with drilling muds into these sediments would result in a large shift in their compositional values; thus, the Atlantic OCS samples also appear to be the most suitable for application of lignin compositions as drilling mud tracers.

The Beaufort Sea, Southern California, and Gulf of Mexico grouping is roughly defined by S/V values between 0·6 and 0·9 and C/V values between 0 and 0·2 (Fig. 1). Compositionally, this grouping remains sufficiently different from the calculated values for lignosulfonates to permit a source and input distinction, and the compositional technique therefore also appears to be applicable in these regions. It is noteworthy that the Beaufort Sea region was judged the least suitable for application of the lignin parameter V as a mud tracer, but is considered more adaptable for use of the compositional parameters approach. It appears that the compositional technique can permit a source distinction even in cases in which a significant lignin (vanillyl phenol) background exists, and the two parameters will therefore be highly complementary.

The group consisting of the Norton Sound samples is defined by S/V values between 0·2 and 0·4 and C/V values between 0·1 and 0·6. The exact degree to which the compositional approach can be employed in this region may vary according to location. Application of this parameter in Norton Sound will result in lower sensitivity than in the other regions.

4.3. Vanillyl Acid/Aldehyde Ratios

Vanillyl acid/aldehyde ratios are another potential tracer parameter for lignosulfonate-containing drilling fluids in OCS sediments. In Sauer *et al.* (1988), ratio values for individual lignosulfonate additives were found to range between 0·45 and 0·57. When lignosulfonates were added to mud formulations, the vanillyl acid/aldehyde ratio decreased with increasing lignosulfonate content and ranged from 0·19 to 0·21. These values are characteristic of relatively unaltered gymnosperm woods (range 0·11–0·15, J. Hedges, personal communication). The exact cause of these differences is not known and, therefore, these variations make the application of this parameter as a tracer somewhat uncertain. Nevertheless, when drilling fluid was added to a marine sediment, a shift toward values characteristic of the drilling fluid formulation was observed (Sauer *et al.*, 1988).

The means and ranges of acid/aldehyde ratios in OCS sediments are shown in Table 4. A ranking of the vanillyl acid/aldehyde ratios in the various OCS regions with respect to the application of this parameter as

TABLE 4
Vanillyl acid/aldehyde ratios (means and ranges) in OCS sediments

OCS region (No. of samples)[a]	Range of acid/aldehyde ratios	$\bar{x} \pm SD$
Georges Bank North Atlantic OCS (3)	0·54–0·63	0·59 ± 0·04
Mid-Atlantic OCS (3)	0·42–1·45	0·79 ± 0·57
Gulf of Mexico OCS (9)	0·09–3·55	0·89 ± 1·03
Southern California OCS (7)	0·41–0·76	0·52 ± 0·12
Beaufort Sea (7)	0·44–0·70	0·57 ± 0·11
Norton Sound (7)	0·35–0·83	0·57 ± 0·15
Lignosulfonates (4)[b]	0·19–0·21	0·20 ± 0·01

SD = standard deviation.
[a]Numbers of samples differ from total analysed because several sediments did not contain vanillyl phenols.
[b]Formulated in muds.

a drilling fluid tracer was not attempted for two reasons: (1) two of the sample sets (Mid-Atlantic and Gulf of Mexico OCS) showed large variability (Table 4), suggesting a significant degree of spatial patchiness, and (2) the remaining four regions showed a remarkable constancy in the ratio, with a very narrow range of mean values (0·52–0·59), rendering any attempt at ranking meaningless. The ratios in the latter four regions (Georges Bank-N. Atlantic OCS, Southern California OCS, Beaufort Sea, Norton Sound) are significantly different from those determined for lignosulfonates in mud formulations, and on this basis this parameter could be applied in these OCS regions. In the Gulf and Mid-Atlantic regions, the sensitivity would vary according to the specific geographic area. Of significant importance in the ultimate use of this parameter is the rate at which the lignosulfonate acid/aldehyde ratio 'weathers' (and thus becomes more similar to the sediment values) once this material is introduced to the marine environment. The longer the lignosulfonate signal is maintained, the more useful this parameter will be as a tracer.

5. SUMMARY AND CONCLUSIONS

The lignin content and composition of sediments from different OCS regions have been examined in relation to the potential application of a cupric oxide oxidation technique as a tracer of drilling mud discharges.

Three lignin parameters—the vanillyl phenol yield (V), the lignin composition (S/V versus C/V plots), and the vanillyl acid/aldehyde ratio—were identified from data generated in another study (Sauer *et al.*, 1988) as the most suitable tracers of lignosulfonates in drilling fluids. The vanillyl phenol yield, lignin composition, and vanillyl acid/aldehyde ratio were used to evaluate sediments in six OCS regions with regard to sensitivity attainable by each tracer parameter. The elevation of the concentration of vanillyl phenols and changes in lignin parameter composition over background in sediments would provide information on the transport and fate of drilling fluids.

Evaluation of sediments from OCS regions by the cupric oxide oxidation method revealed that the concentrations of vanillyl phenol oxidation products were lowest in the Georges Bank and Mid-Atlantic OCS (mean = 0·03, range = 0·0–0·16) and highest in the Beaufort Sea (mean = 0·61, range = 0·31–0·79). A spurious value of 1·18 was measured in one Norton Sound sample. In almost all cases, high values in the total lignin-derived phenol yield, Λ, corresponded to high values in the lignin parameter V. The ranking order of the regions based on mean V was (in order of decreasing sensitivity) as follows: Georges Bank-North Atlantic OCS = Mid-Atlantic OCS < Gulf of Mexico OCS = Norton Sound < Southern California OCS < Beaufort Sea. This ranking does not take into account factors such as spatial patchiness and net depositional and erosional environments, which can influence the ultimate sensitivity attainable in any area.

The lignin compositions (i.e. Hedges plots) of sediments from the Georges Bank-N. Atlantic OCS were the most dissimilar from those of lignosulfonates, and this parameter therefore appears to be most applicable in this region. Lignin compositions of sediments from the Beaufort Sea, Southern California OCS, and Gulf of Mexico OCS were all found to be similar and also easily distinguishable from lignosulfonates. Norton Sound sediment indicated that lignin compositions were the most similar to lignosulfonates, and this tracer approach would be least sensitive in this region. The compositions in Norton Sound were somewhat variable, and thus the sensitivity of this tracer would vary according to area within the Sound.

The vanillyl acid/aldehyde ratios in all regions differed from those of lignosulfonate. This ratio parameter appears to be a suitable tracer of drilling fluids in sediments. A major question in the use of this parameter is the rate at which the ratio that is characteristic of lignosulfonates is altered in the environment.

From the evaluation of these OCS sediments, the cupric oxide oxidation technique appears to be suitable as an environmental monitoring tool for drilling mud discharges in most OCS regions. As discussed previously, for drilling fluids containing lignosulfonate and barite that are discharged into marine receiving waters, lignosulfonate would be the indicator of the drilling fluid additives associated with the finer clay and organic fraction of drilling fluids; whereas barium would be the indicator for those additives that are known to associate with barite in drilling fluids. It may be expected that the fine fraction of drilling fluids may not reach the seabottom near a discharge point and, therefore, this fraction may not be of concern to benthic impact. However, when appraising the fate of the fine particles of drilling fluids in the ocean, one must consider the processes of particle flocculation and the adherence of fine particles to discharged cuttings. As discussed in Ayers *et al.* (1982), finer particles of drilling fluids flocculate when discharged in seawater and as a loosely associated aggregate, the particles increase in fall velocity. Also, the finer fractions of drilling fluids have been observed to adhere to discharged cuttings which fall through the water column at a faster rate than the finer particles alone. Whether the mode of settling of drilling fluid constituents is from flocculated particles or cuttings discharges, lignosulfonate and barium sulfate may have significantly different fates. Lignosulfonate not detected in sediments which are observed to contain drilling fluid barium (the customary tracer of drilling fluid impact) indicates that the constituents of drilling fluids associated with lignosulfonate are not present. The reality of the different fates of drilling fluid constituents has important consequences in interpreting drilling fluid impacts in the marine environment.

ACKNOWLEDGEMENT

We would like to thank the Minerals Management Service (Contract No. 4-6434-10212) for funding the analysis of OCS sediment samples and providing the samples.

REFERENCES

AYERS, R.C., MEEK, R.P., SAUER, T.C. & STEUBNER, D.O. (1982). An environmental study to assess the effect of drilling fluids on water quality parameters during high-rate, high-volume discharges to the ocean. *J. Petrol. Technol.*, **21**, 11–24.

BOEHM, P.D. (1984). Aspects of the saturated hydrocarbon geochemistry of recent sediments in the Georges Bank region. *Org. Geochem.*, **7**, 11–24.

BOEHM, P.D. & REQUEJO, A.G. (1986). Overview of the recent sediment hydrocarbon geochemistry of Atlantic and Gulf Coast outer continental shelf environments. *Estuarine Coastal Shelf Sci.*, **23**, 29–58.

ERTEL, J.R. & HEDGES, J.I. (1984). The lignin component of humic substances: Distribution among soil and sedimentary humic, fulic, fulvic and base-soluble fractions. *Geochim. Cosmochim. Acta*, **48**, 2065–74.

ERTEL, J.R., HEDGES, J.I. & PERDUE, E.M. (1984). Lignin signature of aquatic humic substances. *Science*, **223**, 485–7.

HEDGES, J.I. & ERTEL, J.R. (1982). Characterization of lignin by glass capillary gas chromatography of cupric oxide oxidation products. *Anal. Chem.*, **54**, 174–8.

HEDGES, J.I. & MANN, D.C. (1979a). The characterization of plant tissues by their lignin oxidation products. *Geochim. Cosmochim. Acta*, **43**, 1809–17.

HEDGES, J.I. & MANN, D.C. (1979b). The lignin geochemistry of marine sediments from the southern Washington coast. *Geochim. Cosmochim. Acta*, **43**, 1809–18.

HEDGES, J.I., ERTEL, J.R. & LEOPOLD, E.B. (1982). Lignin geochemistry of a late quaternary sediment core from Lake Washington. *Geochim. Cosmochim. Acta*, **46**, 1869–77.

SAUER, T.C., REQUEJO, A.G., BROWN, J.S., AYERS, R.C. & BOEHM, P.D. (1988). Application of analytical pyrolysis and cupric oxide oxidation to characterization of non-extractable organic constituents in drilling fluids and sediments. In *Proceedings, Symposium on Chemical and Biological Characterization of Municipal Sludges, Sediments, Dredge Spoils and Drilling Muds*, May 20–22, 1986, Cincinnati, Ohio. Environmental Monitoring and Support Laboratory, Environmental Criteria and Assessment Office, US Environmental Protection Agency.

39

Effects of Drilling Fluids on a Shallow Estuarine Ecosystem—I. Characterization and Fate of Discharges

Maurice (Mo) Jones

Enseco Houston, 2400 West Loop South, Houston, Texas 77027-4206, USA

ABSTRACT

To address concerns expressed about the possible effects of drilling mud discharges on shallow, low-energy estuarine ecosystems, a 12-month study was designed to detect alterations in water quality, sediment geochemistry, and the benthic community. A relatively pristine, shallow (1·2 m) estuary (Christmas Bay, Texas) without any drilling activity for the last 30 years was chosen for the study site. After a 3-month baseline study, three stations were selected. At each treatment station (muddy-sand or seagrass community), mesocosms were constructed to enclose a 3·5 m³ water column. Each in situ mesocosm, except the controls, was successively dosed at a mesocosm-specific concentration with 4 field-collected muds (spud, nondispersed, lightly-treated, and heavily-treated lignosulfonate) over a 1·5-month time period to simulate discharges from a drilling rig. Twenty-four hours after each dose, water exchange was allowed until the next treatment. After the last treatment, the enclosures were removed and the site monitored for 6 months. One additional site was similarly dosed with clean sediment from Christmas Bay for comparison. Analysis of the water samples and field data showed that water quality was impacted during the discharges, primarily at the highest dose (1:100), but that elevated levels returned to ambient levels immediately after water exchange was allowed. Barium, from the barite, was used as a tracer during the study to confirm the estimated doses. Barium levels returned to ambient or only slightly elevated levels at the end of the 6-month monitoring period due to sediment deposition, resuspension and

*Present address: ENSR Environmental Laboratory, 2400 West Loop South, Suite 300, Houston, Texas 77027-4206, USA.

bioturbation. Changes in grain size due to the mud treatment were well within the variability due to naturally occurring patchiness.

1. INTRODUCTION

Concerns about the fate and effects of drilling fluids discharged from offshore oil and gas drilling rigs on the marine environment have resulted in numerous studies by governmental agencies, industry and academia. Two international conferences were held in 1975 and 1980 on this topic for publication of these studies (Environmental Protection Agency (EPA), 1975; Anon., 1980). Most marine related studies dealt with the Outer Continental Shelf (OCS) or other high-energy neritic environments. The US Department of the Interior (DOI) reviewed these studies prior to developing policy and noted that the agency was specifically concerned about the lack of information and studies on 'low energy, semi-enclosed bays' where the depositional environment could contribute to any cumulative effects (DOI, 1981). Similarly, the National Marine Pollution Program Office (COPRDM, 1980) found that:

> existing information generated by industry and government sponsored field and laboratory studies has led this panel to the opinion that during exploratory drilling, disposal of conventionally employed drilling muds (excluding those which are petroleum based or have unusually toxic components) pose little threat to the marine environment, except in localized bottom habitats where the deposition of the particulate matter may affect the benthos.

In a more extensive review of the literature, the National Research Council (NRC, 1983) found 'that effects of individual discharges are limited in extent and are confined mainly to the benthic environment. Uncertainties regarding effects still exist for low energy, depositional environments' and 'any additional research on drilling fluids should include acute, sublethal and chronic bioassays using techniques and contaminant exposures that reflect actual discharge and exposure conditions, field studies that take into account inventories and chemical analyses of discharges, and studies of resuspensive transport of particulate contaminants'.

The objectives of this research project were to study the effects of drilling fluids on a shallow, estuarine ecosystem and particularly the benthic community. This paper summarizes the characterization of the drilling fluids, impacts on water quality, and the fate of the drilling fluids.

A second paper will summarize the biological effects of the discharges on the ecosystem, particularly the benthos. Field muds of known composition, after chemical and physical characterization, were tested in the laboratory using acute bioassays (96-h) and elutriation tests to predict probable impacts. The same muds were then discharged into an estuarine ecosystem under carefully controlled conditions and at realistic levels to verify the laboratory results. In addition, clean dredged sediment from the same site were similarly tested and discharged for comparison to determine if any observed effects were due to chemical toxicity or physical alteration of the environment.

2. EXPERIMENTAL DESIGN

Prior to the field study, research, discharge and collecting permits were received from the Texas Railroad Commission (TRC) and Texas Parks and Wildlife (TPW). TRC specifically prohibited the discharge of drilling fluids containing 'free oil' in a shallow estuary. Actively drilling exploratory rigs in a shallow estuary were surveyed for a location without previous drilling activity or diesel oil additions to the drilling fluids. No ideal candidates (using both criteria) were located, and it was decided to simulate the drilling of a well in a nearby estuary. Muds from four depth ranges representing the four most commonly used generic muds on the Outer Continental Shelf (spud, nondispersed, lightly-treated and heavily-treated lignosulfonate) (Ayers *et al.*, 1983) were collected from coastal or offshore wells (Texas, Louisiana (2) and Massachusetts). These muds were representative of a well of approximately 6000 m in depth (a worst-case situation).

2.1. Description of Study Site
Christmas Bay, Texas was chosen as the study site for the field verification study (Fig. 1). Located at the western end of the Galveston Bay system adjacent to the Gulf of Mexico, Christmas Bay (6 km × 3·8 km) had no history of drilling activity for over 30 years. It was believed that this was an important attribute in order to clearly distinguish the possible impacts of this study from previous or simultaneous drilling activity. The bay is semi-enclosed, relatively shallow (1·2 m mean depth), has fairly stable salinity, and receives minimal industrial and municipal inputs. The chemical and biological properties of Christmas Bay have been studied in the past: 1971–3; fishery survey (McEachron *et al.*, 1977); 1972–3, water

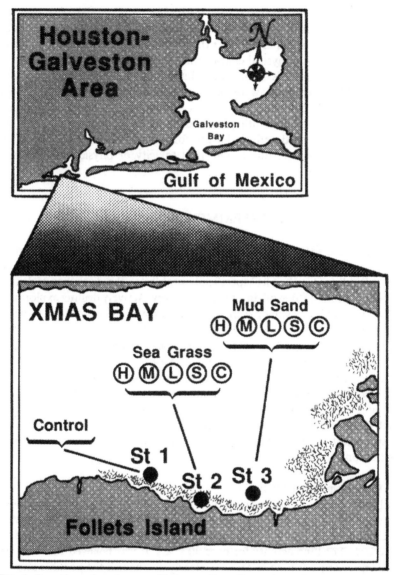

FIG. 1. Location of study area. Christmas Bay, Texas, is at the western edge of the Galveston Bay System and is separated from the Gulf of Mexico by Follets Island (peninsula). Station 2 is located within the seagrass beds (stippled area) and Stations 1 and 3 are in adjacent muddy-sand areas. Legend: H = high; M = medium; L = Low concentration; S = sediment control; C = control mesocosms.

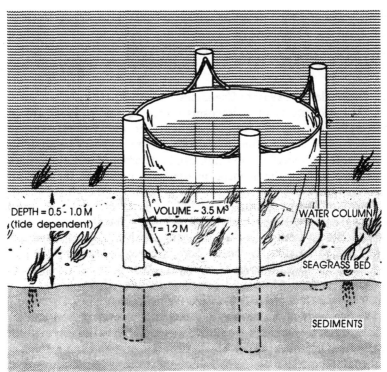

FIG. 2. Schematic of mesocosm used in Christmas Bay research project. Mesocosm is made of polyethylene and supported by plastic pipe. The height of the mesocosm could be adjusted by raising and lowering the top supporting ring.

quality, sediment geochemistry, benthos (US Coast Guard, 1976); 1973–74, zooplankton (Bagnall, 1976); 1976–7, benthos, sediment geochemistry, vegetation (White *et al.*, 1985); 1975–6, seagrass community, crustacea (Penn, 1979), and this study (1981–2).

After the initial selection of Christmas Bay, the bay was surveyed for an area with high benthic diversity. Three stations along the southern shore were selected. Station 1 was a control station in a muddy-sand substrate area; Station 2 was in a *Halodule wrightii* seagrass bed, and Station 3 was in a muddy-sand area. Both Stations 2 and 3 were treated identically for comparison of the macroinfauna; however, Station 3 was destroyed in a winter storm. Stations 1 and 2 were separated by 500 m and Station 1 was slightly upwind of the prevailing wind direction (SE). At Station 2, 5 mesocosms were constructed (Fig. 2) in a line perpendicular to the prevailing wind direction, all at an equal water depth. During the initial survey, these sites within Station 2 were identified as having similar sediment and benthic characteristics and chosen to minimize the interference of benthic and sediment patchiness in the impact assessment.

2.2. Rationale for Exposure Conditions

Enclosed ecosystems or mesocosms (Grice & Reeve, 1982) are ideal systems for studying the effects of pollutants because a natural community can be studied *in situ* while a relatively precise dose can be established. Thus, cause-and-effect relationships can be determined under near natural conditions without the artificial interferences present in the laboratory (e.g. microcosms). These mesocosms are modeled after the Loch Ewe enclosures of Davies & Gamble (1979) and the Den Helder enclosures of Kuiper (1977). Each mesocosm consisted of a polyethylene plastic sleeve anchored at the bottom and supported by a collar raised and lowered by ropes tied to adjacent supports. Total volume was $3 \cdot 5 \, m^3$, but varied depending on the water level. Dosing was adjusted for volume at the time of each treatment. The actual period of the test discharge was 30 min followed by 24 h of restriction within the mesocosm.

The frequency of dosing was modeled after several offshore studies in which discharges had been monitored and found to be discrete events and intermittent in nature (Ray & Meek, 1980; Ayers *et al.*, 1980*a*). The length of the total dispersion period ($1 \cdot 5$ month) is typically less than average for a well of 6000 m depth, and is therefore a worst-case situation in that the biota had less time to recover between doses. The length of exposure to each dose (24 h) is also a worst-case scenario. Ayers *et al.* (1980*b*) found that a high discharge rate under calm sea conditions in 23 m of water and currents of 1–22 cm/s resulted in dispersion of barium and chrome in the upper plume between $0 \cdot 01$ and $0 \cdot 001\%$ of its concentration in the whole mud within 5–100 min of discharge. A 24-h exposure within an enclosed mesocosm should certainly allow all settleable solids to be removed from the water column and settle on the bottom.

The three concentrations chosen for the high ($1:100$ dilution or 10^2 dispersion), medium ($1:1000$ or 10^3), and low doses ($1:10\,000$ or 10^4) for the three drilling fluid treatment mesocosms were based on modifications of a hazard assessment performed by Petrazzuolo (1981) for the US Environmental Protection Agency (EPA). Petrazzuolo presumed that '... benthic impacts of drilling fluids were thought likely to correspond to dispersions of these fluids in the water column. Thus, water column dispersions of 10^4, 10^5, and 10^6 (from Ayers *et al.* (1980*a,b*) and others) were considered to result in sediment levels of drilling fluid solids equivalent to relative dispersions of 10^4, 10^5, and 10^6 in the sediment'. Since no rig discharges have ever been monitored in a shallow, semi-enclosed bay like Christmas Bay, it is uncertain what the actual dispersion achieved would be. Even though current speeds in Christmas Bay often approach those reported by Ayers *et al.* (1980*a,b*), the shallow depth could diminish the dispersion rate. Auble *et al.* (1982) suggested that different and more complex fate and effects behavior in bays and estuaries was likely due to different biota (e.g. seagrass beds), more complex circulation and stratification patterns, and potentially more active resuspension processes. To account for these differences, the 10^4–10^5 dispersion typically observed offshore was decreased by two orders of magnitude (10^2–10^4) for this bay study. Further decreases in the dilution rate below 10^2 were believed to be doses representative of areas subject to physical burial (i.e. beneath the rig). Testing at these concentrations appeared to be unrealistically high.

3. METHODS AND MATERIALS

The sequence of events are described in Fig. 3. Station selection and baseline assessment were performed from April 16 to November 13, 1981 followed by dispersion of the four muds over a 1·5 month period (November 14–December 31, 1981). Dates for dispersal are as follows: spud mud on November 14; nondispersed on November 24; lightly-treated lignosulfonate on December 17; and heavily-treated lignosulfonate on December 31, 1981. Monthly monitoring occurred for 6 months (January 15–July 16, 1982). Each in situ mesocosm, except for the controls and dredged sediment mesocosm, was dosed for 30 min with the appropriate mud for that date at a dosage rate specific to each mesocosm (1:100; 1:1000, or 1:10 000 v/v). For instance, the high dose mesocosm (1:100) received a dose of spud mud on November 14, nondispersed on November 24, *et cetera* in order to mimic the discharges from an active drilling rig. Each mesocosm represents an unknown distance and time from the rig, but a known concentration of drilling fluid wastes.

Drilling fluids were collected from four coastal or offshore drilling rigs during June 18–November 13, 1981. Each mud was collected under the shale shaker during active drilling in an acid-rinsed high density polyethylene pail, chilled to 4 °C, and transported immediately to the laboratory for storage. Daily mud reports from the spud date to day of

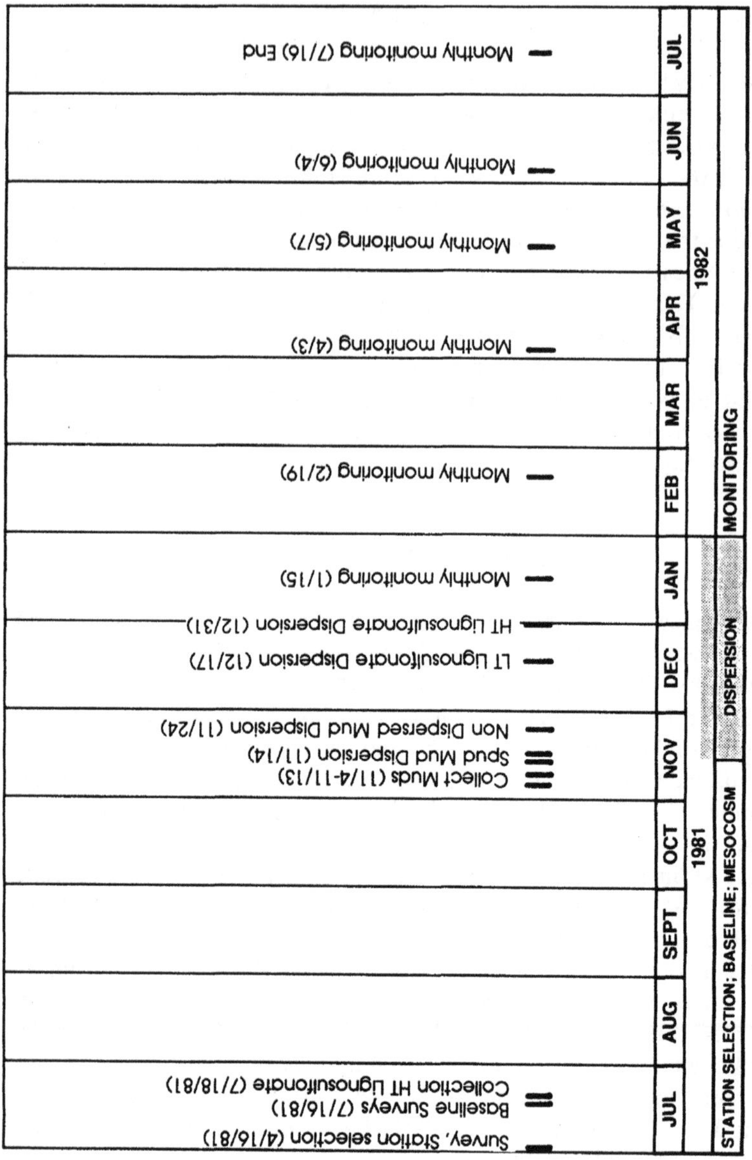

FIG. 3. Sequence of events for drilling fluid research program on shallow estuarine ecosystem (1981–1982).

collection were obtained for determining the product composition and mud properties.

These four muds and a sample of sediment from Station 1 were analysed by X-ray diffraction and X-ray fluorescence (Diano XRD-7/8000) for mineralogical and barium content. In addition, each sample was digested using the EPA Sludge Digestion Procedure (Delfino & Enderson, 1978) with analysis by atomic absorption (AA) spectrophotometry for As, Hg and Pb and inductively-coupled plasma (ICP) spectrophotometry for Al, Cd, Cr, Cu, Fe, Mn, Ni, Sr and Zn. Grain size analysis was done by pipette and a Ladal X-ray gravitational sedimentometer.

In order to predict possible impact on water quality, the four muds and the dredged sediment were mixed with ambient Christmas Bay water using the EPA/COE Elutriation Test Procedure (EPA, 1973a,b). This test involves vigorous mixing of one part sediment to four parts disposal site water (v/v) for 30 min, settling for 1 h, filtration (0·45 μm), and determination of soluble constituents.

Biological impacts were predicted using the EPA Region II Drilling Fluids Protocol that was in use during the research period (Jones & Hulse, 1982). This procedure consists of a 1:4 (v/v) dilution of the mud with $20\,^o/_{oo}$ seawater, mixing for 30 min, 1-h settling, and preparation of three phases (liquid, suspended particulate and solid). The liquid and suspended particulate phases were tested for 96-h with *Mysidopsis bahia* and with *Mercenaria mercenaria* and *Nereis virens* in the solid phase test for 10 days.

Hydrographic properties were measured in the field with a Hydrolab, field turbidity meter, YSI S-C-T meter, and a YSI Dissolved Oxygen Meter. Current speeds were determined with a subsurface neutrally buoyant drogue and water depth with a meter stick. Samples for geochemical, grain size, and benthic analyses were taken in triplicate with a modified Mackin hand corer (Baker *et al.*, 1977). The coring device (53 mm diameter) was designed to penetrate the mat of seagrass blades, rhizomes and roots and to sample this material and the underlying sediments to a depth of 30 cm. The sediment volume (100 cm^3) was kept constant during sampling because the sample volume was visible through the clear Lexan tube, which was calibrated. A relief valve allowed the volume of sediment to be adjusted to 100 cm^3 without disturbing the surficial sediments or seagrass material.

Samples for geochemical and grain size analysis were transported to the lab on ice. Grain size analysis was performed using sieves, a hydrometer, and a Rapid Settling Tube Analyzer. Barium levels were measured with

X-ray fluorescence and confirmed with X-ray diffraction. Replicate samples were analysed separately and the results were pooled.

Water quality samples (4 litres) were grab samples taken below the surface in acid washed glass jars, chilled, and transported to the laboratory immediately for analysis. Water samples were split into two fractions. One fraction was filtered (0·45 μm) and analysed for soluble components while the unfiltered fraction was analysed for total levels. Chemical analyses were performed according to the following methods: NH_3, Sb, As, NO_3-NO_2, N, Se, Vn (EPA, 1979); Ba, Be, Cd, Cr^{6+}, Cr(T), Cu, chlorophyll (A, B, C), Fl, Fe, Hg, Ni, solids, oil/grease, Zn, Co (APHA, 1975); C, Cr^{3+}, phenols, P (ASTM, 1975), and Mn (USGS, 1974).

A QA/QC sample of ambient Christmas Bay water was spiked with a commercially available laboratory spike for heavy metals (ERA Lot No. 8301) and run as a blind sample by the laboratory. Three laboratory standards were also run as blind samples with the X-ray fluorescence samples. The three samples were API standard barite (American Petroleum Institute, Lot No. 8), NBS standard drilling fluid (National Bureau of Standards, Lot No. RDM0101100TWD101S810), and US Geological Survey Standard marine sediment MAG-1 (Flanagan, 1976; Lot No. 27, Position 19).

4. RESULTS

The dates sampled, depth, mud weight, and chemical and physical properties of the four muds are given in Table 1. Mud weight was from 9·0 to 15·6 lbs/gal. Oil content as measured by the API retort was 0% (v/v). Oil and grease measured by the freon extraction and gravimetric procedure indicated a range of 18–161 ppm. At low levels, oil and grease analysis typically gives a higher value than retorting because the freon extracts and measures compounds other than hydrocarbons. The values compare favorably with values reported in Ayers *et al.* (1983). Table 2 lists the chemical product composition for the four muds from the spud date to the day of sampling. The muds became increasingly more complex (6–13 products) in composition with depth. Barite dominated the heavier muds while bentonite clay dominated in the lighter density muds.

Mineralogical composition is listed in Table 3. Barium sulfate levels are less than the reported percentages for barite because barite includes other non-barium minerals. The values are closely correlated. Elemental analysis of the four muds and of clean sediment from Station 2(C) are

TABLE 1

Chemical and physical properties of four drilling fluids used in research program. Properties measured by field and laboratory methods (API, 1980; 8th edn)

Property	Spud	Nondispersed	Lightly-treated lignosulfonate	Heavily-treated lignosulfonate
Date sampled	11/4/81	11/9/81	11/13/81	7/10/81
Depth (ft)	2 570	7 260	12 730	18 122
Weight (lb/gal)	9·0	9·2	9·8	15·6
Viscosity, funnel (s)	38	37	45	41
PV (cP)	7	7	18	16
YP (lb/100 ft^2)	13	12	4	16
10 s/10 m Gels (lb/100 ft^2)	9/19	11/24	2/4	3/30
pH	10·3	11·1	11·5	10·0
API filtrate (ml)	25	18	6·8	14·4
Cake thickness (in)	2/32	2/32	2/32	2/32
Alkalinity, mud (P_m)	0·5	0·6	—	0·3
Alkalinity, filtrate (P_f/M_f)	0·3/0·4	0·3/0·4	1·4/2·4	0·2/0·6
Chlorides (mg/litre)	2 100	3 600	6 500	5 000
Calcium (mg/litre)	trace	20	120	trace
Retort				
Solids, % (v/v)	5	7	10	32
Water, % (v/v)	95	93	90	68
Oil, % (v/v)	0	0	0	0
Lignosulfonate (lb/bbl)	0·06	0·00	2·68	8·7
Cation exchange (meq/100 g)	24·5	24·5	19·0	3·5
Solids, % (w/w)	13·03	16·39	22·59	63·22
Oil/grease, (G/F); (μg/g(dry))	18·0	35·0	20·0	161·0

Maurice (Mo) Jones

TABLE 2

Product composition of four drilling fluids used in research program. Percentage composition based on weight basis of products added from spud date to date of sampling

Product	Spud	Nondispersed	Lightly-treated lignosulfonate	Heavily-treated lignosulfonate
Barite	4·36	25·81	42·58	93·28
Bentonite clay	89·64	65·69	47·22	2·87
Causticized lignite	0·22	3·20	NU	0·13
Sodium hydroxide	4·91	4·35	4·00	0·71
Calcium hydroxide	0·22	0·13	0·01	0·06
Sodium bicarbonate	NU	NU	NU	0·14
Sodium carbonate	0·20	0·64	0·10	0·13
Polyanionic cellulosic polymer	NU	0·10	0·79	NU
Chrome lignosulfonate	NU	NU	1·87	2·90
Pecan hulls	NU	NU	3·05	NU
Sodium acid pyrophosphate	NU	NU	0·02	NU
Tributyl phosphate	NU	NU	0·01	NU
Zinc oxide	NU	NU	0·01	NU
Aluminum stearate	NU	NU	0·01	NU
Polyacrylamide	NU	NU	NU	0·06
Sodium polyacrylate	NU	NU	NU	0·02
Paper	NU	NU	NU	0·25
Lost circulation material	NU	NU	NU	0·07
Sugar cane fiber	NU	NU	NU	0·08
Total percentage	99·55	99·92	99·67	100·02
Total products	6	7	12	13
Total weight (MT)	21	35·4	726	1 150·3

NU = not used: MT = metric ton

TABLE 3

Mineralogical composition of four drilling fluids (with cuttings) used in research program. Analysis by X-ray diffraction in percentage dry weight solids. Range reflects semi-quantitative nature of analytical method

Mineral	Spud	Nondispersed	Lightly-treated lignosulfonate	Heavily-treated lignosulfonate
α Quartz	28–30	20–22	8–10	9–11
Barium sulfate	2·5	1–2	18·9	66·5
Feldspar	4–6	1–3	trace	none
Calcite	trace	2–4	10–12	1–3
Dolomite	none	none	1–3	1–2
Salt (NaCl)	2–4	2–4	2–4	none
Anhydrite	none	none	none	none
Kaolinite	2–4	2–4	none	none
Illite	4–6	4–6	3–5	none
Smectite and mixed layer clays	50–54	50–55	38–42	10–12
Amorphous	1–3	8–10	11–13	8–10
Total percentage (lower range)	93·5	90	91·9	95·5
Total percentage (upper range)	109·5	110	107·9	104·5

given in Table 4. Muds are enriched relative to the sediments in Al, As, Ba, Cd, Cr, Cu, Fe, Mn, Ni, Pb, Sr and Zn, although most enrichment occurs in the heavier weighted or treated muds.

4.1. Elutriation Results

Results of the elutriation tests using Christmas Bay water and the four muds plus a sample of clean sediment are summarized in Table 5. The elutriation test predicted a significant increase above ambient levels for Ba, C, Cr^{3+}, Cr(T), Fe, Mn, N, Oil and Grease, P(T), suspended solids, Sr and Zn for the drilling fluids. Tests with sediment indicated an increase in C(T), Mn, NH_3 and specific conductance. These results are fairly consistent with results reported by Lee *et al.* (1976) in which manganese and ammonia were readily released from most dredged sediments. Other components are released according to their chemical form and sediment matrix. Comparisons of released materials against one single sample of ambient water can be misleading in that the natural variability in water quality is not accurately reflected. Figures 4 and 5 indicate the wide variability observed over 12 months.

TABLE 4

Elemental analysis of four drilling fluids and Christmas Bay dredged material used in research program. Samples digested using EPA Sludge Digestion Procedure and analysis by AA (As, Hg, Pb) or ICP (remaining elements). Barium analysis by X-ray fluorescence. Values are mg/litre by weight

Analyte	Spud	Nondispersed	Lightly-treated lignosulfonate	Heavily-treated lignosulfonate	Dredged material
Al	670	1 200	760	1 800	520
As	1·7	8·3	3·9	9·6	2·0
Ba	8828	14 713	111 227	391 353	345
Cd	<0·019	<0·024	0·070	2·0	<0·018
Cr	1·9	2·8	130	430	0·75
Cu	1·3	1·0	5·4	21	<0·009
Fe	1 600	3 100	1 400	8 100	1 400
Hg	<0·000 2	<0·000 2	<0·000 2	<0·000 2	<0·000 2
Mn	40	54	35	160	32
Ni	1·9	2·9	1·3	8·8	1·0
Pb	2·8	3·2	30	47	3·2
Sr	10	14	58	24	9·9
Zn	23	20	640	260	5·5

4.2. Water Quality Results

Results from the water samples taken in the high dose (1:100) (v/v) mesocosm at Station 2 before, during, 24 h after dispersion but before water exchange, and during the 6 month monitoring period are tabulated in Table 6. Ba, Cr^{6+}, Cr^{3+}, and total Cr were measured as both total and soluble fractions to differentiate between particulate bound and ionic concentrations. The following parameters were significantly increased during (maximum concentration) dispersion of the muds inside of the (1:100 v/v) mesocosm: Al, Ba(T), C, Cr(T), Cr(F), Cr^{3+}(T, F), N, Pb, suspended solids and Zn. At the end of the 24-h settling period prior to water exchange, the following parameters were still elevated above ambient: Al, Ba(T) and suspended solids. All other parameters previously elevated had decreased to ambient levels. Soluble barium (Ba-F) occurred once during the dispersion of the HTL mud. It is uncertain whether this increase truly represents soluble Ba or exceptionally fine particulate matter that passed through the 0·45 µm filter. No Cr^{6+} was detected during any dispersions. Transmittance (%) significantly decreased during the dispersions due to the particulates, but quickly returned to ambient within

TABLE 5

Results of elutriation test with Christmas Bay water, four muds, and dredged sediment (1 : 4 v/v). Results are in mg/litre except for specific conductance (μmhos @ 25 °C)

Parameter	Christmas Bay water	Spud mud, elutriant	ND mud elutriant	LTL mud elutriant	HTL mud elutriant	Sediment elutriant
Al	1·0	0·2	0·5	0·8	0·2	0·1
Ba	<0·2	<0·2	0·9	0·8	1·0	<0·2
Cd	<0·005	<0·005	<0·005	<0·005	<0·005	<0·005
C(T)	24	42	64	710	110	42
Cr^{+6}	<0·03	<0·03	<0·03	<0·03	<0·03	<0·03
Cr^{+3}	<0·03	<0·03	<0·03	<0·03	0·08	<0·03
Cr(T)	<0·03	<0·03	<0·03	<0·03	0·08	<0·03
Cu	<0·02	<0·02	<0·02	<0·02	<0·02	<0·02
Fe	0·02	0·02	0·04	0·11	0·06	<0·02
Mn	<0·02	<0·02	<0·02	0·04	0·80	0·03
Ni	<0·02	<0·02	0·03	<0·02	0·02	<0·02
$N-NH_3$	<0·1	1·0	1·7	0·7	0·1	2·8
N-Kjeldahl	<0·1	2·2	2·8	0·8	0·1	2·8
Oil-Grease	<1	<1	<1	2	2	<1
P(T)	0·04	0·02	0·06	0·08	0·04	0·02
Pb	<0·05	<0·05	<0·05	<0·05	<0·05	<0·05
Solids (D)	27 200	22 200	22 060	25 600	22 900	27 600
Solids (S)	14	14	8	18	4	<1
Sp. Cond.	34 000	29 000	29 000	32 000	30 000	57 000
Sr	8·1	8·1	9·4	16	4·2	8·0
Zn	<0·01	<0·01	<0·01	0·42	<0·01	<0·01

T = total; D = dissolved; S = suspended; ND = nondispersed mud; LTL = lightly treated lignosulfonate; HTL = heavily treated lignosulfonate.

24 h except for the LTL dispersion. Transmittance returned to ambient levels once water exchange was allowed. Chlorophyll A, B, C also decreased during the dispersions and remained at low levels during the first 2 months of the monitoring period (January, February), but returned to ambient or above ambient levels during the last 3 months of the monitoring period. The chlorophyll decrease coincides with the coldest and typically darkest months of the year (Penn, 1979) and it is uncertain whether the decrease is a natural decrease or effect of reduced transmittance. Since transmittance returned to ambient levels at the end of the dispersion period, it is believed that the decrease is normal phytoplankton seasonal variation. The normal decline of the seagrass *Halodule wrightii* during these months may also explain the decrease.

Maurice (Mo) Jones

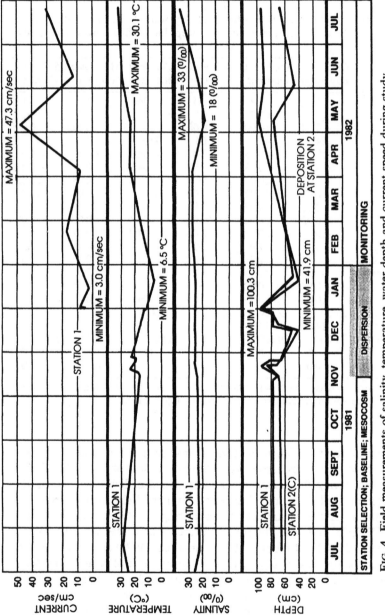

FIG. 4. Field measurements of salinity, temperature, water depth and current speed during study.

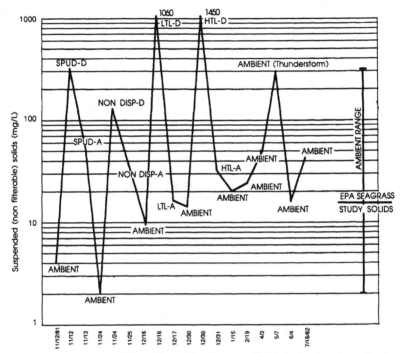

FIG. 5. Suspended (nonfilterable) solids data (mg/litre) during study at Station 2(H) inside of mesocosm. Samples taken during ambient conditions during dispersion (D), and 24 h after (A) dispersion prior to water exchange. Legend: SPUD = spud mud; NON DISP = nondispersed mud; LTL = lightly treated lignosulfonate; HTL = heavily treated lignosulfonate mud.

4.3. Sediment Geochemistry Results

Analysis of the sediment cores for grain size and barium analyses are given in Tables 7 and 8 and graphed in Fig. 6. Barium was chosen as a tracer in order to verify the actual dose calculated. Using the barium enrichment approach described by Boothe & Presley (1983), theoretical barium enrichment factors were calculated for each treatment mesocosm based on the laboratory analysis of the mud, measured background levels, and Petrazzuolo's assumption that water column dispersion rates equal sediment enrichment rates. As can be seen in Table 7, background levels in the bay ranged from 169–345 ppm. This range is consistent with the survey published by White *et al.* (1985) for Christmas and Galveston Bays where a mean of 413 ppm was determined for bay sediments Baker *et al.*

TABLE 6

Water quality at Station 2(H) inside of mesocosm (1:100) before, during, and after dispersion of four drilling muds. All units are mg/litre except for specific conductance (μmhos @ 25°C) and transmittance (%). Range given for last five sampling events

Parameter	Ambient 11/12/81	Dispersion		Ambient 11/24/81	Dispersion		Ambient 12/17/81	Dispersion		Ambient 12/30/81	Dispersion		Monitoring	
		Spud during	Spud after		ND during	ND after		LTL during	LTS after		HTL during	HTL after	1/15/82	2/19–7/16/88
Al	0·15	<0·02	0·26	0·09	<0·02	0·53	0·5	6·0	0·2	1·0	7·0	2·3	0·3	0·4–1·0
Ba-T	<0·2	1·0	1·7	<0·2	1·4	1·5	<0·2	1·1	<0·2	<0·2	16	2·7	<0·2	<0·2
Ba-F	<0·2	<0·2	<0·2	<0·2	<0·2	<0·2	<0·2	0·3	<0·2	<0·2	0·3	<0·2	<0·2	<0·2
Cd	<0·0005	<0·0005	<0·0005	<0·0005	<0·0005	<0·0005	<0·0005	<0·0005	<0·0005	<0·0005	<0·0005	<0·0005	<0·0005	<0·0005–0·003
C-T	28	28	28	30	31	33	27	80	27	24	45	28	33	28–35
CHL-ABC	14	8·3	14	8·3	11	14	2·7	11	8·0	10	0	0	0	0–51
Cr-T	<0·03	<0·03	<0·03	<0·03	<0·03	<0·03	<0·03	1·8	<0·03	<0·03	0·12	<0·03	<0·03	<0·03–0·007
Cr-F	<0·01	<0·03	<0·03	<0·03	<0·03	<0·03	<0·03	0·13	<0·03	<0·03	<0·03	<0·01	<0·03	<0·03
Cr⁶⁺	<0·01	<0·01	<0·01	<0·01	<0·01	<0·01	<0·01	<0·01	<0·01	<0·01	<0·01	<0·01	<0·01	<0·01
Cr⁶⁺-F	<0·03	<0·03	<0·01	<0·01	<0·03	<0·01	<0·01	1·8	<0·03	<0·03	0·12	<0·03	<0·01	<0·03–0·007
Cr³⁺	<0·03	<0·03	<0·03	<0·03	<0·03	<0·03	<0·03	0·13	<0·03	<0·03	<0·03	<0·03	<0·03	<0·03
Cr³⁺-F	<0·03	<0·03	<0·03	<0·03	<0·03	<0·03	<0·01	0·5	<0·01	<0·01	0·6	<0·01	<0·01	<0·01–2·8
N-Kj	<0·01	<0·01	<0·01	<0·01	<0·01	0·1	<0·01	<0·04	<0·04	<0·04	<0·04	<0·04	<0·01	<0·04–0·15
P-T	<0·04	<0·04	<0·04	<0·04	<0·04	<0·04	<0·04	<0·04	<0·04	<0·04	<0·04	<0·04	<0·04	<0·0005
Pb	0·070	<0·0005	<0·0005	<0·0005	<0·0005	<0·0005	<0·0005	<0·0005	<0·0005	<0·0005	0·009	<0·0005	<0·0005	<0·0005
S-Dis	28 200	25 200	25 300	28 400	28 400	28 500	30 000	28 000	28 500	27 200	31 100	30 300	29 000	25 000–36 200
S-Sus	4	326	54	2	130	36	14	1 060	16	14	1 450	32	20	26–308
Sp. Cond.	36 000	35 000	37 000	40 000	40 000	41 000	43 000	NA	NA	34 000	43 000	43 000	39 000	24 000–57 000
Sr	9·0	9·0	9·0	9·0	9·0	9·0	9·0	100	100	811	100	100	100	9·0–14·0
% Tr	95	65	90	98	46	94	94	51	75	94	45	95	100	70–98
Zn	0·11	<0·001	<0·001	<0·001	<0·002	<0·010	0·010	5·2	0·002	<0·01	0·88	0·008	0·004	0·001–0·004

T = total; F = filtered (0·45μ); CHL = chlorophyll A, B, C; S = solids; Sus = suspended; Dis = dissolved; Kj = Kjeldahl; Tr = transmittance; ND = nondispersed; LTL = lightly-treated lignosulfonate; HTL = heavily-treated lignosulfonate.

TABLE 7

Theoretical barium enrichment compared to measured barium enrichment of sediments inside of mesocosms at Station 2 (seagrass). Values are parts per million (ppm) barium as measured by X-ray fluorescence

	High dose (1:100)	Medium dose (1:1 000)	Low dose (1:10 000)	Dredged sediment	Control
Baseline Ba^{2+} (7/16/81)	242	202	181	169	345
Predicted maximum Ba^{2+}	36 010	3 852	516	169	345
Predicted enrichment factor	(148·8×)	(19·1×)	(2·8×)	(1·0×)	(1·0×)
Actual maximum Ba^{2+} (12/31/81)	40 200	2 010	424	460	412
Actual enrichment factor	(166·1×)	(9·9×)	(2·3×)	(2·7×)	(1·2×)
Final measured Ba^{2+} (7/16/82)	678	302	240	255	254

TABLE 8

Grain size analysis of sediments at Station 1 (external control) and within the mesocosms at Station 2 (seagrass station). Mesocosms in Station 2 analysed are high dose (H), dredged sediment treatment (S), and internal control (C). Classes of grain size are expressed in percent by weight and the mean grain size in phi (ϕ).

	Baseline		Dispersion			Monitoring			
	7/16/81	11/14/81	12/31/81	1/15/82	2/19/82	4/3/82	5/7/82	6/4/82	7/16/82
Station 1									
Gravel (%)	0·00	0·03	0·01	0·00	0·04	0·00	0·00	0·01	0·00
Sand (%)	93·56	92·31	93·04	92·79	92·41	94·38	93·65	95·31	93·39
Silt (%)	5·29	4·39	4·69	4·50	4·03	4·89	4·63	3·98	4·52
Clay (%)	1·10	0·91	1·03	1·15	1·81	1·36	1·29	1·01	1·07
Mean (ϕ)	3·36	3·26	3·28	3·31	3·43	3·20	3·38	3·09	3·30
Station 2(H)									
Gravel (%)	0·00	0·00	0·00	0·01	0·04	0·01	0·00	0·00	0·00
Sand (%)	92·58	93·01	87·33	88·51	89·30	93·26	92·91	93·49	93·61
Silt (%)	1·97	1·85	6·01	5·41	5·02	4·72	3·01	2·51	2·01
Clay (%)	5·07	5·14	6·56	6·08	5·39	1·96	3·39	3·17	4·38
Mean (ϕ)	3·64	3·60	4·37	4·25	4·18	3·31	3·37	3·59	3·71
Station 2(S)									
Gravel (%)	0·00	0·00	0·00	0·01	0·00	0·05	0·05	0·00	0·00
Sand (%)	80·81	83·09	91·10	91·19	91·01	93·01	93·61	93·01	92·51
Silt (%)	3·17	3·22	7·41	7·26	7·54	6·07	4·60	4·98	4·72
Clay (%)	15·33	13·19	1·49	1·52	1·45	0·87	1·69	2·06	2·77
Mean (ϕ)	4·52	4·43	3·63	3·59	3·62	3·28	3·30	3·41	3·51
Station 2(C)									
Gravel (%)	0·00	0·10	0·00	0·00	0·00	0·00	0·00	0·00	0·03
Sand (%)	91·00	90·14	90·39	91·03	90·32	94·02	94·11	93·92	93·06
Silt (%)	2·44	2·81	2·39	2·04	2·91	2·36	2·79	2·72	2·90
Clay (%)	6·25	6·35	7·22	6·95	6·79	3·62	3·01	3·36	4·01
Mean (ϕ)	3·75	3·63	3·72	3·70	3·74	3·38	3·31	3·30	3·35

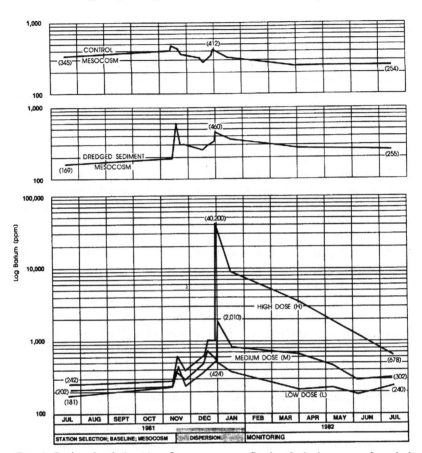

FIG. 6. Barium levels (ppm) at five mesocosms, Station 2, during research period.

(1981) and White *et al.* (1985) found that barium concentrations have a positive correlation with mud (grain size) content. Range for Ba throughout the Galveston Bay system was < 100–1600 ppm. Nearshore Texas shelf sediments were found to have slightly higher mean Ba levels (400–600 ppm).

For the high dose mesocosm (1:100), a theoretical enrichment of 148·8 × background levels was calculated. After all dispersions were made, 40 200 ppm was measured, equal to an enrichment factor of 166·1 × background. The mesocosm apparently was successful in containing the dispersions and no leaks were apparent. The additional barium is possibly explainable by the varying volume of the mesocosm due to tidal flux, which may have resulted in heavier dosing than intended. Within 6 months, the 40 200 ppm was reduced to 678 ppm through resuspension, dilution, or movement of the barium into the deeper sediments.

Barium levels in the medium dose mesocosm (1:1000) were less than predicted (predicted = 3852 ppm; actual = 2010 ppm). This may be a result of losing material through a rip in the plastic sleeve during the same storm that destroyed Station 3. The low dose mesocosm (1:10 000) remained intact apparently because the theoretical enrichment was close to the actual enrichment (516 and 424 ppm, respectively). An enrichment occurred at the dredged sediment mesocosm (1:100) even though none was predicted. Sediments with higher than initially measured barium levels were either used in the treatment or contamination of the mesocosm occurred from resuspended sediments from adjacent mesocosms. The dredged sediment mesocosm was relatively close to station 3 and may have been contaminated by sediments from that station. Some slight elevation was also detected at the Station 2 control, also adjacent to Station 3. The barium levels at the control mesocosm were below the initial ambient levels at the end of the 6-month period.

For grain size analysis, the class ranges of Udden Wentworth, as described by Dyer (1986) were used: gravel = -5 to -1ϕ; sand = $> -1\phi-4\phi$; silt = $4 \cdot 0-8 \cdot 0\phi$; and clay = $8 \cdot 0-15 \cdot 0\phi$. Overall, all stations had between 80 and 95% sand and were considered to be roughly similar in grain size. A detailed grain size analysis indicated that Station 1 had a significantly higher silt fraction; within Station 2 varied; and 2(S) was significantly different (15·33% clay) prior to mud dispersion. During dispersion, Station 1 changed very little while Station 2H had a significant increase in silt and clay fractions due to the drilling mud dispersion. Station 2(S) changed significantly in that the dredged sediment added more sand to the sediment composition, decreasing the levels of clay. Station 2(C) changed very little. During the 6-month monitoring period, both controls remained much the same. Station 2(H) lost the finer particles and had a mean ϕ similar to the baseline mean (3·71 and 3·64ϕ, respectively). Station 2(S) remained significantly coarser than its initial composition due to the dredged sediment addition. The amount of change measured is within the natural areal patchiness initially measured at the two stations.

Data from the quality assurance and quality control program are presented in Table 9 for water analyses and in Fig. 7 for barium analyses of sediments. Agreement between spiked samples and laboratory results was variable. The laboratory reported lower results for Al, Ba and Zn

TABLE 9

QA/QC data for water quality analysis. Christmas Bay seawater was spiked with a commercially available laboratory standard (ERA Lot No. 8301) and run as a blind sample by the laboratory. All values are for total amounts in mg/litre. Analysis by atomic absorption spectrophometry

Element	Lab. ambient value	Lab. spiked value	Certified spike	Anticipated value (ambient + spike)	Agreement
Al	0·7	0·3	0·155	0·855	Low
Ba	<0·2	<0·2	0·220	0·220	Low
Cd	0·0013	0·055	0·045	0·0463	Good
Cr	<0·03	0·48	0·125	0·125	High
Pb	<0·005	0·05	0·057	0·057	Good
Zn	0·004	0·05	0·172	0·176	Low

than anticipated, good agreement for Cd and Pb, and overestimated Cr^{n+}. Agreement between laboratory results and three laboratory standards for sediment barium levels was good over a wide range ($\sim$ 500–500 000 ppm).

5. DISCUSSION

For water quality impacts, results from this study were consistent with results observed on the OCS. Most parameters were affected only for relatively short time periods and quickly returned to ambient levels when water exchange was allowed. Particulate Ba, Cr and Al were the most persistent parameters during the 24-h treatment period. Suspended solids levels reported are in close agreement with solids reported by Ayers *et al.* (1980) for these dilution rates. Suspended solids levels were typically much higher (Fig. 5) in this study than the 15 mg/litre exposure concentration reported by Morton *et al.* (1986) in an EPA seagrass-benthic microcosm study; however, this exposure was intermittent (not constant for 42 days) and resuspension was allowed. Ambient levels of suspended solids (2–308 mg/litre) exceeded treatment levels (130–1450 mg/litre) when storms occurred. Suspended solids, salinity and other parameters in a shallow estuary tend to be highly variable due to storm events, fresh-water input, wind and currents (Sikora & Kjerfve, 1985). The shallow depth does not allow for any stratification and complete mixing was evident (based on temperature and salinity profiles). Water quality did not appear to be significantly different as a result of mud dispersions once water exchange

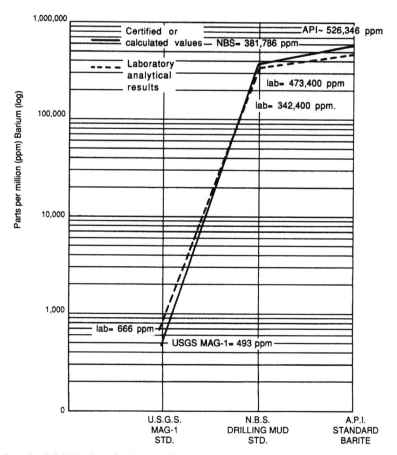

FIG. 7. QA/QC data for X-ray fluorescence analysis of sediments and drilling waste. Legend: A.P.I. = American Petroleum Institute; N.B.S. = National Bureau of Standards; U.S.G.S. = United States Geological Survey.

was allowed in the mesocosms and no persistent effects were detected. The elutriation test tends to be a conservative test in that it predicted more impacts. Elutriation tests using the same dilution ratios as in the field study might be more predictive than the 1:4 ratio used in the laboratory.

Barium enrichment of the sediments was in good agreement with predicted levels when the mesocosm remained intact. Some contamination from Station 3 was evident and suggested that stations should be located further apart. It also suggests that shallow estuaries may not be truly low energy systems, even though mean currents are often less than offshore environments. The frequency and impact of storm events may very well have caused greater resuspension and sediment transport than normally expected. Sediment deposition and scouring was clearly evident from the change in water depths and the unexpected development of deep holes in the study area. The highest level measured (40 200 ppm) was comparable to levels measured by Boothe & Presley (1983) near multi-well production platforms, confirming that the experimental design succeeded in developing a worst case situation.

Effects on the sediment composition decreased much quicker than typically reported for low energy offshore environments (Gettleson, 1980; Schwartz *et al.*, 1980; Gillmer *et al.*, 1985). It is assumed that the recovery rate was influenced by the rapid resuspension of the estuarine sediments and quick dilution of the muds in the sediments. Episodic storm events and weather fronts are significant factors in shallow ecosystems, particularly to water quality and benthic communties (Harper & Hopkins, 1976; Flint, 1985). The presence of a very diverse and abundant biotic community associated with the seagrass beds may have also helped contribute to the faster recovery rate through active bioturbation. Numerous tubes and burrows were observed in the sediment cores, some as deep as 30 cm.

6. CONCLUSIONS

Analysis of the water samples collected before, during and after dispersal of drilling fluid wastes showed an impact on water quality, primarily at the highest dosage level (1 : 100). The elevated levels returned to ambient levels immediately after water exchange was allowed. Episodic weather appeared to be the primary cause of significant variability in ambient water quality. Elutriation tests overestimated the impacts when performed according to protocol.

Barium, from the barite in the drilling fluids, was successfully used as a geochemical tracer and to confirm the estimated doses. Elevated barium levels returned to ambient or slightly above ambient levels for Gulf of Mexico estuarine sediments within 6 months of discharging. Sediment deposition, resuspension and bioturbation are the most probable mechanisms to explain the barium dispersal. Changes in grain size due to the mud or dredged sediment treatments were well within the areal patchiness naturally occurring within the two stations.

Patchiness and the presence of contaminating drilling mud from sediment resuspension somewhat confounded the data analysis, but overall there were similarities in the results of the drilling mud and dredged material dispersal. Both physical and chemical changes were detected, but neither appeared to be very long lasting (> 6 months).

In order to avoid any impact on water quality and sediment geochemistry in a shallow estuary of this type, the data suggest that a minimal dilution of 1:1000 (v/v) is required. This dilution factor could be achieved by predilution before discharging or by controlling discharge rates. This approach has been adopted by EPA and DOI for other shallow areas and would be advisable in this shallow, semi-enclosed ecosystem.

ACKNOWLEDGEMENTS

The author acknowledges funding of this research project by the Halliburton Education Foundation, IMCO Services Division of the Halliburton Company, and the University of Texas School of Public Health. Enseco Houston assisted in the preparation of this manuscript and attendance at the conference. I am particularly indebted to Mike Hulse (Shell Oil Company), Jim Baker, Ph.D. (Rollins Environmental Field Services), and Cynthia Stahl (US Environmental Protection Agency) for their assistance in the field study. Doyle Waller (IMCO Services), Bill Taylor (Halliburton Company), Drs Carl Hacker, Gene Schroder and Ron Harrist (UTSPH), and the staff of Enseco Houston (Dolores Phillips, David Havis) provided continued encouragement and support. Lastly, the assistance of David Giesen made the project possible.

REFERENCES

AMERICAN PUBLIC HEALTH ASSOCIATION (APHA) (1975). *Standard Methods for the Examination of Water and Wastewaters*, 14th edn. APHA, American Waters Works Association, and Water Pollution Control Federation, Washington D.C.

AMERICAN SOCIETY FOR TESTING AND MATERIALS (1975). *Annual Book of Standards*, Part 31. Water ASTM, Philadelphia, PA.

ANON. (1980). Research on the environmental fate and effects of drilling fluids and cuttings, January 21–24, 1980. Lake Buena Vista, FL, Vol I, II. Courtesy Associates, Washington D.C.

AUBLE, G., ANDREWS, A., ELLISON, R., HAMILTON, D. *et al.* (1982). Results of an adaptive environmental assessment modeling workshop concerning potential impacts of drilling muds and cuttings on the marine environment, Sept. 14–18, 1981. Gulf Breeze, FL. US Environmental Protection Agency. EPA 600/9-82-019, 64 pp.

AYERS, R.C., SAUER, T., MEEK, R. & BOWERS, G. (1980a). An environmental study to assess the impact of drilling discharges in the Mid Atlantic I. Quantity and fate of discharges. In *Proc. Symposium on Research on Environmental Fate and Effects of Drilling Fluids and Cuttings.* Lake Buena Vista, FL, Jan. 21–24, 1980, pp. 382–418.

AYERS, R.C., SAUER, T., STEUBNER, D. & MEEK, R. (1980b). An environmental study to assess the effect of drilling fluids on water quality parameters during high rate, high volume discharges to the ocean. In *Proc. Symposium on Research on Environmental Fate and Effects of Drilling Fluids and Cuttings.* Lake Buena Vista, FL, Jan. 21–24, 1980, pp. 351–91.

AYERS, R.C., SAUER, T. & ANDERSON, P. (1983). The generic mud concept for offshore drilling for NPDES permitting. In *Proc., IADC/SPE Drilling Conference.* Society of Petroleum Engineers, Dallas, Texas, pp. 327–34.

BAGNALL, R. (1976). Definition and persistence of an estuarine zooplankton assemblage. Doctoral dissertation, University of Houston, 137 pp.

BAKER, J.H., PUGH, L. & KIMBALL, K. (1977). A simple hand corer for shallow water sampling. *Chesapeake Science,* 18(2), 232–6.

BAKER, J.H., JOBE, W.D., HOWARD, C. & KIMBALL, K. (1981). Benthic biology. In *Ecological Investigations of Petroleum Production Platforms in the Central Gulf of Mexico,* Vol. I, Part 6, ed. C. A. Bedinger. Southwest Research Institute, San Antonio, Texas, 391 pp.

BOOTHE, P.N. & PRESLEY, B.J. (1983). Distribution and behavior of drilling fluids and cuttings around Gulf of Mexico drilling sites. Report to American Petroleum Institute. Texas A&M University, College Station, Texas, 141 pp.

COMMITTEE ON OCEAN POLLUTION RESEARCH, DEVELOPMENT AND MONITORING (COPRDM) (1980). Evaluation panel report on federal programs in environmental impact studies of petroleum in the environment. National Marine Pollution Program Office, NOAA, Rockville, MD, p. 57.

DAVIES, J.M. & GAMBLE, J.C. (1979). Experiments with large enclosed ecosystems. *Phil Trans. R. Soc. Lond. B, Bull. Sci.,* 286, 523–44.

DELFINO, J. & ENDERSON, R. (1978). Comparative study of methods for analysis of total metals in sludge. *Waste and Sewage Works,* 125, 32–48.

DEPARTMENT OF THE INTERIOR (1981). Request for comments on effects from drilling discharges on the Outer Continental Shelf. *Fed. Reg.,* 46(205), 52039, October 23.

DYER, K.R. (1986). *Coastal and Estuarine Sediment Dynamics.* John Wiley. New York, 342 pp.

EPA (1973a). Ocean dumping criteria. *Fed. Reg.,* 38(94), 12872–77.

EPA (1973b). Ocean dumping final criteria. *Fed. Reg.,* 38, 28610–21.

EPA (1979). *Methods for Chemical Analysis of Water and Wastes.* EPA, Washington, D.C.

FLANAGAN, F. (1976). Descriptions and analyses of eight new USGS Rock Standards. Geological Survey Professional Paper 840. US Geological Survey, Washington, D.C., pp. 25–8.

FLINT, W. (1985). Long term estuarine variability and associated biological response. *Estuaries*, **8**(2A), 158–69.

GETTLESON, D.A. (1980). Effects of oil and gas drilling operations on the marine environment. In *Marine Environmental Pollution, Vol. 1, Hydrocarbons*, (ed.) R. Geyer. Elsevier Scientific Publishing Co., New York, pp. 371–411.

GILLMER, R.B., MENZIE, C., MARIANI, G., LEVIN, D., AYERS, R.C. & SAUER, T. (1985). Effects of exploratory drilling discharges on the benthos. In *Wastes in the Ocean, Vol. 4, Energy Wastes in the Ocean*, ed. I. Duedall *et al.* John Wiley, New York, pp. 243–70.

GRICE, G.D. & REEVE, M.R. (Eds) (1982). Introduction and description of experimental ecosystems. In *Marine Mesocosms: Biological and Chemical Research on Experimental Ecosystems*. Springer-Verlag, New York, pp. 1–10.

HARPER, D.E. & HOPKINS, S.H. (1976). The effects of oyster shell dredging on macrobenthic and nektonic organisms in San Antonio Bay, Texas. In *Shell Dredging and its Influence on Gulf Coast Environments*, ed. A. Bouma, Gulf Publishing Company, Houston, Texas, pp. 232–79.

JONES, M. & HULSE, M. (1982). Drilling fluid bioassays and the OCS. *Oil and Gas Journal*, **80**(25), 241–4.

KUIPER, J. (1977). Development of North Sea coastal plankton communities in separate plastic bags under identical conditions. *Mar. Biol.*, **37**, 97–107.

LEE, G.F., LOPEZ, J. & PIWONI, M. (1976). An evaluation of the factors influencing the results of the elutriate test for dredged material. In *Proc. ASCE Conference on Dredging and Its Environmental Impact*. ASCE, New York, pp. 253–99.

MCEACHRON, L., SHAW, C. & MOFFETT, A. (1977). A fishery survey of Christmas, Drum and Bastrop Bays, Brazoria County, Texas. Technical Series, Bulletin 20. Texas Parks and Wildlife Dept., Austin, Texas, 83 pp.

MORTON, D., DUKE, T., MACAULEY, J., CLARK, J., PRICE, W., HENDRICKS, S. *et al.* (1986). Impact of drilling fluids on seagrasses: an experimental community approach. In *Community Toxicity Testing*, ASTM STP 920, ed. J. Cairns. American Society for Testing and Materials, Philadelphia, pp. 199–212.

NATIONAL RESEARCH COUNCIL (1983). *Drilling Discharges on the Marine Environment*. National Academy Press, Washington, D.C., 180 pp.

PENN, G. (1979). Decapod crustacean communities in Texas seagrasses. Doctoral dissertation, University of Houston. 126 pp.

PETRAZZUOLO, G. (1981). An environmental assessment of drilling fluids and cuttings released into the Outer Continental Shelf. Report for EPA Industrial Permits Branch. Technical Resources, Inc., Volumes I and II.

RAY, J. & MEEK, R. (1980). Water column characterization of drilling fluids dispersions from an offshore exploratory well on Tanner Bank. In *Proc. Symposium on Research and in Environmental Fate and Effects of Drilling Fluids and Cuttings*, Lake Buena Vista, FL. Jan. 21–24, pp. 223–58.

SIKORA, W.B. & KJERFVE, B. (1985). Factors influencing the salinity regime of Lake Ponchatrain, Louisiana, a shallow coastal lagoon: analysis of a long-term data set (1946–1982). *Estuaries*, **8**(2A), 170–80.

SCHWARTZ, R.C., DEBEN, W., COLE, F. & BENTSEN, L. (1980). Recovery of the macrobenthos at a dredge site in Yaquina bay, Oregon. In *Contaminants and Sediments, Vol. 2, Analysis, Chemistry and Biology*, ed. R. Baker. Ann Arbor Science, Ann Arbor, MI, pp. 391–408.

US COAST GUARD (1976). Seadock Deepwater Port License Application. Final Environmental Impact Statement. Office of Marine Environment and Systems. US Coast Guard, Washington, D.C., Vols I–IV.

US GEOLOGICAL SURVEY (1974). Methods for collection and analysis of water samples for dissolved minerals and gases. USGS, Denver, CO.

WHITE, W. A., CALNAN, T., MORTON, R., KIMBLE, R., LITTLETON, T., McGOWEN, J. NANCE, H. (1985). Submerged lands of Texas, Galveston-Houston Area: Sediments, Geochemistry, benthic macroinvertebrates. Bureau of Economic Geology, University of Texas, Austin, Texas. 145 pp. + plates.

40

Tainting of Atlantic Cod (*Gadus morhua*) by Petroleum Hydrocarbons

R. J. Ernst[a], W. M. N. Ratnayake[b], T. E. Farquharson[b],
R. G. Ackman[b], W. G. Tidmarsh[c] and J. A. Carter[a]*

[a]*Martec Ltd, 5670 Spring Garden Road, Halifax, Nova Scotia,
Canada B3J 1H6*
[b]*Canadian Institute of Fisheries Technology, Technical University of Nova
Scotia, PO Box 1000, Halifax, Nova Scotia, Canada B3J 2X4*
[c]*Texaco Canada Resources Ltd, Suite 704, 1809 Barrington Street, Halifax,
Nova Scotia, Canada B3J 3K8*

ABSTRACT

Tainting in fish is the development of an atypical flavour caused by natural spoilage or by the assimilation of contaminants into edible tissue. Potential sources of tainting include oil from spills or blowouts, produced water, deck drainage, and the base oil used in drilling mud. This study was concerned with the exposure of cod to the water-soluble fraction (WSF) of three crude oils (Amauligak, Brent and Hibernia) and a drilling mud base oil (Conoco) to determine threshold concentrations which cause tainting. Depuration of hydrocarbons from cod tissue was examined as well.

The hydrocarbon concentration of WSF of the crude oils ranged from 15 to 21 ppm, comprising mainly low boiling aromatic compounds. The concentration of the WSF of the Conoco oil was about 4 ppm, the main components being moderate boiling cyclic and branched alkanes. The threshold concentrations of WSFs of different crude oils imparting a taint after 24 h exposure ranged from 0·1 to 3·4 ppm. The tainting threshold concentration for Conoco base oil was 0·7–1·9 ppm. Exposure of cod to low oil concentrations (0·1–0·2 ppm in WSF) over 3 days also caused tainting. Depuration following tainting generally occurred within 24 h.

These data suggest that tainting of gadid fish due to leaching of base oil from a cuttings pile is unlikely. The study also indicates that tainting of gadid fish exposed to naturally-dispersed oil spills is unlikely.

* To whom correspondence should be addressed.

1. INTRODUCTION

Tainting as a result of exposure to petroleum hydrocarbons has been documented frequently (Mackie *et al.*, 1972; Nitta, 1972; Motohiro & Inoue, 1973; Shenton, 1973; Connell, 1974; Kerkhoff, 1974; Scarratt, 1980); however, tainting is a perceptual problem. Experience, as well as preference, can greatly influence a determination as to whether or not a food is tainted. The importance lies in whether the flavour or odour of the product is altered, not whether a contaminant impairs or improves the flavour.

The threshold level of exposure necessary to cause tainting in fish depends on the chemical composition of the hydrocarbon to which the fish is exposed, and on the species exposed. The principal components of crude and refined oil which cause tainting in fish include the phenols, dibenzothiophenes, naphthenic acids, mercaptans, methylated naphthalenes (GESAMP, 1977), toluene and other alkylbenzenes (Motohiro, 1983). Some of these compounds are polar and some are very volatile; however, most demonstrate some degree of water and lipid solubility. The lipid solubility of these hydrocarbons allows them to be readily transferred into the blood and tissues of organisms. Consequently, it is generally accepted that the proportion of depot fat in the lipids present in a fish is an indication of the susceptibility of that organism to acquisition of a taint (Whittle & Mackie, 1976).

The main goal of this study was to establish a quantitative basis for assessing taint in a representative commercial fish species exposed to various concentrations of water-soluble petroleum hydrocarbons. Such laboratory data can then be used to assess the probability of tainting occurring under conditions where hydrocarbons are released to the environment. The commercial species selected was Atlantic cod (*Gadus morhua*). The test oils were Amauligak, Hibernia and Brent crude oil, and a drilling mud base oil developed by Conoco. These oils represent a range of hydrocarbons which may be discharged as components of cuttings from oil-bearing formations and as base oil residues on cuttings following drilling with an oil-based drilling mud.

2. MATERIALS AND METHODS

Live cod of modest size (typically 700 g) were collected off Halifax, Nova Scotia, and were held in a seawater pool tank (685 m³) at the Dalhousie University Aquatron for 3 months prior to exposure trials. They were fed

TABLE 1
Short-term (24 h) exposure trials

Test oil	Hydrocarbon concentration of exposure water (ppm)		Taste panel evaluation
	Time		
	0 h	24 h	
Brent	0·53	0·13	Tainted ($p < 0.05$)
	0·50	1·05	Tainted ($p < 0.001$)
	2·98	2·48[a]	Tainted ($p < 0.001$)
Amauligak	0·28	0·03	Not tainted
	0·75	0·29	Tainted ($p < 0.05$)
	1·78	2·70	Tainted ($p < 0.001$)
Hibernia	0·21	0·12	Not tainted
	0·42	0·39	Not tainted
	1·41	3·40	Tainted ($p < 0.001$)
Conoco	0·69	0·47	Not tainted
	0·74	0·14	Not tainted
	1·93	1·21	Tainted ($p < 0.001$)

[a]Terminated after 16 h exposure.

a diet of chopped mackerel to satiation on alternate days, and beef liver once a week. All stock water-soluble fractions (WSF) were prepared in a 1100 litre stainless steel vessel equipped with a mechanical stirrer. Seawater and test oil, in a ratio of 99:1 (volume:volume), were mixed for 24 h and the WSF was allowed to settle beneath the oil layer for 48 h.

Exposure and control tanks were plumbed separately to ensure that there was no cross-contamination from the drain system. All tanks were aerated with pure oxygen and dissolved oxygen (DO) levels were monitored regularly. WSF was pumped directly from a bottom drain in the mixing vessel to the exposure tanks as needed. Hydrocarbon concentrations were monitored at regular intervals by measuring the fluorescence of the exposure water. Corresponding water samples taken at the beginning and end of each trial were analysed by gas chromatography (GC). All exposures were conducted under static conditions.

Short-term exposure trials were run with all test oils. Fish were exposed to three different concentrations of hydrocarbons for 24 h (Table 1). At the end of the exposure period, the fish were removed from the exposure tanks and prepared for taste panel assessment.

Depuration trials were conducted with Hibernia and Amauligak crude

oil. Fish were exposed to approximately 2·5 ppm hydrocarbon for periods
of 8 and 24 h. After exposure the fish were transferred to larger tanks
containing clean seawater and were allowed to depurate for 0, 1, 4, 7 and
14 days before taste panel assessment.

Long-term exposure trials were conducted with WSFs of Hibernia and
Amauligak crude oil. Fish were exposed to 0·10 and 0·25 ppm hydrocarbon
for periods of 3 and 7 days.

The hydrocarbon analyses were conducted on a Perkin-Elmer Sigma
3B gas chromatograph equipped with a flame ionization detector (FID)
and a split injection system. The chromatography was done on a DB-1
flexible, fused-silica capillary column (60 m × 0·25 mm I.D.). Conditions
and temperatures were as follows: FID, 280 °C; injector, 280 °C. The
column temperature was programmed as follows: initial temperature 45 °C
held for 15 min, and the oven was then programmed at 13 °C/min up to
280 °C and held for 30 min. The carrier gas (H_2) pressure was 16 psig and
hydrogen and air pressures of the FID were 22 and 30 psig, respectively.

Hydrocarbons in water samples were extracted with hexane using the
microextraction technique of Murray *et al.* (1984) and injected into the
GC without any concentration step. The peak areas were summed and
the concentrations were calculated with respect to an internal standard
(n-C_{21}). Hydrocarbons in fish fillets were extracted by the steam distillation
procedure of Ackman & Noble (1972). The lipid content of selected cod
fillets was determined using the procedure of Bligh and Dyer (1959).

The extended triangle test (Jellinek, 1985) using 14–18 trained taste
panelists was used to detect tainting. From each experimental or control
group, fillets from three or four fish were cut into strips and minced in a
food processor to ensure a homogeneous sample. A subsample of about
25–30 g was formed into a patty, placed in a covered glass petri dish, and
cooked in a microwave oven for 45 s. Panelists were presented with three
samples, two of which were identical, and asked to select the odd sample.
They were also asked to indicate the degree of difference between duplicate
and odd samples and the acceptability of all samples. The taste panel
results were evaluated statistically (Larmond, 1977).

3. RESULTS

3.1. Hydrocarbon Components of the Test Oils

The chromatograms of the four test oils showed n-alkanes from C_{10} to
C_{24}. The major hydrocarbons centred around the n-C_{12}–n-C_{17} alkanes.

Brent crude oil showed high levels of n-alkanes ranging from n-C_{10} to n-C_{21}, while the content of volatile aromatics was low. Amauligak crude showed higher levels of low-boiling hydrocarbons, including the volatile aromatic hydrocarbons. The high wax Hibernia crude oil exhibited the widest range of hydrocarbons, from the low-boiling aromatics to long-chain alkanes up to at least n-C_{34}. The Conoco base oil had only trace levels of the low-boiling aromatic hydrocarbons. The major components eluted between n-C_{10} and n-C_{17}.

3.2. Hydrocarbon Components of the WSFs

The hydrocarbon profiles of the WSFs of the three crude oils were qualitatively similar to each other, but were very different from those of the neat crude oil. Most of the major components in the crude oil were absent or found at very low levels in the WSF, especially the larger alkanes above C_9. The most prominent components present in the WSFs of the three crude oils were benzene, cyclohexane, methylcyclohexane, toluene, xylenes, ethylbenzene, isopropyl- and propylbenzene, ethylmethyl-benzenes, trimethylbenzenes, naphthalene and methylnaphthalenes (Fig. 1). n-Alkanes up to C_{13} and polynuclear aromatic hydrocarbons (PNAH) were detected as very minor components. The major hydrocarbon in all three crude oil WSFs was toluene, followed by benzene. The hydrocarbon profile of the WSF of the Conoco base oil was completely different from those of the three crude oils (Fig. 1). Like the parent oil, the WSF showed the complete absence of low-boiling alkanes and mononuclear aromatics, especially benzene and toluene. The main components were moderate boiling cyclic and branched alkanes. The aromatic components that were identified were ethylmethylbenzene, trimethylbenzenes, tetramethyl-benzenes, naphthalene and methylnaphthalenes, the major components being methylbenzenes and methylnaphthalenes. PNAHs were also detected, but at very low levels.

The hydrocarbon profiles of exposure water (seawater with added stock WSF) were similar to those of the corresponding stock WSFs, indicating no preferential loss of any of the component hydrocarbons after dilution.

3.3. Short-term Exposures

Based on an initial range-finding trial, the exposure levels proposed for the short-term (24 h) exposures were 2·50, 0·50 and 0·25 ppm hydrocarbon. The actual exposure levels, as determined by GC analysis of water samples, are shown in Table 1. During short-term exposures, the temperature rose an average of 3 °C. Dissolved oxygen levels fluctuated between 60 and

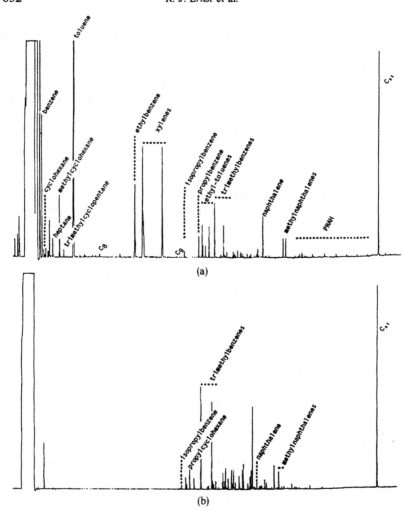

FIG. 1. (a) Gas chromatogram (GC) of Hibernia water-soluble fraction (15 ppm).
(b) GC of Conoco base oil water-soluble fraction (12 ppm).

100% saturation; however, super-saturation occurred in all tanks for
several hours. The fluorescence levels remained fairly constant at the two
lowest exposure levels. More variation was observed at the highest
exposure level.

A significant taint could be detected in all the fish exposed to various
concentrations of Brent crude oil WSF (Table 1). The taste of fish exposed

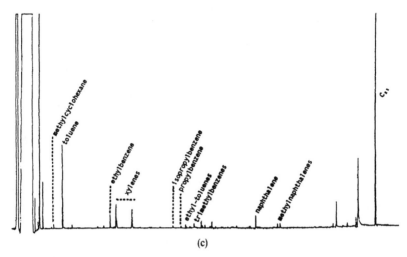

(c)

FIG. 1.—*contd.* (c) GC of cod tissue exposed to Hibernia water-soluble fraction (0·67 ppm). PNAH indicates polynuclear aromatic hydrocarbons. C_{21} is added internal standard.

to a WSF of Amauligak crude oil at concentrations less than 0·3 ppm hydrocarbon were not significantly different from the taste of control fish. Fish exposed to slightly higher concentrations (0·3–2·7 ppm) were significantly tainted. Fish exposed to the two lowest levels of Hibernia crude oil WSF (<0·42 ppm) were not tainted. The highest exposure level (between 1·4 and 3·4 ppm) resulted in significant tainting.

Similarly, the two lowest exposure levels (<0·74 ppm) of Conoco base oil WSF did not result in a significant taint in cod. The highest exposure level (1·21–1·93 ppm) resulted in significant tainting.

Fish exposed for 24 h to Hibernia WSF in the range of 0·5–3 ppm hydrocarbon had about the same lipid levels as fish not exposed to hydrocarbons. These lipid levels (0·65–0·84% lipid wet weight) are within the range found in healthy mature Atlantic cod (Jangaard *et al.*, 1967).

3.4. Depuration Trials

The exposure conditions and taste panel results of the depuration trials are noted in Table 2. All fish were found to be tainted (significant at $p < 0.001$–0·01) after the initial exposure (at 1·5–3·5 ppm hydrocarbon). Most fish completely depurated after 1 day in clean water. One group of fish exposed to Amauligak WSF for 24 h were still tainted after 1 day of

TABLE 2
Depuration trials

Test oil	Duration of exposure (h)	Hydrocarbon concentration of exposure water (ppm)			Taste panel evaluation			
		Time			Time (in clean water)			
		0 h	*8 h*	*24 h*	*0 d*	*1 d*	*4 d*	*7 d*
Amauligak	8	2·68	3·50		Tainted	Not	Not	Not
	8	3·43	2·45		($p < 0.001$)	tainted	tainted	tainted
	24	3·05		2·35	Tainted	Tainted	Not	Not
	24	2·98		1·93	($p < 0.01$)	($p < 0.05$)	tainted	tainted
Hibernia	8	1·50	1·85		Tainted	Not	Not	Not
	8	1·67	2·16		($p < 0.001$)	tainted	tainted	tainted
	24	1·86		3·23	Tainted	Not	Not	Not
	24	1·93		2·55	($p < 0.001$)	tainted	tainted	tainted

depuration ($p < 0.05$), but had completely depurated by the fourth day. Fish from the Hibernia depuration trials were analysed for hydrocarbons after exposure to WSF for 8 and 24 h. They had tissue hydrocarbon levels of 0·67 and 0·48 ppm hydrocarbon, respectively. After 24 h in clean water, hydrocarbon levels in the fillets dropped to trace levels and 0·08 ppm, respectively. The GC hydrocarbon profiles of these fish were similar to the GC profiles of the stock WSF of Hibernia crude oil (Fig. 1). Control fish had a tissue hydrocarbon concentration of 0·19 ppm.

3.5. Long-term Exposures
GC analysis of water samples taken throughout the long-term exposure trials with Amauligak and Hibernia crude oil indicated no significant difference between the actual exposure levels, although two levels were attempted. The results were therefore treated as replicates with an average exposure level of 0·15 ppm hydrocarbon over 3 and 7 day exposures. The exposure conditions and taste panel results are shown in Table 3.

The taste panelists were able to significantly distinguish test fish from controls in one of the 3 day Amauligak exposures and in one each of the 7 day exposures to WSFs of Amauligak and Hibernia crude oil. These data suggest a marginal level of tainting under these exposure conditions.

TABLE 3
Long-term exposure trials

Test oil	Duration of exposure (h)	Hydrocarbon concentration of exposure water (ppm)			Taste panel evaluation
		Time			
		0 h	72 h	168 h	
Amauligak	72	0·30	0·13		Not tainted
	72	0·22	0·15		Tainted ($p < 0.05$)
	168	0·09	0·15	0·13	Tainted ($p < 0.05$)
	168	0·25	0·05	0·05	Not tainted
Hibernia	72	0·10	0·07		Not tainted
	72	0·10	0·10		Not tainted
	168	0·10	0·15	0·23	Not tainted
	168	0·07	0·25	0·09	Tainted ($p < 0.05$)

4. DISCUSSION

The threshold concentration of hydrocarbons (as WSF) which can cause tainting in cod during short-term exposure ranges from 0·1 to 3·4 ppm, depending on the original oil. Lower concentrations over longer exposures (3–7 days) also cause marginally significant tainting. However, depuration of tainted fish occurs relatively quickly after short-term exposure to low levels of petroleum hydrocarbons. The compounds responsible for tainting in cod could include toluene (although not in Conoco base oil) and larger aromatic compounds (some of which are present in the WSF of Conoco base oil).

What are the implications of these data when applied to gadid populations in areas where offshore drilling occurs? The main sources of petroleum hydrocarbons in the marine environment associated with offshore drilling are base oil adhering to drill cuttings, and contaminated deck drainage (Menzie, 1982). During production, produced water from the formation is the main source of petroleum hydrocarbons. The possibility of a spill or blowout exists during any stage of exploration drilling, field development, and production.

Field studies have shown very high levels (151 500 ppm) of Conoco base oil in the cuttings pile beneath drilling rigs near Sable Island (Yunker *et al.*, 1986). However, the amount of leaching of aromatic hydrocarbons and cyclic and branched alkanes into the seawater adjacent to the cuttings pile is unknown. These compounds (making up the WSF) comprise only a very small portion of the original whole base oil. Although our study indicates that low concentrations of Conoco WSF can cause tainting in cod, under natural conditions the volume of water over the cuttings pile in which tainting concentrations of dissolved hydrocarbons occur must be small and variable. The risk of tainting of fish such as cod, which swim and feed in the water column, due to mineral base oil adhering to cuttings, is perceived to be low. The risk of tainting of other fish, such as plaice, which live and feed on the seabed may be higher.

The scale and effect of oil-contaminated deck drainage is poorly documented. Deck drainage could include diesel oil and higher molecular weight compounds associated with lubricants and grease. The contaminant plume near a rig tends to be a very variable surface phenomenon. It is likely that the concentration of dissolved hydrocarbons in the water column beneath a small and sporadic slick is extremely low and not a tainting threat. In any case, most gadids of commercial importance live in deeper water and would not encounter surface contaminant plumes.

Produced water contains the more volatile and soluble hydrocarbons linked to tainting (Lysyj *et al.*, 1981). Although rapid dilution of produced water reduces the volume of a plume which can cause tainting (Law *et al.*, 1982; Menzie, 1982), chronic input under conditions of poor mixing could lead to tainting in fish in the immediate vicinity of a platform. This is, perhaps, an analogue of fish tainting due to refinery effluent.

The remaining potential source of tainting is oil from a spill or blowout. Tainting of finfish and shellfish has been documented for the more extensive coastal oil spills (see Tidmarsh *et al.*, 1985 for a review), but tainting of fish as a result of offshore blowouts has been marginal (Whittle *et al.*, 1978) or negligible (Carter *et al.*, 1985), even though uptake of hydrocarbons into fish tissue has been demonstrated in some of these cases (Zitko *et al.*, 1984).

It is difficult to extrapolate the laboratory data on tainting thresholds to natural oil spill conditions, mainly because the former are expressed as WSF, whereas oil concentration data under slicks are usually expressed as whole oil. However, the risk of tainting associated with offshore blowouts can be put into proper context by examining our mixing ratio (water:oil) and the resultant concentration of the WSF in the stock solution. This simple analysis indicates that a tainting threshold of 0·1 ppm hydrocarbon as WSF requires a whole oil-in-water concentration of greater than 47 ppm. According to McAuliffe (1986), sustained whole oil-in-water concentrations under chemically dispersed oil slicks rarely exceed 50 ppm and those under naturally dispersed slicks are usually less than 1 ppm. This suggests that the risk of tainting of gadid fish due to offshore blowouts is minimal. The risk is further reduced by the fact that potential tainting components, such as toluene, quickly evaporate, even when initially in solution. The behaviour of the fish is a factor as well. Commercially important sizes of gadids live in deeper water most of the time and would not normally encounter hydrocarbon plumes near the surface. Also, our study suggests that cod will show avoidance behaviour when exposed to high hydrocarbon concentrations. This has also been observed in salmon exposed to oil (McAuliffe, 1986).

Factors which could increase the risk of tainting of finfish include any combination of events leading to prolonged contact of lipid-rich fish with the aromatic fractions of oil. For example, the exposure of herring to a high aromatic content oil during a long subsea blowout in shallow coastal waters could be perceived as a worst case scenario for tainting, especially if dispersants are used. However, an encouraging finding of our study is that contaminated fish which can seek clean seawater will probably rapidly depurate.

ACKNOWLEDGEMENT

Funding for this study was provided by the Environmental Studies Research Funds, Canada.

REFERENCES

ACKMAN, R.G. & NOBLE, D. (1972). Steam distillation: a simple technique for recovery of petroleum hydrocarbons from tainted fish. *J. Fish Res. Bd Can.* **30**, 711–14.

BLIGH, E. G. & DYER, W. J. (1959). A rapid method of total lipid extraction and purification. *Can. J. Biochem. Physiol.*, **37**, 911–17.

CARTER, J., MACGREGOR, C., TIDMARSH, G., CHANDLER, P., PARSONS, J. & BONKE, C. (1985). Monitoring programs associated with the Uniacke G-72 gas/condensate blowout on the Scotian Shelf. In *Proceedings of the Eighth Annual Arctic Marine Oilspill Program Technical Seminar*, Edmonton, Technical Services Branch, Environment Canada, Ottawa, Ontario, pp. 403–22.

CONNELL, D.W. (1974). A kerosene-like taint in the sea mullet, *Mugil cephalus* (Linnaeus). I. Composition and environmental occurrence of the tainting substance. *Austr. J. Mar. Fresh. Res.* **25**. 7–24.

GESAMP (1977). IMCO/FAO/UNESCO/WMO/WHO/IAEA/UN/UNEP. Joint group of experts on scientific aspects of marine pollution impact of oil on the marine environment. Report on Studies No. 6, FAO, Rome.

JANGAARD, P.M., BROCKERHOFF, H., BURGHER, R.D. & HOYLE, R.J. (1967). Seasonal changes in general condition and lipid content of cod from inshore waters. *J. Fish Res. Bd Can.*, **26**, 607–12.

JELLINEK, G. 1985. *Sensory Evaluation of Food, Theory and Practice*. Ellis Horwood, Chichester, UK.

KERKHOFF, M. (1974). Oil Pollution of the Shellfish in the Oosterschelde Estuary: December 1973. ICES CM 174/E:13.

LARMOND, E. (1977). Laboratory Methods for Sensory Evaluation of Food. Canadian Department of Agriculture, Publication No. 1637.

LAW, R.J., BLACKMAN, R.A.A. & FILEMAN, T.W. (1982). Surveys of hydrocarbon levels around five North Sea oil production platforms in 1981. ICES Marine Environmental Quality Committee Paper CM 1982/E:14.

LYSYJ, I., PERKINS, G. & FARLOW, J.S. (1981). Effectiveness of offshore produced water treatment. In *Proceedings of the 1981 Oil Spill Conference*, American Petroleum Institute, Washington, D.C., pp. 63–8.

MACKIE, P.R., MCGILL, A.S. & HARDY, R. (1972). Diesel oil contamination of brown trout, *Salmo trutta* (L). *Environ. Pollut.*, **3**, 9–16.

MCAULIFFE, C. D. (1986). Organism exposure to volatile hydrocarbons from untreated and chemically dispersed crude oils in field and laboratory. In *Proceedings of the Ninth Annual Arctic Marine Oilspill Program Technical Seminar*, Edmonton, Technical Services Branch, Environment Canada, Ottawa, Ontario, pp. 497–526.

MENZIE, C.A. (1982). The environmental implications of offshore oil and gas activities. *Environ. Sci. Technol.*, **16**, 454–72.

MOTOHIRO, T. (1983). Tainted fish caused by petroleum compounds—a review. *Water Sci. Technol.*, **15**, 75–83.

MOTOHIRO, T. & INOUE, N. (1973). n-Paraffins in polluted fish by crude oil from "Juliana" wreck. *Bull. Fac. Fish.*, Hokkaido University, **23**(4), 204–8.

MURRAY, D.A.J., LOCKHART, W.L. & WEBSTER, G.R.B. (1984). Analysis of the water-soluble fraction of crude oils and petroleum by gas chromatography. *Oil Petrochem. Pollut.*, **2**, 39–46.

NITTA, T. (1972). Marine pollution in Japan. In *Marine Pollution and Sea Life*, ed. M. Ruivo. Fishing News (Books) Ltd, Surrey, pp. 77–81.

SCARRATT, D.J. (1980). Taste panel assessments and hydrocarbon concentrations in lobsters, clams and mussels following the wreck of the Kurdistan. In *Scientific Studies During the "Kurdistan" Tanker Incident*, Proceedings of a Workshop. June 1979, Bedford Institute of Oceanography, ed. J.H. Vandermeulen, BI-R-80-3, pp. 212–26.

SHENTON, E.H. (1973). An historical review of oil spills along the Maine coast. The Research Institute of the Gulf of Maine (TRIGOM), Publication No. 3.

TIDMARSH, W.G., ERNST, R., ACKMAN, R.G. & FARQUHARSON, T. (1985). Tainting of fishery resources. Environmental Studies Revolving Funds Report No. 021, Ottawa, XIX + 174 pp.

WHITTLE, K.J. & MACKIE, P.R. (1976). Hydrocarbons in marine zooplankton and fish. In *Effects of Pollutants on Aquatic Organisms*, ed. A.P.M. Lockwood.

Cambridge University Press, pp. 85–105.

WHITTLE, K.J., MACKIE, P.R., FARMER, J. & HARDY, R. (1978). The effects of the Ekofisk blowout on hydrocarbon residues in fish and shellfish. In *Proceedings of the Conference on Assessment of Ecological Impacts of Oil Spills*, American Institute of Biological Sciences, pp. 540–59.

YUNKER, M.B., DRINNAN, W. & ENGELHARDT, F.R. (1986). A study of the behavior and fate of the hydrocarbons in sediments resulting from the use of oil-based drilling muds at two East Coast rig sites. In *Northern Hydrocarbon Development Environmental Problem Solving*, ed. M. Lee Lewis. University of Toronto Press, pp. 68–78.

ZITKO, V., BURRIDGE, L.E., WOODSIDE, M. & AKAGI, H. (1984). Low contamination of fish by hydrocarbons from the Uniacke G-72 (Shell Oil, Vinland) wellsite blowout in February 1984. Can. Tech. Rep. Fish. Aquat. Sci., 1305: iii + 43 p.

PART VI
Perspectives on Treatment and Control,
Effects and Monitoring of
Drilling Wastes

41

The Onshore Disposal of Drilling Wastes*

R.J. Donally‡

Nature Conservancy Council, Bowness-on-Windermere, Windermere, Cumbria LA23 3JR, UK

It was suggested to me that I should briefly describe the regulatory arrangements that we have in Great Britain by which control of drilling wastes is achieved, and also to give some account of our experience of landward or onshore petroleum exploration and production in relation to waste control. It is against this background that some of the points which have emerged from the papers presented at the 1988 International Conference on Drilling Wastes will be discussed. It is useful to maintain some perspective on the overall impact of the petroleum industry. This industry appears to be singled out for somewhat special attention in both Great Britain and in other jurisdictions. This summary will acknowledge the fact that the petroleum industry, although in some senses special, is generally no different from other extractive industries.

Firstly, though, and as a point of reference from my own perspective, here is a very brief word about the Nature Conservancy Council (NCC) of the United Kingdom. This is the government body which promotes nature conservation in Great Britain, including the marine environment, and also has a responsibility for geology and physiography. It gives advice to government and all those whose activities affect Britain's wildlife and wild places, for many of which it has direct responsibility. The Council has an official position within the Great Britain government organization, and is constituted by an Act of Parliament. Various other statutes relate to responsibilities which the Nature Conservancy Council has in respect of regulatory, planning and development control. Many of these apply to the petroleum industry.

---

*Editors' note: The text of this paper is an edited version of the conference presentation by R. J. Donally.
‡Present address: Barn Hey, Cartmel, Grange Over Sands, Cumbria LA11 6HD, UK.

The granting of a planning consent by the Mineral Planning Authority in the UK for petroleum activity almost invariably has attached to it a series of environmental conditions. And one of those conditions applies to the wastes derived from drilling activity. The NCC is involved at a very early stage because when an operating licence is granted by the Department of Energy, there is attached to it an advisory note which says, in effect, that the operator would be well advised to consult with the NCC at the earliest possible stage because the Council knows which sites are likely to be of particular ecological interest, and therefore likely to impose very considerable constraints on the petroleum activity. Furthermore, the Council is aware of any need on the part of the local people for some education in respect of the proposed activities. We are all well aware that ignorance of the implications of a particular proposal tends to result in a wish to resist it. The NCC can introduce the operator to the local interest and inform the people of the exact nature of the proposed operations. The NCC can be very helpful at this and even earlier, pre-proposal stages.

The philosophy which should govern responsible environmental regulation is based on the knowledge that man is both a user and polluter of the natural environment. While seeking to make the best overall use of the environment, the quality of the air, land and water must be maintained and in some instances improved. This is important. In other words, conservation of the natural environment is considered a quite appropriate and understandable objective. Effective conservation of the environment is an optimization process, one in which optimal balance is sought between man's use and abuse of the environment. Controversies arise, however, from the, at times, different perspectives on conservation held by the various users. One area which gives rise to controversy is the use of the land environment for the disposal of wastes, a topic in this conference.

This kind of controversy arises mainly from two factors; one is the capability we now have to detect extremely small levels of environmental contaminants. There have been some examples of this in the conference papers, some very sophisticated chemistry has been referenced. A second factor is that our knowledge of the overall impacts of contamination may be inadequate to enable us precisely to decide what may or may not be significant, and therefore unacceptable. It is, for instance, not uncommon that biologists, including those working in an organization like the NCC, will say in reference to a particular environmental concern, 'This is as far as our knowledge allows us to go'. There is a clear need for more factual information about the natural processes of the present global environment. A better understanding is needed about the abilities of man's activities to perturb these processes, and of the consequences of these perturbations. The petroleum industry can play a major role in promoting this under-standing, facilitated through joint consultation with government and local authorities. The relationship of the petroleum industry with the NCC in Britain is a case in point.

Because environmental processes are complex, the required information can only be obtained by scientific research, which has been the thrust of some of the papers presented at this Conference on Drilling Wastes. In addition, the information required for the conservation of the environment must also embrace technological developments. Again there has been some information on this aspect presented at the conference, in the knowledge that this matter embraces many fields, which is something that scientists in the NCC and in other conservation-oriented regulatory bodies have to come to terms with. And as our understanding of environmental processes advances, one can begin to assess with increasing confidence for which reasons particular uses of the environment amount to significant abuse. An important corollary to this linkage to understanding is the fact that too often denial of a particular use of a piece of land is based very much more upon prejudice than upon certain knowledge. The balance needs to be redressed to achieve a stage where one can see clearly what regulatory actions may be required to ameliorate or eliminate the abuse without unnecessary denial or severe curtailment of an industrial activity which is also, in itself, considered beneficial. It is against that kind of background that the production handling disposal of drilling wastes should be viewed.

The following briefly outlines the experience in Great Britain with regard to drilling wastes from onshore operations. Onshore exploration and production takes place on the areas shown on Fig. 1, which have all been licensed for petroleum exploration, and in some instances for production. Any operation can of course only take place once planning consent has been given.

This is an instance where a true perspective is critical. It is interesting that within all those licensed activity areas shown in Fig. 1, the total absolute acreage of all the petroleum industry development including exploration and production sites amounts to only about 1000 acres, that is, the size of a medium-sized agricultural holding in Great Britain. It is part of the perspective exercise, however, to recognize that the individual activity sites, small in themselves, may be competing with important conservation objectives when they are located in sensitive land areas.

The NCC came to a disturbing realization in 1985 when it was realized that the pace of exploration activity was increasing, and that more and

846   R. J. Donally

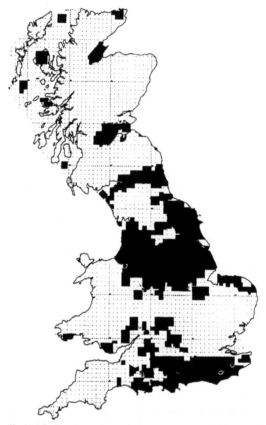

FIG. 1. Areas licensed for petroleum exploration and/or production in Great
Britain, as of March 1988, based upon Ordnance Survey National Grid.

more developments were planned and already taking place in locations
where there was likely to be some impact upon the features for which the
NCC has responsibility. In response, a series of guidelines was produced in
consultation with the industry. These guidelines, the *Nature Conservation
Guidelines for Onshore Oil and Gas Development,** were published in 1986
as an aid to those people in the industry who were perhaps entering the
landward area for the first time. They were produced also for planning
authorities and for the voluntary nature conservation movement in Great

*Editors' note: The guidelines can be obtained by contacting the Nature
Conservancy Council at Northminster House, Peterborough PE1 1UA, UK.

Britain, who can be vociferous and whose influence can be considerable. In general, the guidelines are a tool for public education, targeting these three interest holders.

The results have been remarkable, so that in many parts of Britain there is now a better understanding of the implications of onshore petroleum industry activity so that applications can be dealt with much more speedily, including matters relating to disposal of waste. In Great Britain, most if not all of the wastes derived from well sites are required to be disposed of in licensed tips. Many of these tips or ground fill sites are owned and managed by the local authority, and therefore there is an absolute control over reception and integrity of the particular materials.

The other regulatory authority responsible for determining how and where waste materials might be disposed of are the regional water authorities in England and Wales, and the river purification boards serving a similar function in Scotland. In relation to all of the licensed tips, they ensure the total integrity of each site. Therefore, problems associated with leaching are minimized and monitoring of any leak will be conducted by the regional water authority or the local authority acting on their behalf. At times, there have also been additional monitoring requirements placed on the petroleum industry, which I am convinced have often been unnecessary because of duplication and inherent lack of specificity. In many of these tips, for instance, there has been a cocktail of materials dumped from all kinds of origins, to which the drilling wastes are a relatively small addition.

There is a cost to industry in having to conform to the requirements of conventional waste disposal in Great Britain. There may be useful alternative disposal methods which should be explored. The NCC has had only one experience of land spreading. That was, however, a most unfortunate experience where some of the spent drilling mud and cuttings taken from a well site was side-routed from its licensed pit disposal site through agreement between a local haulage contractor and the farmer land owner who wished to reclaim some land with landfill. Regrettably, this land to be reclaimed was a prime piece of Dorset heath, a lowland heath site and a resource that is rapidly diminishing in Great Britain, very valuable in nature conservation terms. The circumstance under which this heathland was completely destroyed is, however, less of an indictment of alternate waste disposal methods than it is a result of lack of detailed control. Such control over the handling and disposal of wastes must be maintained. Other cases have been examined by the UK Ministry of Agriculture, Fisheries and Food, which advises at the present time that land spreading is not an appropriate route for disposal. There may be a need to revise our views in certain circumstances. Some of the papers at this conference have described new work that is being done in this field, which may force a reevaluation of the issue. At the present time there remain very considerable reservations about the use of this route for disposal, probably because there is such a diminished resource of natural and semi-natural vegetation in Great Britain. More studies on soils and soil chemistry, such as presented in some of the papers at this conference, are informative, not only on the effects of land spreading, but also in relation to reclamation.

The results of studies not dealing specifically with the disposal of drilling wastes may also be useful. For example, the NCC is collaborating on a research program on the disposal of oiled beach material, that is, beach material oiled at the time of an oil pollution incident. The program is being designed jointly with the Marine Pollution Control Unit of the UK Department of Transport which has responsibility for oil pollution matters. This study program will examine disposal of oiled sand in coastal soils, which may be of interest to drilling operators as well as for understanding the effects of disposing some of their materials in this fashion.

The issue of well-site management came out very strongly in a number of the presentations. The view in Great Britain is that the integrity of well sites is vital. For instance, unless precautions are taken in well-site design, problems of ground water pollution may arise from rain-water run-off in areas where the water table is high. It is essential of course that when a good design for a well site and its equipment is accepted by the regulator, that the drilling operation is conducted in accordance with the highest possible required standard. At this point, the responsibility lies primarily with the industry to operate in accordance with its licensing conditions, although compliance with such operational requirements is usually verified through inspections by the appropriate regulatory authority. In the UK, it is the Department of Energy which determines the standards and enforces them in the granting of licences for petroleum exploration and development.

In closing, I would remark that while it is inevitable that at any conference gathering such as this not every detail is novel to everyone, it is the time when we can review, and as appropriate debate, the essential and new features of the theme of the meeting, the control, fate and effects of drilling wastes. This is a rapidly advancing area both onshore and offshore and another updating conference on the topic should be considered in a few years' time.

42

Offshore Drilling Waste Issues*

JAMES P. RAY

Shell Oil Company, PO Box 4320, Houston, Texas 77210, USA

There are two things I would like to present in this overview. First, I would like to back up a little and consider the evolution of what has brought us to this meeting today in the context of the progress we have made in our knowledge of drilling wastes. Then I would like to reflect on some of the papers that have been presented here on offshore matters, and mention some of the things that stand out as being new, interesting, valuable, or usable information. I want to highlight findings that I think add to our knowledge, and have a look to the future for a few topics where more work may be needed.

Back in 1975 at the time of the EPA Drilling Mud Conference in Houston, drilling wastes became a problem of interest. It was the Environmental Protection Agency which decided that this was an interesting area to look at. They knew there must be a problem but they really were not sure what it was. That was a good enough reason to call a conference and try to get everybody together who thought they knew something about drilling mud. On the other hand, the industry did not feel it had a problem, although it also could not substantiate its reasons for thinking this way. So there were two sides which really were not sure of where they were. As a consequence, the 1975 conference presented probably the largest percentage of anecdotal science that I have ever seen at a conference; interesting papers, but really some very subjective, qualitative information. There had evidently been very little research in the US on the environmental side of drilling muds and cuttings. As a matter of fact, some of the major contributions at that particular meeting were by the Canadians who had been very active in trying to address some of the Arctic and fresh water questions. They were the first ones to develop an aquatic white rat, and tested everything with rainbow trout.

---

*Editors' Note: The text of this paper is an edited version of the conference presentation by J. P. Ray.

The US has rapidly caught up and now has its own version of the aquatic white rat, the mysid shrimp. There was also at that time some testing with a limited number of marine invertebrates. Some of the early work by John Sprague on the individual and additive toxicity of drilling mud components was available. We had, however, very little information on the fate of the materials in the environment. Also, in the early 1970s, there was little information on mud composition. All references were to the World Oil Report, listing 600–700 additives and trade names. Most people had little idea of the actual composition of additives. It was even hard for a mud engineer from one company to really know what another company was calling the same product.

Some of the papers presented in the last few days are indicative of the advance from 1975. Apparently, the message back to the industry from the 1975 conference was that there was a great need for information.

In the US, both the industry and the regulatory agencies recognized the lack of critical field and other information, and geared up to become involved in research on drilling fluids. Much was accomplished over that period of 5 years which led to the 1980 drilling fluids conference. It was a meeting jointly sponsored by the US and Canada, much as this 1988 conference, and represented a cross-section of industry and government agencies from the US and Canada working as co-sponsors. There was an important reason behind the 1980 conference other than to just pull together information gathered since 1975. It should be realized that most of the research done in this field is primarily a result of the permitting system. This usually results in a variety of research reports which fall into the category of non-peer reviewed 'gray literature', and which are not readily available to the other researchers or regulators. The purpose of that conference was to provide an opportunity for pulling the many research studies together, subjecting them to peer review, and discussing them in an open forum where the results and interpretations could be challenged.

There were some good papers dealing with the fate of drilling fluids in the 1980 conference. As a matter of fact, since that time, there have not been many new major fate studies because the basic information on dispersion has fairly well stood the test of time. Also, there were a number of platform studies dealing with benthic effects. Many papers focussed on toxicity, some on whole muds, and quite a few on the components of drilling fluids with the intent of finding out which components were the most toxic. This was a period of evolution in toxicity testing, which is continuing, although there are always problems in trying to compare new data with the old because of technique differences. There were also studies on the bioaccumulation of metals, still a current topic.

Also in the 1980 conference there were many good papers dealing with more exotic topics, such as the effects of muds on corals and in arctic and subarctic environments, both Alaskan and Canadian. The first of the mesocosm type studies were presented which were trying to look at whole community responses and recruitment in contaminated sediments.

There were some deficiencies then, as there are now. In the bioassays of the 1980 papers one of the biggest criticisms was of exposure techniques, unrealistic exposure regimes that did not represent what the animal would actually experience in the environment. Because of this, it was very difficult to extrapolate laboratory results to the field environment, which should always be one of our goals so that we can predict impacts. Few data exist on the recovery of impacted environments over time, such as a year or two. Today we are seeing efforts toward assessing recovery. In the late 1970s, bioassays were primarily quick, for the most part 96-h (acute) tests, and few long-term exposures were tested. Further, sublethal effects testing was still fairly limited.

One of the biggest deficiencies in offshore information on the North American side was data on the impact of multiple well discharges, since most of our work was on single well exploratory drilling. Since 1980, however, there have been a lot of data from the North Sea which deal with multiple well platforms. That is very important to us in the US because of the move towards larger and larger platforms, some of them with more than 100 drilling slots for the development of discrete reserve areas, resulting in potentially large quantities of discharges at a single site. Another lack was good data on contamination gradients near platforms, to allow meaningful extrapolation back to the laboratory experiments. We did not know what the levels and gradients for hydrocarbons and metals were in the environment because too many of our previous studies had sampling locations too far from the discharge.

Following the 1980 conference, we once again assessed our progress and current state of knowledge, trying to identify information gaps and to focus on more direct, pointed questions. Much of the research is applied, however, because we are driven by regulation and many of the basic questions end up not being answered.

I would like now to review a few selected papers and mention some key points on how the information may be valuable to us in making environmental decisions. In the fate area, for instance, a paper by Paul Boehm and colleagues concerned a small partitioning study, but a partitioning study that provided basic information for a very important question. The question was, what happens to the diesel and other oil in a water-based mud once it is discharged? Where is the oil going? Into the water or to the bottom? It makes a big difference if you are conducting a hazard assessment and trying to predict what the impact on the environment will be since the dispersion rates and the residence time for exposure to the organisms are totally different between the solid phase or dissolved in the water column. It appears that most of the hydrocarbons went into the water column, which makes a big difference in how we assess potential impact for something like diesel in a water-based mud. This cannot be compared to an oil-based mud with an oil-wet cutting that is taking a trip to the bottom rapidly. Understanding the fate of the materials is very important.

The study that Paul Boothe presented is one of the first with a good tight sampling pattern around a variety of different drilling operations (exploratory, development and production) to take a look at metal contamination levels and the gradients. Both surficial and core samples were taken, to determine how deep into the sediments trace metals penetrate. Another interesting approach taken in that study was the attempt to develop a mass balance of barium distribution. Barium is our most commonly used tracer, present in the largest quantity.

The paper by Erickson, Fowler and Thomas discusses important new analytical approaches. In this study on oil-based muds in the Beaufort Sea, the analyses identified 10 different isoprenoids to be used as labels. The authors were able to follow the signature of a mineral oil associated with the discharges over distance from the drilling locations. In the years to come, we are going to need these kinds of labels so that we can follow individual components of our discharges.

The large monitoring studies of development and production fields in the North Sea have been of considerable interest in North America. The many North Sea studies, summarized here by John Davies of the UK, are important for a number of reasons. They represent the real cumulative effects of large volumes discharged over many years from multiple wells. They also represent a worst-case scenario against which more benign water-based mud systems can be compared. One of Dr Davies' estimates that impressed me was that, in 1986, 29 000 t of oil entering the North Sea came from the offshore oil industry, and upwards of 70–75% of wells in the UK sector of the continental shelf are drilled with oil-based mud. It is going to be interesting to see if the Agreed Facts of the Paris Commission relating to the zones of impact (major deleterious impacts out to 500 m, a transition zone and then a larger measurable-contamination zone) change in the future based on new data.

In that context, Lars-Otto Reiersen of Norway presented an interesting paper that summarized studies in the Norwegian sector of the North Sea. Special emphasis was given to comparison of data sets from 1984 and 1986. There was a lot of interesting debate on that paper, and some controversy over interpretation of data caused by changes in sampling methods, and whether the chemistry in the Reiersen study, or any other study, really support 'cause and effect' conclusions. It is an important question and it needs to be addressed satisfactorily. Further, in this study, and all studies like it, one of the biggest problem areas is the interpretation of benthic data and the separation of natural variability from impact. Do the statistics and all of the exotic but artificial biological indices we use really mean anything? Many are not convinced that they are useful, and it is an area where much care needs to be taken in the confident extrapolation of data. It is a real problem area in environmental monitoring.

Also in the field of benthic work, there is new mesocosm research by Torgeir Bakke and colleagues, presenting research on soft bottom communities brought to the laboratory and exposed to contaminated cuttings. This was a reversal of the usual process approach of going into the environment with recolonization trays to see how contaminated sediments will affect recolonization rate. An interesting difference of opinion arose in this and several other papers. Apparently, researchers are finding different levels of degradation in heavily oiled sediments, in some cases more, and in some only slow degradation. It may be that in the highly contaminated areas, as suggested in the paper by Otto Grahl-Nielsen, there may be a false signal in the naphthalene/dibenzothiophene ratios because there is more biodegradation going on than previously thought. We may be getting undegraded oil from the lower sediment layers that is masking degradation on the surface which would otherwise be evidenced by the ratios.

Studies on recovery from discharged cuttings are appropriate from the biodegradation standpoint, and in more general applied terms for the assessment of impacts. The paper on seagrass impacts off the west coast of Florida by Alan Hart, where complete die off of seagrasses occurred within 300 m of a discharge location, demonstrated recovery starting in about 2 years.

One of the areas in which some of the biggest strides have been made since the last conference is bioaccumulation: the bioavailability of contaminants. There were a number of papers that addressed this subject. Ken Jenkins, for instance, in his work off the coast of California around an exploratory operation, carried out very detailed chemistry, examining the intracellular level of metals in benthic organisms. They documented bioaccumulation in the near field. One of the important results of that study is a first answer to questions on the fate of barium inside the organism, possibly incorporated and sequestered as particulate barium sulfate. This study begs a careful definition of bioaccumulation in relation to toxicity. An organism can carry elevated levels of metals which are not necessarily toxic. It is evidently more than just a straight comparison of how much is in the sediments and how much is in the ground-up organism.

Other studies also reported an uptake of heavy metals. For instance, Morton and Montgomery observed barium and chromium in seagrasses, and Jerry Neff and colleagues examined food web transfer, a very relevant question. They worked with juvenile lobsters and winter flounder which were exposed to both contaminated and uncontaminated sediments and fed contaminated and uncontaminated worms to trace the transfer of metals up the food web. Based on their work, it does not appear that this is happening with the species tested, even with long 99-day exposures; a great improvement over the standard 96-h tests. Another study from these researchers was on the bioavailability of trace metals from barite, which is a controversial issue in the US and in other countries. It is suggested by some people that it is a real problem, although it had not been defined in laboratory or field studies. The paper by Neff in these proceedings has shown, however, that there is only very limited bioavailability of metals of concern, such as copper, barium, mercury and cadmium. Arsenic did not appear to be bioavailable at all.

With regard to bioaccumulation of hydrocarbons, there have been two very interesting papers presented, dealing with tainting, the uptake of petroleum hydrocarbons into fish such that it affects odor and taste. This is an important issue, even if it does not affect the fish biologically. Nonetheless, it affects the perception of the public with regard to the edibility of fish, which is a problem. There still remain questions from the field studies, to date carried out only in the North Sea on migratory fish species. Evidently, when a migratory fish is collected 10 000 m away from a location, there is little certainty about where it was exposed. Also, there is still little information on depuration rates as relates to fish tainting. We do know, however, that in most cases fish metabolize petroleum hydrocarbons at a fairly rapid rate—an interesting and relevant study area for the future. One should keep in mind that this issue is also of major interest to those who work with oil spills, but is shared with the drilling fluids concerns.

There have been several different papers in the area of toxicity testing. Rod Parrish and colleagues from the EPA expanded on the work of John Sprague from many years ago in trying to address the question of additive toxicity of drilling mud components. In the US, there is a concerted attempt to obtain toxicity information on the various additives that are being used in drilling fluids, with the purpose of being able to predict for regulatory purposes what the final toxicity might be. The Parrish paper presents an initial step by characterizing six different additives in different combinations to see if final toxicity could be predicted, and finding that the toxicity formula predicts the component parts to be much more toxic than they really are in the final whole mud. This is obviously an area where more work is needed in the near future.

The Moynihan *et al.* paper focussed on an area that is both innately interesting and controversial, and that is the use of rapid bioassay techniques—Microtox in this particular case, a bacterial bioluminescence test. The Microtox test has been used in a variety of places for effluent testing, especially for refineries and chemical plants. An interesting finding from the studies on drilling fluids is that while the test was repeatable against itself, it did not correlate well with the rainbow trout toxicity data. The paper that Jim O'Reilly and I presented raised the issue of variability problems that we find in commercial laboratory toxicity testing. The main point we made was that if industry is going to be required to do compliance bioassay testing in a pass/fail mode, then either the industry or the government has to establish guidelines and minimum standards for these testing laboratories so that we can have a bit more confidence in the results.

The analytical paper presented by Ted Sauer was very thought provoking because it addresses an area which has long been avoided. The key point was that we should find ways to tag and trace fractions from drilling fluids other than barite. Currently, we have been chasing barite because the rest of the components, such as organics, are not in high enough concentrations to be detectable very far from the rig. Yet organics are usually the compounds of concern; most of the toxicity is from the oils. As proposed in the paper, techniques for lignosulfonates may provide us with methods for following organics in the environment, a research area which must be pursued further to deal with some of the drilling mud issues.

The diesel pill monitoring program is a project in which the US offshore industry has enjoyed a good, cooperative, working relationship with the Environmental Protection Agency. The diesel pill monitoring program presented by Bob Ayers questioned the effectiveness of removing diesel pills following stuck pipe situations, and also discussed the characteristics of that mud, including how toxic it is after you remove the pill. Although there are not good data to address these questions, the cooperative effort between the American Petroleum Institute and the Environmental Protection Agency since about 1986 is a very good first step for the US and has gone some way to resolve the issue.

The third section of this summary focuses on the future. Based on what we know today, what new directions do we need for the future? What information gaps remain to be filled? I have categorized a variety of different areas which need more information for environmental management decisions. But before I go into those, I want to mention two points which Robin Donally made in his summary; that is, firstly, understanding processes is very important, simply because a lot of the information that we do not understand on drilling fluid effects results from a lack of the basic knowledge of processes (for example, resuspension and transport), and, secondly, the impact of other human activities. Drilling discharges from offshore oil and gas in most cases result in fairly small impact zones. There are major impacts occurring in the oceans of a much larger magnitude, whether from pollutants coming in from the coastal zones, or from activities such as commercial fishing. Yet we tend to ignore this in the design and often the interpretations of our monitoring programs.

There are some areas which require more research. While there has been good progress in the bioaccumulation area, there are still some questions to be answered: such as which metals go where, and how much really causes an effect? More field verification of uptake levels is needed. More emphasis should be placed on field studies, in situ work, but carried out with a solid and defensible controlled approach.

Also, biodegradation is a subject very relevant to future interest in the US in the discharge of low-level oily cuttings. If in fact there were little to no biodegradation in heavily oiled sediments, it raises other interesting questions. Can techniques be developed to enhance degradation; such as simple spreading and raking of the bottom to re-aerate the deposit? Alternatives to treatment and mitigation are critically important; for instance, in the Gulf of Mexico, some operators have had to undergo very expensive recovery operations to actually vacuum oil cuttings off the sea floor (e.g. over $2 million for one platform).

A knowledge of transport processes is important in understanding the fate and effects of discharged materials. For example, resuspension and transport are both very important, especially when dealing with the broad variety of oceanographic regimes, ranging from shallow bays to the Continental Shelf area to ice-fast areas. There is still much to be known about vertical migration of pollutants in sediments. There are geochemistry questions to be resolved, dealing with resolubilization of metals in the interstitial pore waters, and the question of bioavailability once the speciation of metals changes.

Clearly, the need for and usefulness of extrapolation is a general category which needs addressing. One of the biggest problems is that so much of the science carried out is in different geographical locations. People seem to think that they cannot use the information if it was not gathered in their own backyard; however, much of the science is basic science. It does not change just because it crosses the border of a region or of a country. If we are going to keep following this pattern of repetition for years and years, it is going to cost tens of millions of dollars, and be a detriment to useful research since funds are ultimately limited. Extrapolation of data needs to be made more definable and palatable to the public, and to those who make the regulatory decisions.

On the subject of treatment technologies, there is a lot to do in the future to make drilling cheaper and also mitigate environmental impact more effectively. For instance, the area of solids control needs further development, to reduce the discharge volume of solids and bulk mud. Many of the contaminants of concern are associated with particulates. Much could be done to control particulate volume, which is an area for more research and development. Further, proper screening methods are needed for a better choice of additives, effective in the job they are supposed to do but of less potential environmental impact. Another area that needs research is predischarge detoxification, to reduce the toxicity of mud systems and make them acceptable for discharge.

With regard to cuttings cleaning technology, it seems that advances are being made to achieve results of around 1% residual oil on cuttings. Better cleaning technology will do much to reduce the concern existing currently in Europe and North America. On the other hand, there has been not nearly enough work with tracer materials to answer some of the fate questions in the environment.

It is incorrect to lump trace metals into one giant pot. Whether the issue is chromium or another metal, it depends on the metal species. In one of the papers which was not accepted for these conference proceedings, and which condemned all of the horrible chemicals in drilling muds, conclusions on studies were based using trace metals originating from the reagent shelf: mercuric chloride, barium chloride, right on down the list, all highly soluble forms of the metals which were then used in an ultimately spurious experiment. Barium chloride cannot be equated to barium sulfate, for instance, based on differences in bioavailability and toxicity.

In the area of benthic ecology, it is evident that we need better techniques to detect change. Currently we kill them and count them, and sort them and name them, and draw fancy dendrograms, but we usually still do not know whether the change was due to seasonal change, water temperature, or because there was a true toxic or physical effect on the sea bottom. This is one example where basic knowledge is essential. An extension of this issue is how small local or regional changes relate to the ecosystem— it is a difficult and big question.

In the bioassay area, both acute and chronic, the key requirement is to make bioassays more realistic. Sooner or later, bioassays should allow direct extrapolation to the environment if the question is one of ecological significance. If the goal is just a screening result to meet a defined standard, the realism is of less importance. Also, rapid bioassays are going to become more and more important. As long as toxicity testing is going to be part of regulatory programs, there need to be quick ways to get results. Such rapid techniques, however, have to be correlated with longer-term tests that we understand a little better in the ecological context.

There are a number of areas which are of concern and remain controversial. For instance, in the US, we are now dealing with a concept called 'Best Available Technology' (BAT); that is, if there is a technology that is better in reducing the pollutants put into the environment, and is economically feasible, it should be used. The decision for such use does not have to relate to the environment, nor to science, but solely to the existence of an affordable technology. This is obviously a very controversial topic with an inherent real danger of affecting a lot of the useful science and research that is funded by the industry and even by government agencies. When you take the reasons for this type of science away, the money will not be made available by the interested agencies. The effectiveness of funding this type of researach without good scientific justification (i.e. basic research) via organizations such as the National Science Foundation is questionable and a cause for concern.

Another but related area which is very important currently and in the future is that of better contact between the research scientist and the regulator. This problem exists even within regulatory agencies, and not just in the US. There is a real communication problem between the laboratory and the head office, and it is something that needs to be resolved.

More specifically, in the US, more properly-designed multiple well studies are needed. But, environmental monitoring has to be both reasonable and scientific. There is concern that monitoring has become a panacea to solve, *a priori*, regulatory uncertainties. As long as we are monitoring, the heat is off. Unfortunately, a lot of these monitoring approaches and programs are open-ended and not efficient or necessarily rational.

This conference has represented a mixed group of disciplines, from the environmental sector, government and industry. One last comment to the scientists in these sectors is that REGULATIONS WILL NOT WAIT FOR THE SCIENCE. The scientists have a challenge in this regard, to seek the answers to the right questions and to give definitive answers or at least some strong guidance to the regulator at early stages. The regulator will not, or cannot, wait and will need to make regulatory decisions anyway, even in the absence of definitive information. It is a sobering thought.

Conversion Factors

The following non-metric units have been used by some of the authors in the preceding papers. A detailed table can be found in the *Supplementary Metric Practice Guide for the Petroleum and Natural Gas Industry and Services*, 4th Edition, 1979, Canadian Petroleum Association, Calgary, Alberta.

Non-metric Unit	SI Units		Other Metric Units and Explanatory Notes	
acre	4·046 856	E + 03 m²	4·046 856	E − 01 ha
barrel (42 US gal)	1·589 873	E − 01 m³		
centipoise	1·0	E − 03 Pa s	1.0	E + 00 mPa s
cubic inch	1·638 706	E − 05 m³		
cubic foot	2·831 685	E − 02 m³		
cubic yard	7·645 549	E − 01 m³		
degree (angle)	1·745 329	E − 02 rad		
degree Centigrade	1·0	E + 00°C	now called degree Celsius	
degree Fahrenheit	5/9 $(t_F - 32) + 273·15 = T_K$			
dram	1·771	E − 03 kg		
foot	3·048	E − 01 m		
gallon (Cdn)	4·546 09	E − 03 m³		
gallon (old imperial)	4·546 092	E − 03 m³		
gallon (US)	3·785 412	E − 03 m³		
inch	2·54	E − 02 m	2·54	E + 00 cm
ounce (avdp)	2·834 952	E − 02 kg		
ounce (fluid)	2·841 308	E − 05 m³		
ounce (fluid US)	2·957 353	E − 05 m³		
ppb	1·0	E − 09		
ppm	1·0	E − 06		
part per thousand (°/₀₀)	1·0	E − 03		
percent (%)	1·0	E − 02		
pound-force per 100 sq.ft	4·788 026	E − 01 Pa		
pound-force per sq.ft	4·788 026	E + 01 Pa		
pound-mass (avdp)	4·535 924	E − 01 kg		
pound-mass per gallon	9·977 637	E + 01 kg/m³		
pound-mass per gallon(US)	1·198 264	E + 02 kg/m³		
pound-mass per sq. ft	4.882 428	E + 00 kg/m²		
psi	6·894 757	E + 03 Pa		
rpm	1·047 198	E − 01 rad/s	1.0	E + 00 r/min
square foot	9·290 304	E − 02 m²		
square yard	8·361 274	E − 01 m²		
ton (short—2000 lb)	9·071 847	E + 02 kg		
ton (long—2240 lb)	1·016 047	E + 03 kg		
yard	9·144	E − 01 m		

Index

Aromatic hydrocarbons, *see*
Hydrocarbon analysis;
Polynuclear aromatic
hydrocarbons; Sediments

Barite, *see* Barium
Barium
levels in sediments, 10–13, 40–3,
47–9, 101–3, 139, 142–6,
148, 150–1, 154, 445, 447–57,
592, 595, 598–600, 809, 811,
815, 817–21
ratio to iron levels, 14, 143–6,
148–51, 154
relationship with hydrocarbon
levels, 51, 53, 103–4
toxicity, 455
tracer in sediments, as, 11, 139, 777,
813
uptake
bioaccumulation, 440, 449, 452,
454–5, 588, 601, 603–4, 624
biological effects, 589, 604–6
exposure system, 441–4, 464
marine animals, in, 448–9, 452,
455–6, 588, 594, 600–4,
620–2
see also Trace elements
Biodegradability test protocols,
687–96

Cadmium
levels in sediments, 20, 103
see also Trace elements
Chromium
levels in sediments, 15–16, 103, 445,
447–57

Chromium — *contd.*
uptake
bioaccumulation, 440, 449, 452,
454–5, 624
exposure system, 441–4, 464
marine animals, in, 448–9, 452,
455–6, 620–2
see also Trace elements
Copper
levels in sediments, 19–20, 103
see also Trace elements

Diversity index, *see* Species diversity
Drill cuttings
dispersion, 25, 47, 49–56, 84, 99,
134, 142, 161, 163, 181, 192,
522, 547, 572–4, 802–3
modeling, 628–33, 635–44
role of ice, 47–53, 181–2
effects, 63, 72, 75, 86, 162, 488–9,
496–8, 523, 540–2
hydrocarbon analysis, in, 33–4, 484
487, 498, 501–3
land spreading, 172, 190, 222,
229–30
mass, 122–5, 127, 522
oil-contaminated
disposal, 160–72, 180–93, 220,
428, 482, 627
drill hole parameters, 126, 128
incineration, 175–6, 191–2
oil content, 61, 121–2, 126, 128,
432–3, 484, 497, 522, 524
particle size, 126, 129, 432–3
toxicity, 162–4, 181, 482, 484,
487–93